RADIATION DOSIMETRY

Instrumentation
and
Methods

Second Edition

RADIATION DOSIMETRY

Instrumentation and Methods

Second Edition

Gad Shani

Department of Nuclear and Biomedical Engineering
Ben Gurion University
Beer Sheva, Israel

CRC Press
Taylor & Francis Group
Boca Raton London New York

CRC Press is an imprint of the
Taylor & Francis Group, an **informa** business

First published 2001 by CRC Press
Taylor & Francis Group
6000 Broken Sound Parkway NW, Suite 300
Boca Raton, FL 33487-2742

Reissued 2018 by CRC Press

© 2001 by Taylor & Francis
CRC Press is an imprint of Taylor & Francis Group, an Informa business

No claim to original U.S. Government works

A Library of Congress record exists under LC control number: 00046855

Publisher's Note
The publisher has gone to great lengths to ensure the quality of this reprint but points out that some imperfections in the original copies may be apparent.

Disclaimer
The publisher has made every effort to trace copyright holders and welcomes correspondence from those they have been unable to contact.

ISBN 13: 978-1-138-10559-1 (hbk)
ISBN 13: 978-1-138-56156-4 (pbk)
ISBN 13: 978-0-203-71070-8 (ebk)

Visit the Taylor & Francis Web site at http://www.taylorandfrancis.com and the CRC Press Web site at http://www.crcpress.com

Preface

This volume is an updated reference book for medical dosimetry. It evolved from the book *Radiation Dosimetry Instrumentation and Methods* (CRC Press, 1991) published 10 years ago, which contains many of the basic facts of radiation dosimetry techniques. The present book contains developments in the last decade, mainly for medical dosimetry. The two books are complementary.

Radiation dosimetry has made great progress in the last decade, mainly because radiation therapy is more widely used. Every medium to large sized hospital has an oncology department with at least one, generally more, linear accelerators for tumor treatment. Radiation dosimetry has become a common need in the medical world. Medical physicist is now a certified profession and is required by the law in every hospital where radiation treatment is given to patients. One of the main tasks of the medical physicist is to provide the physician with an accurate measurement of the dose delivered to the patient. Several measurement methods were developed together with improved calculation methods, correction factors, and Monte Carlo simulation.

It was the intention of the author to assemble this past decade's developments in one short volume. Unfortunately, because of the vast amount of material, only selected information could be included, and many important developments in this field had to be left out.

The book starts with a short introductory chapter where the basic concepts of radiation and dosimetry are defined. The second chapter deals with basic concepts of radiation dosimetry theory. Electron and photon beams are defined and their interaction with the detector material is described with the consequences of that interaction for dosimetry. Theoretical methods for calculating these effects are discussed.

The main tool in medical dosimetry is the ionization chamber. Because of the importance of an accurate dose measurement, several types of ionization chambers were developed for this purpose in the last decade, having different shapes and dimensions and made of various materials. Application of a dosimeter in a measured system causes perturbation; also the dosimeter itself is not an ideal system. The difference between the measured dose and the calculated one requires the use of correction factors. These are calculated using models developed in the last few years. Ion chambers are also used in portable monitors for area survey; some examples are discussed in Chapter 3.

Chapter 4 deals with the new developments in thermoluminescent dosimetry (TLD). The basic concepts were discussed in *Radiation Dosimetry Instrumentation and Methods* and are not repeated here. New developments of TLD in the last decade are shown with scientific results and applications. Other luminescence dosimetry methods and electron spin resonance dosimetry are included in this chapter.

The new development in film dosimetry is mainly radiochromic film. This new type of film has a high spatial resolution and low spectral sensitivity; therefore, it is useful for dose distribution measurement. Again, the basic aspects of film dosimetry are not repeated in Chapter 5.

Some progress in calorimetry dosimetry was made. However, the two materials used for dose measurement are still water and graphite, because of the similarity between these materials' interaction with radiation to that of tissue. Calorimetry dosimetry and chemical dosimetry are discussed in Chapters 6 and 7, respectively.

Solid-state dosimetry has made considerable progress in the last decade. First, the use of diamonds became popular, mainly because diamonds are made of pure carbon and are close to tissue equivalent. Also as a perfect crystal, a diamond makes a good solid-state detector. The development of solid-state devices for electronics found its way to radiation dosimetry too. MOSFET and other devices are small and capable of detecting very low currents when used as solid-state detectors. Another way of using them: radiation effect on the device's performance is an indication of the dose absorbed.

Chapter 9 deals with a new development in radiation dosimetry, a three dimensional dosimeter. Detection of changes taking place in gel is a measure of the dose absorbed. One way to detect changes is by introducing ferrous sulfate to the gel and measuring light absorption following the irradiation, as done with the Fricke dosimeter. Another way is by changing the gel properties with radiation, then using three dimensional nuclear magnetic resonance (NMR) to measure the changes. It seems to give a good solution to the problem of three dimensional dosimetry, done in one measurement. This new dosimetry method is at its early stages of application.

Chapter 10 deals with neutron dosimetry. As is well known, neutron detection cannot be done by direct ionization; as in other kinds of radiation, secondary radiation is used. Neutron measurement is done using all the techniques used for other radiation measurements, i.e., ionization chamber, scintillator (TLD for dosimetry) and solid-state detectors. In addition, methods specific to neutron measurement include activation analysis and track detector. Because neutron interactions with matter are different from interactions with other forms of radiation, it is important for neutron dosimetry to use detectors made of tissue equivalent materials. When gamma radiation is present, pair detectors are used, and the gamma dose is subtracted. When TLD is used, it should contain isotopes with high cross-section for neutron interaction, such as Li^6 for thermal neutrons.

A few new developments in neutron dosimetry came about in the last decade. One of them is the superheated drop detector in which the number of bubbles formed in the liquid is a function of neutron dose. Solid state diodes were developed with polyethylene or boron converters. Gamma dose is subtracted in these detectors by the pulse shape discrimination technique.

Bonner sphere was developed many years ago as an area monitor. In the last decade, better designs were made with better energy response.

The book was written for medical physicists, engineers, and advanced dosimetrists. As mentioned above, it contains only the last decade's developments. If more basic data or methods are sought, the reader should refer to the book *Radiation Dosimetry Instrumentation and Methods*.

The Author

Gad Shani, Ph.D., is a professor in the department of nuclear engineering and the head of the biomedical engineering program at Ben Gurion University, Beer Sheva, Israel. Professor Shani received his B.Sc. degree in electrical engineering and M.Sc. degree in nuclear science at the Technion Israel Institute of Technology in 1964 and 1966, respectively, and his Ph.D. in nuclear engineering from Cornell University, Ithaca, New York. On the faculty at Ben Gurion University since 1970, Dr. Shani was one of the founders of the department of nuclear engineering and the head of the department from 1980 to 1984. Since 1994, he has served as the head of the biomedical engineering program.

Professor Shani is the incumbent of the Davide and Irene Sala Chair in Nuclear Engineering and the head of the Center for Application of Radiation in Medicine. He is a member of the International Euroasian Academy of Science. Professor Shani was a visiting scientist in many laboratories around the world, including the KFA in Germany as a winner of the European Community grant; in Harwell, England as a winner of the British Royal Society grant; at McMaster University in Canada; and at UCLA in the U.S. He spent sabbatical years at the University of California, Santa Barbara, Ohio State University, and Brookhaven National Laboratory (BNL) where he was a collaborator for the last 10 years. Additionally, he spent a few months every year at BNL doing research in the fields of neutron capture therapy, synchrotron radiation, and other methods for cancer treatment with radiation.

Professor Shani's field of scientific activity over the years spans a wide area, including neutron physics, reactor physics, nuclear instrumentation, dosimetry, and medical physics. His professional affiliations include the American Nuclear Society, Israel Nuclear Society (council member), Israel Ecology Society, Israel Society for Medical and Biomedical Engineering (council member), International Federation for Medical and Biomedical Engineering, Israel Society of Medical Physics, Israel Society of Radiation Research, and International Society for Neutron Capture Therapy. He has published about 150 papers in journals and meeting proceedings, authored 4 books, and co-authored 2 books.

Contents

1 Introduction

CONTENTS

I. UNITS AND DEFINITIONS

The energy imparted by ionizing radiation to matter of a given mass is the fundamental quantity of radiation dosimetry. Radiation field can be described by the average number of rays (or particles) per unit area, per unit time at each point. The rays can be in a parallel beam at angle θ to the plane or moving at all directions. In the second case the examined surface should be a sphere. If the examined area is a circle with area Δa (or one quarter of the sphere area) and the number of rays crossing it at time Δt is ΔN, then the flux density is given by

$$\Phi = \frac{\Delta N}{\Delta a \Delta t} \qquad (1.1)$$

and integration over time gives the fluence

$$\Phi = \frac{\Delta N}{\Delta a} \qquad (1.2)$$

Another way to deal with a radiation field is by summing the kinetic energy of all the particles entering the sphere:

$$\Delta E = \sum_i T_i \qquad (1.3)$$

where T_i is the kinetic energy of the ith ray or particle. The intensity is given by

$$I = \frac{\Delta E}{\Delta a \Delta t} \qquad (1.4)$$

If there is more than one kind of ray or particle, the flux density and the energy fluence can be calculated for each separately. If the particles (or rays) have different energies within a range, the flux density will be the integration of the distribution (or spectrum) over the energy range. When a beam of radiation encounters matter, it will be attenuated by the interaction with the matter. The attenuation can be measured by the reduction in number of rays or panicles, or by the reduction of the total beam energy.

The official units used in health physics and dosimetry are those agreed upon by the International Commission on Radiological Units and Measurements (ICRU). [1] In radiation protection the term for the linear energy transfer dependent factor is the quality factor (QF) by which absorbed doses are multiplied to obtain a quantity expression of the irradiation incurred on a common scale. The distribution factor (DF) expresses the nonuniformity effect of the irradiation. The product of the absorbed dose (D) and the two factors above is the dose equivalent

$$DE = D \times QF \times DF \quad (Sv) \qquad (1.5)$$

The units generally used in dosimetry are gray (Gy) for absorbed dose, roentgen (R) for exposure, and curie (Ci) for activity. Definitions of some terms used in dosimetry are listed below:

- Direct ionizing particles—charged particles having sufficient kinetic energy to produce ionization
- Indirect ionizing particles—uncharged particles that can produce ionizing particles
- Ionizing radiation—radiation consisting of directly and indirectly ionizing particles
- Energy imparted by ionizing radiation—the difference between the sum of energies of ionizing particles entering a certain volume and the sum of energies leaving the volume, less the energy spent in increasing any rest mass
- Absorbed dose—the quotient of the energy imparted by ionizing radiation and the mass of this volume:

$$D = \frac{\Delta E}{\Delta m} \qquad (1.6)$$

The units of absorbed dose are 1 Gy = 100 rad where 1 rad = 100 erg/gm.
- Absorbed dose rate—the quotient of the incremental absorbed dose and the absorption time:

$$\dot{D} = \frac{\Delta D}{\Delta t} \qquad (1.7)$$

The units can be Gy/min, Gy/sec, etc.
- Particle fluence—the quotient of the number of panicles ΔN that enter a sphere of area $4\Delta a$ (a test sphere) and the area Δa (the sphere cross-section area):

$$\Phi = \frac{\Delta N}{\Delta a} \qquad (1.8)$$

- Particle flux rate—the incremental particle flux per time interval; ϕ denotes flux distribution with respect to energy, direction, etc.:

$$\Phi = \frac{\Delta \phi}{\Delta t} \qquad (1.9)$$

- Energy fluence—the incremental kinetic energy of all particles entering the sphere of area $4\Delta a$ (cross-section Δa) per cross-section area:

$$F = \frac{\Delta E}{\Delta a} \qquad (1.10)$$

- Energy flux density—the incremental energy fluence per time interval:

$$I = \frac{\Delta F}{\Delta t} \qquad (1.11)$$

- Kerma—the incremental kinetic energy of all charged particles liberated by ionizing particles in a volume element divided by the mass of this volume element:

$$K = \frac{\Delta E}{\Delta m} \qquad (1.12)$$

- Kerma rate—incremental kerma in time interval Δt:

$$\frac{\Delta K}{\Delta t} \qquad (1.13)$$

- Exposure—the ratio between the sum of secondary electrical charge (ions of one sign produced when electrons produced by photons are stopped) in a volume element of air to the mass of that volume:

$$X = \frac{\Delta Q}{\Delta m} \qquad (1.14)$$

The unit of exposure is roentgen: 1 R = 2.58 × 10^{-4} Cb/kg. (This is identical to 1 ESU per 1 cc [0.001293 g] of air.)
- Exposure rate—the incremental exposure in time interval Δt:

$$\frac{\Delta X}{\Delta t} \qquad (1.15)$$

The units are R/sec (or R/min, etc.)
- Mass attenuation coefficient—the property of the material defined by

$$\frac{\mu}{\rho} = \frac{1}{\rho N}\frac{dN}{dl} \qquad (1.16)$$

for indirectly ionizing particles; ρ is the material density, N is the number of particles incident normal to the material, and dN is the number of particles interacting in thickness dl.

- Mass energy transfer coefficient—the property of the material defined by

$$\frac{\mu_k}{\rho} = \frac{1}{\rho E}\frac{dE}{dl} \qquad (1.17)$$

where E is the sum of kinetic energies T_i of indirectly ionizing particles meeting normally on the material of density ρ. dE is the sum of the kinetic energies of all the charged particles liberated in thickness dl. One use of this quantity is the ratio between fluence and kerma:

$$F = K\frac{\mu_k}{\rho} \qquad (1.18)$$

- Mass absorption coefficient—the property of the material defined as

$$\mu_{en} = \frac{\mu_k}{(1 - g)} \qquad (1.19)$$

where g is the part of the energy of the secondary charged particles lost by *bremsstrahlung*.

- Mass stopping power—the property of the material defined as

$$S = \frac{dE_s}{dl} \qquad (1.20)$$

where dE_s is the average energy lost by a charged particle traversing the length dl.

- Linear energy transfer—the energy imparted from charged particles to the medium

$$L = \frac{dE_L}{dl} \qquad (1.21)$$

where dE_L is the average local energy imparted when the particle travels a distance dl.

- Average energy expended in a gas per ion pair formed—

$$W = \frac{E}{N_w} \qquad (1.22)$$

where E is the particle initial energy and N_w is the average number of ion pairs formed by complete stopping of the particle. Activity units are

$$1 \text{ Ci} = 3.7 \times 10^{10} \text{ sec}^{-1}$$

II. ABSORBED DOSE IN TERMS OF EXPOSURE AND STOPPING POWER

When the exposure is 1 R, the energy absorbed in air is 87.7 erg/g. The absorbed dose is

$$D_{air} = 0.877 \, R \text{ rad} \qquad (1.23)$$

where R is the number of roentgens. If the medium is not air, then

$$D_m = 0.877 \, R \, \frac{\mu_{en}/\rho_m}{\mu_{en}/\rho_{air}} = fR \text{ rad} \qquad (1.24)$$

where f is the number of rad per roentgen in the medium.

When the spectrum is continuous, integration should be carried out:

$$D = \int_0^\infty R(E)f(E)\,dE \qquad (1.25)$$

and

$$R = \int_0^{E_{max}} R(E)\,dE \qquad (1.26)$$

The absorbed dose for a charged particle can be expressed in terms of the stopping power. If the stopping power is

$$S(T) = \frac{dT}{dx} \qquad (1.27)$$

and the particle fluence is $\phi(E)$, then using the definition,

$$D = \frac{\Delta E}{\Delta m} \qquad (1.28)$$

$$D = \frac{1}{\rho}\int_0^{T_{max}} S(T)\phi(T)\,dt \qquad (1.29)$$

where ρ is the stopping material density and the charged particles impinge perpendicular to the area.

The mean absorbed dose, D_T, in a specified tissue or organ, T, is given by

$$D_T = \frac{1}{m_T}\int_{m_T} D\,dm \qquad (1.30)$$

where m_T is the mass of the tissue or organ and D is the absorbed dose in the mass element dm. The mean absorbed dose, D_T, in a specified tissue or organ equals the ratio of the energy imparted, ε_T, to the tissue or organ, and m_T, the mass of the tissue or organ.

III. LINEAR ENERGY TRANSFER

Linear energy transfer (LET) denotes the energy lost by a charged particle per unit distance of medium traversed:

$$L = \frac{dE_L}{dl} \qquad (1.31)$$

where dE_L is the average energy locally imparted to the medium. When a nonmonoenergetic radiation interacts with material, there is a distribution of LET. If the distribution of tracks is $T(L)$, then the average LET can be defined as:

$$\overline{L_T} = \int_0^{L_{max}} T(L)L\,dL \qquad (1.32)$$

or dose average

$$\overline{L_D} = \int_{L_{min}}^{L_{max}} D(L)L\,dL \qquad (1.33)$$

Charged particles lose energy by colliding with the atomic electrons and transferring energy to them. This energy can be half the partial energy if the particle is an electron and four times the relative mass between the electron and the particle for heavy particles. The scattered electrons that are δ-rays form their own track, which might branch to a ternary track. δ-rays of energy above 100 eV are generally considered separate particles (in some cases higher energy is taken). The selection of the lower limit of δ-rays affects the LET of the original particle and makes the calculation complicated.

Energy transfer of heavy charged particles (HCP) to nm-size targets have been investigated by Iwanami and Oda [2], taking into account δ-ray generation by HCP as well as associated δ-rays. The energy transfer into the target is mainly due to ionizing collisions of HCP with matter. Secondary electrons generated by ionizing collisions within the target, whose ranges are much larger than the target size, deposit almost all of their energy outside the target. The ionizing collisions generating such secondary electrons are therefore excluded from the energy transferred into the target and are regarded as generating a new electron fluence. The energy of these electrons is greater than the cutoff energy for δ-rays, Δ. Secondary electrons with energy less than Δ dissipate their energy locally at their production site. ICRU [3] defined two kinds of LET: unrestricted and restricted. The unrestricted LET, L_{∞}, is the quotient of dE and dl, where dE is the mean energy lost by a charged particle due to collisions with electrons

in traversing a distance dl; thus,

$$L_{\infty} = \frac{dE}{dl} \qquad (1.34)$$

L_x does not take into account δ-ray production. The restricted LET, L_{Δ}, is the quotient of dE by dl, where dE is the energy lost by a charged particle in traversing a distance dl due to those collisions with electrons in which the energy loss is less than the restricted energy Δ:

$$L_{\Delta} = \left(\frac{dE}{dl}\right)_{\Delta} \qquad (1.35)$$

Where Δ is the cut-off energy for δ-rays and restricted energy of L_{Δ},

A simplified parameter, the event size Y, was suggested by Rossi. [4] It is the ratio between the energy deposited in a small sphere by the primary and secondary particles to the sphere diameter d:

$$Y = \frac{E}{d} \qquad (1.36)$$

The complication here is that Y requires additional parameter d.

A distribution of event size Y can be found as a distribution of the LET or track length. The relation between the distribution of the absorbed dose in Y, $D(Y)$, and the distribution in L, $D(L)$, can be found by examining the relation of a track length within a sphere and the sphere diameter d:

$$D(Y) = \frac{3Y^2}{L^2} \qquad (1.37)$$

where Y is idealized by assuming that the tracks are straight lines, the energy loss is uniform, and Y is independent of d. $Y \le L$ since Y_{max} occurs along the diameter, at which position $Y = L$. It is also possible to write

$$D(Y) = 3Y^2 \int_Y^{L_{max}} \frac{D(L)}{L^3}\,dL \qquad (1.38)$$

A. DOSE-EQUIVALENT QUANTITIES

A quality factor, Q, is introduced to weight the absorbed dose for the biological effectiveness of the charged particles producing the absorbed dose. It is formulated to take account of the relative effectiveness of the different types of ionizing radiation at the low exposure levels encountered in routine radiation protection practice. The quality factor, Q, at a point in tissue is given by

$$Q = \frac{1}{D}\int_L Q(L)D_L\,dL \qquad (1.39)$$

where D is the absorbed dose at that point, D_L is the distribution of D in linear energy transfer L, and $Q(L)$ is the corresponding quality factor at the point of interest. The integration is to be performed over the distribution D_L, due to all charged particles, excluding their secondary electrons.

B. Dose Equivalent

The dose equivalent, H, is the product of Q and D at a point in tissue, where D is the absorbed dose and Q is the quality factor at that point; thus,

$$H = QD \qquad \text{(Sv)} \qquad (1.40)$$

The quantity dose equivalent is defined for routine radiation-protection applications. The dose equivalent, H, at a point is given by

$$H = \int_L Q(L)D_L\,dL \qquad (1.41)$$

where $Q(L)$ is the quality factor for particles with linear energy transfer L and D_L is the spectral distribution, in terms of L, of the absorbed dose at the point.

C. Ambient Dose Equivalent

The ambient dose equivalent, $H^*(d)$, at a point in a radiation field is the dose equivalent that would be produced by the corresponding expanded and aligned field in the ICRU sphere [5] at a depth d on the radius opposing the direction of the aligned field. [6]

For strongly penetrating radiation, a depth of 10 mm is currently recommended. The ambient dose equivalent for this depth is then denoted by $H^*(10)$. For weakly penetrating radiation, depths of 0.07 mm for the skin and 3 mm for the eye are employed, with analogous notation.

Measurement of $H^*(10)$ generally requires that the radiation field be uniform over the dimensions of the instrument and that the instrument have an isotropic response.

D. Directional Dose Equivalent

The directional dose equivalent, $H'(d, \Omega)$ (Sv), at a point in a radiation field is the dose equivalent that would be produced by the corresponding expanded field in the ICRU sphere at a depth d on a radius in a specified direction, Ω.

The ICRU sphere is a 30-cm-diameter tissue-equivalent sphere with a density of 1 g cm^{-3} and a mass composition of 76.2% oxygen, 11.1% carbon, 10.1% hydrogen and 2.6% nitrogen.

IV. DOSIMETRY METHODS

A. Ionization Method

The most widely used method of dosimetry is based on ionization. The number of ion pairs produced is

$$I = \sum_i \int_B^\infty \frac{\varepsilon_i n_i(E)\,d\varepsilon_i}{w(\varepsilon_i)} \qquad (1.42)$$

where B is the lower limit of energy loss and $w_i\,(\varepsilon_i)$ is the energy required for a particle of type i at energy ε to produce an ion pair. Since for many gases w is independent of i and ε, $I = \varepsilon T/w$.

When measurement of the dose at a specific position is required, the detector dimensions must be small compared to the attenuation length of the primary radiation. If this is impossible, the first collision dose in the detector must be the same as in the medium, or, at least, the ratio between the first collision doses in the two materials must be independent of energy. It is always required that the ratio of the stopping powers in the two materials is independent of energy.

B. Chemical Methods

In some systems the chemical composition is changed by the absorbed radiation (including photographic film). If Y is the observed chemical change, then

$$Y = \sum_i \int_0^\infty \varepsilon_i n_i(\varepsilon_i, E)G_i(\varepsilon_i)\,d\varepsilon_i \qquad (1.43)$$

where $G_i\,(\varepsilon_i)$ is the yield per unit energy absorbed. If G is independent of particle type and ε_i, then

$$Y = G\varepsilon T \qquad (1.44)$$

C. Calorimetric Methods

The radiation energy absorbed in the dosimeter changes into thermal energy and raises the dosimetry temperature. The temperature change is given by

$$\Delta T = \frac{1}{c}\sum_i \int_0^\infty \varepsilon_i n_i(\varepsilon_i, E)F_i(\varepsilon_i)\,d\varepsilon_i \qquad (1.45)$$

where $\varepsilon_i n_i\,(\varepsilon_i, E)\,d\varepsilon_i$ is the amount of energy absorbed in a unit mass. $F_i(\varepsilon_i)$ is the fraction of charged particle energy that is degraded to heat. c is the thermal capacity of the substance. $F_i(\varepsilon_i)$ is approximately constant near unity so that

$$\Delta T = \frac{\varepsilon T}{c} \qquad (1.46)$$

D. Thermoluminescence Methods

When radiation is absorbed by an impure crystal, some of the electrons are trapped in the levels created within the forbidden gap. When those electrons are forced by heat to return to the valence band, their energy is emitted as light. The total amount of light emitted is proportional to the dose absorbed in the crystal:

$$L = \sum_i \int_B^x \varepsilon_i n_i(E) \, d\varepsilon_i \qquad (1.47)$$

where L is the total amount of light, ε_i is the light photon energy, and n_i is the number of light photons. B is the lower limit of light detection.

V. Gamma Dosimetry

A. Point Source Dose

If we define dose rate as the energy absorbed per unit volume per unit time, it is given that

$$D' = \mu I(E, r) \qquad (1.48)$$

where $I(E, r)$ is the flux density of energy E at a distance r from a point source. If the point source strength is S, then

$$I(E, r) = \frac{S}{4\pi r^2} \qquad (1.49)$$

when no attenuation in the surrounding material is assumed. With attenuation the flux is

$$I(E, r) = \frac{S}{4\pi r^2} e^{-\mu r} \qquad (1.50)$$

For the dose rate to be in units of energy absorbed per unit time as defined above, the source must be expressed in units of energy

$$S = cE \text{ MeV/s} \qquad (1.51)$$

where c is the source intensity in disintegration per second and E is in MeV. If the source strength c is expressed in curie, then

$$S = 3.7 \times 10^{10} \, cE \qquad (1.52)$$

and the dose rate is

$$D' = 2.96 \times 10^9 \, \mu c E \, \frac{e^{-\mu r}}{r^2} \, \frac{\text{MeV}}{\text{cm}^2/\text{s}} \qquad (1.53)$$

The total dose is obtained by time integration of the dose rate:

$$D = \int D' \, dt \qquad (1.54)$$

or, if the dose rate is constant,

$$D = D't \qquad (1.55)$$

Radioactive isotopes are an exponentially decaying source, so integration must be carried out for at least short-lived isotopes. When the source is other than a point source, the flux must be calculated accordingly. Self absorption should sometimes be included.

For high-dose measurement the following dosimeters are used: calorimeters, alanine/electron spin resonance (ESR) systems, liquid solutions (Fricke, ceric-cerous, dichromate), and polymer systems (polymethyl methacrylate, cellulose triacetate, radiochromic films and optical waveguides).

B. First Collision Dose

When a beam of ionizing radiation meets with a small mass so that the attenuation is small, the dose is referred to as first collision dose. It is expressed in terms of the energy imparted to a unit mass of the material per unit time per unit flux at the incident beam. An expression for the first collision dose for gamma rays of energy E is given by

$$D(E) = 1.602 \times 10^{-8} \sum_i N_i \{\tau_i(E)\varepsilon_{pe}(E)$$
$$+ \sigma_i(E)\varepsilon_c + \kappa_i(E)\varepsilon_{pp}\} \qquad (1.56)$$

where $D(E)$ is in rad/(photon/cm²); N_i is the number of atoms of the ith element per gram of material; $\tau_i(E)$, $\sigma_i(E)$, and $\kappa_i(E)$ are the photoelectric, compton, and pair production cross sections, respectively, in cm²/atom of the ith element; and ε is the average kinetic energy transferred to the electron (or positron) in the effects taking place (pe is photoelectric, c is compton, and pp is pair production). In the photoelectric effect, $\varepsilon_{pe} = E - E_B$, where E_B is the electron binding energy. The kinetic energy transferred to the electron in the compton effect is

$$E_c = \frac{aE(1 - \cos\phi)}{1 + \alpha(1 - \cos\phi)} \qquad (1.57)$$

where $\alpha = E/0.511$ MeV and ϕ is the Compton scattered photon angle.

For the pair production $\varepsilon_{pp} = E - 1.022$. The summation in Equation (1.56) is over all elements in the absorbing material. The factor 1.602×10^{-8} converts MeV/gr to rad. At low energy the main effect is the photoelectric effect. The cross section is decreased when the energy is increased, and it has the Z^5 dependence. At energies above 0.2 MeV, the main interaction is the Compton effect and then the pair production. Both cross sections are relative to the number of electrons per unit volume; hence, the difference between the different materials is small.

VI. BETA DOSIMETRY

There are several methods of calculation of beta dose and different applications of these methods according to the different source geometry. A point source dose rate can be calculated with the Loevinger formula [7]:

$$D'(r) = \frac{KC}{(\mu r)^2}\left\{ c\left[1 - \left(\frac{\mu r}{c}\right)e^{(1 - \mu r/c)} \right] + \mu r e^{(1 - \mu r)} \right\} \tag{1.58}$$

where $D'(r)$ is the beta dose rate in rad per hour at distance r from the point source, r is measured in gr/cm^2, C is the source intensity in curies, c is a parameter dependent on the beta maximum energy (dimensionless), μ is the absorption coefficient in cm^2/gr, and K is a normalization constant

$$K = \frac{1.7 \times 10^5 \rho^2 \mu^3 E_{av}}{[3c^2 - e(c^2 - 1)]} \frac{\text{rad/h}}{\text{curie}} \tag{1.59}$$

where ρ is the absorber density, e is the mathematical e, and E_{av} is the beta average energy. The value of c in air is $3.11e^{-0.55E_{max}}$ and in tissue, $c = 2$ for $0.17 \leq E_{max} < 0.5$ MeV, $c = 1.5$ for $0.5 \leq E_{max} < 1.5$ MeV, and $c = 1$ for $1.5 \leq E_{max} < 3.0$ MeV. μ in air is given by

$$\mu = \frac{16(2 - E_{av}/E_{av}^*)}{(E_{max} - 0.036)^{1.4}} \text{ cm}^2/\text{gr} \tag{1.60}$$

E_{av}^* is called the hypothetical average beta energy per disintegration for a hypothetical forbidden beta disintegration having the same E_{max} as an allowed beta decay transition in the same Z element. For allowed spectra, $E_{av}/E_{av}^* = 1$.

Other simple expressions for beta dose rate calculation are available. In analogy to gamma point source dosimetry, the following equation can be used:

$$D'(r) = 2.14 \times 10^6 \rho^2 \frac{\mu}{\rho} C E_{av} \frac{e^{-\mu r/\rho}}{4\pi r^2} \frac{\text{rad}}{\text{h}} \tag{1.61}$$

for C in curies, E_{av} in MeV, ρ in gr/cm^3, $\mu/\rho = 17E_{max}^{-1.14}$ cm^2/gr, and r in gr/cm^2. Expressions for other source geometries can be found in Fitzgerald et al. [8]

VII. NEUTRON AND HEAVY PARTICLES DOSIMETRY

A. NEUTRON DOSIMETRY

Neutron dosimetry is done by transforming the number density of neutrons (or neutron flux) to dose. This is done by using the equation

$$D'(r,E) = K\phi(r,E)\left[\Sigma_s(E)\frac{AE}{(A + 1)^2} + \Sigma_{n,\gamma}(E)E_\gamma B \right] \tag{1.62}$$

where $D'(r, E)$ is the dose rate in rad/h, K is a conversion factor

$$K = 5.76 \times 10^{-5} \frac{\text{rad}}{\text{h}} \Big/ \frac{\text{Mev}}{\text{cm}^3 \text{s}}$$

$\phi(r, E)$ is the neutron flux in n/cm^2 sec, A is the atomic mass of the target nucleus, E_γ is the radioactive capture gamma-photon energy in MeV, and B is a factor representing the fraction of radioactive capture gamma-photon energy absorbed in the neighborhood of the capture. $A/(A + 1)^2$ is the fraction of incident neutron energy imparted to the recoil nucleus of mass A. Σ_s is the scattering cross section and $\Sigma_{n,\gamma}$ is the (n, γ) reaction cross section. If any other reaction in addition to scattering and radiation capture takes place, the energy transferred to the substance should be included.

The energy transferred to the substance after neutron collision (first collision dose) is given by

$$D(E) = 1.602 \times 10^{-8} \sum_i \sum_j N_i \sigma_{ij}(E)\varepsilon_{ij}(E) \tag{1.63}$$

where N_i is the number of nuclei of type i per gram of substance, σ_{ij} is the cross section of the ith kind of nucleus for the reaction in which particles of type j are produced, and ε_{ij} is the average kinetic energy of the jth particle emitted by the ith nucleus.

In elastic scattering the secondary panicle is the scattered neutron, and for the isotropic case,

$$\varepsilon_{ij} = \frac{2mM_iE}{(m + M_i)^2} \tag{1.64}$$

where m is the neutron mass, M_i is the nucleus mass, and E is the neutron energy.

For an unisitropic scattering, the last expression should be multiplied by $1 - f_{li}(E)$, where $f_{li}(E)$ is given by the expansion of the elastic cross section

$$\sigma_{el}(E, \Theta) = \frac{\sigma_{el,0}(E)}{4\pi} \sum_{l=0}^{\infty} (2l + 1) f_{l,i}(E) P(\cos\Theta)$$

(1.65)

P is the Legendre polynomial.

In the case of nuclear reaction,

$$\varepsilon_{ij}(E) = E + Q_{ij} \quad (Q_{ij} + E > 0) \qquad (1.66)$$

where Q_{ij} is the reaction Q value.

B. HEAVY PARTICLES

The introduction of heavy particles (hadrons) into radiation therapy aims at improving the physical selectivity of the irradiation (e.g., proton beams) or the radiobiological differential effect (e.g., fast neutrons) or both (e.g., heavy ion beams). Each of these therapy modalities requires several types of information; absorbed dose measured in a homogeneous phantom in reference conditions; dose distribution computed at the level of the target volume(s) and the normal tissues at risk; radiation quality from which an evaluation on the RBE could be predicted; and RBE measured on biological systems or derived from clinical observation. The single beam isodoses and thus the dose distributions are similar in neutron and photon therapy. Similar algorithms can then be used for treatment planning and the same rules can be followed for dose specification for prescribing and reporting a treatment. In hadron therapy, the RBE of the different beams raises specific

problems. For fast neutrons, the RBE varies within wide limits (about 2 to 5) depending on the neutron energy spectrum, dose, and biological system. For protons, the RBE values range between smaller limits (about 1.0 to 1.2). A clinical benefit is thus not expected from RBE differences. However, the proton RBE problem cannot be ignored since dose differences of about 5% can be detected clinically in some cases. The situation is most complex with heavy ions since the RBE variations, as a function of particle type and energy, dose, and biological system, are at least as large as for fast neutrons. In addition, the RBE varies with depth. Radiation quality thus has to be taken into account when prescribing and reporting a treatment. This can be done in different ways: description of the method of beam production; computed LET spectra and/or measured microdosimetric spectra at the points clinically relevant; or RBE determination. The most relevant data are those obtained for late tolerance of normal tissues at 2 Gy per fraction ('reference RBE'). Combination of microdosimetric data and experimental RBE values improves the confidence in both sets of data.

VIII. BIOLOGICAL DOSIMETRY

When dose to radiation workers or to patients is a concern, biological dosimetry is the most accurate dosimetry technique. In this way the radiation effect on the human body is measured directly without the intermediary of a technical device. No interpretation of physical or chemical phenomena taking place in the dosimeter is needed, nor is there a need for corrections.

Chromosome aberration analysis is recognized as a valuable dose-assessment method which fills a gap in dosimetric technology. Detection of chromosomal aberrations

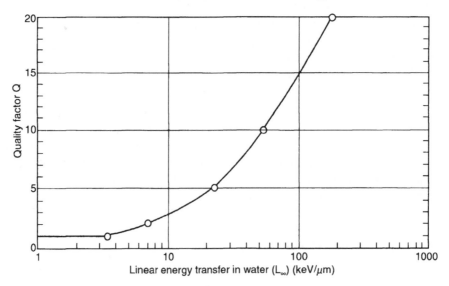

FIGURE 1.1 Quality factor as a function of linear energy transfer in water (L_x).

in the peripheral blood lymphocytes of exposed persons is the most fully developed biological indicator of exposure to ionizing radiation. By using a distribution analysis of the aberrations, it is possible to estimate the proportion of the body exposed and the average dose absorbed by the irradiated fraction.

The influence of the microscopic distribution of the absorbed energy on the detriment is taken into account by the use of the quality factor, Q.

The ICRP [9] recommends the following approximations for the average value of Q:

- X-rays, γ-rays and electrons: 1
- Thermal neutrons: 4.6
- Other neutrons: 20
- Protons and single-charged particles of unknown energy and rest mass $\geq 1 : 10$
- α-particles and multiple-charged particles of unknown energy: 20

The dependence of Q on LET is shown in Figure 1.1.

The RBE varies with the LET such that a hump-shaped response curve is obtained. A generalized curve is shown in Figure 1.2.

In order to produce a dicentric aberration, DNA damage must be induced in the two unreplicated chromosomes involved such that the damaged chromosomes can undergo exchange.

As the dose increases, the contribution of two track-induced dicentrics will also increase. Thus, the dose-response curve for X-ray-induced dicentrics will be a combination of one- and two-track events, with the former being more frequent at low doses and the latter being much more frequent

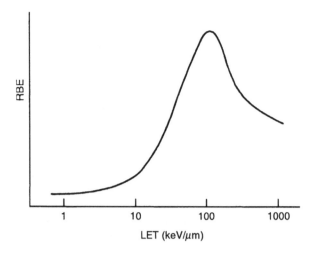

FIGURE 1.2 Generalized relationship between RBE and LET.

at high doses. The dose-response curve is generally assumed to fit the equation

$$Y = \alpha D + \beta D^2 \qquad (1.67)$$

where Y is the yield of dicentrics, D is the dose, α is the linear coefficient, and β is the dose-squared coefficient.

The dose-response curve for low LET radiation (X-rays or γ-rays) will be non-linear and best fit a linear-quadratic model. The dose-response curve for high LET radiation (for example, neutrons, protons, and α-particles) will be linear, or close to linear. RBE increases with increasing LET to a maximum of 100 keV/μm and decreases at higher LET values as a result of overkill.

Figure 1.3 shows a selection of dose-response curves.

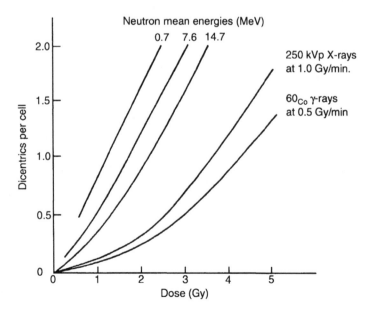

FIGURE 1.3 The relationship between dicentric yield and acute exposure to several types of radiation.

IX. CAVITY THEORY

The dose absorbed in different regions of an inhomogeneous medium is related to the stopping power of the material in the particular region. If the activity is the same and the radiation energy is the same, then the ratio of the doses in the different regions is the reciprocal of the stopping power in these regions. (Most of the investigation and theory was done for X- and γ-rays, which are the actual stopping power of electrons.)

If the cavity is small, most of the electrons are produced in the surrounding material, and the insertion of the cavity to the material does not change the electron spectrum. Bragg-Gray theory [10,11] is based on this assumption. It states that the energy lost by the electrons per unit volume in the cavity is $1/S$ times the energy lost by the gamma ray per unit volume in the solid. S is the stopping-power ratio in the gas and in the solid.

If the energy lost by electrons in crossing a volume is equal to the energy absorbed within the volume, then by the principle above the energy absorbed per unit volume in the gas is $1/S$ times the energy absorbed per unit volume of the solid. The energy absorbed in the gas is JW, where J is the ionization per unit volume of the gas (number of ion pairs produced) and W is the energy dissipated in ion-pair production. The Bragg-Gray relation states that the energy E_s absorbed per unit volume of the solid is

$$E_s = SW_s J \qquad (1.68)$$

By dividing the stopping power by the density, S_m, E_m, and J_m are the same quantities per unit mass:

$$S_m = \frac{\left(\frac{1}{\rho}\frac{dT}{dx}\right)_{solid}}{\left(\frac{1}{\rho}\frac{dT}{dx}\right)_{gas}} \qquad (1.69)$$

If the energy lost in an ion pair production, W, is independent of energy, Laurence [12] derived the expression for ionization in the gas cavity:

$$J = K\int_0^{T_{max}} \lambda \int_0^{T_0} \frac{(dT/dx)_g}{(dT/dx)_s} dT\ dT_0 \qquad (1.70)$$

where K is a constant and λ is the rate of electron production in the wall material per cm³ per gamma ray. T_{max} is the highest electron energy in the initial spectrum. Spencer

and Attix [13] and Burch [14] developed theories in which the secondary electrons were taken into account. In Spencer-Attix theory, a certain energy Δ was fixed, below which secondary electrons were assumed to dissipate all their energy at the site of interaction. Secondary electrons with energy above Δ were included in the fast spectrum. The lower energy limit of energy deposited in the gas was Δ. Spencer and Attix derived an expression for S_m by deriving an expression for the ratio of total to primary electron spectrum and using the Spencer-Fano [15] primary electron spectrum.

In Burch's theory the energy dissipated in the cavity per unit distance is defined as $(dT/dX)_g$ and the ratio of mass energy dissipation in the gas and in the solid $R_{m,T}$ is averaged:

$$\bar{R}_m = \frac{\int_0^{T_{max}} R_{m,T}\, n_{T,g}\, l_{T,g} \left(\frac{dT}{dx}\right)_g' dT}{\int_0^{T_{max}} n_{T,g}\, l_{T,g} \left(\frac{dT}{dx}\right)_g' dT} \qquad (1.71)$$

where $n_{T,g}$ is the number of electrons crossing the cavity, $l_{T,g}$ is the average path length traversed in the gas, and the product $n_{T,g}\, l_{T,g}\, (dT/dX)_g'$ is the electron energy dissipated in the cavity by $n_{T,g} dT$ electrons.

Secondary-electron equilibrium at a point inside a material exists when the quotient

$$\frac{W_{in}^e - W_{out}^e}{\Delta m} \qquad (1.72)$$

disappears for $\Delta m \to 0$. Here W_{in}^e is the sum of the kinetic energies of secondary electrons produced by photon radiation that enter a volume element ΔV containing the point; W_{out}^e is the sum of the kinetic energies of the secondary electrons that leave the volume element; and $\Delta m = \rho \Delta V$ is the mass of the material of density ρ within this volume element. This condition means that the absorbed dose produced at the point of interest is determined solely by the energy balance of the photon radiation entering and leaving the volume element, since the secondary-electron components of the energy balance counterbalance each other. For that reason the ratio of the absorbed doses produced by the same photon fluence at secondary-electron equilibrium in two different substances is equal to the ratio of the mass energy absorption coefficients.

REFERENCES

1. *Radiation Quantities and Units*, ICRU Report 10a, published as National Bureau of Standards (U.S.) Handbook 84, 1962.
2. **Iwanami, S. and Oda, N.**, *Phys. Med. Biol.*, 44, 873, 1999.
3. *Radiation Quantities and Units*, ICRU Report 16, 1970.
4. **Rossi, H. H.**, *Rad. Res.*, 10, 532, 1959.
5. *Radiation Quantities and Units*, ICRU Report 33, 1980.

6. *Quantities and Units in Radiation Protection Dosimetry,* ICRU Report 51, 1993.

7. **Loevinger, R.,** *Radiology,* 62, 74, 1954; 66, 55, 1956.

8. **Fitzgerald, J. J. et al.,** *Mathematical Theory and Radiation Dosimetry,* Gordon & Breach, New York, 1967.

9. **ICRP Report 26,** Pergamon Press, New York, 1977.

10. **Bragg, W. H.,** *Phil. Mag.,* 20, 385, 1910.

11. **Gray, L. H.,** *Proc. Roy. Soc.,* A122, 647, 1929; A156, 578, 1936.

12. **Laurence, G. C.,** *Can. J. Res.,* A15, 67, 1937.

13. **Spencer, L. V. and Attix, F. H.,** *Rad. Res.,* 3, 239, 1955.

14. **Burch, P. R. J.,** *Rad. Res.,* 4, 361, 1955; 6, 79, 1957.

15. **Spencer, L. V. and Fano, U.,** *Phys. Rev.,* 93, 1172, 1954.

2 Theoretical Aspects of Radiation Dosimetry

CONTENTS

I. INTRODUCTION

As the field of radiation dosimetry progressed, in particular for medical purposes, better equipment for more accurate dose measurement was developed. At the same time, theoretical methods for dose calculations and for calculation of correction factors of the various parameters affecting to dosimetry were developed.

Radiation therapy requires that a high dose of radiation is delivered accurately to specific organs. The success or failure of radiation treatment depends on the accuracy of the dose delivered to the tumor. Calibration of the radiation beam is based on complicated measurements and must be supported by accurate calculations and correction factors. Codes of practice were developed to guide medical physicists and physicians toward accurate dose delivery. High-energy electron beams are not monoenergetic but have a certain energy spread. This spread must be taken into consideration when the dose is measured and calculated. The energy spread has an effect on the electron range in the irradiated body. The most probable electron energy in the beam can be calculated. The range is generally obtained by the Harder equation, in which the electron energy is the main parameter.

Photon energy is well known if Cs^{137} and Co^{60} sources are used. When an accelerator is used to produce photons, the spectrum is *bremsstrahlung*, produced by the interaction of an electron beam with a target material. A set of parameters is then used to describe the beam spectrum. The advantage of using a photon beam is that its energy does not change with depth. Low-energy x-rays are generally characterized by the half-value layer (HVL)—the thickness of an absorber which reduces the beam intensity (or its air kerma) to 50% of its value before the attenuation. Aluminum is generally used for x-ray energy below 100 kV and copper is used above 100kV, and there is an overlap energy range below 100 keV.

Before any application of the radiation beam and the dosimeter is done, both beam and dosimeter must calibrated. For the beam calibration an air kerma must be established. The air kerma corresponds to the absorbed dose in air inside the dosimeter chamber cavity. The dosimeter calibration is given by the ratio of the dose to its cavity and the meter reading. These calibrations should be done both at the calibration beam and at the users beam. Following these steps, a series of corrections must be carried out in order to obtain the accurate dose for the patient. Since calibration is generally done in a water phantom, Bragg-Gray theory is used to make the transfer: dose to the cavity—dose to water—dose to patient. Dose correction must be done because of perturbation due to water displacement by the chamber, absorption and scattering in the chamber wall, and the fact that at least three different materials are involved. Energy per ion pair production must be known. The effective location of dose measurement, temperature, and pressure in the chamber must all be taken into consideration for accurate dose evaluation. In this chapter, methods used to calculate these corrections and theoretical ways to obtain an accurate value for the dose are discussed.

Analytical treatments generally begin with a set of coupled integro-differential equations that are prohibitively difficult to solve except under severe approximation. One such approximation uses asymptotic formulas to describe pair production and *bremsstrahlung*, and all other processes are ignored.

The Monte Carlo technique obviously provides a much better way for solving the shower generation problem, not only because all of the fundamental processes can be included, but also because arbitrary geometries can be treated. In addition, other minor processes, such as photoneutron production, can be added as a further generalization.

The most commonly used code for Monte Carlo calculation for dosimetry is the Electron-Gamma Shower (EGS) code. [1] The EGS system of computer codes is a general-purpose package for the Monte Carlo simulation of the coupled transport of electrons and photons in an arbitrary geometry for particles with energies above a few keV up to several TeV. The radiation transport of electrons (+ or −) or photons can be simulated in any element, compound, or mixture. That is, the data preparation package, PEGS4, creates data to be used by EGS4, using cross-section tables for elements 1 through 100. Both photons and charged particles are transported randomly rather than in discrete steps. The following physics processes are taken into account by the EGS4 Code System:

- *Bremsstrahlung* production (excluding the Elwert correction at low energies)
- Positron annihilation in flight and at rest (the annihilation quanta are followed to completion)
- Moliere multiple scattering (i.e., Coulomb scattering from nuclei)
- The reduced angle is sampled from a continuous, rather then discrete, distribution. This is done for arbitrary step sizes, selected randomly, provided that they are not so large or so small as to invalidate the theory.
- Moller (e^-e^-) and Bhabha (e^+e^-) scattering
- Exact, rather than asymptotic, formulas are used.
- Continuous energy loss applied to charged particle tracks between discrete interactions
- Pair production
- Compton scattering
- Coherent (Rayleigh) scattering can be included by means of an option
- Photoelectric effect.

II. ELECTRON DOSIMETRY

Electrons, as they traverse matter, lose energy by two basic processes: collision and radiation. The collision process is one whereby either the atom is left in an excited state or it is ionized. Most of the time the ejected electron, as in the case of ionization, has a small amount of energy that is deposited locally. On occasion, however, an orbital electron is given a significant amount of kinetic energy such that it is regarded as a secondary particle called a delta-ray. Energy loss by radiation (*bremsstrahlung*) is fairly uniformly distributed among secondary photons of all energies from zero up to the energy of the primary particle itself. At low-electron energies the collision loss mechanism dominates, and at high energies the *bremsstrahlung* process is the most important. At some electron energy the two losses are equal, and this energy coincides approximately with the critical energy of the material, a parameter that is used in shower theory for scaling purposes. Therefore, at high energies a large fraction of the electron energy is spent in the production of high-energy photons that, in turn, may interact in the medium. One of three photo-processes dominates, depending on the energy of the photon and the nature of the medium. At high energies, an electron-positron-pair production dominates over Compton scattering, and at some lower energy the reverse is true. The two processes provide a return of energy to the system in the form of electrons which, with repetition of the *bremsstrahlung* process, results in a multiplicative process known as an electromagnetic cascade shower. The third photon process, the photoelectric effect, as well as multiple Coulomb scattering of the electrons by atoms, perturbs the shower to some degree. The latter, coupled with the Compton process, gives rise to a lateral spread. The net effect in the forward (longitudinal) direction is an increase in the number of particles and a decrease in their average energy at each step in the process.

Electron linac fields are usually characterized by the central-axis practical range in water, R_p, and the depth of half-maximum dose, R_{50}, for dosimetry, quality assurance, and treatment planning. The spectral quantity $\langle E_0 \rangle^*$ is introduced, defined as the mean energy of the incident spectral peak and termed the "peak mean energy." An analytical model was constructed by Deasy et al. [2] to demonstrate the predicted relation between polyenergetic spectral shapes and the resulting depth-dose curves. The model shows that, in the absence of electrons at the patient plane with energies outside about $\langle E_0 \rangle^* \pm 0.1 \langle E_0 \rangle^*$, R_p and R_{50} are both determined by $\langle E_0 \rangle^*$.

The two most common range-energy formulas currently used for clinical dosimetry are

$$\langle E_0 \rangle = 2.33 R_{50} \quad \text{MeV} \tag{2.1}$$

and

$$E_{p,0} = (0.22 + 1.98 R_p + 0.0025 R_p^2) \quad \text{MeV} \tag{2.2}$$

where E_0 denotes the energy at the patient plane (typically 100 cm from the source), $\langle E_0 \rangle$ is the mean energy,

$E_{p,0}$ is the most-probable energy of the incident beam, and the range parameters are measured in cm. These formulas imply that R_{50} is determined by the mean energy, whereas R_p is determined by the most-probable energy.

The mathematical technique used here is to relate the dose and dose gradient at depth s to the depth R_p for monoenergetic beams. These relations are then expressed as spectra-averaged quantities. We first write, for s in the linear falloff region,

$$D(s) + (R_p - s)\left(\frac{\partial D(z)}{\partial z}\right)_s = 0 \qquad (2.3)$$

where $D(z)$ is the dose at depth z. Note that Equation (2.3) could be said to define R_p. If $\Phi_0(E_0)$ is defined as the fluence spectrum of electrons at the entrance plane (not the fluence spectrum at depth) and $dE_0 G(z, E_0)$ is defined as the central-axis depth-dose contribution at depth z from electrons of energies between E_0 and $E_0 + dE_0$. That is, $G(z, E_0)$ are depth-dose curves for monoenergetic beams. Also assume that $G(z, E_0)$ all have nearly a common maximum at the polyenergetic beam's depth of maximum dose (d_{max}). The expressions

$$D(z) = \int_{\Delta E} dE_0 \Phi_0(E_0) G(z, E_0) \qquad (2.4)$$

and

$$\left(\frac{\partial D(z)}{\partial z}\right)_s = \int_{\Delta E} dE_0 \Phi_0(E_0)\left(\frac{\partial G(z, E_0)}{\partial z}\right)_s \qquad (2.5)$$

follow. Consistent with the assumption of negligible low-energy contamination, the integrals are taken over only the peak energy region, denoted by ΔE. Equation (2.5) follows since these are definite, well-behaved integrals of well-behaved functions.

An electron-beam dose calculation algorithm has been developed by Keall and Hoban [3] which is based on a superposition of pregenerated Monte Carlo electron track kernels. Electrons are transported through media of varying density and atomic number using electron tracks produced in water. The perturbation of the electron fluence due to each material encountered by the electrons is explicitly accounted for by considering the effect of varying stopping power, scattering power, and radiation yield.

To accurately transport charged particles through a heterogeneous absorbing medium, knowledge of these three parameters of the different components in the medium is required. For an absorbing medium of density ρ,

ICRU Report 35 gives the expression for scattering power T as

$$T = \pi\rho\left(\frac{2r_e Z}{(\tau + 1)\beta^2}\right)^2$$
$$\times \frac{N_A}{M_A}\left[\ln\left(1 + \left(\frac{\theta_m}{\theta_\mu}\right)^2\right) - 1 + \left(1 + \left(\frac{\theta_m}{\theta_\mu}\right)^2\right)^{-1}\right] \qquad (2.6)$$

where r_e is the classical electron radius, $\tau = E/m_e c^2$ is the ratio of the kinetic energy E of the electrons to the rest energy (the rest energy equals the rest mass, m_e, times the speed of light squared, c^2), β is the ratio of the velocity of the electron to c, N_A is Avagadro's constant, M_A is the molar mass of substance A, θ_m is the cutoff angle due to the finite size of the nucleus, and θ_μ is the screening angle. The collision stopping power S_{col} can be calculated from

$$S_{col} = \frac{2\pi\rho r_e^2 m_e c^2 N_A Z}{\beta^2 M_A}\left[\ln\left(\frac{\tau^2(\tau + 2)}{2(I/m_e c^2)^2}\right) + F(\tau) - \delta\right] \qquad (2.7)$$

where I is the mean excitation energy, δ is the density effect correction, and

$$F(\tau) = 1 - \beta^2 + \left[\tau^2/8 - (2\tau + 1)\ln 2\right]/(\tau + 1)^2 \qquad (2.8)$$

For an incident kinetic energy E_0 and radiative stopping power S_{rad}, ICRU 37 uses the following equation to calculate the radiation yield, $Y(E_0)$:

$$Y(E_0) = \frac{1}{E_0}\int_0^{E_0} \frac{S_{rad}(E)}{S_{col}(E) + S_{rad}(E)}\, dE \qquad (2.9)$$

In the Super Monte Carlo (SMC) method, the above equations are used to translate electron transport in water to electron transport in a medium of arbitrary composition.

SMC electron dose distributions are calculated by superimposing the dose contribution from the electron track kernel, which is initiated from every surface voxel within the treatment field. At each surface voxel, the dose from the kernel is determined by transporting each electron step of each electron track. The dose is found by summing the energy deposited in voxel i, j, k from electron track which fall in voxel i, j, k, from all of the tracks

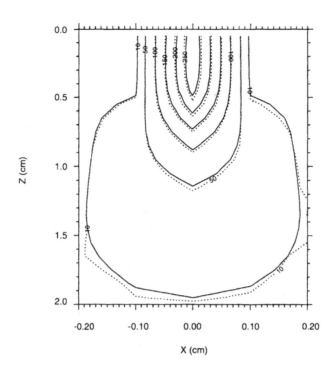

FIGURE 2.1 The dose distributions resulting from a 15-MeV pencil beam incident on a water phantom calculated using Monte Carlo (—) and SMC (---). The numbers on the isodose curves represent dose per incident fluence (pGy cm²). (From Reference [3]. With permission.)

transported from the surface in field XY.

$$D(i, j, k) = \frac{1}{\rho_{i, j, k}} \sum_{XY} \left(\sum_{m=0}^{M} \sum_{l=q}^{q'} \Delta E_{dep, m, l} \right) \qquad (2.10)$$

q, q' are steps in the calculation and m denotes an electron track.

The SMC algorithm can also be used to calculate dose by only taking account of the density of the irradiated medium and not the composition. This option is performed by transporting the electron step by its original length, divided by the density of the irradiated medium, instead of taking full account of the stopping power and scattering power characteristics of the medium. This option saves computation time and is useful when only waterlike media (e.g., soft tissue) are irradiated.

Isodose curves for a 15-MeV pencil beam were calculated using SMC and Monte Carlo methods in a water phantom with $0.1 \times 0.1 \times 0.1$ cm^{-3} voxels. The results are shown in Figure 2.1. Agreement is obtained between the SMC and Monte Carlo dose distributions, as expected, because the electron track kernel is generated in water. Figure 2.1 shows that 3000 electrons give a smooth dose distribution for all but the lowest isodose level and, hence, the neglect of *bremsstrahlung* transport in the SMC calculation is valid. The maximum difference found between the dose in any two corresponding voxels in the dose distribution was 5% of the maximum dose in the distributions.

The Method of Moments was generalized by Larsen et al. [4] to predict the dose deposited by a prescribed source of electrons in a homogeneous medium. The essence of this method is first to determine, directly from the linear Boltzmann equation, the exact mean fluence, mean spatial displacements, and mean-squared spatial displacements as functions of energy, and second to represent the fluence and dose distributions accurately using this information. Unlike the Fermi-Eyges theory, the Method of Moments is not limited to small-angle scattering and small angle of flight, nor does it require that all electrons at any specified depth z have one specified energy $E(z)$. The sole approximation in the Larsen et al. application is that for each electron energy E, the scalar fluence is represented as a spatial Gaussian, whose moments agree with those of the linear Boltzmann solution.

The mean particle motion is along the z-axis. As E decreases from E_0 to 0, $z(E)$ increases from 0 to a finite value, which is less than the electron range unless $T = 0$. Individual electrons may increasingly stray from this mean position as E decreases. Therefore, the variances in particle positions should all increase as E decreases.

Figure 2.2 shows the broad-beam depth doses for the MM-BCSD (BCSD ≡ Boltzmann Continuous Slowing Down), MM-FP (FP ≡ Fokker Planck), and EGS4 simulations. The shapes of the two MM simulations are globally correct. The differences between the Monte Carlo and the MM-BCSD results are due to the assumption of a spatial Gaussian, while the differences between the

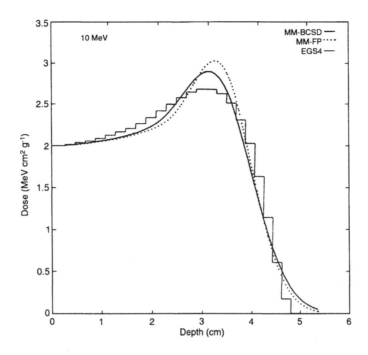

FIGURE 2.2 Depth dose curve for a broad 10-MeV electron beam. (From Reference [4]. With permission.)

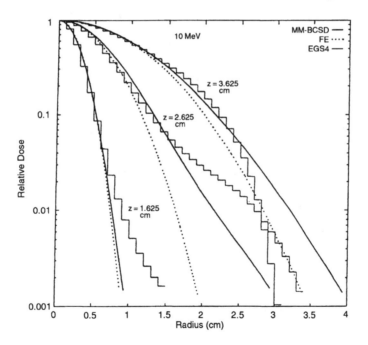

FIGURE 2.3 Radial dose profiles for a 10-MeV electron pencil beam. (From Reference [4]. With permission.)

Monte Carlo and the MM-FP results are due to the spatial Gaussian assumption and the small angle of scattering approximation. For these problems, the omission of large-angle scattering in the MM-FP simulation leads to an error in the depth-dose distribution that is approximately double that of the MM-BCSD simulation. [4]

In Figure 2.3 radial dose profiles are plotted from the MM-BCSD, Fermi-Eyges, and EGS4 simulations at different depths. Near the central axis, the MM-BCSD and

Fermi-Eyges profiles are remarkably similar. Away from the central axis, the MM-BCSD results are consistently greater than the Fermi-Eyges results and agree with the Monte Carlo results over a larger range. The MM-FP profiles are very similar to the other profiles near the central axis; away from the central axis, they are closer to the MM-BCSD results than to the Fermi-Eyges results.

Dose in water at a depth d_{water} is related to the dose in a solid at a corresponding depth d_{med}, provided secondary electron equilibrium exists (normally within a few mm of the surface) and energy spectra at each position are identical [9], by

$$D_{water}(d_{water}) = D_{med}(d_{med})[(S/\rho)_{coll}]_{med}^{water}[\Phi]_{med}^{water}$$

$$(2.11)$$

where $[(\bar{S}/\rho)_{col}]_{med}^{water}$ is the ratio of the mean unrestricted mass collision stopping power in water to that in the solid. $[\Phi]_{med}^{water}$ is the fluence factor, that is, the ratio of the electron fluence in water to that in the solid phantom. Because energy-loss straggling and multiple scattering depend upon the effective atomic number of the phantoms, it is not possible to find corresponding depths where the energy spectra are identical. However, there are corresponding depths where the mean energies are identical. Equation (2.12) is assumed to hold for these depths, which are defined to be the equivalent depths.

It is recommended that the water-equivalent depth be approximated using a density determined from the ratio of R_{50} penetrations by

$$d_{water} = d_{med} \times \rho_{eff} = d_{med}\left(\frac{R_{50}^{water}}{R_{50}^{med}}\right) \qquad (2.12)$$

i.e., that the effective density be given by the ratio of the R_{50} in water to that in the non-water material.

In order for Equation (2.11) to apply, secondary electron equilibrium must hold, which requires that the detector be of minimal mass or be made of material identical to the phantom. Therefore, it is recommended that thick-walled ion chambers (>0.1 g/cm^2) be irradiated in phantoms made of the same material. For example, the Memorial Holt parallel-plate chamber should be used in clear polystyrene and the PTW Markus parallel-plate chamber in PMMA. Thin-walled ion chambers, e.g., the graphite-walled Farmer chamber, can be used in any of the solid phantoms. In the measurement of depth dose, the conversion of ionization to dose is given by Andreo et al. [65]:

$$D_{water}(d_{water}) = N_{gas}Q_{corr}(d_{med})[(\bar{L}/\rho)_{coll}]_{air}^{med}P_{repl}$$
$$\times [(\bar{S}/\rho)_{coll}]_{med}^{water}[\Phi]_{med}^{water} \qquad (2.13)$$

where $Q_{corr}(d_{med})$ is the corrected ionization reading; $[(\bar{L}/\rho)_{coll}]_{air}^{med}$ is the ratio of the mean restricted mass collision stopping power in water to that in air. Percent depth dose is then given by

$$\%D_{water}(d_{water}) =$$
$$\left(\frac{\{Q_{corr}(d_{med}) \times [(\bar{L}/\rho)_{coll}]_{air}^{water} \times [\Phi]_{med}^{water} \times P_{repl}\}}{[...]_{max}}\right) \times 100$$

$$(2.14)$$

where the denominator equals the value of the numerator at the depth-of-maximum dose.

In radiation dosimetry protocols, plastic is allowed as a phantom material for the determination of absorbed dose to water in electron beams. The electron fluence correction factor is needed in conversion of dose measured in plastic to dose in water. There are large discrepancies among recommended values as well as measured values of electron fluence correction factors when polystyrene is used as a phantom material. Using the Monte Carlo technique, Ding et al. [5] have calculated electron fluence correction factors for incident clinical beam energies between 5 and 50 MeV as a function of depth for clear polystyrene, white polystyrene, and PMMA phantom materials.

The dose to water at dose maximum in a water phantom, D_w^w is related to the dose to plastic at dose maximum in a plastic phantom, D_p^p, by

$$D_w^w(d_{max}^w) = D_p^p(d_{max}^p)\left(\frac{\bar{S}}{\rho}\right)_p^w \phi_p^w \quad Gy \qquad (2.15)$$

where $(S/\rho)_p^w$, is the Bragg-Gray ratio of the mean unrestricted mass collision stopping power in water to that in the solid and ϕ_p^w is the electron fluence correction factor defined above. (D_B^A refers to the dose to medium B in a phantom of medium A.) The TG-21 protocol takes $(S/\rho)_p^w$ to be constant with respect to beam energy, with a value of 1.03 for polystyrene and 1.033 for PMMA. It also takes $\phi_{PMMA}^w = 1.0$ for all energies and tabulates $\phi_{polystyrene}^w$ vs. the beam's mean energy on the surface (ranging from 1.039 at 5 MeV to 1.009 at 16 MeV).

The AAPM TG-25 protocol's approach is similar to that of TG-21 except for the dose to water in a water phantom, D_w^w is related to the dose to plastic in a plastic phantom, D_p^p, for arbitrary scaled depths rather than just d_{max}. The relationship is given by:

$$D_w^w(d^w) = D_p^p(d^p)\left(\frac{\bar{S}}{\rho}\right)_p^w \phi_p^w \quad Gy \qquad (2.16)$$

where the depth in water (in cm), d^w, is related to the depth in plastic, d^p, by a depth-scaling factor called the effective density ρ_{eff}:

$$d^w = d^p \rho_{eff} = d^p \frac{R_{50}^w}{R_{50}^p} \quad cm \qquad (2.17)$$

where R_{50}^w and R_{50}^p are the depths (in cm) at which the absorbed dose falls to 50% of its maximum in water and plastic, respectively.

In the AAPM TG-21 protocol, which is based on Spencer-Attix cavity theory, the dose to a medium in a phantom of that medium is given by:

$$D_{med}^{med}(d^{med}) = M^{med}(d^{med})N_{gas}\left(\frac{\bar{L}}{\rho}\right)_a^{med}$$

$$\times (P_{ion}P_{repl}P_{wall})^{med} \text{ Gy} \quad (2.18)$$

where $M^{med}(d^{med})$ is the ion chamber meter reading measured in the medium. d^{med} is the depth of the measurement. $[\bar{L}/\rho]_a^{med}$ is the ratio of the mean restricted mass collision stopping power in the medium to that in air and the P corrections are defined by the AAPM TG-21 protocol and are in principle, dependent on the medium of the phantom. Combining Equation (2.16) and Equation (2.18) gives TG-25's estimate of the dose to water, given a measurement at the scaled depth in a plastic phantom:

$$D_w^w(d^w) = M^p(d^p)N_{gas}\left(\frac{\bar{L}}{\rho}\right)_a^p$$

$$\times (P_{ion}P_{repl}P_{wall})^p \left(\frac{\bar{S}}{\rho}\right)_p^w \phi_p^w \text{ Gy} \quad (2.19)$$

The IAEA TR277 Code of Practice and the NACP protocol have a definition of electron fluence correction factor which is different from that of the AAPM protocols. The IAEA converts the electrometer reading at the ionization maximum in a plastic phantom to the equivalent reading at the ionization maximum in a water phantom using

$$M_u^w = M_u^p h_m \quad C \quad (2.20)$$

where M_u^w and M_u^p are the electrometer readings in water and plastic, respectively, corrected for ion recombination by P_{ion}: D_w^w.

The IAEA gives the absorbed dose to water in a water phantom, D_w^w, at the effective point of measurement by:

$$D_w^w(P_{eff}) = M_u^w N_{gas}\left(\frac{\bar{L}}{\rho}\right)_{air}^w (P_{wall}P_{fl})^w \text{ Gy} \quad (2.21)$$

where in the IAEA notation, $[\bar{L}/\rho]_{air}^w$ is written as $(S_{w,air})_u$, N_{gas} as N_D, $P_{fl}P_{wall}$ as p_u, and P_{eff} is the effective point of measurement.

The overall IAEA equation for assigning dose to water based on a measurement in a plastic phantom is:

$$D_w^w(P_{eff}^w) = M_u^p(z_{max}^p)h_m N_{gas}\left(\frac{\bar{L}}{\rho}\right)_a^w (P_{wall}P_{fl})^w \text{ Gy} \quad (2.22)$$

where z_{max}^p is the depth of the maximum ionization in plastic and p_{eff}^w is at the depth of ionization maximum in the water phantom.

A more rigorous approach gives

$$\left(\frac{\bar{L}}{\rho}\right)_p^w \times \left(\frac{\bar{L}}{\rho}\right)_a^p = \left(\frac{\bar{L}}{\rho}\right)_a^w \quad (2.23)$$

which gives

$$D_w^w(d^w) = M^p(d^p)N_{gas}\left(\frac{\bar{L}}{\rho}\right)_a^w (P_{ion}P_{repl}P_{wall})^p \phi_p^w \text{ Gy} \quad (2.24)$$

where $M^p(d^p)$ is the ion chamber reading measured in a plastic phantom at depth d^p, corresponding to the scaled depth in water at which the dose is desired, and D_w^w is the dose to water in a water phantom. Equation (2.24) can be used to obtain the dose to water in a water phantom directly from the measurement performed in a plastic phantom.

Calculated central-axis depth-dose curves in different phantom materials with an identical incident beam are shown for two beams in Figure 2.4.

Figure 2.5 compares $(\bar{L}, \rho)_w^w$ to $(\bar{L}, \rho)_a^w/(\bar{L}, \rho)_a^p$ and shows that they agree within 0.2% up to the depth of R_{50}. This demonstrates that the depth-scaling procedure, which was defined to give equivalent depths at which mean energies match, also gives depths at which the entire electron spectra are effectively the same since the stopping-power ratios are close to identical. Figure 2.6 shows the calculated Spencer Attix water to plastic stopping power ratio for electron beams.

The accuracy of the Monte Carlo algorithm for fast electron dose calculation, VMC, was demonstrated by Fippel et al. [6], by comparing calculations with measurements performed by a working group of the National Cancer Institute (NCI) of the USA. For both energies investigated, 9 and 20 MeV, the measurements in water are taken to determine the energy spectra of the Varian Clinac 1800 accelerator. In some cases deviations have been observed which could be explained by the incompletely known geometry on the one hand and by inconsistent data on the other.

Most of the commercially available 3D electron beams planning systems use pencil-beam algorithms. Algorithms of this type may produce large errors (up to 20%) when small, dense inhomogeneities are present in an otherwise homogeneous medium. These errors are mainly caused by the semi-infinite-slab approximation of the patient geometry.

The VMC (Voxel Monte Carlo) algorithm can be briefly described as follows. To simulate the head of the accelerator for a definite energy, a fluence distribution in energy $F(E, \beta_0, x_0)$, denoted as the energy spectrum for short, at the phantom surface is necessary, with β_0 and x_0 being the initial direction and position and E being the kinetic energy of the primary electron. For simplicity, we assume that this spectrum is independent of β_0 and x_0; i.e.,

$$F(E, \beta_0, x_0) = F(E, 0, 0) \equiv F(E) \quad (2.25)$$

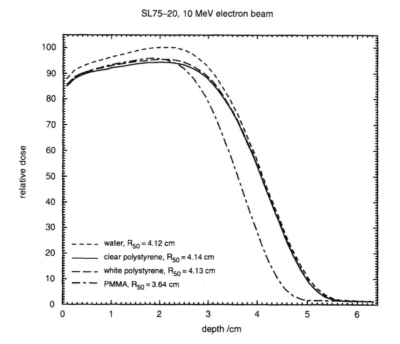

FIGURE 2.4 Calculated central-axis dose curves of a 10-MeV electron beam from an SL75-20 accelerator in different phantom materials. Curves all normalized to the maximum of the water curve (From Reference [5]. With permission.)

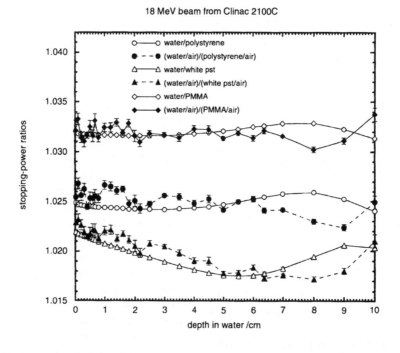

FIGURE 2.5 A comparison between calculated Spencer-Attix stopping-power ratios for an 18-MeV beam ($R_{50} = 7.7$ cm) from a Clinac 2100C. The empty symbols are the calculated $(\bar{L}, \rho)_p^w$ in water and the filled symbols are the calculated $(\bar{L}, \rho)_a^w$ in water divided by the calculated $(\bar{L}, \rho)_a^p$ in plastic where the depths in plastic are scaled to the water-equivalent depth according to Equation (2.17). This shows that $(\bar{L}, \rho)_p^w$ and $(\bar{L}, \rho)_a^w/(\bar{L}, \rho)_a^p$ agree within 0.2%, indicating that the depth-scaling procedure is giving equivalent depths at which the electron spectra are close to identical (From Reference [5]. With permission.)

If this spectrum, the depth dose of a monoenergetic broad beam in water $D_{mono}(E, z)$, and the photon background $D_\gamma(z)$ are known, then the depth-dose curve in water for the accelerator can be calculated by

$$D(z) = D_\gamma(z) + \int dE F(E) D_{mono}(E, z) \qquad (2.26)$$

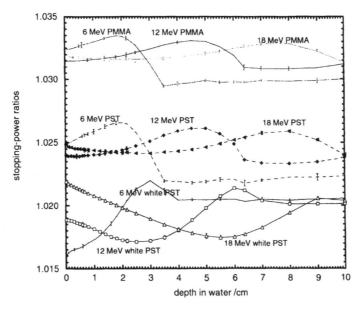

FIGURE 2.6 Calculated Spencer-Attix water to plastic stopping-power ratios for electron beams from the Clinac 2100C as a function of beam energy and depth for PMMA, clear and white polystyrene. The AAPM recommends using energy and depth independent values of 1.03 and 1.033 for polystyrene and PMMA, respectively. (From Reference [5]. With permission.)

Fippel et al. assumed

$$F(E) = C[A(2m + E) + 1/(E_x - E)^3]$$
$$E_{min} \leq E \leq E_p < E_x \qquad (2.27)$$

Here, C is the normalization constant, $m = 0.511$ MeV is the electron mass, $E_{min} = 0.5$ MeV is the minimum energy, and A, E_x, and E_p are free parameters (E_p is the most probable energy and at the same time the maximum energy of the spectrum). To fit these parameters, they precalculated a set of curves $D_{mono}(E, z)$ in the energy range of 0.5 MeV $\leq E \leq 30$ MeV for the large field size (15 cm $\times$ 15 cm) using VMC. From the measured depthdose curves $D_{meas}(z)$ of the Varian Clinac 1800 accelerator in water, the gamma contribution $D_\gamma(z)$ is subtracted. Then, using Equation (2.26), $(D_{meas}(z) - D(z))^2$ is minimized and $F(E)$ is determined. Gauss-Legendre quadrature has been used to perform the integration in Equation (2.26).

Figure 2.7 shows percentage depth doses in water for SSD = 100 cm and SSD = 110 cm as calculated by VMC (with the spectra $F(E)$) and the MDAH (M.D. Anderson Hospital) algorithm compared to measurements. The base data for the pencil-beam algorithm have been created using the water measurements.

A new approach was proposed by Rogers et al. [7] for electron-beam dosimetry under reference conditions. The approach has the following features:

- It uses ion chambers and starts from an absorbed-dose calibration factor for ^{60}Co to be consistent

with the present proposal for the new AAPM photon-beam protocol

- It uses R_{50} to specify the beam quality and the reference depth ($d_{ref} = 0.6 R_{50} - 0.1$ [all quantities in cm])
- It has a formalism which is parallel to the k_Q formalism for photon-beam dosimetry
- It fully accounts for the impact on stopping-power ratios of realistic electron beams
- It allows an easy transition to using primary standards for absorbed dose to water in electron beams when these are available

The equation for dose to water under reference conditions is:

$$D_w^Q = M P_{ion} {}^{P_{gr}^Q} K'_{R50} K_{ecal} N_{D,w}^{60Co} \qquad (2.28)$$

The term P_{gr}^Q is not needed with plane-parallel chambers but corrects for gradient effects with cylindrical chambers and is measured in the user's beam. The parameter k_{ecal} is associated with converting the ^{60}Co absorbed-dose calibration factor into one for an electron beam of quality Q_e and contains most of the chamber-to-chamber variation. The factor k'_{R50} is a function of R_{50} and converts the absorbed-dose calibration factor to that for the electron-beam equality of interest. Two analytical expressions were presented by Rogers that are close to universal expressions for all cylindrical Farmer-like chambers and for well-guarded plane-parallel chambers, respectively. [7]

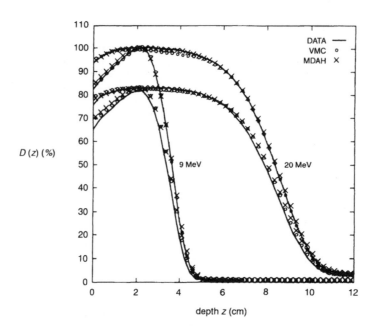

FIGURE 2.7 Depth dose curves in water as calculated by VMC and the MDAH algorithm, compared with measurements for SSD = 100 cm and SSD = 110 cm. (From Reference [6]. With permission.)

The fundamental equations of the k_Q formalism are

$$D_w^Q = MP_{ion}k_Q N_{D,w}^{Q_0} \quad \text{Gy} \tag{2.29}$$

$$N_{D,w}^Q = k_Q N_{D,w}^{Q_0} \tag{2.30}$$

where D_w^Q is the absorbed dose to water (in Gy) at the point of measurement of the ion chamber when it is absent (this is the center of a cylindrical or spherical chamber, or the front of the air cavity in a plane-parallel ion chamber); M is the temperature and pressure-corrected electrometer reading in coulombs (C) or meter units (rdg); P_{ion} accounts for ion chamber collection efficiency not being 100%; $N_{D,w}^{Q_0}$ is the absorbed dose to water calibration factor for an ion chamber placed under reference conditions in a beam of quality Q_0; $N_{D,w}^Q$ is the calibration factor in a beam of duality Q; and k_Q accounts for the variation in the calibration factor between beam quality Q and the reference beam quality Q_0. These equations can be applied to electron or photon beams. In practice, the reference beam quality Q_0 is ^{60}Co. The general equation for k_Q is

$$k_Q = \frac{\left[\left(\frac{\bar{L}}{\rho}\right)_{air}^w P_{wall}P_{fl}P_{gr}^Q P_{cel}\right]_Q}{\left[\left(\frac{\bar{L}}{\rho}\right)_{air}^w P_{wall}P_{fl}P_{gr}^Q P_{cel}\right]_{Q_0}} \tag{2.31}$$

where the numerator and denominator are evaluated for the beam quality Q of interest and the calibration beam quality Q_0, respectively. P_{cel} is a correction for the central electrode if it is made of a material different from the chamber walls. k_Q for electron beams has two components:

$k_{R_{50}}$, which depends on the chamber but is a function only of the beam quality specifier R_{50}, and P_{gr}^Q, which extracts the gradient corrections and which, for a cylindrical chamber, depends on the shape of the particular depth-dose curve being measured; i.e.,

$$k_Q = P_{gr}^Q k_{R_{50}} \tag{2.32}$$

where

$$k_{R_{50}} = \frac{\left[\left(\frac{\bar{L}}{\rho}\right)_{air}^w P_{wall}P_{fl}P_{cel}\right]_{R_{50}}}{\left[\left(\frac{\bar{L}}{\rho}\right)_{air}^w P_{wall}P_{fl}P_{gr}^Q P_{cel}\right]_{^{60}Co}} \tag{2.33}$$

and

$$P_{gr}^Q = I(d_{ref} + 0.5 r_{cav})/I(d_{ref})$$
$$\text{[for cylindrical chambers]} \tag{2.34}$$
$$= 1.0 \quad \text{[for plane-parallel chambers]}$$

where $I(d)$ is the ionization reading of a cylindrical chamber placed with the cylindrical axis at depth d and r_{cav} is the radius of the chamber's cavity in cm.

$$D_w^Q = MP_{ion}P_{gr}^Q k_{R_{50}} N_{D,w}^{^{60}Co} \quad \text{Gy} \tag{2.35}$$

The values of $k_{R_{50}}$ are calculable as a function of R_{50} and depend only on the chamber. For cylindrical chambers, the user must measure P_{gr}^Q in their own electron beam, but for plane-parallel chambers, $P_{gr}^Q = 1.0$ (see Figures 2.8, 2.9, 2.10, and 2.11).

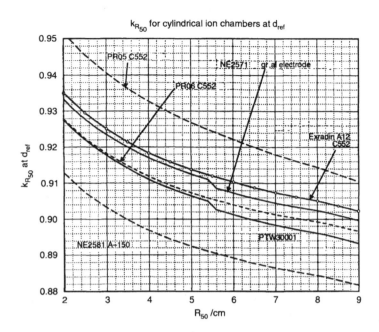

FIGURE 2.8 Calculated values of $k_{R_{50}}$ as a function of R_{50} for several common cylindrical ion chambers. These values can be used with a ^{60}Co absorbed-dose to water calibration factor to assign dose to water at the reference depth $d_{ref} = 0.6\,R_{50} - 0.1$ cm. (From Reference [7]. With permission.)

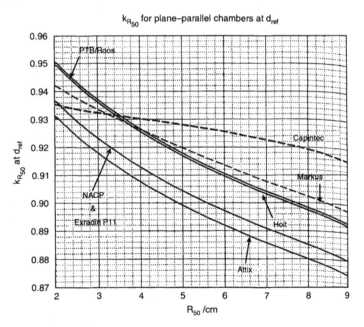

FIGURE 2.9 Calculated values of $k_{R_{50}}$ as a function of R_{50} for several common plane-parallel chambers. These values can be used with a ^{60}Co absorbed-dose to water calibration factor to assign dose to water at the reference depth $d_{ref} = 0.6R_{50} - 0.1$ cm. (From Reference [7]. With permission.)

Once electron-beam absorbed-dose calibration factors are available, one could use the following dose equation:

$$D_w^Q = MP_{ion}\frac{P_{gr}^Q}{P_{gr}^{Q_e}}k'_{R_{50}}\,N_{D,w}^{Q_e} \quad \text{[cylindrical chambers]}$$

$$\tag{2.36}$$

$$= MP_{ion}k'_{R_{50}}N_{D,w}^{Q_e} \quad \text{[plane-parallel chambers]}$$

$$\tag{2.37}$$

The major complication in this procedure is the need to measure P_{gr}^Q correction factors in the user's beam if d_{ref} cylindrical chambers are used.

The recommendations of the Electron Dosimetry Working Party II of the UK Institution of Physics and Engineering in Medicine and Biology [8], consist of a code of practice for electron dosimetry for radiotherapy beams of initial energy from 2 to 50 MeV. The code

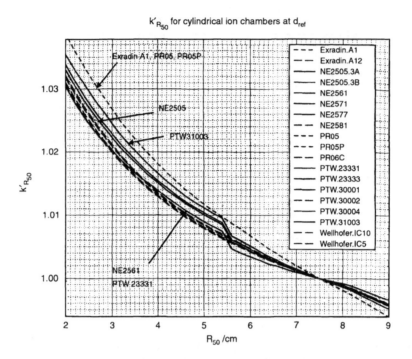

FIGURE 2.10 Calculated values of $k'_{R_{50}}$ as a function of R_{50} for cylindrical ion chambers. These values can be used with a ^{60}Co absorbed-dose to water calibration factor to assign dose to water at the reference depth $d_{ref} = 0.6R_{50} - 0.1$ cm. Note that chambers with aluminum lectodes are shown as solid lines. The upper two curves at smaller values or $k'_{R_{50}}$ represent curves which have cavity diameters less than 6 mm. (From Reference [7]. With permission.)

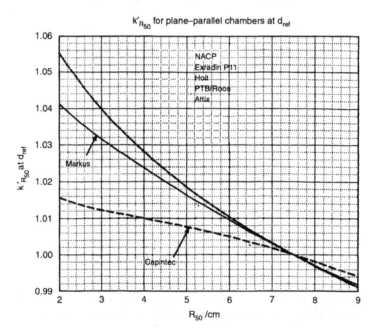

FIGURE 2.11 Calculated values of $k'_{R_{50}}$ as a function of R_{50} for several common plane-parallel chambers. These values can be used to determine absorbed-dose to water at the reference depth $d_{ref} = 0.6R_{50} - 0.1$ cm. Note that the values for the five well-guarded chambers lie on the same line in the figure. (From Reference [7]. With permission.)

is based on the 2 MV (or ^{60}Co) air kerma calibration of the NE 2561/2611 chamber, which is used as the transfer instrument between the national standards laboratory and hospitals in the UK. The code utilizes an $N_{D, air}$ approach. Designated chambers are the NE 2571 (graphite-walled Farmer chamber), to be calibrated against the transfer instrument in a megavoltage photon beam, and three parallel-plate chambers, to be calibrated against the NE 2571 in a higher-energy electron beam.

The current approach to electron dosimetry (*code of practice*) changes the conceptual framework to one based on $N_{D,air}$, a traceable calibration factor for the electron chamber to convert its reading to mean dose to air in the sensitive volume under standard ambient conditions. This enables the calibration of parallel-plate electron chambers in an electron beam. The recommended calibration route retains the NPL-designed secondary-standard chamber as the transfer instrument link between that national standards laboratory and hospital. It also retains a 2 MV/^{60}Co air kerma calibration of the instrument as the basis of the calibration route. Graphite-walled Farmer chambers are still to be calibrated in the hospital against the secondary-standard chamber at 5 cm depth in a phantom in either a ^{60}Co gamma-ray beam or megavoltage x-ray beam, now in terms of $N_{D,air}$. Parallel-plate chambers are then to be calibrated, also in terms of $N_{D,air}$, against a previously calibrated graphite-walled Farmer chamber at a reference depth equal to or nearly equal to the depth of the dose maximum in a sufficiently high-energy electron beam. This is due to the relatively large reported variation in photon beam perturbation factors for parallel-plate chambers. [8]

The National Physical Laboratory (NPL) has introduced a calorimeter-based direct-absorbed-dose-to-water calibration service for megavoltage photon beams, and a dosimetry code of practice based on this was published by IPSM (1991). [66]

The absorbed dose to water under reference conditions in an electron beam should be determined using one of the following designated ionization chambers:

- the NE2571 cylindrical graphite-walled Farmer chamber;
- the NACP-designed parallel-plate chamber (there are two designs of this chamber, differing in their waterproofing arrangements; both are allowed);
- the Markus-designed parallel-plate chamber (PTW/Markus chamber), type 2334;
- the Roos-designed parallel-plate chamber (Roos chamber), type 34001.

The cylindrical chamber is recommended only for use in electron beams where the mean energy at the reference depth ($\bar{E}_z$) is greater than 5 MeV (in practice this typically implies that the mean energy at the surface $\bar{E}_0 \leq 10$ MeV).

A practical ionization chamber does not sample the true electron fluence which is present in the undisturbed phantom at the point corresponding to the geometrical centre of the chamber cavity. The difference between this and the fluence in the cavity depends on a number of factors, one of which is the fluence gradient across the chamber. This gives rise to the concept of the effective point of measurement P_{eff}. A practical definition of P_{eff} is that depth in the medium where the average energy is the same as in the chamber, such that stopping-power ratios as conventionally evaluated can be used.

For cylindrical chambers used in megavoltage photon and electron beams around and beyond the depth of dose maximum, a displacement of 0.6 r toward the radiation source gives a reasonable representation of the experimental data, where r is the internal radius of the cavity.

Each designated electron chamber must be calibrated in terms of $N_{D,air,ch}$, a calibration factor to convert its corrected reading to mean absorbed dose to the air in the sensitive volume. To ensure traceability to national primary standards, this calibration must be via the local standard chamber, i.e., in the UK the NPL-designed secondary-standard chamber.

For the NE2571 chamber, the $N_{D,air,2571}$ calibration is obtained by a comparison against the secondary-standard chamber with both centers at the same depth in a PMMA phantom in a ^{60}Co γ-or megavoltage x-ray beam. For parallel-plate chambers, $N_{D,air,pp}$ is obtained by comparison with an $N_{D,air}$-calibrated NE2571 chamber in a higher-energy electron beam.

The secondary-standard and the NE2571 should be placed with their centers at a depth of 5 cm in a PMMA phantom of at least 20 cm $\times$ 20 cm in area and at least 12 cm thick. They should be irradiated in a ^{60}Co beam using a 10-cm $\times$ 10-cm field at a source-to-surface distance of at least 80 cm (if a ^{60}Co beam is not available). No corrections for ambient temperature and pressure are required, provided these conditions remain stable throughout the measurements.

Ion recombination corrections are negligible in these conditions for ^{60}Co-beam measurements and may be ignored.

The $N_{D,air}$ factor for the NE2571 chamber is given by

$$N_{D,air,2571} = 0.978 \; N_{K,sec}(M_{sec}/M_{2571}) \qquad (2.38)$$

where 0.978 is the combined correction factor $[(1 - g) \times k_{sec}(p_{sec}^\lambda/p_{cyl}^\lambda)_{PMMA}]$. $N_{K,sec}$ is the in-air air-kerma calibration factor for the secondary standard at ^{60}Co γ- or 2-MV x-rays as supplied by the national standards laboratory. The units of $N_{D,air,2571}$ are grays (absorbed dose to air) per chamber reading when $N_{K,sec}$ is in grays (air kerma) per chamber reading. The response of the parallel-plate chamber and the NE2571 should be determined in turn at the reference depth in a high-energy electron beam. The beam should have an $\bar{E}_0$ of preferably 17–22 MeV and no less than 15 MeV in order to maintain the in-scattering correction ($p_{cav,cyl}^e$) for the Farmer chamber within 2% of unity and thus to minimize its uncertainty.

TABLE 2.1
Ratios of the Linear Continuously Slowing Down Range
r_0/ρ^a **as a Function of Electron Energy for Different**
Plastics to the Corresponding Values for Water[b]

$\bar{E}_0$ (MeV)	A−150	Polystyrene	PMMA
0.1	0.879	0.961	0.863
0.2	0.880	0.963	0.864
0.5	0.882	0.964	0.864
1.0	0.886	0.967	0.865
2	0.891	0.970	0.867
5	0.897	0.974	0.870
10	0.902	0.978	0.872
15	0.906	0.983	0.874
20	0.909	0.987	0.876
30	0.914	0.993	0.878
40	0.919	0.999	0.882
50	0.922	1.003	0.883

[a] Note that r_0 conventionally is given in units of g cm^{-2}, while R_t, R_{50}, and R_p generally are given in cm.

[b] The density of plastic may vary from one sample to another. It is therefore recommended that the density is measured and corrections are applied if necessary to the figures given in the table.

Source: From Reference [64]. With permission.

The irradiations should preferably be carried out in a water phantom. It is important to allow phantoms and chambers to reach thermal equilibrium before measurements arc begin.

The readings for each chamber should be corrected for polarity and ion recombination. The $N_{D, air}$ for the parallel-plate chamber is given by

$$N_{D, air, pp} = M_{2571} N_{D, air, 2571}$$
$$p^e_{cav, 2571}(\bar{E}_z)/M_{pp} p^e_{pp}(\bar{E}_z) \qquad (2.39)$$

where M_{2571} and M_{pp} are the readings of the chambers for the same number of mitor units and corrected as mentioned above. $p^e_{cav, 2571}$ is the in-scattering perturbation factor for the NE2571 in this electron beam at this depth (i.e., at the appropriate $\bar{E}_z$).

The densities for A-150, polystyrene, and PMMA were 1.127, 1.06, and 1.19 g cm^{-3}, respectively:

$$\frac{R_{pl}}{R_w} = \frac{(r_0/\rho)_{pl}}{(r_0/\rho)_w} \qquad (2.40)$$

values of r_0/ρ ratios for three materials as shown in Table 2.1.

The effect of a nonstandard SSD on machine output and dose distributions must be assessed in order to assure proper patient treatment. The majority of nonstandard

SSD treatments are at an extended SSD, and as a general rule one should avoid such treatments unless absolutely necessary. Detailed calculations or measurements are required for individual machines and circumstances Khan et al. [9].

The relationship between maximum dose at the nominal (calibration) SSD and the maximum dose at the extended SSD is called the output correction. Two methods of output correction are typically used, namely, the effective SSD method and the virtual SSD method. In the effective SSD method, the dose D'_{max} at an extended SSD′ is related to the dose D_{max} at the nominal SSD by the following inverse-square law relationship:

$$D'_{max} = D_{max} \frac{(SSD_{eff} + d_{max})^2}{(SSD_{eff} + g + d_{max})^2} \qquad (2.41)$$

Where SSD_{eff} is the effective SSD for calibration for the given collimator field size and energy, g equals the difference between SSD′ and SSD, and d_{max} is the depth of maximum dose on the central axis. As stated earlier, the inverse square law relationship may not be exactly followed for the treatment conditions involving low electron energies and large air gaps. In those cases, either the $(Q_0/Q_g)^{1/2}$ vs. g data measured for various field sizes, electron energies, and air gaps may be used directly or a new calibration may be performed for the given treatment conditions.

The use of an extended treatment distance has only minimal effect on the central-axis depth dose and the off-axis ratios. Under the treatment conditions in which the dose rate changes with the effective SSD in accordance with the inverse-square law, the depth-dose distribution can be calculated according to

$$\%D(d, SSD, g) = \%D(d, SSD, 0)$$
$$\times \frac{[(SSD_{eff} + d)/(SSD_{eff} + g + d)]^2}{[(SSD_{eff} + d_{max})/(SSD_{eff} + g + d_{max})]^2} \qquad (2.42)$$

where the depth dose equals 100% at d_{max} for both standard and extended SSD. For typical values of SSD_{eff}, g, d, and d_{max} ($SSD_{eff} = 100$ cm, $d_{max} = 2$ cm, $g < 10$ cm, $d < 15$ cm), this correction will have an insignificant effect (< 1 mm) on the shape of the depth-dose curve (< 1 mm shift on descending portion) so that the use of the standard depth-dose curve should be an acceptable approximation at extended distances.

The mean electron energy at the surface of the phantom is $\bar{E}_0$. The relationship between $\bar{E}_0$ and R_{50} is usually taken to have the general form

$$\bar{E}_0 = C\ R_{50} \qquad (2.43)$$

where C is a constant.

TABLE 2.2

The Relationship Between the Mean Energy at the Phantom Surface, $\bar{E}_0$, and the Half-Value Depths $R_{50,D}$ or $R_{50,I}$ Measured from Broad-Beam Absorbed Dose and Ionization Curves, Respectively, at $SSD = 100$ (From Reference 8 with permission)

$\bar{E}_0$ (MeV)	$R_{50,D}$ (cm)	$R_{50,I}$ (cm)
2	0.7	0.7
3	1.2	1.2
4	1.6	1.6
5	2.1	2.1
6	2.5	2.5
7	3.0	3.0
8	3.4	3.4
9	3.8	3.8
10	4.3	4.3
12	5.1	5.1
14	6.0	5.9
16	6.8	6.7
18	7.8	7.6
20	8.6	8.4
22	9.4	9.2
25	10.7	10.4
30	12.8	12.3

When the R_{50} is taken from a dose distribution which has been obtained with a constant SSD (e.g., $SSD = 100$ cm),

$$\bar{E}_0(\text{MeV}) = 0.818 + 1.935\, R_{50,I} + 0.040(R_{50,I})^2$$
$$(2.43a)$$

for $R_{50,I}$ (depth of 50% of maximum ionization) determined from a depth-ionization curve and

$$\bar{E}_0(\text{MeV}) = 0.656 + 2.059\, R_{50,D} + 0.022\,(R_{50,D})^2$$
$$(2.43b)$$

for the case of $R_{50,D}$ (depth of 50% of maximum dose) determined from a depth-dose curve. As a guide, values are given in Table 2.2 for $\bar{E}_0$ from 2 to 30 MeV.

The absorbed dose to water in grays in the electron beam at the position of the effective point of measurement of the chamber when the chamber is replaced with water is given by

$$D_w(z_{ref,w}) = M^e_{ch,w} N_{D,air,ch}$$
$$s^e_{w/air}(\bar{E}_0, z_{ref,w}) p^e_{ch}(\bar{E}_z) \qquad (2.44)$$

where $M^e_{ch,w}$ is the corrected chamber reading; $s^e_{w/air}(\bar{E}_0, z_{ref,w})$ is the appropriate water-to-air stopping power ratio for an electron beam of mean surface energy $\bar{E}_0$ at depth in water $z_{ref,w}$; and $s^e_{w,air}(\bar{E}_u, z_{red,w}) p^e_{ch}(\bar{E}_z)$ is the

perturbation factor applicable to the particular ionization chamber in an electron beam at a mean energy $\bar{E}_z$ at depth $z_{ref,w}$. The $p^e_{ch}(\bar{E}_z)$ values for the Markus chamber are given by

$$p^e_{Markus} = 1 - 0.039\exp(-0.2816\bar{E}_z) \qquad (2.45)$$

It is generally recommended that water phantoms should be used wherever possible for the experimental procedures involved in the absolute calibration of electron beams. However, it is recognized that in certain situations, particularly for lower-energy electron beams and the use of parallel-plate chambers, it may be more convenient to use solid plastic phantoms. In this case, the depths need to be scaled to water-equivalent depths and the chamber reading needs to be multiplied by an appropriate fluence-ratio correction.

The effective water depth, $z_{w,eff}$, which can be taken as an approximation to the depth of water which is equivalent to a given depth in a plastic phantom, can be obtained from scaling:

$$z_{w,eff} = z_{non-w} C_{pl} \qquad (2.46)$$

The recommended values of the scaling factors, C_{pl}, are given in Reference [8].

The expression to evaluate the absorbed dose to water in this situation is

$$D_w(z_{ref,w}) = M^e_{ch,non-w} h_m N_{D,air,ch} s^e_{w/air}$$
$$(\bar{E}_0, z_{w,eff}) p^e_{ch}(\bar{E}_z) \qquad (2.47)$$

Electron beam absorbed dose distribution in a water phantom is shown in Figure 2.12.

The polarity correction is given by

$$f_{pol} = (|M^+| + |M^-|)/2M \qquad (2.48)$$

where the superscripts $+$ or $-$ indicate the reading (M) with collecting voltage positive and negative, respectively, and M in the denominator is the reading taken with the normal polarity used during measurements.

In the two voltage techniques two ionization chamber readings are taken in the same irradiation conditions, one at the normal collecting voltage (V_1, reading M_1) and one at a lower voltage (V_2, reading M_2). The ratio V_1/V_2 should have a value of 2 or 3. p_{ion}, the recombination correction factor to be applied at the normal collecting voltage, can be obtained from the solutions of the particular expressions for pulsed and pulsed scanned beams. To simplify this, Weinhous and Meli [10] have given quadratic fits to these solutions:

$$p_{ion} = a_0 + a_1(M_1/M_2) + a_2(M_1/M_2)^2 \qquad (2.49)$$

For small corrections ($(p_{ion} - 1) < 0.05$) the theoretical expressions reduce in a first approximation to

$$p_{ion} - 1 = (M_1/M_2 - 1)/(V_1/V_2 - 1) \qquad (2.50)$$

i.e., the percentage correction is simply the percentage change in reading divided by a number equal to one less than the voltage ratio. If the voltage ratio selected, V_1/V_2, equals 2, the percentage recombination correction is equal to the percentage change in reading. Such an approach yields values of p_{ion} accurate to 0.1% for corrections up to 3% and to 0.5% for corrections up to 5%.

For parallel-plate chambers, Boag's analytical formulas may be applied. For pulsed beams p_{ion} can be written as

$$p_{ion} = u/\ln(1 + u) \qquad (2.51)$$

where

$$u = \mu m d^2/V$$

If m is the dose per pulse in centigrays, d is the plate separation in mm, and V is the collecting voltage in volts, then μ is a constant equal to 10.7 V mm^{-2} cGy^{-1}. For cylindrical chambers the effective electrode spacing can be calculated from expressions given by Boag, and for a Farmer-designed chamber, such as the NE2571, the value obtained is 3.36 mm. Thus, the percentage loss of signal due to recombination in a Farmer 2571 chamber is calculated to be approximately 2.4% in a pulsed beam having 0.1 cGy/pulse and using a collecting voltage of around 250 V.

III. PHOTONS DOSIMETRY

The term "medium-energy x-rays" refers to x-rays of half-value layers in the range 0.5–4 mm Cu or equivalently above 8 mm Al covering approximately those generated at tube voltages in the range 160–300 kV. The recommended reference depth in this energy range is 2 cm in a full scatter water phantom. The equation for the calculation of absorbed dose at this reference depth for a given field size is

$$D_{w,z=2} = M N_K k_{ch} \left[\left(\frac{\bar{\mu}_{en}}{\rho} \right)_{w/air} \right]_{z=2,\,\Phi} \qquad (2.52)$$

where $D_{w,z=2}$ is the dose to water in grays at the position of the chamber center at a depth $z = 2$ cm when the chamber is replaced by water. M is the instrument reading obtained with a chamber corrected to standard pressure and temperature (20°C, 1013.25 mbar, and 50% relative humidity); N_K is the chamber calibration factor in grays per scale reading to convert the instrument reading at the beam quality (HVL) concerned to air kerma free in air at the reference point of the chamber with the chamber assembly replaced by air; $[(\mu_{en}/\rho)_{w/air}]_{z=2,\,\Phi}$ is the mass energy absorption coefficient ratio, water to air, averaged over the photon spectrum at a water depth of 2 cm and

field diameter Φ; and k_{ch} is a factor which accounts for the change in the response of the ionization chamber between calibration in air and measurement in a phantom.

The recommended procedure for the intercomparison (in water) of a field instrument with a secondary-standard dosimeter (i.e., the determination of the product $[N_K k_{ch}]_f$ for the field instrument) is given below. The intercomparison should be carried out using the same x-ray facilities and radiation qualities as will subsequently be measured by the field instrument.

Measure the first half-value layer (HVL$_1$) for each radiation quality and determine the secondary-standard calibration factor N_K by interpolation from those given by the standards laboratory. Intercompare the secondary-standard dosimeter and the field instrument by simultaneous irradiation in a water phantom, following the procedures specified below. A "solid water" phantom may be used, provided that a previous, independent intercomparison in this quality range has been made in both the "solid" and natural water phantoms and any necessary corrections applied. A Perspex phantom is not recommended. The phantom should extend outside the beam edges and at least 10 cm beyond the chamber center along the beam axis. The front surface of the phantom should be at the standard source-surface distance (SSD) for the applicator. The field size for the comparison should be 10 cm × 10 cm (at the front surface of the phantom) and the reference points of the two chambers should be at the same depth of 2 cm in water. An appropriate separation of the chamber centers is 3 cm, with each chamber equidistant from the beam axis. If the intercomparison is in water, waterproof sheaths should be used.

The dose in water (indicated here by w) is determined at a depth z of 2 cm. This was chosen because this depth is more relevant in the clinic than that of 5 cm and also because of the rapid reduction of the dose with depth in the kV region. The dose $D_{w,z=2}$ is given by Equation (2.52). The overall correction factor k_{ch} which takes account of various effects, including the modification of the spectrum (primary) by 2 cm of water, the attenuation and scattering by the material (water in this case) in the hole. Figure 2.13 compares the values recommended in the IPEMB CoP with both the original IAEA [64] and the revised IAEA (1993) figures; the dashed line serves to emphasize that this factor was implicitly assumed to be unity in ICRU. [11]

For the determination of the absorbed dose to any medium *med* on the surface of a phantom made of that medium, the following relation [12],

$$D_{med,\,z=0} = M N_K B_{med} \left[\left(\frac{\bar{\mu}_{en}}{\rho} \right)_{med,\,air} \right]_{air} \qquad (2.53)$$

is used, where M is the in-air chamber reading corrected for temperature and pressure, N_K is the air-kerma calibration

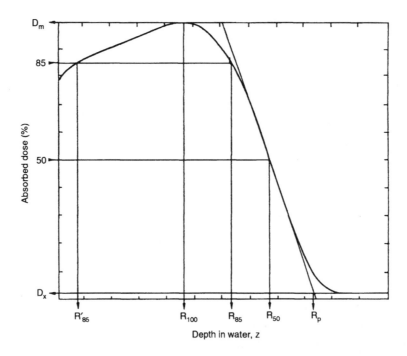

FIGURE 2.12 An electron beam absorbed dose distribution in a water phantom: D_m is the maximum absorbed dose; D_x I is the absorbed dose due to *bremsstrahlung*; R_{100} is the depth of dose maximum; R_{85} is the therapeutic range (it is here assumed that $R_t = R_{85}$; the depth at which the therapeutic interval intersects the depth dose curve near the skin entrance is designated by R'_{85}); R_{50} is the depth for 50% absorbed dose; and R_p is the practical range.

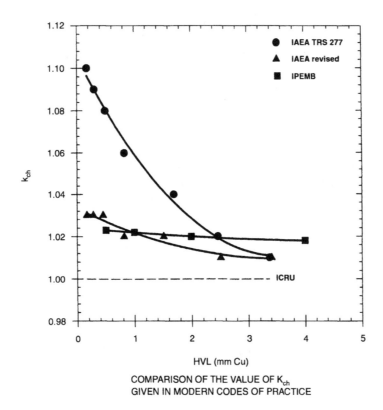

FIGURE 2.13 Comparison of the value of K_{ch} given in modern codes of practice (From Reference [11]. With permission.)

factor, B_{med} is the backscatter factor defined as the ratio of the collision kerma to the phantom material at a point on the beam axis at the surface of a full scatter phantom, to the collision kerma to the same material at the same point in the primary beam with no phantom present, and $[(\bar{\mu}_{en}/\rho)_{med, air}]_{air}$ is the ratio of the mass energy- absorption coefficients for the phantom material to air averaged over the primary photon spectrum. An advantage of the above formalism [Equation (2.53)] over the previous formalism is that by changing the definition of the backscatter factor from air-kerma ratio to water-kerma ratio, the mass energy–absorption coefficient ratio, water to air, becomes independent of phantom scatter. Thus, it is field size–independent.

Equation (2.53) determines kerma to the medium at $z = 0$ rather than absorbed dose to the medium, as the latter is highly affected by electron contamination and electron build-up. In general, charged particle equilibrium (CPE) does not exist on the surface. The dose to the medium at a depth where CPE has just been established is a more meaningful quantity for clinical radiotherapy and its value is practically the same as the surface kerma.

To determine the absorbed dose to any medium, *med*, at a reference depth (e.g., 2 cm) in that medium using an ionization chamber, the relation

$$D_{med, z=2} = MN_K P_{Q, cham} P_s \left[\left(\frac{\bar{\mu}_{en}}{\rho} \right)_{med, air} \right]_{z=2} \quad (2.54)$$

can be used where M is the in-phantom chamber reading corrected for temperature and pressure, N_K the air kerma calibration factor, $P_{Q, cham}$ the overall correction factor which accounts for the effects of the change in beam quality between chamber calibration and measurement and the perturba-tion due to the introduction of the chamber in the medium, P_s the water proofing sleeve correction factor, and $[(\bar{\mu}_{en}/\rho)_{med, air}]_{z=2}$ the mass energy–absorption coefficients for the medium to air averaged over the spectrum present in undisturbed medium at the depth of the chamber center.

The percentage depth dose (*PDD*) is defined as

$$PDD(z) = \frac{D_{med, z}}{D_{med}(max)} \times 100\% \quad (2.55)$$

where $D_{med}(max)$ is the maximum dose in the phantom, which is usually taken as the dose on the surface $D_{med, z=0}$ for kilovoltage x-ray beams. For convenience, 100% is ignored in the following derivations. Therefore,

$$PDD(z) = D_{med, z}/D_{med, z=0} \quad (2.56)$$

Substituting Equation (2.54) into Equation (2.56),

$$PDD(z) = \frac{M_z [P_{Q, cham}]_z \left[\left(\frac{\bar{\mu}_{en}}{\rho} \right)_{w, air} \right]_z}{M_{z=0} [P_{Q, cham}]_{z=0} \left[\left(\frac{\bar{\mu}_{en}}{\rho} \right)_{w, air} \right]_{z=0}}$$

$$= \frac{M_z}{M_{z=0}} C_z \quad (2.57)$$

where

$$C_z = \frac{[P_{Q, cham}]_z \left[\left(\frac{\bar{\mu}_{en}}{\rho} \right)_{w, air} \right]_z}{[P_{Q, cham}]_{z=0} \left[\left(\frac{\bar{\mu}_{en}}{\rho} \right)_{w, air} \right]_{z=0}} \quad (2.58)$$

is the *PDD* correction factor for an ion chamber in water. As P_s is almost independent of phantom depth, it has been omitted in Equation (2.58). C_z is detector-dependent and may deviate from unity significantly. So far there has been no study on C_z in the literature. [12]

One can then relate the dose anywhere in the phantom to the dose on the surface or at the reference depth using the PDD data; i.e.,

$$D_{med, z} = D_{med, z=0} PDD(z) \quad (2.59)$$

based on the "in-air" method or

$$D_{med, z} = D_{med, z=ref} PDD(z)/PDD(z = ref) \quad (2.60)$$

based on the "in-phantom" method.
To study the consistency of the two methods, Ma et al. [12] calculated the ratio R of the doses at the same depth determined by Equations (2.59) and (2.60):

$$R = \frac{D_{med, z=0}}{D_{med, z=2}} PDD(z = ref) \quad (2.61)$$

If the ratio R is unity, the "in-air" and "in-phantom" procedures are consistent with each other. It is interesting to see that no matter at what depth we compare the dose values, the consistency remains the same and is determined by the accuracy of the dose at the reference points ($z = 0$ and $z = ref$) and of the *PDD* at the ref depth. If both the dose on the surface and at the ref depth are accurately known, Equation (2.61) becomes truly a test for the *PDD* data. By substituting Equations (2.53), (2.54), and (2.57) into Equation (2.61), we have

$$R = \frac{M_{air}B_{med}\left[\left(\frac{\bar{\mu}_{en}}{\rho}\right)_{med, air}\right]_{air}}{M_{z=ref}P_{Q, cham}\left[\left(\frac{\bar{\mu}_{en}}{\rho}\right)_{med, air}\right]_{z=ref}} PDD(z=ref)$$

(2.62)

The parallelism between existing air kerma formalisms based on the "absorbed dose to air" chamber factor, N_D (or N_{gas}), and the simpler approach based on calibrations in terms of absorbed dose to water, N_w, was described by Andreo.[13] The importance of avoiding steps performed by the user that introduce avoidable uncertainties in the dosimetric procedure was emphasized.

A simple relation can be given as the basis of the formalism, based on the calibration of an ionization chamber at Standard Dosimetry Laboratories in terms of absorbed dose to water, D_w, at a certain radiation quality, Q_0 (usually ^{60}Co gamma-rays):

$$D_{w, Q_0} = M_{Q_0}N_{w, Q_0}$$

(2.63)

This expression is based directly on the definition of the calibration factor, N_w. The main advantage of Equation (2.63) is that N_w is obtained with better accuracy than if it is derived through successive steps from the air kerma formalism. It has the simplicity that the absorbed dose to water at another photon beam with quality Q will be obtained with a relation of the type

$$D_{w, Q_0} = M_Q N_{w, Q_0} k_Q$$

(2.64)

where M_Q is the electrometer reading in the user's photon beam (corrected for influence quantities of a different nature, such as temperature, pressure, humidity, etc.) with the ionization chamber positioned in a water phantom, and N_w is the absorbed dose to water calibration factor supplied by the Standard Dosimetry Laboratory for the reference quality Q_0. The factor k_Q, corrects the calibration factor for differences in beam quality between the laboratory and the user. Other advantages of the method are the use of only one quantity (absorbed dose to water) along the dosimetry chain and very similar experimental conditions both at calibration and during user's measurements.

Two steps are generally required to obtain D_w from measurements in a water phantom with a chamber having a calibration factor N_K. In the first step, the (free in air) air kerma, k_{air, Q_0} is related to the mean absorbed dose D_{air, Q_0} inside the air cavity of the user's ionization chamber at the calibration quality Q_0 using

$$\bar{D}_{air, Q_0} = k_{air, Q_0}(1 - g)k_{att}k_m k_{cel}$$

(2.65)

where k_{att} is a factor that takes into account the attenuation (absorption and scattering) of ^{60}Co-gamma-rays in the chamber material during the calibration in a ^{60}Co beam; k_m takes into account the deviations from air equivalence of the chamber wall and build-up cap material; k_{cel} accounts for similar deviations in the material of the central electrode; and g is the fraction of the energy of the photon-produced charged particles expended in radiative processes (mainly *bremsstrahlung* in air). Specifically, k_m takes into account the lack of equivalence to air of the materials in the ionization chamber and therefore should include k_{cel}. This would allow unambiguous comparisons between experimental and calculational procedures.

An "absorbed dose to air" chamber factor, N_D, is defined as

$$\bar{D}_{air, Q_0} = N_{D, Q_0}M_{Q_0}$$

(2.66)

where M_{Q_0} is the meter reading of the ionization chamber system (corrected for influence quantities) in the calibration beam.

The equations above yield

$$N_{D, Q_0} = N_{K, Q_0}(1 - g)k_{att}k_m k_{cel}$$

(2.67)

The basic assumption in the "absorbed dose to air" chamber factor formalism is that N_{D, Q_0} is also valid at the user's beam quality Q; i.e., $N_{D, Q_0} = N_{D, Q}$. The mean absorbed dose inside the air cavity, $\bar{D}_{air, Q_0}$, can be written as

$$\bar{D}_{air, Q_0} = \frac{M_Q}{V \rho_{air}} \frac{W_Q}{e}$$

(2.68)

where M_Q is the electric charge produced in the ionization chamber, V is the cavity volume and W/e is the mean energy expended in air per ion formed divided by the charge of the electron. The above assumption holds only if $W_{Q_0} = W_Q$ and implies that N_D is a function only of the mass of air inside the cavity ($V \rho_{air}$) and therefore is a constant of the chamber.

In the second step, the Bragg-Gray equation is used to determine the absorbed dose to water at the effective point of measurement, P_{eff}, from the mean absorbed dose to air $\bar{D}_{air, Q}$. Alternatively, a displacement factor P_{displ} can be used and D_w is determined at the position of the geometrical center of the chamber using

$$D_{w, Q} = \bar{D}_{air, Q_0}(s_{w, air})_Q p_Q = N_D M_Q(s_{w, air})_Q p_Q$$

(2.69)

where $(s_{w, air})_Q$ is the water to air stopping-power ratio and p_Q is a perturbation factor to correct for all the different contributions to the departure from ideal Bragg-Gray conditions, when an extended non-water-equivalent walled detector is introduced in the medium, all of them given at the user's beam quality Q.

When an ionization chamber has independent calibration factors both in terms of air kerma N_K (or exposure N_x) and in terms of absorbed dose to water N_w at the same reference beam quality Q_0, it is possible to obtain a relationship between both calibration factors. Equating the absorbed dose to water at a quality Q reference point, obtained using the "absorbed dose to air" chamber factor formalism above, with that used in Equation (2.64) results in

$$N_{w,Q_0} = N_{k,Q_0}(1-g)k_{att}k_m k_{cel}(s_{w,air})_{Q_0}P_{Q_0} \qquad (2.70)$$

and the factor k_Q in Equation (2.64) becomes

$$k_Q = \frac{(S_{w,air})_Q}{(S_{w,air})_{Q_0}}\frac{p_Q}{p_{Q_0}}\frac{W_Q}{W_{Q_0}} \qquad (2.71)$$

which explicitly includes the quality dependence of W; the meanings of the factors $S_{w,air} P_Q$ and W have been given above. [13]

A quality index can be used to define the absorbed dose to water if its energy dependence is known.

Three parameters, HVL, mean energy at a depth of 2 cm in water, and the ratio of the doses at depths of 2 and 5 cm in water (D_2/D_5) were investigated by Rosser [14] as functions of $[(\bar{\mu}_{en}/\rho)_{w/a}]_{z=2,\phi}$. The quality index that best defines $[(\bar{\mu}_{en}/\rho)_{w/a}]_{z=2,\phi}$ is the ratio of doses at depths of 2 and 5 cm in water.

$[(\bar{\mu}_{en}/\rho)_{w/a}]_{z=2,\phi}$ varies by 5.8% between 0.47 and 4 mm Cu HVL. Therefore, the energy-dependent factor of the absorbed dose to water will be taken as being the same as that for $[(\bar{\mu}_{en}/\rho)_{w/a}]_{z=2,\phi}$.

There are two practical problems associated with this quality index. First, it is not easy to measure the absorbed dose to water at a depth in water. The ratio of doses at water depths of 2 and 5 cm (D_2/D_5) is given by:

$$\frac{D_2}{D_5} = \frac{(MN_K k_{ch}[(\bar{\mu}_{en}/\rho)_{w/a}])_2}{(MN_K k_{ch}[(\bar{\mu}_{en}/\rho)_{w/a}])_5} \qquad (2.72)$$

For general dosimetry, k_{depth} is negligible. Therefore, Equation (2.72) reduces to:

$$\frac{D_2}{D_5} = \frac{M_2}{M_5}k_{depth} \qquad (2.73)$$

thus showing that D_2/D_5 may be a practical quality index.

The absorbed-dose calibration service from National Physical Laboratory (NPL) is based on a primary-standard calorimeter that measures absorbed dose to graphite. Secondary-standard dosimeters are calibrated in absorbed dose to water in a ^{60}Co gamma-ray beam and in x-ray beams over a range of generating potentials from 4 MV

to 19 MV. Two methods were used by Burns [15] to convert the calibrations of working-standard ionization chambers from absorbed dose to graphite into absorbed dose to water. One method involved the use of published interaction data for photons and secondary electrons and required a knowledge of the chamber construction. The second method involved the calculation of the ratio of absorbed dose in graphite and water phantoms irradiated consecutively in the same photon beam using the photon-fluence scaling theorem. The two methods were in agreement to 0.1%.

There are three stages to the calibration procedure of NPL:

1. Working-standard ionization chambers in a large graphite phantom are calibrated in absorbed dose to graphite, N_c, by comparison with the primary-standard graphite calorimeter.
2. The calibration factors of the working-standard chambers in absorbed dose to graphite, N_c, are converted into calibration factors in absorbed dose to water, N_w, with the chambers in water, using conversion factors derived from a program of work.
3. Secondary-standard chambers are calibrated in absorbed dose to water by comparison with the working-standard chambers in a water phantom.

A group of three NE2561 ionization chambers is kept at NPL and these chambers are used as working standards. These are recalibrated once a year in a graphite phantom by direct comparison with the calorimeter over the whole range of energies. NE2561 chambers were chosen as working standards because of their known long-term stability.

The calibration of the working standards in comparison with the calorimeter is in terms of absorbed dose to graphite, whereas the quantity required for the calibration of secondary-standard dosimeters for use in radiotherapy is absorbed dose to water. The subscript following a symbol indicates the particular medium to which the physical quantity represented by the symbol applies: c = graphite, w = water, and P = Perspex. The formula used to calculate the ratio of calibration factors N_w/N_c for a working-standard ionization chamber exposed in water and graphite to a beam of high-energy photons is given by Equation (2.74). The point of measurement is taken as the center of the chamber.

$$\frac{N_w}{N_c} = \frac{k_w}{k_c}\frac{p_w}{p_c}\frac{s_w}{s_c}\frac{(\bar{\mu}_{en}/\rho)_w}{(\bar{\mu}_{en}/\rho)_c}\frac{\beta_w}{\beta_c} \qquad (2.74)$$

where k is the photon energy-fluence rate at the center of the chamber collecting volume divided by the ionization current, p corrects for the replacement of the medium by the chamber cavity and wall, s corrects for the replacement of the medium by the chamber stem, $\bar{\mu}_{en}/\rho$ is the mean mass energy-absorption coefficient for the photons at the point of measurement, and β is the absorbed dose at the point of measurement divided by the collision kerma at the same point.

For a graphite-walled chamber which is fitted with a Perspex waterproof sheath when the chamber is in water, k_w/k_c is given by

$$\frac{k_w}{k_c} = \left(\alpha_{wall} + \alpha_{sheath} \frac{(\mu_{en}/\rho)_p (L/\rho)_c}{(\mu_{en}/\rho)_c (L/\rho)_p} \right.$$
$$\left. + \alpha_{wall} \frac{(\bar{\mu}_{en}/\rho)_w (L/\rho)_c}{(\mu_{en}/\rho)_c (L/\rho)_w} \right)^{-1} \qquad (2.75)$$

where α is the fraction of ionization in the cavity arising from photon interactions in the component indicated by the subscript and L/ρ is the mean restricted mass collision stopping power for secondary electrons.

The absorbed dose calibration factor is defined by $N_m = D_m/Q_m$, where D_m is the absorbed dose to the medium at a point corresponding to the center of the chamber with the chamber replaced by the medium, and Q_m is the charge produced by the ionization in the cavity of the chamber. Hence, the ratio of calibration factors of the chamber in water and graphite is given by

$$\frac{N_w}{N_c} = \frac{D_w/Q_w}{D_c/Q_c} = \frac{D_w}{D_c} \frac{Q_c}{Q_w} \qquad (2.76)$$

To show how the photon-fluence scaling theorem is applied, it is necessary to first convert the ratio of absorbed doses into a ratio of photon energy fluences. The relationship between absorbed dose D_m and photon energy fluence Ψ_m at a point in a medium m irradiated by a photon beam under conditions of transient electron equilibrium is

$$D_m = \Psi_m (\bar{\mu}_{en}/\rho)_m \beta_m \qquad (2.77)$$

Hence, the ratio of absorbed dose at given points in the two media is

$$\frac{D_w}{D_c} = \frac{\Psi_w}{\Psi_c} \frac{(\bar{\mu}_{en}/\rho)_w}{(\bar{\mu}_{en}/\rho)_c} \frac{\beta_w}{\beta_c} \qquad (2.78)$$

Thus, from Equation (2.76),

$$\frac{N_w}{N_c} = \frac{\Psi_w}{\Psi_c} \frac{Q_c}{Q_w} \frac{(\bar{\mu}_{en}/\rho)_w}{(\bar{\mu}_{en}/\rho)_c} \frac{\beta_w}{\beta_c} \qquad (2.79)$$

Values of N_w/N_c for the working-standard chambers derived from the two conversion methods are in excellent agreement with each other over the range of beam energies covered by the NPL calibration service, as shown in Figure 2.14.

A dose calculation algorithm has been developed by Chui et al. [16] for photon beams with intensity modulation generated by dynamic jaw or multileaf collimations. The correction factors are used to account for the effect of the intensity distribution of a beam and are calculated, at

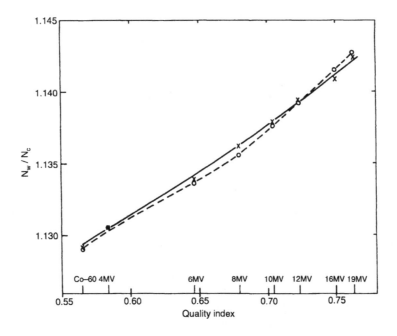

FIGURE 2.14 Comparison of N_w/N_c for NE2561 chamber derived by both methods. Crosses and continuous line are derived from dose-ratio method; circles and dashed line are derived from cavity-ionization theory method. (From Reference [15]. With permission.)

each point, as a ratio of the dose resulting from an "idealized" dynamic-collimated field to that from the corresponding "idealized" static-collimated field. The dose due to an idealized dynamic-collimated field is computed as

$$D'_{dynamic}(x, y, d)$$
$$= \int_w \int_h \phi(x', y')k(x - x', y - y', d)dx'dy' \quad (2.80)$$

where $\phi(x, y)$ is the fluence distribution of the dynamic collimated and $k(x, y, d)$ is the pencil-beam kernel at depth d in the medium.

Similarly, doses due to a corresponding idealized static-collimated field are calculated as

$$D'_{static}(x, y, d)$$
$$= \int_w \int_h U(x', y')k(x - x', y - y', d)\, dx'dy' \quad (2.81)$$

where $U(x, y)$ is a step function describing the uniform fluence distribution of a static-collimated field and the limits of integration are the same as that for the dynamic-collimated field in Equation (2.80).

The dose from a clinical, dynamic-collimated field, $D_{dynamic}$, is finally calculated by applying the correction factors to doses calculated for a clinical static-collimated filed D_{static}:

$$D_{dynamic}(x, y, d) = D_{static}(x, y, d)\left[\frac{D'_{dynamic}(x, y, d)}{D'_{static}(x, y, d)}\right]$$
$$(2.82)$$

The dose, D_{static}, due to a clinical, static-collimated field may be calculated by a number of calculation models based primarily on conventionally acquired beam data. For example, it can be calculated as

$$D_{static}(x, y, d) = MU \times OF_{med}(w \times h) \cdot TMR(w \times h; d)$$
$$\times OCR(w \times h; x, y, d) \quad (2.83)$$

The term MU is the number of monitor units, or beam-on time. The term $OF_{med}(w \times h)$ represents the output factor at a depth, typically d_{max}, in the medium for a field of size $w \times h$, where the field dimensions w and h are the same as the integration limits used in the convolutions described above. The *TMR* term is the tissue maximum ratio measured along the central axis of the beam, and d is the equivalent depth which accounts for patient inhomogeneity correction. The *OCR* term accounts for off-axis corrections which includes effects such as "horns," variation of beam quality with off-axis distance, and penumbra region near the edge of the field.

The relationship between $\%dd(10)_x$ and the Spencer-Attix water to air restricted stopping power ratio $(\bar{L}/\rho)_{air}^{water}$ is [17]:

$$(\bar{L}/\rho)_{air}^{water} = 1.2676 - 0.002224(\%dd(10)) \quad (2.84)$$

Yang et al. [18] have shown that it is inappropriate just to use a simple $1/r^2$ correction to convert depth-dose curves calculated for parallel beams into those for a fixed SSD. It is well known that there are scatter corrections needed as well as the $1/r^2$ corrections. Yang et al.'s data suggest changes in the value of $\%dd(10)_x$ of up to nearly 2% at low energies but no significant change for beams with $\%dd(10)_x$, greater than 80%.

Figure 2.15 presents the revised data along with the revised fit to the data for the *bremsstrahlung* beams. The revised relationship is:

$$(\bar{L}/\rho)_{air}^{water} = 1.275 - 0.00231(\%dd(10)_x) \quad (2.85)$$

A convolution/superposition based-method was developed by Liu et al. [19] to calculate dose distributions and wedge factors in photon treatment fields generated by dynamic wedges.

A dose distribution $D(x, y, z)$ as a convolution or superposition between terma $T(x, y, z)$ and photon dose kernel $K(x, y, z)$ was calculated as follows:

$$D(x, y, z) = \iiint T(x', y', z')$$
$$\times K(x - x', y - y', z - z') \times dx'dy'dz'$$
$$(2.86)$$

Terma is a product of photon energy fluence and the mass attenuation coefficient, which represents the total energy released per unit mass by photons at their initial interactions. Dose kernel describes the distribution of relative energy deposition per unit volume around the photon initial interaction site. Thus, a convolution or superposition between the two functions in Equation (2.86) yields a three-dimensional dose distribution in the irradiated volume.

The total terma was calculated from the contribution of both photon sources; i.e.,

$$T(x, y, z) = T_p(x, y, z) + T_e(x, y, z) \quad (2.87)$$

in which $T_p(x, y, z)$ and $T_e(x, y, z)$ are terma from the primary photon source and the extra-focal photon source, respectively. Both $T_p(x, y, z)$ and $T_e(x, y, z)$ were calculated as

$$T_{p,e}(x, y, z) = \sum_E \phi_{p,e}(E, x, y, z) \times E \times [\mu(E)/\rho]$$
$$(2.88)$$

in which $\phi_{p,e}(E, x, y, z)$ is the fluence distribution of the primary or the extra-focal sources; $\mu(E)/\rho$ is the mass

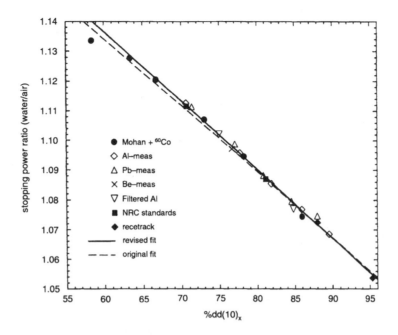

FIGURE 2.15 Calculated Spencer-Attix water to air stopping-power ratios (based on ICRU 37 stopping powers) vs. $\%dd(10)_x$. The solid straight line is the linear fit [Equation (2.85)] to the revised data for all the *bremsstrahlung* beams, whereas the dashed line is the fit to the original data. (From Reference [17]. With permission.)

attenuation coefficient of photons at energy E. The $\phi_{p,e}(E, x, y, z)$ was obtained from the results of the Monte Carlo simulation.

The total photon fluence in the dynamic field after the entire treatment being delivered can be calculated as

$$\phi(x, y) = \int s(t) \times \phi_s(x, y, t)\, dt \qquad (2.89)$$

in which $s(t)$ describes the machine output per unit time at time t; $\phi_s(x, y, t)$ is the photon fluence per unit machine output in the component static field at position (x, y) and time t.

Techniques for reducing computation time in 3D photon dose calculations were addressed by Aspradakis and Redpath [20] with specific emphasis given to the convolution/superposition approach. A single polyenergetic superposition model calculating absorbed dose per incident photon fluence (Gy cm^2) was developed in terms of terma and a total energy deposition kernel (a total point spread function). A novel approach was devised for reducing calculation time. The method, named the CF method, was based on the use of a conventional, fast model for the generation of 3D dose distributions on a fine dose matrix. Superposition calculations were carried out on a coarse matrix and calculation speed was increased simply by reducing the number of calculations. A set of correction factors was derived on the coarse grid from the ratio of the dose values from superposition to

those from the conventional algorithm. These were interpolated onto the fine matrix and used to modify the dose calculation from the conventional algorithm. The method was tested in a worst-case example where large dose gradients were present and in a clinically relevant irradiation geometry. It is shown that the time required for the generation of a 3D matrix with superposition can be reduced by at least a factor of 100 with no significant loss in accuracy.

The distribution of terma was calculated analytically (in voxels of 0.5 cm in all dimensions) as a product of the mass attenuation coefficient and the energy fluence at each depth. The kernels represent the energy deposited in a large water medium from a monoenergetic pencil beam of photons that experience their first interaction at the center of a large water medium. They were stored on a spherical coordinate system and their use from the dose deposition point of view was based on the reciprocity exhibited between photon interaction and dose deposition sites. Dose at $\bar{r}$ in water calculated with the superposition method is expressed by

$$D(\bar{r}) = \int_E \iiint_{volume} T_E(\bar{s}) h_{\rho_{water}}(E, \bar{r} - \bar{s})\, d^3 s\, dE \qquad (2.90)$$

where T_E is the terma differential in energy and h is the (monoenergetic) energy deposition kernel generated in water.

The code of practice for the determination of absorbed dose for x-rays below 300 kV [21] has been approved by the IPEMB and introduces the following changes to the previous codes:

1. The determination of absorbed dose is based on the air kerma determination (exposure measurement) method.
2. An air kerma calibration factor for the ionization chamber is used.
3. The use of the F (rad/roentgen) conversion factor is abandoned and replaced by the ratio of the mass-energy absorption coefficients of water and air for converting absorbed dose to air to absorbed dose to water. Perturbation and other correction factors are incorporated in the equations.
4. New backscatter factors are recommended.
5. Three separate energy ranges are defined, with specific procedures for each range. These ranges are:

   (a) 0.5 to 4 mm Cu HVL: for this range calibration at 2 cm depth in water with a thimble ion chamber is recommended;
   (b) 1.0 to 8.0 mm Al HVL: for this range calibration in air with a cylindrical ion chamber and the use of tabulated values of the backscatter factor are recommended;
   (c) 0.035 to 1.0 mm Al HVL: for this range calibration on the surface of a phantom with a parallel-plate ionization chamber is recommended. [21]

The reading of each instrument should be corrected to a chamber temperature of 20°C and an ambient air pressure of 1013.25 mbar. Ion recombination effects are generally small for continuous radiation. If the secondary-standard dosimeter has been calibrated at the United Kingdom National Physical Laboratory (NPL), this correction is included in the calibration factor provided. A series of at least three exposures should be given to the two chambers (series a) and the ratio M_s/M_f of the readings of the two chambers should be calculated for each exposure, where M_s and M_f are the instrument readings of the secondary standard and field instrument, respectively. The chambers should then be interchanged and the readings repeated (series b). The whole procedure should be repeated at least once so that each series contains a total of at least six exposures. The standard error of the mean of either series of ratios should not exceed 0.5%. The standard error of the mean of all the ratios calculated (series a and b together) should normally be less than 1%. The true ratio of the instrument readings

is given by

$$[M_s/M_f]_t = \sqrt{[M_s/M_f]_a [M_s/M_f]_b} \qquad (2.91)$$

where $(M_s/M_f)_a$ is the mean of the ratios M_s/M_f for series a, $(M_s/M_f)_b$ is the corresponding mean ratio for series b, and $(M_s/M_f)_t$ is the true ratio of the instruments' readings. By equating the absorbed doses according to equation (2.52), the field instrument calibration factor product $(N_K k_{ch})_f$ is given by

$$(N_k k_{ch})_f = N_{K,s} k_{ch,s} (M_s/M_f)_t \qquad (2.92)$$

The field instrument calibration factor $N_{k,f}$ is given by

$$N_{k,f} = N_{K,s} (M_s/M_f)_t \qquad (2.93)$$

The term "very low-energy x-rays" refers to x-rays of half-value layers in the range 0.035–1.0 mm Al, covering approximately those generated at tube voltages in the range 8–50 kV. The values of k_{ch} are not known for these energies.

In the case of low- and medium-energy x-radiation, the determination of the absorbed dose is based on treating the measuring instrument as an exposure meter by making use of the following general expression [21]:

$$D_w = X(W/e)(\bar{\mu}_{en}/\rho)_{w/air} \qquad (2.94)$$

where D_w is the absorbed dose to water, X is the exposure at the point at which the dose is required, W/e is the quotient of the average energy expended to produce an ion pair in air by the electronic charge, and $(\bar{\mu}_{en}/\rho)_{w/air}$ is the ratio of the mass energy absorption coefficients of water to air averaged over the photon spectrum at the point of measurement. Air kerma has replaced exposure as the preferred quantity. The air kerma k_{air} is related to the exposure X by

$$K_{air} = X(W/e)(1/(1 - g)) \qquad (2.95)$$

where g is the fraction of energy of secondary charged particles lost to *bremsstrahlung* in air. It is assumed that $g = 0$; therefore, $\mu_{tr} = \mu_{en}$, and $K = K_{col}$, and $D_w = K_{col,w}$. Equation (2.94), expressed in terms of air kerma, becomes

$$D_w = K_{air}(\bar{\mu}_{en}/\rho)_{w/air} \qquad (2.96)$$

$$K_{air} = MN_K \qquad (2.97)$$

where N_K is the air-kerma calibration factor for the chamber, free in air at the particular HVL in question. Its purpose is to convert the reading to air kerma at the point corresponding to the center of the chamber in the absence of the chamber, under standard ambient conditions; M is the chamber reading in the user's radiation beam.

In medium-energy x-rays

$$K_{air,hole} = MN_K k_Q \qquad (2.98)$$

and, with the perturbation factor P_{dis},

$$K_{air,z} = MN_k k_Q P_{dis} \qquad (2.99)$$

where $K_{air,z}$ is the air kerma at the position of the chamber center (at a depth of 2 cm) in the undisturbed medium.

The air kerma free in air, $K_{air,air}$, is converted into water kerma free in air, $K_{w,air}$, by multiplying by the water-to-air mass-energy absorption coefficient ratio for the HVL in question (note that this quantity pertains to the primary beam only and is therefore independent of field size):

$$K_{w,air} = K_{air,air}[(\bar{\mu}_{en}/\rho)_{w/air}]_{air} \qquad (2.100)$$

where $(\bar{\mu}_{en}/\rho)_{w/air}$ is averaged over the photon energy spectrum in air.

The dependence of $(\bar{\mu}_{en}/\rho)_{w/air}$ on HVL is shown in Figure 2.16.

The recommended secondary-standard dosimeter for medium-energy x-ray qualities is either an NE2561, NE2611, or NE2571 ionization chamber connected to an electrometer of secondary-standard quality [21].

The recommended field instrument is any thimble ionization chamber with a volume of less than 1.0 cm³ for which $[N_K k_{ch}]$ varies smoothly and by less than 5% over the energy range of interest, connected to any suitable electrometer. No materials with an atomic number greater than that of aluminum may be used in the vicinity of the cavity, and the chamber must be vented to the atmosphere.

The preferred chambers are the NE2505/3A, NE2571, NE2581, PTW30002, or any chambers of the "Farmer" type constructed with either a carbon wall and aluminum central electrode or with both electrodes of conducting plastic that is approximately air-equivalent. Chambers with graphite-coated nylon or PMMA walls may be vulnerable to sudden changes of response at low energies and should be used only with appropriate precautions.

The recommended secondary-standard dosimeters for low-energy x-ray qualities are NE2561, NE2611, and NE2571 ionization chambers connected to an electrometer of secondary-standard quality. The dosimeter should be calibrated in terms of air kerma, at appropriate radiation qualities, at a standards laboratory. The ionization chamber and electrometer should preferably be calibrated together.

The recommended secondary-standard dosimeter for very low-energy x-rays qualities is a PTW23342 (0.02 cm³) or PTW23344 (0.2 cm³) soft-x-ray ionization chamber

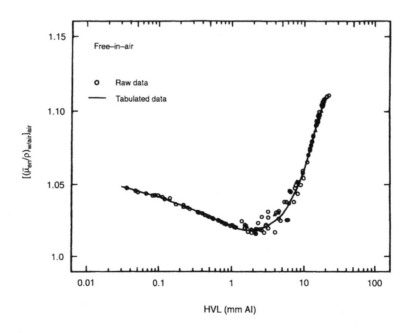

FIGURE 2.16 Mass energy absorption coefficient ratio, water to air, as a function of HVL, corresponding to the in-air (primary) spectra. (From Reference [21]. With permission.)

connected to an electrometer of secondary-standard quality.

The recommended field instruments are parallel-plate ionization chambers with volumes in the range 0.02–0.8 cm³.

A COMPARISON OF THE IAEA 1987 AND AAPM 1983 PROTOCOL

The IAEA 1987 protocol is an international protocol which has made a number of improvements over the AAPM 1983 protocol for calibration of high-energy photon and electron beams. [22] For photons, the IAEA protocol gives results which are in good agreement with the AAPM protocol; on average, the IAEA results, are 0.6% smaller than the AAPM results, while discrepancies between the two are in the range of -0.4% to -1.2%. For 10-MeV electrons also, the IAEA protocol gives results which are in excellent agreement with the AAPM protocol; on average, the IAEA results, are 0.3% smaller than the AAPM results, while discrepancies between the two are in the range of -1.0% to $+0.5\%$.

Following the formalism developed by Loevinger, the AAPM protocol introduced the concept of the cavity gas calibration factor N_{gas} for ionization chambers, as dose to the gas in the chamber divided by the electrometer reading corrected for temperature, pressure, and ionization collection efficiency; i.e.,

$$N_{gas} = D_{gas}A_{ion}M^{-1} \qquad (2.101)$$

where D_{gas} is the mean absorbed dose in the cavity gas, M is the electrometer reading corrected for temperature and pressure, and A_{ion} is the ion collection efficiency in the Standards Laboratory beam.

N_{gas}/A_{ion} is a constant for a chamber and the AAPM protocol recommends that N_{gas} be determined from the ^{60}Co exposure calibration factor N_x using the following expression:

$$N_{gas} = N_x$$
$$\times \frac{k(W/e)A_{ion}A_{wall}\beta_{wall}}{\alpha(\bar{L}/\rho)_{gas}^{wall}(\bar{\mu}_{en}/\rho)_{wall}^{air} + (1-\alpha)(\bar{L}/\rho)_{gas}^{cap}(\bar{\mu}_{en}/\rho)_{cap}^{air}}$$
$$(2.102)$$

where k is a constant equal to the charge produced in air per unit mass per unit exposure (2.58×10^{-4}C kg^{-1}R^{-1}); W/e is the quotient of the average energy expended to produce an ion pair in air by the electronic charge; A_{wall} is a factor that accounts for attenuation and multiple scattering of the primary photons in the chamber wall and the buildup cap; β_{wall} is the ratio of absorbed dose to collision part of kerma; $(\bar{L}/\rho)_{gas}^{med}$ is the ratio of the mean restricted

collision mass stopping power of a medium to that of the gas in the cavity of the chamber averaged over the electron spectrum; $(\bar{L}/\rho)_{med}^{air}$ is the ratio of the average mass energy absorption coefficient for air and medium averaged over the photon spectrum; α is the fraction of ionization due to electrons from the chamber wall; and $(1-\alpha)$ is the fraction of ionization due to electrons coming from the buildup cap.

The equation for N_{gas} in the AAPM used to modify the equation for N and revised values for W/e results in cancellation of errors such that the revised values of N_{gas} are not different from the AAPM 1983 protocol values by more than 0.5%. The AAPM Radiation Therapy Committee, and the Radiological Physics Center in Houston, Texas recommended that the AAPM 1983 protocol should be used in its original form as presented in 1983.

The IAEA protocol defines an absorbed dose-to-air chamber factor N_D which has the same meaning as N_{gas} in the AAPM protocol. In the IAEA protocol, N_D is given by

$$N_D = N_k(1 - g)k_{att}k_m \qquad (2.103)$$

where N_k is the air-kerma calibration factor, g is the fraction of energy of the secondary charged particles that is converted to *bremsstrahlung* in air at the calibration beam, k_{att} accounts for absorption and scatter of the primary photons in the chamber wall and buildup cap, and k accounts for the lack of air equivalence of the chamber wall and buildup cap material at the ^{60}Co calibration beam and is given by

$$k_m = \alpha S_{air,wall}(\mu_{en}/\rho)_{wall,air}$$
$$+ (1-\alpha)S_{air,wall}(\mu_{en}/\rho)_{cap,air} \qquad (2.104)$$

where $S_{air,wall}$ is the ratio of averaged stopping powers for air-to-wall material (the average is taken over the total electron energy spectrum at the point measurement in accordance with the Spencer-Attix theory); $S_{air,cap}$ has the same meaning as $S_{air,wall}$ except that the wall material is replaced by buildup cap material; $(\mu_{en}/\rho)_{wall,air}$ is the ratio of the mass energy absorption coefficient of wall to that of air; $(\mu_{en}/\rho)_{cap,air}$ is the ratio of the mass energy absorption coefficient of buildup cap to that of air; α is the fraction of ionization inside an ion chamber due to electrons arising in the chamber wall; and $1 - \alpha$ is the fraction of ionization inside an ion chamber due to electrons arising in the buildup cap.

It should be noted that the IAEA protocol uses the air kerma calibration factor N_k instead of the exposure calibration factor N_x; this is in compliance with the new recommendations of the standards laboratories across the world. The relationship between N_x and N_k is given by

$$(W/e)N_x = N_k(1 - g) \qquad (2.105)$$

The AAPM protocol recommends the use of water, PMMA, or polystyrene as phantom materials for calibration of high-energy photon and electron beams. According to this protocol, the absorbed dose-to-plastic $D^{plastic}$ in the user beam quality at a reference depth d_0 is given by

$$D^{plastic}(d_0) = MN_{gas}(L/\rho)^{plastic}_{gas} P_{ion}P_{wall}P_{repl} \qquad (2.106)$$

where M is the electrometer reading corrected for temperature and pressure from an ionization chamber with its center at a depth of d_0; P_{ion} is a factor that corrects for the ionization recombination loss; P_{wall} is a correction factor that accounts for differences in composition between the chamber wall and the medium; and P_{repl} is a replacement correction which accounts for the change in photon and electron fluence that occurs because of replacement of the phantom material by chamber wall and cavity.

Absorbed dose to water, D^{water}, is then calculated from

$$D^{water}(d_0) = D^{plastic}(d_0)(\bar{\mu}_{en}/\rho)^{water}_{plastic} \; ESC \text{ for photons} \qquad (2.107)$$

and

$$D^{water}(d_0) = D^{plastic}(d_0)(\bar{\mu}_{en}/\rho)^{water}_{plastic} \Phi^{water}_{plastic} \text{ for electrons} \qquad (2.108)$$

where $(\bar{\mu}_{en}/\rho)^{water}_{plastic}$ is the ratio of the average mass energy absorption coefficient for water to that for plastic; ESC is a correction for the fractional increase in scattered photons which occurs in polystyrene and PMMA phantoms compared to that of a water phantom; $(\bar{S}/\rho)^{water}_{plastic}$ is the ratio of the average unrestricted mass collision stopping power of water to that of plastic; and $\Phi^{water}_{plastic}$ is the ratio of electron fluence at d_{max} in water to that at d_{max} in plastic.

The IAEA protocol recommends water as the phantom material for photon beams. However, for electron beams with average energy $E_0 < 10$ MeV, both water and plastic phantoms are recommended. The absorbed dose to water in the user's beam at a reference depth of d_0 is given by

$$D_w(d_0) = M_u N_D(S_{w,air})_u P_u P_{cel}h_m \qquad (2.109)$$

where N_D is the absorbed dose-to-air chamber factor and is given by

$$N_D = N_x(W/e)k_{att}k_m \qquad (2.110)$$

M_u in Equation (2.109) denotes the electrometer reading corrected for temperature, pressure, humidity, and ionization recombination loss for an ionization chamber with its center at a depth of d_0 plus a fraction f of the inner radius of the chamber where $f = 0.5$ for electron beams and $f = 0.75$ for x-ray beams under investigation. $(S_{w,air})_u$ is the ratio of the average stopping power for water to air. The stopping powers are of the Spencer-Attix type. P_u is a perturbation correction to the Bragg-Gray equation and is a product of two factors which account for the lack of water equivalence of the chamber wall and the perturbation of electron fluence that occurs because of the replacement of water by chamber wall and cavity. P_{cel} is a factor that accounts for the central electrode correction; h_m is defined for electron beams only and has the same meaning as $\Phi^{water}_{plastic}$ of the AAPM protocol. It is a measured quantity and is the ratio of the signal at the ionization maximum on the central axis in water to that in a plastic phantom at the ionization maximum.

$$\frac{D^{IAEA}_w(d_0)}{D^{AAPM}_w(d_0)}$$

$$= \frac{M_u}{MP_{ion}} \frac{N_D}{N_{gas}} \frac{(S_{w,air})_n}{(L/\rho)^{water}_{gas}} \frac{P_u}{P_{wall}P_{repl}} \frac{P_{cel}}{1} \qquad (2.111)$$

The results of the absorbed dose intercomparison in water for all the chambers employed in this study and exposed to 4 and 25 MV photon beams are given in Table 2.3.

An equation similar to Equation (2.111) but applicable only to electron beams can be written:

$$\frac{D^{IAEA}_w(d_0)}{D^{APM}_w(d_0)} = \frac{M_u}{MP_{ion}} \frac{N_D}{N_{gas}} \frac{(S_{w,air})_u}{(\bar{L}/\rho)^{plastic}_{gas}(\bar{S}/\rho)^{water}_{plastic}}$$

$$\times \frac{P_u}{P_{wall}P_{repl}} \frac{h_m}{\Phi^{water}_{plastic}} \frac{P_{cel}}{1} \qquad (2.112)$$

As in the case of photons, the values of M_u, M, and P_{ion} are related to each other by the ratio of percent depth doses at $d_0 + 0.5r$ and d_0.

The AAPM protocol recommends that $\bar{E}_0$ the mean electron energy at the surface of the phantom, should be determined from the following expression:

$$\bar{E}_0 = 2.33, \quad MeVcm^{-1} \times d_{50} cm \qquad (2.113)$$

where d_{50} in the above equation is the depth in water at which an ionization chamber reading is reduced to 50% of its maximum value. If plastic phantoms are used, then

TABLE 2.3
Absorbed Dose Intercomparison for 4- and 25-MV Photon Beams in a Water Phantom

Chamber	Photon Energy (MV)	$\dfrac{M_u}{MP_{ion}}$	$\dfrac{N_D}{N_{gas}}$	$\dfrac{(S_{w,air})_u}{(\bar{L}/\rho)_{air}^{water}}$	$\dfrac{P_u}{P_{wall}}$	$\dfrac{1}{P_{repl}}$	$\dfrac{D_{re}^{IAEA}(d_0)}{D_{ie}^{AAPM}(d_0)}$
PTW	4	0.99	1.005	0.994	1.000	1.008	0.996
Capintec	4	0.99	1.006	0.994	0.998	1.008	0.996
NEL	4	0.99	1.008	0.994	0.997	1.008	0.996
PTW	25	0.995	1.005	0.988	1.001	1.005	0.994
Capintec	25	0.99	1.006	0.988	0.999	1.006	0.988
NEL	25	0.99	1.008	0.988	1.000	1.006	0.991

Note: The values of TPR_{10}^{20} for 4- and 25-MV x-rays are 0.64 and 0.80, respectively.

Source: From Reference [22]. With permission.

d_{50} should be scaled appropriately. The AAPM provides these scale factors.

IV. STOPPING POWER

The stopping power of a material is defined as the average energy loss per unit path length that charged particles suffer as the result of Coulomb interactions with electrons and atomic nuclei when traversing the material. For protons and alpha particles, the predominant contribution to the total stopping power comes from the electronic stopping power, $-(dE/dx)_{el}$, also collision stopping power S_{col}, due to inelastic collisions with electrons. A smaller contribution comes from the nuclear stopping power, $-(dE/dx)_{nuc}$ or S_{nc}, due to elastic Coulomb collisions in which recoil energy is imparted to atoms.

The mass electronic stopping power is defined in terms of the inelastic scattering cross sections $d\sigma_{in}(W, T)/dW$ for collisions with atomic electrons:

$$\frac{1}{\rho}S_{col}(T) = NZ\int_0^{W_m} W\frac{d\sigma_{in}}{dW}\,dW \qquad (2.114)$$

where T is the initial kinetic energy and W is the energy loss of the incident particle (projectile). The upper limit of integration, W_m, is the largest possible energy loss in an inelastic collision with an atomic electron. N is the number of atoms (or molecules) per gram of material, and Z is the number of electrons per atom (or molecule).

The mass nuclear stopping power is defined in terms of the elastic scattering cross section $d\sigma_{el}(\theta, T)d\Omega$ for collisions with atoms:

$$\frac{1}{\rho}S_{nuc}(T) = 2\pi N\int_0^{W_m} W(\theta, T)\frac{d\sigma_{el}}{d\Omega}\sin\theta\,d\theta \qquad (2.115)$$

where θ is the deflection angle (in the center-of-mass system) and $W(\theta,T)$ is the recoil energy received by the

target atom. The number of atoms per gram of material is

$$N = N_A/M_A = (uA)^{-1} \qquad (2.116)$$

where N_A is the Avogadro constant, M_A is the molar mass in g mol^{-1}, A is the relative atomic (or molecular) mass (sometimes denoted by A_r), and u is the atomic mass unit (1/12 of the mass of an atom of the nuclide ^{12}C).

To calculate the penetration, diffusion, and slowing down of charged particles in bulk matter, one must utilize—in principle—the complete set of differential cross sections for energy losses and angular deflections in inelastic and elastic Coulomb collisions. In the continuous-slowing-down approximation (csda), energy-loss fluctuations are neglected and charged particles are assumed to lose their energy continuously along their tracks at a rate given by the stopping power. The csda range, calculated by integrating the reciprocal of the total stopping power with respect to energy, is a very close approximation to the average path length traveled by a charged particle in the course of slowing down to rest.

In the straight-ahead approximation, the angular deflections due to multiple elastic scattering are neglected, and charged particle tracks are assumed to be rectilinear. For protons and alpha particles, this is a good approximation except near the ends of their tracks.

Stopping powers pertain to the loss of energy by the incident charged particles, whereas in radiation dosimetry, one is often more interested in the spatial pattern of energy deposition in some target region (such as an organ, cell, or cell nucleus). A large fraction of the energy lost by protons or alpha particles along their tracks is converted to kinetic energy of secondary electrons (delta-rays), and the transport of these electrons through the medium influences the spatial pattern of energy deposition. In some dosimetry calculations, the transport of energy by secondary photons (fluorescence radiation or *bremsstrahlung*) or by recoil nuclei may also have to be taken into account.

The formula for the mass collision stopping power for a heavy charged particle can be written in the form

$$\frac{1}{\rho}S_{col} = -(1/\rho)(dE/dx)_{el} = \frac{4\pi r_e^2 mc^2}{\beta^2}\frac{1}{u}\frac{Z}{A}z^2 L(\beta)$$

(2.117)

where $r_e = e^2/mc^2$ is the classical electron radius, mc^2 is the electron rest energy, u is the atomic mass unit, β is the particle velocity in units of the velocity of light, Z and A are the atomic number and relative atomic mass of the target atom, and z is the charge number of the projectile. With standard numerical values for the various constants, one finds that $4\pi r_e^2 mc^2/u$ has the value 0.307075 MeV cm^2 g^{-1}.

The quantity L is called the stopping number. The factors preceding the stopping number take into account the gross features of the energy-loss process, whereas L takes into account the fine details. It is convenient to express the stopping number as the sum of three terms:

$$L(\beta) = L_0(\beta) + zL_1(\beta) + z^2 L_2(\beta) \qquad (2.118)$$

The first term is given by

$$L_0(\beta) = \frac{1}{2}\ln\left(\frac{2mc^2\beta^2 W_m}{1-\beta^2}\right) - \beta^2 - \ln I - \frac{C}{Z} - \frac{\delta}{2}$$

(2.119)

where I is the mean excitation energy of the medium, C/Z is the shell correction, and $\delta/2$ is the density-effect correction. W_m is the largest possible energy loss in a single collision with a free electron, given by

$$W_m = \frac{2mc^2\beta^2}{1-\beta^2}$$
$$\times [1 + 2(m/M)(1-\beta^2)^{-1/2} + (m/M)^2]^{-1}$$

(2.120)

where m/M is the ratio of the electron mass to the mass of the incident particle and mc^2 is the electron rest energy (0.511 MeV). If the factor in square brackets in Equation (2.120) is set equal to unity, the maximum energy transfer for protons is overestimated by only 0.1% at 1 MeV and 0.23% at 1000 MeV. In the non-relativistic limit, $W_m \sim 2 mv^2 = 4(m/M)T$.

The mean excitation energy I is a quantity independent of the properties of the projectile and depends only on the properties of the medium. For the Thomas-Fermi model of the atom, it is proportional to the atomic number, $I = I_0 Z$, with I_0 approximately equal to 10 eV.

For gases, the mean excitation energy can be obtained from the expression

$$\ln I = \int_0^\infty \frac{df}{dE}\ln E\, dE \Big/ \int_0^\infty \frac{df}{dE}\, dE \qquad (2.121)$$

where df/dE is the density of optical dipole oscillator strength per unit excitation energy E above the ground state. The oscillator strength is proportional to the photo-absorption cross section, for which abundant experimental data are available.

For materials in the condensed phase, the analogous formula for the mean excitation energy is

$$\ln I = \frac{2}{\pi\omega_p^2}\int_0^\infty \omega Im[-1/\varepsilon(\omega)]\ln(h\omega)\,d\omega \qquad (2.122)$$

where $\varepsilon(\omega)$ is the complex-valued dielectric response function and

$$h\omega_p = (4\pi h^2 e^2 n_e/m)^{1/2} = 28.816\rho(\rho Z/A)^{1/2}$$

(2.123)

is the plasma energy (in eV). e is the charge of the electron, n_e is the number of electrons per unit volume, and p is the density.

The selection of stopping power ratios water/air for clinical electron beams is normally based on the use of the so-called $S_{w,air}(\bar{E}_0, z)$ method, where $\bar{E}_0$ is the mean energy of the incident electron beam at the phantom surface and z is the depth of measurement. Stopping power ratios are determined for monoenergetic electron beams, usually by means of Monte Carlo calculations, and $\bar{E}_0$ provides the link between the clinical electron beam and the energy of the electrons used for the calculation. In a sense, $\bar{E}_0$ provides an approximate indication of the electron beam quality.

The relationship between the mean electron energy at the surface of a phantom, $\bar{E}_0$, and the depth of the 50%-absorbed dose in water, R_{50}, receives special attention.

All dosimetry protocols use the 2.33 approximation; i.e., $\bar{E}_0 = 2.33\, R_{50}$.

A detailed analysis of R_{50} values in terms of relative depth-dose distributions shows very good agreement at high energies. However, at lower energies, in the range most commonly used in radiotherapy with electron beams, discrepancies in R_{50} increase with decreasing energy, differences being close to 2% at 5 MeV.

Figure 2.17 compares results obtained with the two Monte Carlo codes, EGS4 and ITS3, showing that, although depth-dose distributions in water are similar, discrepancies in energy deposition close to the depth of the maximum (R_{100}) yield different estimates of R_{50} for monoenergetic beams. [23]

The depth of the effective point of measurement of the ionization chamber is used to select the stopping power ratio, at the same depth, of a monoenergetic beam with energy equal to $\bar{E}_0$. In order to verify the validity of the $s_{w,air}(\bar{E}_0, z)$ method, electron beams with energies close to 10 MeV and with varying energy and angular spread, which in most cases exceed the conditions existing for clinical beams, were analyzed. [23]

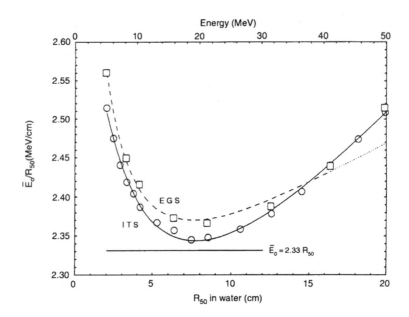

FIGURE 2.17 A comparison of $\bar{E}_0/R_{50}$, vs. R_{50} (and approximate energy) for monoenergetic beams obtained with two Monte Carlo codes (circles: ITS3; squares: EGS4). The dashed is $\bar{E}_0/R_{50}$, reproducing the EGS data in the energy range of 5–40 MeV. The solid line is a fit to the ITS data, which reproduces the values in the range of 5–50 MeV. (From Reference [23] With permission.)

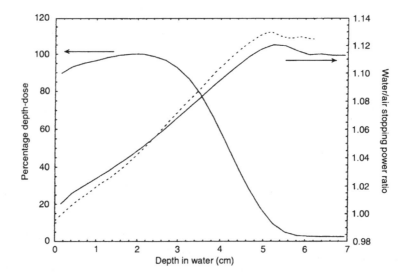

FIGURE 2.18 Monte Carlo calculated depth-dose and water/air stopping power ratio distributions (solid lines) for a "realistic" electron beam, including the energy and angular spread of electrons and contaminant photons produced in the beam-defining system. (From Reference [23]. With permission.)

Depth-dose data and water/air stopping power ratio distributions for the "realistic" beam are shown in Figure 2.18, together with $s_{w,air}$ values for a monoenergetic electron beam of 9.8 MeV (based on $\bar{E}_0 = 2.33\ R_{50}$). It can be seen that the $s_{w,air}$ value at the reference depth computed for the "realistic" beam differs by less than 0.5% from the $s_{w,air}(\bar{E}_0, Z_{ref})$ value determined according to most dosimetry protocols, differences being larger at shallow depths. From the depth-dose curve, $R_{50} = 4.2$ cm is obtained, which, together with the "2.33 approximation,"

yields $\bar{E}_0 = 9.8$ McV. Stopping power ratios for a monoenergetic beam of 9.8 MeV, as recommended by most protocols, are shown by the dashed line in Figure 2.18.

Collimated electron beams of about 5, 10, and 20 MeV, having passed through single thick scattering foils, air, and water, were simulated as a full configuration (i.e., without splitting the simulation at the phantom surface), and the above procedure was repeated. Results for the full calculations of stopping power ratios are shown in Figure 2.19. At high energies, it can be observed that,

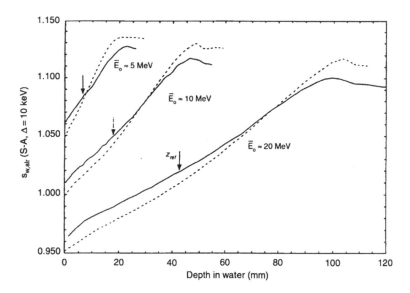

FIGURE 2.19 Full Monte Carlo simulations (solid lines) of collimated electron beams with energies of 5, 10, and 20 MeV having passed through single scattering lead foils, air, and water, compared with the $s_{w,air}(\bar{E}_0 = 2.33\ R_{50}, z)$ method (dashed lines) used in dosimetry protocols. (From Reference [23]. With permission.)

mainly because of the *bremsstrahlung* contamination produced in the scattering foils, discrepancies at the reference depth can be slightly larger than 1% but will not generally exceed this limit in practical cases.

From a linear fit of results simulating different therapy machines with scattering foils, an expression is provided to correct for the *bremsstrahlung* contamination, the stopping power ratio at the reference depth obtained with $s_{w,air}(\bar{E}_0, z_{ref})$:

$$s_{w,air} = (0.9952 + 0.0032\ D_x)s_{w,air}(\bar{E}_0, z_{ref}) \quad (2.124)$$

where D_x is the percentage depth dose in the *bremsstrahlung* tail, 2 cm past the practical range R_p.

The total stopping power of a medium is the average rate at which charged particles lose energy at any point along their tracks in the medium. It is customary to separate total stopping power into two components: the collision stopping power, which is the average energy loss per unit path length due to inelastic Coulomb collisions with atomic electrons of the medium resulting in ionization and excitation; and the radiative stopping power, which is the average energy loss per unit pathlength due to the emission of *bremsstrahlung*.

The following initial energy distribution of electrons set in motion in a medium by a beam of monoenergetic photons of energy $E_0 = h\nu$ is proposed: [24]

$$\frac{dN(T)}{dT} = \frac{d}{dT}\left(\frac{\sigma(T)}{\rho}\right) + 2\frac{d}{dT}\left(\frac{\kappa(T)}{\rho}\right) + \delta(T - E_b)\frac{\tau E_0}{\rho}$$
$$(2.125)$$

where E_0 is the binding energy of the K-shell electron. This expression gives the number of electrons set in

motion in a medium through Compton collision, pair production, and photoelectric interaction with initial kinetic energies in the range of T to $T + dT$. In Equation (2.125), positrons are assumed to behave essentially like electrons, and triplet production has not been included since it is one order of magnitude less in the energy range of interest in radiation therapy. One may concentrate on the first two terms in Equation (2.125).

For free and stationary electrons, the differential Compton cross section $d\sigma(T)/dT$ takes the form

$$\frac{d}{dT}\left(\frac{\sigma(T)}{\rho}\right) = \frac{N_e \pi r_0^2}{\alpha E_0}$$

$$\times \left(2 - \frac{2T}{\alpha(E_0 - T)} + \frac{T^2}{\alpha^2(E_0 - T)^2} + \frac{T^2}{E_0(E_0 - T)}\right)$$
$$\text{for } T \in [0, T_{max}] \quad (2.126)$$

where r_0 denotes the classical electron radius, N_e is the number of electrons per gram of the medium, $\mu_0 = m_e c^2$ is the rest mass of the electron, $\alpha = E_0/\mu_0$, and $T_{max} = E_0 \times 2\alpha/(1 + 2\alpha)$.

As pointed out above, Equation (2.126) describes the case of free and stationary electrons. However, atomic electrons are neither free nor stationary. To take this fact into account, Equation (2.126) should be modified as follows:

$$\frac{d}{dT}\left(\frac{\sigma_{inc}(T)}{\rho}\right) = \frac{d}{dT}\left(\frac{\sigma(T)}{\rho}\right)\frac{S(x, Z)}{Z} \quad (2.127)$$

where $S(x, Z)$ denotes the incoherent scattering function.

In order to take electrons and positrons into account separately, Tome and Palta used of the following initial energy; distributions of electrons and positrons set in motion in a medium by a monoenergetic photon beam, [24]

$$\frac{dN_-(T)}{dT} = \frac{d}{dT}\left(\frac{\sigma_{inc}(T)}{\rho}\right) + \frac{d}{dT}\left(\frac{\kappa(T)}{\rho}\right)$$
$$+ \delta(T - E_b)\frac{\tau(E_0)}{\rho} \tag{2.128}$$

$$\frac{dN_-(T)}{dT} = \frac{d}{dT}\left(\frac{\kappa(T)}{\rho}\right)$$

When the continuous slowing down approximation (CSDA) is used, the electron spectrum "seen" in the medium is given by:

$$\frac{d\Phi_-^{T_0}(T)}{dT} = \frac{N}{S_{tot}^-(T)} \tag{2.129}$$

where N is the total number of electrons set in motion per gram. If, for convenience, $N = 1$; then

$$\frac{d\Phi_-^{T_0}(T)}{dT} = \frac{1}{S_{tot}^-(T)} \tag{2.130}$$

Using results of Spencer and Fano, Spencer and Attix proposed the following approximate integral equation for the description of the electron spectrum generated by a monoenergetic source that takes the buildup of secondary electrons generated in knock-on collisions into account; i.e., when CSDA is not assumed to hold,

$$\frac{d\Phi_-^{T_0, \delta}(T)}{dT} = \frac{1}{S_{tot}^-(T)}$$
$$\times \left[1 + \int_{2T}^{T_0} K_-^*(T', T)\frac{d\Phi_-^{T_0, \delta}(T')}{dT'} dT'\right] \tag{2.131}$$

where (d/dT) $\Phi_-^{T_0, \delta}(T)$ denotes the electron spectrum "seen" in the medium arising from electrons with an initial kinetic energy T_0, including the production of fast secondary electrons. Equation (2.130) is clearly the zeroth-order approximation to the integral equation (Equation 2.131). Substitution of the zeroth-order approximation into Equation (2.131) yields the following first-order approximation to (d/dT) $\Phi_-^{T_0, \delta}(T)$,

$$\frac{d\Phi_-^{T_0, \delta}(T)}{dT} = \frac{1}{S_{tot}^-(T)} \times \left[1 + \int_{2T}^{T_0} \frac{K^*(T', T)}{S_{tot}^-(T')} dT'\right] \tag{2.132}$$

Calculations of stopping power ratios, water to air, for the determination of absorbed dose to water in clinical proton beams using ionization chamber measurements have been undertaken by Medin and Andreo [25] using the Monte Carlo method. A computer code to simulate the transport of protons in water (PETRA) has been used to calculate $s_{e, air}$-data under different degrees of complexity, ranging from values based on primary protons only to data including secondary electrons and high-energy secondary protons produced in nonelastic nuclear collisions.

The influence on the depth-dose distribution from the different particles can be seen in Figure 2.20 for two different energies. It should be noted that the distribution labelled "primary protons" includes only the contribution from energy losses below the chosen value of a cut-off energy Δ.

V. ENERGY FOR ION PAIR PRODUCTION

Dosimetry of charged particles is generally done by measuring the ionization yield produced in a gas-filled ionization chamber. For the absorbed dose determination, the mean energy per ion pair production, W, or its differential value, ω, is needed. W values for protons in TE gas was calculated and measured by Grosswendt and Baek. [26] Characterizing by N the number density of target molecules, by $\sigma_t(T)$ the sum of the cross section $\sigma_{ex}(T)$ for the charge-exchange cycle and the weighted cross section $\sigma_i(T)$ for ionization with respect to protons and neutral hydrogen projectiles, and by $(dT/dx)_{T'}$ the linear total stopping power at energy T', the ionization yield is given by

$$N(T) = N\int_t^T \frac{\sigma_t(T')f(T')}{(dT/dx)_{T'}} dT' \tag{2.133}$$

Here, $f(T)$ is a factor to include in the ionization yield contribution by secondary electrons:

$$f(T) = 1 + \frac{\sigma_{ex}(T)}{\sigma_t(T)} \frac{E_{ex}}{W_e(E_{ex})}$$
$$+ \frac{\sigma_i(T)}{\sigma_t(T)}\int_I^{E_{max}} q(T, E) \frac{E}{W_e(E)} dE \tag{2.134}$$

where $E_{ex} = T/1836$ is the energy of electrons set free during the charge-exchange cycle, W_e is the electron W value, $q(T, E)$ is the spectral distribution of secondary electrons due to direct ionizing collisions, and $E_{max} = T/459.0$ is the maximum electron energy in the classical limit.

The reciprocal of the integrand of Equation (2.133), divided by the number density of target molecules N, is equal to the differential value ω.

Figures 2.21 and 2.22 summarize our calculated absolute W values of protons slowed down in TE gas or air as

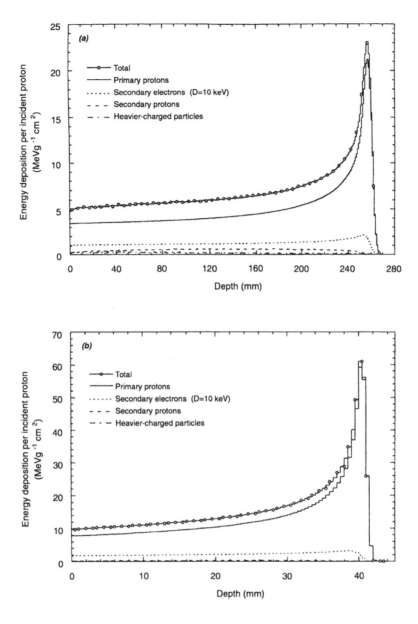

FIGURE 2.20 Monte Carlo–calculated depth-dose distributions in water for monoenergetic proton beams calculated with the code PETRA. Proton Monte Carlo transport cut-off energy, 1.0 MeV and $\Delta = 10$ keV, 10^5 histories, (a) 200 MeV, (b) 70 MeV. (From Reference [25]. With permission.)

a function of energy T, in comparison with the evaluated experimental data (note that the experimental values at T greater than a few MeV are actually ω values). The steep decrease of the calculated or measured data with increasing energy, which is typical in the low-energy region, is obvious from both figures. In the case of the calculated data, it is followed by a more or less clear structure of $W(T)$ up to energies of about 1 MeV and an almost constant or only slightly decreasing value at higher energies. At lower energies, this behavior of $W(T)$ is due to the competition between direct proton or hydrogen impact ionization and of the ionization caused by charge-changing effects, and, at higher energies, it is caused by the ionization-yield con-

tribution produced by secondary electrons. The dependence of W for protons on T in dry air is shown in Figure 2.23.

The mean energy W, expended per ion pair formed, has been determined experimentally by Waibel and Willems [28] for protons completely stopped in methane-based tissue-equivalent gas (TE gas) and its constituents, methane, carbon dioxide, and nitrogen, covering the energy range from 1 to 100 keV.

The degradation of low-energy electrons in a methane-based TE gas was studied experimentally by Waibel and Grosswendt [29] by ionization chamber experiments and theoretically by Monte Carlo electron transport simulation in the energy range between 25 eV and 5 keV.

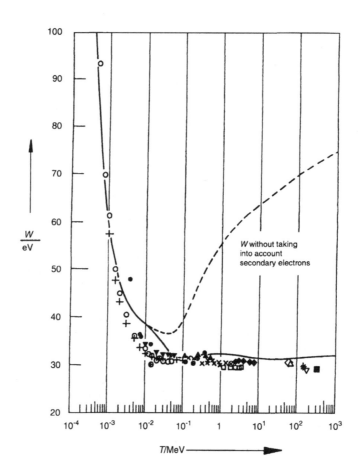

FIGURE 2.21 Dependence of W for protons slowed down in *TE* gas as a function of the initial energy T (—); Grosswendt and Baek results of the complete analytical model: (- - -); model calculations without taking into account the ionization-yield contribution of secondary electrons; experimental data. (From Reference [26]. With permission.)

The electron beam entered the chamber perpendicular to one plate through a set of diaphragms and its current, i_0, was measured in vacuum using a Faraday cup. The primary energy, T_0, was determined each time, applying the retarding field method.

The mass stopping power can be determined from the derivative, di_+/dp at $p = 0$ of the pressure-dependent ion current $i_+(p)$ if the conversion factor from ionization to energy, the w value, is known and some precautions are taken.

The mass stopping power of the gas for a primary electron with energy, T_0, then becomes

$$S/\rho = \omega(T_0) \lim_{z\rho p \to 0} d$$
$$\times \{[i_+(z\rho p)i_0] I[n_{tot}/n_e(T, z\rho)]\}/d(z\rho) \quad (2.135)$$

The secondary electron distribution has been approximated by

$$f(T, E_s) \cong (E_s^2 + E_{so}^2)^{-1} \quad \text{with} \quad E_{so} = 8\,\text{eV}$$
$$E_s \leq T/2 \quad (2.136)$$

Figure 2.24 shows the mean free path for electrons as a function of energy. Figure 2.25 shows the ratios of cross section s for ionization and excitation for inelastic electron scattering in TE(CH$_4$) gas.

The experimental W values are approximated by the following analytical expression between 35 eV and 5 keV with deviations of less than 0.5%:

$$W(eV) \approx 29.4 + 146.2\,[T(eV) - 13.61]^{-0.749} \quad (2.137)$$

Figure 2.26 presents W values vs. electron energy.

VI. ELECTRON BACKSCATTERING

Scattering probability of low-energy electrons is high, and a problem is therefore imposed when low-energy electron dose is measured. The parallel-plane ionization chambers are designed with a thick supporting body at the back of the sensitive volume. The backscattered electrons into the ion chamber (in-scattering) contribute a significant dose to the measurement. The scattering from the detector material is different than back-scattering from water. In-scattering was found in cylindrical chambers also, 3% for Farmer chamber for electrons of energy below 8 MeV.

The exponential empirical expression proposed by Klevenhagen et al. [30, 31] is

$$EBS = A - B[\exp(-kZ)] \quad (2.138)$$

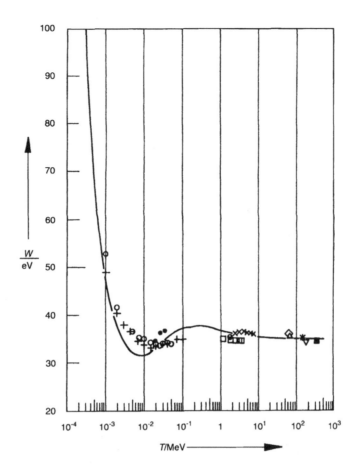

FIGURE 2.22 Dependence of W for protons slowed down in air as a function of the initial energy T(—); Grosswendt and Baek results of the complete analytical model. (From Reference [26]. With permission.)

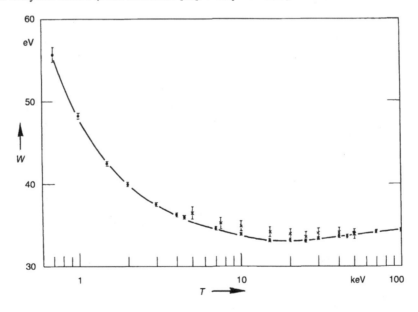

FIGURE 2.23 Energy dependence of the experimental W values for protons in dry air. (From Reference [27]. With permission.)

where the equation parameters A, B, and k are defined in Table 2.4 as a function of electron energy at the scatterer surface, and Z is the atomic number of the scatterer.

The effect of the thickness of tissue-equivalent material on the backscattering of low-energy x-rays has been studied by Lanzon and Sorell [32] by direct measurement for x-ray beams with first-half-value thicknesses in the

TABLE 2.4
List of Parameters for Equation (2.138)

Beam Energy E_m (MeV)	A	B	$k \times 10^{-2}$
3	1.932	1.037	1.7
4	1.945	1.035	1.5
6	1.960	1.033	1.15
8	1.970	1.031	0.85
10	1.980	1.029	0.68
12	1.985	1.027	0.60
14	1.990	1.025	0.53
16	1.994	1.023	0.45
18	1.998	1.021	0.39
20	2.000	1.019	0.33

Source: From Reference [31]. With permission.

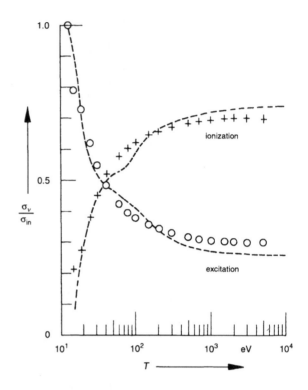

FIGURE 2.25 Ratios of the cross section σ_v for impact ionization and excitation to the cross section σ_{in} for inelastic scattering of electrons in $TE(CH_4)$ gas as a function of energy T, compared with those for water vapor according to Paretzke. [68] The ratios in $TE(CH_4)$ gas are characterized by the symbols, and those in water are characterized by the broken curves. (From Reference [29]. With permission.)

Nuclear Enterprises 2577 0.2-cm³ thimble ionization chamber with a 0.36-mm thick graphite cap. The x-ray generator was a Philips RT-100 superficial therapy machine. The ionization chamber was positioned with its center at the surface of the water phantom, and a lead disk of thickness 1.5 mm was mounted on the transport carriage. The surface ionization was measured as the lead disk traversed from its depth to the lower edge of the ionization chamber.

The surface dose measurements are represented as relative to the surface dose for full backscatter conditions. The quantity backscatter reduction factor $BSRF(z)$ is defined as this ratio:

$$BSRF(z) = D_0(z)/D_0(z_{fullBS}) \qquad (2.139)$$

where $D_0(z)$ is the dose on the beam axis at the surface of a phantom of thickness z and $D_0(z_{fullBS})$ is the dose on the beam axis at the surface of a phantom which provides full backscatter.

The results for the 8-mm Al HVT beam for the range of field sizes are shown in Figure 2.27 and the results for the 70-mm square field size for the range of beam qualities

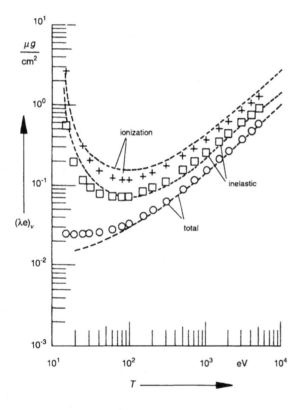

FIGURE 2.24 Mean free path lengths $(\lambda\rho)_v$ of electrons in $TE(CH_4)$ gas as a function of energy T, compared with those in water vapor of Paretzke. [67] The mean free path lengths for total scattering, inelastic scattering, and ionization impact in TE (CH_4) gas are characterized by the symbols, and those in water are characterized by the broken curves. (From Reference [29]. With permission.)

range 0.5 mm to 0.8 mm Al and for filed sizes ranging from 15 mm in diameter to 70 mm square at 100 mm ssd. Measurements were made in a water phantom using a

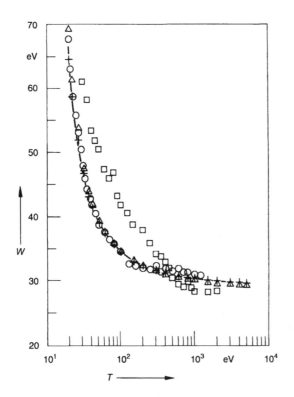

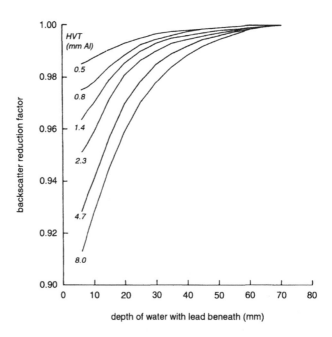

FIGURE 2.26 Energy dependence of the W values for electrons stopped in methane-based, TE gas: $+$, experimental results; O, experimental data from Combecher [69]; $\square$, from Smith and Booz [70]; Δ, Monte Carlo results; full curve, fitting function of Equation 2.137 [29]. With permission.

FIGURE 2.28 The dependence of the surface dose on beam quality and the depth of water with lead beneath for a 70-mm-square field. (From Reference [32]. With permission.)

are shown in Figure 2.28. The results show that the contribution to surface dose by photons backscattered from the water medium below any depth z is greater than that from photons backscattered from the lead material placed at depth z.

For a given depth of water with lead beneath, the backscatter reduction factor decreases with increasing beam energy and field size. An increase in energy implies an increase in the penetration of the photons and a greater relative contribution to the surface dose by photons backscattered from below any given depth. The decrease in the backscatter reduction factor with increasing field size indicates that the contribution to the central-axis surface dose by photons backscattered from the lead material at a depth z relative to that from photons backscattered from the water medium below depth z decreases with increasing lateral distance from the central axis of the beam. The backscatter reduction factors have been parametrized to condense the data for presentation and to enable such data to be readily generated for use elsewhere. The data is well described by the following exponential function:

$$BSRF(z) = 1 - e^{-(az + b)} \qquad (2.140)$$

The parameters a and b are functions of the beam area and energy. The values of the parameters are derived from the experimental data by a least-squares best fit method; see Lanzon and Sorell. [32]

Monte Carlo calculations of the dose backscatter factor for monoenergetic electrons was done by Cho and Reece. [33] There has been growing interest in beta emitters for

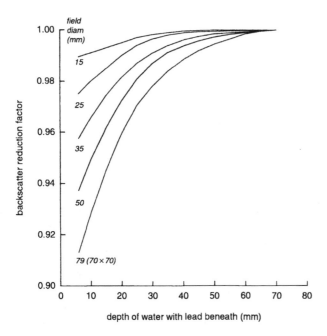

FIGURE 2.27 The dependence of the surface dose on field size and the depth of water with lead beneath for 8-mm Al HVT x-rays. (From Reference [32]. With permission.)

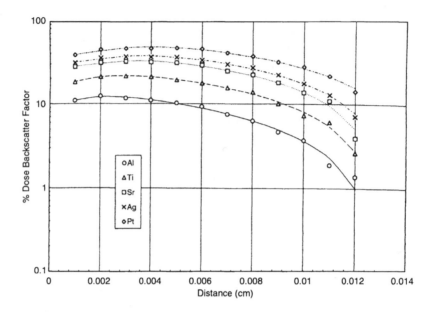

FIGURE 2.29 Dose backscatter factor depth profile for $E_0 = 0.1$ MeV. Note, in Figures 2.29 to 2.31, that the statistical errors in the data are comparable with the size of the symbols. (From Reference [33]. With permission.)

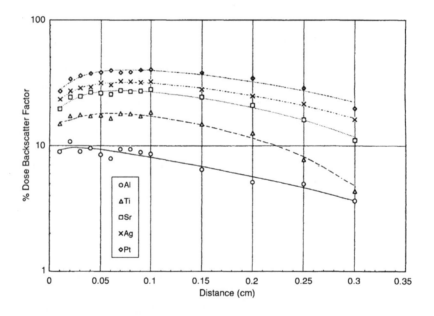

FIGURE 2.30 Dose backscatter factor depth profile for $E_0 = 1.0$ MeV. (From Reference [33]. With permission.)

therapeutic uses, especially in connection with so-called endovascular (or intravascular) brachytherapy. Since accurate dose estimation is necessary for the success of such applications, some problems in beta-ray dosimetry need further study. Among these problems is the effect of electron backscattering on dose, which has significance not only for accurate dose estimation but also for new source design. In that study, an empirical measure of electron backscattering, known as the dose backscatter factor, was calculated using EGS4 Monte Carlo calculations for monoenergetic electrons and various scattering materials. Electron energies were 0.1, 0.5, 1.0, 2.0, and 3.0 MeV in

combination with Al ($Z = 13$), Ti ($Z = 22$), Sr ($Z = 38$), Ag ($Z = 47$), and Pt ($Z = 78$) scatterers. The dose backscatter factor ranged from 10% to 60%, depending on electron energy and material, and was found to increase with the atomic number Z by a log ($Z + 1$) relationship. Results are shown in Figures 2.29 to 2.31.

In Figure 2.32, dose backscatter factors at 0.001-cm depth for 0.1 MeV and 0.01-cm depth for other energies are plotted against log ($Z + 1$) of the scatterers used in this study. Data were fitted well by straight lines in a semi-log scale, showing that the dose backscatter factor is proportional to log ($Z + 1$).

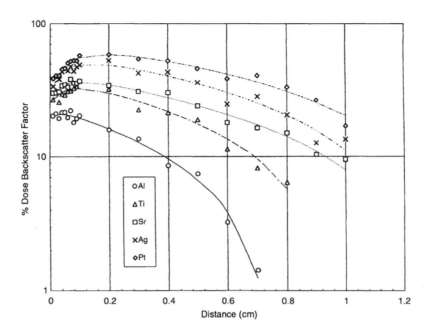

FIGURE 2.31 Dose backscatter factor depth profile for $E_0 = 3.0$ MeV. (From reference [33]. With permission.)

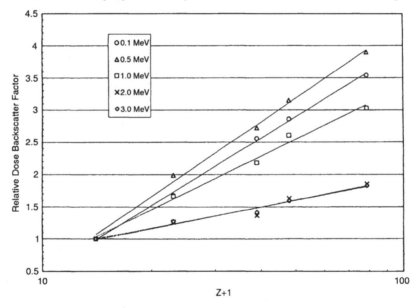

FIGURE 2.32 Log $(Z + 1)$ dependence of the dose backscatter factor data for each electron energy. (From Reference [33]. With permission.)

Beta-ray dose backscatter factors with respect to soft tissue were measured by Nunes et al. [34] using an extrapolation chamber. The beta-ray dose backscatter factor is a measure of the change effected in absorbed dose to a soft-tissue medium when part of the medium is replaced by a material other than soft tissue (i.e., a scatterer); the source is located at the boundary between the two media.

The variation of backscatter factor with distance from the boundary is well represented analytically by sums of exponentials. Therefore, the rate of decrease of backscatter factor with distance can be specified by a relaxation length, defined as the depth through which the backscatter factor is reduced by $1/e$, where e is the base of the natural logarithm. With a ^{32}P planar source, relaxation lengths in Mylar are 588 mg/cm^2 and 238 mg/cm^2 for bismuth and aluminum scatterers, respectively.

The backscatter factor is given by $B = D_i/D_h - 1$, where D_h is the dose measured when both slabs are soft tissue; this is the "homogeneous dose," and D_i is the dose measured when the replaceable slab is substituted with a scatterer—a material that is not soft tissue; this is the "inhomogeneous dose." B is usually multiplied by 100 and referred to as a percentage. D_i/D_h is called the dose ratio;

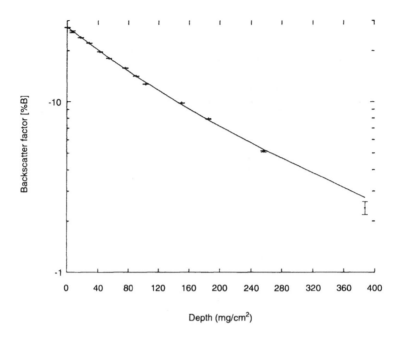

FIGURE 2.33 Backscatter factor depth profile for a ^{32}P distributed source located at an air/Mylar interface. Symbols denote experimental data and the solid curve is a fit to these data. (From Reference [34]. With permission.)

if it is larger than unity, it is a dose enhancement factor, and if it is smaller than unity, it is a dose reduction factor.

The extrapolation chamber, model EIC-I, is 35 cm long, including the connectors. The stem is 29.2 cm long and the head, about 2.3 cm long, with a diameter of 3.81 cm. It weighs 150 g. The air gap spacing is continuously variable, from 0.3 mm to 4.5 mm; the space is between the entrance window, which is one electrode, and the collecting electrode. The entrance window is made of graphite-coated polypropylene and is 0.2 mg/cm² thick. The collecting electrode and guard ring are made of A-150 (Shonka) plastic, which is a soft-tissue equivalent material with an effective atomic number of 5.49. The sensitive region of the chamber is a right circular cylinder with a diameter of 1.05 cm and height equivalent to the interelectrode spacing. Phosphorus-32 point source, with a diameter about 2 mm, was prepared by depositing 5 μl of stock solution onto a Mylar substrate and allowing it to evaporate in air. The activity of the source at the time of preparation was about 92.5 MBq. Both point and distributed or planar sources were supported by, and covered with, Mylar, 0.35 mg/cm² thick.

The variation of backscatter factor with distance from an interface where a ^{32}P source is located is found to be well represented by sums of exponentials. Specifically,

$$\%B(x) = \sum_i A_i e^{-x/v_i} \quad \text{for solid material interfaces and}$$

$$-\%B(x) = \sum_i A_i e^{-x/v_i} \quad \text{for air interfaces.}$$

The air interface depth profile, Figure 2.33, demonstrates that the dose reduction, which is largest at the interface, falls off in roughly two stages. Close to the boundary, the dose reduction decreases sharply; at larger distances, the falloff is less rapid.

The aluminum and bismuth interface depth profiles, Figure 2.34 for aluminum, exhibit gentle maxima that are located away from the interface.

VII. DOSIMETER PERTURBATION

Perturbation effects are defined as departures from ideal large-detector or Bragg-Gray cavity behavior. Perturbations are involved in the determination of the absorbed dose to a medium irradiated by the photon and electron beams used in external-beam radiotherapy. Kilovoltage x-ray beams and electron beams were discussed by Nahum. [35,36] Many correction factors are involved in radiation dosimetry, including, for an ionization chamber, corrections for differences between the temperature, pressure, and humidity of the air in an ion chamber at measurement and at calibration, the recombination of ions before they can contribute to the measured charge, and possible polarity effects.

Assuming the absorbed dose to the sensitive material of the detector, D_{det}, is known, one can generally write

$$D_{med} = f D_{det} \tag{2.141}$$

where D_{med} is the desired quantity, the absorbed dose in the undisturbed medium, it is the task of cavity theory to evaluate the factor f. In general, there will be a perturbation

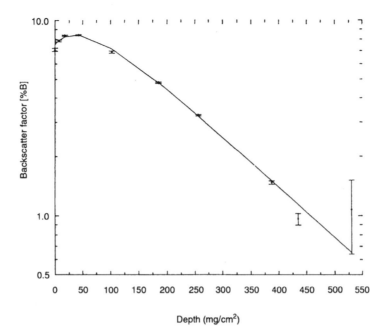

FIGURE 2.34 Backscatter factor depth profile for a ^{32}P point source located at an aluminum/Mylar interface. Symbols denote experimental data, and the solid curve is a fit to these data. (From Reference [34]. With permission.)

due to differences in atomic composition, density, or both between the detector material and the undisturbed medium, and to correct for this, a perturbation correction factor, p, is introduced.

In the case of a photon detector, p can be defined by

$$D_{med} = \overline{D}_{det}(\overline{\mu}_{en}/\rho)_{med,det} \, p \qquad (2.142)$$

where it is the photon (energy) fluence that is perturbed.

In the case of a Bragg-Gray cavity or electron detector, p is defined by

$$D(z)_{med} = \overline{D}_{def} s_{med,det} p \qquad (2.143)$$

where it is the electron fluence that is perturbed.

A simple rearrangement yields

$$p = D(z)_{med}/\overline{D}_{det} s_{med,det} P \qquad (2.144)$$

The quantities $D_{med}(z)$, D_{det} and $s_{med,det}$ can be expressed as a cavity integral using the Spencer-Attix formulation. The ratio D_{med}/D_{det} can be written as

$$\frac{D_{med}(z)}{D_{det}}$$

$$= \frac{\int_{\Delta}(\Phi_E)^z_{med}(L_{\Delta}/\rho)_{med}\,dE + [\Phi(\Delta)^z_{med}(S(\Delta)/\rho)_{med}\Delta]}{\int_{\Delta}(\overline{\Phi_E})_{det}(L_{\Delta}/\rho)_{det}\,dE + [\overline{\Phi(\Delta)}_{det}(S(\Delta)/\rho)_{det}\Delta]}$$

$$(2.145)$$

where Φ_E is the standard nomenclature for the electron fluence differential in energy E, with the subscripts *med* and *det* referring to its value at a reference point z in the undisturbed medium and its value averaged over the sensitive detectors volume, respectively. The other quantities have their usual meaning. The term outside the integral is the dose due to track ends below energy Δ. The expression for p is

$$p = \frac{\int_{\Delta}(\Phi_E)^z_{med}(L_{\Delta}/\rho)_{det}\,dE + [\Phi(\Delta)^z_{med}(S(\Delta)/\rho)_{det}\Delta]}{\int_{\Delta}(\overline{\Phi_E})_{det}(L_{\Delta}/\rho)_{det}\,dE + [\overline{\Phi(\Delta)}_{det}(S(\Delta)/\rho)_{det}\Delta]}$$

$$(2.146)$$

This is equivalent to writing

$$p = (D_{det})_{BG}/(\overline{D}_{det})_{meas} \qquad (2.147)$$

where $(D_{det})_{BG}$ is the detector dose that would result from ideal Bragg-Gray behavior an $(D_{det})_{meas}$ is the actual detector dose.

p is unity if Φ_E in the medium at z is identical to that avenged over the detector. In the case of the difference being only one of magnitude and not of spectral shape, i.e., if

$$(\overline{\Phi_E})_{det} = k(\Phi_E)^z_{med} \qquad (2.148)$$

where k is a constant for all E, then

$$p = \Phi^z_{E_{med}}/\overline{\Phi}_{det} \qquad (2.149)$$

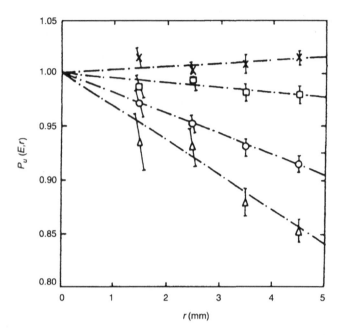

FIGURE 2.35 The dependence of the mono-energetic perturbation correction factor on the chamber radius r for monoenergetic photon beams with energies E = 20 (Δ), 30 ($\bigcirc$), 75 ($\times$), and 200 keV ($\square$). (From Reference [35]. With permission.)

The perturbation (displacement) correction factor for chamber radius r in beam energy E was defined as

$$p_{dis}(E,r) = K(E,0)/K(E,r) \qquad (2.150)$$

where the $K(E,0)$ were obtained by extrapolation of the $K(E,r)$ to zero radius. This was done by a weighted least-squares fit of the straight line to the $1/K(E,r)$ values as a function of r. The kerma $K(E,r)$ was computed by integrating the photon fluence in the cavity.

Figure 2.35, taken from Seuntjens et al. [37], shows that $p_{dis}(E,r)$ is appreciably below unity at 20 and 30 keV, which can be explained by the virtually scatter-free attenuation of the beam at these energies. At higher energies the deviations from unity are very small. These $p_{dis}(E,r)$ were then integrated over the primary photon fluence spectrum to yield $p_{dis}(r)$ for different x-ray spectra. Their results are shown in Figure 2.36 as a function of HVL in both mm of copper and mm of aluminium. For the range 2 mm Al < HVL < 5 mm Cu and for the Farmer chamber (r = 3.5 mm), the factor p_{dis} lay between 0.99 and 1.01. Thus, the authors concluded that the original assumption made by ICRU 23 [77] that the perturbation due to displacement could be neglected was justified. They were unable to find any support in their calculations for the p_u values given by IAEA [64], which exceed unity by several percent.

The overall perturbation factor is denoted by p_u', it is now given by

$$p_u' = (K_{air}^{med}/K_{air}^{fs})(M^{fs}/M^{med}) \qquad (2.151)$$

where fs indicates free space.

For a cylinder of height t and radius R irradiated parallel to the axis, i.e., a plane-parallel chamber, the Harder expression can be written in terms of a perturbation factor p as: [38]

$$p = 1 - 0.13(t/R)\sqrt{tT_m} \qquad (2.152)$$

where T_m is the linear electron scattering power, $d\bar{\theta}^2/ds$, for the medium surrounding the cavity. Harder [71] also gave an expression for the more practically useful geometry of a long cylinder irradiated perpendicular to its axis, i.e., a cylindrical or thimble chamber. This can be written as

$$p = 1 - 0.040\sqrt{rT_m} \qquad (2.153)$$

where r has been used for the chamber radius to distinguish it from R for the case of a coin-shaped cavity.

The dependence of the perturbation correction factor on the radius of the chamber is

$$p_{w,R} = 1 + kr \qquad (2.154)$$

where r is the cavity radius and k is energy dependent. Harders theoretical treatment [Equation (2.153)] had predicted a dependence on $\sqrt{r}$. Figure 2.37 shows their experimental results, at mean energy at depth $\bar{E}$ = 2.5 MeV, as a function of cavity radius.

Figure 2.38a shows how the perturbation is confined to the edges of a coin-shaped cavity. The guard ring in the NACP chamber was designed to prevent the signal due to this edge perturbation being collected.

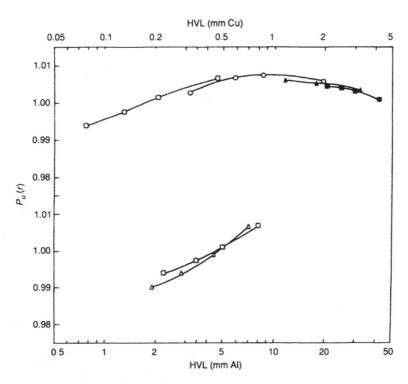

FIGURE 2.36 Variation of the perturbation correction factor p_u with HVL (mm Al) for radiation qualities with tube potential 100 (Δ) and 120 ($\square$) and with HVL (mm Cu) for 120 ($\square$), 150 (O), 200 ($\blacktriangle$) and 250 kV ($\blacksquare$). The factor is independent of the wall material and central electrode. (From Reference [35]. With permission.)

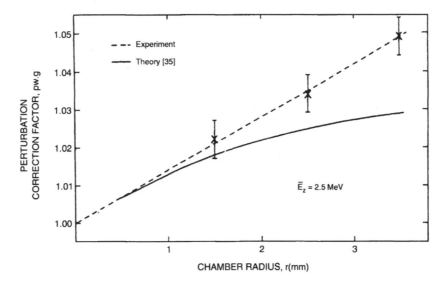

FIGURE 2.37 Perturbation correction factors $p_{w,g}$ for cylindrical chambers for electron radiation ($\bar{E}_z = 2.5$ MeV) as a function of the chamber radius. Experimental values were obtained from measurements using plane-parallel and cylindrical chambers. Theoretical values calculated according to Harder. (From Reference [35]. With permission.)

In a series of measurements, Roos [39] compared the responses of the NACP, PTW/Markus, and PTW/Schul (M23346) designs with the Roos chamber (PTW 34001) which has a 4-mm guard. The latter chamber was specifically designed to exhibit negligible perturbation in low-energy electron beams. It was found that there was negligible difference between the responses of the Roos and NACP chambers. However, for the other two chambers

investigated, the PTW/Marklis and PTW/Schulz models, the perturbation factor began to decrease below unity for energies of $\bar{E}_z \sim 10$ MeV and below (Figure 2.39). The ratio of the response of the PTW/Markus and NACP chambers that Roos obtained was in excellent agreement with the measurements of Van der Piaetsen et al. [40]; the latter's fit to the P_{MARKUS}/P_{NACP} ratio is shown as the full line in Figure 2.39.

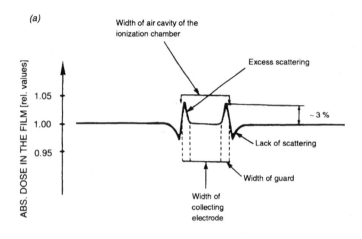

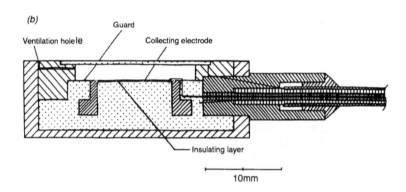

FIGURE 2.38 (a) Film measurements across the front surface of an air cavity in a PMMA phantom at the depth of maximum dose in an $E_0 = 6$ MeV electron beam: the effect of the guard ring in reducing the perturbation is clearly demonstrated. (b) A design for a plane-parallel chamber (the "NACP" chamber) which follows the specification set out in the NACP [72] protocol. The very narrow (2-mm) air gap and the presence of the guard ring minimize perturbation effects. The collecting electrode is very thin (<0.1 mm) and is mounted on a thin insulating layer (0.2 mm) in order to give a negligible polarity effect. The front wall (0.5 mm thick to enable measurements to be made at small depths) and back wall are made of one single material (in this case, graphite). (From Reference [35]. With permission.)

If backscatter plays a role in the response of certain plane-parallel chambers, then this should be taken into account in the perturbation factor. It is suggested, therefore, that a total or overall perturbation factor p_Q for plane-parallel chambers should be written as the product of the cavity factor and two separate wall factors:

$$p_Q = p_{cav} p_{wall}^{\delta} p_{wall}^{sc} \qquad (2.155)$$

where the component of the wall correction due to back scatter effects has been written as p_{wall}^{sc}. Naturally, this expression can also be applied to cylindrical chambers.

VIII.   DOSIMETRY FOR BRACHYTHERAPY

In brachytherapy treatment planning, dose distributions are most commonly calculated from point-source approximations where correction factors are applied to account for the geometrical shape of a non-pointlike source, the radial

dependence and angular anisotropy of the photon absorption and scatter in the source, encapsulation, and the surrounding medium. The functions describing the radial dependence of the total dose distribution, both primary and scatter, for a pointlike source are usually polynomial functions of the radial distance from the center of the source and along the perpendicular bisector to the long axis of the source.

A model based on the EGS4 Monte Carlo system was developed by Luxton [41] for calculating dose rate to water from an embedded low-energy brachytherapy source, given measurement data of dose rate to water within a water-substitute solid phantom for a source of given strength. The EGS4-based model was used to calculate point source dose rate distributions per unit source strength for water and for several species of solid phantoms. The Monte Carlo system was used to calculate dose rate to a thin spherical shell of water contained within the solid phantom at various distances centered on the source. Correction factors were calculated for polymethylmethacrylate

(PMMA or acrylic), solid water (WTI), and RW-I, a material optimized for low-energy dosimetry, with photon spectra from Pd-103 and from two commercial models of 1-125 seed used as input. For model 6711 1-125 seeds at 1 cm in PMMA and WTI, the calculated ratios of dose rate to water to dose rate to water in the solid phantom are 0.893 and 1.038, respectively. Atomic composition of phantom materials is given in Table 2.5.

The dose rate in an arbitrary medium is written in the point source model as

$$D_{med,point}(r) \;=\; \Lambda_{med} S_K \frac{g(r)}{(r/r_0)^2} \qquad (2.156)$$

TABLE 2.5
Percent Atomic Composition (by weight) and Densities of Phantoms

Atomic Species	Water	PMMA	WT1	RW-1
H	11.2	8.0	8.1	13.2
C	...	60.0	67.2	79.4
N	...	...	2.4	...
O	88.8	32.0	19.9	3.8
Ca	...	...	2.3	2.7
Mg	...	...	...	0.9
Cl	...	...	0.1	...
Density (g cm^{-3})	1.00	1.17	1.015	0.97

Source: From Reference [41]. With permission.

---

where $D_{med,point}(r)$ is the dose rate to the medium in cGy h^{-1} at the radial distance r in cm from the center of a point source; $r_0 = 1$ cm is the reference distance; S_K is the air-kerma strength of the source in units of cGy cm^2 h^{-1}, numerically equal to the air-kerma strength in the recommended unit $U = 1 \ \mu$Gy m^{2}h^{-1}; and Λ_{med} is the dose rate constant for the source in that medium, defined here as the dose rate to the medium at 1 cm from the center of the point source per U of air-kerma strength. The units of Λ_{med} are cGy h^{-1} U^{-1}. Λ_{med} will be understood to refer to a 1 cm distance along the transverse axis when reference is made to any physical nonpoint source that possesses cylindrical symmetry. $g(r)$ is defined by this equation as the dimensionless radial dose function for that source/medium combination, normalized to $g(r) = 1$ at $r = r_0 = 1$ cm.

The dose rate to infinitesimal water substance in the medium may be written as

$$D_{med}^{w}(r) \;=\; f_{med}^{w}(r) D_{med,point}(r) \qquad (2.157)$$

where $f_{med}^{w}(r)$ is the dimensionless ratio of dose rate at distance r from the point source to water as compared to the substance of the medium. The Monte Carlo method was used to calculate dose rate distribution for $D_{med}^{w}(r)$ as well as for $D_{med,point}(r)$ because it is conventional and clinically more pertinent to express dose rate and dose rate distributions in terms of water substance, whether in water or in a water substitute phantom. $D_{med}^{w}(r)$ distributions were calculated per unit primary photon emission per unit

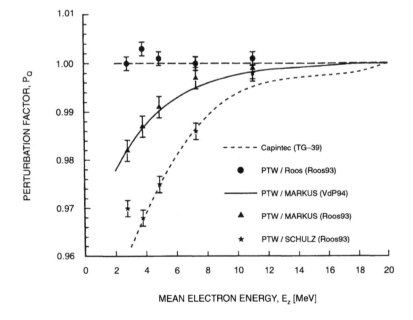

FIGURE 2.39 The variation of the overall perturbation factor p_Q for several different plane-parallel chambers in common use, relative to the NACP chamber, indicated by the dashed line drawn at $p_Q = 1.000$. All the measurements were made at the depth of dose maximum and normalized to the quotient test chamber/NACP in a high-energy electron beam. The full line is a fit to three separate measurement series on different accelerators using the PTW/Markus chamber. [40] The individual data points are measurements on three different PTW designs taken from Roos [39] and re-normalized so that $p_Q = 1.000$ for the NACP chamber; the dashed curve is for the Capintee-PS-003 chamber as given by AAPM. (From Reference [35]. With permission.)

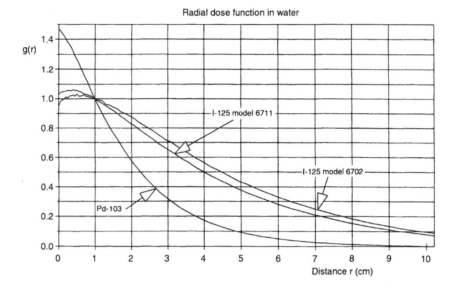

FIGURE 2.40 Calculated radial dose functions in water for point sources of ^{103}Pd and two models of ^{125}I seed. (From Reference [41]. With permission.)

time. The magnitude of primary photon emission per unit time is proportional to S_K for each brachytherapy source.

The radial dose distribution in water for I^{125} and Pd^{103} is shown in Figure 2.40. The dose rate distributions per unit activity to water in the solid phantoms are obtained from the $g(r)$ for the homogeneous media by multiplication by the product of corresponding ratios. The results are given in Figure 2.41.

Computer algorithms for electron-binding correction to Compton scattering and for detailed simulation of K-edge characteristic x-ray production were incorporated into EGS4 unix version 2.0 by Wang and Sloboda. [42] Based on detailed modelling of the internal structures of sources, the modified version was used to calculate dose rate constants, radial dose functions, and anisotropy functions on the long axis for an ^{125}I model 6711 source, ^{169}Yb Type 5 and Type 8 sources, and a stainless steel clad (SS) ^{192}Ir source. The geometry of these sources is cylindrically symmetric.

Following the dose calculation formalism proposed originally by the Interstitial Collaborative Working Group (ICWG) for interstitial brachytherapy sources, and expanded to all brachytherapy sources by TG-43, dose rate in medium at distance r (cm) from the source center and angle θ relative to its long axis is expressed as

$$\dot{D}(r, \theta) = S_k \Lambda \frac{G(r, \theta)}{G(1, \pi/2)} F(r, \theta) g(r) \qquad (2.158)$$

where S_k is the air-kerma strength of the source, Λ is the dose rate constant, $G(r, \theta)$ is the geometry factor, $F(r, \theta)$ is the anisotropy function, and $g(r)$ is the radial dose function. With S_k and $\dot{D}(r, \theta)$ calculated, Λ, $g(r)$, and $F(r, \theta)$ can be formulated as follows:

$$\Lambda = \frac{\dot{D}(1, \pi/2)}{S_k} \qquad (2.159)$$

where $\dot{D}(1, \pi/2)$ is the dose rate at 1 cm from the source center on the transverse axis, and

$$g(r) = \frac{\dot{D}(r, \pi/2) \times G(1, \pi/2)}{\dot{D}(1, \pi/2) \times G(r, \pi/2)} \qquad (2.160)$$

where $\dot{D}(r, \pi/2)$ is the dose rate at distance r on the transverse axis and $G(1, \pi/2)$ and $G(r, \pi/2)$ are geometry factors for the source at 1 cm and at a distance r on the transverse axis. Finally

$$F(r, \theta) = \frac{\dot{D}(r, \theta) G(r, \pi/2)}{\dot{D}(r, \pi/2) G(r, \theta)} \qquad (2.161)$$

The electron binding correction added to EGSU was carried out for both total and differential Compton scattering cross sections. Data for the total cross sections and for the incoherent scattering function were taken from photon cross-section library DLC-99.

To explore the influence of the binding effect of Compton scattering, the modified and original EGS4 codes were used to calculate dose distributions for a realistic ^{125}I 6711 source in unbounded water phantoms. For simplicity, call the doses evaluated in the two cases bound (Compton) dose and free (Compton) dose. Figure 2.42 shows ratios of bound-to-free doses on the transverse axis, delivered by the source in water. At distances less than or equal to 7 cm, the influence of the binding correction is small, less than or equal to 1%. However, the bound dose increases by a factor of 1.02 to 1.04 at distances from 8 to 12 cm and by a factor of up to 1.14 at further distances from 13 to 20 cm.

Figure 2.43 compares the radial dose function $g(r)$ calculated for an ^{125}I Model 6711 source in unbounded water with corresponding Monte Carlo results obtained by Williamson (using his photon transport code'3"4),

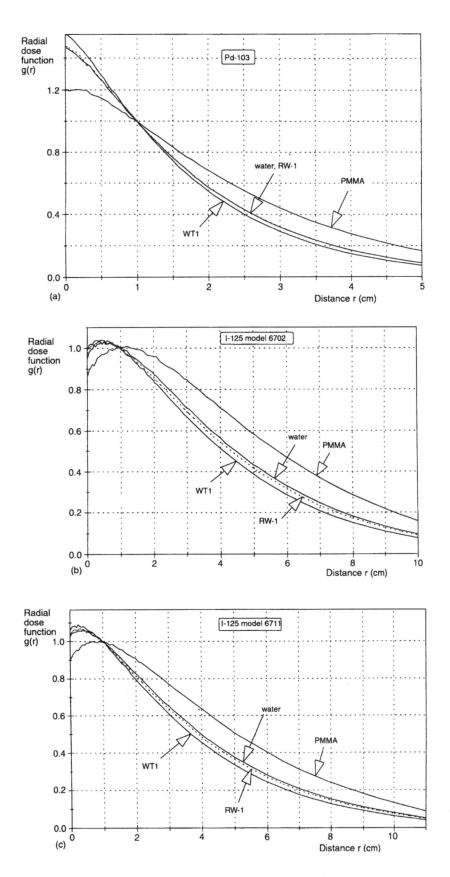

FIGURE 2.41 Calculated radial dose functions for PMMA, WTI, RW-1, and water, respectively, from point sources of (a) [103]Pd, (b) model ^{125}I, and (c) model 6702[125]I. (From Reference [41]. With permission.)

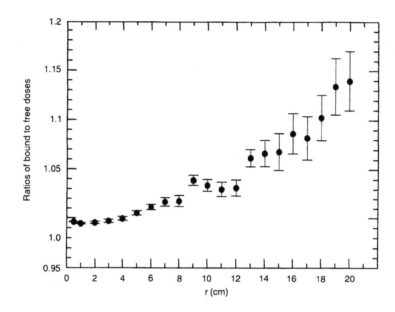

FIGURE 2.42 Influence of binding correction for Compton scattering on calculated doses for an [125]I 6711 source in water. The ratio of doses with and without the correction is plotted as a function of distance on transverse axis. (From Reference [42]. With permission.)

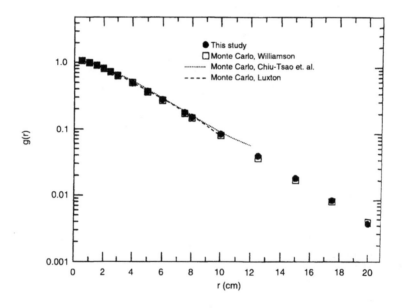

FIGURE 2.43 Calculated radial dose function for an [125]I Model 6711 source in unbounded water compared with Monte Carlo results. (From Reference [42]. With permission.)

by Chiu-Tsao et al. (using a MORSE Monte Carlo code), and by Luxton (using an EGS4 code and assuming a point source model).

In order to describe the decrease of dose with distance, ionization chamber measurements were performed by Venselaar et al. [43] for the ^{60}Co, ^{192}Ir and ^{137}Cs sources in a water tank. Source and chamber could be moved independently. Dimensions of the water tank were changed to simulate different patient sizes. As an independent check of the results, Monte Carlo (MC) calculations with the EGS-4 code system were performed in a simulated water

phantom. Using these results, a mathematical model was used to describe the dose at large distances (10–60 cm), which can be applied in addition to the existing calculation models for the short range (<10 cm).

The decrease of the dose around a brachytherapy point source is described mainly by the inverse square law. Within the first few centimeters, a calculation of the dose with only the inverse square law correction is accurate within a few percent for medium- and high-energy sources. However, deviations occur and these are due to the additional contribution to the dose from scattered

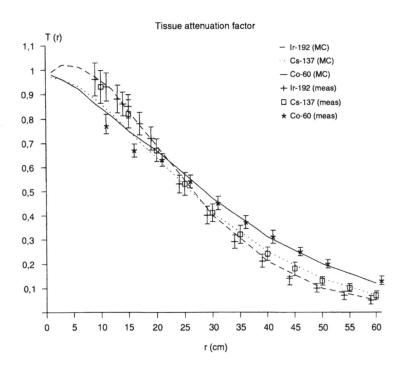

FIGURE 2.44 Results of the measurements and the calculations for an "infinitely" sized phantom, presented together for ^{192}Ir, ^{137}Cs, and ^{60}Co. Results in the range of 10–60 cm obtained by measurement are presented as points; the error bars indicate the estimated inaccuracy (1 sd) of the data points. Lines are used for the representation of the Monte Carlo results in the range of 1–60 cm. (From Reference [43]. With permission.)

radiation and from absorption of radiation. At 10 cm, deviations from the inverse square law may be as large as 20%, depending on the energy of the emitted radiation. The scatter and absorption effects are usually taken into account in a combined "tissue attenuation factor," $T(r)$, in order to calculate the dose rate $\dot{D}(r)$ as a function of the distance r from the assumed point source:

$$\dot{D}(r) = S_K \times (\mu_{en}/\rho)_a^w \times T(r)/r^2 \qquad (2.162)$$

When S is taken to account for the air-kerma strength of the source expressed in the units μGy m^2 h^{-1} and r in cm, the dose rate $\dot{D}(r)$ is in cGy h^{-1}. For $(\mu_{en}/\rho)_a^w$, the average mass energy absorption coefficient, the value 1.112 may be inserted for the sources ^{137}Cs, ^{192}Ir, and ^{60}Co. $T(r)$ in this equation is traditionally written as the ratio of exposure in water to exposure in air or as a ratio of dose in water to water kerma in free space.

For monoenergetic radiation, an expression can be written on a more fundamentally physical basis: $T(r) = B(r) \times \exp(-\mu r)$, in which $B(r)$ can be interpreted as a buildup factor and μ as the linear attenuation coefficient in water. In many cases the isotopes used as brachytherapy sources have a complex spectrum. B as a function of r has to be parametrized, so the advantage as a physically acceptable model is limited. The mathematical expression used is

$$T(r) = [1 + ka \times (\mu r)^{kb}] \times \exp(-\mu r) \qquad (2.163)$$

Tripathy and Shanta [44] have further developed this approach and presented a model with $T(r) = (a_1 + a_2 r) \cdot E_{av}^{(b1 + b2 \cdot r)}$. Its validity is reported again for the same isotopes in the range up to 30 cm.

The measurement equipment consisted of a water tank with dimensions $100 \times 50 \times 35$ cm^3, in which a source holder and an ionization chamber could be moved independently. Absolute dose was measured with two ionization chambers, one calibrated chamber with a sensitive volume of 0.6 cm^3 (graphite, NE 2505/3) and one with a 35-cm^3 volume (NE 2530). The chambers were connected to a NE Ionex 2500/3 electrometer.

In Figure 2.44 the results of the measurements and the calculations for an "infinitely" sized phantom are presented together for the three isotopes. Points are used for the measurements in the range of 10–60 cm. Error bars indicate the estimated inaccuracy (1 sd) of the data points. Lines are used for the representation of the Monte Carlo results in the range of 1–60 cm.

A dose calculation algorithm for brachytherapy was presented by Russell and Ahnesj [45] that reduced errors in absolute dose calculation and facilitates new techniques for modelling heterogeneity effects from tissues, internal shields, and superficially positioned sources. The algorithm is based on Monte Carlo simulations for specific source and applicator combinations. The dose is scored separately, in absolute units, for the primary and different categories of scatter according to the photon scatter generation.

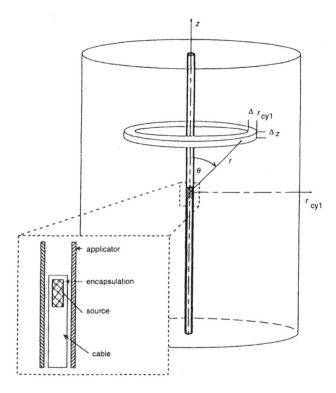

FIGURE 2.45 The cylindrical scoring geometry for the Monte Carlo calculations. The water or air phantom is divided into a number of rings of thickness Δz and radial width Δr_{cyl}. The dose to points within the voxel is estimated as the total energy absorbed by the ring divided by the mass of the ring. The details of the source geometry are shown in the inset. Each shaded area is treated as a separate region of medium in the Monte Carlo calculations. The scoring volume is centered in a large volume to simulate infinite-phantom conditions. (From Reference [45]. With permission.)

Radial dose distributions for the primary dose and the total scatter dose are parametrized using functions based on simple one-dimensional transport theory. The fitted radial parameters are functions of the angle to the long axis of the source to account for the anisotropy of the dose distribution.

$[D/R]_c^{w, pos}$ denotes the calculated dose per radiant energy to a point at the position *pos* in a large water phantom. The calculated dose rate $\dot{D}_c^{w,pos}$ to this point from a single radioactive source is then

$$\dot{D}_c^{w, pos} = \dot{K}_m^{a, ref}\frac{(D/R)_c^{w, pos}}{(K/R)_c^{a, ref}}e^{-\lambda T} \qquad (2.164)$$

where the exponential corrects for the decay of the source during the time T and λ is the source-specific radioactive decay constant. The calculated kerma-per-radiant energy in air at the reference point, $(K/R)_c^{a, ref}$, is used together with the measured air-kerma rate in air, $\dot{K}_m^{a, ref}$, at the same reference point to determine the absolute dose rate. Since $(K/R)_c^{a, ref}$ includes the effects of attenuation and scatter in air, the measured reference air-kerma rate must also

include these effects. This procedure eliminates the need for correction factor (and associated errors) for the attenuation and scatter in air.

The dose to a point at radial distance r and angle θ is often defined as

$$\dot{D}(r, \theta) = S_K\Lambda_0\frac{G(r, \theta)}{G(r_0, \theta_0)}F(r, \theta)g(r)e^{-\lambda T} \qquad (2.165)$$

where $S_K = K_m^{a, ref}Cr_{100}^2$, the reference air-kerma strength at radial distance $r_{100} = 100$ cm in free space (C is a correction factor for converting kerma in air to kerma in free space by accounting for attenuation and scattering in air); (r_0, θ_0) is a standard point located 1 cm along the perpendicular bisector $(1, \pi/2)$; $\Lambda_0 = D(r_0, \theta_0)/S_K = D(r_0, \theta_0)/K_m^{a, ref}Cr_{100}^2$ is the specific dose rate constant; $F(r, \theta)$ is the anisotropy distribution describing the angular dependence of the absorption and scatter in the source and surrounding material; $G(r, \theta)$ is the $1/r^2$ geometry distribution including effects for the spatial distribution of the source; $g(r)$ is the radial dose profile along the perpendicular bisector of the source; and $e^{-\lambda T}$ is the correction for exponential decay between the date of source calibration and the treatment date. Equation (2.164) may be rewritten as

$$\begin{aligned} D_c^{w, pos} = &\ K_m^{a, ref}Cr_{100}^2\frac{[D/R]_c^{w, (r_0, \theta_0)}}{[K/R]_c^{a, ref}Cr_{100}^2} \\ &\times \frac{[D/R]_c^{w, pos}}{[D/R]_c^{w, (r_0, \theta_0)}}e^{-\lambda T} \end{aligned} \qquad (2.166)$$

and by comparing Equations (2.165) and (2.166), one finds that

$$\Lambda_0 = \frac{[D/R]_c^{w, (r_0, \theta_0)}}{[K/R]_c^{a, ref}Cr_{100}^2} \qquad (2.167)$$

and

$$\frac{G(r, \theta)}{G(r_0, \theta_0)}F(r, \theta)g(r) = \frac{[D/R]_c^{w, pos}}{[D/R]_c^{w, (r_0, \theta_0)}} \qquad (2.168)$$

where all the effects of the source, encapsulation, and applicator on the dose distribution are implicit in the Monte Carlo dose deposition kernel.

The total dose per radiant energy is the sum of the primary and scatter dose contributions (all generations).

$$[D/R]_c^{w, pos} = [D/R]_{c, prim}^{w, pos} + [D/R]_{c, scat}^{w, pos} \qquad (2.169)$$

The EGS4/PRESTA computer code was used together with a cylindrical scoring geometry to generate dose distributions in a large homogeneous water phantom. The source description and calculation model are shown in Figure 2.45.

An example of the dose deposition for a Monte Carlo simulation of the ^{192}Ir source, described above, centered in a nylon catheter with an outer diameter of 1.7 mm and

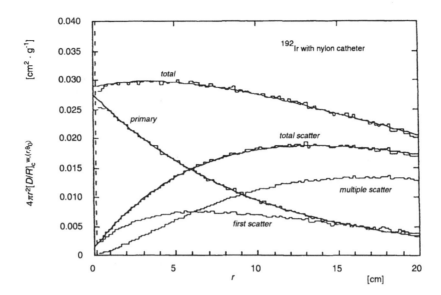

FIGURE 2.46 Monte Carlo simulation (histogram lines) and parametrization (smooth lines) for ^{192}Ir source and nylon catheter combination (from Reference [45] with permission).

a 0.2-mm wall thickness, is shown in Figure 2.46. The dose per radiant energy, with the $(4\pi r^2)^{-1}$ factor removed, and perpendicular to the bisector of the source $\theta_0 = \pi/2$ out to a radial distance of 20 cm, is shown for the primary, first-scatter and multiple-scatter dose fractions separately, as well as for the total and the all-scatter dose. The all-scatter dose has a maximum, approximately one photon mean free path from the source axis, where the build-up of scattered photons has saturated, and then it decreases as a function of distance from the source with the overall attenuation of the primary photon fluence. The parametrizations for the total, primary, and scatter doses are also shown in Figure 2.46 for the primary and scatter doses.

Watanabe et al. [46] have performed computational and experimental dosimetry of the Henschke applicator with respect to high-dose rate ^{192}Ir brachytherapy using a GAM-MAMED remote afterloader. Monte Carlo simulations were performed using the MCNP code. The computational models included the detailed geometry of ^{192}Ir source, tandem tube, and shielded ovoid. The measured dose rates were corrected for the dependence of TLD sensitivity on the distance of measurement points from the source.

Tandem and ovoid schematics are given in Figures 2.47a and 2.47b.

The dose rate at point P at 0.25 cm in water from the transverse bisector of a straight catheter with an active pepping source (Nucletron microSelectron HDR source) with a dwell length of 2 cm was calculated by Wong et al. [47] using Monte Carlo code MCNP 4-A.

The dose calculation formalism used by the Plato brachytherapy planning system (BPS) version 13.X is

$$\dot{D}(r, \theta) = S_K \left(\frac{\mu_{en}}{\rho}\right)_{air}^{tiss} \frac{1}{r^2} \varphi(r) F(\theta) \qquad (2.170)$$

(a) [mm]

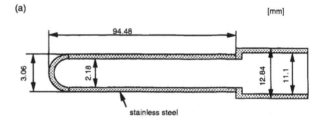

(b)

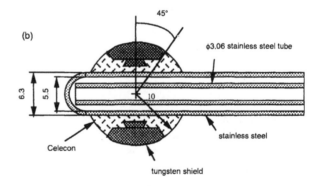

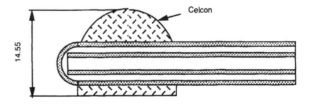

FIGURE 2.47 Longitudinal section of applicators. Dimensions are in mm. (a) Tandem applicator. (b) Ovoid applicator, planes through and midway between shield centers. (From Reference [46]. With permission.)

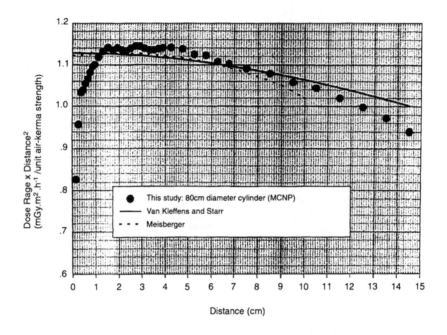

FIGURE 2.48 Dose rate times distance^2 (μGy m^2 h^{-1}) in a water phantom (80-cm diameter) per unit air-kerma strength (1 U = 1 μGr m^2 h^{-1}) vs. distance from the source center. All results are absolute with no cross-normalization. (From Reference [47]. With permission.)

where S_k is the air kerma source strength (Gy m^{-2} h^{-1}); $(\mu_{en}/\rho)_{air}^{tiss}$ is the ratio of the mean mass energy absorption coefficient in tissue and air, having a value of 1.11 for the ^{192}Ir source; $\varphi(r)$ is the modified Van Kleffens and Star correction for absorption and scattering in water; $F(\theta)$ is the anisotropy function; and $1/r^2$ is the geometric factor. Equation (2.170) can be modified to the AAPM TG-43 formalism: [47]

$$\dot{D}(r, \theta) = S_k \Lambda \frac{1}{r^2} g(r) F(r, \theta) \qquad (2.171)$$

where Λ is the dose-rate constant for the source and surrounding medium (cGy h^{-1} U^{-1}) (1 U = 1 cGy cm^{-2} h^{-1}) for the microSelectron HDR source; the value of Λ in water is 1.115 cGy h^{-1} U^{-1}; $g(r)$ is the radial dose function; and $F(r, \theta)$ is the anisotropy factor. In AAPM TG-43, the recommended value for Λ in water is 1.12, which is a rounded-off value of 1.115. Strictly, the TG-43 formalism uses a line source model for the geometric factor, but the point source model ($1/r^2$) is retained here for ease of comparison [47].

Figure 2.48 shows the variation of the radial dose (dose rate times distance) with the distance from the source.

Figure 2.49 shows the angular dose profiles at distances of 0.25 cm, 0.5 cm, and 1 cm from the source center. The profiles show significant variation at distances of 0.25 cm and 0.5 cm from the source center compared with the BPS's anisotropy function.

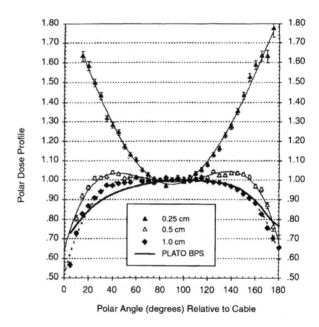

FIGURE 2.49 Monte Carlo calculated polar dose profiles in water at a distance of 0.25, 0.5, and 1.0 cm from the source center (normalized to 1.0 at 90°) vs. the polar angle relative to the source's cable. The BSP dose profile shows large errors with the Monte Carlo results at distances of 0.25 and 0.5 cm from the source center. (From Reference [47]. With permission.)

IX. PROTON DOSIMETRY

Protons have a well-defined range in homogenous tissue and deposit most of their energy at the end of their range in the so called Bragg peak. Single incident proton beam can be modulated to produce a flat high-dose region, the spread-out Bragg peak, at any depth of the body with a lower dose in entrance region. No dose is deposited past the distal border of the spread-out Bragg peak. With this characteristic depth-dose curve, protons have a considerable potential for sparing normal tissue surrounding a tumor.

Two dosimetry protocols have been published: one by the American Association of Physicists in Medicine TG 20 [75] and one by the European Clinical Heavy Particle Dosimetry Group (ECHED) [76], both providing a practical procedure for the determination of absorbed dose to tissue or water, in proton beams, using ionization chambers calibrated in terms of air kerma in a ^{60}Co beam. Both protocols give an uncertainty of about 4% on this absorbed dose determination. Because protons undergo little scattering in low-Z materials, it is generally assumed that perturbations caused by the non-water equivalence of ionization chambers must be very small.

In proton beam dosimetry the use of N_k-calibrated ionization is well established. This is the procedure recommended by the ECHED dosimetry protocol for therapeutic proton beams when no calorimetric calibration of the ion chamber is available. The American TG-20 protocol also recommended this procedure as an alternative to calorimetry or Faraday-cup-based dosimetry methods.

A method which reduces the final uncertainity in the measured dose to water, is to calibrate the ionization chamber in terms of absorbed dose to water.

Ionization chamber dosimetry of proton beams using cylindrical and plane-parallel chambers was discussed by Medin et al. [48]

The N_D formalism in the IAEA Code of Practice for photon and electron dosimetry [64] can be extended to proton beams. The absorbed dose to water at the user's beam quality, $D_{w,Q}$, at the reference point of the ionization chamber P_{eff} is generally given by

$$D_{w,Q}(P_{eff}) = M_Q N_{D,Q_0}[(W_{air})_{Q_0}/(W_{air})_{Q_0}](S_{w,air})_Q p_Q$$

(2.172)

where M_Q is the reading of the electrometer plus ionization chamber at the user's beam quality corrected for influence quantities (temperature, pressure, humidity, saturation, etc.), $(W_{air})_Q/(W_{air})_{Q_0}$ is the ratio of the mean energy required to produce an ion pair in the user's proton beam quality Q and in the calibration quality Q_0, and $(s_{w,air})_Q$ is the water to air stopping power ratio in the user's beam. The factor p_Q contains the product of different perturbation correction factors of the ion chamber at the user's beam quality, which generally consists of the factors p_{wall}, $p_{cel-gbl}$, and P_{cav}. These correct, respectively, for the lack of water equivalence of the chamber wall, the influence of the central electrode of the chamber both during calibration and in-phantom measurements,

and the perturbation of the fluence of secondary electrons due to differences in the scattering properties between the air cavity and water. If the concept of effective point of measurement is not used and the determination of absorbed dose is referred to the center of the ionization chamber, a fourth factor P_{displ} has to be included in P_Q to take into account the replacement of water by the detector. N_{D,Q_0} is the absorbed dose to air chamber factor at the beam quality used by the dosimetry laboratory, which is defined by the well-known relation. [64]

$$N_{D,Q_0} = N_{K,Q_0}(1-g)k_{air}k_m$$

(2.173)

In the N_w formalism the absorbed dose to water at the center of the chamber is given by the relationship

$$D_{w,Q} = M_Q N_{w,Q_0} k_Q$$

(2.174)

where N_{w,Q_0} is obtained at the standard laboratory from the knowledge of the absorbed dose to water at the point of measurement in water for the calibration quality,

$$N_{w,Q_0} = D_{w,Q_0}/M_{Q_0}$$

(2.175)

The factor k_Q in Equation (2.174) corrects for the difference in beam quality at the standard laboratory and at the user's facility and should ideally be determined experimentally at the same quality as the user's beam. When no experimental data are available, k_Q can be calculated according to the expression

$$k_Q = [(s_{w,air})_Q/(s_{w,air})_{Q_0}][(W_{air})_Q/(W_{air})_{Q_0}](P_Q/P_{Q_0})$$

(2.176)

It should be noted that the chamber-dependent correction factors k_{att} and k_m are not included in the definition of k_Q.

Another potential advantage in the use of a calibration factor in terms of absorbed dose to water is that the product of $s_{w,air}$ and W_{air}, referred to as ω, can be determined using calorimetry with a much higher accuracy than that obtained when the two factors are considered independently, yielding

$$k_Q = (w_Q/w_{Q_0})(p_Q/p_{Q_0})$$

(2.177)

The expression for N_D of a plane-parallel ionization chamber can be written as

$$N_{D,x} = N_{D,ref}M_{ref}p_{cav,ref}p_{cel-gbl,ref}/M_x p_{cav,x}p_{wall,x}p_{cel-glb,x}$$

(2.178)

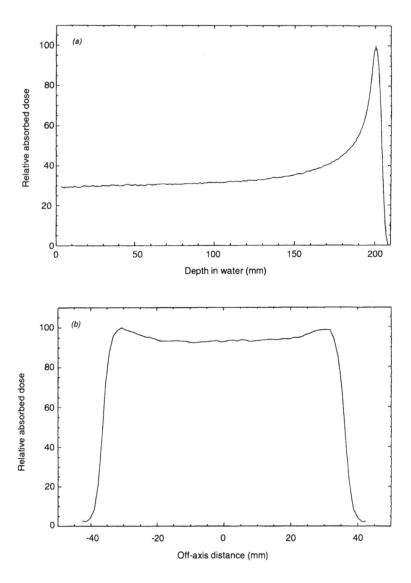

FIGURE 2.50 (a) The central axis depth-ionization distribution of the 170-MeV unmodulated proton beam used in this investigation, measured with a silicon diode detector. (b) A transverse scan of the proton field at the depth of 29 mm where the measurements with the ionization chambers were performed. (From Reference [48]. With permission.)

where the numerator corresponds to the D_w determination using the reference chamber (usually cylindrical) and stopping power ratios cancel out. M_{ref} and M_x are the average ratios of the readings of the two chambers to those of an external monitor to take into account possible accelerator output fluctuations.

Figure 2.50 shows the 170-MeV unmodulated proton depth-ionization distribution, measured with a silicon diode detector (Figure 2.50a), together with a transverse scan of the proton field at the depth where the ion chamber measurements were performed. It should be pointed out that the increase in dose at the lateral edges of the field (shown in Figure 2.50b) arises from protons scattered in the collimator and is not related to the design of the flattening filters.

Proton pencil beams in water, in a format suitable for treatment-planning algorithms and covering the radiother-

apy energy range (50–250 MeV), have been calculated by Carlsson et al. [49] using a modified version of the Monte Carlo code PTRAN.

A general assumption in the straggling theories of heavy charged particles is that the distribution of ranges is described by a Gaussian function $\varphi(\xi)$; this is justified for heavy charged panicles, since hard collisions are rare and *bremsstrahlung* generation is negligible. The distribution of the ranges is described by

$$\varphi(\xi) = \frac{1}{\sqrt{2\pi}\sigma(z)} e^{\frac{-(\xi - r_0)^2}{2\sigma^2(z)}} \tag{2.179}$$

The broad-beam central-axis depth-dose distribution, $D(z)$, is then calculated by convolving the specific energy

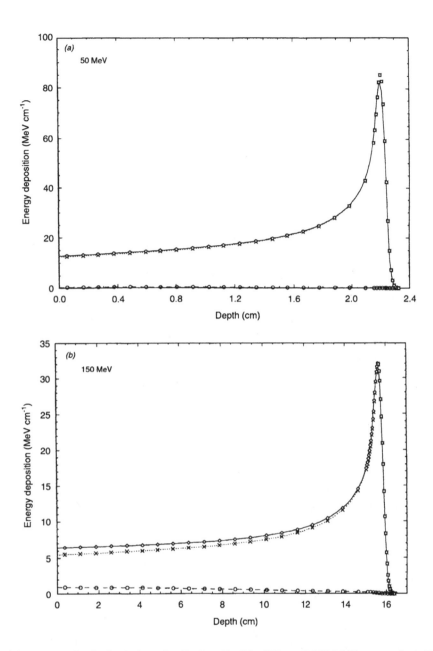

FIGURE 2.51 Broad-beam central-axis depth-dose distributions for 50-, 150-, and 250-MeV protons from PTRAN and Carlsson et al. analytical model. The total depth-dose distribution (solid line, PTRAN, ◇, analytical) as well as the individual contributions from Colomb (short dashes, PATRAN; ×, analytical) and inelastic nuclear interactions (long dashes, FTRAN; O, analytical) are shown. (From Reference [49]. With permission.)

deposition distribution, $d(z)$, in the csda with the range straggling distribution, $\varphi(\xi)$, according to

$$D(z) = \int \frac{d(\xi - z)}{\sqrt{2\pi}\sigma(z)} e^{\frac{-(\xi - r_0)^2}{2\sigma^2(z)}} d\xi \qquad (2.180)$$

where $d(\xi - z)$ is the specific energy deposition at the distance $(\xi - z)$ from the end of the depth-dose curve, $\sigma(z)$ is the standard deviation in position of the protons at the depth z, and r_0 is the csda range. Broad-beam central axis depth-dose distribution for 50 and 150 MeV protons is shown in Figure 2.51.

Proton depth-dose distribution was discussed by Palmans and Verhaegen. [50] The extended use of proton beams in clinical radiotherapy has increased the need to investigate the accuracy of dosimetry for this type of beam. As for photon and electron beams, Monte Carlo simulations are a useful tool in the study of proton dosimetry. The existing proton Monte Carlo code PTRAN, developed for dosimetry purposes, is designed for transport of protons in homogeneous water only. In clinical proton dosimetry, as well as in treatment conditions, several other materials can be present, such as plastic phantoms, plastic modulator wheels, and several materials in ionization

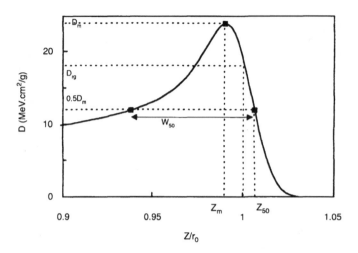

FIGURE 2.52 Parameters z_m, z_{50}, w_{50}, D_m, and D_{rg}, defined in the text, used by Berger [51] and by Palmans and Verhaegen [50] to characterize the Bragg peak of the depth-dose distribution (numerical values are for 200-MeV monoenergetic protons). (From Reference [50]. With permission.)

chambers. Investigation of protons transport in other media started from the PTRAN code and implemented proton transport in other materials, including heterogeneous systems composed of different materials. With this extended code, calculations of depth-dose distributions for some low-Z materials are performed and compared with those obtained for water. The results show that for plastics (PMMA, polystyrene, and A-150), the depth-dose characteristics are comparable to those of water. For graphite, air, and aluminum, larger differences are observed.

The original code PTRAN, developed and described extensively by Berger [52], simulates monodirectional monoenergetic proton beams in homogeneous water without transport of secondary particles. The depth doses obtained from the calculations agreed well with those for three experimental proton beams with energies above 150 MeV. The Bragg peak was characterized by a number of parameters of which a limited number, used in our study, are shown in Figure 2.52. These parameters are the depth z_m at which the maximum dose D_m occurs; D_{rg}, the dose at the CSDA range r_0 (taken from ICRU report 49 [78]; w_{50}, the peak width at 50% of the maximum dose; and z_{50}, the depth distal to the Bragg peak where the dose amounts to 50% of the maximum dose. All depths and widths are expressed relative to r_0. The entrance dose D_0 was also calculated. In Berger [52] it is shown that the radial dose distributions obtained with PTRAN agree with theoretical calculations based on Molliere's theory. Palmans and Verhaegen's calculations show that the root mean square (rms) values of the radial-dose distributions also agree with those obtained with the Fermi-Eyges method except at the end of the range, where the Fermi-Eyges method gives a steeper decrease. This is because the latter method does not include energy loss straggling.

An analytical approximation of the Bragg curve for therapeutic proton beam was discussed by Bortfeld. [53] The model is valid for proton energies between about 10 and 200 MeV. Its main four constituents are:

1. a power-law relationship describing the range-energy dependency;
2. a linear model for the fluence reduction due to nonelastic nuclear interactions, assuming local deposition of a fraction of the released energy;
3. a Gaussian approximation of the range-straggling distribution; and
4. a representation of the energy spectrum of poly-energetic beams by a Gaussian with a linear "tail."

Based on these assumptions, the Bragg curve can be described in closed form using a simple combination of Gaussians and parabolic cylinder functions.

Consider an initially monoenergetic broad proton beam along the z axis, impinging on a homogeneous medium at $z = 0$. The energy fluence, Ψ, at depth z in the medium can be written in the form:

$$\Psi(z) = \Phi(z)E(z) \qquad (2.181)$$

where $\Phi(z)$ is the particle fluence, i.e., the number of protons per cm^2, and $E(z)$ is the remaining energy at depth z. The total energy released in the medium per unit mass (the "terma," T) at depth z is then:

$$T(z) = -\frac{1}{\rho}\frac{d\Psi}{dz} = -\frac{1}{\rho}\left(\Phi(z)\frac{dE(z)}{dz} + \frac{d\Phi(z)}{dz}E(z)\right) \qquad (2.182)$$

where ρ is the mass density of the medium. The first term in the brackets represents the reduction of energy of the

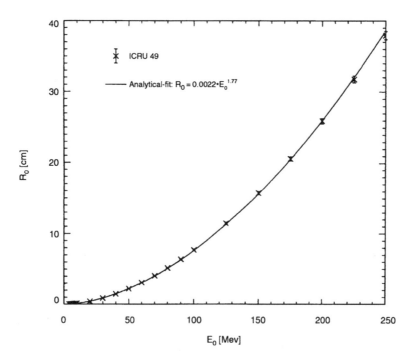

FIGURE 2.53 Range-energy relationship according to ICRU 49 and analytical fit with Equation (2.184). The error bars represent a relative error of the ICRU data of $\pm 1.5\%$. (From Reference [53]. With permission.)

protons during their passage through matter. The "lost" energy is transferred mainly to atomic electrons. The range of these secondary electrons is negligible for our purposes. This is due to the fact that only a relatively small energy is transferred to each electron, which, in turn, is due to the small ratio of electron and proton mass. Therefore, the terma corresponding with the first term of Equation (2.182) produces an absorbed dose which is equal to the terma.

The total absorbed dose, $\hat{D}(z)$, is consequently given by

$$\hat{D}(z) = -\frac{1}{\rho}\left(\Phi(z)\frac{dE(z)}{dz} + \gamma\frac{d\Phi(z)}{dz}E(z)\right) \quad (2.183)$$

In order to determine this depth-dose curve, only the functional relations $E(z)$ and $\Phi(z)$ are needed. These are obtained from the known range-energy relationship and the probability of nonelastic nuclear interactions, respectively.

The relationship between the initial energy $E(z = 0) = E_0$ and the range $z = R_0$ in the medium is given approximately by

$$R_0 = \alpha E_0^p \quad (2.184)$$

With $p = 1.5$, this relationship is known as Geiger's rule, which is valid for protons with energies up to about 10 MeV. For energies between 10 and 250 MeV, the exponent p increases to $p \approx 1.8$. The factor α is approximately proportional to the square root of the effective atomic mass of the absorbing medium, $\sqrt{A_{eff}}$ (Bragg-Kleeman rule).

It is also inversely proportional to the mass density of the medium (see Figure 2.53).

The depth-dose distribution is calculated by [53]

$$\hat{D}(z) =$$
$$\begin{cases} \Phi_0\dfrac{(R_0 - z)^{(1/p)-1} + (\beta + \gamma\beta p)(R_0 - z)^{1/p}}{\rho p \alpha^{1/p}(1 + \beta R_0)} & \text{for } z < R_0 \\[2ex] 0 & \text{for } z < R_0 \end{cases}$$

$$(2.185)$$

which is of the form

$$\hat{D}(z) = \hat{D}_1(z) + \hat{D}_2(z)$$
$$= a_1(R_0 - z)^{1/p - 1} + a_2(R_0 - z)^{1/p} \quad (2.186)$$

The first term, $\hat{D}_1(z)$, is the dose contribution from those protons that have no nuclear interactions. It is proportional to the (non-nuclear) stopping power and exhibits to some degree the form of a Bragg curve, as it increases monotonically from $z = 0$ to $z = R_0$ and has a peak at R_0. However, due to the neglect of range straggling, the peak is unrealistically sharp and there is a singularity at $z = R_0$. The second term, $\hat{D}_1(z)$, represents the dose delivered by the relatively small fraction of protons that have nuclear interactions. It decreases monotonically and is zero at $z = R_0$. Note that $\hat{D}_1(z)$ comprises the dose resulting not only from nuclear but also non-nuclear interactions that take place before the nuclear collision.

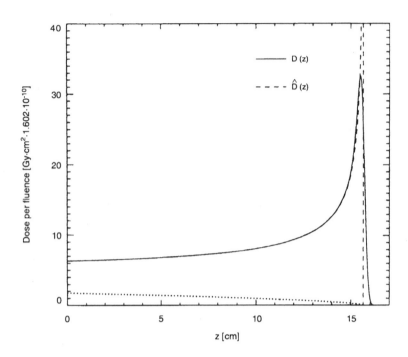

FIGURE 2.54 Bragg curves with and without consideration of straggling for 150-MeV protons in water. The dotted line at the bottom is the dose contribution from the fraction of protons that have nuclear interactions, i.e., $D_2(z)$ or $\hat{D}_2(z)$ (these are indistinguishable within the resolution of the figure). (From Reference [53]. With permission.)

Equation (2.185) gives $\hat{D}$ in units of MeV/g if ρ is given in g/cm³. To obtain $\hat{D}$ in Gy, one needs to multiply by the factor $10^9 e / C = 1.602 \times 10^{-10}$, where e is the elementary charge.

The dose delivered right after the protons have lost the energy $E_0 - E$ may be written as $\hat{D}_1(\bar{z}(E, E_0))$. The dose $D_1(z)$ at the actual depth z is then obtained by folding the Gaussian depth straggling into $\hat{D}_1$ by means of:

$$D_1(z) = \langle \hat{D}_1 \rangle(z) = \int_0^{R_0} \hat{D}_1(\bar{z}) \frac{e^{-(z-\bar{z})^2/2\sigma_z^2(\bar{z})}}{\sqrt{2\pi}\sigma_z(\bar{z})} d\bar{z}$$

(2.187)

The calculation of $D_2(z)$, i.e., the consideration of straggling for the fraction of protons that have nuclear interactions, is less straightforward but also less critical, because these protons contribute a smaller and smoother amount to the total dose.

One can write $D(z) = D_1(z) + D_2(z)$ as the convolution integral [53]

$$D(z) = \langle \hat{D} \rangle(z) = \frac{1}{\sqrt{2\pi}\sigma} \int_{-\infty}^{R_0} \hat{D}(\bar{z}) e^{-(z-\bar{z})^2/2\sigma^2} d\bar{z}$$

(2.188)

The resulting Bragg curves with and without consideration of straggling are shown in Figure 2.54.

An attempt to include density heterogeneity effects in the analytical dose calculation models for proton-treatment planning using a dynamic beam delivery system was made by Schaffner et al. [54] Different specialized analytical dose calculations have been developed which attempt to model the effects of density heterogeneities in the patient's body on the dose. Their accuracy has been evaluated by a comparison with Monte Carlo calculated dose distributions in the case of a simple geometrical density interface parallel to the beam and typical anatomical situations. A specialized ray-casting model which takes range dilution effects (broadening of the spectrum of proton ranges) into account has been found to produce results of good accuracy.

The dose calculation of Schaffner et al. consists of a superposition of individual scanned pencil beams. The description of the physical pencil beam uses calculated look-up tables of the depth-dose curve and the depth-width relation of a proton beam in water. The depth-dose curve is characterized by the nominal beam energy and the width of the initial energy spectrum (momentum band). The dose deposited by a single proton of known initial energy is derived from the Bethe-Bloch equation.

The formula for the dose deposited at a position (x, y, z) by a pencil beam along the z-axis and positioned at (x_0, y_0) is given by

$$D(x, y, z) = N_{p^+} ID(wer) \frac{1}{2\pi\sigma_x\sigma_y}$$
$$\times \exp[-(x_0 - x)^2/2\sigma_x^2] \exp[-(y_0 - y)^2/2\sigma_y^2] \quad (2.189)$$

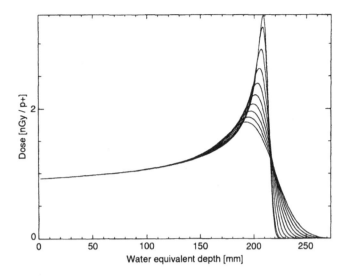

FIGURE 2.55 Depth-dose curve for a 177-MeV beam with a momentum band of 1.1% and with a relative range dilution of 0–9% in steps of 1%. (From Reference [54]. With permission.)

where

$$wer = wer(x_0, y_0, z)$$
$$\sigma_i = \sigma_i(wer(x_0, y_0, z)) \quad \text{with} \quad i = x, y.$$

ID is the integral dose, interpolated from the depth-dose look-up table; N_{p^+} is the number of protons in the beam spot; *wer* is the total water-equivalent range (range shifter plates plus patient) along the central axis of the beam; and σ_x, σ_y are the standard deviations of the Gauasians in *x* and *y* directions. They include all contributions to the beam width, i.e., initial phase space, Coulomb scattering in the patient and range shifter and its propagation in the air gap between range shifter and patient.

The resulting depth-dose curves for a relative range dilution from 0 to 9% standard deviation are shown in Figure 2.55. The influence of range dilution on the shape of the Bragg curve is very high in the peak region, while there are almost no changes in the plateau. This underlines the importance of the peak region and justifies their approach of calculating the range dilution only for protons stopping in the point of interest.

Utilization of air-filled ionization chambers with ^{60}Co-based reference calibrations in proton dosimetry requires application of water to air stopping power ratios and the mean energy required to produce an ion pair (*W* or *w*). Proton dosimetry protocols recommend the use of calorimetry as the absorbed dose standard. Siebers et al. [55] used calorimetry in conjunction with an ionization chamber with ^{60}Co reference calibrations to deduce the proton *w* value in the entrance region of a 250-MeV proton beam: 34.2 ± 0.5 eV.

For high-energy charged particles (protons), the differential value *w* is used since only a small fraction of the particle energy is deposited in the ion chamber gas cavity.

For photon and electron beams, the stopping power ratios and *W* values are known to within a few tenths of 1%.

The dose delivered to the water calorimeter was determined from [55]:

$$D_{c, water} = c\Delta T(1 + D_T) \qquad (2.190)$$

with $D_{c, water}$ being the absorbed dose to water, *c* the specific heat of water, ΔT the change in water temperature measured using the thermistors, and D_T the thermal defect of water. Water temperature changes were monitored using a Wheatstone bridge circuit. ΔT is expressed in terms of the Wheatstone bridge deflection ΔV_u and the thermistor calibration factor Θ as

$$\Delta T = \Theta\Delta V_u \qquad (2.191)$$

To evaluate ionization dosimetry, measurements were made in a geometry similar to that used for the calorimeter measurements through use of the dummy calorimeter. In place of the thermistors, a Farmer-type PTW ionization chamber (model W30001, volume 0.6 cm³, PMMA walls, aluminum central electrode) was inserted into a waterproof PMMA cap which was securely mounted in the core region. Ionization charge measurements were made for runs of length equal to that of the water calorimeter runs.

The ionization chamber calibration factor in the proton beam was determined from the calorimeter and ion chamber responses by applying the recommendations of the AAPM and ECHED code of practice.

$$N_{D_p} = \frac{D_{c, water}}{MP_{ion}C(IS, att)_p} \qquad (2.192)$$

where N_{D_p} is the proton calibration factor, $D_{c,water}$ is the dose measured using the calorimeter [Equation (2.190)], and M is the charge collected in the ionization chamber, corrected to reference temperature and pressure conditions. P_{ion} corrects for recombination in the ionization chamber and $C(IS,att)_p$ corrects for inverse square and attenuation (depth-dose) variations due to minor geometric differences between the calorimeter and the dummy calorimeter. The proton calibration factor for an ionization chamber with a ^{60}Co reference calibration is

$$N_{D_P} = N_{gas}(w_p/W_\gamma)_{gas}(\bar{S},\rho)_{gas}^{water}\, p_p \qquad (2.193)$$

To determine w_p, the calorimeter-based calibration factor [Equation (2.192)] is set equal to the calibration factor determined from the ^{60}Co-based calibration. $(w_p/W_\gamma)_{gas}$ is solved to reveal

$$(w_p/W_\gamma)_{gas} = (w_p/W_\gamma)_{air}$$

$$= \frac{D_{c,water}}{MN_{gas}(\bar{S},\rho)_{gas}^{water}\, p_p P_{ion}C(IS,att)_p}$$

$$\hspace{6cm}(2.194)$$

In Equation (2.194), the subscript *air* refers to dry air, whereas *gas* denotes humid air. The equivalence $(w_p/W_\gamma)_{gas} = (w_p/W_\gamma)_{air}$ can be made by assuming that humidity corrections to w values are independent of irradiation modality.

X.   CAVITY THEORY

Cavity theory is used to relate the radiation dose deposited in the cavity (the sensitive volume of the detector) to that in the surrounding medium. The dose to the cavity depends on the size, atomic composition, and density of the cavity and the surrounding medium. The size of the cavity is defined relative to the range of the electrons set in motion. A cavity is considered small when the range of the electrons entering the cavity is much greater than the cavity dimensions. The electron spectrum within a small cavity is solely determined by the medium surrounding the cavity. The ratio of absorbed dose in the cavity to that in the surrounding medium is given by the Bragg-Gray or Spencer-Attix theory. When the cavity dimensions are many times larger than the range of the most energetic electrons, the electron spectrum within the cavity is determined by the cavity material. A cavity whose dimensions are comparable to the range of electrons entering the cavity has a spectrum within the cavity that is partially determined by the medium and partially determined by the cavity material. Burlin proposed a general cavity theory to include all cavity sizes

A detector will give a signal proportional to the absorbed dose in the sensitive detector material, which in general differs from the medium. One can write

$$D_{med} = fD_{det} \qquad (2.195)$$

where D_{det} is the dose to the radiation-sensitive material, D_{med} is the dose to the medium, and f is the ratio of the two doses. In case of photon radiation, the factor f can be given either by the mass stopping power ratio $s_{med,det}$, if the detector is small compared to the electron range (Bragg-Gray cavity), or by the ratio of the mass energy absorption coefficients $(\mu_{en}/\rho)_{med,det}$, if the detector is large compared to the electron range. For both categories Burlin's theory yielded expressions combining both $s_{med,det}$ and $(\mu_{en}/\rho)_{med,det}$.

According to Bragg-Gray theory, Equation 2.195 becomes

$$D_{med} = D_{air}s_{med,air}$$

where

$$D_{air} = J_g(W/e)$$

$s_{med,air}$ is the mass stopping power ratio, J_g is the charge collected per unit mass of air, and W/e is the energy required to create one ion pair in air. When the incident photon energy decreases, the energy absorbed in air in the chamber increases and will finally invalidate the Bragg-Gray condition.

The validity of the Bragg-Gray cavity theory in photon radiation dosimetry for photon energies from 10 keV to 10 MeV has been investigated by Ma and Nahum [56] quantitatively. The ratio F_{air} of the absorbed dose resulting from photon interactions in an air cavity D_{air}^{PA} to that in air under the condition of charged-particle equilibrium D_{air}^{CPE} has been used as a parameter to determine if the air cavity can be classified as a Bragg-Gray cavity. Burlin general cavity theory seriously overestimates the departure from Bragg-Gray behavior. For clinical photon beams, the dose ratio F_{air} is 0.29 for a 150-kVp beam and 0.27 for a 240-kVp beam, compared to 0.006 for a ^{60}Co-beam, if the cavity is placed at a depth of 5 cm in water. The study confirms that typical air-filled ionization chambers cannot be considered to be Bragg-Gray cavities for low- and medium-energy photon radiation.

For a given absorbed dose to the medium, D_{med}, the absorbed dose to the sensitive detector material in the case of a small detector, D_{det}^S, is in general different from that for a large detector, D_{det}^L. By substituting $s_{med,det}$ and $(\mu_{en}/\rho)_{med,det}$ into Equation (2.195), in turn, one obtains

$$D_{det}^L/D_{det}^S = s_{med,det}(\mu_{en}/\rho)_{det,med} \qquad (2.196)$$

Figure 2.56 shows how the dose ratio, D_{det}^L/D_{det}^S, calculated using Equation (2.196), varies with energy when a monoenergetic photon beam is incident on water and

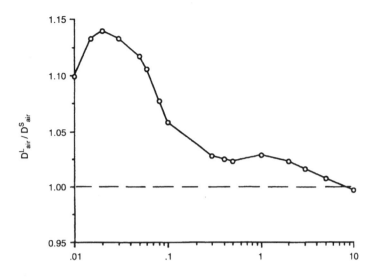

FIGURE 2.56 The variation with the incident photon energy of the dose ratio, D_{air}^L/D_{air}^S at a depth of 8 cm in water calculated using Equation 2.196. The mass energy absorption coefficient ratios and the mass stopping power ratios were calculated using Monte Carlo techniques. (From Reference [56]. With permission.)

the (air-filled) detector is placed at a depth of 8 cm in water (chosen to be greater than the maximum possible CSDA electron range at all energies). The value of D_{air}^S is usually smaller than that of D_{air}^L and the difference between D_{air}^S, D_{air}^L increases as the incident photon energy decreases. This means that significant dosimetric errors could be introduced by using such detectors at lower photon energies.

Suppose that an air cavity (i.e., without walls) is placed at a depth in water where charged-particle equilibrium (CPE) has been established. The mean absorbed dose in the air cavity, D_{air}, can be divided into two parts; i.e.,

$$D_{air} = D_{air}^{EW} + D_{air}^{PA} \qquad (2.197)$$

where D_{air}^{EW} is the dose resulting from the electrons generated by photon interactions in the surrounding water and D_{air}^{PA} is the dose resulting from photon interactions in the air cavity. If $D_{air}^{PA} \ll D_{air}$ and the disturbance by the detector of the electron fluence present in the undisturbed water is negligible, then the absorbed dose in the air cavity depends solely upon the electron fluence present at the depth of interest in water (the condition for Bragg-Gray theory to apply). If D_{air}^{PA} is close to D_{air}^{CPE}, the absorbed dose in air under the CPE condition, then the absorbed dose in the air cavity depends mainly upon the photon fluence present at the depth of interest in water (the condition for large detectors). For the intermediate case, if one still uses the Bragg-Gray cavity theory, errors will occur, the severity of which can be judged from Figure 2.56.

Let the quantity $F_{air} = D_{air}^{PA}/D_{air}^{CPE}$ be a measure of the validity of the Bragg-Gray cavity theory at different

photon beam energies; the condition for the Bragg-Gray cavity theory to apply is $F_{air} \to 0$. For a monoenergetic photon beam of energy E, we have

$$D_{air}^{CPE} = \Phi E(\mu_{en}/\rho)_{air} \qquad (2.198)$$

where Φ is the photon fluence and $(\mu_{en}/\rho)_{air}$ is the mass energy absorption coefficient of air at energy E. The dose ratio, F_{air}, for a monoenergetic photon beam of energy E can then be expressed as

$$E_{air}(E) = D_{air}^{PA} / \Phi_E E(\mu_{en}/\rho)_{air} \qquad (2.199)$$

Suppose that one has a photon spectrum and wishes to calculate the corresponding quantity, F_{air}^{spec}, for the spectrum. This can be done by integrating $F_{air}(E)$ over the photon spectrum according to [56]

$$F_{air}^{spec} = \int_0^{E_{max}} F_{air}(E)\Phi_E E(\mu_{en}(E)/\rho)_{air} dE /$$

$$\int_0^{E_{max}} \Phi_E E(\mu_{en}(E)/\rho)_{air} dE$$

where Φ_E is the photon fluence, differential in energy, and $(\mu_{en}(E)/\rho)_{air}$ is the mass energy absorption coefficient for air at energy E.

The Burlin general cavity relation can be written as

$$D_{det} = [ds_{det,med} + (1-d)(\mu_{en}/\rho)_{det,med}]D_{med}$$

$$(2.200)$$

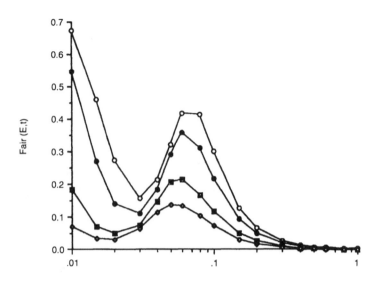

FIGURE 2.57 The variation with the incident photon energy, E, of $F_{air}(E, t)$ for an air cavity of thickness t and 6-mm diameter in vacuum irradiated by monoenergetic photon beams. The Monte Carlo calculational uncertainty is smaller than 0.5%. $\Diamond$, $t = 0.5$ mm; $\blacksquare$, $t = 1$ mm; $\bullet$, $t = 3$ mm; $\bigcirc$, $t = 6$ mm. (From Reference [56]. With permission.)

where d is a parameter related to the cavity size that approaches unity for small cavities and zero for large ones. Burlin expresses $(1 - d)$ as

$$(1 - d) = (\beta L + e^{-\beta L} - 1)/\beta L \qquad (2.201)$$

where L is equal to four times the cavity volume, V, divided by its surface area, with β satisfying

$$e^{-\beta L} = 0.01 \qquad (2.202)$$

That is, the maximum depth of electron penetration, T, is to be arbitrarily taken as the depth to which only 1% of the electrons can travel.

Suppose the *CPE* condition has been established and the values of $s_{det, med}$ and $(\mu_{en}/\rho)_{det, med}$ are known. Then one can write

$$D_{med} = s_{med, det}D_{det}^S = (\mu_{en}/\rho)_{med, det}D_{det}^L \qquad (2.203)$$

By combining Equation (2.203) with equation (2.197), one obtains [56]

$$D_{air}/D_{med} = (D_{air}^{EW}/D_{air}^S)S_{air, med}$$
$$+ (D_{air}^{PA}/D_{air}^L)(\mu_{en}/\rho)_{air, med} \qquad (2.204)$$

By comparing Equations (2.200) and (2.204) and noting that $D_{air}^L = D_{air}^{CPE}$, Ma and Nahum arrived at

$$F_{air} = D_{air}^{PA}/D_{air}^{CPE} = 1 - d \qquad (2.205)$$

$$D_{air}^{EW}/D_{air}^S = d \qquad (2.206)$$

It is worth noting that, since D_{air} is in general different from either D_{air}^S or D_{air}^L, it is evident that [56]

$$(D_{air}^{EW}/D_{air}^S) + (D_{air}^{PA}/D_{air}^L) \neq 1 \qquad (2.207)$$

This is ignored in the Burlin theory.

Figure 2.57 shows the ratio $F_{air}(E, t)$ of the absorbed dose in an air cavity of thickness t and 6-mm diameter in vacuum to that in air under the condition of *CPE* for monoenergetic photon beams. Above about 0.1 MeV, the dose ratio $F_{air}(E, t)$ decreases rapidly with the incident photon energy and becomes negligible for high photon energies (smaller than 0.01 for photon energies above 500 keV). The $F_{air}(E, t)$ curves exhibit maxima between 50 and 80 keV [F_{air} (80 keV, 6 mm) = 0.41, F_{air} (60 keV, 3 mm) = 0.37, F_{air} (60 keV, 1 mm) = 0.22, and $F_{air} \times$ (50 keV, 0.5 mm) =0.14]. The fraction of the photon energy transferred to Compton electrons decreases steadily with decreasing photon energy, whereas, for photoelectrons, this fraction is essentially unity. As a consequence of more secondary electrons resulting from Compton interactions than from photoelectric interactions, the total average energy transfer actually decreases for photon energies greater than 20 keV, increasing again only for photon energies greater than 60 keV.

Comparing F_{air} values with the $(1 - d)$ values in Burlin theory, we note the following. In Figure 2.58, the values of $(1 - d)$ normalized to $F_{air}(E)$ are presented as a function of photon energy for monoenergetic photon beams. The Monte Carlo calculated results of the dose ratio, $F_{air}(E)$, show a much smaller fraction of the absorbed dose in the

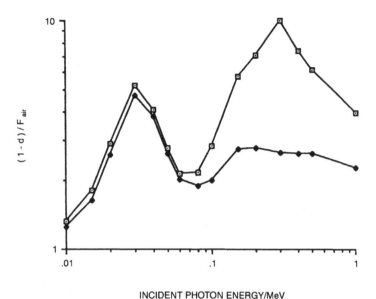

INCIDENT PHOTON ENERGY/MeV

FIGURE 2.58 Values of the ratio $(1 - d)/F_{air}(E)$ for an air cavity of 6-mm thickness and 6-mm diameter in vacuum; $(1 - d)$ was calculated using Equation (2.201) with $L = 4$ mm. For Burlin's $(1 - d)$, β was calculated using Equation (2.202) with $T = 0.95$ R_{SCDA}. For Janssens et al.'s [73] $(1 - d)$, a constant of 0.04 was used in place of 0.01 in Equation (2.202). □, Burlin; ◆, Janssens et al. (From Reference [56]. With permission.)

air cavity resulting from the direct photon interactions with the air in the cavity compared to that predicted by Burlin theory. Using 0.04 instead of 0.01 in Equation (2.202) only slightly reduces the difference between $F_{air}(E)$ and $(1 - d)$.

The Burlin theory ignores all secondary-electron scattering effects which results in large discrepancies in dose to the cavity compared with the experimental results in high atomic number media. Kearsley [57] proposed a new general cavity theory which includes secondary-electron scattering at the cavity boundary. The Kearsley theory showed excellent agreement with experimental results for ^{60}Co γ-rays but poor correlation for 10-MV x-rays. The Kearsley theory has numerous parameters and the magnitude of the input parameters is arbitrary; therefore, the dose to the cavity depends on the choice of parameters. Haider et al. [58] have developed a new cavity theory which includes secondary-electron backscattering from the medium into the cavity. The strength of this proposed theory is that it contains few parameters and a methodical way of determining the magnitude of the parameters experimentally. The theory gives better agreement with experimental results in lithium fluoride thermoluminescence dosimeters for ^{60}Co γ-rays and 10-MV x-rays in aluminum, copper, and lead than do the Burlin and Kearsley cavity theories.

Haider et al. assume that the Compton interaction is the dominant radiation interaction and, thus, the applicable energy range of the theory is from 500-kV to 20-MV x-rays. The electron density is proportional to the energy loss cross section of electron stopping power which is mainly determined by the excitation energy of the molecules and the Compton scatter cross section. When the Compton process is dominant and the difference of mean excitation energies per electron is small, then the stopping power ratio averaged over the cavity volume is equal to the ratio of electron densities, independent of cavity size. The total electron fluence in the cavity $(\Phi_{e,g,T}(z))$ of thickness t is divided into three groups: [58]

1. the electron fluence that originated in the cavity, including the backscattered electrons generated by it in the cavity traveling in the opposite direction $(\Phi_{e,g}(z))$;
2. the electron fluence that originated in the front-wall medium, including the backscattered electrons generated by it in the cavity traveling in the opposite direction $(\Phi_{e,w}(z))$; and
3. the electron fluence in the cavity resulting from the difference in the backscattering coefficients of the cavity medium and the back-wall medium $(\Phi_{e,m,b}(z))$.

In the following, w is referred to as the front-wall medium, g as the cavity medium, m as the back-wall medium, and b as the backscattering coefficient. The electron fluence in the cavity is given by

$$\Phi_{e,g,T}(z) = \Phi_{e,g}(z) + \Phi_{e,w}(z) + \Phi_{e,m,b}(z)$$

$$(2.208)$$

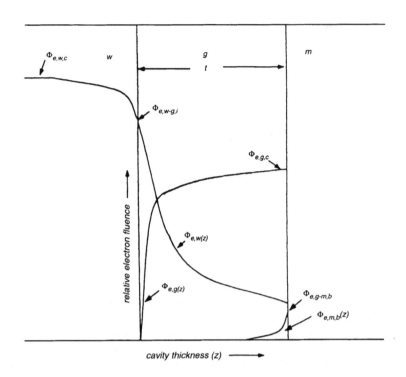

FIGURE 2.59 Illustration of cavity parameter. $\Phi_{e,g}(z)$ is the electron fluence that originated in the cavity (g), including its backscatter. $\Phi_{e,w}(z)$ is the electron fluence spectrum that originated in the front-wall medium (w) and its backscatter as it crosses the cavity. $\Phi_{e,m,b}(z)$ is the electron fluence spectrum in the cavity resulting from the difference in the backscattering coefficient of the cavity (g) and the back-wall medium (m). (From Reference [58]. With permission.)

Therefore, the dose to the cavity is given by

$$
\begin{aligned}
D_g &= \Phi_{e,g,T}(z)\langle S/\rho \rangle_g \\
&= (\Phi_{e,g}(z) + \Phi_{e,w}(z) + \Phi_{e,m,T}(z))\langle S/\rho \rangle_g
\end{aligned}
$$

$$(2.209)$$

where $\langle S/\rho \rangle$ is the mean mass collision stopping power. In the absence of the cavity, the dose to the wall is given by

$$
D_w = \Phi_{e,w,c}(z)\langle S/\rho \rangle_w \qquad (2.210)
$$

where the subscript c denotes the equilibrium fluence in the wall medium (i.e., under charged particle equilibrium (CPE), where dose is equal to the collision part of the kinetic energy released in medium A (kerma). The ratio f of the average dose to the cavity (g) and to the front wall medium (w) is given by

$$
f(z) = \frac{\Phi_{e,g}(z) + \Phi_{e,w}(z) + \Phi_{e,m,b}(z)}{\Phi_{e,w,c}}\langle S/\rho \rangle_w^g
$$

$$(2.211)$$

where $\langle S/\rho \rangle_w^g$ is the ratio of the mean mass collision stopping powers of medium g and of medium w. $\Phi_{e,g}$, $\Phi_{e,w}$, and $\Phi_{e,m,b}$ are functions of the distance z from the front cavity interface (Figure 2.59). Each term can be represented by the product of a distance-independent electron fluence term at the interface and a distance-dependent

weighting factor. Therefore, [58]

$$
\begin{aligned}
f(z) = {} & \frac{C_1(z)\Phi_{e,w,c} + C_1(z)\Phi_{e,w-g,i} + C_3(z)\Phi_{e,g-m,b}}{\Phi_{e,wg}} \\
& \times \langle S/\rho \rangle_w^g
\end{aligned}
$$

$$(2.212)$$

where $\Phi_{e,g,c}$ is the equilibrium cavity fluence, $\Phi_{e,w-g,i}$ is the electron fluence at the front wall-cavity (w-g) interface, $\Phi_{e,g-m,b}$ is the electron fluence at the back wall-cavity (g − m) interface, and C_1, C_2, and C_3 are dependent weighting factors: C_1 is the ratio of the electron fluence generated in the cavity medium at a distance z from the front wall-cavity interface and the equilibrium fluence in the cavity medium; C_2 is the ratio of the electron fluence generated in the front wall medium at a distance z from the wall-cavity interface (Figure 2.59) and the electron fluence at the front wall-cavity interface; and C_3 is the ratio of the backscattered electron fluence at a distance $(t - z)$ from the back wall-cavity interface to the back-scattered electron fluence at the cavity-back wall interface.

If the cavity and the wall materials are irradiated by the same photon fluence separately, then the dose ratio between the cavity and the wall under *CPE* is related by the ratio of the average mass energy absorption coefficients. Therefore,

$$
\frac{D_g}{D_w} = \langle \mu_{en}/\rho \rangle_w^g \qquad (2.213)
$$

where $\langle \mu_{en}/\rho \rangle_w^g$ is the mean mass energy absorption coefficient. The dose ratio between the cavity and the wall is also related by

$$\frac{D_g}{D_w} = \langle \mu_{en}/\rho \rangle_w^g \qquad (2.214)$$

using the relationship of Equation 2.210:

$$\Phi_{e,w,c} = \Phi_{e,w,c} \langle \mu_{en}/\rho \rangle_w^g \langle S/\rho \rangle_g^w \qquad (2.215)$$

Let F be the fraction of the total number of electrons per unit area which crosses a reference plane r from left to right in a medium under *CPE* and let b be the fractional number of incident electrons that are backscattered from the reference plane r traveling from right to left. In a homogeneous medium, the total number of electrons or the equilibrium fluence $\Phi_{e,c}$ which includes the backscattered electrons traveling in the opposite direction is given by

$$\Phi_{e,c} = F\Phi_{e,c}(1 + b) \qquad (2.216)$$

The term $F\Phi_{e,c}$ is the total number of electrons per unit area which crosses a reference plane r from left to right and the team $Fb\Phi_{e,c}$ is the total number of backscattered electrons crossing the same reference plane from right to left. If the medium to the left of the reference plane r is w and to the right of the reference plane is g, then the electron fluence at the interface of $w - g$ ($\Phi_{e,w-g,i}$) is given by [58]

$$\Phi_{e,w-g,i} = F_w \Phi_{e,w,c}(1 + b_g) \qquad (2.217)$$

where F_w is the fractional forward electron fluence originating in the medium w and b_g is the probability of backscattering of the medium g. If medium g of thickness t is sandwiched between medium w and medium m, then the relative backscattered electron fluence between medium g and medium m at the interface $g - m$ ($\Phi_{e,w-g,i}$) is given by

$$\Phi_{e,g-m,b} = F_{e,g-m,i} \Phi_{e,g-m,i}(b_m - b_g) \qquad (2.218)$$

where $F_{e,g-m,i}$ is the fractional forward electron fluence of the total electron fluence $\Phi_{e,g-m,i}$ arriving at the interface $g - m$; b_m is the backscattering coefficient of the medium m; and $b_m - b_g$ is the difference in the backscattering coefficient of medium g and medium m. Electron backscattering within the medium g is already accounted for in $\Phi_{e,g}$, and $\Phi_{e,w}$. $\Phi_{e,g-m,i}$ consists of $\Phi_{e,g}$ and $\Phi_{e,w}$, evaluated at cavity thickness t given by

$$F_{e,g-m,i}\Phi_{e,g-m,i} = F_g\Phi_{e,g}(t) + F_w\Phi_{e,w}(t) \qquad (2.219)$$

where F_g is the fractional forward electron fluence originating in the cavity medium and F_w is the fractional forward electron fluence originating in the front-wall medium. $\Phi_{e,g}(t)$ and $\Phi_{e,w}(t)$ can be expressed by a product of cavity size–independent electron fluence terms and cavity size–dependent weighting factors. Therefore,

$$F_{e,g-m,i}\Phi_{e,g-m,i} = d_4 F_g \Phi_{e,g,c} + d_5 F_w \Phi_{e,w-g,l} \qquad (2.220)$$

where d_4 and d_5 are the fractions of the electron fluences $\Phi_{e,g,c}$ and $\Phi_{e,w-g,l}$, respectively, that arrive at the interlace $g - m$ of cavity thickness t.

When the cavity size is very large compared with the range of the electrons, the above equations lead to the ratio of the mass-energy absorption coefficients of cavity and medium as in the Burlin and the Kearsley cavity theories. When the cavity size is very small compared with the range of the electrons and the front-wall and back-wall mediums are identical, f reduces to the Bragg-Gray theory. However, when the front and the back walls are not identical, f does not reduce to the Bragg-Gray theory. Since the contributions from the front-wall, cavity, and back-wall mediums are calculated separately, all three media could be of different atomic composition. [58]

The weighting factors d_4 and d_5 are given by

$$d_4 = (1 - d_5) = 1 - \exp(-\beta_a t) \qquad (2.221)$$

where β_a is the effective electron absorption coefficient (cm^2 g^{-1}) in the cavity. [58]

XI. ELEMENTS OF MICRODOSIMETRY

Radiation dose in conventional dosimetry is a macroscopic concept. Target volumes are many orders of magnitude greater than the individual cellular entities which make up tissue. The dose to a macroscopic multicellular volume is obtained by the summation of the total energy deposited by multiple radiation tracks over the volume divided by the mass of that volume. Microdosimetry is the study of radiation energy deposition within microscopic volumes, where "microscopic" encompasses sensitive target volumes ranging from the diameter of a cell (typically 20 mm) down to the diameter of the DNA molecule (2 μm). Although microdosimetry is concerned with the same concept of energy deposition per unit mass as dosimetry, the difference in size of the target volume of interest introduces stochastic effects which are negligible in conventional dosimetry. The magnitude and importance of stochastic fluctuations in the target volumes depend greatly on the target diameter, on the energy and linear energy transfer (LET) of the particles, and on the relative number of particles, i.e., the magnitude of the radition dose.

The fluctuations of energy deposition are incorporated in the stochastic quantities of a subdiscipline of radiological physics that has been termed *microdosimetry*. By focusing on actual distributions of absorbed energy rather than on their expected (mean) values, microdosimetry has demonstrated the complexity of the problem of finding a numerical index of radiation quality. On the other hand, a realistic view of the pattern of energy absorption in irradiated matter is essential for an understanding of the mechanisms responsible for radiation effects, not only in radiobiology but also in such fields as radiation chemistry and solid-state dosimetry, and microdosimetry is required in theoretical approaches to these subjects. Considerable effort has therefore been expended in the measurement and calculation of microdosimetric spectra. Once appropriate microdosimetric data were obtained, there were efforts to employ them in radiobiology, radiation chemistry, and solid-state research. Such activities ran parallel with further efforts in instrument design and an important fundamental analysis of microdosimetry. [59]

The microdosimetric quantity that was first formulated bears a relation to LET and it was, in fact, determined in g scheme to derive $t(L)$ by measurements of energy deposition in a microscopic region termed the *site*. In a spherical proportional counter made of tissue-equivalent (TE) plastic, the energy that is deposited by an event (i.e., the passage of a charged particle) in a TE gas volume of density ρ and diameter d is a good approximation to the energy that would be deposited in unit density tissue within a site of diameter $d\rho$. Since in the proportional counters utilized, ρ is typically of the order of 10^{-5}, it is possible to simulate very small tissue regions with counters of convenient size. The quantity measured by these devices was originally termed the "event size." In a later change, this became the lineal energy y, which has also been given a slightly different definition: [59]

$$y = \varepsilon/l \qquad (2.222)$$

where ε is the energy deposited in an event and l is the mean chord length in the tissue region simulated. (Originally the "event size" Y was defined as ε/d, with d being the diameter of the sphere. The newer definition of y permits generalization to other shapes. In the case of the sphere, $l = 2d/3$.) This can be changed readily from a few tenths to at least several tens of micrometers by a change of gas pressure. Utilizing modern low-noise preamplifiers, it is possible to measure most of the pulses corresponding to a single ionization. (Because of the statistics of avalanche formation in the proportional counter, initial ionizations result in a range of pulse heights). Pulse-height analysis permits determination of the probability (density) distribution $f(y)$, i.e., the probability that the lineal energy of an event is equal to y. A typical, widely used, TE-counters design is illustrated in Figure 2.60. An example for ^{60}Co y radiation is given in Figure 2.61.

The quantity termed the specific energy z (the original name and symbol were "local energy density, Z"), which applies to any number of events, is the stochastic analogue

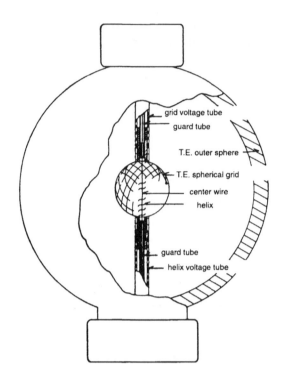

FIGURE 2.60 Typical spherical proportional counter for microdosimetry. The electrically grounded shell is molded from tissue-equivalent (TE) plastic and the insulators are made of Lucite. The spherical grid (TE) serves to largely eliminate the wall effect that causes separate events in the simulated tissue regions to coincide in the counter. The helix reduces variations of the electric field along the center wire. These electrodes are of stainless steel. The counter is traversed by TE gas at a pressure of the order of 10^3 Pa. The potential of the center wire is several hundred volts positive and the potential of the helix is about 20% of this. The spherical grid is at negative potential. An internal α or soft x-ray source serves to calibrate the counter. (From Reference [59]. With permission.)

of the absorbed dose and defined by

$$z = \varepsilon/m \qquad (2.223)$$

where ε is the energy absorbed in a mass m in the receptor. D, the average value of the absorbed dose in m, is equal to $\bar{z}$, the average value of z.

The specific energy z that is produced by an event can be expressed in terms of y because the mean chord length in a convex body is $4V/S$ where V is the volume and S is the surface. Accordingly,

$$z = 4y/(S \times \rho) \qquad (2.224)$$

where ρ is the density. For a unit density sphere,

$$z = 0.204\, y/d^2 \qquad (2.225)$$

when z is expressed in gray, y in keV/μm, and d in μm. Because of the statistical independence of events, a

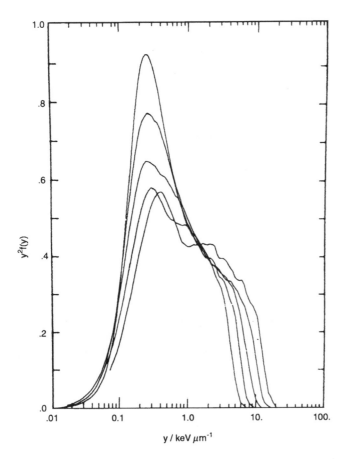

FIGURE 2.61 Microdosimetric spectra for ^{60}Co gamma radiation. The spectra represent, from top to bottom, the following site diameters: 8, 4, 2, 1, and 0.5 nm. $yf(y)$ is the relative fraction of the absorbed dose in y. In this plot, multiplication of the ordinate by another factor y makes the area under the curve proportional to the logarithmic interval in the abscissa. (From Reference [59]. With permission.)

knowledge of $f(y)$ permits calculations of $f(z,D)$, the probability (density) distribution for any value of the absorbed dose D. This calculation involves z_F, the frequency-averaged specific energy in single events:

$$z_F = \int_0^\infty z f_1(z) dz \left(\int_0^\infty f_1(z) \, dz \right)^{-1} \qquad (2.226)$$

From this, one can obtain the mean number of events n at dose D:

$$n = D / z_F \qquad (2.227)$$

At sufficiently small values of D, there is usually no event in a site and very rarely more than one. $f(z, D)$ then consists of a Dirac delta function $\delta(z)$ and the single-event distribution $f_1(z)$. This distribution has the same shape as $f(y)$ and any reduction in D merely increases the relative contribution of the delta function and reduces the fraction of sites that receive an invariant spectrum $f_1(z)$ of ener-

gies deposited in single events. At large values of D, when $n \gg 1$, multiple events ultimately cause $f(z, D)$ to assume a bell-shaped curve about D. [59]

The product between the point-pair distance distribution and the measure of the site is termed the proximity function of that domain. Thus, one has a proximity function of energy deposition

$$t(r) = (\text{total energy}) p_{track}(r) \qquad (2.228)$$

and a proximity function of the sensitive matrix

$$s(r) = (\text{site volume}) P_{site}(r) \qquad (2.229)$$

$t(r)dr$ can also be defined as the expected energy imparted to a shell of radius r and thickness dr centered at a randomly chosen transfer point.

Proximity functions $t(x)$ can be calculated from Monte Carlo generated particle tracks. For monoenergetic particles (energy T) one has

$$t(r)\Delta r = E\left(\frac{1}{T} \sum_{i,k} e_i e_k \right) \qquad (2.230)$$

where i and k refer to energy transfers, e_i and e_k, separated by a distance between r and $r + \Delta r$. The expectation value refers to an ensemble of tracks. In simple cases $t(r)$ can be calculated directly. For a track segment with constant LET, L (no radial extension of the track)

$$t(r) \, dr = 2L \, dr \qquad (2.231)$$

For a homogeneous distribution of energy transfers in a medium of density exposed to dose D:

$$t(r) = 4\pi r^2 \rho D \qquad (2.232)$$

Equations (2.231) and (2.232) follow directly from the definition of $t(r)$. $t(r)$ can not increase faster than r^2 for any three-dimensional structure. The proximity function can also be obtained directly from a series of microdosimetric measurements of $z_D(r)$ in spherical sites of diameter d: [59]

$$t(r) = \frac{md^2}{3dr^2}\left[\frac{1}{r^2}\left(\frac{d}{dr} r^4 z_D(r) \right) - r z_D(r) \right] \qquad (2.233)$$

Microdosimetric concepts in radioimmunotherapy were discussed by Humm et al. [60] There are three principal areas where microdosimetry has been applied: radiation protection; high LET radiotherapy, e.g., neutron therapy; and incorporated radionuclides. In this latter category the importance of microdosimetry to the radiobiology of radiolabeled antibodies is becoming increasingly recognized. The objective of microdosimetry is the complete characterization of energy deposition

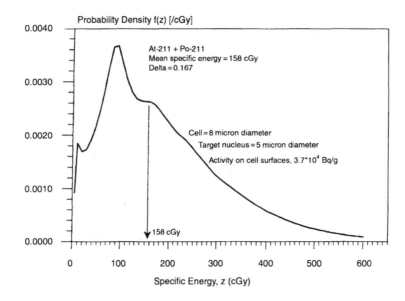

FIGURE 2.62 Distribution of α-particle dose calculated by the Fourier convolution method to tumor cell nuclei from ^{211}At and its daughter ^{211}Pa for a 3.7×10^4 Bq/g specific activity after complete decay. Cells were assumed to be tightly packed and all the activity distributed uniformly on the cell surface. Cell and cell nuclei are of 8 and 5 μm in diameter, respectively. The mean specific energy is 159 cGy. (From Reference [60]. With permission.)

within all target volumes throughout the tissue of interest. The importance and relevance of this pursuit will depend upon the properties of the radionuclide emissions and the spatial distribution of the radionuclide relative to the target volumes.

When the distribution of radiolabeled antibody is non-uniform, techniques of dose-averaging over volumes greater in size than the individual target volumes can become inadequate predictors of the biological effect.

The most fundamental parameter in microdosimetry is the energy deposition ε_1, from a single event (track intersection) with a target volume. Given a specific irradiation geometry, target size and shape, and particle type, the entire probability distribution of energy deposits $f(\varepsilon_1)$ within the target volume from single traversals is referred to as a single-event energy deposition spectrum. A more commonly used quantity is the single-event specific energy z_1. The quantity z_1 is the energy deposition ε_1 divided by the mass of the target volume m. The quantity z_1 has been calculated and also measured, using a Rossi proportional counter, for a number of radiations. For high-dose radiation fields, of concern for therapy, one is primarily interested in multiple-event spectra, where the stochastics of individual target hits is convoluted with the hit probability distribution. The microdosimetric quantity, specific energy z, used to denote the stochastic energy deposition per unit mass from multiple track traversals, is the microdosimeter analogue of absorbed dose. Indeed, the frequency mean $\langle z_F \rangle$ of a multiple-event specific energy spectrum $f(z)$ is, under most circumstances, equal to the absorbed dose. However, identical z_F resulting from differing specific energy spectra do not imply equitoxicity.

This notion is of immense importance for radioimmunotherapy, since microdosimetry can predict widely differing tissue toxicities resulting from identical average tissue doses. [60]

The full three-dimensional distribution of activity as a function of time is required for the exact evaluation of a microdosimetric spectrum. Such spectra have been calculated from theoretical distributions.

An alternative approach is the point dose summation methods by Monte Carlo or other methods. A distribution of sources (which may be located extracellular, on the cell surface, or intracellular) may be simulated or obtained from digitized images of autoradiographs. Each source decay is simulated, with an energy and direction of the emission chosen. For a-particles the tracks are assumed to be straight. If the line intersects a biological target, the specific energy z deposited is determined by [60]

$$z = \frac{1}{m} \int_{t_1}^{t_2} \frac{dE}{dx} dx \qquad (2.234)$$

where m is the target mass, dE/dx is the energy deposited per unit track length, and t_1 and t_2 are the entrance and exit coordinates of the track through the target. If the track ends in the target, t_2 is the coordinate of the end of range of the particle. If a track begins in the target, t_1 is zero.

The total distribution of specific energy $f(z)$ for the cellular targets represents a complete description of the physical dose deposition throughout the target volume. These physical data can be combined with a biological cell inactivation model to estimate the fraction of cell survivors. Such inactivation models evaluate the fraction

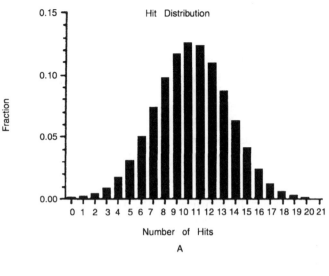

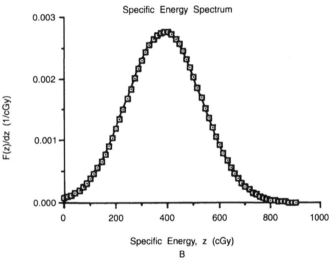

FIGURE 2.63 Hit distribution (a) and specific energy spectrum (b) for the cell nucleus obtained by the Monte Carlo method, assuming a uniform extracellular specific activity of 7.4×10^4 Bq/g of ^{211}At after complete decay of parent and daughter, within 25 decays per cell surface. The stopping power table of Janni for protons, scaled for α-particles, was used to determine energy deposition. Cell and cell nuclear diameters of 10 and 7.5 μm, respectively, were used. Resultant distributions are based on 1000 iterations, and the data is smoothed. The average number of hits is 10.3, and the average specific energy is 388 cGy with a standard deviation of 144 cGy. (From Reference [60]. With permission.)

of cell survivors within each energy deposition bin and then perform a weighted sum of these surviving fractions over the cell populations.

Two methods are employed for the calculation of microdosimetric spectra for internal radionuclides: the Fourier convolution technique developed by Roesch [61] and applied to problems with radiolabeled antibodies by Fisher [74], and the application of full Monte Carlo simulation by Humm [60] and by Roeske. [62] Examples of specific energy spectra calculated by both methods are illustrated in Figures 2.62 and 2.63.

Figure 2.62 is a specific energy spectrum calculated by the Fourier convolution method for a population of tightly packed cells, diameter 8 μm, nuclear diameter 5 μm, uniformly labeled on the cell surface, after the complete

decay of 3.7×10^4 Bq/g of ^{211}At with ^{211}Po daughter. The mean specific energy to the nucleus is 158 cGy, and the fraction of cell nuclei receiving zero dose (delta) is 0.17.

Figure 2.63 is an a-particle hit and specific energy spectrum calculated by the Monte Carlo method for a uniform 7.4×10^4 Bq/g extracellular distribution of ^{211}At labeled antibody with 25 radiolabeled antibodies bound per cell surface. The cell and cell nuclear diameters are 10 μm and 7.5 μm, respectively. These are only two examples of many different types of calculations that are possible. The shape of microdosimetric spectra depends on a number of parameters: the shape and size of the target volume, the geometry of the source distribution relative to the targets, the energy emission spectrum of the radionuclide, etc.

The Medical Internal Radiation Dose (MIRD) Committee of the Society of Nuclear Medicine has provided guidance on methods for calculating radiation-absorbed dose estimates since 1968. The MIRD Primer gives a complete explanation of the schema, which is a series of general equations adaptable for use with either simple or complex anatomical and kinetic models. By definition, the absorbed dose is the energy absorbed from ionizing radiation per unit mass of tissue. Because absorbed dose from internally distributed radionuclides is never completely uniform, the MIRD equations give the average, or mean, absorbed dose to a volume of tissue. [63]

Microdosimetric techniques that account for statistical aspects of particle track structures and energy distribution patterns in microscopic volumes can be used to express energy deposition in tissues from materials labeled with alpha-particle or Auger-electron emitters, particularly those incorporated within cells. The equation for calculating the absorbed dose may be written in various forms, depending on available information, such as: [63]

$$\bar{D}(r_k \leftarrow r_h) = \tilde{A}_h \sum_i \Delta_i \phi_i((r_k \leftarrow r_h)/m_k) \qquad (2.235)$$

where $\bar{D}(r_k \leftarrow r_h)$ is the mean absorbed dose in a target region r_k from activity in a source region r_h; $\tilde{A}_h$ is the cumulated activity (time integral of activity over the time interval of interest) in the source; Δ_i is the mean energy emitted by a radionuclide per nuclear transition; $\phi_i(r_k \leftarrow r_h)$ is the absorbed fraction (fraction of energy emitted in region r_h that is absorbed in region r_k); and m_k is the mass of the target r_k. The absorbed fraction divided by the mass may be represented by $\phi_i(r_k \leftarrow r_h)$, the specific absorbed fraction. The total mean absorbed dose in a target region is calculated by summing the doses from all source regions to the target.

Each type of radiation emitted by a radionuclide is characterized by its own mean energy per particle $\bar{E}_i$ and its own intensity or number of particles emitted per transition n_i. The mean energy emitted per transition, Δ_i, is

equal to $k n_i \bar{E}_i$, where k is a constant that depends on the units used for the terms in Equation (2.235).

The absorbed fraction varies with the type and energy of the radiation, the type of material through which the radiation passes, and geometric configuration and composition of the source and the target. If the amount of energy imparted to any target other than the source is so insignificant as to have little effect on the absorbed dose, the radiation is considered to be nonpenetrating.

The product of Δ and Φ is a constant for a given radionuclide and a given source-target combination, a value designated by the MIRD Committee as the S value. The mean absorbed dose equation can thus be written as

$$\bar{D}(r_k \leftarrow r_h) = \sum_h \tilde{A}_h S(r_k \leftarrow r_h) \qquad (2.236)$$

where

$$S(r_k \leftarrow r_h) = \sum_i \Delta_i \Phi_i(r_k \leftarrow r_h)$$

The cumulated activity $\tilde{A}_h$ represents the total number of nuclear transformations occurring during the time of interest in the source region r_h and may be expressed in units of microcurie hours, Becquerel seconds, or an appropriate multiple of these units. A compilation of cumulated activities for various radionuclides or radioactive compounds has not been published by the MIRD Committee because the source regions differ for each radiolabeled material, and the source regions and their cumulated activities often change as new research results become available.

Autoradiographic studies have clearly shown nonuniform distribution of radiolabeled monoclonal antibodies in tumors. The MIRD schema can be used to estimate absorbed doses for nonuniform distributions if the necessary data are obtained. The limitation is in the lack of an adequate model rather than in the schema. As the volume for which the absorbed dose is calculated becomes smaller, the nonuniformity of dose within that volume also becomes smaller. [63]

REFERENCES

1. **Nelson, R. W. et al.,** *The EGS4 Code System,* SLAC-265, 1985.
2. **Deasy, J. O. et al.,** *Med. Phys.,* 21, 1369, 1994.
3. **Keall, P. J. and Hoban, P. W.,** *Med. Phys.,* 23, 2023, 1996.
4. **Larsen, E. W. et al.,** *Med. Phys.,* 24, 111, 1997.
5. **Ding, G. X. et al.,** *Med. Phys.,* 24, 161, 1997.
6. **Fippel, M. et al.,** *Phys. Med. Biol.,* 42, 501, 1997.
7. **Rogers D. W. et al.,** *Med. Phys.,* 25, 310, 1998.
8. **IPEMB,** Thwaites, D. I. (Chair) et al., *Phys. Med. Biol.,* 41, 2557, 1996.
9. **Khan, F. M. et al.,** *Med. Phys.,* 18, 73, 1991.
10. **Weinhous, M. S. and Meli, J. A.,** *Med. Phys.,* 11, 846, 1984.
11. **Nahum, A. E.,** in *AAPM Proc. No. 11, Kilovolt X-Ray Beam Dosimetry for Radiotherapy and Radiobiology,* 1997, 7.
12. **Ma, C. M. et al.,** in *AAPM Proc. No. 11, Kilovolt X-Ray Beam Dosimetry for Radiotherapy and Radiobiology,* 1997, 69.
13. **Andreo, P.,** *Phys. Med. Biol.,* 37, 2189, 1992.
14. **Rosser, K. E.,** *Phys. Med. Biol.,* 43, 587, 1998.
15. **Burns, J. E.,** *Phys. Med. Biol.,* 39, 1555, 1994.
16. **Chui, C. S. et al.,** *Med. Phys.,* 21, 1237, 1994.

17. **Rogers, D. W. O. and Yang, C. L.**, *Med. Phys.*, 26, 538, 1999.
18. **Yang, C. L. et al.**, *Med. Phys.*, 25, 1085, 1998.
19. **Liu, H. H. et al.**, *Med. Phys.*, 25, 56, 1998.
20. **Aspradakis, M. M. and Redpath, A.**, *Phys. Med. Biol.*, 42, 1475, 1997.
21. **IPEMB**, Klevenhagen, S. C. (Chair) et al. *Phys. Med. Biol.*, 41, 2605, 1996.
22. **Huq, M. S. and Nath, R.**, *Med. Phys.*, 18, 26, 1991.
23. *The Use of Plane Parallel Ionization Chambers in High Energy Electron and Photon Beams*, IAEA Tech. Rep., Series No. 381, 1997.
24. **Tome, W. A. and Palta, J. R.**, *Med. Phys.*, 25, 758, 1998.
25. **Medin, J. and Andreo, P.**, *Phys. Med. Biol.*, 42, 89, 1997.
26. **Grosswendt, B. and Baek, W. Y.**, *Phys. Med. Biol.*, 43, 325, 1998.
27. **Willems, G. and Waibel, E.**, *Phy. Med. Biol.*, 37, 2319, 1992.
28. **Waibel, E. and Willems, G.**, *Phys. Med. Biol.*, 37, 249, 1992.
29. **Waibel, E. and Grosswendt, B.**, *Phys. Med. Biol.*, 37, 1127, 1992.
30. **Klevenhagen, S. et al.**, *Phys. Med. Biol.*, 27, 363, 1982.
31. **Klevenhagen, S.**, *Phys. Med. Biol.*, 36, 1013, 1991.
32. **Lanzon, P. J. and Sorell, G. C.**, *Phys. Med. Biol.*, 38, 1137, 1993.
33. **Cho, S. H. and Reece, W. D.**, *Phys. Med. Biol.*, 44, 13, 1999.
34. **Nunes, J. et al.**, *Med. Phys.*, 20, 223, 1993.
35. **Nahum, A. E.**, *Phys. Med. Biol.*, 41, 1531, 1996.
36. **Nahum, A. E.**, *Phys. Med. Biol.*, 41, 1957, 1966.
37. **Seuntjens, J. et al.**, *Phys. Med. Biol.*, 33, 1171, 1988.
38. **Svenson, H. and Brahme, A.**, *Recent Advances in Electron and Photon Dosimetry*, Plenum, New York, 87, 1986.
39. **Roos, M.**, *The State of the Art in Plane Parallel Chamber Hardware*, 1993. (unpublished report see ref. 35)
40. **Van der Plaetsen, A. et al.**, *Med. Phys.*, 21, 37, 1994.
41. **Luxton, G.**, *Med. Phys.*, 21, 631, 1994.
42. **Wang, R. and Sloboda, R. S.**, *Med. Phys.*, 23, 1459, 1996.
43. **Venselaar, J. L. M. et al.**, *Med. Phys.*, 23, 537, 1996.
44. **Tripathy, U. B. and Shanta, A.**, *Med. Phys.*, 12, 88, 1985.
45. **Russell, K. R. and Ahnesjo, A.**, *Phys. Med. Biol.*, 41, 1007, 1996.
46. **Watanabe, Y. et al.**, *Med. Phys.*, 25, 736, 1998.
47. **Wong, T. et al.**, *Phys. Med. Biol.*, 44, 357, 1999.
48. **Medin, J. et al.**, *Phys. Med. Biol.*, 40, 1161, 1995.
49. **Carlsson, A. K. et al.**, *Phys. Med. Biol.*, 42, 1033, 1997.
50. **Palmans, H. and Verhaegen, F.**, *Phys. Med. Biol.*, 42, 1175, 1997.
51. **Berger, M. J.**, *NISTIR Report*, 5226, 1993.
52. **Berger, M. J.**, *NISTIR Report*, 5113, 1993.
53. **Bortfeld, T.**, *Med. Phys.*, 24, 2024, 1997.
54. **Schaffner, B. et al.**, *Phys. Med. Biol.*, 44, 27, 1999.
55. **Siebers, J. V. et al.**, *Phys. Med. Biol.*, 40, 1339, 1995.
56. **Ma, C-M. and Nahum, A. E.**, *Phys. Med. Biol.*, 36, 413, 1991.
57. **Kearsley, E.**, *Phys. Med. Biol.*, 29, 1179, 1984.
58. **Haider, J. A. et al.**, *Phys. Med. Biol.*, 42, 491, 1997.
59. **Rossi, H. H. and Zaider, M.**, *Med. Phys.*, 18, 1085, 1991.
60. **Humm, J. L. et al.**, *Med. Phys.*, 20, 535, 1993.
61. **Roesch, W. C.**, *Rad. Res.*, 70, 494, 1977.
62. **Roeske, C. et. al.**, *J. Nuc. Med.*, 31, 788, 1990.
63. **Watson, E. E. et al.**, *Med. Phys.*, 20, 511, 1993.
64. **IAEA Tech. Rep.**, Series No. 277, 1987, and Proc. of a consultants meeting, 1993.
65. **Andrew, P. et al.**, *Med. Phys.*, 11, 874, 1984.
66. **Institute of Physical Sciences in Medicine (IPSM)**, *Phy. Med. Biol.*, 36, 1027, 1991.
67. **Paretzke, H. G. et al.**, *Proc. 4th Symp. on Microdosimetry*, EUR511, 22, 1973.
68. **Paretzke, H. G. and Bericht**, 24/88, 1988.
69. **Combecher, D.**, *Rad. Res.*, 84, 189, 1980.
70. **Smith, B. G. R. and Booz, J.**, *Proc. 6th Symp. on Microdosimetry* EUR6064, 759, 1972.
71. **Harder, D.**, *Biophysics*, 5, 157, 1968.
72. **Nordic Association of Clinical Physics (NACP)**, *Acta Radiol. Oncol.*, 2nd vol., 1981.
73. **Janssens, A. et al.**, *Phys. Med. Biol.*, 19, 619, 1974.
74. **Fisher, D. R.**, *4th Int. Symp. Radiopharm. Dosim., Cont.*, 85 1113, 466, 1986.
75. **AAPM Task Group 20**, *AAPM Report* 16, 1986.
76. **Vyncikier, S. et al.**, *Radiother. Oncol.*, 20 53, 1991, 32, 174, 1994.
77. **ICRU 23**, *Measurement of Absorbed Dose in Phantom*, 1973.
78. **ICRU 49**, *Stopping Powers and Ranges for Protons and Alpha Particles*, 1973.

3 Ionization Chamber Dosimetry

CONTENTS

I. INTRODUCTION

The most commonly used dosimeters for dose measurements and calibration of photon and electron beams are exposure-calibrated ionization chambers. One disadvantage of ionization chambers for electron-beam dosimetry is that the ionization-to-dose conversion factors, such as stopping-power ratios and perturbation factors, depend strongly upon electron energy and depth in the phantom.

The ionization chamber should have low leakage current as well as negligible stem and cable effects. Irradiation of the chamber's stem, collecting electrode, or cable by high-energy electrons can lead to emission of secondary electrons or collection of electrons that could produce a spurious ionization current.

Both cylindrical and plane-parallel ionization chambers have been used for determining central-axis depth-dose curves for electron beams. For thin plane-parallel chambers, it is generally accepted that the effective point of measurement is the front surface of the collecting volume. For a cylindrical chamber, the effective point of measurement is displaced toward the source from the center of the chamber.

The following procedure is recommended for determining relative depth-dose values in a homogeneous phantom using an ionization chamber. [1]

- Leakage: Leakage should be less than 0.1% of maximum signal.
- Stability: Variation in sensitivity should not be greater than 1%.
- Data collection
- Polarity
- Ion recombination
- Incorporate the above corrections of polarity and ion recombination to determine the corrected ionization charge $Q_{corr.}$
- Depth correction
- From the plot Q_{corr} vs. d, obtain the depth of 50% ionization R_{50} and practical range R_p, and then determine the mean incident energy $\bar{E}_0$ and mean energy $\bar{E}_d$, as a function of depth d.

$$\bar{E}_d = \bar{E}_0(1 - d/R_p)$$

- Determine the mean restricted mass collision stopping power ratio, $(\bar{L}/\rho)_{air}^{water}$, for the mean

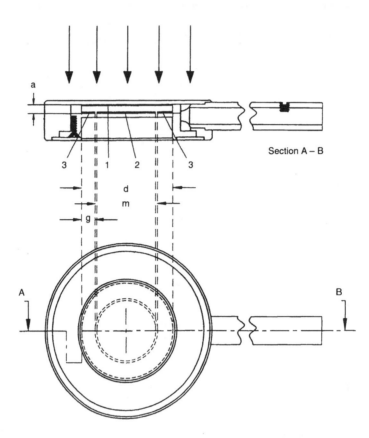

Section A – B

FIGURE 3.1 Diagram of a well-designed plane-parallel ionization chamber. Indicated in the diagram are height a and diameter d of the air cavity (1), diameter m of the collecting electrode (2), and width g of the guard ring (3). (From reference [8]. With permission.)

incident electron beam energy at each corrected depth of measurement.

• Determine chamber replacement factor.

The central axis percent depth dose, %D, is given by

$$\%D(d) \;=\; 100\,\frac{D_{med}(d)}{D_{med}(d_{max})}$$

Plane-parallel chambers are usually characterized by the following constructional details. The air volume is a disc-shaped right circular cylinder, one flat face of which constitutes the entrance window. The inside surface of the entrance window is electrically conducting and forms the outer electrode. The inner electrode is a conducting circular disc inset in the body insulator which forms the other flat face of the cylinder opposite to the entrance window. The sensitive volume is that fraction of the total air volume through which the lines of electrical force between the inner and outer electrodes pass. The inner and outer electrodes are mounted in a supporting block of material (the chamber body) to which the connecting cable is attached. The cable usually exits the body in a direction parallel to the entrance window. The sensitive volume is typically between 0.05 and 0.5 cm³. The polarizing potential is applied to the outer electrode, and the signal charge is

collected from the inner electrode. There is usually a third electrode surface between the other two electrodes which is not connected electrically to either of them but which is designed to be held at the same potential as the inner electrode. If the chamber assembly is fully guarded, this third electrode will be present in the air volume as a ring around the inner electrode. Plane-parallel chambers for electron radiation have the following typical dimensions: the entrance window thickness is 1 mm or less; the distance between the inner and outer electrode is 2 mm or less; and the diameter of the inner (collecting) electrode is 20 mm or less.

For measurements in electron beams of energies below $\bar{E}_0 = 10$ MeV, plane-parallel ionization chambers are recommended, and they must be used below $\bar{E}_0 = 5$ MeV. However, plane-parallel chambers are also suitable for use above $\bar{E}_0 = 10$ MeV. The chambers should preferably be designed for measurements in water, and the construction should be as homogeneous and water equivalent as possible; i.e., mass stopping powers and linear scattering powers should be similar to those of water. It is especially important to consider backscattering effects from the rear wall of the chamber. Chambers designed for measurements in solid phantoms should, accordingly, be as phantom-equivalent as possible. Some chambers, however, have a design that includes several materials, resulting in a significant departure from homogeneity. In these cases, there is no simple rule for the selection of phantom material. One of the main advantages of plane-parallel chambers for electron beam dosimetry is the possibility of minimizing in-scattering perturbation effects, P_{cav}. Plane-parallel ionization chambers may be designed so that the chamber samples the electron fluence incident through the front window, the contribution of electrons entering through the side walls being negligible. This design justifies considering the effective point of measurement P_{eff} to be at the center of the front surface of the air cavity. For practical purposes, it is also convenient to choose the reference point of the chamber at the same position. In order to fulfill the requirements concerning P_{cav} and P_{eff} within a reasonable approximation, plane-parallel chambers must have a flat cavity; i.e., the ratio of cavity diameter and cavity depth must be large (of the order of ten), the cavity height should not exceed 2 mm, and the collecting electrode should be surrounded by a guard electrode having a width not smaller than 1.5 times the cavity height. Such dimensions are reported to sufficiently reduce the in-scattering perturbation effect. [2] Furthermore, the diameter of the collecting electrode should not exceed 20 mm in order to reduce the influence of radial nonuniformities of the beam profile. The thickness of the front window should be restricted to 1 mm at most to make measurements at shallow depths possible. It is also necessary for the air cavity to be vented so that it will equilibrate rapidly with the external pressure and temperature.

An overview of desirable plane-parallel chamber properties is compiled in Table 3.1. Characteristics of several commercial parallel chambers are shown in Table 3.2.

The mean energy at the phantom surface, $\bar{E}_0$, is required for the evaluation of quantities and parameters

TABLE 3.1
Desirable Properties for a Plane-Parallel Chamber in Electron Radiation

Chamber Dimensions	
Front window thickness	≤ 1 mm
Collecting electrode diameter	≤ 20 mm
Ratio of guard ring width to cavity height	≥ 1.5
Cavity height	≤ 2 mm
In-scattering perturbation effect, p_{cav}	$<1\%$
Wall and backscattering perturbation effects, p_{wall}	$<1\%$
Polarity effect	$<1\%$
Leakage current	$<10^{-14}$ A
Long-term stability	$\pm 0.5\%$[a]

Source: From reference [8]. With permission.

used in the formalism and mainly affects the choice of stopping power ratios of water to air, $s_{w,air}$, at the reference depth, namely $s_{w,air}(\bar{E}_0, z_{ref})$. For dosimetry purposes it has become customary to specify the quality of electron beams in terms of the mean energy at the surface of the phantom, $\bar{E}_0$, determined from empirical relationships between electron energy and the half-value depth in water, R_{50}. [3] The recommendation is to determine $\bar{E}_0$ by using the energy-range relationship [4]

$$\bar{E}_0[\text{MeV}] = CR_{50}$$

where $C = 2.33$ MeV cm^{-1} and R_{50} in cm in water is obtained from a depth-dose distribution measured with constant source-chamber distance.

A photon beam is characterized by the ratio of absorbed doses at depths of 20 and 10 cm for a constant source-detector distance and a 10-cm × 10-cm field at the plane of the chamber, TPR^{20}_{10}. Attention should be paid to the use of certain solid-state detectors (some types of diodes and diamonds) to measure depth-dose distributions for the purpose of deriving TPR^{20}_{10}. Significant discrepancies between the reading of some detectors and that of a reference ionization chamber at different depths have been reported, which in certain cases might result in absorbed dose deviations above 1% (see Figure 3.2). [5] A solid-state detector whose response has been regularly verified against a reference detector (ion chamber) should be selected for these measurements.

The formalism for the determination of the absorbed dose to water in high-energy photon and electron beams is based on the use of an ionization chamber having a calibration factor in terms of air kerma, N_K. The reference quality Q_0 is usually gamma rays from a ^{60}Co source. From the chamber's N_K value, the absorbed dose to air chamber factor at the reference quality, N_{D,air,Q_0}, is obtained.

The formalism is based on the assumptions that the volume of air in the sensitive region of the chamber cavity and the average energy required to produce an ion pair, W_{air}, are identical in the user's beam quality Q and at the calibration quality Q_0. Therefore, $N_{D,air,Q} = N_{D,air,Q_0} = N_{D,air}$.

The absorbed dose to water $D_{w,Q}$ in the user's beam of quality Q, when the effective point of measurement of the ionization chamber P_{eff} is positioned at the reference depth, is given by

$$D_{w,Q}(P_{eff}) = M_Q N_{D,air}(s_{w,air})_Q P_Q \quad (3.1)$$

where M_Q is the reading of the ionization chamber and electrometer system in the user's beam, corrected for influence quantities (temperature, pressure, humidity, and saturation); $N_{D,air}$ is the absorbed dose to air chamber factor; $(s_{w,air})_Q$ is the stopping power ratio, water to air, in the user's beam; and p_Q is an overall perturbation factor for

TABLE 3.2
Characteristics of Plane-Parallel Chamber Types (as stated by manufacturers)

Chamber	Materials	Window Thickness	Electrode Spacing (mm)	Collecting Electrode Diameter (mm)	Guard Ring Width (mm)	Polarity Effect[a] (%)	Leakage Current[a] (A)	Long-Term Stability[a] (%)	Recommended Phantom Material
Vinten 631 (Pitman 631)	Aluminized mylar foil window, graphited mylar foil electrode, styrene co-polymer, rear wall, PMMA body	1 mg/cm² 0.006 mm	2 × 1 cavity height 2	20	3	<0.2 larger reported [73]			PMMA (phantom integrated)
NACP01 (Scanditronix) Calcam-1 (Dosetek)	Graphite window, graphited rexolite electrode, graphite body (back wall), rexolite housing	90 mg/cm² 0.5 mm	2	10	3	<0.5 larger reported [73]	$<10^{-14}$	±0.5	Polystyrene, graphite, water (with waterproof housing)
NACP02 (Scanditronix) Calcam-2 (Dosetek)	Mylar foil and graphite window, graphited rexolite electrode, graphite body (back wall), reloxite housing	104 mg/cm² 0.6 mm	2	10	3	<0.5 larger reported [74]	$<10^{-14}$	±0.5	Water, PMMA
Markus chamber PTW 23 343 NA 30-329 NE 2534	Graphite polyethylene foil window, graphited polystyrene collector, PMMA body, PMMA cap	102 mg/cm² 0.9 mm (incl. cap)	2	5.3	0.2	<0.5 larger reported [75]	2×10^{-16} larger reported		Water, PMMA
Holt chamber (Memorial) NA 30-404	Graphited polystyrene wall and electrode, polystyrene body	416 mg/cm² 4 mm	2	25	5	<1 larger reported [76]	10^{-15}	±0.5	Polystyrene (phantom integrated)
PS-033 (Capintec)	Aluminized mylar foil window, carbon impregnated air equivalent plastic electrode, polystyrene body	0.5 mg/cm² 0.004 mm	2.4	16.2	2.5		$<10^{-14}$	±1	Polystyrene

Chamber	Wall and electrodes	Window				Leakage	Phantom
Exradin 11	Conducting plastic wall and electrodes, model Pll: polystyrene equivalent; model A11: C552, air equivalent; model T11: A 150, tissue equivalent	P11: 104 mg/cm² 1 mm	2	20	5.1	<10⁻¹⁵	P11: poly-styrene, water
Roos chamber PTB FK6 PTW 34001 Wellhöfer IC40	PMMA, graphited electrodes	118 mg/cm² 1 mm	2	16	4	<10⁻¹⁴	Water, PMMA
Attix chamber RMI 449	Kapton conductive film window, graphited polyethylene collector, solid water body	4.8 mg/cm² 0.025 mm	1 0.7 reported	12.7	13.5	3 × 10⁻¹⁴	Solid water
Schulz chamber PTW 23346	Graphited polymide foil electrodes, PMMA window and body	60 mg/cm² 0.5	2 ×1.5 cavity height: 5	19	3	<10⁻¹⁴	Water
Memorial pipe rectangular	Graphited mylar window, graphite layer on polystyrene collector	5 mg/cm² 0.03 mm	2.5	11.3 × 2.5	5		Polystyrene
Memorial pipe circular	Graphited mylar window, graphite layer on polystyrene collector	5 mg/cm² 0.03 mm	2.5	10	5		Polystyrene

Note the leakage values in scientific notation: $<10^{-15}$, $<10^{-14}$, 3×10^{-14}, $<10^{-14}$.

Additional column (between value and leakage): <0.5, ≤0.3, <0.5.

ᵃ Blanks correspond to no information available.

Source: From Reference [8]. With permission.

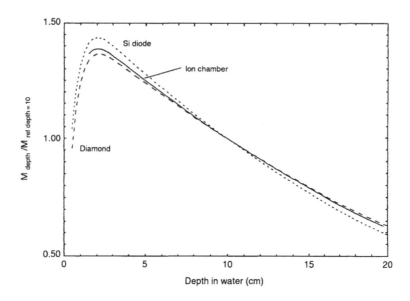

FIGURE 3.2 Relative depth-dose distributions obtained with a silicon diode, a diamond detector, and an ionization chamber (NE-2561) in water using a 10-MV clinical photon beam. Results are normalized to unity at a depth of 10 cm. (From Reference [8]. With permission.)

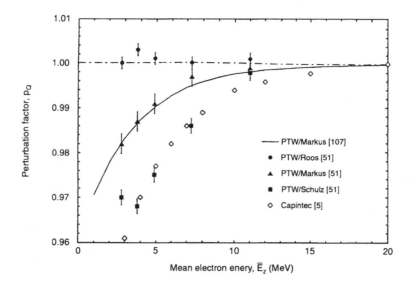

FIGURE 3.3 Variation of the perturbation factor p_Q for several different plane-parallel chambers in common use, relative to the NACP chamber, indicated by the dashed line drawn at $p_Q = 1.00$. All the measurements were made at the depth of dose maximum and normalized to the quotient test chamber/NACP in a high-energy electron beam. The full line is a fit to three separate measurement series on different accelerators using the P7W/Markus chamber. The filled data points are measurements on three different PTW designs and renormalized so that $p_Q = 1$ for the NACP chamber; the unfilled symbols are for the Capintec-PS-033 chamber. (From Reference [8]. With permission.)

the ionization chamber. p_Q in Equation (3.1) replaces the perturbation factor p_u. The factor p_Q is the product of the various perturbation correction factors for the ion chamber at quality Q, namely, p_{wall}, p_{cav} and p_{cel}. The first two factors correct for the lack of phantom equivalence of the chamber wall and for the perturbation of the electron fluence due to differences in the scattering properties between the air cavity and the phantom, respectively. The factor p_{cel} corrects for the effect of the central electrode

of the chamber during in-phantom measurements. Relative values of p_Q are shown in Figure 3.3.

The various steps between the calibration of ionization chambers at the standards laboratories and the determination of absorbed dose to water at hospitals introduce undesirable uncertainties into the realization of D_w. [6] Different factors are involved in the dosimetric chain that starts with the quantity K_{air}, measured in air using a ^{60}Co beam, and ends with the absorbed dose to water measured

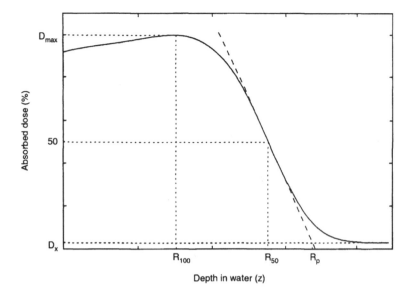

FIGURE 3.4 An electron beam depth-dose distribution in water showing the various range parameters.

in water D_w, using clinical beams. Uncertainties in the chain arise mainly from conversions performed by the user at the hospital, for example, the k_m and k_{att} factors. The transfer of N_K (calibration factor) to $N_{D,air}$ (or N_{gas}) means that in practice the starting point of the calibration of clinical beams already involves a considerable uncertainty. Procedures to determine absorbed dose to water using methods to measure appropriate base or derived quantities have considerably improved at the Accredited Dosimetry Laboratories. The required conversion and perturbation factors are available at some laboratories. These developments support a change in the quantity used to calibrate ionization chambers, i.e., N_K, replacing it by a calibration factor in terms of absorbed dose to water. Electron beam depth close distribution in water is shown in Figure 3.4.

The absorbed dose to water at the reference point of the chamber in a phantom irradiated by a beam of reference quality Q_0 is given by the simple relationship

$$D_{w,Q_0} = M_{Q_0} N_{D,w,Q_0} \qquad (3.2)$$

where N_{D,w,Q_0} is obtained at the standards laboratory from the knowledge of the standard quantity absorbed dose to water at the point of measurement in water for the calibration quality Q_0. A practical approach in common use is to provide users with N_{D,w,Q_0}, i.e., calibration at the reference quality ^{60}Co, and to apply beam quality correction factors k_Q for other beam qualities. For beams other than the reference quality, the absorbed dose to water is given by

$$D_{w,Q} = M_Q N_{D,w,Q_0} k_Q \qquad (3.3)$$

where the factor k_Q corrects for the difference between the reference beam quality Q_0 and the actual quality being used, Q. The value of k_Q should ideally be determined experimentally at the same quality as the user's beam, although this is seldom achievable.

A general expression which includes the ratio $(W_{air})_Q/(W_{air})_{Q_0}$ has been given in Reference [7]:

$$k_Q = \frac{(s_{w,air})_Q}{(s_{w,air})_{Q_0}} \frac{(W_{air})_Q}{(W_{air})_{Q_0}} \frac{p_Q}{p_{Q_0}} \qquad (3.4)$$

which can be used in any type of therapeutic beam. The overall perturbation factors p_Q and p_{Q_0} include p_{wall}, p_{cav}, p_{cel}, and p_{dis} as the reference point of a cylindrical ion chamber in the center of the cavity volume. In therapeutic electron and photon beams, the general assumption of $(W_{air})_Q = (W_{air})_{Q_0}$ yields the better known equation for k_Q:

$$k_Q = \frac{(s_{w,air})_Q}{(s_{w,air})_{Q_0}} \frac{p_Q}{p_{Q_0}} \qquad (3.5)$$

which depends only on ratios of stopping power ratios and perturbation factors.

The connection between the $N_{D,air}$- and the $N_{D,w}$-based formalisms is established by the relationship

$$N_{D,w,Q} = N_{D,air}(s_{w,air})_Q P_Q \qquad (3.6)$$

When a plane-parallel ionization chamber has a calibration factor in terms of absorbed dose to water at a reference quality Q_0, N_{D,w,Q_0}, and the reference point positioned at z_{ref}, the absorbed dose to water at the reference depth is given by

$$D_{w,Q}(z_{ref}) = M_Q N_{D,w,Q_0} K_Q \qquad (3.7)$$

TABLE 3.3
Recommended Values of p_{wall}^{pp} for Plane-Parallel Chambers in Various Phantom Materials Irradiated by A ^{60}Co Gamma-Ray Beam

Chamber type	Phantom Material			
	Water	PMMA	Polystyrene	Graphite
Capintec PS-033	0.989	0.982	0.960[a]	0.989
Exradin P11	—	1.000[a]	—	—
Holt Pancake	—	—	1.000[a]	—
NACP	1.024	1.012	1.000	1.027
PTW/Markus M23343	1.009	1.006	0.979	1.013
PTW/Schulz M23346	1.001	1.001	0.992	1.009
Roos FK-6	1.003[b]	—	—	—

Source: From Reference [8]. With permission.

TABLE 3.4
Ratios of Stopping Powers ($s_{med,\ air}$) and Mass Energy Absorption Coefficients ($(\mu_{en}/\rho)_{med,air}$) for ^{60}Co Gamma-Rays

Chamber Wall or Phantom Material	$s_{med,air}$	$(\mu_{en}/\rho)_{med,air}$
Water	1.133	1.112
PMMA	1.102	1.081
Polystyrene	1.110	1.078
Graphite	1.002	1.001
A-150	1.142	1.101
C-552	0.995	1.001
Delrin	1.080	1.068
Nylon	1.142	1.098

Note: (To evaluate $(\mu_{en}/\rho)_{med_1,med_2}$, use $((\mu_{en}/\rho)_{med_1,air}/(\mu_{en}/\rho)_{med_2,air})$.

Source: From Reference [8]. With permission.

or, more explicitly,

$$D_{w,Q}(z_{ref}) = M_Q N_{D,w,Q_0} \frac{(s_{w,air})_Q}{(s_{w,air})_{Q_0}} \frac{(p_{cav}p_{wall})_Q}{(p_{cav}p_{wall})_{Q_0}} \quad (3.8)$$

The following steps and practical details, common to the different methods to determine $N_{D,air}^{pp}$ for plane-parallel ionization chambers, are recommended: [8]

1. The long term stability and the leakage current of the ionization chambers and the electrometer should be checked before the measurements for the determination of $N_{D,air}^{pp}$.

2. Sufficient time should be allowed for the ionization chamber, build-up cap, and phantom material to reach equilibrium with the room temperature and for the electrometer to stabilize.

TABLE 3.5
Recommended Values of k_{att},k_m, Needed for ^{60}Co In-Air Calibration of Parallel-Plate Chambers

Chamber	Build-up Cap Material	$k_{att}k_m$
Capintec PS-033	polystyrene	1.012[b]
Exradin P11	polystyrene	0.973
Holt Pancake	polystyrene	0.980
PTW Markus	PMMA	0.985
NACP	graphite	0.975
Roos FK-6	PMMA or water	N/A

Note: (The build-up caps are of the material as shown in Column 2 and the additional build-up cap thickness is 0.5 g cm^{-2}, except for the Holt chamber, which comes with a build-up cap.)

Source: From Reference [8]. With permission.

3. The time during which the chambers and the build-up cap material are handled with bare hands should be minimized.

4. The reference chamber (when applicable) should be positioned with its recommended orientation in the radiation beam (the reference orientation should be stated in the calibration certificate).

5. The ambient air pressure and the temperature in the air or in the phantom, depending on the type of measurement, should be measured before, during, and after the irradiation measurements, and corrections should be applied to the reading of the detectors for any variations. In the case of using a phantom, the temperature should be measured inside the phantom material.

6. Before the first reading of each series of measurements is taken, the chambers placed inside

the phantom or equipped with the build-up cap should be pre-irradiated with 2 to 5 Gy to achieve charge equilibrium.

7. For each chamber, five independent determinations of $N_{D,air}^{pp}$ should preferably be made and the mean value should be used.

8. The $N_{D,air}^{ref}$ for the reference chamber is determined from

$$N_{D,air}^{ref} = N_K^{ref}(1 - g)k_{att}k_m k_{cel} \qquad (3.9)$$

where N_K^{ref} is the air-kerma calibration factor of the reference chamber provided by the standards laboratory, $(1 - g) = 0.997$ (for ^{60}Co gamma-rays).

Parameters related to several ionization chambers are listed in Tables 3.3, 3.4, and 3.5.

II.   CORRECTION FACTORS

Ionization chambers calibrated with either a ^{60}Co gamma source or in high-energy electron beam plane-parallel and cylindrical chambers are compared in the calibration correction and must therefore be made to obtain the right dose.

The main purpose of the correction factors is to allow calibration procedures more widely practicable than those based on the use of an electron beam of sufficiently high energy. The condition required to apply the correction factors is that the characteristics of the chambers to be calibrated are the same as those reported for the chambers considered in the investigation.

A. PLANE-PARALLEL IONIZATION CHAMBERS

Correction factors for calibration of plane-parallel ionization chambers with a ^{60}Co gamma-ray beam were investigated by Laitano et al. [9] Correction factors for free-in-air calibration were measured as follows. The plane-parallel chambers were irradiated free in air, in a ^{60}Co gamma-ray beam of given size with a build-up disc and with their air cavity center at a point where the air kerma, K_{air}, was known at that point the absorbed dose to the air in the cavity of the plane-parallel chamber with its build-up disc is given by

$$(D_{air})_{cav} = K_{air}(1 - g)(k_{pp})_{A_{1,2}} = M_{c,pp}M_{D,pp} \qquad (3.10)$$

where $M_{c,pp}$ is the plane-parallel chamber reading at the calibration conditions considered; $(k_{pp})_{A_{1,2}}$, is the plane-parallel chamber correction factor for the conditions A_1 and A_2, and $N_{D,pp}$ is the plane-parallel chamber calibration factor (details in Figure 3.5). The reading $M_{c,pp}$ is corrected for ambient para-meters, for ion recombination, and for polarity effect. For a plane-parallel chamber, the correction factors $(k_{pp})_{A_{1,2}}$ can be obtained from Equation (3.10) if the $N_{D,pp}$ factor of that chamber is already known.

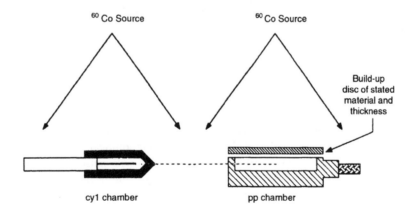

BUILD-UP DISC MATERIAL AND THICKNESS				
Condition		A_1		A_2
CHAMBER				
NACP 02	Graphite	(0.52 gcm⁻²)	PMMA	(0.43 gcm⁻²)
Markus	Graphite	(0.52 gcm⁻²)	PMMA	(0.52 gcm⁻²)
Schulz	Graphite	(0.52 gcm⁻²)	PMMA	(0.45 gcm⁻²)
Capintec	Polystyrene	(0.51 gcm⁻²)	PMMA	(0.52 gcm⁻²)

FIGURE 3.5 Experimental conditions A_1 and A_2: free-in-air calibration, build-up material density: 1.73 g cm⁻³ (graphite), 1.18 g cm⁻³ (PMMA), and 1.04 g cm⁻³ (polystyrene). The disc diameter was equal to the chamber external diameter. In the Schulz chamber characterized by a thin backwall, two discs were used on the front and rear of the chamber, respectively. (From Reference [9]. With permission.)

By equating the expressions of the absorbed dose relevant the plane-parallel and to the cylindrical chamber, the factor $N_{D,pp}$ can be obtained by

$$N_{D,pp} = \frac{M_{u,cyl}N_{D,cyl}(p_{u,cyl})_{eb}}{M_{c,pp}} \qquad (3.11)$$

where $M_{u,cyl}$ and $M_{u,pp}$ are the corrected readings, at the electron-beam irradiation conditions, of the reference cylindrical chamber and of the plane-parallel chamber, respectively. $(p_{u,cyl})_{eb}$ refers to the cylindrical chamber and is the correction factor due to the electron fluence perturbation. $N_{D,cyl}$ is the calibration factor of the reference cylindrical chamber, given by the expression:

$$N_{D,cyl} = \frac{K_{air}(1-g)k_m k_{att}}{M_{c,cyl}} \qquad (3.12)$$

where $M_{c,cyl}$ is the corrected reading of the reference cylindrical chamber in the ^{60}Co gamma-ray calibration condition, and k_m and k_{att} are the factors that take into account the lack of air equivalence of ionization chamber material and the attenuation and scattering of the ^{60}Co photons in the chambers wall, respectively. From Equation (3.10) one finds

$$(k_{pp})_{A_{1,2}} = \frac{M_{c,pp}N_{D,pp}}{K_{air}(1-g)} \qquad (3.13)$$

Combining these equations, one obtains

$$(k_{pp})_{A_{1,2}} = \frac{M_{c,pp}M_{u,cyl}}{M_{c,cyl}M_{u,pp}} k_m k_{att}(p_{u,cyl})_{eb} \qquad (3.14)$$

Correction factors for calibration at d_{max} in PMMA and in polystyrene phantoms have been found as follows. The plane-parallel chambers were irradiated in a ^{60}Co gamma beam with the center of their air cavity at d_{max} in PMMA and polystyrene phantoms. The K_{air} value at the position of the plane-parallel chamber center in the absence of chamber and phantom was known. At this reference point the ^{60}Co beam size was 10.5×10.5 cm^2. In this experimental condition, an expression similar to Equation (3.10) holds:

$$K_{air}(1-g)(k_{pp})_{B_{1,2}} = M_{c,pp}N_{D,pp} \qquad (3.15)$$

Here the chamber reading $M_{c,pp}$ now includes the additional contribution due to the radiation scattered by the phantom into the chamber cavity, and $(k_{pp})_{B_{1,2}}$ is the plane-parallel chamber correction factor for the conditions B_1 and B_2. [9] From Equation (3.15) one finds

$$(k_{pp})_{B_{1,2}} = \frac{M_{c,pp}N_{D,pp}}{K_{air}(1-g)} \qquad (3.16)$$

Combining with Equations (3.11) and (3.12) will give

$$(k_{pp})_{B_{1,2}} = \frac{M_{c,pp}M_{u,cyl}}{M_{c,cyl}M_{u,pp}} k_m k_{att}(p_{u,cyl})_{eb} \qquad (3.17)$$

For correction factors for in-phantom calibration at a given depth, the correction factors of the plane-parallel chambers were determined in phantoms of different materials and at a given depth where the absorbed dose to the phantom medium was known.

The correction factor can be expressed as

$$(p_{u,pp})_{C_{1,2,3,4}} = \frac{M_{c,cyl}N_{D,cyl}(p_{u,cyl})_c}{M_{c,pp}N_{D,pp}} \qquad (3.18)$$

$$(p_{u,pp})_{C_{1,2,3,4}} = \frac{M_{c,cyl}M_{u,pp}(p_{u,cyl})_c}{M_{c,pp}N_{u,cyl}(p_{u,cyl})_{eb}} \qquad (3.19)$$

The calibration depth was 5 cm in PMMA, polystyrene, and water (see Figure 3.6).

Plane-parallel ionization chambers are often calibrated at the clinic against a cylindrical chamber in a high-energy electron beam. If the perturbation due to the presence of the plane-parallel chamber in a water phantom at the calibration quality (i.e., ^{60}Co) is known, a straightforward dose-to-water calibration is possible. If the perturbation as a function of photon-beam quality is known, the plane-parallel chamber might be used for all radiation qualities. [10]

A number of uncertainties may be added to the $N_{D,pp}$ factor for the plane-parallel chamber:

1. Any uncertainties in the clinic's cylindrical chamber calibration are transferred to the plane-parallel dosimetry.
2. The two chambers must be irradiated at slightly different positions or at the same position but at different times. In the former case an inhomogeneous dose distribution within the field will cause a difference in the dose delivered, and in the latter case a very stable dose-monitoring device is needed to avoid influence of accelerator output fluctuations.
3. Most cylindrical chambers are not of a homogeneous design, but they all have a central electrode made of aluminium. A p_{cel} factor must then be introduced into the calculations to take into account the disturbance of the electrode in the electron beam relative to the ^{60}Co-gamma-ray beam. The size of the p_{cel} factor for commonly used cylindrical chambers is being debated and has been the subject of some recent papers. [10]
4. If a plastic phantom is used, which is often the case, charge build-up in the plastic may

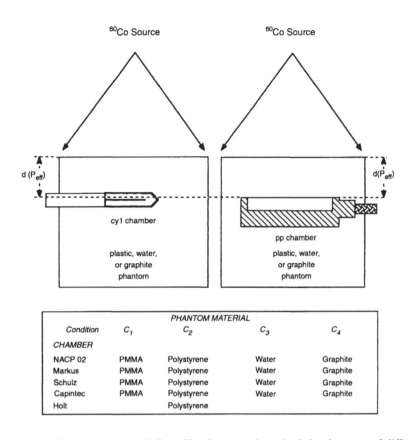

		PHANTOM MATERIAL		
Condition	C_1	C_2	C_3	C_4
CHAMBER				
NACP 02	PMMA	Polystyrene	Water	Graphite
Markus	PMMA	Polystyrene	Water	Graphite
Schulz	PMMA	Polystyrene	Water	Graphite
Capintec	PMMA	Polystyrene	Water	Graphite
Holt		Polystyrene		

FIGURE 3.6 Experimental conditions $C_1, C_2, C_3,$ and C_4: calibration at a given depth in phantoms of different materials (PMMA, polystyrene, water, and graphite). The calibration depth d was 4.45 g cm^{-2} in graphite and 5 cm in other phantoms. With regard to the cylindrical chamber position, the figure refers to measurements in water, polystyrene, and PMMA, for which the P_{eff} approach was used. For measurements in the graphite phantom, the cylindrical chamber was positioned with its center at depth d, then correcting the reading by the P_{repl} factor. (From Reference [9]. With permission.)

introduce a disturbance in the measurement. The h_m factor—the ratio of the measured signal at dose maximum in water to that measured in the phantom material—may be not only phantom material–dependent but also chamber type–dependent, thus yielding different h_m factors for the plane-parallel and cylindrical chambers.

The cylindrical and plane-parallel chambers were calibrated by Nystrom and Karlsson [10] against the Fricke dosimeter in a high-energy electron beam and in four different photon beams, including ^{60}Co γ-rays. Similar plane-parallel chambers were included in the study (Figure 3.7).

The $p_{u,pp}$ factors are shown in Figure 3.8; the P_u values for the two NACP chambers are plotted against TPR_{10}^{20} for the photon beams. The full lines indicate the p_u values for polystyrene and graphite, with chamber wall thickness equal to that of the entrance window of the NACP chambers and the walls of the cylindrical chambers, i.e., 0.5 mm.

In some situations, especially for the calibration of low-energy electron beams, where parallel-plate chambers are recommended for the determination of absorbed dose at the reference point, solid plastic phantoms may, for

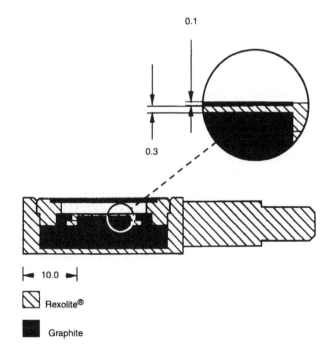

FIGURE 3.7 Principal view of the NACP chamber. (From Reference [10]. With permission.)

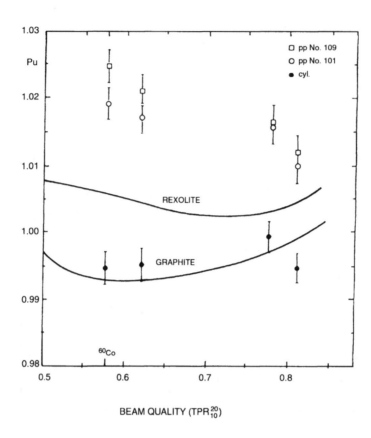

FIGURE 3.8 The *Pu,pp* factors as functions of photon beam quality for the investigated chambers. Error bars indicate ±SD. The full line indicates *Pu,cyl* factors for polystyrene and graphite derived with wall thickness of 0.5 mm, the thickness at the entrance window of the NACP chambers and wall thickness of the chambers investigated. From Reference [10]. With permission.

practical reasons, be used. Corrections are then required to obtain the absorbed dose to water at a reference point situated at an equivalent depth in water. [11]

Scaling factors have often been based on R_{csda}, the linear continuous-slowing-down range; this quantity, however, does not include differences in multiple scattering between the two materials, which might play an important role at low electron energies, where plastic phantoms are more commonly used.

An additional correction factor is needed to account for differences in the magnitude of the primary electron fluence in water and in the solid material at equivalent depths due to the different scattering properties of the phantom materials. The AAPM protocol (1994) [14] provides an electron fluence factor, denoted by the symbol ϕ_{med}^{water}, evaluated analytically from the Fermi-Eyges theory of multiple scattering. To determine this fluence factor experimentally a perturbation free chamber is needed.

IAEA TRS-277 [4] used the factor h_m, first suggested by NACP (1981) [4a] and defined according to IAEA as the ratio of the signal measured using a plane-parallel chamber at the ionization maximum on the central axis in a water phantom to that measured (at the same source to phantom surface distance) for the same accelerator monitor setting at the ionization maximum in other phantom

materials. The new IAEA code of practice, IAEA TRS-381, [8] assumes that the h_m-factor is approximately equal to ϕ_{med}^{water}. The equality between the factors ϕ_{med}^{water} and h_m is based on the assumption of a perturbation free chamber, and any departure from these assumptions would make the set of data dependent on the chamber used. [11]

According to the IAEA TRS-381 protocol, the absorbed dose to water when measurements are performed in a solid plastic phantom is obtained using the equation

$$D_{w,Q} = M_{Q,plastic} N_{D,air} (s_{w,air})_Q p_Q h_m \qquad (3.20)$$

where $N_{D,air}$ is the absorbed-dose-to-air chamber factor. From the quantity determined with this factor, $\bar{D}_{air}$, the absorbed dose to water at a point D_w is derived by the application of the Bragg-Gray principle. $M_{Q,plastic}$ is the ionization measured in the solid plastic phantom, corrected for influence quantities (p, T, recombination, etc.). $(s_{w,air})_Q$ is the stopping power ratio for water to air, p_Q is the overall perturbation factor including both cavity (replacement) perturbation and wall effects, and h_m is the correction factor to account for differences in the magnitude of the primary electron fluence in water and in the solid material at equivalent reference depths, due to the different scattering properties of the phantom materials.

Both IAEA TRS-381 and the new UK-protocol (IPEMB) [12] define h_m as the ratio of electron fluences at equivalent depths in the water and plastic phantoms. If a criterion is applied where perturbation factors in water and plastic differ, the experimental determination of the factor h_m should strictly be modified according to

$$h_m^{exp} = \phi_{med}^{water} p_{Q.med}/p_{Q,w} \quad (3.21)$$

where the ratio of perturbation factors for the chamber used in the measurements appears explicitly for the two media. ϕ_{med}^{water} is the ratio of the electron fluences as noted by AAPM Task Group 21. [13]

AAPM Task Group 39 [14] has presented a different approach. Instead of calculating the dose to water directly, these protocols determine first the dose to the plastic phantom using the equation

$$D_{med} = M N_{gas}^{pp}(\bar{L}/\rho)_{gas}^{med} P_{ion} P_{repl} P_{wall, plastic} \quad (3.22)$$

where M is the electrometer reading in the solid plastic phantom, corrected for pressure and temperature. N_{gas}^{pp} is the cavity gas calibration factor, $(\bar{L}/\rho)$ is the mean restricted mass stopping power of plastic to gas, P_{ion} the correction for ion recombination, P_{repl} the perturbation replacement factor, and $P_{wall,plastic}$ the perturbation wall factor in the plastic medium.

The dose to water is then obtained in a second step by scaling the electron fluence in water and the solid phantom by the relation [11]

$$D_{med}(d_{water}) = D_{med}(d_{med})[(\bar{S}/\rho)_{coll}]_{med}^{water} \phi_{med}^{water} \quad (3.23)$$

where ϕ_{med}^{water} is the fluence factor, i.e., the ratio of electron fluence in water to that in the solid phantom at equivalent depths, and $(\bar{S}/\rho)_{coll}$ is now the mean unrestricted mass stopping power ratio.

Figure 3.9 shows the measured P_{wall} factors as a function of energy for the simulated Attix, Roos, and NACP chambers placed in phantoms of aluminum (Figure 3.9a) and polyethylene (Figure 3.9b). These materials are not typical for clinical dosimetry but have atomic numbers that differ significantly from the chamber wall materials and thus illustrate wall effects more clearly. [11]

Figure 3.10 shows the P_{wall}-factors for graphite, PMMA, clear polystyrene and Plastic Water™. These phantom materials are closer to the wall materials with respect to atomic number and scattering effects and, therefore, the P_{wall}-factors are also closer to unity. The NACP chamber is recommended to be used with a PMMA phantom. At low electron energies where plastic phantoms are normally used, it can be seen in Figure 3.10 that the P_{wall}-factor is 0.99. The Roos chamber has a P_{wall}-factor equal to unity in a PMMA phantom. Figure 3.10 also shows that the P_{wall}-factor for the Attix chamber in PMMA is close

to 1.00, as the backscatter in PMMA and Solid Water™ are similar. In a clear polystyrene phantom, the P_{wall}-factors are below unity for all chambers, especially for the NACP chamber, where a P_{wall}-factor of 0.97 was obtained at low energies. Also, the Attix and Roos chambers have P_{wall}-factors below 0.99 at low energies.

B. k_Q FACTORS

k_Q, the beam quality correction factor, corrects the absorbed dose-to-water calibration factor $N_{D,w}$ in a reference beam of quality Q_0 to that in a user's beam of quality Q_1. It was discussed by Vatnitsky et al. [15]

If an ionization chamber with a ^{60}Co absorbed dose-to-water calibration (reference beam, quality Q_0) is used for measurements in a beam with quality Q_1, the absorbed dose-to-water $D_w(Q_1)$ is given by (rewriting Equation 3.3)

$$D_w(Q_1) = M_{Q_1} \times N_{D,w,\gamma} \times k_{Q_\gamma} \quad (3.24)$$

where M_Q is the charge collected in the ionization chamber in the beam Q_1 corrected to reference temperature and pressure conditions and for ionic recombination; $N_{D,w,\gamma}$ is the absorbed dose-to-water calibration factor in the ^{60}Co beam; and k_{Q_γ} corrects the reference absorbed dose-to-water calibration factor $N_{D,w,\gamma}$ to the beam with quality Q_1. The subscript γ is placed on k_Q to specify that the reference beam is the ^{60}Co beam. The term "quality" is used to specify the radiological properties of the beam.

For photons, the quality is specified in terms of the TPR_{10}^{20} and, for electrons, in terms of the depth of 50% dose in water. The beam quality specifier for a proton beam is the effective energy of the protons at the calibration point.

Water calorimetry was used as the absolute dose standard to measure k_{Q_γ}, employing the water calorimeter developed by R. Schulz. [16]

The dose delivered to the water for irradiation within a reference beam of quality Q_0 (^{60}Co beam) was determined from

$$D_{water}(Q_0, cal) = c \times \theta \times \Delta V_0 \times (1 + D_T) \quad (3.25)$$

where c is the specific heat of water at the calorimeter operation temperature, θ is the calibration factor obtained from the thermistor calibrations, ΔV_0 is the bridge deflection, and D_T is the thermal defect.

Vatnitsky et al. determined k_{Q_γ} by treating the calorimeter response to the irradiation as a deflection of the Wheatstone bridge. When the water calorimeter was replaced with a water phantom, the absorbed dose to water measured by an ionization chamber in the same beam Q_0 was given by

$$D_{water}(Q_0, ion) = N_{D,w,\gamma} \times M_0 \quad (3.26)$$

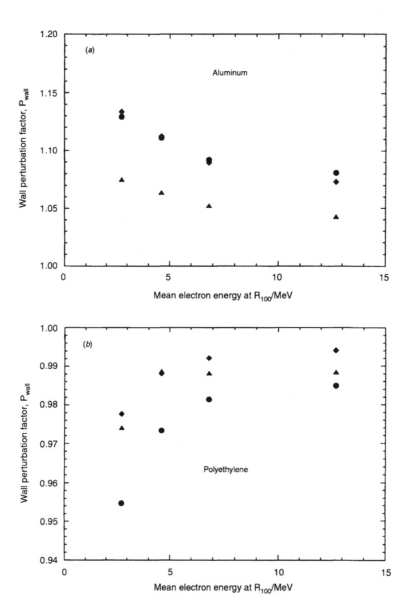

FIGURE 3.9 Wall perturbation factors at the depth of ionization maximum for the simulated Attix (◆), NACP (•), and Roos (▲) chambers as a function of the mean electron energy; (a) phantom material: aluminum, (b) phantom material: polyethylene. (From Reference [11]. With permission.)

where $N_{D,w,\gamma}$ is the ionization chamber absorbed-dose-to-water calibration factor in a reference ^{60}Co beam Q_0 and M_0 is the charge collected in the ionization chamber, corrected to the reference temperature and pressure and for recombination.

Using the following equation, the dose measured by the calorimeter was then related to the dose measured in the dummy calorimeter with an ionization chamber:

$$D_{water}(Q_0, cal) = D_{water}(Q_0, ion) \times C_0(IS, Att)$$
$$(3.27)$$

Because of the slight geometrical differences between water and dummy calorimeters, a correction factor for

inverse square and attenuation $-C_0(IS, Att)$ was applied to the dummy calorimeter data in a ^{60}Co beam. The beam quality correction factor, k_{Q_γ}, relates calibration factors in a beam with a quality Q_1 to that in a beam with a quality Q_0. For measurements in a proton beam with a quality Q_1, Equations (3.25)–(3.27) were rewritten as

$$D_{water}(Q_1, cal) = c \times \theta \times \Delta V_1 \times (1 + D_T) \quad (3.28)$$

$$D_{water}(Q_1, ion) = N_{D,w,\gamma} \times M_1 \times k_{Q_\gamma} \quad (3.29)$$

$$D_{water}(Q_1, cal) = D_{water}(Q_1, ion) \times C_1(IS) \quad (3.30)$$

where $C_1(IS)$ includes only inverse square corrections, since attenuation corrections in the plateau region and at

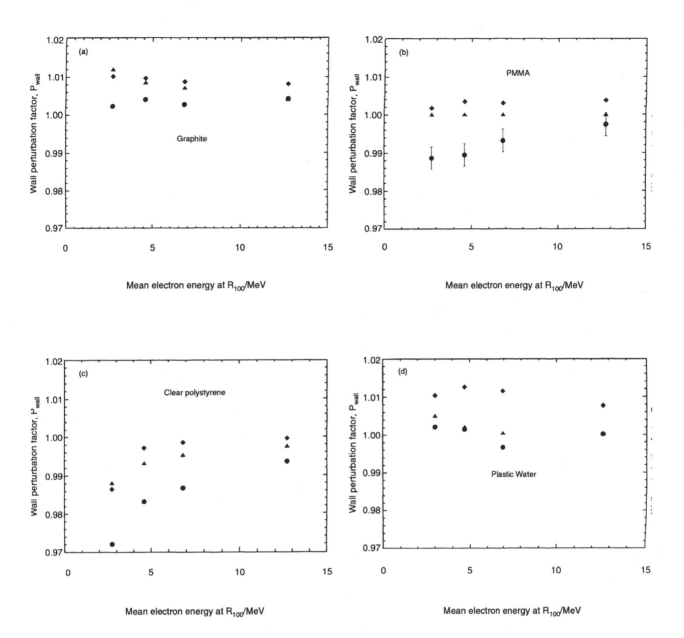

FIGURE 3.10 Wall perturbation factors at the depth of ionization maximum for the simulated Attix (♦), NACP (•), and Roos (▲) chambers as a function of the mean electron energy; (a) phantom material: graphite, (b) phantom material: PMMA, (c) phantom material: polysterene, (d) phantom material: Plastic Water™. (From Reference [11]. With permission.)

the spread-out Bragg peak of the proton beam are very small to be counted. Equations (3.25)–(3.30) were solved for $k_{Q\gamma}$ to reveal [15]

$$k_{Q\gamma} = \frac{\Delta V_1}{\Delta V_0} \times \frac{M_0}{M_1} \times \frac{C_0(IS, Att)}{C_1(IS)} \quad (3.31)$$

To provide a comparison with measured values, the proton beam quality correction factors $k_{Q\gamma}$ were calculated for both the PTW and Capintec chambers, employing equations discussed elsewhere. The $k_{Q\gamma}$ values were derived from ratios of absorbed doses to water in two beams using standard dosimetry protocols. Absorbed dose

to water in a proton beam was calculated using an assumption that the protons interact only with the gas of the ionization chamber and wall effects are ignored. The mass stopping power ratio of water to air was derived for the effective energy of protons at the measurement depth. This energy was determined using the European Hearg Particle Dosimetry Group (ECHED) [17] recommendations. Absorbed dose to water in a reference ^{60}Co beam was calculated using the TG 21 protocol. The resulting equation is given by

$$k_{Q\gamma} = \frac{W_{pr,air}}{W_{\gamma,air}} \times \frac{[(\bar{S}/\rho)_{air}^w]_{Q_1}}{[(\bar{L}/\rho)_{air}^w]_{Q_0}} \times \frac{[p_{air}^w]_{Q_1}}{[p_{air}^w]_{Q_0}} \quad (3.32)$$

TABLE 3.6
A Comparison of Measured and Calculated Proton K_{Q_r} Values

	250 MeV		155 MeV	
	Calculated	Measured	Calculated	Measured
PTW W30001	1.020 ± 0.045	1.032 ± 0.012	1.022 ± 0.046	1.040 ± 0.018
Capintec PR-06	1.032 ± 0.047	1.039 ± 0.011	1.034 ± 0.046	1.051 ± 0.019

Note: Uncertainties (one standard deviation) given include both statistical (type A) and systematic (type B) values.

Source: From Reference [15]. With permission.

TABLE 3.7
A Comparison of Absorbed Doses to Water Obtained with a PTW Ion Chamber Using Different Calibration Methods in a 155-MeV Range Modulated Proton Beam

$D_w(k_{Q_\gamma})$—calculated k_{Q_γ}	2.21 ± 0.11 Gy
$D_w(k_{Q_\gamma})$—measured k_{Q_γ}	2.23 ± 0.06 Gy
$D_w(k_{Q_p})$—calculated k_{Q_p}	2.21 ± 0.05 Gy
$D_w(N_{D_p})$	2.20 ± 0.11 Gy

Note: Uncertainties (one standard deviation) given include both statistical (type A) and systematic (type B) values.

Source: From Reference [15]. With permission.

where $W_{pr,air}$ and $W_{\gamma,air}$ are the mean energies needed to produce one ion pair in air for protons and ^{60}Co gamma rays; $[(\bar{S}/\rho)^w_{air}]_{Q_1}$ is the stopping power ratio of water to air for a proton beam with quality Q_1 at the measurement point; $[(\bar{L}/\rho)^w_{air}]_{Q_0}$ is the stopping power ratio of water to air for the reference ^{60}Co beam; and $[p^w_{air}]_{Q_1}$ and $[p^w_{air}]_{Q_0}$ are the perturbation factors. Using the formalism presented by Loevinger, the perturbation factor $[p^w_{air}]_{Q_1}$ for the reference ^{60}Co beam was calculated using data from the TG 21 protocol:

$$[p^w_{air}]_{Q_0} = p_{repl} \times p_{wall} \qquad (3.33)$$

where P_{repl} is the replacement correction factor and P_{wall} is the wall correction factor.

When beams Q_0 and Q_1 are proton beams, the ratio of absorbed doses to water (the proton beam quality correction factor k_{Q_p}) is simply the ratio of the stopping power ratios at the two energies:

$$k_{Q_p} = \frac{[(\bar{S}/\rho)^w_{air}]_{Q_1}}{[(\bar{L}/\rho)^w_{air}]_{Q_0}} \qquad (3.34)$$

The proton W-values for air and perturbation factors cancel since they are energy independent. As in Equation (3.32),

the effective energy of protons at calibration depth was used to select the stopping power ratios. [15]

Measured k_{Q_γ} values for the Capintec PR-06 and the PTW W30001 ion chambers are shown in Table 3.6. A comparison of the results obtained with the different calibration methods is presented in Table 3.7.

C. Cylindrical Ionization Chambers

Correction factors were derived by Seuntjens [18] for a cylindrical NE2571 ionization chamber, for absorbed dose determinations in medium-energy x-rays. These new correction factors are proposed as weighted mean values of the factors derived from two methods.

Let M_u denote the electrometer reading for measurements in the phantom. N_K is the air-kerma calibration factor and $(\bar{\mu}_{en}/\rho)_{w,air}$ is the ratio of the mean mass energy absorption coefficients, water to air, both averaged over the spectral energy fluence distribution of the photons at the point of measurement in the water phantom for the radiation quality of interest.

In ionometry the absorbed dose can be derived from

$$D_w = M_u N_K (\bar{\mu}_{en}/\rho)_{w,air} k_\alpha k_{st} p_{disp} \qquad (3.35)$$

The total correction $k = k_\alpha k_{st} p_{disp}$ (to be compared with $k_u p_u$ in the IAEA code) comprises three components: k_α, a correction factor for the energy and angular dependence of the response of the ionization chamber within the water phantom; k_{st}, a correction factor for the influence of the stem on the photon radiation field, both free in air and in phantom; and p_{disp}, a correction factor taking into account the perturbation of the photon radiation field by the chamber volume displacing the phantom material, which is defined as the volume limited by the outer dimensions of the chambers, i.e., the sensitive air cavity and the chamber wall but without the stem.

The correction factor k_α is derived as follows. The response of the ionization chamber at the reference point in water is defined as the ratio $R_w = M_u/K_{air,w}$, where M_u is the meter reading in water and $K_{air,w}$ is the air kerma at the reference point in the water phantom. If the energy-angular response free in air is $R(E,\theta)$, defined as the ratio of the meter reading of the ionization chamber in air (for photons of energy E and angle θ between the chamber axis and the incoming photon beam) and the air kerma at the same position, the response R_w of the chamber at the reference position in the water phantom can be calculated from [18]

$$R_w = \frac{\int_0^{E_{max}} \int_\Omega \frac{\mu_{en,air}}{\rho}(E) R(E,\theta) \Phi_{E,\Omega}(E,\theta) E\, dE\, d\Omega}{\int_0^{E_{max}} \int_\Omega \frac{\mu_{en,air}}{\rho}(E) \Phi_{E,\Omega}(E,\theta) E\, dE\, d\Omega}$$

(3.36)

where $\Phi_{E,\Omega}(E,\theta)$ is the double differential energy–angular photon fluence distribution at the reference point in the water, $(\mu_{en,air}/\rho)(E)$ is the mass energy absorption coefficient for air, and $d\Omega = 2\pi\sin\theta\, d\theta$.

The energy-angular response of the chamber thimble resulting from the angular measurements was fitted with polynomes both with respect to the angle axis (between 0 and 90°) and with respect to the energy axis (from 30 to 170 keV). The total correction factor for the energy-angular dependence of the response is given by

$$k_\alpha = R_{air}/R_w$$

(3.37)

where R_{air} is the response free in air for the same radiation quality. It can be seen that R_{air} is given by

$$R_{air} = \frac{\int_0^{E_{max}} \frac{\mu_{en,air}}{\rho}(E) R_a(E) \Phi_E(E) E\, dE}{\int_0^{E_{max}} \frac{\mu_{en,air}}{\rho}(E) \Phi_E(E) E\, dE}$$

(3.38)

where $R_a(E)$ represents the measured response free in air for photons perpendicular to the chamber axis (i.e., $R_a(E) = R(E,90)$) and $\Phi_E(E)$ is the primary photon fluence spectrum, differential in energy, free in air. [18]

The behavior of the correction factors k_α is completely different from that of k_u. It always exceeds unity and is maximal in the region (at 80 keV mean energy) where the contribution of scattered radiation, which is ignored in the evaluation of k_u, is maximal. However, the factor k_α is close to unity ($1 \leq k_\alpha \leq 1.015$), due to the small number of photons at solid angles close to $\theta = 0$.

When an ionization chamber is calibrated free in air, its measured response is increased compared with the situation without its stem. In order to undo the effect of the stem on the chamber response free in air, a correction factor smaller than unity must be applied to the response.

Suppose the reading of the chamber in air with the stem is $M_{s,a}$ and the reading of the chamber using an additional dummy stem of similar material is $M_{ss,a}$; then the correction factor required on the response to undo the stem effect free in air is given by [18]

$$k_{s,a} \approx M_{s,a}/M_{ss,a}$$

(3.39)

When the chamber is used for absorbed dose measurements in a water phantom, the presence of the stem influences the number of scattered photons reaching the chamber. A similar correction factor $k_{s,w}$ can be defined for stem effect in the phantom. The total stem correction for an ionization chamber is given by

$$k_{st} = k_{s,w}/k_{s,a}$$

(3.40)

The free-in-air and the in-phantom stem effects are cumulative, and the resulting correction factor therefore exceeds unity by up to 1.5% at a mean energy of 50 keV. It should be pointed out that, due to the asymmetric sensitivity of the ionization chamber on both sides of the thimble, the correction factor may be overestimated by up to 0.5% at 30 keV mean photon energy according to Monte Carlo calculations by Ma and Nahum. [19]

The displacement correction factor is calculated as

$$p_{disp} = K_{air,w}/K'_{air,w}$$

(3.41)

where $k_{air,w}$ is the air kerma at the reference point in the phantom and $K'_{air,w}$ is the air kerma in the center of the air-filled cavity at reference depth in the water phantom.

Figure 3.11 shows the results in comparison with the previously published p_u correction factors.

The overall correction factor k as a result of the comparison water calorimeter with ionization chamber measurements is plotted in Figure 3.12. The error bars shown correspond with the total uncertainty as discussed above (on a 1σ level). Also shown in Figure 3.12 (by the full curve) are the results of the determination of the overall correction factor for the NE2571 ionization chamber.

D. OTHER CORRECTION FACTORS

The air-kerma calibration factors for cylindrical or end-window parallel-plate ionization chambers are usually determined free in air, and the x-ray machine output is stated as the air-kerma rate free in air, which, when multiplied with the appropriate backscatter factor, gives the air-kerma rate on the surface of a phantom or patient. For end-window chambers, especially when they are used for measurements of small fields or low x-ray energies, the air-kerma calibration factors may also be determined with the chamber embedded in a tissue-equivalent phantom.

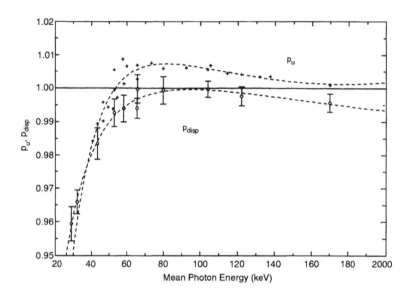

FIGURE 3.11 Correction factor $p_{disp}(0)$ for the volume displaced by the chamber calculated for a cylindrical air cavity with its center at 5 cm in a water phantom, its axis perpendicular to the radiation beam (a parallel beam of 15 cm diameter), and its dimensions equal to the outer dimensions of the NE2571 chamber (i.e. 2.5-cm length and 0.35-cm radius). For comparison the former results $p_u(+)$[20] are shown. (From Reference [18]. With permission.)

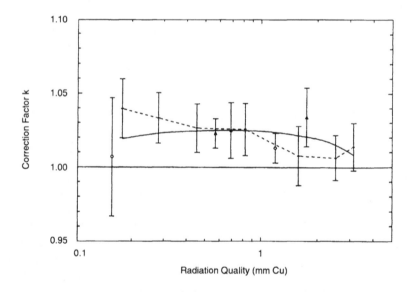

FIGURE 3.12 Overall correction factor k for the NE2571 chamber derived from the comparison of water calorimetry with ionization chamber dosimetry (broken curve) and as the result of the independent determination $k = k_a k_{st} p_{disp}$ by a combination of experimental investigations and Monte Carlo calculations (full curve). The results of the work of Mattsson (1985) [17a] and Kubo (1985) [17b] are shown by (O) and the (▲), respectively. (From Reference [18]. With permission.)

[21] This results in field-size dependent air-kerma in-air calibration factors but obviates the requirement for knowledge of backscatter factors when determining the air-kerma rate on the surface of a phantom.

The body of these end-window chambers acts as a mini-phantom which, at low and medium photon energies, provides an inherent non-negligible scattering component, making the measurement of exposure in air or air kerma in free air with such a chamber problematic, especially when radiation field sizes on the order of chamber cross-sectional dimensions are used. [21]

The x-ray tube output, expressed as the air-kerma rate $\dot{K}_{air}^p(B)$ free in air, is usually obtained by multiplying the appropriate N_K (air-kerma calibration factor) provided by the standards laboratory by the ionization chamber signal M (measured in air per unit time for arbitrary field size B and corrected for temperature, pressure, and recombination losses); i.e.,

$$\dot{K}^p_{air}(B) = MN_k \qquad (3.42)$$

For measurements with the cylindrical chamber, Equation (3.42) is valid irrespective of field size B, provided that the chamber stem effect, on the order of 1% for rectangular fields with high elongation factors, is accounted for. With the phantom-embedded end-window chamber, on the other hand, the air-kerma rate in free air calculated from Equation (3.42) is correct only for field sizes B identical to field size A, which was used during the calibration procedure in the standards laboratory. For arbitrary field size B, the air-kerma rate in free air $K^p_{air}(B)$ can be determined with the phantom-embedded end-window chamber using the following relationship: [21]

$$\dot{K}^p_{air}(B) = M(B)N_K(A)\frac{BSF_{air}(A)}{BSF_{air}(B)} \qquad (3.43)$$

where $M(B)$ is the chamber reading, corrected for temperature, pressure, and recombination losses, and measured with field size B; $N_K(A)$ is the in-phantom air-kerma calibration factor provided for field A by the standards laboratory; and $BSF_{air}(A)/BSF_{air}(B)$ is the ratio of air kerma–based backscatter factors for fields A and B.

For a given field size Φ, the backscatter factor $BSF_{air}(\phi)$ is defined as the ratio of the air-kerma rate on the surface of the phantom $K^s_{air}(\phi)$ to the air-kerma rate free in air $K^p_{air}(\phi)$; i.e.,

$$BSF_{air}(\phi) = \frac{\dot{K}^s_{air}(\phi)}{\dot{K}^p_{air}(\phi)} \qquad (3.44)$$

This is in contrast to water kerma-based backscatter factor $BSF_w(\phi)$, which is defined as the ratio of the water-kerma rate on the phantom surface $K^s_w(\phi)$ to the water-kerma rate free in air $K^p_w(\phi)$, or

$$BSF_w(\phi) = \frac{K^s_w(\phi)}{K^p_w(\phi)} \qquad (3.45)$$

For a given field size ϕ, the two backscatter factors $BSF_{air}(\phi)$ and $BSF_w(\phi)$ are related through the following expression:

$$\frac{BSF_w(\phi)}{BSF_{air}(\phi)} = \frac{[(\bar{\mu}_{ab}/\rho)^w_{air}]_{S,\phi}}{[(\bar{\mu}_{ab}/\rho)^w_{air}]_p} \qquad (3.46)$$

where $[(\bar{\mu}_{ab}/\rho)^w_{air}]_{S,\phi}$ is the ratio of mass-energy absorption coefficients for water and air, averaged over the photon energy fluence spectrum present on the phantom surface S

for field ϕ, and $[(\bar{\mu}_{ab}/\rho)^w_{air}]_p$ is the ratio of mass-energy absorption coefficients for water and air evaluated over the primary x-ray energy fluence spectrum in air. [21]

The results for an x-ray effective energy of 34 keV (120kVp) are presented in Figure 3.13. Angle $\alpha = 0°$ corresponds to normal incidence, with the beam impinging directly onto the polarizing electrode, and $\alpha = 180°$ corresponds to normal incidence with the beam traversing the body of the chamber.

In superficial and orthovoltage radiotherapy, the air-kerma rate determined free in air is only of secondary importance. It is the dose rate on patient's surface obtained for a given field size B which plays the most important role in dose delivery and radiation dosimetry.

It is shown in Figure 3.14a that the chamber response measured in phantom depends on beam energy and field size in an expected fashion. In Figure 3.14b the chamber response measured in air follows the expected behavior for fields above 2.5×2.5 cm^2, showing that for a given beam energy, the chamber response is essentially independent of field size. For a given beam energy and fields smaller than 2.5×2.5 cm^2, however, the chamber exhibits a field size dependence, with the in-air chamber response decreasing with field size, indicating that the scatter from the chamber body has an effect on the chamber reading, the effect diminishing with a decreasing field size.

Seuntjens and Verhaegen [22] examined the depth and field size dependence of the overall correction factor k_{ch} for in-phantom dose determinations in orthovoltage x-ray beams. The overall correction factor is composed of three contributions: a contribution from the angular dependence of the chamber response free in air, derived from the measured directional response of the NE2571 for different energies combined with Monte Carlo calculations; a displacement effect; and a stem effect. The displacement effect and stem effect are both calculated using the Monte Carlo method for different field sizes and depths. The results show a variation of, at most, 2.2% at the lowest photon energies (29.8-keV average photon energy) when going from 2 to 5 cm for a small circular 20-cm^2 field. In the medium-energy range (≥ 100 kV), variations are limited to, at most, 1.5% for 120 kV–150 kV when comparing the most extreme variations in field size and depth (i.e., 2-cm depth; 20-cm^2 area compared to 5-cm depth; 200-cm^2 area). Depth variations affect the overall correction factor most significantly by hardening of the photon fluence spectrum, whereas field diameter variations affect the factor by increase or decrease of contributions of photon scattering.

Absorbed dose to water is determined using the equation

$$D_w = MN_K\left(\frac{\bar{\mu}_{en}}{\rho}\right)_{w,air} k_{ch} \qquad (3.47)$$

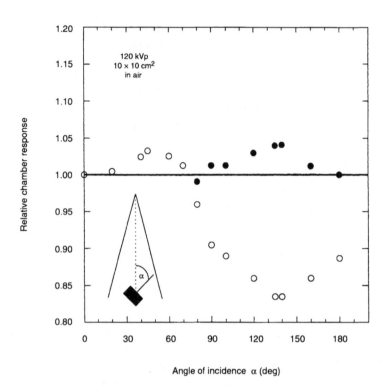

FIGURE 3.13 End-window chamber response vs. the angle α of the incident beam; raw data (open circles) and data corrected for chamber attenuation (solid circles). (From Reference [21]. With permission.)

where M represents the chamber reading at depth z in water corrected for temperature, pressure and recombination; $(\mu_{en}/\rho)_{w,air}$ represents the mass energy absorption coefficient ratio of water to air averaged over the photon fluence spectrum at the point of measurement at a depth in water; and k_{ch} represents the overall correction factor, incorporating all changes in chamber response between the calibration in air at the standards laboratory and the in-phantom measurement in the user's beam. Using Monte Carlo simulations, there are three practical methods to infer an overall correction factor for medium-energy x-ray beams. The first method is the direct calculation of k_{ch} as the quotient of the in-water to in-air kerma ratio and the in-water to in-air chamber reading ratio in a correlated sampling process, including electron transport, as performed by Ma and Nahum: [23]

$$k_{ch} = \frac{K_{air}^{w}}{K_{air}^{a}} \bigg/ \frac{D_{cav}^{w}}{D_{cav}^{a}} \qquad (3.48)$$

where K_{air}^{w} and K_{air}^{a} represent the air-kerma in phantom at the point of measurement and the air-kerma free in air, respectively, and D_{cav}^{w} and D_{cav}^{a} (calculated) represent chamber doses in phantom and free in air, respectively. The disadvantage of this method is the enormous computing

time required to generate acceptable statistical uncertainties for the in-phantom calculations.

A second method employs the same concept but uses measurements of the chamber readings in phantom (M^{w}) to free in air (M^{a}) as opposed to the Monte Carlo calculations of these quantities such as in the previous method. When deriving absolute values of the overall correction factor k_{ch}, this method suffers from problems associated with the exact knowledge of the low-energy tail of the photon fluence spectrum that significantly contributes to calculated air kerma free in air, but not to the calculation of kerma at depth in the phantom. However, the variation of the overall correction factor for a different field size (d') and/or depth (z') relative to the value at the standard conditions (d, z) can be derived from: [22]

$$\frac{k_{ch}(z',d')}{k_{ch}(z,d)} = \frac{K_{air}^{w}(z',d')}{K_{air}^{w}(z,d)} \frac{M^{w}(z,d)}{M^{w}(z',d')} \qquad (3.49)$$

A third successful method of inferring the overall correction factor is based on an experimental study of the combined angular-energy (or directional) response of a cylindrical chamber, a Monte Carlo calculation of the photon fluence distribution differential in energy and angle at the reference depth in the phantom, a separate Monte Carlo calculation of the effect of the displaced volume, and a measurement of the stem effect in air and in phantom.

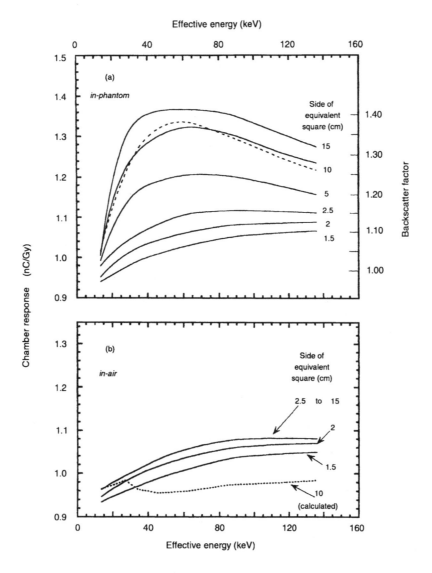

FIGURE 3.14 Parallel-plate end-window chamber response (solid curves) normalized to an in-air dose of 1 Gy in the effective beam energy range from 10 to 140 keV for various square fields in the range from 1.5 × 1.5 cm² to 15 × 15 cm². Part (a): chamber response in phantom; part (b): chamber response in air. The dashed curve of (a) represents the backscatter factor for a 10 × 10-cm² field; the dotted curve of (b) represents the chamber response calculated with Equation (3.43) from the measured data given for a 10 × 10-cm² field in (a). (From Reference [21]. With permission.)

The values shown in Figure 3.15 are obtained representing the overall correction factor for the two depths and for field areas varying between 20 and 200 cm². The error bars shown in this figure represent the combined statistical errors of the Monte Carlo calculations only and disregard any systematic effects.

Absorbed dose values determined with the commonly applied NACP and PTW/Markus parallel-plate chambers and the cylindrical NE2571 Farmer chamber were compared by Van dell Plaetsen et al. [24] to values obtained with ferrous sulphate dosimetry in a number of electron beams. For the ionometry with the parallel-plate chambers, the dose-to-air chamber factor N_D (or N_{gas}) was derived

from a ^{60}Co-beam calibration free in air with an additional buildup layer of 0.54 g cm^{-2} graphite, as proposed by the protocol for electron dosimetry published by the Netherlands Commission on Radiation Dosimetry. The behavior of the fluence perturbation correction factor p_f vs. the mean electron energy at depth was deduced for the flat PTW/Markus and cylindrical NE2571 chamber by comparison with the NACP chamber, for which p_f was assumed unity. The results show a small but significant energy dependence of p_f for the PTW/Markus chamber.

For cylindrical chambers, N_D (or N_{gas}) can be derived directly from the air-kerma calibration factor N_K, determined free in air against the national standard in a beam

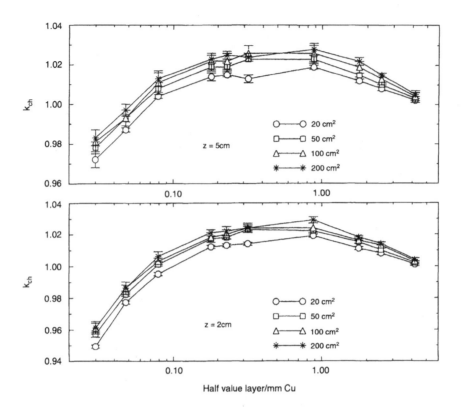

FIGURE 3.15 Overall chamber correction factor k_{ch} for the NE2571 ion chamber, representing the combination of all three components. Upper panel represents data at 5-cm and lower panel at 2-cm depth. (From Reference [22]. With permission.)

of ^{60}Co gamma-rays, using the expression

$$N_D = N_K(1 - g)k_{att}\,k_m k_{cel} \qquad (3.50)$$

Here k_m is a correction for the nonair equivalence of the chamber wall and buildup materials and k_{cel} is a correction for the nonair equivalence of the central electrode. k_{att} represents the correction for attenuation and scatter of the photon beam in the ionization chamber during calibration.

With the known N_D factor, the absorbed dose to water in the electron beam at the effective point of measurement in the phantom material can be determined with the equation

$$D_w(P_{eff}) = M N_D s_{w,\,air} p_{wall} p_d p_f p_{cel} \qquad (3.51)$$

In this expression p_{wall} represents a correction for the difference in composition between the chamber and phantom materials, p_d is a correction for the difference in ionization at the effective point of measurement and the depth at which the absorbed dose is stated, and P_{cel} is a correction for the difference in composition between the central electrode and the phantom material. The fluence perturbation correction factor p_f accounts for the imbalance between the number of electrons scattered from adjacent phantom material into the air cavity and the number of electrons scattered by the air cavity.

The fluence perturbation factor p_f of the parallel-plate PTW/Markus chamber is presented in Figure 3.16. In this figure p_f is given as a function of $\bar{E}_z$, the mean energy at the depth of the effective point of measurement, which is the depth of dose maximum for the electron beams.

Sometimes there are significant differences in temperature between the chamber and measurement phantom. To obtain reliable ionization data, the temperature of the air in the chamber must be allowed to equilibrate with the measuring phantom. The air temperature inside a thimble of a Farmer-type ion chamber was measured by Tailor et al. [25] as a function of time for various phantom materials (air, water, and plastic). In each case, temperature difference is plotted vs. time on a semi-log scale. Distinct differences between the different types of media in the thermal equilibration are noted and discussed. Negligible differences were seen between conductive or nonconductive thimble material and between heating and cooling rates. The thermal-equilibration curves are verified by heat-diffusion theory. Further, the temperature measurements are confirmed by ionization measurements in a plastic medium.

Let the time after zero time be denoted as t, the temperature difference between the "chamber" and the phantom be ΔT, the initial temperature difference be ΔT_0, and the initial phantom temperature be T_0. Data are presented as semi-log plots of the normalized temperature difference, $\Delta T/\Delta T_0$, vs. time.

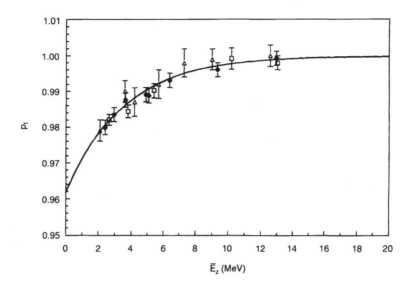

FIGURE 3.16 The fluence perturbation correction factor p_f vs. the mean electron energy at depth $\bar{E}_z$ for the PTW/Markus chamber. The error bars represent 1 s.d. The full line is the result of a fit to the data of the functional form $p_f = 1 - A \exp(B \bar{E}_z)$ with the values of the coefficients $A = 0.039$ and $B = -0.2816 \text{MeV}^{-1}$. (From Reference [24]. With permission.)

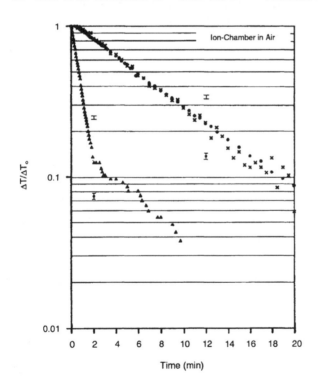

FIGURE 3.17 Cooling/warming curves for an ion chamber in air. No build-up cap (▲), build-up cap warming (O). (From Reference [25]. With permission.)

The data in Figure 3.17 represent thermal equilibration of a Farmer-type ion chamber in air. The solid-circle symbols represent warming of the chamber and build-up cap in air. The initial shoulder (width = 0.5 min) represents a delay in heat transmission through the 0.46-cm Lucite wall of the build-up cap. The shoulder is followed by a single-exponential fall-off; ionization measurements were

made in a ^{60}Co-beam with a similar ion chamber and build-up cap cooling in air. The change in ionization readings was then used to deduce the change in temperature of the air inside the thimble using the ideal gas law. The cross symbols in Figure 3.17 represent the ionization data converted to temperature-change data. The ionization and the direct temperature measurements are in excellent agreement. The solid-triangle symbols represent cooling of the chamber with a bare thimble in air. The absence of the shoulder in these data indicates a negligible delay in heat transmission through the thimble because of negligible wall thickness (0.036-cm nylon). As expected, temperature equilibration is much faster because there is no build-up cap. These data show more than a one-component exponential fall-off. The initial fall-off, until ΔT reaches close to 10% of ΔT_0, is a fast component, whereas the latter fall-off is much slower and perhaps arises from continuation of heat transfer from the stem of the chamber. The stem, because of insulating contents and large heat capacity, takes longer to equilibrate with the surrounding air.

The data in Figure 3.18 represent thermal equilibration of an ion chamber in a water phantom. The solid circle and cross symbols represent cooling and warming data, respectively, when the chamber, along with its build-up cap, are inserted into the water phantom. The two sets of data corresponding to warming and cooling are in good agreement within measurement error. The initial shoulder, representing a delay in heat transmission through the 0.46-cm Lucite wall of the build-up cap, is followed by a single-component exponential fall-off similar to that of the chamber with build-up cap in Figure 3.17. Equilibration is about five times faster in water than in air. The solid-triangle symbols represent the data for insertion of the ion chamber's bare thimble into a build-up cap already

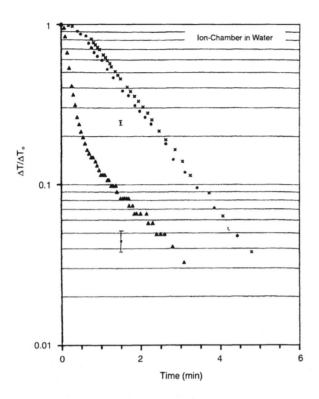

FIGURE 3.18 Cooling/warming curves for an ion chamber in a water phantom. No build-up cap (▲), lucite build-up cap cooling (X) and warming (O). (From Reference [25]. With permission.)

at thermal equilibrium in the water phantom. The initial shoulder is absent, as exhibited previously by the bare-thimble data in Figure 3.17. The second exponential component is again probably attributable to the influence of the stem. [25]

TG-21 states that c-chambers should be avoided for electron-beam energies below 10 MeV. [26] From the Dutch protocol, a limit of about 6 MeV can be deduced. Two criteria are given for the choice of a pp-chamber: a "position criterion" related to the poor definition of the effective point of measurement of c-chambers, quoted to be between 0.3 and 0.7 of the inner chamber radius in front of the chamber center, and an "energy criterion" related to the consistency of dose determination with both chamber types. At least two more concerns with c-chambers at low electron energies become obvious from a look at a typical low-energy electron depth dose curve: the effective point of measurement, usually derived at the steep part of the depth dose curve, may be different near the peak dose and may be dependent on energy as well as position; and, even assuming that the effective point of measurement would be known exactly, the meaning of a measured maximum chamber response remains uncertain because of the relatively large chamber diameter which results in an underestimation of the true maximum dose at positions around peak-dose depth rather than at the descending part of the depth dose, where a good average will be derived. In addition, electron fluence will

be disturbed in an uncontrollable way. TG-21 ignores the effect of the central electrode in c-chambers which may be distinct at low-electron energies. [26]

In Figure 3.19 measurements are compiled after modification, where necessary, to allow for plotting them over the effective energy at depth of measurement. There appears to be a drop of $P_{repl,pp}$ for the Markus chamber from unity at high energies by about 3 to 5% at the low-energy end (Figure 3.19, open symbols). Interestingly, this drop results in cases where the comparison is made to c-chambers by the same (insufficient) methodology. Wittkamper et al. have derived $P_{repl,pp}$ from a comparison to Fricke dosimetry. Their results agree with ours, also derived with Fricke dosimetry, and both show a clearly different result (Figure 3.19, solid symbols). The deviation of $P_{repl,pp}$ from unity at low energies is within measurement accuracy, although there is some indication of a deviation from unity, as shown by the experimental points. Van der Plaetsen et al. [24] found the mean value for this chamber to be 0.8% higher than unity over the full-energy range in question, which would confirm a relative difference between the Markus and the NACP chambers. [26]

E. MONTE CARLO CALCULATED CORRECTION

Ion-chamber responses were calculated for graphite, PMMA, and aluminum-walled ion chambers free in air in ^{60}Co beams and 200-keV beams for a graphite chamber by Rogers. [27] The EGS4 Monte Carlo system was used with various electron step-size algorithms, in particular the PRESTA algorithm, the much simpler ESTEPE constraint on the energy loss per step, and in combination. Contrary to previous reports, it was found that there are variations in the calculated ion-chamber response of up to 3% in ^{60}Co beams and up to 8% in the 200-keV beam.

According to Spencer-Attix cavity theory, the dose to the air in the cavity per unit air kerma at the midpoint of the chamber in the absence of the chamber is given by

$$\frac{D_{air}K_{wall}}{K_{col,air}} = \left(\frac{\overline{L}}{\rho}\right)_{wall}^{air}\left(\frac{\overline{\mu_{en}}}{\rho}\right)_{air}^{wall} = k_m \qquad (3.52)$$

where D_{air} is the dose to the air, $K_{col,air}$ is the air-collision kerma at the center of the ion chamber in the absence of the chamber, K_{wall} is the correction for wall attenuation and scatter, $(\overline{L}/\rho)_{wall}^{air}$ is the Spencer-Attix stopping-power ratio for an energy threshold of 10 keV, and $(\overline{\mu_{en}}/\rho)_{air}^{wall}$ is the ratio of mass-energy coefficients for the wall to air. This expression is the fundamental equation used for primary standards of air kerma in ^{60}Co beams. The right-hand side can be evaluated with negligible statistical uncertainty. Table 3.8 and Figure 3.20 show calculated values of k_m.

TABLE 3.8
Values in ^{60}Co Beams of k_m, $(\bar{L}/\rho)_{wall}^{air}$, and $(\bar{\mu}_{en}/\rho)_{air}^{wall}$

	$\left(\dfrac{\mu_{en}}{\rho}\right)_{air}^{wall}$		$\left(\dfrac{\bar{L}}{\rho}\right)_{wall}^{air}$		k_m	
	1.25 MeV	^{60}Co	1.25 MeV	^{60}Co	1.25 MeV	^{60}Co
Graphite	1.0006	1.0009	0.9993	0.9979	0.9999	0.9988
PMMA	1.0805	1.0800	0.9080	0.9075	0.9811	0.9801
Aluminum	0.9615	0.9629	1.1602	1.1606	1.1156	1.1176

Note: Calculated using EGS4 photon data sets and the electron stopping powers of ICRU Report 37. The stopping-power ratios were calculated for electron spectra in the center of small mini-phantoms of the appropriate material. The values of $(\bar{\mu}_{en}/\rho)_{air}$ are 5.337×10^{-12} and 4.554×10^{-12} Gy cm^2 for the 1.25-MeV photons and the ^{60}Co beam, respectively. *Bremsstrahlung* losses were 0.32% for graphite and PMMA and 0.7% for aluminum. The one-standard-deviation statistical uncertainties in the Monte Carlo calculations were less than 0.1% in all cases.

Source: From Reference [27]. With permission.

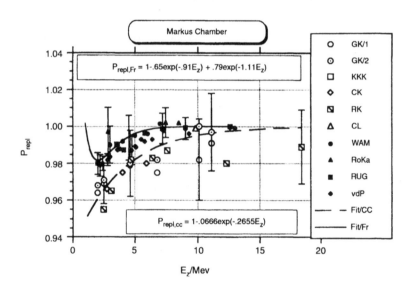

FIGURE 3.19 Compilation of $P_{repl,pp}$ values from the literature. (From Reference [26]. With permission.)

Ma and Nahum [28] calculated the stem-effect correction factors for the NE2561 and NE2571 ionization chambers calibrated in air and used at a depth in a phantom irradiated by medium-energy x-ray beams. The in-air and in-phantom stem effect correction factors were calculated as the ratios of the absorbed dose in the air cavity of an ionization chamber with and without a chamber stem. The "global" stem correction factor was then calculated as the ratio of the in-phantom correction factor to the in-air correction factor. The results show that, in general, the chamber stem increases the chamber response in air but decreases it in phantom.

The global stem effect correction factor, $k_{stem,global}$, is defined as the ratio of the in-water stem correction factor, $k_{stem,water}$, to the in-air stem correction factor, $k_{stem,air}$;

i.e.,

$$k_{stem,\, global} = k_{stem,\, water}/k_{stem,\, air} \qquad (3.53)$$

The variation of the stem-effect correction with SSD for an NE2561 chamber irradiated at 3-cm depth in water by a point source of a 100-kV photon beam is shown in Figure 3.21.

The magnitude of the stem effect depends strongly on the beam field size. As is shown in Figure 3.22, for an NE2561 chamber irradiated at 3-cm depth in water by a point source of a 100-kV photon beam with a 50-cm SSD, the stem correction factor increases with field radius (R) and stabilizes for $R \geq 4$ cm. The global factor $k_{stem,global}$ for an NE2561 chamber is 1.034 ± 0.002 for $R = 5$ cm but decreases to 1.013 ± 0.002 for $R = 1$cm.

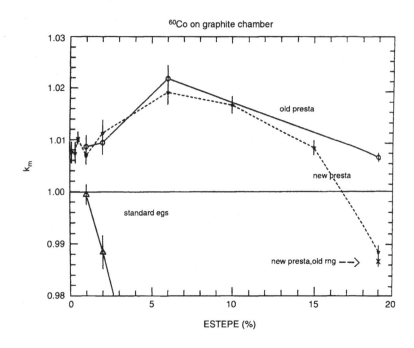

FIGURE 3.20 Variation in calculated value of k_m as a function of maximum continuous energy loss per electron step (ESTEPE in %) for a graphite-walled ion chamber irradiated by a parallel beam of 1.25-MeV photons. The horizontal line corresponds to the predictions of Spencer-Attix cavity theory given in Table 3.8. Default step-size algorithm results are shown with ESTEPE = 19%. (From Reference [27]. With permission.)

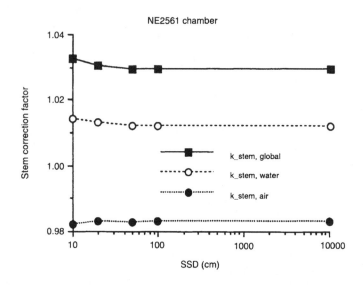

FIGURE 3.21 The variation of the stem correction factors, $k_{stem,air}$ and $k_{stem,global}$, with SSD for an NB2561 chamber used at 3-cm depth in water irradiated by a point source of 100-kV photons. The field size was 3 cm in radius. The statistical uncertainty is within 0.1% for $k_{stem,air}$ and about 0.2% for $k_{stem,water}$ and $k_{stem,global}$. (From Reference [28]. With permission.)

There is a slight dependence of the stem effect on the depth of the chamber center in the phantom. Figure 3.23 shows that, for an NE2561 chamber irradiated at various depths in water by a parallel 70-kV photon beam with a 100-cm² field, $k_{stem,water}$ increases with depth for $d < 2$ cm. The variation of $k_{stem,water}$ becomes negligible for larger depths. The statistical uncertainty in the calculated stem-effect corrections was about 0.2%.

III.   BUILD-UP CAP AND BUILD-UP REGION

The absorbed dose to a point within a phantom can be divided into two components: a part due to primary radiation and a part carried by photons scattered in the treatment head reaching the point of interest. The use of head scatter may be incorporated into different sets of dosimetric calculation models in external-beam radiotherapy.

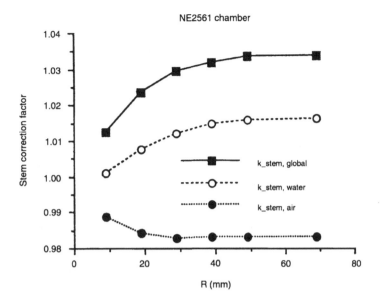

FIGURE 3.22 The variation of the stem correction factors, $k_{stem,air}$, $k_{stem,water}$, and $k_{stem,global}$ with field size for an NE2561 chamber used at 3-cm depth in water irradiated by a point source of 100-kV photons with a 50-cm SSD. The statistical uncertainty is within 0.1% for $k_{stem,air}$ and about 0.2% for $k_{stem,water}$ and $k_{stem,global}$. (From Reference [28]. With permission.)

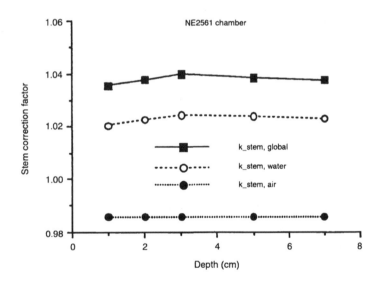

FIGURE 3.23 The variation of the stem correction factors, $k_{stem,air}$, $k_{stem,water}$, and $k_{stem,global}$ with the depth of the chamber center in water for an NE2561 chamber irradiated by a parallel beam of 70-kV photons with a 100-cm² field. The statistical uncertainty is within 0.1% for $k_{stem,air}$ and about 0.2% for $k_{stem,water}$ and $k_{stem,global}$. (From Reference [28]. With permission.)

The suitability of high-Z materials as build-up caps for head-scatter measurements has been investigated by Weber et al. [29] Build-up caps are often used to enable characterization of fields too small for a mini-phantom. The results show that the use of lead and brass build-up caps produces normalized head-scatter data slightly different from graphite build-up caps for large fields at high photon energies. At lower energies, however, no significant differences are found. The intercomparison between the two different plastic mini-phantoms and graphite caps shows no differences.

The determination of the head-scatter factor, S_c, is usually done by in-air measurements with sufficient material surrounding the detector to prevent contaminating secondary particles from reaching the detector volume and to provide enough charged particles for signal strength.

Head-scatter factors may also be derived from phantom measurements. [30] However, two main methods have evolved for the measurements of head scatter. The first method incorporates build-up caps of tissue-equivalent materials such as different plastics or graphite. When higher-energy beams became available, it became evident that materials of low density resulted in caps of substantial dimensions as compared with small field sizes. As a result, interest turned to materials of higher density, e.g., aluminum, brass, or lead.

The caps were used on a 0.6-cm³ BF-type ionization chamber (NE 2571, Nuclear Enterprise, UK) with a graphite wall and connected to an electrometer. The ionization chamber was operated at +360 V. Another ionization chamber used was a 0.12-cm³ RK chamber (RK 83-05, Scanditronix, Sweden) connected to the electrometer and operated at +360 V. The influence of the direction of the ionization chamber, i.e., parallel or perpendicular to the radiation beam, was investigated in order to quantify any influence of stem and cable effects on the measurements. Both ionization chambers, equipped with a build-up cap, were checked for these effects at 4 and 18 MV for the BF-type and at 4 MV for the RK ionization chamber.

The normalized head-scatter factor, or output factor in air, OF_{air}, in the terminology of the treatment planning system is defined as

$$OF_{air} = \frac{(\Psi/M)|^{test}}{(\Psi/M)|^{calib}} \qquad (3.54)$$

where $(\Psi/M)_{calib}$ is the energy fluence per monitor unit on the central axis in the calibration situation, i.e., 10 cm × 10 cm, and $(\Psi/M)_{test}$ is the energy fluence per monitor unit in the beam of interest.

A problem arises when the outer dimensions of the build-up caps or the mini-phantom approach the field size at the isocenter. The normalized head-scatter factor is valid only if the amount of scattering material surrounding the detector, and contributing to the signal, is kept constant, i.e., is not obscured by the radiation field. For very small field sizes, this is not possible and the measurements must be performed at an extended Source Chamber Distance (SCD). Other investigators have shown that the head-scatter factor is independent of SCD. Figures 3.24 and 3.25, indicate that this is not the case at higher energies.

If the detector is moved away from the isocenter, the solid angle of the flattening filter and thus the amount of

TABLE 3.9
Physical Data for the Build-up Caps

Material	Density (g cm⁻³)	Wall Thickness (mm)/(g cm⁻²)	
		Low-energy (4.6 MV)	High-energy (10.18 MV)
Graphite	1.863	11.3/2.1	36.0/6.7
Brass	8.455	2.5/2.1	7.9/6.7
Lead	11.200	1.9/2.1	6.0/6.7

Source: From Reference [29]. With permission.

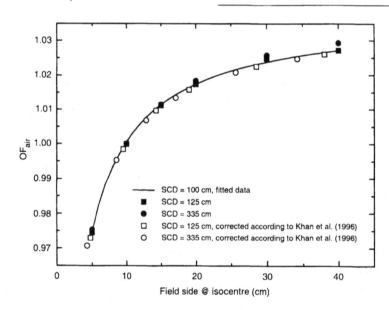

FIGURE 3.24 Head-scatter measurements at 4 MV with brass build-up cap measured at SCD = 100, 125, and 335 cm. The data at SCD = 100 cm have been fitted to a polynomial. Open symbols are corrected values according to Khan el al. [31] (From Reference [29]. With permission.)

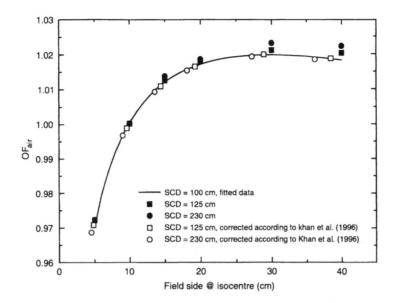

FIGURE 3.25 Head-scatter measurements at 18 MV with brass build-up cap measured at SCD = 100, 125, and 230 cm. The data at SCD = 100 cm have been fitted to a polynomial. Open symbols are corrected values according to Khan el al. [31] (From Reference [29]. With permission.)

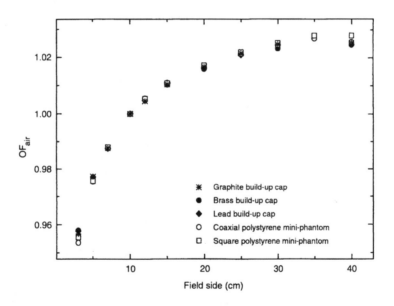

FIGURE 3.26 Head-scatter measurements for a CLINAC 600C, 4 MV. (From Reference [29]. With permission.)

filter, as viewed by the detector, decreases. [31] However, as data are normalized to a 10 cm × 10 cm field, the smaller fields may be covered by measuring at extended distances without introducing any large errors. Measured head-scatter data at extended distances have been recalculated to SCD = 100 cm with an equivalent field size according to Khan et al. [31], assuming the flattening filter to be the single largest source of head scatter. Good agreement with the data measured at SCD = 100 cm was found (see Figures 3.24 and 3.25). [29]

The measured output factors for two beam qualities are shown in Figures 3.26 and 3.27. At lower energies, no significant differences were found between the different build-up caps and mini-phantoms. However, as the energy increases, the differences between the low- and high-Z materials increase and the largest discrepancies were found at 18 MV for large field sizes.

The effect of build-up cap materials on the response of an ionization chamber to ^{60}Co gamma-rays was measured by Rocha et al. [32]

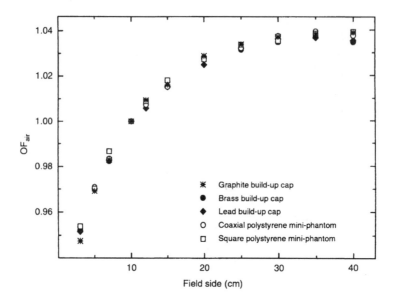

FIGURE 3.27 Head-scatter measurements for a Philips SL15/MLC, 6 MV. (From Reference [29]. With permission.)

The equations relating to the exposure calibration factor (N_X) for the cavity absorbed dose calibration factor (N_{gas} or N_D) are given by

AAPM:

$$N_{gas} =$$

$$\frac{N_X k(W/e)_{gas} A_{ion} A_{wall} K_{humid}^{-1}}{[\alpha(\bar{L}/\rho)_{wall,gas}(\overline{\mu_{en}}/\rho)_{air,wall} + (1-\alpha)(\bar{L}/\rho)_{cap,gas}(\overline{\mu_{en}}/\rho)_{air,cap}]}$$

$$(3.55)$$

SEPM: $$N_D = N_X(W/e)k_{as}k_m/(k_s)_c(k_{st})_c \quad (3.56)$$

IAEA: $$N_D = N_X(W/e)k_{att}k_m \quad (3.57)$$

$$k_m = \alpha s_{air,wall}(\overline{\mu_{en}}/\rho)_{wall,air} + (1-\alpha)s_{air,cap}(\overline{\mu_{en}}/\rho)_{cap,air}$$

$$(3.58)$$

where $(k_s)c$ and $(k_{st})c$ are correction factors to allow for lack of saturation of charge collection and for stem irradiation at the calibration quality, respectively. The notation used in the above equations is in accordance with the appropriate protocols. A comparison of Equations (3.55), (3.56), and (3.57) shows that (a) the correction factor for the attenuation and scattering in the ion chamber wall and build-up cap is

$$k_{att} \equiv k_{as} \equiv A_{wall}$$

and (b) the non-equivalence to air is taken into account by the k_m factor or by the denominator of Equation (3.55).

The attenuation and scattering correction in the chamber and build-up cap in the calibration beam can be calculated from the equation [33]

$$k_{att} = 1 - \gamma t \quad (3.59)$$

where t is the total wall thickness in g cm^{-2} and γ is the attenuation and scattering fraction per wall thickness unit.

The measurements were performed with the ÖFS ionization chamber model TK 01 (secondary standard therapy level) with a 0.5-mm thick Delrin $(CH_2O)_n$ wall, with its inner side coated with a mixture of graphite (15%)/alumina (85%) powder and epoxy resin in the ratio 1 (powder): 3 (resin). The whole thickness was estimated to be 10^{-2} mm. The central electrode, build-up cap, and stem were also made of Delrin.

Caps with different thicknesses were made of Delrin, PMMA, graphite, C-552 (air-equivalent plastic), A-150 (tissue-equivalent plastic), and aluminum (see inset in Figure 3.28).

The mean measured charge was plotted as a function of the total thickness (wall and build-up cap) for each material. The results are presented in Figure 3.29. The k_{att} values were obtained by extrapolating the plots of Figure 3.29 to zero wall thickness and correcting for the center of electron production (CEP). This correction was done by taking into account the overestimation of the beam attenuation of the zero-wall correction.

The CEP was determined to be 105 cm of air for ^{60}Co, corresponding to 0.136 g cm^{-2}. Assuming that the cap materials and air are nearly alike, and taking the experimental values of the slope, the correction for the CEP can be obtained from the expression

$$1 + (slope)(CEP). \quad (3.60)$$

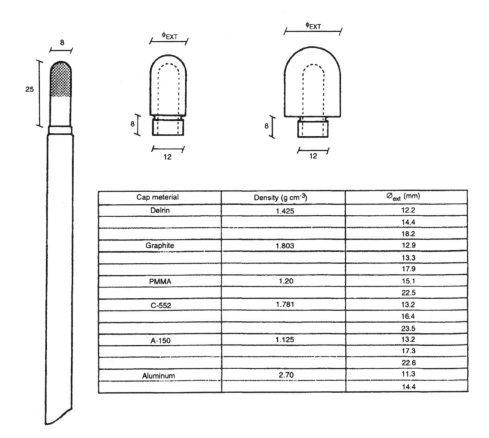

Cap meterial	Density (g cm⁻³)	$\varnothing_{ext}$ (mm)
Delrin	1.425	12.2
		14.4
		18.2
Graphite	1.803	12.9
		13.3
		17.9
PMMA	1.20	15.1
		22.5
C-552	1.781	13.2
		16.4
		23.5
A-150	1.125	13.2
		17.3
		22.6
Aluminum	2.70	11.3
		14.4

FIGURE 3.28 Drawing of the chamber and build-up caps. All the dimensions are in mm. (From Reference [32]. With permission.)

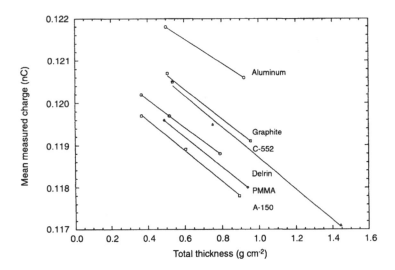

FIGURE 3.29 Mean measured charge as a function of the total thickness for each material. (From Reference [32]. With permission.)

The mean value of k_{att} for a total thickness of 0.5 g cm⁻² was 0.9894 ± 0.0004, independent of the build-up cap material for the low-density materials used in this experiment. [32]

The values of k_m (normalized to the original cap material, Delrin) were obtained by the following equation:

$$k_m = [R(t)/k_{att}(t)]_m/[R(t)/k_{att}(t)]_{delrin} \quad (3.61)$$

where $R(t)$ is the measured charge corrected to reference conditions for a particular value of the wall thickness t. Values of k_m given by AAPM (1983) and SEFM and CAEA (1987) are given in Table 3.10.

From Figure 3.30 one can verify that the theoretical values of k_m are correlated with the mean excitation energy (I), which is the only material-dependent term in the expression used for the calculation of k_m. The variation of k_m as a function of *I* is only 1.7% for the low-*Z* materi-als com-monly used in ion chambers but differ by more than 4% for aluminum compared with graphite, for example. [32]

Perturbation effects are caused in the ionization chamber used in the buildup region because no charged particle equilibrium exists there. This is mainly due to electron fluence through the chamber side wall. More accurate dose measurements in this region can be done with the extrap-olation chamber.

It is generally accepted that parallel-plate extrapola-tion chambers provide the most accurate means of mea-suring dose at the surface and in the buildup region of megavoltage photon beams. An assumption is made that the measurement position is at the inside face of the entrance window of the chamber. For a parallel-plate chamber of finite electrode separation, this assumption is not valid since part of the chamber ionization current is due to electrons emitted from the adjacent side wall in an extrapolation chamber; on the other hand, the chamber signal per unit-collecting volume is determined for differ-ent electrode separations and extrapolated to zero electrode separation, thus eliminating the side-wall contribution. A parallel-plate ion chamber was designed by Rawlinson et al. [34] to study photon build-up. The study has shown that the side-wall error is primarily dependent on the ratio of the electrode separation to the wall diameter, as well as on the wall density and wall angle. Based on these findings, the design of a fixed-separation chamber is described which reads to within about 1% of the correct dose.

A cross section through the special ion chamber con-structed for the study is shown in Figure 3.31. The col-lector and guard electrodes were formed from a 2-mm plate of polymethylmethacrylate (PMMA) coated with a thin (approximately 100 microns) layer of Aquadag colloi-dal graphite and mounted on a base of A-150 conducting plastic. A series of fine concentric grooves of diameters

0.5,1,2,4, and 7.5 cm cut into the surface of the PMMA plate defined the edge between the collector electrode and guard electrode; changes in the collector diameter could be made by scribing an insulating gap through the graphite at one of the circles and by shorting an existing gap. The center wire of a Belden type 9239 low-noise coaxial cable fixed in a fine hole through the center of the PMMA plate pro-vided the electrical connection to the collector. The outer conductor of this cable was connected to the A-150 base and through it to the guard electrode. The front window consisted of 1.6 mg cm^{-2} of aluminized polyethylene terephthalate (Mylar) stretched over a circular PMMA frame of 9.5-cm diameter and held by a PMMA retaining ring. The frame fitted over an annular PMMA spacer of density 1.17 g cm^{-3} which determined both the electrode separation and the wall diameter of the chamber.

TABLE 3.10
Values of k_m as Calculated by the Protocols

Cap Material	AAPM (1983)	SEFM & IAEA (1987)
Delrin	0.982	0.989
Graphite	0.985	0.993
PMMA	0.981	0.986
C-552	0.989	0.996
A-150	0.974	0.979
Aluminum[a]	1.029	1.037

Note: The α value was taken as 0.59. The differences in k_m between the AAPM, SEFM, and IAEA protocols are mainly due to differ-ences in stopping-power ratios.

[a] Stopping power and: absorption coefficient from Rogers *et al.* (1985).

Source: From Reference [32]. With permission.

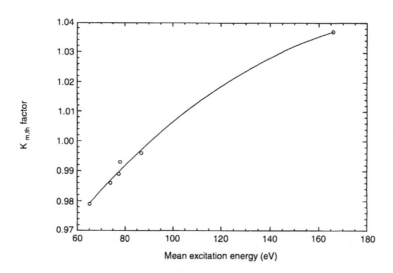

FIGURE 3.30 Theoretical k_m values as a function of mean excitation energy. (From Reference [32]. With permission.)

TABLE 3.11
Characteristics of Commercial Chambers

Parameter	Capintec Model PS-033	PTW ("Markus") Model 30-329
Wall:		
Material, density	C552, $\rho = 1.76$ g/cm^3	PMMA, $\rho \approx 1.17$ g/cm^3
Diameter	17.4 to 23.9 mm	5.7 mm
Angle	$-65°$ (approx.)	0°
Collector:		
Material	C552	Graphited PMMA
Diameter	14.9 mm	5.4 mm
Electrode separation	2.4 mm	2.0 mm
Window:		
Material	Aluminized Mylar	Graphited polyethylene
Thickness	0.5 mg/cm^2	2.7 mg/cm^2

Source: From Reference [34]. With permission.

A150 plastic

PMMA

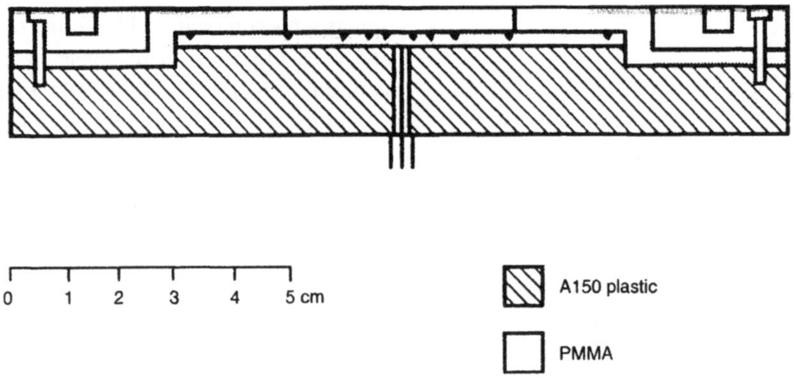

FIGURE 3.31 Cross section through the variable geometry ion chamber. (From Reference [34]. With permission.)

Figures 3.32a and b show the results obtained using the variable-geometry chamber to measure the surface-to-maximum dose ratio in the 10×10-cm^2 cobalt beam. Figure 3a shows, for a chamber of fixed collector diameter, that the chamber response increases approximately linearly with electrode separation and that the rate of increase depends strongly on the wall diameter (or collector edge-wall distance).

In Figure 3.32b these slopes and those obtained for other collector diameters are plotted as a function of collector edge-wall distance. The over-response decreases nonlinearly with collector edge-wall distance, and it also depends on the collector diameter. Trends similar to those shown in Figures 3.32a and b are observed at 6 and 18 MV, though the magnitude of the over-response is smaller at the higher energies for the same chamber geometry. The surface dose for a 10×10 cm^2 field was measured to be $13.9 \pm 0.5\%$ at 6 MV and $9.7\% \pm 0.4\%$ at 18 MV.

The three curves in Figure 3.33 yield chamber over-responses which are within $\pm 1.5\%$ of the original measured data. The results show that the over-response of a chamber of given wall diameter is virtually independent of the relative magnitude of its collector diameter and its collector-edge wall distance.

A chamber whose wall diameter is 10 mm and electrode separation is 3 mm will, from Figure 3.33, have an over-response at the surface of a cobalt beam of 15.3% and therefore a shift, or "effective window thickness," of 29 mg/cm^2 (in addition to the actual window thickness of 1.6 mg/cm^2). This example shows that there is little point in having a parallel-plate ion chamber for buildup measurements with an ultrathin entrance window if it also has a large side-wall contribution which will substantially increase the effective window thickness. [34]

From a least-squares linear fit to the initial part of the ^{60}Co data of Figure 3.33, a fixed-separation parallel-plate ion chamber with PMMA walls whose wall diameter is about 30 times its electrode separation will provide acceptable accuracy for routine measurements in the buildup region of beams of ^{60}Co or greater energy.

The effects of angled walls and low-density walls were investigated. The results show that when the wall has a

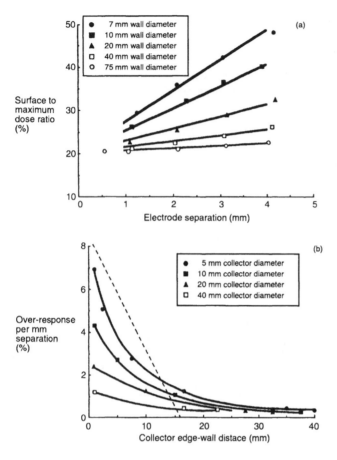

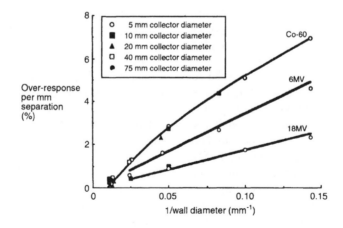

positive angulation, the chamber over-response decreases substantially. [34]

The results of the experiments to determine the effect of the density of the wall material are summarized in Table 3.12. It can be seen that the density of the wall material has an important impact on the size of the chamber over-response. Balsa wood walls, for example, reduce the over-response to approximately 30% of that expected for PMMA walls.

One possible design of a fixed-separation ion chamber that conforms to the criteria discussed above is shown in Figure 3.34. The chamber employs sloped PMMA walls of average diameter 24 mm, an electrode separation of 2 mm, a collector diameter of 10 mm, and a guard width of 5 mm. The window diameter is 27 mm. A flange of inner diameter 35 mm projecting 0.5 mm above the entrance plane of the chamber is employed to protect the window.

Fixed-separation plane-parallel ionization chambers have been shown by Gerbi and Khan [35] to overestimate the dose in the build-up region of normally incident high-energy photon beams. These ionization chambers exhibit an even greater over-response in the build-up region of obliquely incident photon beams. This over-response at oblique incidence is greatest at the surface of the phantom and increases with increasing angle of beam incidence. In addition, the magnitude of the over-response depends on field size, beam energy, and chamber construction. This study showed that plane-parallel ionization chambers can over-respond by a factor of more than 2.3 at the phantom surface for obliquely incident high-energy photon fields.

A model 30-360 extrapolation chamber manufactured by PTW-Freiburg was the standard of comparison for all measurements, with an electric field strength of 30 V/mm of plate separation. This polarizing voltage produced a current within 1% of saturation. For each extrapolation chamber data point, readings were taken at electrode separations of 0.5, 1.0, and 2.5 mm at both a positive and

FIGURE 3.32 Results obtained using variable geometry chamber to measure the surface-to-maximum ratios of the 10 × 10 cm² cobalt beam. (a) Surface-to-maximum ratios as a function of electrode separation for different wall diameters and a collector diameter of 5 mm. (b) Percent over-response per mm of electrode separation as a function of collector edge-wall distance for different collector diameters. The dashed line gives the dependence on collector edge-wall distance. (From Reference [34]. With permission.)

FIGURE 3.33 Percent over-response per mm of electrode separation as a function of the inverse of the wall diameter for a PMMA-walled chanber at the surface of 10 × 10-cm² beams of ⁶⁰Co, 6 and 18 MV. (From Reference [34]. With permission.)

negative chamber bias. These readings were corrected for the polarity effect.

Build-up curves in polystyrene for 5×5 and 20×20 cm^2 6-MV x-ray fields angled at 60° and 75° with no blocking tray in the beam are shown in Figure 3.35. The %DD values in the graphs are obtained by dividing the ionization charge collected at the slant depth along the central ray of the beam by the ionization charge collected with the same detector at the reference depth of a normally incident beam. The relative response as a function of depth along the central axis of the beam for the Memorial chamber and TLD powder is also shown in Figure 3.35.

The magnitude of the over-response at the phantom surface for several fixed-separation plane-parallel chambers as a function of beam angulation for a 5×5-cm^2 field is shown in Figure 3.36. Relative response for these figures is defined as the %DD at the surface for the test chamber for a beam incident at $\theta°$ divided by the %DD for the extrapolation chamber, 0-mm plate separation, at the surface for a beam incident obliquely at the same angle. The response of the chambers at 0°, 30°, 45°, 60°, and 75° relative to the response of the extrapolation chamber is shown in Figure 3.36a for 6-MV x-rays. Based

on this data, the relative response of the chambers at 0° ranged from 1.6 to as high as 2.3 times greater than the response of the extrapolation chamber. The amount of over-response increases gradually with increasing angle, reaching a maximum at about 45°. At beam angles greater than 45°, the amount of chamber over-response begins to decline. At a beam angulation of 75°—the highest angle of oblique incidence investigated—the chamber over-response ranges from 1.4 to 1.8. In comparison to other chambers, the Attix chamber shows a relatively small amount of over-response at 0°—a factor of about 1.12—with little variation up to a beam angulation of 75°. Angular chamber response at the surface for 24-MV x-rays is shown in Figure 3.36b. Three of the chambers show approximately a 1.25 over-response at 0°, while the Markus chamber over-responds by a factor of approximately 1.4. This over-response increases dramatically with beam angulation, reaching a maximum between 1.7 and 1.95 at 60°. From 60° to 75°, chamber response declines for the Markus and the Memorial (circular) chamber but remains constant for the Capintec and the Memorial (rectangular) chamber. By comparison, the Attix chamber shows an over-response of about 1.05 at 0°, with a maximum over-response of ~1.2 at 75°. Again, the amount of over-response of the Attix chamber is very similar to that exhibited by the extrapolation chamber with a 2.5-mm plate separation. [35]

From the data shown, electron fluence across a plane-parallel ionization chamber is disrupted less when the chamber separation is small, and when the active volume is isolated from the in-scattering effects of the side walls of the chamber. This is not only true when high-energy photons are normally incident upon the chamber, but also when these beams are obliquely incident upon the chamber.

IV. CHARGE COLLECTION, ION RECOMBINATION, AND SATURATION

The output current, i, from an ionization chamber is reduced from the saturation current, i_s, by initial recombination, back-diffusion to electrodes, and volume recombination.

TABLE 3.12
Effect of Density of Chamber Walls (^{60}Co; 10 × 10-cm^2 field at 80 cm; 5-mm-diameter collector)

| | | | Over-response | |
Wall Density	Wall Diameter (mm)	Electrode Separation (mm)	For Indicated density	For Density = 1.17
0.22	12	3.5	5.8%	17.0%
0.22	22	4.1	3.7%	10.3%
0.22	12	2.0	2.2%	9.7%
0.22	22	2.0	1.3%	5.0%
2.15	12	3.1	26.5%	15.1%
2.15	22	3.1	14.0%	7.8%

Source: From Reference [34]. With permission.

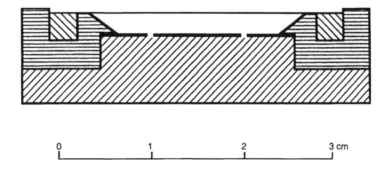

FIGURE 3.34 A suggested design of a fixed-separation parallel-plate ion chamber giving acceptably small side-wall error. (From Reference [34]. With permission.)

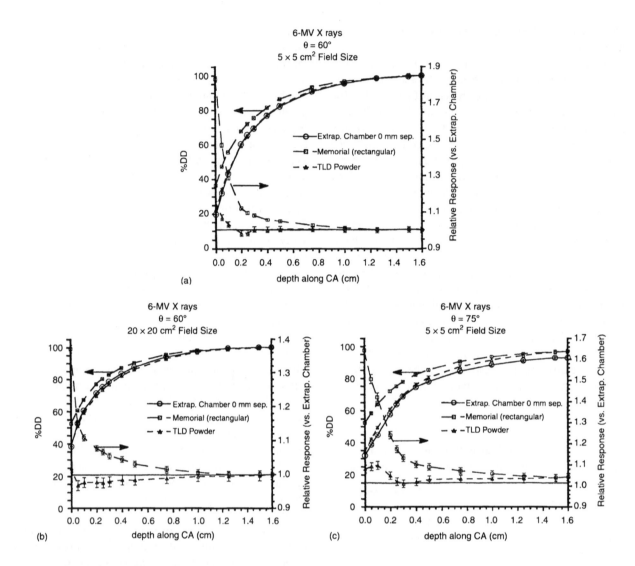

FIGURE 3.35 Typical 6-MV x-ray buildup curves in polystyrene for (a) 5×5 cm² 60°, (b) 20×20 cm² 60°, and (c) 5×5 cm² 75° angled beams. The data is taken at 100-cm SSD with no blocking trays in the beam. %DD on the y axis is normalized to the dose at the reference depth ($t_0 = 1.6$ cm) for a normally incident beam. Depth along *CA* represents the slant depth in polystyrene from the phantom surface to the point of measurement of the plane-parallel chamber. Relative response vs. the extrapolation chamber is shown on the right axis. (From Reference [35]. With permission.)

If the loss due to these phenomena is small, the collection efficiency, $f(= i/i_s)$, can be expressed as

$$f = f_i f_d f_v$$

where $f_i, f_d,$ and f_v are collection efficiencies corresponding to initial recombination, back-diffusion, and volume recombination, respectively.

The two main ways in which ions recombine in an air-filled ionization chamber are by "initial recombination" and "general recombination," the total recombination being the sum of these two effects. Initial recombination takes place when positive and negative ions formed in the same collision event meet and recombine along their initial track. This is a dose rate independent effect, unless

the ion density becomes so great that the field strength responsible for ion collection is reduced. The initial recombination is small with radiation encountered in clinical use.

As irradiation continues, the initial trade structure in the chamber cavity is lost by drift or diffusion of ions and the recombination effect evolves into what is called general recombination. In this process ions produced by different ionizing particles along distinct tracks encounter each other and recombine as they move toward the two electrodes. This effect depends on ion density and, thus, on the dose per pulse and is dominant in pulsed radiation.

Measurements concerning the collection efficiency were made by Takata and Matiullah [36] for a parallel-plate ionization chamber which has a variable space

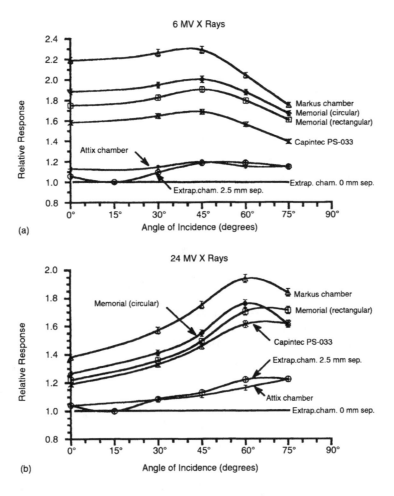

FIGURE 3.36 The angular response of the various chambers relative to the response of the extrapolation chamber with 0-mm plate separation. Measurements were done for (a) 6-MV and (b) 24-MV x-rays in a 3 × 5-cm² field. (From Reference [35]. With permission.)

between the polarizing electrode and collector. To do so, the chamber was exposed to ^{60}Co γ-rays at several different exposure rates and $m = (\alpha/eK_+K_-)^{1/2}$, where α is the recombination coefficient, e is the charge per ion, and K_+ and K_- are the mobilities of positive and negative ions, respectively. It was deduced from the collection efficiencies, which were determined at various applied voltages. It was found that m depends upon the lifetime of ions in the chamber (i.e., it increases from 17.1 to 18.5 MVA$^{-1/2}$ m$^{-1/2}$, with a decrease in the ratio of the voltage to the square of the chamber space from 600 to 20 kVm^{-2}).

Figure 3.37 shows the collection efficiencies of the chamber for a fixed chamber space of 15 mm at different exposure rates as a function of $U/d^2q^{1/2}$ (U is the applied voltage, d is the chamber space, and q is the ion charge). They are normalized to each saturation current and only the values in the range of $U/d^2q^{1/2}$ less than 1000 MVA$^{-1/2}$ m$^{-1/2}$ are plotted. Theoretical values, calculated with various equations for $\beta = 3.99$, $\mu = -0.163$, and $m = 17.5$ MVA$^{-1/2}$ m$^{-1/2}$ are also shown in the figure.

Values for m were calculated from experimental data on $f(= i/i_s)$ using Greening's equation both with and with-

out taking the initial recombination and diffusion losses into consideration. Figure 3.38 shows the results obtained without taking these losses into consideration (i.e., assuming $f_i = f_d = 1$).

The procedure recommended by radiation dosimetry protocols for determining the collection efficiency f of an ionization chamber assumes the predominance of general recombination and ignores other charge-loss mechanisms such as initial recombination and ionic diffusion. [37] For continuous radiation beams, general recombination theory predicts that f can be determined from a linear relationship between $1/Q$ and $1/V^2$ in the near saturation region ($f > 0.7$), where Q is the measured charge and V is the applied chamber potential. Measurements with Farmer-type cylindrical ionization chambers exposed to ^{60}Co gamma-rays reveal that the assumed linear relationship between $1/Q$ and $1/V^2$ breaks down in the extreme near-saturation region ($f > 0.99$) where Q increases with V at a rate exceeding the predictions of general recombination theory. [37]

Boag and Wilson [38] developed the following relationship for the charge-collection efficiency $f_g(V)$ for general recombination as a function of applied potential V in

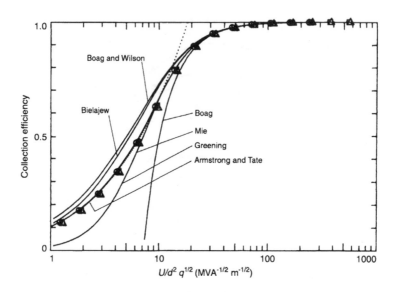

FIGURE 3.37 Diagram showing the collection efficiency, f, measured for a chamber space of 15 mm at exposure rates of Δ, 3.6; $\square$, 31; and O, 310 $\mu Ckg^{-1}s^{-2}$. Saturation curves, f_v, calculated with various equations for $\beta = 3.99$, $\mu = -0.163$, and $m = 17.5$ $MVA^{-1/2}$ $m^{-1/2}$ are also represented. (From Reference [36]. With permission.)

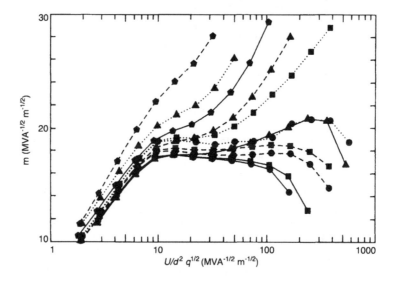

FIGURE 3.38 Values of m obtained without taking into account the initial recombination and diffusion losses. Chamber spaces: $\blacklozenge$, 2; $\blacktriangle$, 5; $\blacksquare$, 15; and $\bullet$, 30 mm. Exposure rates: ..., 3.6; ---, 31; and —, 310 (145 for $\bullet$) μ Ckg^{-1} s^{-1}. (From Reference [36]. With permission.)

an ionization chamber containing an electronegative gas and irradiated by a continuous radiation field with a constant dose rate:

$$f_g(V) = \frac{Q(V)}{Q_{sat}} = \frac{1}{1 + \Lambda_g/V^2} \qquad (3.62)$$

where Λ_g contains chamber and air parameters and is proportional to the dose rate of the continuous beam. The relationship was found to be valid in the near-saturation region ($f_g \geq 0.7$) and may also be written in the following

form:

$$\frac{1}{Q} = \frac{1}{Q_{sat}} + \frac{\lambda_g}{V^2} \qquad (3.63)$$

where Λ_g and λ_g are related through $\lambda_g = \Lambda_g/Q_{sat}$, making λ_g independent of the dose rate.

A simplified version of this approach is the so-called two-voltage technique, in which the collected charge is measured at only two voltages, V_H and V_L, assuming that Equation (3.63) is valid in the region spanned by the two voltage points. If Q_H and Q_L are the charges measured at V_H and V_L, respectively, then $f_g(V_H)$ can be written as: [37]

$$f_g(V_H) = \frac{Q_H}{Q_{sat}} = \frac{Q_H/Q_L - (V_H/V_L)^2}{1 - (V_H/V_L)^2} \quad (3.64)$$

In the AAPM-TG21 protocol which deals with the calibration of high-energy photon and electron beams, Equation (3.64) is further simplified by using $V_H = 2V_L$ (typically, $V_H = 300$ V and $V_L = 150$ V), so that the collection efficiency for general recombination in continuous radiation beams $f_g(V_H)$ may be found from

$$f_g(V_H) = \frac{Q_H}{Q_{sat}} = \frac{4}{3} - \frac{Q_H}{3Q_L} \quad (3.65)$$

Jaffe's theory of initial recombination reduces to a problem of simple Brownian motion under the influence of two forces: the Coulomb attraction between oppositely charged ions produced in the same charged-particle ion track and the applied electric field in an ionization chamber. Jaffe found that the collection efficiency for initial recombination f_i in an ion chamber with a collecting field normal to the ion track of E (V/cm) could be expressed as follows:

$$f_i = \frac{Q}{Q_{sat}} = \frac{1}{1 + gh(x)} \quad (3.66)$$

where g is a constant and

$$h(x) = e^x \left(\frac{i\pi}{2}\right) H_0^{(1)}(ix) \quad (3.67)$$

with $x \propto E^2 = (V/d)^2$. For large x, Equation (3.67) has the asymptotic approximation $h(x) \to \sqrt{\pi/2x}$, which is accurate enough to permit, for large polarizing potentials, Equation (3.66) to be written in the following form:

$$f_i = \frac{Q}{Q_{sat}} = \frac{1}{1 + \Lambda_i/V} \quad (3.68)$$

where Λ_i is a parameter incorporating various chamber and gas parameters and is independent of the dose rate. Ritz and Attix [39] showed that for initial recombination, one should find that

$$\frac{1}{Q} = \frac{1}{Q_{sat}} + \frac{\lambda_i}{V} \quad (3.69)$$

Collection efficiencies for ion loss due to initial recombination and back-diffusion were measured by

Takata [40] for several humidities using a parallel-plate cavity ionization chamber irradiated with ^{60}Co γ-rays. It was shown, from measurements in a range of inverse electric field strengths from 0.05 to 14 mm V^{-1}, that initial recombination took place both in clusters and columns of ions produced along the path of the secondary electrons ejected by the γ-rays. The ion loss due to recombination in clusters was found to increase with humidity, but that in columns did not. Effects of ion-clustering reactions on recombination may be reduced after longer periods of ion drift, when recombination in columns takes place.

Figure 3.39 shows the collection efficiencies measured for relative humidities of 1.5 ± 0.3, 30.7 ± 0.5, and $74.2 \pm 0.7\%$, respectively. These are indicated as functions of $1/E$ with the applied voltage U as a parameter. Solid lines in the figures are obtained by the least-squares method for each result at different applied voltages. They were all measured at temperatures in a range from $21.6°C$ to $21.8°C$. The fluctuation of the temperature during each measurement for one saturation curve was less than $0.04°C$. The pressure in the chamber during measurements was in the range of 991–1018 hPa, and the average was 1006 hPa. Values of the ordinate at the intercepts correspond to the ion loss due to back-diffusion. Arrows at the ordinate show theoretical values for each applied voltage at $T = 21.7°C$.

Figure 3.40 shows the results of calculations for initial recombination in clusters (solid lines) and for columns (broken lines). The values of the parameter h (ion clusters merging parameter) for solid lines A, B, C, and D were ∞, 5.7, 2.85, and 5.7, respectively. The value of v_0 (initial number of ion pairs in a cluster) was 1.5 for line D. The value of h was 5.7 and $\bar{N}_0$ was 16.7 and 33.4 mm^{-1} for broken lines E and F, respectively.

Ion-recombination corrections for plane-parallel and thimble chambers in electron and photon radiation was discussed by Havercroft and Klevenhagen. [41] The aim of the work was to investigate the collection efficiency of several popular ionization chambers used in high-energy electron and photon dosimetry. The recombination effect was evaluated in plane-parallel-type ion chambers (Markus, NACP, Calcam, and Vinten-631) and in a thimble-type chamber (NE2571). The chambers' response was studied in a pulsed accelerator beam for both photons and electrons, and in continuous radiation from a ^{60}Co machine. Two main conclusions from the work may be drawn here. First for the Farmer NE2571 0.6-cm^3 chamber used here, the correction factors required are on the order of 0.1% for ^{60}Co radiation, 0.5% for 5- and 8-MV x-ray and 10- and 12-MeV electrons, and 0.9% for 15-MV x-ray and 15- and 18-MeV electrons, all at typical clinical dose rates of 200–400 cGy min^{-1} (0.017 cGy pulse^{-1}). Second, the recombination effect for the plane-parallel chambers was found to be smaller than for the thimble

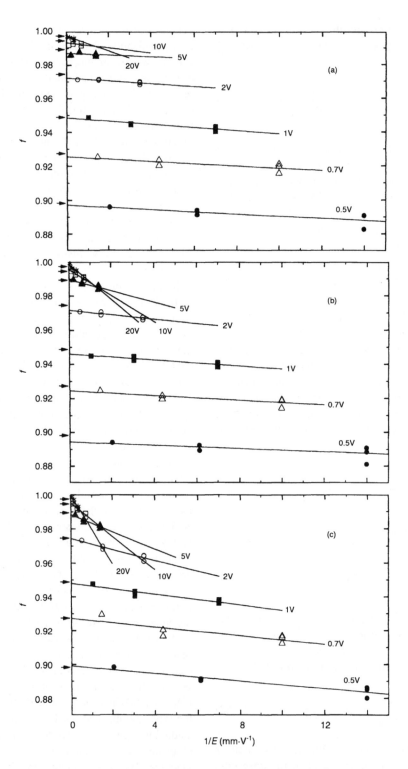

FIGURE 3.39 Collection efficiencies measured at humidities of (a) 1.5%, (b) 31%, and (c) 74%. Solid lines were obtained by the least-squares method for data measured at the applied voltage indicated in the figure. Data from left to right, for each applied voltage, correspond to chamber distance 1, 3, and 7 mm, respectively. (From Reference [40]. With permission.)

chamber, with values of $< 0.1\%$ in continuous radiation and 0.2–0.4% in pulsed radiation.

It is convenient to express the collection efficiency (for pulsed radiation), f, in terms of q, the charge collected per pulse, i.e., that which is being measured by the ion-ization chamber. The expression for collection efficiency is as follows: [42]

$$f = v/(\exp\,(v) - 1) \qquad (3.70)$$

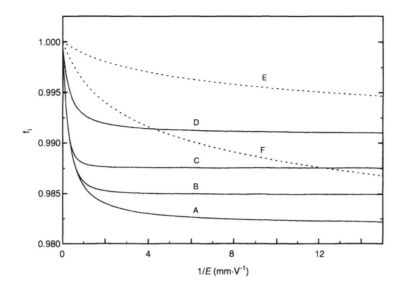

FIGURE 3.40 Calculated collection efficiencies for initial recombination in clusters (solid lines) and in columns (broken lines). The value of the parameter h is 5.7 in general but ∞ and 2.85 for lines A and C, respectively. The value of m was set equal to 3 except for line D, where it was 1.5. The value of $\bar{N}_0$ was 16.7 mm^{-1} and 33.4 mm^{-1} for broken lines B and F, respectively. (From Reference [40]. With permission.)

where

$$v = \mu d^2 q / V$$

and

$$\mu = [(\alpha/e)/(k_1 + k_2)]$$

Where α is the ionic recombination coefficient, e is the electronic charge, k_1 and k_2 are the mobilities of the positive and negative ions, respectively, q is the initial charge density of the positive and negative ions collected by the ion chamber during irradiation, d is the effective electrode spacing (for plane-parallel chambers, this is the actual electrode spacing, but for cylindrical or spherical chambers the equivalent spacing must be calculated), and V is the collecting voltage.

Burns and Rosser (1990) [40a] employed the expansion of $\exp(v)$ to simplify Equation (3.70), neglecting all terms after v^2 and assuming that the dose per pulse does not exceed 0.2 cGy with linear accelerators. Thus, the formula for collection efficiency, in a form convenient for numerical evaluation, becomes

$$f = 1 \Big/ \left(1 + \frac{1}{2}v\right) \tag{3.71}$$

The correction factor to compensate for lost (recombined) ions, for general recombination in pulsed radiation,

is the reciprocal of the collection efficiency

$$p_{ion(gen)} = 1/f = 1 + \frac{1}{2}v \tag{3.72}$$

The equivalent electrode separation for the farmer-type chamber was determined using Boag's expressions. Due to the chamber shape, it was considered to consist of two parts, a cylindrical main volume and a hemispherical section at the tip (see Figure 3.41). The equivalent plate separations for the cylindrical and hemispherical chamber sections, with weighting calculated as a percentage of the total volume, were found for this chamber to be $d_{cyl} = 2.984$ mm, weighting 86% and $d_{hem} = 5.168$ mm, weighting 14%. Substituting these values, along with the polarizing potential $V = -244$ V and $\mu = 3.02 \times 10^{10}$ mC^{-1} V (ICRU 1982), into Equation (3.70), the following values for v were obtained:

$$v_{cyl} = 1102q \qquad v_{hem} = 3306q$$

where the units of q are Cm^{-3}.

By applying the given weightings, one obtains

$$v = 1411q$$

The charge, q, can be expressed in terms of the dose per pulse, D_p, by utilizing the calculated chamber volume of 6.95×10^{-7} m^3 and calibration factor of

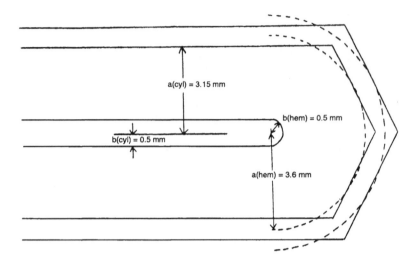

FIGURE 3.41 Calculation of equivalent plate separation for NE2571 chamber using an equivalent hemisphere. The chamber is considered in two sections: a cylinder and a hemisphere, equivalent in volume to the conical end. (From Reference [41]. With permission.)

4.48×10^9 cGy C^{-1}:

$$v = 0.453 D_p$$

where D_p is the dose (cGy) to water per pulse, corrected for temperature and pressure, as measured by the dosimeter. Combining these equations, the recombination correction factor's dependence on pulse density used in a pulsed beam is found: [41]

$$p_{ion(gen)} = 1 + 0.23 D_p \qquad (3.73)$$

for an NE2570/1 (Farmer-type) dosimeter with a standard polarizing potential of approximately -250 V, where D_p is in centigrays.

For continuous radiation, the field-ionization chambers used in electron dosimetry are calibrated against a reference chamber, usually in a local ^{60}Co beam and hence in low-dose-rate continuous radiation in which the initial recombination is dominant. Boag's theory applies also to continuous radiation and provides an equation for collection efficiency, f, which is as follows:

$$f = 1 - \eta^2 \qquad (3.74)$$

where

$$\eta^2 = (m^2/6) d^4 \dot{q} / V^2$$

d is the electrode separation (effective for a thimble chamber), V is the polarizing potential, m is a constant depending on the nature of the gas (2.01×10^7 V m$^{-1/2}$

s$^{1/2}$ C$^{-1/2}$ for air), and $\dot{q}$ is the rate of charge collected per unit volume of the gas (C m^{-3} s^{-1}).

Employing the same derivation method as in the case of pulsed radiation, a correction for general recombination in continuous radiation was obtained for a Farmer chamber:

$$p_{ion(gen)} = 1/f = 1/(1 - \eta^2) \cong 1 + \eta^2 \quad \text{for } \eta^2 \ll 1$$
$$(3.75)$$

Therefore

$$p_{ion(gen)} = 1 + 1.0 \times 10^{-6} \dot{D} = 1.00015 \qquad (3.76)$$
$$\text{for} \quad \dot{D} = 150 \, \text{cGy min}^{-1}$$

This is a very small correction and difficult to verify experimentally. [41]

The correction factor for ion recombination for a chamber irradiated in ^{60}Co radiation was obtained using the two-voltage technique in conjunction with the analytical expression of Almond [43]

$$p_{ion(tot)} = [(V_1^2/V_2^2) - 1]/(V_1^2/V_2^2) - (Q_1/Q_2) \qquad (3.77)$$

where Q_1 is the charge collected at the normal operating voltage V_1 and Q_2 is the charge collected at the lower bias voltage V_2.

Table 3.13 shows correction factors measured experimentally for plane-parallel chambers. With factors between 1.002 and 1.004, it is clear that the experimentally determined correction is larger than the theoretically

TABLE 3.13
Experimentally Determined Recombination Correction Factors for Plane-Parallel Chambers in Electron Beams (using electrometer settings of −244 V and −61 V)

Chamber	4 MeV	6 MeV	8 MeV	10 MeV	12 MeV	15 MeV	18 MeV
Markus	1.0025	—	—	1.0029	—	—	1.0043
NACP	1.0036	1.0031	1.0033	1.0031	1.0031	1.0045	1.0045
Calcam	1.0031	—	—	1.0024	—	—	1.0043
Vinten-631	1.0017	—	—	1.0024	—	—	1.0025

Source: From Reference [41]. With permission.

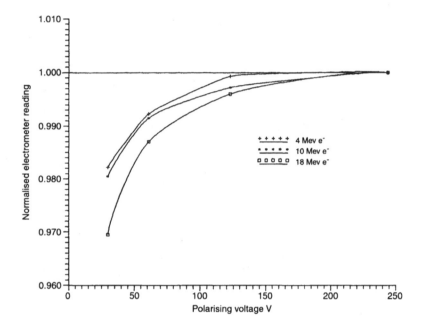

FIGURE 3.42 Saturation curves for Markus chamber in electron beams, i.e., variation of electrometer reading with polarizing voltage. (From Reference [41]. With permission.)

derived value. The latter was for general recombination only; however, measured values of initial recombination were too small to give recombination of experiment and theory.

The other major point to note is a slight energy dependence of the recombination correction factor. By looking at the saturation curve for each chamber individually, it can be seen that the 18-MeV curves consistently fall off faster than the 10-MeV and then the 4-MeV curves, as the polarizing voltage decreases. An example is shown in Figure 3.42. At present, no explanation of this phenomenon has been suggested. [41]

The effect of recombination in plane-parallel chambers was found to be smaller than in thimble chambers, particularly for pulsed radiation.

General ion recombination has been studied by Geleijns et al. [44] under irradiation conditions relevant for diagnostic radiology and with four different ionization

chambers. Recombination was estimated by measuring the collected charge at various collecting potentials of the ionization chamber. An alternative approach is to vary the exposure or exposure rate over a wide range at a constant collecting potential.

For an ionization chamber exposed to continuous radiation, a linear relation between $1/f$ and $1/V^2$ was derived by Boag under the condition that $f > 0.7$:

$$\frac{1}{f} = 1 + \left(\frac{\alpha}{6ek_1k_2}\right)d^4q\left(\frac{1}{V^2}\right) \quad (3.78)$$

where α is the recombination coefficient, q is the charge liberated in the ionization chamber and escaping initial recombination per unit volume and per unit time, and e is the elementary charge. The mean numerical value of the coefficient $\alpha/6ek_1k_2$ in air is estimated by Boag as $(6.73 \pm 0.80) \times 10^{13}$ s m^{-1} C^{-1} V^2.

For an ionization chamber exposed to pulsed radiation, the collection efficiency f for pulsed radiation has been derived by Boag as

$$f = (1/u) \ln(1 + u) \qquad (3.79)$$

where u is $\mu rd^2/V$, r is the charge liberated and escaping initial recombination per unit volume and per pulse, and μ [$\mu = \alpha/e(k_1 + k_2)$] is a constant that depends only on properties of the ions in the ionization chamber. In this study, a value for μ of 3.19×10^{10} mVC^{-1} was used. At relatively high collecting potentials and short pulses, the inverse of the collection efficiency ($1/f$) can be approximated by a linear equation:

$$\frac{1}{f} = 1 + \frac{u}{2} = 1 + \left(\frac{\mu rd^2}{2V}\right) \qquad (3.80)$$

Thus, under these conditions, a linear relation between $1/f$ and $1/V$ is derived.

Examples of general recombination measured with the Keithley 600-cm^3 ionization chamber at two collecting potentials are given in Figures 3.43a and b. These figures illustrate recombination due to continuous radiation and

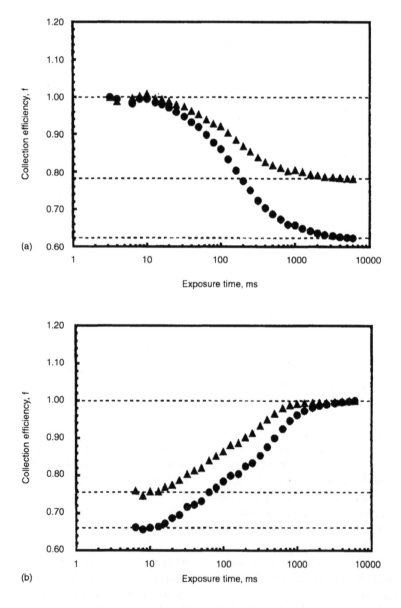

FIGURE 3.43 The collection efficiency of the 600-cm^3 ionization chamber. (a) Measurements at a constant exposure rate of about 3.0×10^{-3} mC kg^{-1} s^{-1} (0.011 R s^{-1}); and (b) measurements at a constant exposure (pulse) of about 0.74×10^{-3} mC kg^{-1} (2.9 $\times$ 10^{-3} R). Collecting potential -200 V: ● and -300 V: ▲. (From Reference [44]. With permission.)

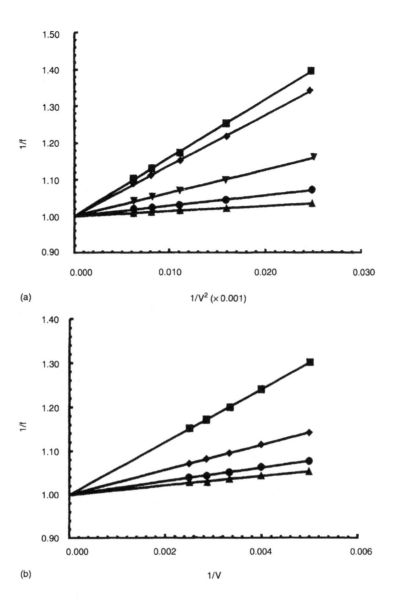

FIGURE 3.44 Recombination measurements at different collecting potentials and different exposure rates or exposures. (a) NE 35-cm^3 chamber and continuous radiation model. ▲: 0.27 mC kg^{-1} s^{-1} (1.0 R s^{-1}); •: 0.56 mC kg^{-1} s^{-1} (2.2 R s^{-1}); ▼: 1.2 mC kg^{-1} s^{-1} (4.6 R s^{-1}); ◆: 2.4 mC kg^{-1} s^{-1} (9.4 R s^{-1}); and ■: 2.8 mC kg^{-1} s^{-1} (11 R s^{-1}). (b) Keithley 600-cm^3 chamber and pulsed radiation model. ▲: 0.050 × 10^{-3} mC kg^{-1} (0.19 × 10^{-3} R), •: 0.095 × 10^{-3} mC kg^{-1} (0.37 × 10^{-3} R): ◆: 0.19 × 10^{-3} mC kg^{-1} (0.75 × 10^{-3} R); and ■: 0.41 × 10^{-3} mC kg^{-1} (1.6 × 10^{-3} R). (From Reference [44]. With permission.)

due to pulsed radiation. Collecting potentials of −200 and −300 V were used. Under these conditions, ion transit times for this ionization chamber are about 100–200 ms.

In Figure 3.44, measurement results are summarized for two ionization chambers. The NE 35-cm^3 ionization chamber is exposed to several exposure rates between 0.27 mC kg^{-1}s^{-1} (1.0 Rs^{-1}) and 2.8 mCkg^{-1} s^{-1} (11 Rs^{-1}) at exposure times that correspond with the continuous radiation model (Figure 3.44a). The Keithley 600-cm^3 ionization chamber is exposed to pulses of 0.50 × 10^{-4} mC kg^{-1} (0.19 × 10^{-3} R) to 4.1 × 10^{-4} mC kg^{-1} (1.6 × 10^{-3} R) at exposure times, corresponding with the pulsed radiation model (Figure 3.44b). During the measurements the collecting potential varied from −200 to

−400 V. Linear extrapolation of charge values measured at different collecting potentials yielded the initial charge from which, subsequently, the collection efficiency was calculated.

Ion recombination and polarity effect of ionization chambers in kilovoltage x-ray exposure measurements was investigated by Das and Akber. [45] Exposure measurements with ionization chambers are dependent on the correction factors related to the beam energy (k_e), temperature and pressure (k_{tp}), ionization recombination (P_{ion}), and polarity (k_{pol}) effects. Six different chambers commonly used in diagnostic radiology were investigated for the P_{ion}, and k_{pol} at various exposure rates by changing the tube voltage, beam current, exposure time, and distance.

The radiation exposure from an x-ray tube can be measured by an ion chamber as

$$\text{Exposure } (X) = MN_X K \qquad (3.81)$$

$$K = k_e k_{tp} k_{pol} P_{ion} \qquad (3.82)$$

where X is exposure measured in C/kg, M is meter reading in C, N_X is the calibration factor (R/C) of an ion chamber for standard beam energy from a national or accredited dosimetry calibration laboratory (ADCL), K is a composite correction factor, k_e is the energy correction factor, k_{tp} is the correction for the air density due to the temperature and pressure, k_{pol} is correction for the polarity, and P_{ion} is the ion recombination correction factor. An ADCL provides values of N_X for each user's beam quality condition for an ion chamber and electrometer combination. However, the dose rate might be a variable and the ADCL may not be able to match the user's exposure condition. Even though for routine use, a single factor might be useful, scientifically, separation of each component of the calibration factor individually as shown in Equation (3.82) is desired. In such situations, an ADCL would prefer to provide individual factors for each beam condition. Ideally, ionization chambers should be independent of an electrometer such that chamber factors can be provided by an ADCL very similar to the factors for chambers used in radiation therapy applications. If a chamber is calibrated by an ADCL with a polarity that is used by a user, k_{pol} is 1.00 and further correction is not needed. However, if calibration and measurement polarities are different, k_{pol} is needed for accurate exposure measurement.

For an x-ray machine, the exposure is dependent upon the applied voltage, V; beam current, mA (milliampere); time, s (seconds); and distance, d. A rule of thumb expression for exposure can be given as: [45]

$$\exp(R) = \frac{kV^n mAs}{d^2} \qquad (3.83)$$

where k and n are constants for a machine. The typical values of k and n are 1.5 and 2.5, respectively, in the kilovoltage range. Hence, the polarity and ion recombination effects in general may depend on these parameters.

For the continuous radiation beam, the ion-recombination correction P_{ion} can be calculated by the half-voltage method as suggested by Attix. If Q_1 and Q_2 are the charges collected at the chamber potential of V and $V/2$, respectively, P_{ion} is calculated as

$$P_{ion} = \left(\frac{4}{3} - \frac{Q_1}{Q_2} \right)^{-1} \qquad (3.84)$$

The polarity effect depends strongly on the medium between the plates of chamber and is the highest for air ($\leq 30\%$) and the lowest ($< 0.3\%$) for dielectric medium. If the charges collected at negative and positive potentials are denoted by Q_- and Q_+, respectively, then k_{pol} is defined as the absolute value of the ratio.

$$k_{pol} = \left| \frac{Q_-}{Q_+} \right| \qquad (3.85)$$

The actual measured charge Q for the measurement of exposure is the absolute value of mean of the two charges:

$$Q = \left| \frac{Q_+ + Q_-}{2} \right| \qquad (3.86)$$

For the measurements, if only a single polarity is used, the error in Q will be

$$\% error = \left(\frac{k_{pol} - 1}{k_{pol} + 1} \right) \times 100 \quad \text{for } Q_- \qquad (3.87)$$

$$\% error = \left(\frac{1 - k_{pol}}{1 + k_{pol}} \right) \times 100 \quad \text{for } Q_+ \qquad (3.88)$$

The P_{ion} and k_{pol} were measured with various exposure factors kVp, mA, s, and d for all chambers in the diagnostic energy range by changing one parameter at a time while keeping the others constant.

Figure 3.45 shows the variation of the P_{ion} vs. x-ray tube voltage in mammography range. Data are shown for various ion chambers at a focus-to-chamber distance (FCD) of 25 cm and 50 cm in Figure 3.45a and b, respectively. The value of P_{ion} is relatively independent of the tube voltage for small-volume (≤ 150 cm^3) ion chambers within 1%; however, for the large-volume ion chambers, the magnitude of P_{ion} is much higher and rises steadily with the tube potential. In general, P_{ion} is greater for the higher exposures. The slope of curve for the 180-cm^3 chamber was anomalously higher. Independent repeated measurements with this chamber showed the same results within $\pm 1.5\%$.

The effect of beam current for a given exposure (fixed kVp and mAs) is shown in Figures 3.46a and b for the P_{ion} and k_{pol}, respectively, for the 125-kVp and 0.1-s exposure typically used in diagnostic radiology. The value of P_{ion} rises linearly with beam current for most chambers, but the effect is more pronounced for the large-volume ion chambers. For small-volume chambers, the P_{ion} is constant within 1.0%. It suggests that P_{ion} is significant only at high-exposure rate measurements with large chambers. The polarity effect, on the other hand, is constant over a wide range of beam current. The magnitude of the k_{pol} is

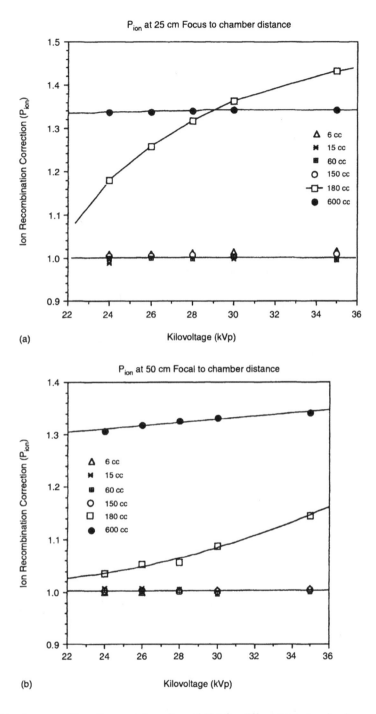

FIGURE 3.45 Ion recombination correction, P_{ion}, vs. tube voltage (kVp) for different ion chamber in mammography range at a focal to chamber distance (FCD) of 25 cm (a) and 50 cm (b). (From Reference [45]. With permission.)

nearly constant within ±4% with respect to the beam current for all chambers, except for the 600-cm³ ion chamber, which has the value of 1.16. This indicates that k_{pol} is independent of the exposure conditions and depends only on the chamber characteristics.

P_{ion} is higher for higher kilovoltage and rises with exposure time for a 600-cm³ chamber. In contrast, the 180-cm³ chamber shows the reverse effect. The polarity effect has

a similar trend, as noted for the P_{ion}, except that the slope is higher for higher exposure time.

For low exposure and scatter radiation measurements such as those used in diagnostic radiology, the large-volume ion chambers are used for a better signal; however, when large-volume chambers are used, the ion recombination and polarity effect become significant and must be counted for accurate exposure measurements. In general,

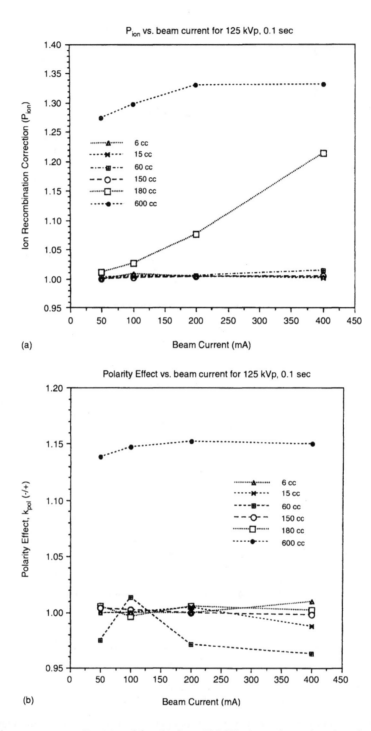

FIGURE 3.46 Effect of beam current on P_{ion} (a) and k_{pol} (b) for a 125-kVp beam for various ion chambers and a fixed exposure time of 0.1 s. (From Reference [45]. With permission.)

the k_{pol} depends on the area and thickness of the electrodes, window thickness, the potential across the electrodes, and the depth of chamber in phantom. It is shown that k_{pol} is significant for air exposure measurements with large volume chambers.

Volume recombination parameter in ionization chambers was investigated by Boutillon. [46] The parameter m^2 governing the volume recombination in ionization chambers was measured under conditions which allow strict application of the basic theory. The method consists of measuring the ratio of ionization currents I_{V_1} and I_{V_2} obtained at two given voltages V_1 and V_2 as a function of I_{V_1}. The value of m^2 is derived from a linear extrapolation to zero current. Several pairs of voltages (V_1, V_2) were used. The value of m^2 obtained in that work is 3.97×10^{14} s m^{-1} C^{-1} V^2 with a relative uncertainty of 1.7%.

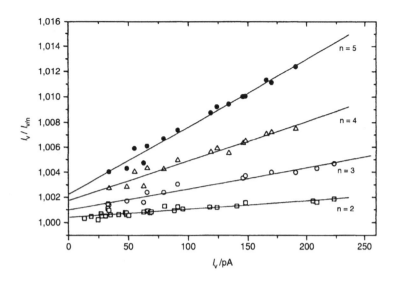

FIGURE 3.47 Ratio of the ionization currents I_V and $I_{V/n}$, measured in chamber B with voltages V and V/n, respectively, as a function of I_V, for several values of n (100–250 kV). (From Reference [46]. With permission.)

The dependence of m^2 on atmospheric conditions has also been investigated.

The various processes by which ions recombine in an ionization chamber have been reviewed by Boag. [42] Near saturation, the ratio between the saturation current I_s and the ionization current I_V, measured at voltage V, can be expressed, neglecting terms of higher order, by the basic equation

$$I_s/I_V = 1 + A/V + m^2(g/V^2)I_s \qquad (3.89)$$

with

$$m^2 = \alpha/ek_+k_-$$

where A is a constant depending on the chamber type, α is the recombination coefficient under continuous irradiation, e is the electron charge, k_+ and k_- are the mobilities of the positive and negative ions, and g is a factor depending on the chamber geometry. The first variable term on the right-hand side of Equation (3.89) describes the initial recombination and diffusion. The second term describes the volume recombination.

The effect of the free electrons on the recombination can be neglected for the low air-kerma rates.

The geometric factor g is given by Boag [47] for a parallel-plate free-air chamber as

$$g = L^2/2\pi D \qquad (3.90)$$

where L is the plate separation and D is the length of the collecting plate in the direction of the photon beam. This relation is based on the assumption of a sharp Gaussian distribution of the ionization around the beam axis.

Two parallel-plate free-air chambers were used by Boutillon for the determination of m^2, one of them in a low-energy x-ray beam and the other in a medium-energy x-ray beam.

Figure 3.47 shows the data for chamber B. The experimental values for the ratio $I_V/I_{V/n}$ have been corrected for space charge according to Boag [42] and for variations in the atmospheric conditions. The slope of each curve, divided by $(n^2 - 1)g/V^2$ gives an estimate of m^2. The correction applied for space charge amounts to about 0.01% for an ion-recombination loss of 1%.

Ion recombination corrections for the NACP parallel-plate chamber in a pulsed electron beam was studied by Burns and McEwen. [48] The NACP electron chamber is one of three parallel-plate chambers recommended for use in the UK. Measurements with this chamber type have indicated a problem in determining the recombination correction. This is due to a variation of the ionization current I with polarizing voltage V, which deviates from the accepted Boag theory. It is shown that there is a chamber-dependent threshold voltage below which the NACP chamber follows the Boag theory. Above this voltage the chamber should be used with caution, although it is still possible to correct for the dependence of the chamber response on the dose per pulse. The existence of such deviations from theory demonstrates the usefulness of the $1/I$ against $1/V$ plot and the limitations of the Boag two-voltage analysis.

Since the component of initial recombination k_{init} is independent of the dose per pulse, k_{ion} will tend to k_{init} as the dose per pulse, and hence u, tends to zero. Thus, the practical measurement is better represented by

$$k_{ion} \approx k_{init} + \frac{u}{2} \qquad (3.91)$$

where the value of $k_{ion}(V_u)$ at the normal user polarizing voltage V_u is

$$k_{ion}(V_u) \approx k_{init}(V_u) + \frac{\mu d^2 \rho_{air}}{2V_u(W_{air}/e)} D_{air,p} \quad (3.92)$$

It follows that if $k_{ion}(V_u)$ is measured using the $1/I$ against $1/V$ method at each of a series of doses per pulse $D_{air,p}$, then a plot of $k_{ion}(V_u)$ against $D_{air,p}$ should be approximately a straight line with intercept $k_{ion}(V_u)$ and gradient $C_{gen}(V_u)$ defined by

$$C_{gen}(V_u) = \frac{\mu d^2 \rho_{air}}{2V_u(W_{air}/e)} \quad (3.93)$$

The parameter $C_{gen}(V_u)$ describes the effect of general recombination on a chamber response and is referred to here as the coefficient of general recombination.

A typical plot is shown in Figure 3.48 for chamber N3009 at a moderate $D_{air,p}$ of 0.43 mGy. The y-axis is normalized by the best estimate of the saturation current I_s, as determined by a fit using the full Boag equation. The solid line is a simple linear fit to the data. The value of the intercept given by the linear fit is within 0.01% of the best estimate of I_s using the Boag fit, thereby justifying the linear approximation at this dose per pulse.

For all four Scanditronix chambers, the value of $k_{ion}(100\ V)$ evaluated as above was measured as a function of $D_{air,p}$. According to Equation (3.92), this dependence should be approximately linear, as was found to be the case for all chambers (including the Calcam chambers). Figure 3.49 shows the results for chamber N3701. The solid line shows a linear fit. The intercept of the linear fit is taken to be the value of the correction for initial recombination $k_{init}(100\ V)$. The gradient is the coefficient of general recombination $C_{gen}(100\ V)$. As one would expect, the value of $k_{ion}(100\ V)$ is dependent only on the dose per pulse and not the electron energy.

It was found that for all the chambers tested (including the Scanditronix chambers), the close agreement with the Boag theory does not extend to voltages significantly beyond 100 V.

A measurement of $C_{gen}(V_u)$ allows one to extract a value for the effective plate separation d_{eff} for each chamber consistent with Equation (3.93). (This is not necessarily the actual plate separation d.) For a polarizing voltage of 100 V and assuming the values μ equal to 3.02×10^{10} V m C^{-1} (ICRU 1982), (W_{air}/e) equal to 33.97 JC^{-1}, and ρ_{air} equal to 1.205 kg m^{-3} (at standard temperature and pressure), Equation (3.93) gives

$$d_{eff} = \sqrt{\frac{2V_u(W_{air}/e)C_{gen}(V_u)}{\mu\rho_{air}}}$$

$$= 4.32 \times 10^{-4}\sqrt{C_{gen}(100V)} \quad (3.94)$$

where $C_{gen}(100\ V)$ is in Gy^{-1} and d_{eff} is given in m.

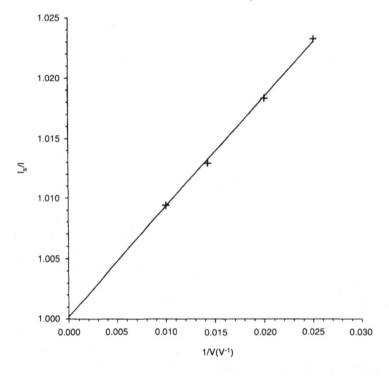

FIGURE 3.48 The inverse of the chamber current I as a function of the inverse of the polarizing voltage V for chamber N3009 at a dose per pulse of 0.43 mGy to air. The chamber current is normalized by the saturation current I_s as determined by a full Boag analysis. The solid line is a linear fit to the data. (From Reference [48]. With permission.)

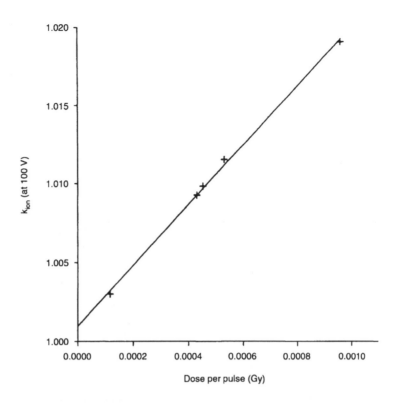

FIGURE 3.49 The total recombination correction k_m as a function of the absorbed dose to air per pulse for chamber N3701 at a polarizing voltage of 100 V. The correction at each dose per pulse was evaluated using a Boag fit over the polarizing voltage range 40–100 V. The solid line is a linear fit to the data. (From Reference [48]. With permission.)

Ion-recombination correction factor k_{sat} for spherical ion chambers irradiated by continuous photon beams was investigated by Piermattei et al. [49] When large-volume ionization chambers are used, the ion-recombination correction factor k_{sat} has to be determined. Three spherical ion chambers with volumes ranging from 30 to 10^4 cm³ have been irradiated by photons of a ^{192}Ir source to determine the k_{sat} factors. The k_{sat} values for large-volume ionization chambers obtained by considering the general ion recombination as predominant (Almond's approach) are in disagreement with the results obtained using methods that consider both initial and general ion-recombination contributions (Niatel's approach). Such disagreement can reach 0.7% when high currents are measured for a high-activity source calibration in terms of reference air-kerma rate. A "two-voltage" method, independent of the voltage ratio given by a dosimetry system, is proposed for practical dosimetry of continuous x- and gamma-radiation beams. In the case where the Almond approach is utilized, the voltage ratio V_1/V_2 should be less than 2 instead of Almond's limit of $V_1/V_2 < 5$.

For continuous radiation, such as x- and gamma-ray beams of high dose rate, generally used in external radiotherapy treatments, the general recombination dominates the total recombination losses. In this case, the trend of the reciprocal of the observed ionization current, i, against

the voltage, V, can be described by

$$1/i = 1/i_{sat} + b/V^2 \qquad (3.95)$$

with b a constant that summarizes the recombination coefficient, the electron charge, and ion mobilities. The ion-recombination correction factor k_{sat}, used in several dosimetric protocols, is the reciprocal of the collection efficiency:

$$k_{sat} = 1/f = i_{sat}/i$$
$$f = i_1/i_{sat} \qquad (3.96)$$

Almond [50] and Weinhous and Meli [51], using Equations (3.95) and (3.96), obtained for k_{sat} the expression

$$k_{sat} = [(V_1/V_2)^2 - 1]/[(V_1/V_2)^2 - (i_1/i_2)] \qquad (3.97)$$

with i_1 and i_2 the currents obtained using voltages V_1 (the normal operating bias voltage) and V_2 ($V_1 > V_2$), respectively. Equation (3.97), known as a "two-voltage" technique, can be written replacing the currents with the collected charge Q_1 and Q_2.

A new "two-voltage" method to determine the $p(i_{sat}, V)$ factor was proposed. If V_1 and V_2, with $V_1/V_2 = n > 1$, are the only two voltages given by the dosimetry system and i_1 and i_2 are the relative ionization currents, then [49]

$$p(i_{sat}, V_1) = a/V_1 + bi_{sat}/V_1^2 \qquad (3.98)$$

$$p(i_{sat}, V_2 = V_1/n) = na/V_1 + n^2 bi_{sat}/V_1^2 \qquad (3.99)$$

and the difference between Equations (3.98) and (3.99) can be approximated for $i_1 \cong i_{sat}$ by

$$(i_1 - i_2)/i_1 \cong (n - 1)a/V_1 + (n^2 - 1)bi_1/V_1^2 \qquad (3.100)$$

If the experimental data $(i_1 - i_2)i_1$ as a function of i_1 can be fittted by a straight line, then the ordinate value A for $i_1 = 0$, and the slope value B of the line are given by

$$A = (n - 1)a/V_1$$
$$B = (n^2 - 1)b/V_1^2$$

Then $k_{sat}(i_{sat}, V_1)$ can be evaluated as

$$k_{sat}(i_{sat}, V_1) = 1 + p(i_{sat}, V_1) = 1 + a/V_1 + bi_{sat}/V_1^2$$
$$(3.101)$$

which, near the saturation condition $i_1 \cong i_{sat}$, can be rewritten as

$$k_{sat}(i_{sat}, V_1) = 1 + A/(n - 1) + Bi_1/(n^2 - 1)$$
$$(3.102)$$

The k_{sat} values for the ion chambers examined, obtained using the three methods, are reported in Figure 3.50 as a function of the i_1 values and the air-kerma rates. The arrows indicate the maximum air-kerma rate values related to minimum source-chamber distances used in the K_r calibration of ^{192}Ir (740 GBq) when multiple measurements (at different source-chamber distances) are made. In particular, the k_{sat} average values, obtained using Equation (3.102) with varying V_2 values, are reported with the maximum variations shown by bars; these data are in good agreement with the results obtained by applying the Niatel approach. [49]

The collection efficiency of a 5.7-cm-diameter spherical ionization chamber was measured by Biggs and Nogueira. [52] Due to its large volume, the collection efficiency is low when placed in the primary beam at isocenter and, hence, the correction factor is large. By measuring the primary fluence with the machine operating at a low dose rate and at a distance where the collection efficiency is high, a measure of the dose with this chamber at isocenter can be obtained by extrapolating back to the isocenter using the inverse square law.

Although Boag's theory was derived for parallel-plate geometry, he extended the calculation to cylindrical and spherical chambers by providing suitable correction factors.

The collection efficiency of this chamber was measured using three approaches. The first approach was to use the conventional two-voltage technique based on Boag's theory. In this case measurements were made at two voltages ($V_1 = 2V_2$, V_1 being the standard operating voltage) and the collection efficiency was calculated from the formalism outlined below. The second approach was to measure the collection efficiency using the voltage extrapolation technique. Here, the voltage across the chamber was varied between 50 V and 1000 V and the inverse of the collected charge, normalized to unity at the standard collection voltage, was plotted against the inverse of the voltage. The collection efficiency was then derived by extrapolating this plot to $1/V \rightarrow 0$.

Geleijns et al. [53] showed that for large chambers, up to 600 cm³, recombination effects due to transit-time effects can be quite significant.

The chamber used by Biggs and Nogueira was a Shonka (Model A5, Exradin, Lisle, IL 60532) 100-cm³ spherical ionization chamber with a 5.7-cm inner diameter. The leakage current for this chamber is less than 1 fA. A cubic, PMMA build-up cap with a minimum thickness of 1 cm was made for this chamber. This shape was chosen to provide build-up for the chamber in the primary beam and to allow for easy incorporation into a lead-shielding arrangement with adequate mechanical protection during the scattered radiation measurements. The build-up was found to be adequate for the 4-MV primary beam, but additional build-up was necessary for the 10-MV beam.

Measurements were also taken with a 0.6-cm³ Farmer-type chamber at 100% and 50% bias at the same distances as the Shonka chamber. Since the efficiency of a Farmer chamber when calibrating at isocenter is generally >98%, this chamber acts as a useful reference.

Geleijns et al. gave the formula for the transit time for ions in an ionization chamber as

$$\tau_i = \frac{d^2}{(Vk_i)} \qquad (3.103)$$

where k_i is the ion mobility, V is the collecting potential, and d is the effective electrode separation for a spherical chamber given by

$$d = K_{sph}(a - b) \qquad (3.104)$$

Here, a and b are the radii of the air volume and central electrode, respectively, and K_{sph} is a constant, given by

$$K_{sph} = \sqrt{\frac{1}{3}} \left\{ \frac{a}{b} + 1 + \frac{b}{a} \right\} \qquad (3.105)$$

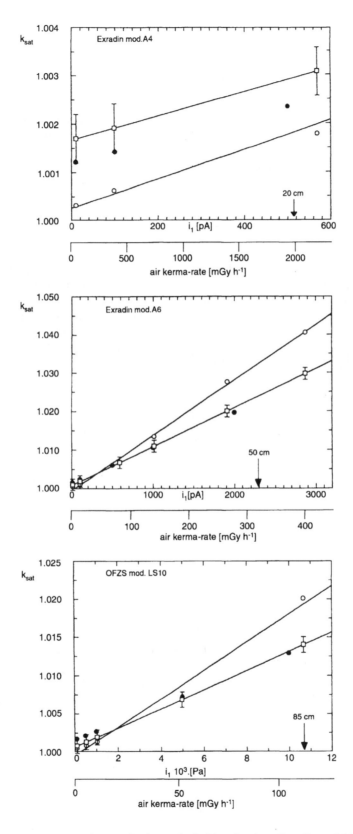

FIGURE 3.50 A comparison of the k_{sat} values obtained for three spherical ion chambers. Exradin model A4 (with $V_1 = 380$ V), Exradin model A6 (with $V_1 = 380$ V), and OFZS model LS10 (with $V_1 = 1000$ V) as a function of the ionization current, i_1. Results obtained using Almond (1981) (O) (with $V_1/V_2 \cong 3.5$) and Equation (3.102) (□) (the bars show the maximum variation obtained changing the V_1/V_2 ratio) are compared with the results obtained by applying the Niatel (1967) method (●). The arrows show the air-kerma rate values related to minimum source-chamber distances used in the ^{192}Ir (740 GBq) calibration. (From Reference [49]. With permission.)

For this particular chamber, $a = 2.84$ cm and $b = 0.3$ cm, giving a value of $K_{sph} = 1.88$ and $d = 4.78$ cm. Using the value for k_i of 1.58×10^{-4} m²s⁻¹V⁻¹ given by Geleijns et al., one obtains a value for τ_i of approximately 0.05 s.

TWO-VOLTAGE TECHNIQUE:

The ion-collection efficiency was calculated using the two-voltage technique based on the Boag formalism. One first takes the ratio, r:

$$r = \frac{Q_{100\%}}{Q_{50\%}} \qquad (3.106)$$

where $Q_{100\%}$ corresponds to the charge collected at 100% bias and $Q_{50\%}$ corresponds to the charge collected at 50% bias. One then calculates the value of u that satisfies the equation

$$r = \frac{2\ln(1 + u)}{\ln(1 + 2u)} \qquad (3.107)$$

and the ion-collection efficiency f is then given by

$$f = \frac{1}{u}\ln(1 + u) \qquad (3.108)$$

GELEIJNS' METHOD:

In this method, which is based on the Boag theory, a plot is made of the inverse of the charge measured by the Shonka chamber vs. the inverse of the charge measured by the Farmer chamber. The various points correspond to measurements made at different distances from the source; the line is a least-squares fit to the data with an r^2 value of 0.9999. This linearity derives from an approximation to the Boag formula. The chamber collection efficiency f at any distance, is then given by

$$f = S \times \frac{Q}{Q_{mom}} \qquad (3.109)$$

where S is the slope of the plot, Q is the charge measured by the Shonka chamber, and Q_{mon} is the charge measured by the Farmer chamber.

Figure 3.51 shows the charge measured by the Farmer chamber at 4 MV multiplied by the square of the distance from the source plotted against the source distance.

The general collection efficiency in pulsed radiation was studied by Johansson et al. [54] for isooctane (C_8H_{18}) and tetramethylsilane ($Si(CH_3)_4$). These two liquids were used as sensitive media in a parallel-plate liquid ionization chamber with a 1-mm sensitive layer. Measurements were carried out using 20-MV photon radiation from a linear

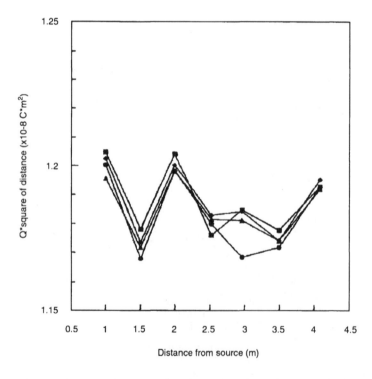

FIGURE 3.51 Variation of the charge measured by the Farmer chamber multiplied by the square of the distance from the target for 10-MV x-rays. This function, which is constant to within ±1.3% over the distance from 1 to 4.5 m, verifies the applicability of the inverse-square law for this beam over that distance. (■) 100 MU/min; (◆) 200 MU/min; (▲) 400 MU/min; (●) 600 MU/min. (From Reference [52]. With permission.)

accelerator with a pulse repetition frequency of 30 pulses/s and a pulse length of 3.5 μs. The aim of the work was to examine whether the theoretical equation for general collection efficiency for gases in pulsed radiation, derived by Boag (1969), can be used to calculate the general collection efficiency for isooctane and tetramethylsilane.

The body of the parallel-plate liquid ionization chamber used for the experiments was made of Rexolite®, a styrene copolymer with density 1.05 g cm^{-1}, and the electrodes were made of pure graphite. The chamber had cylindrical symmetry without any guard.

The theory neglects the space-charge screening effect and assumes that recombination loss during the radiation pulse is negligible; i.e., the pulse length is short. Furthermore, it assumes that the densities of the ions of different sign are equal; i.e., the negative charge is carried by negative ions, and that diffusion loss can be neglected.

For two ions of opposite sign at a separation distance R and approaching each other with a velocity v, the recombination rate constant α in a nonpolar molecular liquid is given by [55]

$$\alpha = \frac{e(k_1 + k_2)}{\varepsilon_0 \varepsilon} \left[1 + \frac{r_c D}{R^2 v} \exp\left(-\frac{r_c}{R}\right) - \exp\left(-\frac{r_c}{R}\right) \right]^{-1}$$

$$(3.110)$$

where r_c is the Onsager escape distance [56], the separation distance at which the Coulombic potential energy is equal to the diffusion kinetic energy of the ions; D is the sum of the diffusion coefficients of the ions; ε_0 is the permittivity of free space; and ε is the permittivity of the dielectric liquid in the chamber ($\varepsilon = 1.94$ for isooctane and $\varepsilon = 1.84$ for tetramethylsilane). For dielectric media with a low permittivity, the ratio r_c/R is large and the recom-

bination rate constant can be approximated by the equation derived by Debye:

$$\alpha = \frac{e(k_1 + k_2)}{\varepsilon_0 \varepsilon}$$

$$(3.111)$$

Using the Debye equation for μ in the Boag equation gives

$$\mu = \frac{1}{\varepsilon_0 \varepsilon}$$

$$(3.112)$$

μ is thus a characteristic for the general recombination rate in the sensitive media in the ionization chamber, determined by the permittivity of the dielectric liquid.

The relation between the ionization current in a dielectric liquid and the electric field strength has been shown by several authors, experimentally and by computer simulations, to follow a linear equation in a limited interval of the electric field strength, where the general recombination can be ignored. The linear relation is given by

$$i = (c_1 + c_2 E)\dot{D}$$

$$(3.113)$$

where i is the ionization current, c_1 is a constant characteristic of the dielectric liquid and determined by the probability for initial recombination escape due to diffusion, c_2 is a constant characteristic of the dielectric liquid and determined by the probability for initial recombination escape due to the external electric field strength, E is the external electric field strength, and $\dot{D}$ is the dose rate. Figure 3.52 shows schematically the relation between the

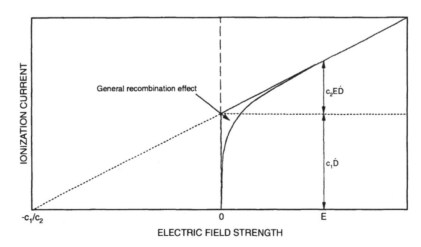

FIGURE 3.52 The characteristic relation of the ionization current from a dielectric liquid and the electric field strength in a liquid ionization chamber. The ionization current is the sum of the two components $c_1\dot{D} + c_2 E\dot{D}$, where c_1 is the fraction of ions that escape initial recombination due to diffusion, c_2 is the fraction of ions that escape initial recombination due to the external electric field strength, E is the electric field strength, and $\dot{D}$ is the dose rate. (From Reference [54]. With permission.)

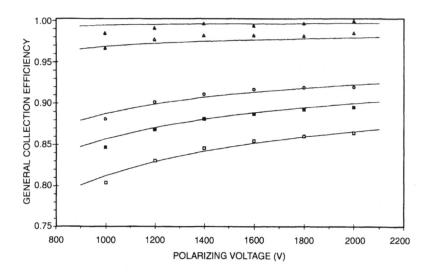

FIGURE 3.53 The experimental and theoretical general collection efficiencies for isooctane, with a liquid-layer thickness of 1 mm, as a function of applied polarizing voltage for different pulse doses. Experimental general collection efficiency at 1.9 mGy per pulse (□); 1.4 mGy per pulse (■); 0.65 mGy per pulse (O); 0.27 mGy per pulse (△); and 0.06 mGy per pulse (▲). The theoretical general collection efficiency is represented by a solid line for each pulse dose. (From Reference [54]. With permission.)

ionization current of a dielectric liquid and the applied electric field strength for constant dose rate. For low electric field strengths, the ionization current is reduced from the linear relation because of general recombination ion loss.

The theoretical and experimental general collection efficiencies are shown as a function of the polarizing voltage for different pulse doses in Figure 3.53 for isooctane and tetramethylsilane, respectively. The differences between the theoretical and experimental general collection efficiencies of the two liquids were within ±1% for general collection efficiencies down to 80% for isooctane and 75% for tetramethylsilane, respectively, using the permittivity of each of the liquids and electric field strengths exceeding 10^6 V m^{-1}. The maximum difference between the theoretical and experimental general collection efficiencies occurs for the lowest electric field strengths, 1000 V for isooctane and 500 V for tetramethylsilane, and for the highest pulse doses used in the experiments, 1.9 mGy per pulse. For isooctane, the maximum difference was less than 1%, and for tetramethylsilane the maximum difference was about 3%.

V. DETECTOR WALL EFFECT

Plane-parallel ionization chambers are recommended for the dosimetry of electron beams with mean energies below about 10 MeV. It is recommended to calibrate plane-parallel ionization chambers relative to a reference cylindrical ionization chamber in a phantom irradiated by a high-energy electron beam.

Because commercial plane-parallel chambers are not homogeneous in composition, calculations of k_{att}, and particularly k_m, will have large uncertainties.

The experimental method applied for the determination of $(k_{att}k_m)$ consists of comparing the reading of a plane-parallel chamber with that of a reference cylindrical chamber both in air in a ^{60}Co gamma-ray beam and in a phantom irradiated by a high-energy electron beam. The wall correction factor for a plane-parallel ionization chamber was measured by Wittkämper et al. [57]

For a plane-parallel chamber, it can be shown that

$$(k_{att}k_m)_{pp} = (k_{att}k_m k_{cal})_{cyl}(M_{cyl}/M_{pp})$$
$$\times (N_k)_{cyl}/(N_k)_{pp}((p_f)_{cyl}/(p_f)_{pp}) \quad (3.114)$$

where the subscripts *cyl* and *pp* refer to the cylindrical and plane-parallel chamber, respectively. It is assumed that p_{wall} and p_{cel} are unity in the electron beam.

By performing measurements in a photon beam with the same chambers as used in the high-energy electron beam, values for p_{wall} can be obtained from a similar equation: [57]

$$(p_{wall})_{pp} = (p_{wall}p_{cel})_{cyl}(M_{cyl}/M_{pp})_x$$
$$\times (M_{pp}/M_{cyl})_e((p_f)_{pp}/(p_f)_{cyl}) \quad (3.115)$$

where subscripts *X* and *e* refer to the photon beam and the high-energy electron beam, respectively. By positioning both chambers with their effective point of measurement at the same depth, the displacement correction will be unity.

The results of the $(k_{att}k_m)$ determinations of the four NACP chambers and four PTW/Markus chambers are summarized in Table 3.14.

TABLE 3.14

($k_{att}k_m$) Values of Different NACP and PTW/Markus Ionization Chambers

Ionization chamber	Type	Serial Number	($k_{att}k_m$)
NACP	01	06–02	0.978
	01	10–01	0.982
	01	10–02	0.980
	02	12–09	0.978
Dosetek	01	08–09	0.984
		weighted average	0.980
PTW/Markus	M23343	212	0.993
		373	0.992
		421	0.996
		923	0.992
		weighted average	0.993

Source: From Reference [57]. With permission.

The average results of the p_{wall} determinations for the three type-01 NACP chambers and one PTW/Markus chamber are summarized in Figure 3.54, presented as a function of the quality index of the photon beam.

To calibrate a megavoltage therapy beam using an ionization chamber, it is necessary to know the fraction of the ionization arising in the chamber wall when this is made of a material different than the medium. A method for measuring the ionization fraction produced by electrons arising in the chamber wall (α) was presented by Kappas et al. [58] The method uses three measurements at the same point in a medium in order to calculate α. These measurements are made using the examined chamber with and without a build-up cap and one reference chamber of wall material equivalent to the medium.

In the AAPM protocol for an absorbed dose in high-energy photon beams, the general relationship between the dose to the gas (air) of the chamber and the dose to the medium that replaces the chamber when it is removed is given as

$$D_{med} = \left(\frac{L}{\rho}\right)_{gas}^{med} D_{gas} \qquad (3.116)$$

where $(L/\rho)_{gas}^{med}$ is the ratio of the mean, restricted collision mass stopping power of the phantom material to that of the chamber gas and D_{gas} is given by the formula

$$D_{gas} = MN_{gas}P_{ion}P_{repl}P_{wall} \qquad (3.117)$$

where M is the electrometer reading, N_{gas} is the dose to the gas to the chamber per electrometer reading, P_{ion} is a factor that corrects for the ionization collection efficiency, P_{repl} is a correction factor for the replacement of the point of measurement from the geometrical center of the chamber, and P_{wall} is a wall correction factor that takes into account the fact that the chamber wall is usually made of different material than that of the dosimetry phantom. P_{wall} is equal to unity when the chamber wall and the medium are of the same composition or in the case of electron beams. Also, P_{wall} is given by a semiempirical expression

$$P_{wall} = \alpha\left(\frac{L}{\rho}\right)_{med}^{wall}\left(\frac{\mu_{en}}{\rho}\right)_{wall}^{med} + (1 - \alpha) \qquad (3.118)$$

when the chamber wall is of a composition different to the medium. In the above formula, α gives the fraction of the total ionization produced by electrons arising in the chamber wall and $1-\alpha$ gives the fraction of the total ionization produced by electrons arising in the dosimetry phantom. $(\mu_{en}/\rho)_{wall}^{med}$ is the ratio of the mean mass energy absorption coefficient for the dosimetry phantom (medium) to that of the chamber wall.

The method used to extract α from the three dose measurements at the same point is based on the fact that the absorbed dose in medium, D_{med}, is independent of the measuring device (i.e., chamber). In particular, the dose D_{med} given by each chamber is as follows:

$$D_{med} = M(w)N_{gas}(w)\left(\frac{L}{\rho}\right)_{gas}^{med}P_{ion}P_{repl}(w)$$
$$\times \left[\alpha \cdot \left(\frac{L}{\rho}\right)_{med}^{wall}\left(\frac{\mu_{en}}{\rho}\right)_{wall}^{med} + (1 - \alpha)\right] \qquad (3.119)$$

$$D_{med} = M(w')N_{gas}(w')\left(\frac{L}{\rho}\right)_{gas}^{med}P_{ion}P_{repl}(w')$$
$$\times \left(\frac{L}{\rho}\right)_{med}^{wall}\left(\frac{\mu_{en}}{\rho}\right)_{wall}^{med} \qquad (3.120)$$

$$D_{med} = M(m)N_{gas}(m)\left(\frac{L}{\rho}\right)_{gas}^{med}P_{ion}P_{repl}(m) \qquad (3.121)$$

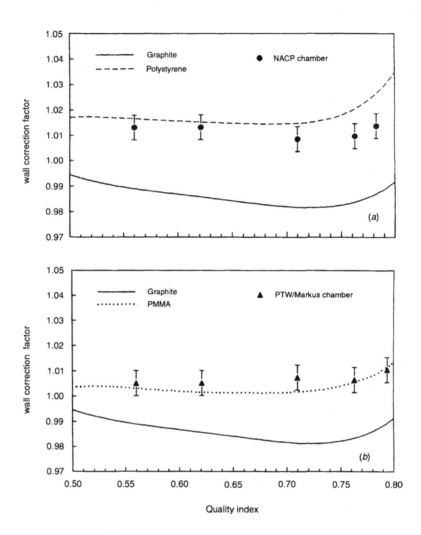

FIGURE 3.54 The wall correction factor, p_{wall}, for (a) the NACP chamber and (b) the PTW/Markus chamber in water, as a function of the quality index of the photon beam. Also indicated are p_{wall} values calculated for a homogeneous PMMA, polystyrene, or graphite chamber. (From Reference [57]. With permission.)

where W is the chamber wall composition, α unknown w' as w with build-up cap, $\alpha = 1$ and m is the chamber wall material equivalent to the medium, $\alpha = 0$.

Factors N_{gas} and P_{repl} are dependent on the individual chamber. Finally,

$$\alpha = \frac{1 - M(m)/M(w)}{1 - M(m)/M(w')} \quad (3.122)$$

The fraction of ionization due to the wall (α) is independent of the wall material, even if it is aluminum. This is very well demonstrated in Figures 3.55a and b, where all four combinations of wall + cap give results for α, depending only on the beam energy and wall thickness and not on the wall material.

Wall attenuation and scatter corrections for ion chambers were measured and calculated by Rogers and Bielajew. [59] Using the EGS4 system shows that Monte Carlo–calculated A_{wall} factors predict relative variations in

detector response with wall thickness which agree with all available experimental data within a statistical uncertainty of less than 0.1%. Calculated correction factors for use in exposure and air-kerma standards are different by up to 1% from those obtained by extrapolating these same measurements.

The measured ionization from a chamber with wall thickness t is proportional to $R(t)$, the absorbed dose to the gas in the cavity. For walls that are thick enough to establish charged particle equilibrium in the chamber's cavity, and assuming the normal tenets of cavity theory hold, one has

$$R(t) = K_{col,air} s_{air,wall} (\bar{\mu}_{en}/\rho)^{wall}_{air} A_{wall}(t) A_{oth} \quad (3.123)$$

where $K_{col,air}$ is the collision kerma in air at the geometric center of the cavity in the absence of the chamber, $s_{air,wall}$ is the stopping-power ratio, $(\mu_{en}/\rho)^{wall}_{air}$ is the ratio of spectrum-averaged mass-energy absorption coefficients in the

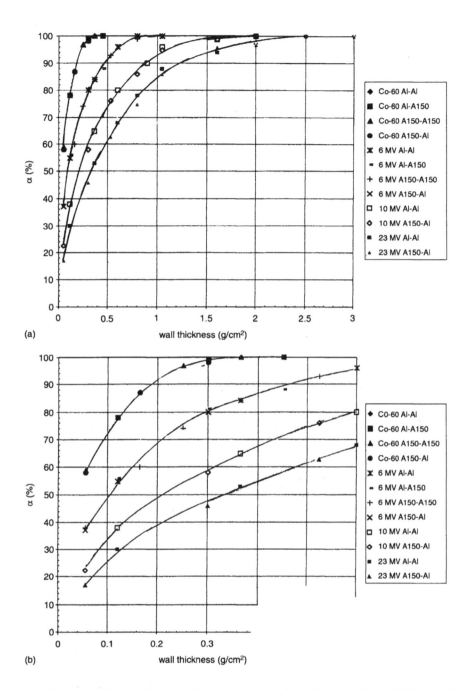

FIGURES 3.55 (a) A graph representing the fraction of ionization due to the wall (α) vs. the wall thickness for beam energies of ^{60}Co, 6, 10, and 23 MV and for combinations of wall-cap material Al-Al, Al-A150, A150-A150, and A150-Al, respectively. (The line is fit by eye to the data.) (b) The same graph as in (a) but for an expanded wall-thickness scale between 0 and 1 g/cm^2. (The line is fit by eye to the data.) (From Reference [58]. With permission.)

wall to those in the air, $A_{wall}(t)$ is the wall attenuation correction factor for the particular wall thickness, t, and A_{oth} groups several other small correction factors which are taken as independent of the wall thickness (stem, electrode, and field non-uniformity effects).

Since A_{wall} corrects for both attenuation (which decreases the response) and scattering (which increases the response), it can be either greater or less than unity. However, attenuation usually dominates so that A_{wall} is less than unity.

The value of A_{wall} is determined by scoring

$$A_{wall} = A_{sc}A_{at} \qquad (3.124)$$

where

$$A_{sc} = \sum_i (r_i^0 + r_i^1)\left(\sum_i r_i^0\right)^{-1} \qquad (3.125)$$

$$A_{at} = \sum_i (r_i^0)\left(\sum_i r_i^0 e^{+d_i}\right)^{-1} \qquad (3.126)$$

TABLE 3.15
Wall Materials Used in the Measurements

Material	Composition (Z of constituent: fraction by weight)	Density (kg m^{-3})^a
Graphite, C	C: 1.00	1.69×10^3
Polymethylmethacrylate, PMMA $(C_5H_8O_2)_n$	H:0.0805, C:0.5998, O:0.3196	1.17×10^3
Polyethylene $(C_2H_4)_n$	H:0.1437, C:0.8563	0.95×10^3
Clear polystyrene $(C_8H_8)_n$	H:0.0774, C:0.9226	1.04×10^3
White polystyrene $(C_8H_8)_n$ + TiO$_2$		1.06×10^3
Solid Water-457™	H:0.081, C:0.672, N:0.024, O:0.199, Cl:0.001, Ca:0.023	1.04×10^3
Plastic Water™	H:0.0926, C:0.6282, N:0.010, O:0.1794, Cl:0.0096, Ca:0.0795, Br:0.0003	1.02×10^3

Source: From Reference [60]. With permission.

r_i^0 is the energy deposited by electrons generated by the ith primary photon interaction, r_i^1 is the energy deposited by electrons generated from the second- and higher-order scattered photons that arise from the ith primary photon, and d_i is the number of mean free paths in the chamber to the point of interaction of the ith primary photon.

The AAPM protocol (TG39) includes a cavity replacement factor p_{repl} that differs from unity for some chambers but assumes that the wall perturbation factor, p_{wall}, may be taken as unity. The perturbation of the wall has been determined by Nilsson et al. [60], using a large plane-parallel ionization chamber with exchangeable front and back walls. The results show that in many commercial chambers there is an energy-dependent p_{wall} factor, mainly due to differences in backscatter from the often thick chamber body as compared to the phantom material. Backscatter in common phantom and chamber materials may differ by as much as 2% at low electron energies. The front walls are often thin, resulting in negligible perturbation, but the 0.5-mm front wall of graphite in the NACP chamber was found to increase the response by 0.7% in a PMMA phantom.

The BPPC-1 chamber has a negligible polarity effect. Thus, all measurements were made with the same polarity using a field strength of 50 V mm^{-1}. The leakage current was less than 0.01 pA, which is less than 0.01% of the ionization current obtained in the measurements.

Plane-parallel chambers commercially available often have composite backscatter walls with thin electrodes of a material different from the chamber body. Nilsson et al. measurements simulating the NACP (Scanditronix

NACP-02) and the Attix chambers (Gammex/RMI model 449) were made both with and without the composite electrodes in a PMMA phantom. The Roos chamber (PTW type 34001 or Wellhofer IC40), with walls made of PMMA, was simulated by just using the BPPC-1 chamber in the phantom, and the ionization obtained with this combination was used as a reference. The measuring geometries of the different simulated chambers are indicated in Figure 3.56.

If measurements are not performed in a PMMA phantom, the 1-mm front and back walls of PMMA in the Roos chamber may have an effect on the response of the chamber. Some measurements were thus performed simulating the Roos chamber in a phantom of polystyrene and compared to the use of a homogeneous polystyrene chamber in a polystyrene phantom. The experiments were simulated using the EGS4 Monte Carlo system.

Calculations were performed assuming a monoenergetic parallel electron beam with an energy equal to the mean energy at the surface obtained from depth-dose distributions in water and calculated according to IAEA.

The results of the variation of the backscatter factor with the mean electron energy at R_{100} are plotted in Figure 3.57. The data are normalized to backscatter from PMMA. Solid Water appears to be equivalent to PMMA with respect to backscatter within the experimental uncertainties. Graphite has a backscatter factor about 0.5–0.6% higher than PMMA, showing a slight increase with decreasing energy. The backscatter factors of the other materials show a stronger energy dependence. White and clear polystyrene have similar backscatter factors, around

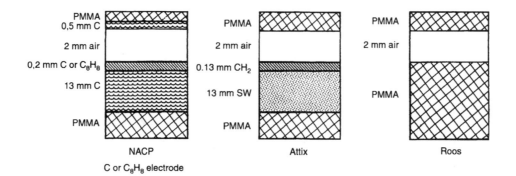

FIGURE 3.56 The geometries used in the simulation of the NACP, Attix, and Roos plane-parallel chambers. (From Reference [60]. With permission.)

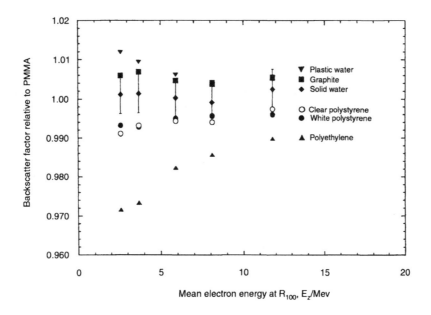

FIGURE 3.57 Backscatter factors relative to PMMA as a function of the mean electron energy, E_z, at R_{100} for different low-atomic-number materials. (From Reference [60]. With permission.)

1% lower than that of PMMA for low electron energies at $E_z \approx 3$ MeV. Polyethylene has a significantly lower backscatter factor, which decreases from 0.99 at $E_z = 12$ MeV to 0.97 at $E_z = 3$ MeV. Plastic Water, on the other hand, has a 0.5–1% higher backscatter factor as compared to PMMA and is more graphite-equivalent from this point of view.

In Figure 3.58 the experimental results for some materials are compared with the backscatter factors obtained by Hunt et al. [61], together with data obtained from the fit proposed by Klevenhagen [114]. The backscatter factors for polystyrene show very good agreement with the results obtained in both references, while differences can be observed for graphite.

Figure 3.59 shows a comparison between the measured and Monte Carlo values of the electron backscatter coefficient calculated using the EGS4 code. The agreement is generally within the statistical uncertainties, indicating that

the EGS4 code can be used for relative backscatter determinations in low-atomic-number materials where the backscatter component is low.

VI. CHAMBER POLARITY EFFECT

The polarity effect is a phenomenon encountered when using ionization chambers for electron measurements in which the measured readings vary significantly depending upon whether the bias applied to the chamber is positive or negative. TG-25 recommends correcting all ionization chamber readings if polarity effects greater than 1% are found. In order to correct for these effects, the readings must be taken at full positive and full negative bias voltages. In addition, it has been suggested that some ionization chambers require an adjustment period of several minutes in order for the readings to stabilize. The net effect

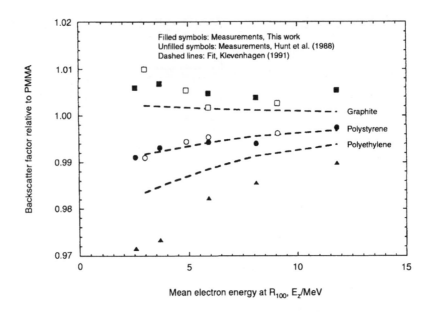

FIGURE 3.58 A comparison of backscatter factors for graphite, polystyrene, and polyethylene as a function of the mean electron energy E_z at R_{100} obtained in the present measurements (filled symbols), in measurements by Hunt el al. (unfilled symbols), and using the fit suggested by Klevenhagen (dashed lines). (From Reference [60]. With permission.)

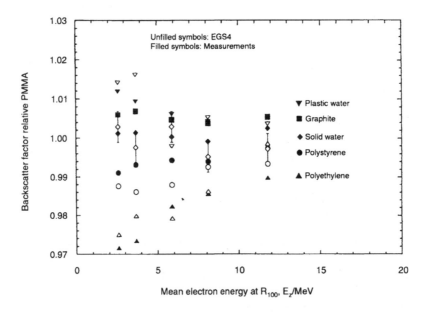

FIGURE 3.59 A comparison of backscatter factors relative to PMMA as a function of the mean electron energy E_z at R_{100} obtained by measurements (filled symbols) and by EGS4 Monte Carlo calculations (unfilled symbols). (From Reference [60]. With permission.)

is an approximate doubling of the time required to take measurements.

Reversing the polarity of the collecting voltage applied to some flat ionization chambers may change the value of the readings for photon beams or electron beams. Consequently, double polarity measurements are recommended for most of the plane-parallel ionization chambers. The average of the absolute value of the charges collected with a positive and then a negative polarity is considered close to the true value. This effect is related to the balance between electrons stopped in or knocked away from the collecting electrode.

Aget and Rosenwald [62] have recorded the electrometer readings Q^+ and Q^- corresponding, respectively, to a positive and negative bias voltage. The voltage is applied to the electrode of the ionization chamber which is not the collector, i.e., the front window for the plane-parallel chambers and the external electrode for the cylindrical

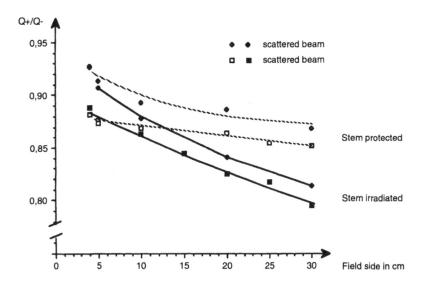

FIGURE 3.60 Variation of Q^+/Q^- vs. field size for the flat 0.03-cm^3 chamber. Nominal electron energy is 9 MeV, depth is 2 cm, and SSD is 100 cm. Continuous lines are for stem and cable irradiated, dashed lines for stem and cable protected. (From Reference [62]. With permission.)

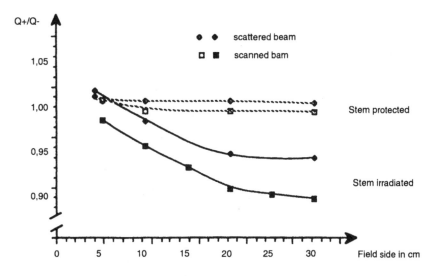

FIGURE 3.61 Variation of Q^+/Q^- vs. field size for the cylindrical 0.2-cm^3 chamber. Nominal electron energy is 9 MeV, depth is 2 cm, and SSD is 100 cm. Continuous lines are for stem and cable irradiated, dashed lines for stem and cable protected. (From Reference [62]. With permission.)

chambers. The Q^+/Q^- ratio is used to estimate the magnitude of the effect obtained in reversing the bias.

Curves in Figures 3.60 and 3.61 show, respectively, for the flat and the cylindrical chamber, the variation of the ratio Q^+/Q^- vs. the field size. In both figures, it can be seen that the ratio Q^+/Q^- is significantly smaller than 1 and that the polarity effect is maximum for larger field sizes. The effect is slightly larger for the scanned beam than for the scattering foil beam. In the scanned beam, it reaches as much as 21% for the flat chamber and 11% for the cylindrical one. It is about 4% less for the scattering foil beam. The explanation for the difference between the two beams is not obvious, but it is likely to be due to the difference in time structure rather than to the difference in energy or angular spectrum.

It is clear that the polarity effect is mostly related to the average energy $\bar{E}_z$ at the point of measurement. However, for the same $\bar{E}_z$ obtained from different combinations of $\bar{E}_0$ and depths, differences are observed which indicate that the angular and energy distributions of the electrons should also be taken into account. The maximum of the polarity effect has been found at depths where energy is around 2 MeV (Figure 3.62). This effect is larger for lower incident energies: $Q^+/Q^- = 0.66$ for $\bar{E}_0 = 3.5$ MeV, 0.73 for $\bar{E}_0 = 9$ MeV, and 0.77 for $\bar{E}_0 = 19$ MeV for the 30 × 30-cm^2 field.

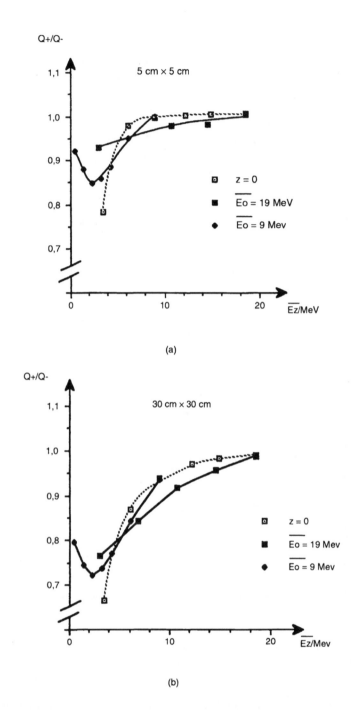

FIGURE 3.62 Q^+/Q^- ratio as a function of E_z. The continuous line is for surface measurements, changing the nominal incident energy $\bar{E}_0$. The dashed lines are for incident energies of 9 and 19 MeV and different depths. In all cases the 0.03-cm³ flat chamber was used in the scattering foil beam. (From Reference [62]. With permission.)

It was found by Havercroft and Klevenhagen [63] that the NACP, Markus, and Vinten chambers require a correction on the order of 0.2% in the energy range between 4.5 MeV and 18 MeV. Results have been published in the past where the polarity effect is on the order of 1%–3%, increasing to 4.5% in depth or to 20% for small plane-parallel chambers (Gerbi and Khan). [64] More physical explanation of the polarity can be found in Reference [65].

A well-constructed chamber should have a polarity effect not exceeding 1%. If a correction is made for this effect, the generally accepted approach is to average the reading taken with the negative (Q^-) and positive (Q^+) polarities, to obtain the mean true ionization charge value from

$$Q_{true} = (|Q^+| + |Q^-|)/2 \qquad (3.127)$$

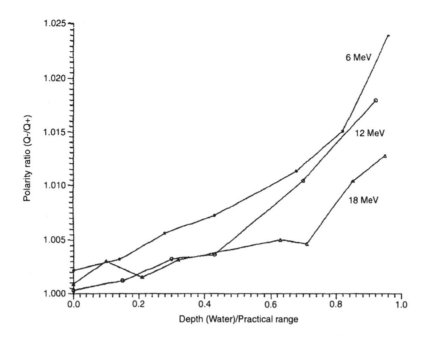

FIGURE 3.63 Ratio Q^-/Q^+ vs. chamber position in depth of water for the Calcam chamber measured in 6-, 12-, and 18-MeV (nominal) energy electron beams. (From Reference [63]. With permission.)

The correction factor for readings taken at negative bias is given by

$$f_{pol(-)} = (|Q^+| + |Q^-|)/(2Q^-) \qquad (3.128)$$

Figure 3.63 shows data for a Calcam chamber obtained in this study at three different electron energies (6, 12, and 18 MeV) where the energy is, in this case, the incident (nominal) beam energy at the phantom surface. The ratio (Q^-/Q^+) was calculated for readings taken at increasing depths from the phantom surface down to approximately $z/R_p = 0.9$.

The polarity effects of four commercially available ionization chambers were characterized by Williams and Agarwal [66] and correction factors as a function of mean energy at depth were tabulated. These included a Farmer-type chamber, two parallel-plate chambers, and one cylindrical chamber used in a scanning water phantom dosimetry system. Polarity effects were measured at representative depths along the depth dose curves of 6, 9, 12, 16, and 20 MeV electron beams. The term "polarity error" was introduced and defined as the error which is present if polarity effects are ignored. Polarity errors for the four ionization chambers studied were shown to monotonically decrease with increasing mean energy at depth and were largely independent of the energy of the incident electron beam. Only at very low energies, that is, very near the end of the practical range, did the correction factors for beams of different incident energy diverge. Three of the four chambers studied had correction factors which were independent of field size, to within ±1/2%. One chamber showed an increase in correction factor with increasing field size, which was shown to be mainly due to stem and cable irradiation.

The polarity error is plotted as a function of mean energy at depth for the four ionization chambers analyzed in this study and are shown in Figure 3.64. These data were fit to the third-order polynomial and the resulting curve or "characteristic curves" are also shown.

Van Dyk and MacDonald [65] demonstrated that the polarity effect is due in part to the lack of equilibrium in the number of electrons entering and leaving the collecting volume. For monoenergetic electron beams, the net deposition of charge is positive at the surface from the ejection of secondary charged particles and negative at depth where the electrons stop in the material. The charge deposition from electrons stopping in the medium influences the magnitude of the collecting charge. This, in turn, is dependent on the polarity of the voltage applied to the collecting electrode. Nilsson and Montelius [67] found that the polarity effect is dependent on sidewall material, chamber geometry, electron contamination, and the angular distribution of the electron fluence. Gerbi and Khan [64] found that the polarity effect was small (approximately 1%) up to a depth of d_{max} but increased up to 4.5% at greater depths near the practical range of the electron beam.

Two plane-parallel ionization chambers (PTW 23343) and two cylindrical chambers (NEL 2571) were used by Ramsey et al. [68] to compare the magnitude of the polarity effect on different types of ionization chambers as well as individual chambers. The effective volumes of the parallel-plane chambers and cylindrical chambers were 0.045 cm³

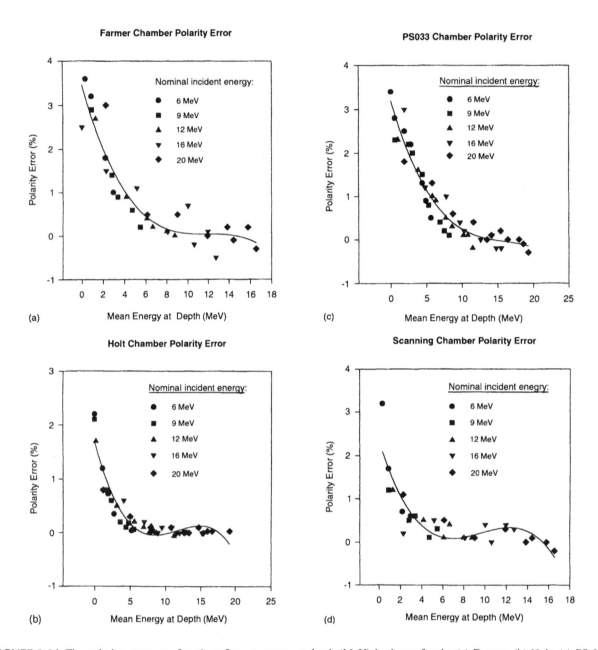

FIGURE 3.64 The polarity error as a function of mean energy at depth (MeV) is shown for the (a) Farmer, (b) Holt, (c) PS-033, and (d) scanning chambers. The characteristic curve resulting from a third-order polynomial fit to the data is also shown for each chamber. Data for the first three chambers were taken using a 15×15-cm^2 square field in a polystyrene phantom with 100-cm source-to-surface distance. Data for the scanning chamber was taken using an identical setup in a water phantom. (From Reference [66]. With permission.)

and 0.6 cm^3, respectively. The N_{gas} for each of the chambers is 4.90×10^8 (PTW 430), 4.75×10^8 (PTW 1663), 4.04×10^7 (NEL 484), and 4.01×10^7 Gy/C (NEL 1626).

The polarity error is defined as the fractional error between the reading taken at positive bias and the correct reading

$$polarity\ error = \frac{|Q_+| - \frac{|Q_+| + |Q_-|}{2}}{\frac{|Q_+| + |Q_-|}{2}} = \frac{|Q_+| - |Q_-|}{|Q_+| + |Q_-|}$$

(3.129)

In order to compare the polarity error for the different ionization chambers, the measured data was fit using a non-linear regression of the model

$$P.E. = \frac{1}{A + B \times (\bar{E}_d)^c}$$

(3.130)

where $P.E.$ = polarity error, E_d = mean energy at depth, and A, B, C = fitting parameters.

This model resulted in a better correlation coefficient than either a second- or third-order polynomial fit. The model uses the Levenberg-Marquardt method to combine the steepest descent method and a Taylor series–based method for non-linear optimization. Although this model will approach zero polarity error at large mean energies, the model adequately fits in the region of lower mean energies where the polarity error is most pronounced.

VII. DETECTOR PERTURBATION

Two effects perturb the electron fluence compared to that at the same position in the undisturbed medium. One of them is due to an imbalance between the number of electrons that come into the gas cavity through the side walls and the number of electrons already in the cavity that leave it. This is due to the fact that the average angular divergence of the former is greater than that of the latter. This has become known as in-scattering and will cause the fluence in the cavity to increase. The other effect is that the fluence of electrons entering the front wall of a plane-parallel cavity is hardly affected by the scattering in the air, whereas in the medium the fluence continues to build up, owing to scattering. This effect results in a lower fluence in the gas cavity than in the medium at the same depth. Thus, the two effects act in opposite directions. [69]

The second, or fluence build-up, effect is generally taken care of by referring the reading of the chamber to an effective point of measurement which differs from that of the chamber center.

The first effect exceeds 3% for Farmer-type chambers for $\bar{E}_z$ below 8 MeV and is one of the principal reasons why plane-parallel chambers are recommended in low-energy electron beams.

In coin-shaped gas cavities, the in-scattering perturbation in the electron fluence is confined to a region close to the edges of the cavity. Therefore, by constructing the chamber with a guard ring of sufficient width, it is possible to exclude the contribution to the ionization signal from this edge region.

Backscatter plays a role in the response of certain plane-parallel chambers. It should be part of any theoretical treatment of the variation of the overall perturbation factor p_Q with electron energy and with chamber material. The overall perturbation factor p_Q will be written as the product of two factors:

$$p_Q = p_{cav} p_{wall} \qquad (3.131)$$

where p_{cav} refers exclusively to the perturbation due to the air cavity, i.e., predominantly in-scattering effects, and p_{wall} refers to any effects due to the composition of the chamber wall, which could include possible backscatter effects.

A series of chambers with different guard ring widths, each with the same cavity height of 2 mm, were compared to each other in a 6-MeV beam (Figure 3.65). It can be seen that for widths greater than 3 mm, the chamber perturbation remained constant. p_{cav} is unity for all guard ring widths greater than 3 mm.

Values for the fluence perturbation correction factor were determined by Wittkamper et al. [71], in a number of electron beams for the commonly applied PTW/Markus plane-parallel chamber and a cylindrical NE 2571 farmer chamber. These data were determined relative to the

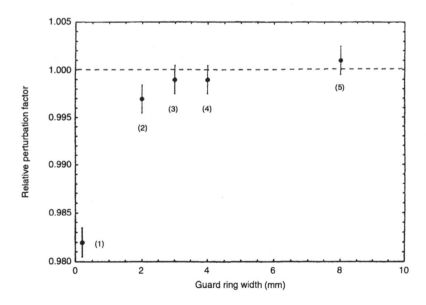

FIGURE 3.65 Experimental study of the effect of guard ring width on the perturbation factor for various plane-parallel chamber designs. The points labelled (1) and (3) are for the PTW/Markus chamber and the NACP chamber, respectively. Points (2), (4), and (5), represent specially constructed chambers of the PTW/Roos design with guard ring widths of 2, 4, and 8 mm, respectively. The measurements were made at the depth of dose maximum in a 6-MeV beam against a chamber with an ultrabroad (8-mm) guard ring, with the normalization of the response quotients obtained in a 20 MeV beam. [70] (From Reference [8]. With permission.)

TABLE 3.16
Characteristics of the Reference and Plane-Parallel Ionization Chambers and Build-Up Caps

Chamber Type		NE 2571 Cylindrical 0.6 cm³ Farmer		NACP Plane-Parallel	PTW/Markus Plane-Parallel
Cavity	Length	24 mm	Plate distance	2 mm	2 mm
Dimensions	Diameter	6.3 mm	Diameter	16.4 mm	6 mm
Chamber wall	Material	graphite	Front wall material	Graphite	Graphited polyethylene
	Thickness	65 mg cm⁻²	Thickness	90 mg cm⁻²	2.3 mg cm⁻²
			Side wall material	Polystyrene	Graphited PMMA
Electrode	Material	aluminum	Material	Graphited polystyrene	Graphited polystyrene
	Diameter	1 mm	Diameter	10 mm	5.4 mm
Build-up cap	Material	PMMA	Material	Graphite	Graphite
	Thickness	558 mg cm⁻²	Thickness	540 mg cm⁻²	540 mg cm⁻²
Guard ring			Width	3.2 mm	0.7 mm

NE, Nuclear Enterprises Ltd.; NACP, Scanditronix, Dosetek (Calcam); PTW, Physikalisch Technische Werkstatte.

Source: From Reference [71]. With permission.

NACP plane-parallel chamber as a function of the mean energy at depth E_z.

In recent codes of practice the formalism to determine the absorbed dose in high-energy electron beams is based on an air kerma calibration of the ionization chamber and application of the Bragg-Gray equation. This leads to the following expression:

$$D_w = MN_K(1-g)\Pi k_i s_{w,air}\Pi p_i \qquad (3.132)$$

where D_w is the absorbed dose to water in the electron beam at the depth of the effective point of measurement, M is the instrument reading corrected for temperature, pressure, recombination, polarity and humidity, N_K is the air kerma calibration factor, g is the fraction of the energy of secondary-charged particles converted to *bremsstrahlung* in air at the calibration quality, and Πk_i is the product of correction factors to be applied for the determination of the absorbed dose to the air in the ionization chamber cavity from the air kerma in the photon calibration beam in the absence of the chamber:

$$\Pi k_i = k_{att}k_m k_{cel}k_{stem} \qquad (3.133)$$

$s_{w,air}$ is the restricted stopping-power ratio of water to air and Πp_i is the product of correction factors to be applied to the measurements in the water phantom at the user's quality:

$$\Pi p_i = p_{wall}p_d p_{cel}p_f \qquad (3.134)$$

The wall and central electrode corrections p_{wall} and p_{cel} are negligible for cylindrical ionization chambers commonly applied in electron beam dosimetry. If measurements are performed using the effective point of measurement, the displacement correction p_d can also be omitted, and the only correction factor in the formalism to be applied during phantom measurements in electron beams is the fluence perturbation correction factor p_f.

For cylindrical chambers p_f was determined experimentally by Johansson et al. [72] for a number of beam energies. p_f is usually given as a function of the mean electron energy at depth z, $\bar{E}_z$.

The plane-parallel chambers are not calibrated in terms of air kerma in a ⁶⁰Co gamma-ray beam but are calibrated in a high-energy electron beam against a cylindrical chamber. In this way, the product

$$N_K\Pi k_i\Pi p_i \qquad (3.135)$$

is determined. The fluence perturbation correction factor of plane-parallel chambers in the calibration electron beam is thus incorporated in the calibration factor.

By applying Equations (3.132), (3.133), and (3.134) twice, it follows that

$$p_{f,x} = \frac{(k_{att}k_m k_{cel})_{ref}}{(k_{att}k_m k_{cel})_x}\frac{(N_K)_{ref}}{(N_K)_x}\frac{(p_{wall}p_d p_{cel}p_f)}{(p_{wall}p_d p_{cel})_x} \qquad (3.136)$$

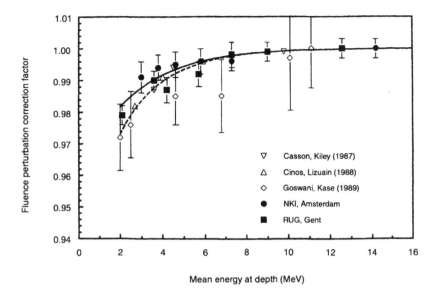

FIGURE 3.66 Fluence perturbation correction values for the PTW/Markus plane-parallel chamber as a function of the mean energy at depth. The results are normalized to the NACP chambers. The error bars represent one standard deviation. The solid curve represents a fit through the data points of this work and is adopted in the NCS Code of Practice (1989). The broken curve gives a fit through all data points in the figure. (From Reference [71]. With permission.)

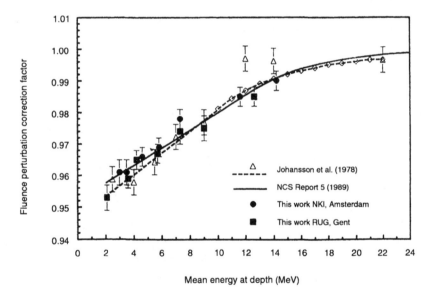

FIGURE 3.67 Fluence perturbation correction values for the NE2571 cylindrical Farmer-type chamber as a function of the mean energy at depth. The results are normalized to the NACP chambers. The error bars represent one standard deviation. The solid curve represents a fit through the data points of this work and is adopted in the NCS Code of Practice (1989). (From Reference [71]. With permission.)

where the subscripts x and *ref* refer to the chamber with unknown fluence perturbation correction values and to the reference chamber, respectively.

In Figure 3.66 the resulting p_f values are given for the PTW/Markus chambers as a function of the mean energy at the depth of the effective point of measurement, which is the depth of dose maximum.

In Figure 3.67 p_f values for the 0.6-cm³ NE2571 cylindrical chambers are given as a function of the mean energy

at depth. In the same figure, the data derived from Johansson et al. [72] are depicted as a function of $\bar{E}_z$.

The p_f values reported in this work are obtained by averaging the results of at least five individual measurement series. The error bars indicated in Figures 3.66 and 3.67 are one standard deviation, derived by quadratic summation of the statistical uncertainties of the ionization chamber readings in the electron beams M_x and M_{ref} and at the air-kerma calibrations $N_{k,x}$ and $N_{N,ref}$. A contribution to the overall

TABLE 3.17
Detector Characteristics

| | | Entrance wall | | | | |
Type	Model	Material	Thickness (g/cm²)	Gap (cm)	Guard (cm)	Diameter (cm)
NACP	...	Graphite	0.09	0.2	0.3	1.0
ATTIX	449	Kapton	0.0048	0.1	1.35	1.27
MARKUS	30–329	Polyethylene	0.0023	0.2	0.02	0.53
FARMER	30–351	Acrylic	0.059	0.25	...	...

Source: From Reference [73]. With permission.

uncertainty of random and systematic errors in the k_i and p_i correction factors of the reference and second chamber in Equation (3.136) was not taken into account.

The solid curve in Figure 3.66, which is given by

$$p_f = 1 - 0.041 e^{-0.40 \bar{E}_z} \qquad (3.137)$$

is an exponential fit to the data determined in this work, weighted with the one standard uncertainty.

A second exponential term was added to make a best fit to all data given in Figure 3.66. This curve is the broken curve and is given by

$$p_f = 1 - 0.041 e^{-0.40 \bar{E}_z} - 0.036 e^{-0.72 \bar{E}_z} \qquad (3.138)$$

Figure 3.67 shows that for the cylindrical 0.6-cm³ ionization chamber, good agreement exists between Wittkämper et al. p_f values and those determined by Johansson and colleagues, while Wittkämper et al. [71] $\bar{E}_z$ values are computed using the Andreo and Brahme tables. A partial linear and partial exponential fit was applied to the data determined in our work, which is depicted in Figure 3.67 by the solid curve. The analytical expression of the fit is given by

$$p_f = 0.952 + 0.0028 \, \bar{E}_z \quad \text{for } \bar{E}_z < 12 \, \text{MeV} \quad (3.139)$$

$$p_f = 1 - 0.18 e^{-0.21 \bar{E}_z} \qquad \text{for } \bar{E}_z \geq 12 \, \text{MeV} \quad (3.140)$$

In electron beam dosimetry, the perturbation effect in the medium by the ionization chamber cavity is accounted for by introducing a replacement correction factor, p_{repl}. Another perturbation correction factor, denoted as p_{wall}, is due to the materials of the walls of the parallel-plate chamber differing from the phantom material. Because of the difficulties in separating these two components, Reft and Kuchnir [73] measured the overall perturbation factor, $p_q = p_{repl}p_{wall}$. A distinct advantage of parallel-plate ionization chambers over cylindrical chambers is that, p_q has been shown to be close to unity at the standard calibration depth, d_{max}.

At the same depth in the phantom, they equate the percentage depth dose (PDD) measured with the NACP chamber to that measured with the other chambers, to derive an equation for the overall perturbation factor:

$$PDD(d, E_d) = [Q_{NACP} \times (L/\rho)_{air}^{H_2O}]_{d_{max}}^d$$

$$= [Q_{ch} \times (L/\rho)_{air}^{H_2O} (p_q)_{ch}]_{d_{max}}^d \quad (3.141)$$

Therefore,

$$(p_q)_{d,ch} = [(p_q)_{d_{max},ch}] \times [Q_{NACP}]_{d_{max}}^d [Q_{ch}]_{d_{max}}^d \quad (3.142)$$

where $[Q]_{d_{max}}^d$ is the ratio of the corrected measured charges at depths d and d_{max}, respectively, and $[(L/\rho)_{air}^{H_2O}]_{d_{max}}^d$ represents the restricted mass collision stopping power ratio of water to air evaluated at depth d divided by its value at depth d_{max}. In this derivation they have assumed that p_q is unity for the NACP chamber and the restricted stopping power ratios cancel because the measurements are performed at the same depths in the phantom. For the parallel-plate chambers, p_q was taken as 1.00 at d_{max} with one exception, the Markus chamber. For this chamber there are experimental data showing that, at energies below 12 MeV, p_q measured at d_{max} decreases with decreasing energy.

Using the NACP chamber as the reference detector, p_q was obtained for the other three ionization chambers. The data as summarized in Table 3.18 are obtained by using Equation (3.142) and show that for the depths slightly greater than d_{ref}, p_q is unity for the parallel-plate chambers within the experimental uncertainties.

For the Markus chambers, results show that p_q increases significantly with an increasing dose gradient. This behavior is illustrated in Figure 3.68. There are a number of studies showing that this chamber over-responds ($p_q < 1.00$) at d for low-energy electrons. The results reported here show that it under-responds ($p_q > 1.00$) in regions beyond d_{max}. The response of the Markus chamber is reduced in the decreasing portion of the depth ionization

TABLE 3.18
Summary of Data Used to obtain p_q for the Other Ionization Chambers Using Equation (3.142) and the NACP Chamber as the Reference Detector

E (MeV)	d (cm)	Ed (MeV)	$\Delta PDD/\Delta d$ (%/mm)	$[Q]^d_{d_{max}}$				p_q			
				NACP	Attix	Markus	Farmer	Attix	Markus	Farmer	
22	5.3	10.9	0.4	0.8913	0.8929	0.8993	0.9033	0.998	0.991	0.981	(0.983)
22	6.4	8.8	0.8	0.8150	0.8177	0.8203	0.8326	0.997	0.994	0.973	(0.977)
18	5.3	7.0	1.0	0.8850	0.8834	0.8881	0.8993	1.002	0.996	0.968	(0.972)
18	5.8	6.1	1.3	0.8264	0.8244	0.8251	0.8382	1.002	1.002	0.970	(0.969)
22	9.0	3.6	2.3	0.4184	0.4232	0.4239	0.4350	0.989	0.987	0.956	(0.961)
18	7.6	2.7	3.1	0.4018	0.4075	0.3931	0.4175	0.986	1.022	0.947	(0.958)
15	6.1	2.5	3.8	0.4948	0.4876	0.4660	0.5079	1.015	1.062	0.951	(0.957)
12	4.9	1.9	4.5	0.4796	0.4679	0.4398	...	1.025	1.085[a]	...	

[a] This value was obtained using 0.985 for $(P_q)_{d\,max}$ from TG 39.

The values in parentheses for the Farmer chamber are taken from Table VIII in the TG-21 protocol.

Source: From Reference [73]. With permission.

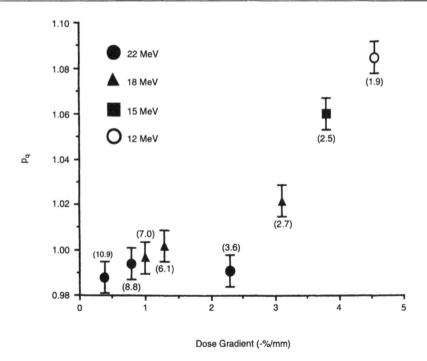

FIGURE 3.68 A plot of p_q vs. the dose gradient for the Markus chamber. The values in parentheses are the energies at depth. (From Reference [73]. With permission.)

curve. This behavior is summarized in Figure 3.69; the dose gradient is shown in parentheses.

VIII. DISPLACEMENT CORRECTION

For measurements free in air as commonly employed in calibration situations, the variation in the radiation field from a point source is governed by the inverse square law. Inside a phantom, the effective point of measurement can be displaced relative to the free-in-air condition due to the replacement of phantom material by the cavity of the ionization chamber. A correction has to be made for changes in attenuation and scattering due to this replacement. This can be achieved in terms of a radial displacement using an effective point of measurement or by taking the geometrical center of the chamber as the measuring point and applying a displacement correction factor. Such a factor is essentially the same as the perturbation factor.

Displacement corrections for in-phantom measurements with ionization chambers for the purpose of mam-

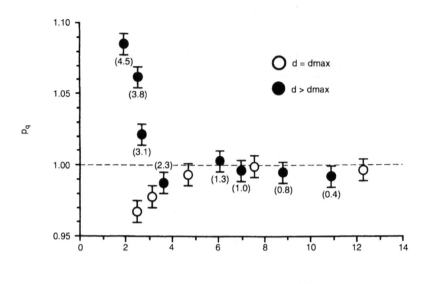

FIGURE 3.69 A plot of p_q vs. energy at d_{max} and at the depth for the Markus chamber. The values in parentheses are the dose gradients in % per mm. (From Reference [73]. With permission.)

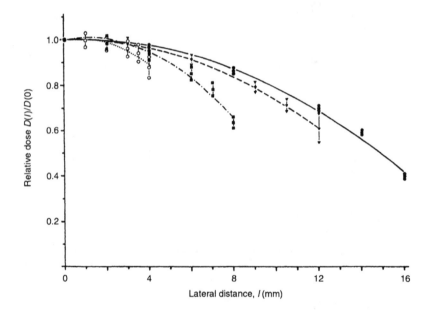

FIGURE 3.70 Lateral dose relative to dose in the center of the cavity for calculations inside spherical cavities of different radii ($\bigcirc$, 4 mm; $\blacksquare$, 8 mm; $\blacktriangledown$, 12 mm; and $\bullet$, 16 mm) placed with their geometric centers at 30-mm depth in PMMA phantoms. Fits are made according to Equation 3.143. (From Reference [74]. With permission.)

mography are large and represent a major correction to consider for dose determinations. Experimental data on displacement corrections depend to a large degree on the model used to extrapolate to zero cavity radius. Calculations of displacement correction factors using a Monte Carlo code for different cavity shapes were carried out by Zoetelief et al. [74]

For photons employed in mammography, experimental displacement correction factors obtained by linear extrapolation to zero cavity radius differ considerably from

unity; e.g., k_d is about 0.7 for a 0.6-cm^3 Baldwin-Farmer (Nuclear Enterprises Ltd., UK, type BF2571) ionization chamber. [75] Non-linear extrapolation to zero cavity radius for mammography radiation qualities would lead to different displacement correction factors and their associated errors. Some of the experimental data at different cavity radii suggest a nonlinear dependence of the reading on cavity size. There is, however, a lack of extrapolation models. In addition, for photons as employed in mammography, information is not available for phantom materials other than polymethylmethacrylate (PMMA) and is restricted to cylindrical and spherical ionization chambers. There is only limited knowledge of the dependence of displacement correction factors on depth in phantom. [74]

Calculated dose values at points along the central lateral axis (a line through the center of the cavity parallel to the x-axis) are given in Figure 3.70 for the spherical cavities located at 30-mm depth in the PMMA phantom. A decrease in calculated dose values with increasing lateral distance, l, from the center of the cavity is found. This is due to an increase in the amount of phantom material traversed by the photons with increasing lateral distance from the center of the cavity. Due to partial lack of side scatter, the calculated dose values at the maximum lateral distance are still somewhat higher (up to about 10%) than the dose at the same depth in the homogeneous phantom. Also shown in the figure are fits using STAR-PAC, a least-squares program for nonlinear functions [76], according to

$$f_r(l) = 1 + dl - bl^2 \qquad (3.143)$$

The results of calculations of dose in the cavities relative to dose at 30-mm depth in the homogeneous phantoms, $(D(r)/D(0))$, for spherical cavities of different radii placed with their centers at 30-mm depth inside the PMMA, A-150 plastic, and EVA/28 phantoms are shown in Figure 3.71 as a function of cavity radius.

It is evident from Figure 3.71 that the exponential fits are superior.

When parallel-plate chambers are used for dosimetry in electron fields, the AAPM dosimetry protocol recommends a value of 1.0 for the replacement correction factor, $P_{repl,pp,E}$, until further data become available. $P_{repl,pp,E}$ for five commercially available parallel-plate chambers was measured by Reft and Kuchnir [77] as a function of electron energy from a nominal value of 5.5 to 22 MeV by comparison with a cylindrical chamber whose $P_{repl,cyl,E}$ was obtained from data in the protocol. It was found that for three of the chambers, $P_{repl,pp,E}$ is independent of energy, consistent with unity within one or two standard deviations (s.d.). For the fourth chamber, $P_{repl,pp,E}$ is similarly consistent with one above 10 MeV but decreases at lower energies, while for the fifth one it shows a systematic drop with decreasing energy.

For equal dose to the homogeneous medium at the effective point of measurement, the following relation holds:

$$N_{gas,pp} M_{pp} P_{repl,pp} P_{wall,pp} = N_{gas,cyl} M_{cyl} P_{repl,cyl} P_{wall,cyl}$$

$$(3.144)$$

where M is the measured charge corrected for temperature, pressure, polarity effect, and ion collection efficiency. P_{repl}

and P_{wall} are the replacement and wall correction factors, and pp and cyl refer to the parallel-plate and cylindrical chambers, respectively. The effect of chamber polarity is evaluated from the ratio of charges measured with positive and negative polarizing voltages. For measurements in a photon field, the AAPM protocol considers $P_{repl,pp,x}$ to equal 1, and when a cylindrical chamber is used at d_{max}, $P_{repl,cyl,x} = 1$, also. With the exception of the NACP chamber, which has a graphite entrance wall, all measurements were taken in a phantom matched to the chamber wall material. Therefore, $P_{wall} = 1.0$. Equation (3.144) then reduces to:

$$N_{gas,pp} = N_{gas,cyl}[Mx]_{pp}^{cyl} \qquad (3.145)$$

where the notation $[M_X]_{pp}^{cyl}$ is used to represent the ratio $(M_x)_{cyl}/(M_x)_{pp}$. For the NACP chamber, $P_{wall} = 1$ was used; it varied from 0.995 to 0.997 over the energy range investigated. In the case of the 22-MeV electron beam irradiation, values of $P_{repl,cyl,E}$ are available only for the Farmer chamber. The Exradin cylindrical chamber, therefore, was not used with the electron beams. Following recommendations of the AAPM protocol, $P_{wall} = 1$ for all chambers and $P_{repl,pp,E} = 1$ at high electron energies. Referring to Equation (3.144),

$$N_{gas,pp} = N_{gas,cyl}[P_{repl,E}]_{pp}^{cyl}[M_E]_{pp}^{cyl} \qquad (3.146)$$

Using Equation (3.144) and (3.145), Reft and Kuchnir found that for the matched pair of chambers in homogeneous media,

$$[P_{repl,E}]_{cyl}^{pp} = ([M_E]_{pp}^{cyl}/[Mx]_{pp}^{cyl})P_{wall,pp,X} \qquad (3.147)$$

where E and X refer to electron and photon fields, respectively. All of the measurements were made in phantom with a source-to-surface distance of 100 cm and the effective point of measurement of each chamber placed at d_{max}. A 10×10-cm^2 field defined at the phantom surface was used at nominal electron energies of 5.5, 6, 9, 12, 15, 18, and 22 MeV from a Cl 2500 linear accelerator.

Results for P_{repl} as a function of electron energy for the five chambers are summarized in Table 3.19.

As seen in Figure 3.72, the experimentally determined values of $P_{repl,pp,E}$ for the Holt, NACP, and Exradin chambers are constant over the energy range investigated. The respective average values are 0.985, 0.980, and 0.989 with a standard deviation of 0.5% and an overall uncertainty of 1% (63% confidence level).

The TG-21 protocol analyzes the replacement correction factor in two components: (a) gradient correction and (b) electron fluence correction. The gradient correction, according to the protocol, arises because, on the descending portion of a depth-dose curve, the proximal surface of a cylindrical cavity intercepts an electron fluence that is

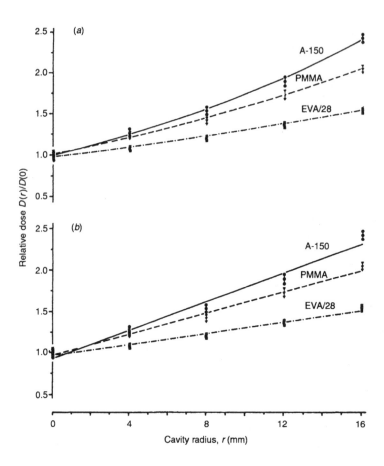

FIGURE 3.71 Absorbed dose calculated for spherical cavities with radii of 4, 8, 12, and 16 mm, placed with their geometric centers at 30-mm depth inside the PMMA phantom, relative to dose in the homogeneous phantom, $D(r)/D(0)$, as a function of cavity radius. (a) Exponential fits; (b) linear fits. (From Reference [74]. With permission.)

more intense than that at the measurement depth when the chamber is removed.

The protocol assumes that P_{repl} for photons, under the calibration conditions, is mainly due to the dose gradient and that electron fluence corrections are not required since there is a transient electronic equilibrium at the center of the chamber. In the case of electron beams, the gradient correction is ignored as the chamber is positioned at the depth of maximum dose, but an electron fluence correction is used as a function of beam energy and cavity diameter. For the plane-parallel chamber, P_{repl} is assumed to be unity for both photon- and electron-beam calibration. The protocol does not address the problem of measuring dose distributions with ionization chambers that require replacement corrections. [78]

In an idealized case where a parallel beam of electrons (or secondary electrons generated by photons) enters the chamber from a direction perpendicular to the chamber axis, it has been shown theoretically that the displacement in the effective point of measurement is given by $0.85r$ ($= 8r/3\pi$). Experimental data show that the actual values lie between 0 and $0.75r$, depending upon beam energy, in the energy range of medium energy x-rays to 42-MV

x-rays. In the megavoltage range of photon beams, the measurements of Johansson et al. [72] show good agreement between depth-dose curves measured with plane-parallel and cylindrical chambers if a constant shift correction of $0.75r$ is used for the cylindrical chambers.

The TG-21 protocol equates P_{repl} or "gradient correction" to photon attenuation corrections, calculated by Cunningham and Sontag [79] for cylindrical phantoms of the same size as the chamber that displaces the medium. These factors are used to correct for change in energy fluence of photons at the center of the chamber compared to the energy fluence that would exist at that point in the absence of the chamber. Under conditions of transient electronic equilibrium, the change in photon energy fluence is equated to the change in electron fluence, which is the basis of P_{repl} corrections.

P_{repl} can then be thought of as the ratio of dose to the medium at the depth of central axis of chamber to that at the depth of the effective point of measurement. Mathematically: [78]

$$P_{repl}(d) = D(d)/D(d - 0.75r) \qquad (3.148)$$

TABLE 3.19
Summary of Results for P_{repl}

Nominal Energy	E_0 (MeV)	E_z (MeV)	$P_{repl,cyl}$ (TG #21)	Chamber Type	$[M]_{pp}^{cyl}$	$[P_{repl,E}]_{cyl}^{pp}$	$P_{repl,pp,E}$
E-(MeV)					E		
5.5	4.7	2.5	0.957	N	3.404	1.022	0.978
				M	10.34	0.998	0.955
				H	0.641	1.032	0.988
				C	1.175	1.044	0.999
				E	1.027	1.027	0.983
6	5.2	3.1	0.959	N	3.411	1.024	0.982
				M	10.42	1.006	0.965
				H	0.641	1.032	0.990
				C	1.167	1.036	0.994
				E	1.024	1.024	0.982
9	8.0	4.7	0.964	N	3.385	1.016	0.979
				M	10.56	1.019	0.982
				H	0.637	1.026	0.989
				C	1.186	1.053	1.015
				E	1.027	1.027	0.990
12	11.0	6.4	0.970	N	3.349	1.005	0.975
				M	10.49	1.013	0.983
				H	0.628	1.011	0.981
				C	1.168	1.037	1.006
				E	1.021	1.021	0.990
15	14.0	7.6	0.974	N	3.356	1.007	0.981
				M	10.49	1.013	0.987
				H	0.627	1.010	0.983
				C	1.174	1.043	1.016
				E	1.018	1.018	0.992
18	17.0	12.3	0.986	N	3.306	0.992	0.978
				M	10.30	0.994	0.980
				H	0.617	0.994	0.980
				C	1.168	1.037	1.022
				E	1.001	1.001	0.987
22	20.6	18.4	0.994	N	3.303	0.991	0.985
				M	10.31	0.995	0.989
				H	0.616	0.992	0.986
				C	1.168	1.037	1.031
				E	1.005	1.005	0.999
X-MV					X		
4	...	...	1.000	N	3.332	...	...
				M	10.36	...	...
				H	0.621	...	...
				C	1.126	...	...
				E	1.000	...	...

Note: The values for $[P_{repl}]_{cyl}^{pp}$ in column 7 were obtained by dividing the electron charge ratio, $[M_E]_{pp}^{cyl}$, by the photon ratio, $[Mx]_{pp}^{cyl}$, according to Equation (3.147), for $X = 4$ MV. The last column is the product of values in columns 4 and 7. N = NACP; M = Markus; H = Holt; C = Capintec; E=Exradin.

Source: From Reference [77]. With permission.

where D is the percent depth dose and d is the depth at which the chamber central axis is located. P_{repl} values are shown in Table 3.20.

Ma and Nahum [80] investigated the displacement-effect corrections for ionization chambers calibrated in air in terms of air kerma and used at a depth in a phantom irradiated by medium energy x-ray beams for tube potentials between 100 and 300 kV. The Monte Carlo code system was used to simulate the coupled transport of photons and electrons.

The air kerma at a point, usually the center of the air cavity of the chamber on an air cavity axis, that has a volume defined by the outer dimensions of the ionization chamber and the stem in the water phantom can be written as:

$$K'_{air,u} = M_u N_K k_Q \qquad (3.149)$$

The air kerma at the depth of the chamber center in water in the absence of the chamber can be written as

$$K_{air,u} = K'_{air,u} p_{dis} = (M_u N_K) k_Q p_{dis} \quad (3.150)$$

where p_{dis} is the displacement correction factor that accounts for the effect of the displacement of water by the ionization chamber. Equation (3.150) defines p_{dis} as a ratio of air kermas.

The air kerma $K_{air,u}$ can be converted to the water kerma at the same point in water, $K_{w,u}$, through the ratio of the mass energy absorption coefficients for water to air averaged over the photon fluence at the same point in the undisturbed phantom, $(\bar{\mu}_{en}/\rho)_{w,air}$.

In order to obtain the displacement correction factor, p_{dis}, one must first calculate air kermas $K_{air,u}$ and $K'_{air,u}$.

$$K'_{w,u} = K'_{air,u}(\bar{\mu}_{en}/\rho)'_{w,air} \qquad (3.151)$$

and

$$K_{w,u} = K_{air,u}(\bar{\mu}_{en}/\rho)_{w,air} \qquad (3.152)$$

where $(\bar{\mu}_{en}/\rho)'_{w,air}$ and $(\bar{\mu}_{en}/\rho)_{w,air}$ are averaged over the photon fluence' and primary and scattered, respectively. In general, the difference between $(\bar{\mu}_{en}/\rho)'_{w,air}$ and $(\bar{\mu}_{en}/\rho)_{w,air}$ is very small and therefore, according to Equation (3.150), we have

$$p_{dis} = K_{air,u}/K'_{air,u} \approx K_{w,u}/K'_{w,u} \quad (3.153)$$

TABLE 3.20
P_{repl} factors for photon beams

		k (mm^{-1})	
Energy	a	b	c
^{60}Co	0.0040	0.0043	0.0030
5 MV	0.0040	0.0043	0.0026
8 MV	0.0031	0.0037	0.0022
16 MV	0.0024	0.0027	0.0020
42 MV	0.0025	0.0026	0.0016

Note: The measurement point was assumed to be at the central axis of chamber, located at a depth of 5 cm for ^{60}Co and 10 cm for other beams. $P_{repl} = 1 - kr$, where r is chamber radius in mm; *a*: from measured data of Johansson et al; *b*: from Equation (3.148), using BJR 17 depth-dose data; *c*: from AAPM TG-21 protocol.

Source: From Reference [78]. With permission.

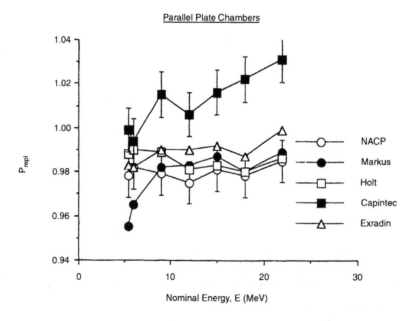

Parallel Plate Chambers

○ NACP
● Markus
□ Holt
■ Capintec
△ Exradin

FIGURE 3.72 Experimental values of $P_{repl,pp,E}$ for the five parallel-plate chambers studied in this work. Error bars are $\pm 1\%$. (From Reference [77]. With permission.)

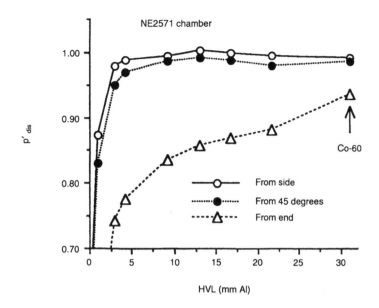

FIGURE 3.73 The displacement corrections for an NE2571 chamber, calculated using the simple attenuation and scattering method as ratios of the water kerma in a water cylinder of 7.4-mm diameter and 26-mm length, $K_{w,u}$, to the water kerma in a low-density (the same as that of air) water cavity, $K'_{w,u}$, for x-ray beams incident from either the side or the end of the water cylinder, or at a 45° angle (note that the quality of the ^{60}Co beam does not correspond to the HVL value). The statistical uncertainty in the calculated kerma ratios is about 0.2%. (From Reference [80]. With permission.)

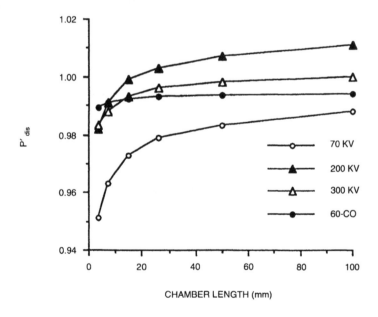

FIGURE 3.74 The variation of the displacement correction factor with chamber length for a chamber of 7.4-mm outer diameter, calculated using the simple attenuation and scattering method for 70, 200, and 300-kV photons and ^{60}Co gamma-rays incident from the side (90°). The statistical uncertainty in the calculated kerma ratios is about 0.1%. (From Reference [80]. With permission.)

Figure 3.73 gives the p'_{dis} values, calculated using the simple attenuation and scattering method for a stemless NE2571 chamber of 7.4-mm outer diameter and 26-mm length for photon beams incident from different angles. p'_{dis} varies with both photon energy and incident angle. It decreases rapidly with photon energy for low-energy photons.

Figure 3.74 shows the dependence of p'_{dis} on chamber length for photons incident at a 90° angle. At 200 kV, p'_{dis} is 1.011 for 100-mm length but decreases to 1.003 for the dimensions of an NE2571 chamber. For ^{60}Co gamma-rays, p'_{dis} increases with chamber length and stabilizes at about 0.994 ± 0.001 for chamber length greater than 50 mm. Further calculations show that for a 7.4-mm diameter and

7.4-mm long cylinder, p'_{dis} is 0.991, while p'_{dis} decreases to 0.988 if the beam is incident on the end wall of the cylinder (i.e., 0°) rather than on the side wall (i.e., 90°). For a cylinder of 7.4-mm diameter and 7.4-mm length, p'_{dis} values are smaller than unity for photon energies throughout the medium-energy range.

IX. IONIZATION CHAMBER CALIBRATION

Several methods for the calibration of plane-parallel chambers have been reported in the literature. These are shown in Figure 3.75 and can be described as follows. [81]

1. Calibration with high-energy electrons in phantom. The location of the point of measurement is taken as the center of the cylindrical chamber and the inner surface (of the wall that is proximal to the source) of the plane-parallel chamber, respectively. The point for each chamber is placed at d_{max} for the ionization curve, as determined by measurement for the electron beam being used, and the field size is measured at the phantom surface Figure 3.75a.

2. "In-air" calibration of a plane-parallel chamber using a ^{60}Co beam with a fixed source-to-detector distance (SDD) and field size (FS),

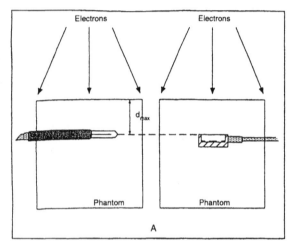

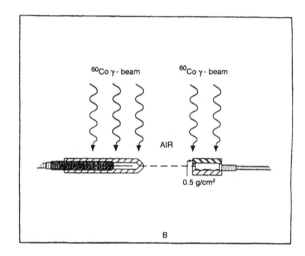

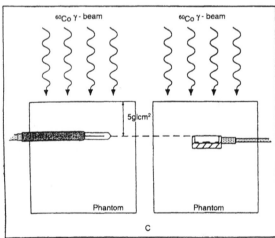

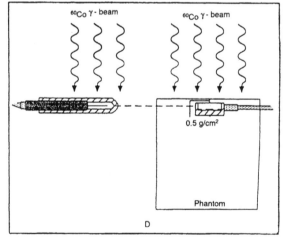

FIGURE 3.75 Schematic illustration of the irradiation geometries used for making calibrations of parallel-plate chambers with cylindrical chambers. (A) Calibration with 20-MeV electrons. The cylindrical chamber is aligned with the midpoint of its collecting volume, located on the beam axis at d_{max}, and its axis of rotation, perpendicular to the beam axis. The midpoint of the inner surface of the front wall of the plane-parallel chamber is located on the beam axis, and the chamber walls are perpendicular to that axis. (B) "In-air" calibration of a parallel-plate chamber using a ^{60}Co beam. Point of measurement taken as the chamber center. (C) Calibration "in phantom at depth," with a ^{60}Co beam. Both chambers are placed in phantoms at depth of 5 g/cm². Point of measurement taken as the center of the cylindrical chamber and the inner surface of the proximal electrode of the plane-parallel chamber. (D) Calibration "in phantom at dose maximum," with a ^{60}Co beam. The parallel-plate chamber was placed in the phantom with full backscatter at d_{max} (determined by measurement). Point of measurement taken as the chamber center. In all cases, the cylindrical chamber is shown on the left and the plane-parallel chamber is on the right. (From Reference [81]. With permission.)

TABLE 3.21

Recommended Values of $N_{gas}^{pp}/(N \times A_{ion})^{pp}$ and $N_{gas}^{pp}/(N_K \times A_{ion})^{pp}$ for Five Commercially Available Plane-Parallel Chambers

Chamber	Build-up[a] material or phantom material	A_{wall}	K_{comp} and P_{wall}^{pp}	$N_{wall}^{pp}/(N_x A_{ion})^{pp}$ (10^{-3} Gy/R)	$N_{gas}^{pp}/(N_K A_{ion})^{pp}$
Capintec PS-033	Polysty	1.0022	0.960	8.84	1.005
Exradin P-11	Polysty	1.0016	1.000	8.48	0.964
Holt	Polysty	1.0097	1.000	8.55	0.972
NACP	Graphite	1.0016	1.027	8.45	0.961
PTW-Markus	Acrylic	1.0030	1.000	8.59	0.977

Factors to be applied in the calibration and use of plane-parallel chambers.

[a] The build-up material of thickness 0.5 g/cm^2 is added to the front face of the chamber, except for the Holt in which the build-up material is inherent in the design.

Source: From Reference [81]. With permission.

measured at the SDD, with the point of measurement being taken as the chamber center, and with build-up cap required for both chambers (Figure 3.75b).

3. Calibration "in phantom at depth," with a ^{60}Co beam, both chambers being placed in phantoms at depth $d = 5$ g/cm^2. The point of measurement is taken as the center of the cylindrical chamber and at the inner surface (of the wall that is proximal to the source) of the plane-parallel chamber Figure 3.75c.

4. Calibration "in phantom at dose maximum," with a ^{60}Co beam. The plane-parallel chamber is placed in the phantom at d_{max} with full back-scatter (determined by measurement). The point of measurement is taken as the chamber center. The SDD and field size are the same as for the "in-air" method. The comparison is made with a cylindrical chamber irradiated in air (Figure 3.75d).

5. Method (d) is essentially the same as method (b), the two being related by the backscatter factor. There are only three independent methods, (a), (b), and (c), and these are described in this report; method (d) is not recommended. Method (a) employs high-energy electrons, method (b) makes use of a ^{60}Co gamma-ray beam with the chambers located at a point in air, and method (c) also employs a ^{60}Co beam but with the chambers placed at depth in a phantom.

N_{gas} for a plane-parallel chamber may be determined as follows. Using the highest electron-beam energy available and the cylindrical chamber for which N_{gas} is known, determine the response per monitor unit at d_{max}. Next, place the plane-parallel chamber into the same dosimetry phantom, taking care to position the inner surface of its proximal electrode at the depth of the central axis of the cylindrical chamber, and determine its response per monitor unit. The cavity gas calibration factor for the plane-parallel chamber is given by [82]

$$N_{gas}^{pp} = (MN_{gas}P_{ion}P_{repl})^{cyl}/(MP_{ion})^{pp} \qquad (3.154)$$

where the terms in the numerator apply to the cylindrical chamber and those in the denominator apply to the plane-parallel chamber. N_{gas} is the dose to the gas in the chamber per electrometer reading (Gy/C or Gy/scale division) M, P_{ion} is the factor that corrects for ionization recombination losses, and P_{repl} is the replacement correction factor. M is the electrometer reading (C or scale division), and it represents the measured average ionization for positive and negative polarities, in Coulombs or scale division, corrected to air density at 22°C, 760 mm Hg, but not corrected for relative humidity, assuming it to be typical of laboratory conditions (50% ± 25%).

The highest electron-beam energy available must be high enough to make the P_{repl} value for the cylindrical chamber no smaller than 0.98. For a typical Farmer-type cylindrical chamber of inner diameter 6.3 mm, an electron beam is required, with a mean energy of at least 10 MeV at d_{max}, the depth of the measurement required. The mean energy at depth is related approximately to the mean incident energy.

Electron-beam diameters should be large enough to provide complete in-scattering to the beam axis. Conservatively, this requires a beam diameter of twice the range R of the electrons in the phantom medium. Thus, a beam diameter (cm) numerically equal to the incident electron energy (MeV) is ample, assuming unit density.

The quantity N_{gas}^{cyl} in Equation (3.154) is derived from the NIST calibration value of N_x^{cyl} or N_K^{cyl} for the chamber.

Equation (3.155) relates N_{gas}^{cyl} to N_x^{cyl}, based on the conditions present during the NIST calibration of the chamber in a ^{60}Co gamma-ray beam:

$$N_{gas}^{cyl} = \frac{N_x^{cyl} k (W/e)_{gas} A_{ion} A_{wall} \beta_{wall}}{(\bar{L}/\rho)_{gas}^{wall} (\bar{\mu}_{en}/\rho)_{wall}^{air} K_{comp}^{cyl}} \quad (3.155)$$

where N_x^{cyl} is stated in R/C, R/scale division, Ckg^{-1}/C, or Ckg^{-1}/scale division; $k = 2.58 \times 10^{-4} \, CKg^{-1}R^{-1}$ or unity if exposure is stated in Ckg^{-1}; and $(W/e)_{gas} = 33.7$ J/C. A_{ion} is the ion-collection efficiency, at the time of calibration. A_{wall} is the correction factor for attenuation and scattering of gamma-rays in the chamber wall and buildup cap. β_{wall} is the quotient of absorbed dose by the collision kerma in the chamber wall; $\beta_{wall} = 1.005$ for ^{60}Co gamma-rays. $(L/\rho)_{gas}^{wall}$ is the mean restricted mass stopping power ratio for the chamber wall material, relative to the gas (ambient air) inside. $(\mu_{en}/\rho)_{wall}^{air}$ is the mean mass energy absorption coefficient ratio for dry air relative to the chamber wall material. K_{comp} corrects for the composite nature of the chamber and build-up cap. For a cylindrical chamber and buildup cap, it is given by [81]

$$K_{comp}^{cyl} = \frac{\alpha(\bar{L}/\rho)_{gas}^{wall}(\bar{\mu}_{en}/\rho)_{wall}^{air} + (1-\alpha)(\bar{L}/\rho)_{gas}^{cap}(\bar{\mu}_{en}/\rho)_{cap}^{air}}{(\bar{L}/\rho)_{gas}^{wall}(\bar{\mu}_{en}/\rho)_{wall}^{air}}$$

$$(3.156)$$

where α is the fraction of ionization due to electrons arising from photon interactions in the chamber wall and $(1-\alpha)$ is the fraction of ionization due to electrons arising from photon interactions in the build-up material.

If the NIST calibration of the cylindrical chamber is stated as an air-kerma calibration factor, N_K in Gy/C, the value of N_X, in R/C for use in Equation (3.155), can be obtained from

$$N_x = N_K(1-g)/(W/e)_{air} \quad (3.157)$$

where g is the average fraction of secondary electron kinetic energy that is spent in *bremsstrahlung* production. NIST takes g as 0.0032 and $(W/e)_{air}$ for dry air as 33.97 J/C for ^{60}Co gamma-rays. Thus, for these values, $N_x = 113.7N_K$. Recommended values of $N_{gas}/N_x A_{ion}$ for the cylindrical chambers commonly used can be found in Gastorf et al. [83] $N_{gas}/N_K A_{ion}$ values can be determined using Equation (3.157).

In the ^{60}Co in-air method, the intercomparison between the plane-parallel chamber and the NIST calibrated spherical or cylindrical chamber is performed in a ^{60}Co gamma-ray beam in air. An exposure or air-kerma calibration factor N_x^{pp} or N_K^{pp} will be obtained for the plane-parallel chamber by direct comparison with the spherical or cylindrical

chamber and N_{gas}^{pp} is obtained from given $N_{gas}^{pp}/(N_x A_{ion})^{pp}$ values. The following procedures should be followed: [81]

1. The plane-parallel chamber should be submitted to the ADCL for calibration with the necessary dose build-up material in place, unless the proximal wall is already thick enough, as is the case for the Holt chamber. The added build-up material should have the same outer diameter as the chamber and be of the material specific in Table 3.21; its thickness should be 0.5 g/cm^2 to ensure charged-particle equilibrium, exclude electro-contamination that may be present in the ^{60}Co beam, and match the calculated A_{wall} values given in Table 3.21.

2. The ^{60}Co beam should be 10 × 10 cm^2 at the chamber measurement location, at a distance of at least 80 cm from the source.

3. The correct alignment of the plane-parallel chamber for the intercomparison is with the *midpoint* of its ion-collecting volume located on the beam axis at the measurement location and its flat chamber walls perpendicular to that axis. This is the point of measurement for in-air calibrations only and is required by the ion chamber theory used to extract N_{gas}. Similarly, the use-provided build-up cap material must be in place for the in-air calibration only.

In the ^{60}Co in-phantom method, the intercomparison between the plane-parallel chamber and the NIST-calibrated cylindrical chamber is performed in a ^{60}Co gamma-ray beam at a depth of 5 g/cm^2 in a phantom of material selected to match that of the plane-parallel chamber. N_{gas}^{pp} is related to the known value of N_{gas}^{cyl} by

$$(M/A_{ion})^{pp}(N_{gas}P_{wall})^{pp} = (M/A_{ion})^{cyl}(N_{gas}P_{repl}P_{wall})^{cyl}$$

$$(3.157)$$

where M^{cyl} and M^{pp} are, respectively, the meter readings of the cylindrical and plane-parallel chambers under the above conditions, corrected for temperature and pressure, and M^{pp} is the average of each polarity. A_{ion} is the ion-collection efficiency at the time of calibration; N_{gas}^{cyl} is obtained from the NIST calibration value of N_x^{cyl} or N_K^{cyl}; P_{repl} is the correction factor for the replacement of phantom material by the cavity of the ionization chamber [P_{repl} is taken as unity in the 1983 AAPM protocol for plane-parallel chambers, so it does not appear on the left-hand side of Equation (3.157)]; and P_{wall} is the correction factor to account for the chamber material being different from that of the phantom.

P_{wall} for the plane-parallel chamber is difficult to determine. In the ideal case, where the plane-parallel chamber is made of only a single material that is identically matched by the phantom medium, P_{wall} equals unity. However, the construction of plane-parallel chambers usually employs more than one material. Thus, the phantom medium should be selected to match whichever material is the most important contributor of the secondary electrons that produce the measured ionization.

If the chamber has a very thin front wall (for which α is nearly zero), its influence can be ignored because practically all of the electrons then originate in the phantom medium. That medium should be selected to match the principal material of which the thicker rear wall is constructed. Since secondary electrons are projected preferentially in the forward hemisphere by ^{60}Co gamma-ray interactions, the electron backscattering ability of the rear wall (a strong function of atomic number) will be the most important influence on the ionization. If the chamber has a thicker front wall, for α which is significantly greater than zero, the phantom medium should be made of that same material. However, if the back wall differs appreciably in atomic number, this in-phantom procedure should not be expected to yield a satisfactory result, due to the difference in electron backscattering from the back wall in comparison with the phantom medium.

Once N_{gas}^{pp} has been calculated, the determination of absorbed dose for the user's beam will proceed according to TG-21. Since these chambers are designed primarily for use with electron beams, calibration of those beams only will be presented. When the chamber is placed in a suitable phantom (medium), the dose to the medium will be given by

$$D_{med} = M N_{gas}^{pp} (\bar{L}/\rho)_{gas}^{med} P_{ion} P_{repl} \qquad (3.158)$$

where M is the electrometer reading (Coulombs or scale division corrected to 22°C, 760 mm Hg). $(\bar{L}/\rho)_{gas}^{med}$ is the ratio of the mean, restricted collision mass stopping power of the phantom material to that of the chamber gas (ambient air). P_{ion} is the factor that corrects for ionization recombination losses that occur at the time of calibration of the user's electron beam.

In the 1983 AAPM protocol, the replacement correction factor P_{repl} is taken as unity for plane-parallel chambers, irrespective of electron-beam energy. Two of the commercial chambers (Capintec and Markus) show clear experimental evidence that P_{repl} decreases with decreasing electron energy. The Netherlands Commission on Radiation Dosimetry, in their Code of Practice of the Dosimetry of High-Energy Electron Beams, considered this situation for the Markus chambers and recommended the following equation for P_{repl} (designated p_f in their protocol):

$$P_{repl} = 1 - 0.041 e^{-0.4\bar{E}_z} \qquad (3.159)$$

For five commonly used parallel-plate ion chambers, Monte Carlo calculations of the wall attenuation and scatter correction factors ($K_{wall} = A_{wall}^{-1}$ or K_{att}^{-1}) and the correction for nonhomogeneous composition of the chamber (K_{comp} or related to k_m) were presented by Rogers. [84] The chambers were assumed to have 0.5 g/cm^2 build-up caps made of the predominant material in each chamber. These correction factors are needed if air-kerma calibration factors are used to deduce the chamber's cavity-gas calibration factor, N_{gas}. The scatter from the material around the cavity more than compensates for the attenuation in the front wall and, hence, K_{wall} values are less than unity. Thin collectors or insulators behind the collecting volume can have a major effect because electron backscattering depends strongly on material.

The AAPM's TG-21 dosimetry protocol recommends calibrating parallel-plate ion chambers by establishing the dose in a high-energy electron beam using a calibrated cylindrical ion chamber and then using this dose to establish the calibration of the parallel-plate chamber. By equating the dose measured at d_{max} in an electron beam using cylindrical and parallel-plate ion chambers, one finds by following TG-21 that

$$K_{wall} = N_{gas, cyl} \frac{M^{cyl} P_{repl}^{cyl} P_{wall}^{cyl}}{M^{pp} P_{repl}^{pp} P_{wall}^{pp}} (Gy/C) \qquad (3.160)$$

where the charge readings from ion chambers are represented by M and are understood to be corrected for temperature, pressure, ion recombination, and polarity effects.

If an electron beam is not to be used, there are two options: base the calibration on an in-air air-kerma calibration factor, $N_K [N_K = N_X(W/e_{air}/(1-\bar{g}))]$, or base the calibration on an in-phantom absorbed-dose calibration factor, N_D, or equivalent.

If the in-air route is chosen, N_{gas} is determined from

$$N_{gas} = \frac{N_K(1-\bar{g})}{(\bar{L}/\rho)_{air}^{wall} (\mu_{en}/\rho)_{wall}^{air} K_{wall} K_{comp} K_{ion}^c} (GyC^{-1}) \qquad (3.161)$$

where N_K is the air-kerma calibration factor, $\bar{g}$ is the fraction of the electron's energy lost by radiative processes, $K_{wall} (= A_{wall}^{-1})$ corrects for the lack of charged-particle equilibrium in the chamber walls (caused mostly by photon attenuation and scatter), K_{ion}^c corrects for any charge-collection losses in the calibration beam, and K_{comp} corrects for any components in the chamber made of materials different from the wall material.

If the calibration of the parallel-plate chamber is done starting from an absorbed-dose measurement in a phantom,

TABLE 3.22

Calculated Values of $K_{wall} = A_{att}^{-1} = k_{att}^{-1}, k_m$, and K_{comp} for the 5 Parallel-Plate Chambers, Each with a 0.5-g/cm^2 Build-Up Cap of the Chamber's Predominant Material Added to the Front Face (except for the Holt chamber)

Chamber	Cap Material	K_{wall}	$A_{wall} = k_{att}$	k_m	K_{comp}	$k_{att}k_m^c$
Capintec	polystnoy	0.9978(4)	1.0022(4)[a]	1.020(4)[a]	0.952(4)	1.022(4)
Exradin P[11b]	polystnoy	0.9984(3)	1.0016(3)	0.971(4)	1.000(4)	0.973(4)
Holt[b,d]	polystnoy	0.9904(4)	1.0097(4)	0.974(5)	0.997(5)	0.983(5)
Markus[b]	PMMA	0.9970(6)	1.0030(6)	0.982(4)	1.000(4)	0.985(4)
NACP	graphite	0.9984(2)	1.0016(2)	0.973(3)	1.027(3)	0.975(3)

Note: The collector and/or insulators were included in the calculations. All cases are for a point source with a ^{60}Co spectrum at a distance of 80 cm from the front face of the build-up cap. For in-air measurement, the point of measurement is the midpoint of the air cavity. One-standard-deviation statistical uncertainties in the last digit are shown in brackets. Estimated systematic uncertainties are 1% for k_m and K_{comp} and 0.05% for K_{wall} or A_{wall}.

[a] Results for the Capintec chamber using a radius of 1. 6 cm are 0.1% less, i.e., insignificantly different. This means the approximation of a square detector by a circular one has no effect on these results. Recall that stem effect corrections must be used with this detector.

[b] No insulator.

[c] Note: $N_{gas}/N_K = k_{att}k_m(1 - g)$ where $g = 0.003$. Alternatively, $N_{gas}/N_X = (W/e)k_{att}k_m$ [J/C] for N_X in C/kg or $N_{gas}/N_x = 8.764 \times 10^{-3} k_{att}k_m$ [Gy/R] for N_X in R/C.

[d] Holt chamber results are for a beam area of 100 cm^2.

Source: From Reference [84]. With permission.

one has [84]

$$N_{gas} = \frac{N_D}{[P_{wall}P_{repl}(\bar{L}/\rho)_{air}^{water}]^{60}Co}(Gy/C) \quad (3.162)$$

where N_D is the absorbed dose to the phantom material per unit charge from the parallel-plate chamber (corrected for ion recombination and polarity effects), P_{wall} corrects for the replacement of the phantom material by the material in the ion chamber, and P_{repl} corrects for the effects of the cavity on the electron fluence in the phantom at the point of measurement. This equation is the basis of many proposals and measurements related to determining N_{gas} for parallel-plate ion chambers, where N_D is usually determined using a calibrated cylindrical ion chamber in the same phantom. In most cases, the phantom material is the same as the material of the parallel-plate chamber and, hence, P_{wall} is taken as unity, as is P_{repl}. Note, however, that in a ^{60}Co beam for an ion chamber in a phantom or free in air with a build-up cap of the phantom material, P_{wall} and K_{comp} correct for the same physical effect. K_{comp} corrects the chamber response for the fact that not all of the chamber is made of the same material as the build-up cap.

Results for the standard configurations are shown in Table 3.22. The K_{wall} factors are all less than unity, which

means that photon scatter is more important than attenuation. With the exception of the Holt chamber, which comes in a slab-phantom, the chambers have a K_{wall} value within 0.1% of 0.998. This should be compared to a typical value of 1.010 for a cylindrical thimble chamber with a 0.5 g/cm^2 wall, in which case the photon attenuation dominates the scatter. The value of 0.9904 for the Holt chamber is 0.3% greater than the previous value published using this computer code. This may be attributed mainly to improved statistical precision (0.04% here vs. 0.2% previously) and somewhat (0.1%) to the fact that the current calculations are for a 100-cm^2 beam, whereas the previous ones were for a broad beam.

The effects of using different buildup caps have been investigated for the Markus, NACP, and Capintec chambers. The results in Table 3.23 show some results. Effects of the build-up caps on the calculated values of k_m for heterogeneous chambers do not follow what is expected from a naive theory, in which the k_m value of the chamber is assumed to behave as the k_m value of the cap. [84]

It was shown by Rogers [85] that basing clinical dosimetry on absorbed-dose calibration factors N_D leads to considerable simplification and reduced uncertainty in dose measurement. A quantity k_Q is defined which relates an absorbed-dose calibration factor in a beam of quality Q_0 to that in a beam of quality Q. For 38 cylindrical ion chambers, two sets of values were presented for N_D/N_X

TABLE 3.23
Calculated Effect on k_m and K_{wall} of Changing the Material of the 0.5-g/cm² Build-Up Cap for the NACP, Markus, and Capintec Chambers

Chamber	k_m	K_{wall}	$k_{att}k_m$
Capintec			
polyst cap	1.020(4)	0.9978(4)	1.022(4)
PMMA cap	1.008(3)	0.9981(4)	1.010(3)
graphite cap	1.018(3)	0.9964(4)	1.022(3)
Markus			
polyst cap	1.000(4)	0.9972(4)	1.003(4)
PMMA cap	0.982(4)	0.9970(6)	0.985(4)
graphite cap	0.984(3)	0.9949(4)	0.989(3)
NACP			
polyst cap	0.984(3)	1.0003(3)	0.982(3)
PMMA cap	0.977(5)	1.0007(3)	0.976(5)
graphite cap	0.974(3)	0.9984(2)	0.975(3)

Source: From Reference [84]. With permission.

and N_{gas}/N_D, and for k_Q for photon beams with beam quality specified by the TPR_{10}^{20} ratio. One set is based on TG-21's protocol to allow the new formalism to be used while maintaining equivalence to the TG-21 protocol. To demonstrate the magnitude of the overall error in the TG-21 protocol, the other set uses corrected versions of the TG-21 equations and the more consistent physical data of the IAEA Code of Practice.

Given N_D^Q, the absorbed dose-to-water calibration factor for an ion chamber in a beam of quality Q, and under reference conditions of field size and depth:

$$D_{water}^Q = MP_{ion}N_D^Q \qquad (3.163)$$

where D_{water}^Q is the absorbed dose to water at the location of the center of the ion chamber when it is absent; the ion chamber reading M has been corrected to reference conditions of temperature and pressure; and P_{ion} corrects for lack of complete charge collection in the user's beam and must be measured for each beam quality. This definition of N_D^Q is analogous to the definition of the exposure calibration factor N_X given by $X = MN_x$, where X is the exposure at the location of the center of the ion chamber (except that in the absorbed-dose equation, the P_{ion} correction is made explicit because its effects are not negligible, unlike the normal case for exposure measurements). The following dose equation from the AAPM protocol also applies:

$$D_{water}^Q = MP_{ion}N_{gas}[P_{wall}P_{repl}(\bar{L}/\rho)_{air}^{water}]_Q \qquad (3.164)$$

In the case where the absorbed-dose calibration factor N_D^Q is known in some other beam quality Q_0, let k_Q be

a factor which accounts for changes in the calibration factor between beam quality Q_0 and Q; i.e.,

$$N_D^Q = k_Q N_D^{Q_0} \qquad (Gy/c) \qquad (3.165)$$

and from Equation (3.163):

$$D_{water}^Q = MP_{ion}k_Q N_D^{Q_0} \qquad (Gy) \qquad (3.166)$$

Equations (3.163) and (3.164) give

$$N_D^Q = N_{gas}[P_{wall}P_{repl}^0(\bar{L}/\rho)_{air}^{water}]_Q \qquad (Gy/c) \qquad (3.167)$$

Eliminating N_{gas} from this equation by using the analogous equation for a beam of quality Q_0 allows one to solve for k_Q:

$$k_Q = \frac{[P_{wall}P_{repl}(\bar{L}/\rho)_{air}^{water}]_Q}{[P_{wall}P_{repl}(\bar{L}/\rho)_{air}^{water}]_{Q_0}} \qquad (3.168)$$

The quantity k_Q relates the absorbed-dose calibration factor in a beam of quality Q to that for a beam of quality Q_0 in the same medium. From the user's point of view, this is the simplest possible approach since it uses a correction factor k_Q which is conceptually very straightforward yet takes into account all of the physics.

A computer program, called PROT, has been written to calculate k_Q and other protocol-related quantities for photon beams as a function of beam quality Q for an arbitrary ion chamber in a phantom of water, PMMA, or polystyrene. The program calculates k_Q using the equations and data of the original TG-21 protocol or using the TG-21 equations with the more consistent physical data of the IAEA Code.

Conversion from N_X to N_D (in a ^{60}Co beam) gives

$$\frac{N_D}{N_X} = \frac{N_{gas}}{N_X}[P_{wall}P_{repl}(\bar{L}/\rho)_{air}^{water}]_{^{60}Co} \qquad (Gy/R) \qquad (3.169)$$

where N_X is in R/C [divide N_D/NX by 2.58×10^{-4} (C/kg)/R to get the result in Gy/(C/kg), for use with N_X kg^{-1} when using the SI unit for exposure C/kg]. The k_Q formalism can be applied to cases of electron beams as well as photon beams. Conceptually, it loses its simplicity somewhat because different quantities are involved in the ratio for k_Q for electron and photon beams and because k_Q becomes a function of both the beam quality Q and the depth of dose maximum.

To investigate the relative merits of using an absorbed dose calibration–based approach vs. an in-air calibration factor–based approach, one must first define the optimal approach based on the in-air calibration factors. This can

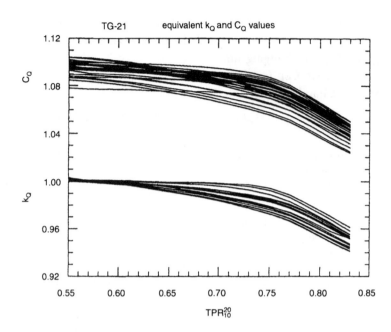

FIGURE 3.76 Plots vs. TPR_{10}^{20} of TG-21 equivalent values of k_Q and C_Q values for the chambers discussed by Rogers. There is a reduction in chamber-to-chamber variation obtained by using the k_Q approach, although the functional shape for a given detector is the same in each case. (From Reference [85]. With permission.)

best be done by starting from an air-kerma calibration factor $N_K[= N_X(W/e)_{air}/(1-\bar{g})]$ and defining a single factor, C_Q, to relate the ion-chamber readings in phantom M to the dose to the medium; i.e., C_Q is defined such that:

$$D_{med} = M P_{ion} N_K C_Q \quad \text{(Gy)} \quad (3.170)$$

This is usually accompanied by requiring use of one of a few specified types of ion chambers. Using Equation (3.164) for the dose to the medium and substituting the relationship between N_K and N_X gives:

$$C_Q = \frac{(1-\bar{g})[P_{wall} P_{repl}(\bar{L}/\rho)_{air}^{med}]_Q}{[(\bar{L}/\rho)_{air}^{wall}(\mu_{en}/\rho)_{wall}^{air} K_{wall} K_{comp} K_{ion}^C]_{^{60}Co}} \quad (3.171)$$

One of the advantages of using this approach rather than the N_{gas} and N_X approach is that C_Q is within about 10% of unity.

The C_Q approach was introduced to make use of an in-air calibration as simple as possible and analogous to the k_Q procedure in practical terms. This bypasses the equivalent, now traditional approach of the AAPM TG-21 protocol where Equation (3.170) can be written:

$$D_{med} = M P_{ion} N_{gas} [P_{wall} P_{repl}(\bar{L}/\rho)_{air}^{med}]_Q \quad \text{(Gy)} \quad (3.172)$$

with a corrected equation for N_{gas} given by

$$N_{gas} = \frac{N_K(1-\bar{g})}{(\bar{L}/\rho)_{air}^{wall}(\bar{\mu}_{en}/\rho)_{wall}^{air} K K_{ion}^C} \quad \text{(Gyc}^{-1}) \quad (3.173)$$

where N_{gas} has the standard meaning but has been written in terms of the air-kerma calibration factor. In using this approach, the complexities of using an in-air calibration factor are handled within the factor N_{gas}.

Figure 3.77 shows the size of the differences between the AAPM value of P_{repl} and the effective value of P_{repl} used by the IAEA Code for a cavity diameter typical of a Farmer-like chamber. In photon beams, this is the only difference between the physics in the AAPM TG-21 protocol and IAEA Code of Practice (ignoring aluminum electrode differences, after corrections, and using the same data sets).

Note that the IAEA's effective P_{repl} for ^{60}Co is not on the curve for the TPR_{10}^{20} values commonly found for ^{60}Co beams because the offset for ^{60}Co beams is different than for all other beams. The net effect is that k_Q values for the IAEA Code do not go to unity TPR_{10}^{20} for values near 0.57, which is a typical value for a ^{60}Co beam. [85]

Figure 3.78 presents a comparison of various P_{repl}, correction factors as a function of cavity diameter in a ^{60}Co beam. The upper curve is that from the TG-21 protocol. The long dashed and solid curves are those given by the IAEA offsets in the effective point of measurement, as defined explicitly for a ^{60}Co beam (long dash) or assuming a general beam with TPR_{10}^{20} of 0.57 (solid). The final dashed-dot curve represents the values implied by the original data of Johanson et al. [72], on which the IAEA's offsets are based. These are significant differences which deserve further research.

Figure 3.79 presents k_Q curves for a more restricted situation, namely for Farmer-like chambers. In this case

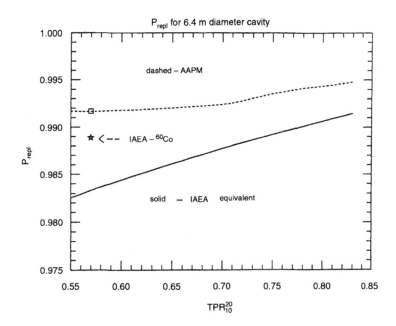

FIGURE 3.77 Comparison of the AAPM TG-21 values of P_{repl} and the value of the equivalent correction in the IAEA Code of Practice for a 6.4-mm diameter Farmer-like chamber as a function of beam quality, specified as TPR_{10}^{20}. The symbols show the values specified for ^{60}Co beams. (From Reference [85]. With permission.)

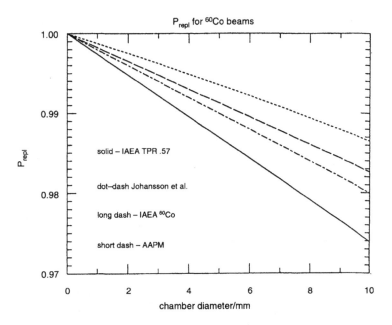

FIGURE 3.78 Comparison, as a function of chamber diameter, of the effective values of the replacement correction factor given by a variety of approaches. The upper short-dash curve is that given by TG-21. The long-dash curve is equivalent to the offset recommended by the IAEA for a ^{60}Co beam, whereas the solid curve is the same quantity for a beam with a TPR_{10}^{20} value of 0.57. The dot-dash curve is based directly on the data of Johansson *et al.*, upon which the IAEA approach was based. (From Reference [85]. With permission.)

there is only a ±0.7% variation about a "neutral" median value. This is given accurately by any of the PMMA-walled Farmer chambers, which are what make up the dark central group in Figure 3.79. Alternatively, the formula

$$K_Q(Q) = 1.0 - 0.0215(TPR - 0.57)$$
$$- 0.736(TPR - 0.57)^2 \qquad (3.174)$$

reproduces the data for a water-walled (i.e., wall-less) chamber with a diameter of 6.4 mm to within 0.25% and is within 1% of all the IAEA-based data.

Using the protocol of the American Association of Physicists in Medicine (AAPM) for the determination of absorbed dose from high energy, the calibration factor N_{gas} for the PTW M23343 plane-parallel ionization chamber has been determined by Murali et al. [86] in ^{60}Co and high-energy electron beams. Two cylindrical chambers of different wall materials and known values of N_{gas} have been used as reference chambers. Results show that the N_{gas} value for the plane-parallel chamber determined

in the ^{60}Co beam is about 2% higher than the value obtained in the high-energy electron beams. The discrepancy is attributed to not taking into account a wall-correction factor for the plane-parallel chamber while calibrating in the ^{60}Co beam. N_{gas} is calculated using the relation

$$N_{gas} = (MN_{gas}P_{ion}P_{repl}P_{wall})^{C}/(MP_{ion}P_{repl}P_{wall})^{P}$$

(3.175)

where c represents a cylindrical chamber; p represents a plane-parallel chamber; M is the meter reading; P_{ion} the correction factor for the ion recombination; P_{repl} is the fluence correction factor (which depends on the dimensions of the chamber); and p_{wall} is the correction factor for the non-air equivalence of the wall material. P_{wall} and P_{repl} for the plane-parallel chamber are assumed to be unity (AAPM 1983).

The results for the N_{gas} values obtained with ^{60}Co and the electron beams are presented in Table 3.24. It is seen that the N_{gas} value obtained with ^{60}Co is 2.16% higher than that obtained with the electron beams. The increase can be recognized in both sets of measurements.

A set of dosimetric measurements was made by Huq et al. [87] with Farmer-type PTW and Capintec ionization chambers in solid water, PMMA, and polystyrene phantoms and exposed to a 4-MV photon beam from a Varian Clinac 4S at Yale, a 10-MV photon beam, 6-and 15-MeV electron beams, and a 25-MV photon beam. The results are as follow.

TABLE 3.24
N_{gas} Values

	Set I with PTW (Gy C⁻¹)	Set II with NE (Gy C⁻¹)
^{60}Co	0.4817	0.4812
at $d = 5$ cm	(2.16)	(2.06)
Electrons (15 MeV)	0.4707	0.4710
$d_{max} = 2.2$ cm	(−0.17)	(−0.11)
Electrons (20 MeV)	0.4721	0.4722
$d_{max} = 2.2$ cm	(0.13)	(0.15)

Note: Values in parentheses show the percentage difference of the N_{gas} value from the mean value (0.4715 Gy C⁻¹) obtained with electron beams.

Source: From Reference [86]. With permission.

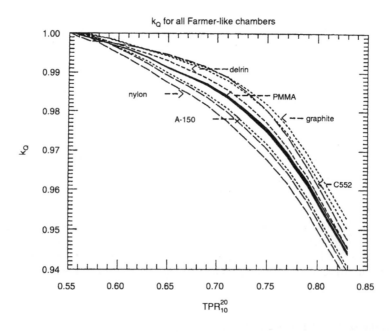

FIGURE 3.79 Calculated k_Q factors for all 17 Farmer-like chambers discussed by Rogers. The IAEA data set and AAPM values of P_{repl} were used in the calculations. Chambers of the same material are shown by the same type of curve. (From Reference [85]. With permission.)

TABLE 3.25
Absorbed Dose Intercomparisons for 4, 10, and 25-MV Photon Beams in a Solid Water Phantom

Chamber	TPR_{10}^{20}	$\dfrac{M_u}{M}$	$\dfrac{(s_{w,air})_u}{(\bar{L}/\rho)_{air}^{water}}$	$\dfrac{p_u}{P_{wall}}$	$\dfrac{\%DD_{P\,eff}^{P}}{P_{repl}}$	$\dfrac{D_w^{IAEA}(d)}{D_w^{AAPM}(d)}$
PTW	0.63	0.988(0.99)[a]	0.995	1.000	0.996	0.996(0.99)
Capintec	0.63	0.987(0.99)[a]	0.995	0.996	0.994	0.991(0.99)
PTW	0.73	0.992	0.989	1.001	0.998	0.993
Capintec	0.73	0.991	0.989	0.998	0.997	0.990
PTW	0.78	0.995(0.995)[a]	0.989	1.001	1.000	0.995(0.99)
Capintec	0.78	0.992(0.99)[a]	0.989	0.999	0.997	0.991(0.98)

Average 0.993 ± 0.002(0.994)

Note: N_D/N_{gas} has a value of 1.005 for the PTW chamber and 1.006 for the Capintec chamber, and P_{cel} has a value of unity for both chambers. The 4, 10, and 25-MV data are shown in rows having TPR_{10}^{20} values equal to 0.63, 0.73, and 0.78, respectively. [87]

[a] Values given in parentheses are from theoretical intercomparison in Reference [22] of Chapter 2.

Source: From Reference [87]. With permission.

The ratio of the electrometer readings corrected for ionization-collection efficiency is equal to the ratio of percent depth doses at $d_o + fr$ and d_o, i.e.,

$$\frac{M_u}{M} = \frac{\%DD(d_0 + f_r)}{\%DD(d_0)} \qquad (3.176)$$

where M_u, and M are electrometer readings corrected for ionization collection efficiency at $d_0 + fr$ and d_0, respectively. d_o is a reference depth, f is a displacement factor ($f = 0.5$ for electrons and 0.75 for protons), and r is the inner radius of the chapter. The above ratios for both photons and electrons were obtained from dosimetry data in clinical use.

The ion chambers used for dose intercomparisons were all 0.6-cm^3 Farmer-type chambers having different wall, central electrode, and build-up cap materials.

For electron beams, all measurements were taken with a source-surface distance (SSD) of 100 cm and a field size of 10×10 cm at the phantom surface.

The basic equations that are used by both protocols for the determination of absorbed dose to water are as follows:

1. AAPM protocol
   • photons,

$$D^{water}(d_0) = M N_{gas}(\bar{L}/\rho)_{gas}^{plastic} P_{ion} P_{wall} P_{repl}$$
$$\times (\bar{\mu}_{en}/\rho)_{plastic}^{water} ESC \qquad (3.177)$$

   • electrons,

$$D^{water}(d_0) = M N_{gas}(\bar{L}/\rho)_{gas}^{plastic} P_{ion} P_{wall} P_{repl}$$
$$\times (\bar{S}/\rho)_{plastic}^{water} \phi_{plastic}^{water} \qquad (3.178)$$

2. IAEA protocol
   • photons and electrons,

$$D_w(d_0) = M_u N_D (S_{w,air})_u P_u P_{cel} h_m \qquad (3.179)$$

The results of the IAEA absorbed dose to solid water are compared with those of AAPM in Table 3.25.

Results of absorbed doses to solid water for 6-and 15-MeV electron beams are given in Table 3.26 for a PTW chamber. Measurements were performed by using both chambers in solid water, PMMA, and polystyrene phantoms.

N_{gas} was determined by Reft et al. [88] for six commercially available parallel-plate ionization chambers using in-phantom and an in-air ^{60}Co irradiation and an in-phantom irradiation with a high-energy electron beam. The chamber characteristics for the six parallel-plate chambers and three cylindrical chambers are listed in Tables 3.27 and 3.28, respectively. All parallel-plate chambers used were designed to produce minimal perturbation in the electron field when placed in a homogeneous medium.

A 22-MeV electron beam with a 15×15-cm^2 (0.15×0.15-m^2) field at a source to surface distance of 1 m was used for all the electron irradiations. The cylindrical chamber was placed with the axis of symmetry at d_{max} and perpendicular to the beam, and the parallel-plate chamber was positioned perpendicular to the beam axis, with the center of the inner surface of the front wall at d_{max}.

For the same exposure in air from the ^{60}Co beam for a pair of plane-parallel and cylindrical chambers at the same point of measurement, one can write

$$M_{air,pp,Co}N_{x,pp} = M_{air,cyl,Co}N_{x,cyl} \qquad (3.180)$$

where the measured charge is corrected for temperature, pressure, polarity effect, and ion-collection efficiency. In the revised notation of the AAPM protocol by Rogers, and the letter of clarification by Schulz et al. N_{gas} and N_x are related by

$$N_x(W/e) = N_{gas} K_{wall} K_{comp} K_{ion} \overline{(L/\rho)}_{air}^{wall} (\bar{\mu}_{en}/\rho)_{wall}^{air}$$
$$(3.181)$$

TABLE 3.26
Absorbed Dose Intercomparisons in Electron Beams (PTW Chamber)

Energy (MeV)	Phantom	$\dfrac{(S_{w,air})_u}{(\bar{L}/\rho)_{air}^{med} \times (\bar{S}/\rho)_{med}^{water}}$	$\dfrac{P_u}{P_{repl}}$	$\dfrac{M_u}{M}$	$\dfrac{h_m}{\phi_{med}^{water}}$	$\dfrac{D_w^{IAEA}(d_0)}{D_w^{AAPM}(d_0)}$
6	Solid water	0.996	0.998	0.990	1.000	0.997 (0.987)[a]
15	Solid water	0.999	0.999	1.000	1.000	1.010 (1.005)[a]
6	PMMA	0.993	0.998	0.994	1.000	0.999 (0.989)[a]
15	PMMA	0.998	0.999	1.000	1.000	1.010 (1.007)[a]
6	Poly	0.993	0.998	0.993	0.977	0.973 (0.965)[a]
15	Poly	0.994	1.000	1.000	0.999	1.005 (1.007)[a]

Average 0.999 ± 0.014 ($0.993 \pm L0.017$)

Note: For this chamber P_{cel} has a value of 1.008 and N_D/N_{gas} as has a value of 1.005.

[a] Values given in parentheses are from the theoretical intercomparison.

Source: From Reference [87]. With permission.

TABLE 3.27
Physical Characteristics of the Parallel-Plate Chambers Used by Reft et al.

Chamber Type	Model	Guard Width (10^{-3} m)	Sensitive air-volume Radius (10^{-3} m)	Sensitive air-volume Depth (10^{-3} m)	Entrance Window Material	Entrance Window Thickness (kg/m²)	Buildup Cap Radius (10^{-3} m)	Buildup Cap Thickness (kg/m²)	Buildup Cap Material
NACP	...	3.0	5.0	2.0	Graphite	0.900	15	4.10	Graphite
Markus	30–329	0.2	2.7	2.0	Polyethylene	0.023	15	5.00	Acrylic
Holt	30–404	5.0	12.5	2.0	Polystyrene (Carbon-coated)	4.200	...	...	...
Capintec	PS-033	0.5	8.1	2.4	Mylar (Aluminized)	0.005		5.00	Polystyrene
Exradin	P-11	5.1	10.0	2.0	Polystyrene-equiv. (Conducting plastic)	1.180	22.5	4.00	Polystyrene
Attix	449	13.5	6.35	1.0	Conductive Kapton	0.048	30.0	4.95	Solid Water (RMI)

Source: From Reference [88]. With permission.

for any one chamber, where K_{wall} corrects for the lack of charged particle equilibrium in the chamber walls, K_{ion} corrects for ion-collection efficiency, K_{comp} corrects for any compositional material differences between the chamber wall and build-up cap, $\bar{L}/\rho$ is the mean restricted collision mass stopping power, $\bar{\mu}_{en}/\rho$ is the mean mass energy absorption coefficient, and W/e is the mean energy expended per unit charge in dry air. Rearranging Equation (3.181) and using Equation (3.180),

$$N_{gas,pp} = (W/e)N_{x,cyl}[M_{air,Co}]_{pp}^{cyl}[(\overline{L/\rho})_{wall}^{air}(\bar{\mu}_{en}/\rho)_{air}^{wall}]_{pp}/$$
$$[K_{wall}K_{comp}K_{ion}]_{pp} \qquad (3.182)$$

If the product of terms $K_{wall}K_{comp}$ for the parallel-plate chambers is known, $N_{gas,pp}$ can be obtained from the ratio of charge measurements and dosimetric quantities given in TG-21.

The parallel-plate and cylindrical chamber of like material were positioned in the phantom with their respective points of measurement at a depth of 5.0 g/cm² (50 kg/m²) and exposed to a ^{60}Co beam under identical irradiation conditions. A 10×10-cm² field defined at a source-to-detector distance of 1 m was used for the exposures.

Results for the determination of $N_{gas,pp}$ using the three calibration procedures are presented in Table 3.29.

TABLE 3.28

Physical Characteristics of the Cylindrical Chambers Used by Reft et al.

Chamber Type Model	Inner Radius (10^{-2} m)	Inner Axial length (10^{-2} m)	Entrance Window Thickness Material	Entrance Window Thickness (kg/m²)	Build-up cap Thickness Material	Build-up cap Thickness (kg/m²)	N_{gas} 10^7 Gy/C
Farmer 30–35[a] PTW	0.305	2.20	Acrylic	0.59	Acrylic	5.30	4.57
Farmer N30–351[b] PTW	0.304	2.20	Graphite	0.96	Graphite	4.89	4.62
Farmer 2571[c] NEL	0.315	2.25	Graphite	0.65	Graphite	4.89	3.99

[a,b] Farmer-type chamber available from Nuclear Associates, Carle Place, NY 11514.

[c] Farmer-type chamber available from Nuclear Enterprises America, Fairfield, NJ 07006.

Source: From Reference [88]. With permission.

TABLE 3.29

Summary of Results for $N_{gas,pp}$ ($\times$ 10^7 Gy/C) Using the Three Different Calibration Methods

Chamber	22E $[M]_{pp}^{cyl}$	22E $N_{gas,pp}$	⁶⁰Co (Phantom) $[M]_{pp}^{cyl}$	⁶⁰Co (Phantom) $P_{wall,pp}$	⁶⁰Co (Phantom) $N_{gas,pp}$	⁶⁰Co (Air) $[M]_{pp}^{cyl}$	⁶⁰Co (Air) $K_{wall}K_{comp}$	⁶⁰Co (Air) $N_{gas,pp}$	$\langle N_{gas} \rangle$
NACP	2.9900	13.530	3.0176	1.027	13.380	3.0000	1.0250	13.49	13.5 ± 0.1
Markus	10.280	46.700	10.3140	1.000	46.920	1.0340	0.9970	47.25	47.0 ± 0.3
Holt	0.6157	2.797	0.6258	1.000	2.807	0.6199	0.9904	2.821	2.81 ± 0.01
Capintec	1.2120	5.506	1.1719	0.960	5.475	1.1640	0.9579	5.476	5.49 ± 0.02
Exradin	0.9326	4.236	0.9518	1.000	4.269	0.9400	0.9984	4.243	4.25 ± 0.02
Attix	5.1680	23.480	5.2790	1.015	23.640	0.5223	1.0060	23.61	23.6 ± 0.1

Source: From Reference [88]. With permission.

The question whether results of measurements made in different phantom materials using different ionization chambers are consistent with each other was investigated by Huq et al. [89] Two 0.6-cc Farmer-type chambers, a PTW N23333, NA 30-351, NA 30-352, and a Capintec PR 06G chamber were used. They consisted of very different walls, build-up caps, and central electrode materials. The wall and build-up cap material of the PTW chamber was made of PMMA, and the central electrode was made of aluminum. By contrast, the Capintec chamber had an air-equivalent wall, polystyrene build-up cap, and an air-equivalent plastic central electrode.

Absorbed dose at a point in a water phantom was determined from measurements in a nonwater (or water) phantom using the following expression:

$$D_{wat}(d_{wat}) = N_{gas}Q_{corr}(d_{med})(\bar{L}/\rho)_{air}^{med}(\bar{S}/\rho)_{med}^{wat}\phi_{med}^{wat}P_{repl}$$

(3.183)

where $D_{wat}(d_{wat})$ is the dose in water at a depth d_{wat}; N_{gas} is the dose to the gas in the chamber per unit electrometer reading corrected for temperature, pressure, and

ion collection efficiency; $Q_{corr}(d_{med})$ is the ion-chamber reading corrected for polarity and ion recombination at the effective depth d_{med} in the medium; $(\bar{L}/\rho)_{air}^{med}$ is the ratio of mean restricted collision stopping powers of *med* to *air*; $(\bar{S}/\rho)_{med}^{wat}$ is the ratio of average unrestricted collisional stopping powers of *wat* to *med*; ϕ_{med}^{wat} is the difference in electron fluence between nonwater and water phantoms at water equivalent depths; and P_{repl} is a replacement correction that accounts for the change in photon and electron-beam fluence that occurs because of the replacement of phantom material by the air cavity.

If measurements are made in a nonwater-equivalent phantom, then the protocol recommends that the depth in the nonwater-equivalent phantom be scaled to that of the water-equivalent phantom by the following expression:

$$d_{water} = d_{med}\rho_{eff} = d_{med}\frac{R_{50}^{wat}}{R_{50}^{med}}$$

(3.184)

If cylindrical chambers are used for depth-ionization measurements, then replacement corrections P_{repl} are needed to convert depth-ionization data to depth-dose data.

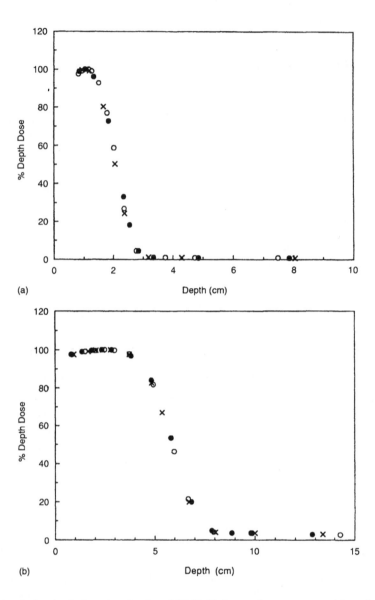

FIGURE 3.80 Central axis relative depth-dose data for 6- and 15-MeV electron beams measured by using a PTW chamber in solid water, PMMA, and clear polystyrene phantoms; (a) 6-MeV and (b) 15-MeV electron beams. Solid circles are data for solid water; open circles are for clear polystyrene; and crosses are for PMMA phantom. (From Reference [89]. With permission.)

Figure 3.80 shows the relative depth-dose data for 6- and 15-MeV electron beams generated by using a PTW chamber in solid water, PMMA, and clear polystyrene phantoms, using TG-25 except for the *bremsstrahlung* tail section of the curves. Similar data for the Capintec chamber are given in Figure 3.81. The figures clearly show that the depth-dose data measured in a solid water phantom are essentially indistinguishable from those obtained from measurements made in the PMMA and clear polystyrene phantoms at both energies. Thus, if one measures depth ionization in a solid water, PMMA, or clear polystyrene phantom and follows the recommendations of the TG-25 protocol to convert relative depth ionization to relative depth dose, then no significant differences will be observed between the relative depth doses obtained from

measurements made in the solid water, PMMA, and clear polystyrene phantom.

The AAPM TG-39 protocol has proposed three different methods of calibrating plane-parallel ionization chambers, i.e., in-phantom irradiation with a high-energy electron beam and in-phantom and in-air ^{60}Co irradiation. To verify the consistency of the three methods, Almond et al. [90] have measured N_{gas}^{pp} values using each of these techniques for the five most commonly used plane-parallel chambers considered by the protocol. Their results demonstrate that the measured N_{gas}^{pp} values for the three different methods for any of the chambers agree to within ±0.6%.

The five plane-parallel chambers considered by the AAPM TG-39 protocol were used and designated as

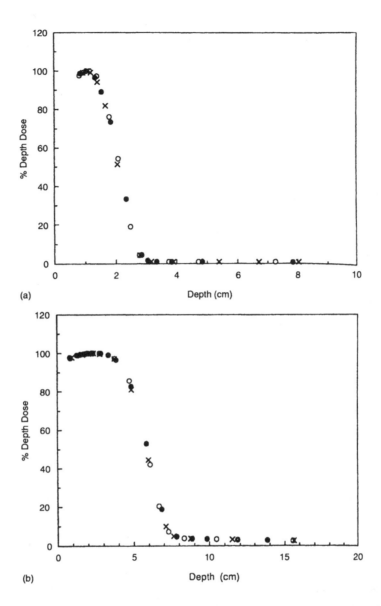

FIGURE 3.81 Central axis relative depth-dose data for 6- and 15-MeV electron beams measured by using a Capintec chamber in solid water, PMMA, and clear polystyrene phantoms; (a) 6-MeV and (b) 15-MeV electron beams. Solid circles are data for solid water; open circles are for clear polystyrene; and crosses are for PMMA phantom. (From Reference [89]. With permission.)

"Capintec," "Exradin," "Holt," "NACP," and "PTW-Markus." Their characteristics and individual parameters are listed in Table 3.30.

The comparisons were made against a PTW graphite cylindrical Farmer-type chamber, number N30001G, with an ADCL exposure factor $N_x = 5.455(R/nC)$ and a cavity-gas calibration factor $N_{gas}^{cyl} = 4.653(R/nC)$, which was calculated by using $N_{gas}^{cy}/(N_x A_{ion})^{cyl} = 8.53 \times 10^{-3}$ The experimental setup is schematically shown in Figure 3.82.

Intercomparison between the plane-parallel chamber and the PTW Farmer cylindrical chamber was performed in a ^{60}Co beam in air using a 10×10-cm^2 (0.10×0.10 m^2) field defined at 0.80 m. The experimental setup is shown in Figure 3.83. Under these conditions, the exposure calibration factor for the plane-parallel chamber is given by

$$N_x^{pp} = \frac{N_x^{cyl} M^{cyl}}{M^{pp}} \qquad (3.185)$$

where M^{cyl} and M^{pp} are, respectively, the meter readings of the cylindrical and plane-parallel chambers, corrected for temperature, pressure, and polarity; N_x^{cyl} and N_x^{pp}, are, respectively, the exposure calibration factors for ^{60}Co radiation of the cylindrical and plane-parallel chambers. N_{gas}^{pp} was then calculated from the ratios of $N_{gas}^{pp} = (N_x A_{ion})^{pp}$, given by the AAPM TG-39 protocol.

The intercomparison between the plane-parallel chamber and the PTW Farmer chamber was performed in

TABLE 3.30

Physical Characteristics of Plane-Parallel Ionization Chambers Used by Almond et al.

Chamber	Wall: Material Thickness (cm)/(mg/cm²)	Collector: Material/ Diameter (cm)	Guard Ring Radial Width (cm)	Electrode Spacing (cm)	Nominal Volume (cm³)	Recommended Build-Up/Thickness (cm)/dia.(cm)
Capintec	Aluminized polyethylene 0.001/0.5	Conducting air-equivalent plastic 1.6	Conducting air-equivalent plastic 0.05	0.24	0.5	Polystyrene/ 0.48/3.0
Exradin	Conducting Polystyrene 0.1/104	Conducting polystyrene 2.0	Conducting, polystyrene 0.5	0.2	0.6	Polystyrene/ 0.48/4.5
Holt	Polystyrene 0.4/418	Graphite polystyrene 2.5	Graphite polystyrene 0.5	0.2	1.0	No additional
NACP	Graphite and Mylar 0.5 and 0.008/93	Graphite/1.0	Graphite/0.3	0.2	0.16	Graphite/0.3/3.0
PTW-Markus	Graphite Polyethylene 0.0025/2.3	Graphite PMMA 0.54	Graphite PMMA 0.07	0.2	0.055	Acrylic/0.42/3.0

Source: From Reference [90]. With permission.

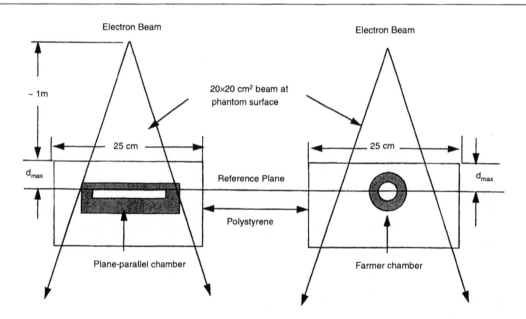

FIGURE 3.82 Experimental setup for the electron-beam method. (From Reference [90]. With permission.)

a ^{60}Co beam at a depth of 5 g/cm² (50 kg/m²) in a phantom of a material selected to match that of the plane-parallel chamber as described in the AAPM TG-39 protocol. The experimental setup is shown in Figure 3.84. Under these conditions, N_{gas}^{pp} is given by

$$N_{gas}^{pp} = \frac{(MN_{gas}P_{ion}P_{repl}P_{wall})^{cyl}}{(MP_{ion}P_{wall})^{pp}} \qquad (3.186)$$

where M^{cyl} and M^{pp} are, respectively, the meter readings

of the cylindrical and plane-parallel chambers under the above conditions, corrected for temperature, pressure, and polarity; P_{ion} is the factor that corrects the ionization recombination measured by the two-potential method; and P_{repl} is the correction factor for the replacement of phantom material by the cavity of the cylindrical ionization chambers. P_{repl} is taken as unity in the AAPM TG-21 protocol for the plane-parallel chambers, so it does not appear in the above equation. P_{wall} is the correction factor for the chamber materials being different from that of the phantom.

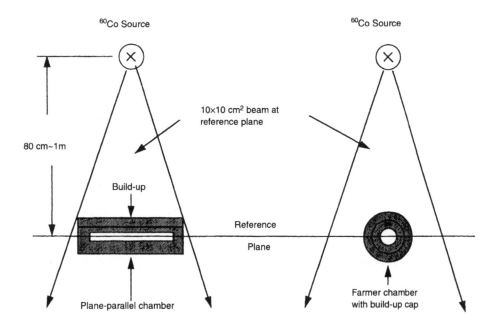

FIGURE 3.83 Experimental setup for the ^{60}Co in-air method. (From Reference [90]. With permission.)

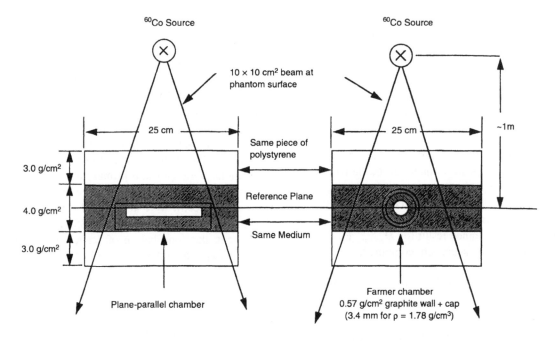

FIGURE 3.84 Experimental setup for the ^{60}Co in-phantom method. (From Reference [90]. With permission.)

The N_{gas}^{pp} values obtained for each chamber by using three different calibration methods are shown in Table 3.31a. [90]

Calibration of ^{192}Ir PDR brachytherapy source was performed by Reynaert et al. [91] in a water phantom at short distances (1.0, 2.5, and 5.0 cm) using an NE2571 Farmer-type ion chamber. To convert the measured air-kerma rate in water to dose rate to water, a conversion factor (*CF*) was

calculated by adapting the medium-energy x-ray dosimetry protocol for a point-source geometry. The *CF* was calculated by Monte Carlo simulations, where the source-ionization chamber geometry was modeled accurately. In a second method, a combination of Monte Carlo simulations and measurements of the air-kerma rate in water (at 1.0, 2.5, and 5.0-cm distance) and in air (1-m distance) was used to determine the *CF*. The following values were

TABLE 3.31
(a) Summary of Results for N_{gas}^{pp}, Using Three Different Calibration Methods.
(b) Percentage Difference from the Average Values of N_{gas}^{pp}

(a)

Chamber	Electron Beam $N_{gas}(10^{+7}$ Gy/C)	^{60}Co in Air $N_{gas}(10^{-7}$ Gy/C)	^{60}Co in phant. $N_{gas}(10^{+7}$ Gy/C)	Average $N_{gas}(10^{+7}$ Gy/C) $\pm$ s.d.
Capintec	5.973	5.947	5.906	5.942 ± 0.6%
Exradin	4.614	4.591	4.649	4.618 ± 0.6%
Holt	2.740	2.755	2.766	2.754 ± 0.5%
NACP	14.990	14.820	14.920	14.910 ± 0.6%
PTW-Markus	48.510	48.929	49.074	48.838 ± 0.6%
Uncertainty	2.1%	2.6%	2.3%	

(b)

Chamber	Electron beam	^{60}Co in air	^{60}Co in phantom
Capintec	+0.5	+0.1	−0.6
Exradin	−0.1	−0.6	+0.7
Holt	−0.5	+0.0	+0.4
NACP	+0.5	−0.6	+0.1
PTW-Markus	−0.7	+0.2	+0.5

Source: From Reference [90]. With permission.

obtained (medium-energy x-ray protocol): CF (1 cm) = 1.458; CF (2.5 cm) = 1.162; CF (5.0 cm) = 1.112 (1σ = 0.7% for the three distances of interest).

A conversion factor (CF) relates the dose rate to water to the measured air-kerma rate in the phantom

$$\dot{D}_w(r) = CF \times \dot{K}_{det}(r) \qquad (3.187)$$

where $\dot{K}_{det}$ is the air-kerma rate measured by the ion chamber (cavity plus wall), which can be derived from the chamber reading. An in-water calibration has the advantage that one is measuring geometry and scatter conditions in the correct beam.

To convert the obtained result for $D_{det,phot}$ to $K_{det,phot}$, the two following expressions are combined:

$$D_{gas} = N_{K,^{60}Co}k_{att}k_m(1-g)M \quad \text{and} \quad K_{det} = N_{k,^{192}Ir}M \qquad (3.188)$$

where the air-kerma calibration factor for ^{60}Co was used in the first expression and that for the ^{192}Ir radiation quality (without build-up cap) in the second; k_m is a correction for the non-air equivalence of the chamber wall and the build-up cap; and k_{att} is a correction for attenuation in the wall and the build-up cap. M is the chamber reading. Then

$$K_{det,phot} = D_{det,phot}\frac{N_{k,^{92}Ir}}{N_{k,^{60}Co}k_{att}k_m(1-g)} \qquad (3.189)$$

The conversion from $D_{det,phot}$ to $K_{det,phot}$ introduces an extra uncertainty on the CF obtained with this approach.

The conversion factor CF is obtained directly by dividing $D_{w,phot}$ by $K_{det,phot}$.

For the calibration factor with build-up cap for the ^{192}Ir radiation quality obtained by the calibration procedure, Reynaert et al. found $N_k(^{192}Ir, air)/N_k(^{60}Co) = 0.988$ and $N_k(^{192}Ir, water)/N_k(^{60}Co) = 0.987$ at the three distances of interest in the water phantom.

To determine the scatter contribution to the total kerma, the photon fluence spectra obtained in the EGS4 simulations were split into primary and scattered photon spectra and multiplied by $4\pi r^2$ to remove the r^{-2} dependence. The spectra were normalized per emitted photon. These results are plotted in Figure 3.85. At distances larger than 6 cm, the contribution of scattered photons to the kerma is larger than that of the primaries.

CF_H can be defined by the following expression

$$\dot{D}_w(r) = N_K M \times CF_H \qquad (3.190)$$

where N_K is the calibration factor for kerma in ^{60}Co and the index H denotes the high-energy photon dosimetry protocol. CF_H can be defined as the conversion factor obtained from the high-energy photon protocol, or

$$CF_H = \frac{N_d}{N_k}\left(\frac{\bar{S}}{\rho}\right)_{w,air} p_{wall}p_{cel}p_dp_n \qquad (3.191)$$

The behavior of a chamber during measurements, with variation of response with beam quality, was measured by National Physical Laboratory (NPL).[92] The results, normalized to 1 mm Cu HVL, are shown in Figures 3.86 and 3.87. As can be seen from these figures, the air-kerma

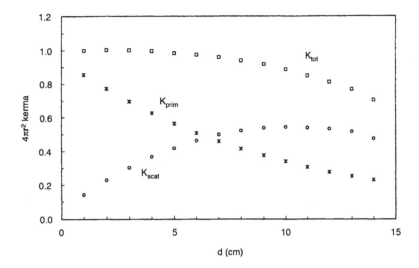

FIGURE 3.85 Calculated total kerma is split into separate contributions of scattered and primary photons. The kerma are multiplied by $4\pi r^2$. At distances larger than 6 cm, the scatter contribution is larger than that of the primaries. The results are normalized at 1 cm. This figure can be compared with the results obtained by Russell and Ahnesjö. [91a] (From Reference [91]. With permission.)

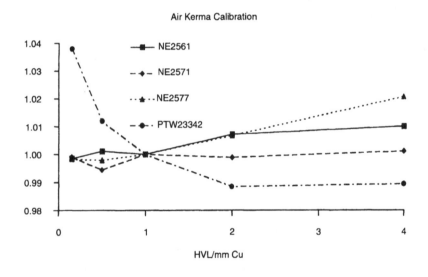

FIGURE 3.86 The variation of the chamber calibration factor with beam quality. (From Reference [92]. With permission.)

calibration factors for the NE2561 and NE2571 vary by less than 1% between 0.15 mm Cu and 4 mm Cu. The responses of the (short) NE2577 and PTW Grenz ray chambers are slightly less flat, varying by 2% and 4%, respectively. If the NACP electron chamber is included, the variation becomes much greater (25%).

For Farmer types of chamber, the variation in response with HVL has been extracted from data measured during routine calibration. This is shown in Figure 3.88, normalized to 3.5 mm Cu.

As can be seen, the variation of response for the NE2581 is greater than might be expected. Although this graph is plotted over a greater range (down to 0.068 mm Cu), there is still about a 15% difference between 1.0 and 0.15 mm Cu, with significant variation between individual

chambers. Experience suggests that this may also be typical of chambers of the Farmer type which use either conducting or graphite-coated plastic outer electrodes.

Experimental investigation of a simple design of a plane-parallel electron chamber, which has very thin layers of copper (0.018 or 0.035 mm) as conducting material was done by Ma et al.[93] The results show that the C_e factors (proportional to the product of water/air stopping power ratio and perturbation factor) for converting the in-phantom air-kerma-calibrated chamber reading to the absorbed dose to water are nearly constant for incident electron energies between 4 and 11 MeV for prototype chambers with 0.018-mm-thick copper layers and between 4 and 15 MeV for chambers with 0.035-mm-thick copper layers.

Air Kerma Calibration

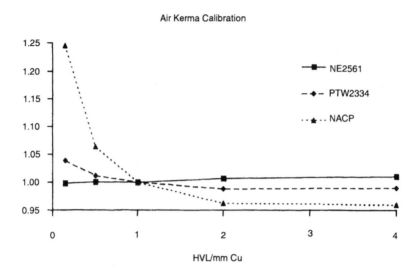

FIGURE 3.87 The variation of the chamber calibration factor with beam quality. (From Reference [92]. With permission.)

Air Kerma Calibration

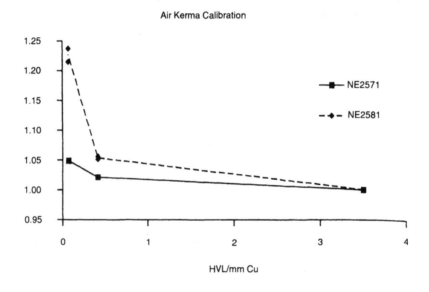

FIGURE 3.88 The variation of the chamber calibration factor with beam quality. (From Reference [92]. With permission.)

The results show that the C_e factors for converting the in-phantom air-kerma-calibrated chamber reading, to the absorbed dose to water, are nearly constant for incident electron energy between 4 and 11 MeV, for prototype chambers with 0.018 mm thick copper layers, and between 4 and 15 MeV for chambers with 0.035 mm thick copper layers. The C_e factors are proportional to the product of water/air stopping-power ratio and perturbation factor. When the designated chamber is irradiated by an electron beam at a depth d in a water phantom the absorbed dose to water is

$$D_w = N_f C_e E_e \qquad (3.192)$$

The variation of C_e, with energy for a particular electron chamber, depends on both the electron fluence–per-

turbation correction factor p_u^e and the stopping power ratio for the phantom material to air, $S_{med,air}$. The conversion factor C_e can be written as

$$C_e = \{k_{ch}^s k_{att}^s (1 - g) / k_s\}(P_{cal}^s / P_{cal}^e)S_{water,air}P_u^e$$

$$(3.193)$$

where k_{ch}^s is a correction factor for a secondary-standard NE2561 chamber to account for the lack of air equivalence of the chamber wall, electrode, and build-up cap, and k_{att}^s is a correction factor for the effect of absorption and scattering in the chamber wall, electrode, and build-up cap at the calibration of the chamber irradiated in air by a ^{60}Co beam or a 2-MV x-ray beam. g is the fraction of the energy of the charged particles lost to *bremsstrahlung*. The perturbation factor, p_{cal}, corrects for the effect of

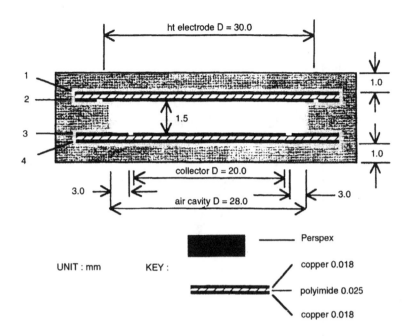

FIGURE 3.89 A cross-sectional view of the NPL plane-parallel chamber with 0.018-mm-thick copper layers. The thickness of each copper layer (1,2,3, or 4) is 0.018 mm and the thickness of the polymide layer is 0.025 mm. The diameter of the collecting electrode is 20 mm and the width of the guard ring is 3 mm. The diameter of the air cavity is 28 mm and the thickness is 1.5 mm. The thickness of the front and back Perspex wall is 1 mm. For a 0.035-mm-thick copper-layered version, the geometry is exactly the same, except that the copper layers are 0.035 mm thick. (From Reference [93]. With permission.)

non-medium-equivalent wall material and central electrode and for the deviation of the effective point of measurement of the chamber from the chamber center during the intercomparison of a secondary-standard NE2561 chamber and an electron chamber.

The geometrical details of the NPL chambers are shown in Figure 3.89.

The conversion factor C_e for a particular electron chamber varies with both the electron fluence–perturbation correction factor p_u^e and the stopping power ratio for water to air, $S_{water, air}$. Relative values of C_e are shown in Figure 3.90.

X. LIQUID DIELECTRIC IONIZATION CHAMBERS

Ionization chambers filled with a dielectric liquid instead of a gas have many basic advantages. The sensitive volume of a liquid-ionization chamber (LIC) can be made much smaller than that of a gas-ionization chamber and still provide the same ionization current. A high spatial resolution, not possible with a gas chamber, can thereby be achieved. The stopping power ratio for the dielectric liquid to water varies by only a few percent in the range 1–50 MeV, compared to about 20 percent for air to water in the same energy range (Figure 3.91).

Two dielectric liquids, isooctane and tetramethylsilane (TMS), have been tested by Wickman and Nystrom [94] as sensitive media in parallel-plate ionization chambers.

The ionization chambers use graphite electrodes and have a high spatial resolution due to a coin-shaped ionization volume of 2 mm³ (diameter 3 mm, height 0.3 mm). The detectors show a calibration stability within ±1% over several years. In neither of the liquids did general recombination exceed 2% in pulsed radiation fields and dose rates up to 7 Gy min⁻¹.

The mobility of ions is reported to be extremely high. The collision stopping power ratio to water is found to be almost constant (within about ±0.2%) in the electron energy range of 0.01–50 MeV, while isooctane has a moderate variation of ±2% (Figure 3.92).

The leakage current is a rough measure of the amount of impurities in a dielectric liquid. Obviously, there are some initial chemical exchanges due to liquid and chamber wall interactions, as a new ionization chamber has to be rinsed several times by emptying and filling it with fresh liquid during the first weeks before a low and stable residual current is established. A typical residual current at room temperature that can be achieved at a polarization potential of 900 V (3.0 MV m⁻¹) is about 5×10^{-14} A in isooctane, as well as in TMS. This current corresponds to a resistivity of about 3×10^{14} Ωm. The resistivity reported for highly purified hydrocarbon liquids is 3 to 30 times higher. The residual current is temperature-dependent, as well as field strength–dependent. The leakage current is slightly increased directly after the polarization voltage has been connected but stabilizes within a few minutes. When a

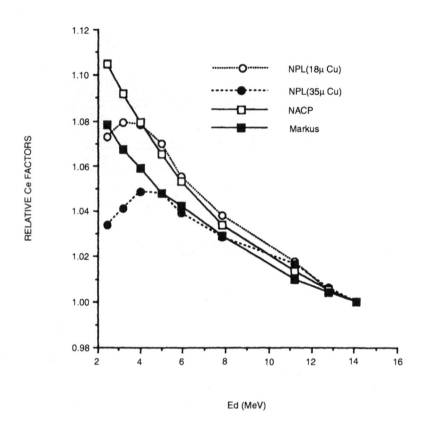

FIGURE 3.90 Relative C_e factors for the plane-parallel chambers used by Ma et al. The values normalized to the values for $E_d = 14.4$ MeV. (From Reference [93]. With permission.)

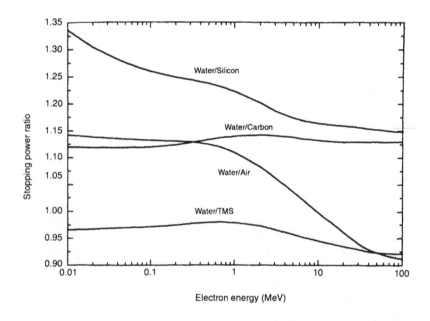

FIGURE 3.91 Stopping power ratio to water of the dielectric liquid TMS and other materials commonly used as sensitive media for dosimeters. (From reference [99]. With permission.)

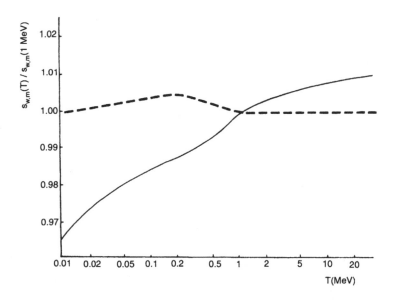

FIGURE 3.92 Mass collision stopping power ratios, water to isooctane (—) and TMS (- -), normalized to 1.00 at the electron energy $T = 1$ MeV. (From Reference [94]. With permission.)

chamber has been rinsed and a low current has been established, this will typically remain stable for years and appears to be unaffected by normal use of the chamber. A residual current of 5×10^{-14} A corresponds to a dose rate from ionization radiation of ≈ 0.2 mGy min^{-1} in isooctane and ≈ 0.13 mGy min^{-1} in TMS for the type of chamber used in this investigation.

The basic chamber design is illustrated in Figure 3.93. Strong emphasis during the development of the chamber has been placed on the creation of a small, robust, versatile, and easy-to-handle detector with a stable response that can be used as simply as any gas-ionization chamber for routine calibration and measurements in radiation therapy fields. The chamber is watertight and can be used with the same liquid for years.

The electrodes are made of thin discs of pure graphite. In the earlier chamber designs, thin layers of evaporated beryllium were used. The reason for the exchange was that beryllium reacts with the liquid in a way not yet understood, causing a slow and constant drop in the detector sensitivity. No significant change in calibration stability has been found when graphite discs are used as electrodes.

An air bubble has been introduced in order to compensate for the approximately four times greater volume change with temperature in the liquid compared to the chamber body materials. The liquid container has been divided into two compartments; the air bubble is placed in the outer part, and a "bubble lock" between the two compartments is introduced in order to prevent air from coming between the electrodes.

The distances between the ions formed in the track of an ionizing particle are much shorter than in the gas.

The ion recombination in the liquid is much more pronounced and must often be corrected for. In the track of a highly energetic particle, liquid ions are formed in clusters containing one or a few ion pairs, and there is a significant initial recombination of the ions in their original clusters. The amount of ions liberated per 100 eV of absorbed energy at zero external electric field is commonly defined as G_{fi}^0. The intra-cluster recombination is affected by the temperature and by the external electric field. The G_{fi}^0 value at room temperature varies from approximately 0.1 to 1 for the various dielectric liquids. For "impurity-free" 2, 2, 4-trimethylpentane (isooctane) and TMS, G_{fi}^0 is 0.33 and 0.74, respectively. [94]

The free-ion yield, G_{fi}, is a measure of the probability that the electrons formed by the ionization radiation have sufficient energy to escape from the electric field of the parent positive ions.

In the measurements of the collection efficiency, the absorbed dose in water was simultaneously monitored by a small plane-parallel air-ionization chamber with a well-known and small general recombination that was corrected for. The results are presented in Figures 3.94. It is seen from the figures that the general recombination in isooctane is noticeably dependent on the pulse repetition rate when the polarizing voltage is 500 V. This effect decreases when the voltage is increased to 900 V and is, at that voltage, almost negligible for pulse rates under 100 pulses per second. This pulse rate effect depends on the fact that the transport time of some of the ions is longer than the pulse interval; this means that new ions are injected by the following pulse before these slow-motion

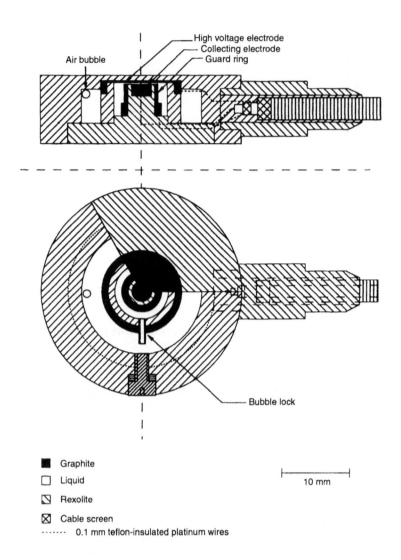

FIGURE 3.93 Cross-sectional views of the chambers used. (From Reference [94]. With permission.)

ions have been cleared from the liquid. This then causes an increased recombination when the pulse repetition rate increases. The ion transport in the chamber has been studied by the use of a DC-amplifier with high sensitivity and a fairly good response time.

Chemical decomposition of the liquid by radiation-induced bond fissions and an uncontrolled recombination of the fragments to form other molecules than the original compound can affect the calibration.

One kg of 224-trimethylpentane contains 5.3×10^{24} molecules, and the absorbed energy of 1 J creates 3.2×10^{16} affected molecules. From these figures, the relative number of affected molecules per Gy of absorbed dose in the liquid is $(3.2 \times 10^{16})/(5.3 \times 10^{24})$, or $6 \times 10^{-6}\%$. Most hydrocarbons have their G-values within a factor of three of that of isooctane. The dependence of G_{fi} on the field strength is shown in Figure 3.95.

The absorbed dose to liquid, D_{liq}, at a field strength E is given by

$$D_{liq} = Q_{liq}[C_{sat}m_{liq}G_{fi}(E)] \qquad (3.194)$$

The absorbed dose to water can then be calculated from

$$D_w = D_{liq}f \qquad (3.195)$$

where D_w is the absorbed dose to water, D_{liq} is the absorbed dose to liquid, $G_{fi}(E)$ is the free-ion yield per absorbed energy at field strength E, C_{sat} is the correction factor for general recombination, m_{liq} is the mass of active liquid volume, and Q_{liq}, is the collected charge in liquid. The factor f accounts for different cavity-theory effects in the conversion from dose in the liquid layer of the chamber to dose in water. In electron beams, $f = S_{W,liq}^{SA}P_u$, where $S_{W,liq}^{SA}$ is the Spencer-Attix stopping power ratio of water to liquid with a cut-off energy corresponding to the energy of electrons with a range corresponding to the average distance across the liquid layer, i.e., 0.02 g cm^{-2}. P_u corrects for any electron fluence change due to the Rexolite-graphite wall and should be close to 1, as the materials are thin and are very nearly "water equivalent."

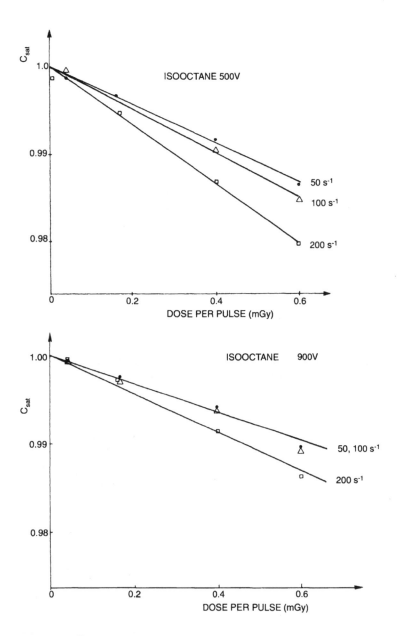

FIGURE 3.94 Collection efficiency C_{sat} with pulse dose at pulse radiation (pulse length ~3 μs) and different pulse repetition frequency and polarization voltages. (From Reference [94]. With permission.)

In photon beams the factor f is dependent on the energy and on the physical dimensions of the cavity and wall. At very low energies, the major contribution to the energy deposition in the liquid is from electrons released by photon interactions in the liquid. In this case the factor f can be approximated by

$$\frac{1}{f} = (\mu_{en}/\rho)_W^{liq} \qquad (3.196)$$

At high energies, on the other hand, where the liquid layer and walls are thin compared with the secondary electron ranges, only a small fraction of the electrons in the liquid is generated in the liquid or in the wall. In this case

the sensitive volume behaves like a Bragg-Gray cavity:

$$\frac{1}{f} = s_W^{liq} \qquad (3.197)$$

In the megavoltage range, however, none of these situations is approximated very accurately. Burlin and Chan (94a) suggested a weighting factor, d, related to cavity size for the intermediate situation in which the cavity behaves neither as a Bragg-Gray cavity nor as a large cavity:

$$\frac{1}{f} = ds_W^{liq} + (1 - d)(\mu_{en}/\rho)_W^{liq} \qquad (3.198)$$

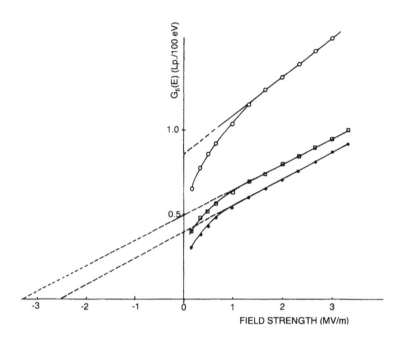

FIGURE 3.95 The amount of free-ion pairs per 100 eV of absorbed energy G_{fi} at different field strengths in TMS (○) and isooctane (□) irradiated with ^{60}Co photons. Isooctane irradiated with 190-kV x-rays (●). (From Reference [94]. With permission.)

Here d is the fraction of the electrons in the liquid generated by photon interactions outside the liquid.

The first term of Equation (3.198) then becomes

$$d[\alpha(\mu_{en}/\rho)_w^{wall} s_{wall}^{liq} + (1 - \alpha)S_w^{liq}] \qquad (3.199)$$

where α is the fraction of the outside-cavity electrons generated in the wall. The fact that the liquid-ionization chamber in the present work has a front wall consisting of two materials, graphite and Rexolite, demands a further split of Expression (3.199):

$$d\{\alpha[\beta(\mu_{en}/\rho)_w^{gr} s_{gr}^{liq} + (1 - \beta)(\mu_{en}/\rho)_w^{rex} s_{rex}^{liq}]$$
$$+ (1 - a)s_w^{liq}\} \qquad (3.200)$$

Here β is the fraction of the wall-generated electrons originating from the graphite and $(1 - \beta)$ is the fraction of those generated in the Rexolite. An approximate expression for the f-factor would thus be [94]

$$\frac{1}{f} = d\{\alpha[\beta(\mu_{en}/\rho)_w^{gr} s_{gr}^{liq} + (1 - \beta)(\mu_{en}/\rho)_w^{rex} s_{rex}^{liq}]$$
$$+ (1 - \alpha)s_w^{liq}\} \qquad (3.201)$$

The general collection efficiency has been measured by Johansson and Wickman [95] in liquid isooctane (C_8H_{18}) and tetramethylsilane ($Si(CH_3)_4$) as the sensitive media in a parallel-plate ionization chamber, with an electrode distance of 1 mm, intended for photon and electron dosimetry applications. The measurements were made at potential differences of 50, 100, 200, and 500 V. Measurements were performed for each liquid and electric field strength, with the decay rate of ^{99m}Tc used as the dose-rate reference. The maximum dose rate was about 150 mGy min^{-1} in each experiment.

If the general collection efficiency, defined as the ratio of the ionization current to the saturation current, is denoted by f, the ionization current–polarizing voltage relation derived by Mie [96] can be expressed as

$$f = \frac{1}{1 + \xi^2\{1 - [(4 - \lambda)/10](1 - f)\}^2}$$
$$\xi^2 = \frac{1}{6}m^2\frac{d^4q}{U^2}$$
$$m = \left(\frac{\alpha}{ek_1 k_2}\right)^{1/2} \qquad (3.202)$$
$$\lambda = \frac{e(k_1 + k_2)}{\alpha\varepsilon_r\varepsilon_0}$$

where k_1 and k_2 are the mobilities of the ions; ξ is the recombination rate constant; e is the electronic charge; d is the distance between the collecting electrodes; q is the amount of charge liberated by the radiation and escaping initial recombination per unit volume and time in the medium; U is the polarizing voltage of the ionization chamber; ε_r is the permittivity of the medium in the chamber; and ε_0 is the permittivity of free space. The recombination

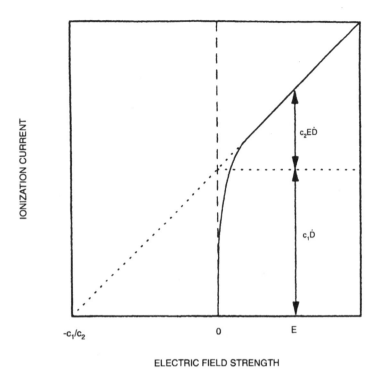

FIGURE 3.96 The characteristic relation of the ionization current from a dielectric liquid and the electric field strength in a liquid-ionization chamber. The ionization current is the sum of the two components $c_1\dot{D} + c_2E\dot{D}$, where c_1 is the fraction of ions that escape initial recombination due to diffusion, c_2 is the fraction of ions that escape initial recombination due to the external electric field strength, E is the electric field strength, and $\dot{D}$ is the dose rate. (Fom Reference [95]. With permission.)

rate constant and the ion mobilities are specific constants of the sensitive medium.

If the ionization chamber contains air at normal pressure and temperature with a λ value of ≈ 3.56, the factor in the square brackets in Equation (3.202) can be ignored for $0.7 < f < 1$. The equation for the general collection efficiency is then reduced to

$$f = \frac{1}{1 + \xi^2} \qquad (3.203)$$

When using an ionization chamber with a dielectric liquid instead of a gas as the sensitive medium, the ion-production density is increased by a factor of about 300. The ionization current never achieves saturation with respect to the initial recombination. The characteristic relation between the ionization current from an irradiated dielectric liquid and the electric field strength is shown in Figure 3.96. The ionization current can be described as the sum of two components. The first term consists of the product $c_1\dot{D}$, and the second term is the product $c_2E\dot{D}$, where c_1 and c_2 are constants characteristic of the liquid, E is the electric field strength, and $\dot{D}$ is the dose rate.

The ions can escape initial recombination by diffusion or by the combined effect of diffusion and the influence of the external electric field. Under constant irradiation conditions and with negligible general recombination, the ionization current will increase linearly with electric field strength. In Figure 3.96, the constants c_1 and c_2 represent the fractions of the ionization current that escape initial recombination due to diffusion and due to the external electric field, respectively.

The body of the parallel-plate liquid-ionization chamber used by Johansson and Wickman for the experiments was made of Rexolite®, a styrene copolymer with a density of 1.05 g cm^{-1}, and the electrodes were made of pure graphite. The chamber had cylindrical symmetry without any guard. It was irradiated towards the plane frontal side during the experiments. The thickness of the chamber wall in the irradiation direction was 1 mm, consisting of 0.7 mm Rexolite® and 0.3 mm graphite. The sensitive volume had a diameter of 3 mm and thickness of 1 mm, and was filled through a hole with diameter of 0.3 mm in the center of the second electrode. The total sensitive volume of the chamber was 7.07×10^{-9} m^3. The dielectric liquids used were isooctane (C_8H_{18}; Merck, isooctane analysis grade 99.5%) and tetramethylsilane (TMS; Si(CH$_3$)$_4$; Merck, NMR calibration grade 99.7%). The liquids were used without any further purification.

The relative dose-rate distribution in the liquid layer at the irradiation geometry was calculated by the function

$$\Phi = \mu t \int_{\mu t}^{\infty} \frac{e^{-y}}{y^2}dy - (\mu t + \mu_s h) \int_{\mu t + \mu_s h}^{\infty} \frac{e^{-y}}{y^2}dy$$

$$(3.204)$$

TABLE 3.32
Experimentally Determined m Values and the Corresponding Mobilities of the Ions (k_1, k_2) and the Recombination Rate Constants (α) for Isooctane and Tetramethylsilane

	m ($s^{1/2}C^{-1/2}$ V $m^{-1/2}$)	k_1 (m^2 s^{-1} V^{-1})	k_2 (m^2 s^{-1} V^{-1})	α (m^3 s^{-1})
Isooctane	2.0×10^9	2.9×10^{-8}	2.9×10^{-8}	5.4×10^{-16}
Tetramethylsilane	1.4×10^9	5.3×10^{-8}	9.0×10^{-8}	1.4×10^{-15}

Source: From Reference [95]. With permission.

where Φ is the relative photon fluence, μ the linear attenuation coefficient of the different materials, t is the thickness of the different materials, μ_s is the linear attenuation coefficient, and h is the thickness of the ^{99m}Tc solution.

For the theoretical general collection efficiency, a mean q value for the sensitive volume was used. This mean value was calculated as

$$\bar{q} = q_f \frac{i_{theor}}{V_{liquid}} \qquad (3.205)$$

where $\bar{q}$ is the mean value of the amount of charge liberated per unit volume and time in the liquid by the radiation and escaping initial recombination, q_f is an experimentally determined correction factor, i_{theor} is the theoretical ionization current in the absence of general recombination, and V_{liquid} is the liquid volume in the ionization chamber. The general collection efficiency derived by Mie, Equation (3.202), also contains the parameter λ. For two ions of opposite sign at a separation distance R and approaching each other with a velocity, v the recombination rate constant α is given by

$$\alpha = \frac{e(k_1 + k_2)}{\varepsilon_0 \varepsilon_r} \left[1 + \frac{r_c D}{R^2 v} \exp\left(-\frac{r_c}{R}\right) - \exp\left(-\frac{r_c}{R}\right) \right]^{-1} \qquad (3.206)$$

where r_c, the Onsager escape distance [56], is the separation distance at which the Coulombic potential energy is equal to the diffusion kinetic energy of the ions and D is the sum of the diffusion coefficients of the ions. For dielectric media with a low permittivity, the ratio r_c/R is large and the recombination rate constant can be approximated by the equation (Debye 1942)

$$\alpha = \frac{e(k_1 + k_2)}{\varepsilon_0 \varepsilon_r} \qquad (3.207)$$

This approximation of the recombination rate constant leads to a λ value of 1 for dielectric liquids. [95]

m (in Equation 3.202) is a constant purely dependent on the recombination rate constant and the mobilities of the ions in the medium. From the experimental results, the specific m values for isooctane and tetramethylsilane were empirically determined by an iterative method, using the values for the electrode distance and the polarizing voltages and by assuming $q_f = 1$. Measured values of m, k, k_2, and α are given in Table 3.32.

From the definition of the m value (Equation (3.202)) and the approximation of the recombination rate constant (Equation (3.206)), a relation between the mobilities of the ions of different signs can be derived as

$$k_2 = \frac{1}{m^2 \varepsilon_r \varepsilon_0 - (1/k_1)} \qquad (3.208)$$

The experimentally determined general collection efficiencies are well described by the theoretical general collection efficiency equation derived by Mie (Figures 3.97 and 3.98). However, probably as a result of the approximate space charge correction, the theoretical expression shows a lower general collection efficiency for all of the experiments. The maximum deviation is less than 1% of the saturation current at a general collection efficiency of 60%. For general collection efficiencies down to 90%, the simplified equation proposed by Greening [95a] can also be used, showing a deviation from the experimental general collection efficiency of less than 1% of the saturation current. [95]

Two new liquid-ionization chamber designs, consisting of cylindrical and plane-parallel configurations, were presented by Wickman et al. [97] They are designed to be suitable for high-precision measurements of absorbed dose to water at dose rates and photon energies typical for LDR intermediate photon energy brachytherapy sources. The chambers have a sensitive liquid–layer thickness of 1 mm and sensitive volumes of 7 mm^3 (plane-parallel) and 20 mm^3 (cylindrical). The liquids used as sensitive media in the chambers are either isooctane (C_8H_{18}), tetramethylsilane ($Si(CH_3)_4$) or mixtures of these two liquids in the approximate proportions 2 to 1.

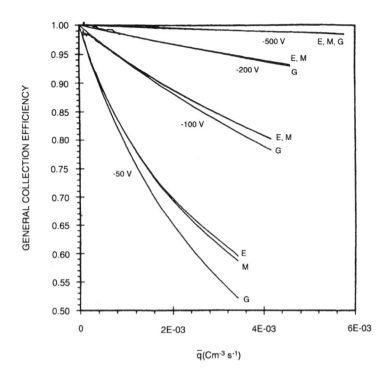

FIGURE 3.97 Experimental (E) and theoretical (Mie (M); Greening (G)) general collection efficiency in the parallel-plate liquid-ionization chamber, filled with isooctane and operated at the polarizing voltages 50, 100, 200, and 500 V, as a function of $\bar{q}$, the mean value of the amount of charge liberated by irradiation and escaping initial recombination. (From Reference [95]. With permission.)

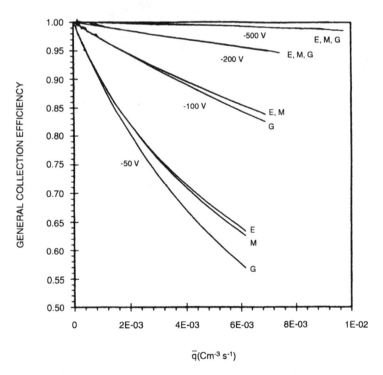

FIGURE 3.98 Experimental (E) and theoretical (Mie (M); Greening (G)) general collection efficiency in the parallel-plate liquid-ionization chamber, filled with tetramethylsilane and operated at the polarizing voltages 50, 100, 200, and 500 V, as a function of $\bar{q}$ the mean value of the amount of charge liberated by irradiation and escaping initial recombination (From Reference [95]. With permission.

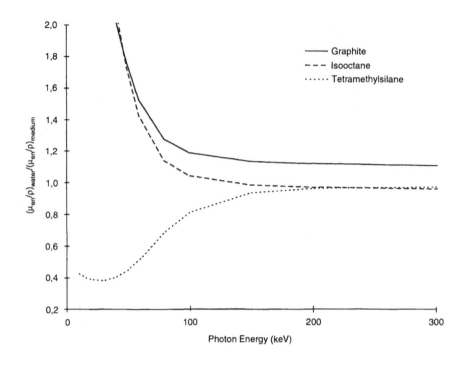

FIGURE 3.99 Mass energy–absorption coefficient ratio to water for: graphite (—), isooctane (----), and TMS (.....). (From Reference [97]. With permission.)

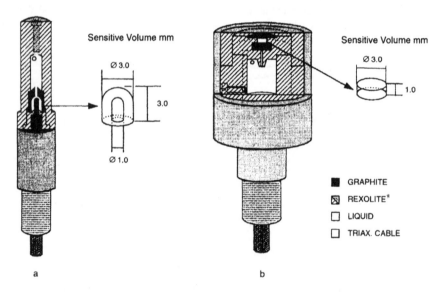

FIGURE 3.100 Cross-sectional views showing the construction details of the two chamber types tested: (a) cylindrical; (b) plane-parallel. (From Reference [97]. With permission.)

The two new LIC designs, consisting of cylindrical and plane-parallel configurations, were tested (Figures 3.100a and b). The intended function of this testing was to prevent leakage through the insulator from reaching the charge-collecting electrode. The high yield of ions in LICs, however, makes the influence from such leakage less critical than in air-filled ion chambers. Furthermore, the introduction of a guard electrode in an LIC of small size makes it difficult to manufacture. Moreover, the guard electrode introduces fringes in the electric field that defines the active ionization volume. Omission of the guard electrode implies that the ionization volume has to be filled through a thin (Ø 0.3-mm) duct in one of the chamber electrodes.

The body-construction material is a styrene-copolymer (Rexolite®), a material tested in a variety of LIC designs for many years and shown to have excellent insulator properties, even after contact with the liquids isooctane and TMS for years and after an accumulated absorbed dose of 10 kGy. The chamber electrodes are made of pure

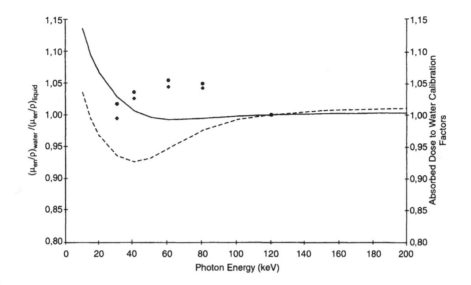

FIGURE 3.101 Water-to-liquid mass energy–absorption coefficient ratios for two TMS/isooctane mixtures, 30/70 (—) and 35/65 (----). Absorbed dose to water calibration factors for cylindrical LICs filled with TMS/isooctane mixtures, 35/65 (•) and 40/60 (♦). Each data set normalized at 120 keV. (From Reference [97]. With permission.)

graphite and the internal electrical connections are made by 0.1-mm Teflon-insulated Pt-wires. Three cylindrical and two plane-parallel chambers were used for the test. The liquids used as sensitive media in the experimental work were isooctane (Merck, isooctane analysis grade 99.5%) and tetramethylsilane (TMS; Merck, NMR calibration grade 99.7%) or mixtures of these two liquids. No further purification was done before their use.

The calculated values for this absorption ratio for two mixture fractions of TMS and isooctane are plotted in Figure 3.101, calculated with the Bragg rule using mass energy–absorption coefficients given by Hubbell and Seltzer. [98]

At low dose rates the uncertainty from the background current corrections will increase. The chambers tested in this project having a liquid-layer thickness of 1 mm and a polarizing voltage of 300 V were optimized for absorbed dose measurements in water around LDR brachytherapy sources. For the current LIC specifications, the dose-rate range within which the accuracy can be predicted to be better than ±1% using a high-quality electrometer and an ionization charge-collecting time of approximately 60 s is approximately 0.1 to 100 mGy min^{-1}. At 0.1 mGy min^{-1} the typical background current at room temperature is about 20% of the ionization current, and at 100 mGy min^{-1} the general recombination is still less than 2%. For a specific LIC, the optimum dose-rate range can be moved slightly up or down by changing the polarizing voltage. The most efficient way, however, to change the optimal dose-rate range is by a change of the liquid-layer thickness. Chambers of the current design but having a liquid-layer thickness of 0.3 mm have been tested in high dose-rate fields (a few Gy/min) and have been found to work

properly with respect to their leakage current at a polarizing voltage of approximately 600 V. An LIC operating at these specifications has an optimum dose-rate range approximately three orders of magnitude higher than the presented LICs. [97]

A liquid-ionization chamber design optimized for high spatial resolution was used by Dasu et al. [99] for measurements of dose distributions in radiation fields intended for stereotactic radiosurgery (SRS). The work was focused mainly on the properties of this detector in radiation fields from linear accelerators for clinical radiotherapy (pulsed radiation with dose rates from approximately 0.5 to 5 Gy/min and beam diameters down to 8 mm). The narrow beams used in stereotactic radiosurgery require detectors with small sizes in order to provide a good spatial resolution.

The LIC used for that work (Figure 3.102) is a plane parallel, liquid-filled, unguarded ionization chamber developed at Umeå University. Its body is made of Rexolite®, a styrene copolymer with density 1.05 g cm^3. The sensitive volume with diameter of 3 mm and thickness of 0.3 mm is placed behind a 1-mm wall of 0.7-mm Rexolite® and 0.3 mm graphite. It is filled through a 0.3-mm hole in the center of the collecting electrode. The dielectric liquid used is tetramethylsilane (TMS; Si(CH$_3$)$_4$; Merck, NMR calibration grade 99.7%).

With a liquid-layer thickness of 1 mm and operated at a polarization voltage of 300 V, excessively high general recombination can be expected at the dose rates commonly used in accelerator applications. This problem can be eliminated by using a thinner liquid layer (0.3 mm) and a much higher electric field strength. For all the measurements, a polarizing voltage of 900 V was applied to the LIC.

All the materials surrounding the sensitive volume of the LIC (graphite and polystyrene) are approximately water-like in density and atomic number. Therefore, the

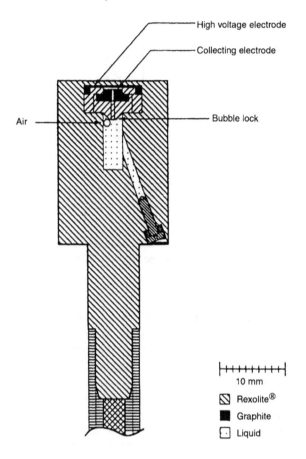

High voltage electrode

Collecting electrode

Air

Bubble lock

10 mm

◈ Rexolite®
■ Graphite
⊡ Liquid

FIGURE 3.102 Unguarded liquid-ionization chamber used by Dasu et al. (From Reference [99]. With permission.)

disturbance of the radiation field by the chamber in water is expected to be negligible.

The output factor (OF) was measured as the ratio of the dose per monitor unit for each collimator to that of the $100 \times 100\text{-mm}^2$ rectangular field. The dose rate for the calibration condition in the measuring point was 3 Gy min^{-1}.

The dose-rate dependence of the LIC response for dose rates up to 6 Gy min^{-1} is presented in Figure 3.103. The uncertainty in the measurement is estimated to be about ±0.5%. These results show a loss of collection efficiency of less than 0.5% at the highest dose rate investigated.

The leakage current of the LIC during the entire study was quite stable, in the range of 1.1–1.5 pA. This current corresponds to the ionization current at a dose rate of about 3–5 mGy min^{-1}; therefore, the correction and error introduced by the leakage current are negligible.

The results of OF measurements for rectangular fields are presented in Figure 3.104 for field sizes from 30 to 200 mm.

As can be seen in Figure 3.105, when the field size decreases from 20 to 10 mm, the LIC response comes closer to that of the unshielded diode. For these very small fields (smaller than the range of secondary electrons generated by the 6-MV photons), the electron spectrum at the center of the beam is changed due to the lack of in-scattered low-energy electrons.

A plane-parallel ionization chamber having a sensitive volume of 2 mm^3 and using the dielectric liquid tetramethylsilane as the sensitive medium instead of air was described by Wickman and Holmstrom. [100] In the design of the chamber, special attention was given to the factors that can

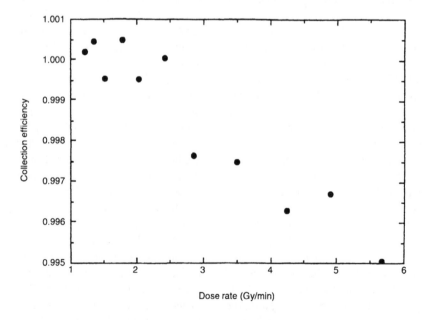

FIGURE 3.103 Dose-rate dependence of the LIC. (From Reference [99]. With permission.)

cause unwanted currents in the cable, stem, or the chamber dielectric material. Experimental results showed that, despite the extremely small ionization volume in the liquid-ionization chamber, the polarity effect never exceeds a few tenths of a percent in field positions.

The liquid-ionization chamber used is shown in Figure 3.106. Special attention has been given to charge deposition and induced leakage effects in order to minimize the contribution to the measured signal from charges not originating in the chamber-sensitive volume. The cable connected to the chamber is a solid Teflon-insulated triaxial cable, where the central conductor, including its insulator, has been made thin ($\phi = 1$ mm). The inner dielectric has received a "low-noise" treatment by the use of a thin film of conducting plastic. This type of cable design has been shown to give the best overall preformance for use with ionization chambers. The cable has been specially made. The inner screen of the cable forms an unbroken screen for the central cable conductor until it meets the chamber guard ring. This measure has been taken in order to minimize dielectric leakage currents reaching the collecting electrode or the central conductor. The liquid used, tetramethylsilane, has a density of ≈ 0.7 g/cm^3, and the net ionization charge collected at a polarization voltage of 900 V is approximately 300 times as high as in an air cavity of the same volume. For all the measurements, the chambers are imbedded in a 40×40-cm lucite phantom, machined to get a close fit with the chambers. When the bias of the chambers is reversed, at least 5 min are allowed for the chamber and cable dielectrics to settle.

Figure 3.107 shows the polarity effect at measurements with the liquid-ionization chamber in a 10×10-cm electron field with electrons having an incident energy of 9 MeV.

Sensitivity, general recombination, and reproducibility of liquid-filled parallel-plate ionization chambers for

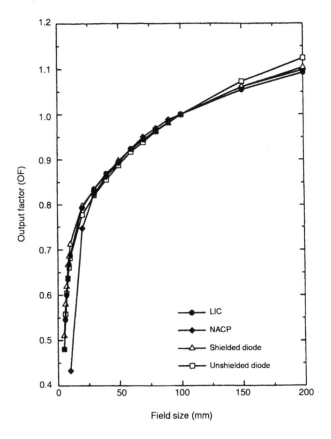

FIGURE 3.104 Output factors for square fields. (From Reference [99]. With permission.)

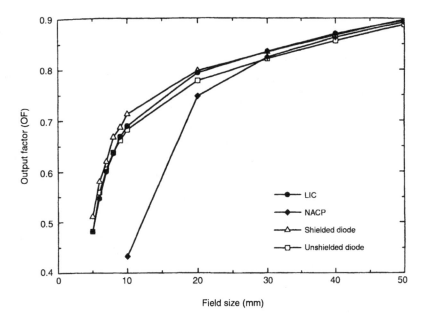

FIGURE 3.105 Output factors for small square fields. (From Reference [99]. With permission.)

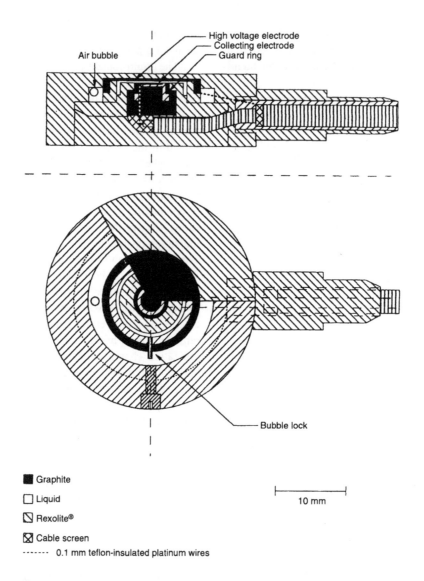

FIGURE 3.106 Cross-sectional view of the liquid ionization chamber. (From Reference [100]. With permission.)

dosimetry in low-dose-rate brachytherapy radiation fields have been evaluated by Johansson et al. [101] Two different dielectrics liquids, isooctane (C_8H_{18}) and tetramethyl-silane ($Si(CH_3)_4$), have been used as sensitive media in chambers having a coin-shaped sensitive volume of 3 mm in diameter and 1-mm thickness. An electric field strength of 300 kV m^{-1} was found to be optimal with respect to sensitivity, leakage current, and general recombination.

The density and chemical composition of the dielectric liquids commonly used do not significantly perturb the radiation field in tissue-like materials, and liquid-ionization chambers can be made as simple and rugged as air-ionization chambers.

A high electric field strength reduces the magnitude of general recombination, but it also reduces the absorbed dose-rate resolution, due to increased magnitude and variation of the leakage current. A low electric field strength, on the other hand, increases the absorbed dose-rate resolution because of

reduced leakage current but decreases the ion-collection efficiency. If the magnitude and variation of the general recombination increase, this will counteract the gain in the absorbed dose-rate resolution due to the reduced leakage current. Different liquids will have different magnitudes and variations of the ionization and leakage currents.

The thickness of the liquid layer limits the geometric resolution, and this can be a significant factor in the measurement of the absorbed dose-rate distribution near brachytherapy sources. When the thickness of the liquid layer is increased, the ion transport time increases, thereby increasing the general recombination at a given electric field strength. On the other hand, the leakage current, which depends on the resistivity of the liquid, will decrease linearly with the thickness. These facts imply that the chamber properties at low dose rates are improved when the layer thickness is increased, due to a better ionization current to leakage current ratio. [101]

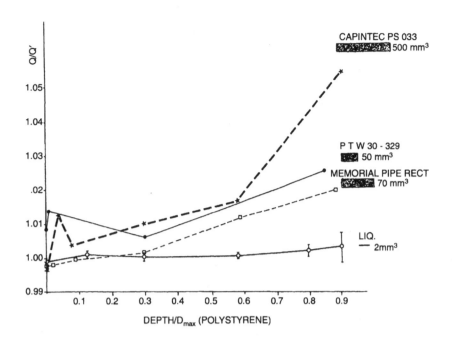

FIGURE 3.107 The polarity effect in three commercially available plane-parallel ionization chambers and the liquid-ionization chamber at different depths in polystyrene irradiated by 9-MeV electrons. (From Reference [100]. With permission.)

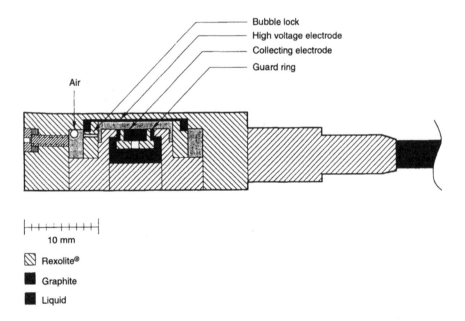

FIGURE 3.108 A cross-sectional view of the liquid-ionization chamber. The figure illustrates the electrodes, liquid, and guard ring. (From Reference [101]. With permission.)

The liquid-ionization chambers used by Johansson et al. (Figure 3.108) are of parallel-plate type. The body is made from Rexolite®, a styrene copolymer with density 1.05 g cm^{-3} and the electrodes made from pure graphite. The liquids used are tetramethylsilane (TMS; Merck, NMR calibration grade 99.7%) and isooctane (Merck, isooctane analysis grade 99.5%). The liquids were used without any further purification. TMS and isooctane responses as a function of electric field is shown in Figure 3.109. A typical response of the ionization chamber is shown in Figure 3.109.

Figure 3.110 shows the general recombination for TMS at the electric field strengths of 25, 50, 100, and 200 kV m^{-1}. The curves for the isooctane chamber are very similar to those for the TMS chamber but lie somewhat higher.

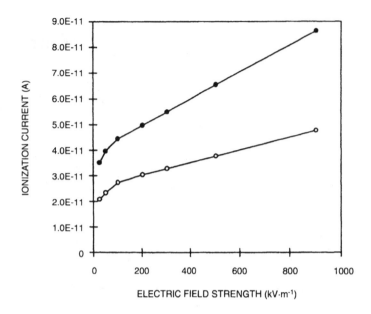

FIGURE 3.109 Typical responses for TMS (●) and isooctane (○) as a function of electric field strength. (From Reference [101]. With permission.)

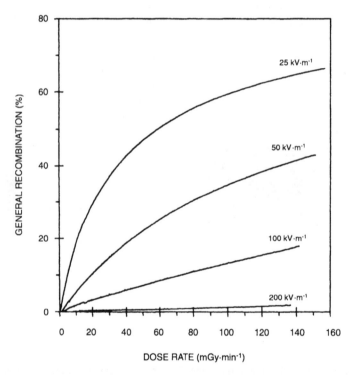

FIGURE 3.110 General recombination for the TMS liquid-ionization chamber, with 1-mm liquid thickness, as a function of dose rate. The graph shows the results at the electric field strengths of 25 kV m^{-1}, 50 kV m^{-1}, 100 kV m^{-1}, and 200 kV m^{-1}. The values are expressed as small dots, which, due to the large number of dots, gives the impression of solid curves. (From Reference [101]. With permission.)

XI. TECHNICAL DATA FOR SOME IONIZATION CHAMBERS

A number of commercially available ionization chamber dosimeters are generally used in radiation oncology therapy dosimetry. Most of them are man-ufactured by a small number of manufacturers. Generally, the ionization chambers are divided into two geometric shapes: cylindrical (thimble) and flat or parallel-plane ionization chambers. A few examples are described here; the data is as provided by the manufacturers.

TABLE 3.33
PTW 23342 Ion Chamber Dimensions
Figure 3.111

Volume:	0.02 cm^3
Response:	3×10^{-10} C/Gy
Leakage:	$\pm 1 \times 10^{-14}$ A
Polarizing voltage:	max. 300 V
Cable leakage:	1×10^{-12} C/(Gy $\times$ cm)
Wall material:	PE(CH$_2$)$_n$
Membrane thickness:	0.03 mm
Area density:	2.5 mg/cm^2
Electrode:	Graphite;
	3 mm $\varnothing$
Range of temperature:	$+10°$C .... $+40°$C
Range of relative humidity:	20% ... 75%
Ion collection time:	150V:0.05ms
	300V:0.03ms

Ionization chamber type PTW 23342 is a small soft x-ray chamber of volume 0.02 cm^3. It is the standard chamber for measurements in skin therapy. It can be used in air or in the depth of solid phantoms (e.g., phantom type T2962). The usual calibration is 15 kV to 70 kV. This chamber has a very flat energy dependence from 10 to 100 kV.

Source: From Reference [102]. With permission.

TABLE 3.34
PTW 23344 Ion Chamber Dimensions
Figure 3.112

Volume:	0.2 cm^3
Response:	7×10^{-9} C/Gy
Leakage:	$\pm 1 \times 10^{-14}$ A
Polarizing voltage:	max. 500 V
Cable leakage:	1×10^{-12} C/(Gy $\times$ cm)
Wall material:	PE(CH$_2$)$_n$
Membrane thickness:	0.03 mm
Area density:	2.5 mg/cm^2
Electrode:	Graphite with amber,
	graphite-coated;
	3 mm $\varnothing$
Range of temperature:	$+10°$C .... $+40°$C
Range of relative humidity:	20% ... 75%
Ion collection time:	300V:0.05ms
	400V:0.04ms
	500V:0.03ms

Ionization chamber type PTW 23344 is a big soft x-ray chamber of volume 0.2 cm^3.

Source: From Reference [102]. With permission.

Exradin [103] designs and manufactures ionization chambers for a wide range of dosimetric measurements. The chambers find applications in the fields of radiation therapy, diagnostic radiology, and radiation protection and

TABLE 3.35
PTW 23323 Ion Chamber Dimensions
Figure 3.113

Volume:	0.10 cm^3
Response:	3.8×10^{-9} C/Gy
Leakage:	$\pm 4 \times 10^{-15}$ A
Polarizing voltage:	max. 500 V
Cable leakage:	1×10^{-12} C/(Gy $\times$ cm)
Wall material:	PMMA(C$_5$H$_8$O$_2$)$_n$
Wall density:	1.18 mg/cm^3
Wall thickness:	1.75 mm
Area density:	210 mg/cm^2
Electrode:	Aluminum, graphite-coated;
	0.8 mm $\varnothing$; 10 mm long
Range of temperature:	$+10°$C .... $+40°$C
Range of relative humidity:	20% ... 75%
Ion collection time:	300V:0.05ms
	400V:0.04ms
	500V:0.03ms

Ionization chamber type PTW 23323 is a micro chamber of volume 0.1 cm^3.

The ionization chamber type 23323 has been designed for use with therapy dosimeters as a standard sealed chamber. It is watertight and equipped with a flexible stem, so it can be placed inside the patient's rectum during afterloading treatment. The usual calibration is for 280 kV x-ray and ^{137}Cs for afterloading use, so ^{192}Ir calibration can be obtained by interpolation, or 140 kV to ^{60}Co for general purpose use. The measuring volume is closed to the surrounding air.

Source: From Reference [102]. With permission.

research. Applications for which Exradin chambers are suited include beam calibration, quality assurance, dose assessment, depth-dose studies, mixed field dosimetry, monitoring, background measurements, and a variety of research studies.

All Exradin chambers are the three-terminal type (guarded design) and employ homogeneous construction. The chamber electrodes are made entirely from one of three conducting Shonka plastics except for the magnesium chambers, which utilize pure magnesium.

The 0.6-cm^3 2571A ionization chamber was developed by NE Technology [104] from the original Farmer design. It is constructed from a thin-walled, high-purity graphite thimble and aluminum electrode, with a detachable build-up cap which also protects the thimble. Careful guarding of the signal conductors and cable has ensured low post-irradiation leakage. The 2571A is used throughout the world for measuring dose distributions in photon or electron beam dosimetry.

The 0.2-cm^3 2577C chamber is only one third the length of the 2571A but otherwise identical. The improved spatial resolution offers greater precision in beam profiles

TABLE 3.36
PTW 31002 Ion Chamber Dimensions
Figure 3.114

Volume:	0.125 cm³
Response:	4×10^{-9} C/Gy
Leakage:	$\pm 4 \times 10^{-15}$ A
Polarizing voltage:	max. 500 V
Cable leakage:	10^{-12} C/(Gy × cm)
Wall material:	PMMA($C_5H_8O_2$)$_n$
Wall density:	1.18 mg/cm³
Wall thickness:	0.70 mm
Area density:	84 mg/cm²
Electrode:	Aluminum;
	1 mm ∅; 5 mm long
Range of temperature:	+10°C .... +40°C
Range of relative humidity:	20% ... 75%
Ion collection time:	300V:0.15ms
	400V:0.10ms
	500V:0.08ms

Ionization chamber type 31002 is a semiflex tube chamber of volume 0.125 cm³.

The 0.125-cm³ ionization chamber type 31002 is designed for measurements in the useful beam of high-energy photon or electron fields. The chamber is watertight and used mainly for relative measurements with a water phantom or air scanner for characterization of the radiation fields of therapy accelerators and teletherapy cobalt sources. The measuring volume is open to the surrounding air via cable and connector. The measuring volume is approximately spherical, resulting in a flat angular response over an angle of 160° and a uniform spatial resolution during phantom measurements along all three axes. The chamber has a short rigid stem for mounting and flexible connection cable.

Source: From Reference [102]. With permission.

TABLE 3.37
PTW 31003 Ion Chamber Dimensions
Figure 3.115

Volume:	0.3 cm³
Response:	1×10^{-8} C/Gy
Leakage:	$\pm 4 \times 10^{-15}$ A
Polarizing voltage:	max. 500 V
Cable leakage:	1×10^{-12} C/(Gy × cm)
Wall material:	PMMA($C_5H_8O_2$)$_n$
Wall density:	1.18 mg/cm³
Wall thickness:	0.75 mm
Area density:	90 mg/cm²
Electrode:	Aluminum, graphite-coated;
	1.5 mm ∅; 14.25 mm long
Range of temperature:	+10°C .... +40°C
Range of relative humidity:	20% ... 75%
Ion collection time:	300V:0.10ms
	400V:0.08ms
	500V:0.06ms

Ionization chamber type 31003 is a flex tube chamber of volume 0.3 cm³.

The ionization chamber type 31003 is designed for measurements in the useful beam of high-energy photon or electron fields. The chamber is watertight and used mainly for relative measurements with a water phantom or air scanner for characterization of the radiation fields of therapy accelerators and teletherapy cobalt sources. The measuring volume is open to the surrounding air via cable and connector. The chamber has a short rigid stem for mounting and a flexible connection cable. This chamber is type-tested by the PTB Braunschweig. The design of the chamber is similar to the 0.125-cm³ chamber type 31002 but with a larger volume for higher response.

Source: From Reference [102]. With permission.

and iso-dose contour measurements. The 0.6-cm³ 2581A ionization chamber has similar applications, using a tissue-equivalent inner electrode and thimble. It is particularly suitable for routine checking of the output from x-ray, ⁶⁰Co, and linac units.

The secondary Standard System [104] forms an essential measurement link between the primary standard held at a national laboratory and dosimeters in routine use. Exceptionally high design qualities, a flat energy response, and proven stability are mandatory for acceptance for calibration by the UK National Physical Laboratory. Certification by the laboratory is recognized in many other countries. The 0.3-cm³ ionization chamber type 2611A incorporates a graphite cap, aluminum electrode, and amber insulator with a guarded stem, resulting in negligible leakage and soakage effects. It is fitted with a TNC connector and is used with a standard 2590 Ionex Dosemaster. The x-ray energy range of 32 to 285 kV is extended to 2 MV using build-up cap 2565/NPL and to 35 MV using Perspex intercomparison phantom 2566A. The 10-m cable

is terminated by a TNC connector. A carrying case is included. The 2562A stability check source provides a constant dose rate at a nominal 0.22 Gy/min, allowing the system stability to be confirmed during the recommended three-year interval between calibrations.

The shallow cylindrical shape of the 2534D MARKUS electron beam chamber reduces uncertainties due to energy loss and scattering in dosimetric measurements, while the small volume (0.055 cm³) minimizes the uncertainty in dose-distribution measurements due to non-uniformity of the beam. The 2534D is supplied with 1 m of cable, Perspex phantom plates, waterproof cover, and source adaptor.

The applications of the 2535A and 2535B extrapolation chambers include dosimetric measurements and investigations of beta sources, calibrating the dose rates encountered in beta dosimetry, and investigating electron build-up from beams in the MV range. The chamber depth is adjustable by a micrometer screw permitting responses to be extrapolated to zero volume and extrapolation to "zero" weight. The 2535A is an air-equivalent chamber

TABLE 3.38
PTW 30001 Ion Chamber Dimensions
Figure 3.116

Volume:	0.6 cm^3
Response:	2×10^{-8} C/Gy
Leakage:	$\pm 4 \times 10^{-15}$ A
Polarizing voltage:	max. 500 V
Cable leakage:	10^{-12} C/(Gy $\times$ cm)
Wall material:	PMMA($C_5H_8O_2)_n$
Wall density:	1.18 mg/cm^3
Wall thickness:	0.45 mm
Area density:	53 mg/cm^2
Electrode:	Aluminum;
	1 mm $\varnothing$; 21.2 mm long
Range of temperature:	+10°C .... +40°C
Range of relative humidity:	20% ... 75%
Ion collection time:	300V:0.18ms
	400V:0.14ms
	500V:0.11ms

Ionization chamber type 30001 is a Farmer chamber of volume 0.6 cm^3.

The 0.6-cm^3 ionization chamber type 30001 is the standard chamber for absolute dosimetry for use with therapy dosimeters. This chamber is of rugged construction and equipped with an acrylic cap and an aluminum central electrode. It is fully guarded up to the measuring volume.

The outer dimensions are fully compatible with the Farmer chambers of other manufacturers.

Source: From Reference [102]. With permission.

TABLE 3.39
PTW 30002 Ion Chamber Dimensions
Figure 3.117

Volume:	0.6 cm^3
Response:	2×10^{-8} C/Gy
Leakage:	$\pm 4 \times 10^{-15}$ A
Polarizing voltage:	max. 500 V
Cable leakage:	10^{-12} C/(Gy $\times$ cm)
Wall material:	Graphite
Wall density:	1.85 mg/cm^3
Wall thickness:	0.45 mm
Area density:	83 mg/cm^2
Electrode:	Graphite;
	1 mm $\varnothing$; 20.7 mm long
Range of temperature:	+10°C .... +40°C
Range of relative humidity:	20% ... 75%
Ion collection time:	300V:0.18ms
	400V:0.14ms
	500V:0.11ms

Ionization chamber type 30002 is a graphite Farmer chamber of volume 0.6 cm^3.

The graphite 0.6-cm^3 ionization chamber type 30002 is the chamber for absolute dosimetry for use with therapy dosimeters where a completely graphite-built chamber is required. This chamber is of delicate construction since it is equipped with a graphite cap and a graphite central electrode. It should be handled with care. The chamber is fully guarded up to the measuring volume.

The outer dimensions are fully compatible with the Farmer chambers of other manufacturers.

Source: From Reference [102]. With permission.

with brass connectors. The 2535B collector electrode and guard ring are constructed from tissue-equivalent material. A 1-m cable is included as standard.

Data regarding NE-Technology is summarized in Table 3.46.

Quality conversion factor K_Q for some of the chambers is shown in Figure 3.124.

XII. PORTABLE IONIZATION CHAMBERS

Many radiation monitors are used in laboratories around the world. A few examples are mentioned here. The Eberline Model ASP-1, Analog Smart Portable [105], is a microcomputer-based portable radiation-measuring instrument designed to operate with most Eberline radiation detectors. The ASP-1 retains the simplicity of a meter display while adding the capabilities of a computer. The computer compensates for detector resolving (dead) time, provides an alarm if the radiation is higher than the detector is capable of measuring, provides various meter response times, and permits a large number of operating ranges. The ASP-1 displays both detection rate (mR/h, etc.) and integrated amount (mR, etc.).

The ASP-1 is designed to accommodate a wide variety of detectors, radiation units, and ranges with a minimum of actual changes to the instrument. The user can readily make these changes to meet new requirements as they occur.

The full-scale count rate for the ASP-1 is between approximately 20 counts per minute (cpm) and 3 million cpm, true count rate from the detector. With dead-time correction, the meter indication may be equivalent to as high as 5 million cpm.

The input sensitivity is adjustable from 1 to 50 mV. When the optional pulse-height analyzer (PHA) is installed, the threshold range is 1 to 20 mV and the window is adjustable from 0 to 20 mV, always referenced to the adjustable threshold.

The detector is powered by the high-voltage power supply. When the detector reacts to radiation, a pulse is coupled to the amplifier. If the amplified pulse is larger than the discriminator setting (and smaller than the window limit, if PHA is in use), the signal is sent to the microcomputer. The computer senses the various switch settings, computes the value and outputs a code to the digital-to-analog (D/A) converter. The resulting analog signal drives the meter.

TABLE 3.40
PTW 23343 Ion Chamber Dimensions
Figure 3.118

Volume:	0.055 cm³
Response:	1×10^{-9} C/Gy
Leakage:	$\pm 2 \times 10^{-15}$ A
Polarizing voltage:	max. 300 V
Guard potential :	max. 100 mV
Cable leakage:	3.5×10^{-12} C/(Gy × cm)
Wall material:	CH_2 Polyethylene
Membrane thickness:	0.03 mm
Area thickness:	2.3 mg/cm²
Electrode:	Acrylic, graphite-coated;
	5.3 mm Ø
Range of temperature:	+15°C .... +40°C
Range of relative humidity:	20% ... 75%
Ion collection time:	150V:0.20ms
	300V:0.10ms

Ionization chamber type 23343 is a Markus chamber of volume 0.055 cm³.

The Markus chamber is the first chamber in the world specially designed for electron dosimetry. Its small measuring volume and watertight construction make this chamber ideal for measurements in a water phantom, giving a good spatial resolution. The diaphragm front allows measuring in the build-up region of the electron field up to a depth of virtually zero. The measuring volume is open to the surrounding air via the connecting cable. This chamber can be delivered calibrated for absolute electron dosimetry.

Source: From Reference [102]. With permission.

TABLE 3.41
Farmer-Type Chamber Model A12 Dimensions
Figure 3.119

• Farmer chamber geometry • Axially symmetric design • Homogeneous construction • Complete guarding
• Waterproof construction
• Collecting volume: 0.651 cm³ • Collector diameter: 1.0 mm
• Wall thickness: 0.5 mm • Wall, collector, and guard material: Model A12 - Shonka air-equivalent plastic C552 • Maximum polarizing potential: greater than 1000 V • Gas flow capability
• Integrally flexible, low-noise cable • Separable minimum scatter stem • Matching ^{60}Co cap • Handsome, rugged wooden case • Inherent leakage currents: less than 10^{-15} amperes •
Cable: 50 Ohms, 29 pF/f, 2 m long • Signal connector: choice of coaxial and triaxial BNC plugs and jacks and triaxial BNC plug

Source: From Reference [103]. With permission.

TABLE 3.42
Spokas Thimble Chamber Models A2, M2, P2, T2 Dimensions
Figure 3.120

• Axially symmetric design • High collection efficiency • Homogeneous construction • Complete guarding • Waterproof construction • Gas flow capability • Integrally flexible, low-noise cable • Separable minimum scatter stem • Matching equilibrium caps • Handsome, rugged wooden case
• Collecting volume: 0.5 cm³ • Collector diameter, 4.6 mm
• Wall thickness: 1.0 mm • Wall, collector and guard material: Model A1–Shonka air-equivalent plastic C552; Model M1–magnesium; Model P1–polystyrene equivalent; Model T1–Shonka tissue-equivalent plastic A 150
• Maximum polarizing potential: greater than 1000 V • Inherent leakage currents: less than 10^{-15} ampere • Cable: 50 Ohms, 29 pF/f, 2 m long • Signal connector: choice of coaxial and triaxial BNC plugs and jacks and triaxial TNC plug • Nominal calibration factor for ^{60}Co: 6.0 R/nC

Source: From Reference [103]. With permission.

TABLE 3.43
Miniature Shonka Thimble Chamber Models A1, M1, T1 Dimensions
Figure 3.121

• Axially symmetric design • Small gas volume • Homogeneous construction • Complete guarding • Waterproof construction • Gas flow capability • Integrally flexible, low-noise cable • Separable minimum scatter stem • Matching equilibrium caps • Handsome, rugged wooden case
• Collecting volume: 0.05 cm³ • Collector diameter: 1.5 mm • Wall thickness: 1.0 mm • Wall, collector, and guard material: Model A1–Shonka air-equivalent plastic C552; Model M1–magnesium; Model T1–Shonka tissue-equivalent plastic A150
• Maximum polarizing potential: greater than 1000 V • Inherent leakage currents: less than 10^{-15} amperes • Cable: 50 Ohms, 29 pP/f, 2 m long • Signal connector, choice of coaxial and triaxial BNC plugs and jacks and triaxial TNC plug • Nominal calibration factor for ^{60}Co: 6.0 R/nC

From Reference [103]. With permission.

The microcomputer system consists of computer, address latch, and program chip. The onboard clock oscillator runs at 6 MHz and the program updates the meter reading every one-half second. A block diagram of the ASP-1 system is shown in Figure 3.125.

The RAM R-200 is a portable gamma meter designed for measuring wide-range gamma radiation fields. It is lightweight (approximately 500 g), in small dimensions (80 × 35 × 130 mm), with emphasis on ease of operation

TABLE 3.44
Xradin Ionization Chambers, Spokas Planar Chamber Models A11, P11, T11 Dimensions Figure 3.122

• Planar geometry • Homogeneous construction • Complete guarding • Waterproof construction • Gas flow capability • Low mass collector • Integrally flexible, low-noise cable • Separable minimum scatter stem • Handsome, rugged wooden case
• Collecting volume: 0.62 cm³ • Collector diameter: 20.0 mm • Window thickness: 1.0 mm • Body, collector, and guard material: Model A11–Shonka air-equivalent plastic C552; Model P11–Polystyrene equivalent plastic D400; Model T11–Shonka tissue-equivalent plastic A 150
• Maximum polarizing potential: greater than 1000 V • Inherent leakage currents: less than 10^{-15} amperes • Cable: 50 Ohms, 29 pF/f, 2 m long • Signal connector: choice of coaxial and triaxial BNC plugs and jacks and triaxial TNC plug • Nominal calibration factor for ^{60}Co :5.5 R/nC

Source: From Reference [103]. With permission.

TABLE 3.45
Thin-Window Chamber Models A11TW, P11TW, T11TW Dimensions Figure 3.123

• Planar geometry • Homogeneous construction • Complete guarding • Thin stretched conductive window • Waterproof with cover over window • Low mass collector • Gas flow capability • Integrally flexible, low-noise cable • Separable minimum scatter stem • Matching equilibrium caps • Handsome, rugged wooden case
• Collecting volume: 0.94 cm³ • Collector diameter: 20.0 mm • Window: Conductive Kapton film, 3.86 mg/cm³ • Body, collector, and guard material: Model A11TW–Shonka air-equivalent plastic C552; Model P11TW–Polystyrene-equivalent plastic D400; Model T11TW–Shonka tissue-equivalent plastic A150

Source: From Reference [103]. With permission.

and ergonomic structure. [106] The RAM R-200 meter contains an internal detector for gamma field measurements which covers a wide dynamic range of 0.1 μSv h^{-1} to 1 Sv h^{-1}. The meter includes two energy-compensated GM tubes, high-voltage regulated power supply, signal processing electronics, and embedded microprocessor circuitry with dedicated software for data processing and display. A microprocessor-controlled auto ranging switch determines the appropriate GM tube to be used according to the dose rate. Additional functions include accumulated dose measurement, malfunction detection and provision of appropriate alarm, auto recognition of external probes,

and three serial RS-232 communications channels. The RAM R-200 supports the logging of beta-gamma surface contamination activity and dose-accumulation and dose-rate measurements into its internal battery-backed-up memory. The logged data includes location (provided by an external GPS or a bar-code reader) and acquisition time. The sophisticated software enables smooth analog and digital display and fast response when abrupt changes in the dose rate measurements occurs. The meter readout is displayed on a large easy-to-read custom-designed Liquid Crystal Display (LCD).

The RAM R-200 meter's ability to concurrently measure and analyze radiation fields from both internal and external probes enables dose-accumulated and dose-rate alarms from the internal detector while the instrument measures and displays readings from the external probe. This feature improves the operator's safety.

A dedicated integrated circuitry was designed and implemented in order to achieve very low power consumption, yielding prolonged system operability over 100 hours, using a standard 9-V alkaline battery. The RAM R-200 probes can be operated using the RAM R-200 meter or by direct connection to a PC via the serial port. Testability and calibration of the meter and external probes are performed directly from a PC. This concept for testability significantly simplifies maintenance and operation.

The RAM R-200 system includes three external probes: RG-40 for a high-range gamma field covering a dose rate range of 1 mSv h^{-1} to 100 Sv h^{-1}, and RG-12 and RG-10 end-window GM probes for beta/gamma contamination detection and measurement. Similar to the meter, these probes are designed to withstand vibrations, shocks, and extreme temperature conditions.

All RAM R-200 external probes include an internal microcontroller circuitry, self-regulated power supply for the internal electronics, high-voltage power supply, and GM output signal processing electronics. The external probe microcontroller performs data processing, enables RS-232 serial communication, and executes a continual built-in test for detection of malfunction conditions to increase reliability and easy maintenance.

After connecting the external probes to the meter, the latter performs automatic probe identification. No further calibration or operator's action is required.

Figure 3.126 shows the main board layout; Figure 3.127 shows a block diagram of the internal detector board; Figure 3.128 shows a block diagram of the external detector for high dose rate monitoring; Figure 3.129 shows a block diagram of the CPU board; Figure 3.130 shows the energy response of the internal detector; and Figure 3.131 shows the dose rate response of the R-200 detector.

The RAM R-200 meter and the RG-40 probe were tested over a wide range of gamma fields. The intrinsic error of the whole effective measurement range was less than ±10%. The results meet international standard

TABLE 3.46
Charecteristics of the NE-Technology Therapy Dosimetry Chambers

	2571A		2577C		2581A		2532/3C		2536/3C	
Sensitivity	4.0 rad/nC	40 Gy/μC	11.7 rad/nC	117 Gy/μC	5.2 rad/nC	52 Gy/μC	105 rad/nC	1050 Gy/μC	7.9 rad/nC	79 Gy/μC
Energy Range	Photons	Electrons	Photons	Electrons	Photons	Electrons	Photons 7.5–100kV		Photons 7.5–100kV	
	0.05–35MV	5–35 MeV	0.05–35MV	5–35 MeV	0.1–35MV	5–35 MeV				
Max Dose Rate	3.5 krad/min	35 Gy/min	3.5 krad/min	35 Gy/min	44.0 krad/min	440 Gy/min	25 krad/sec	250 Gy/sec	4.8 krad/sec	48 Gy/sec
99% Eff. 250V	22 mrad/pulse	0.22 mGy/pulse	22 mrad/pulse	0.22 mGy/pulse	88 mrad/pulse	0.88 mGy/pulse	400 mrad/pulse	4 mGy/pulse	180 mrad/pulse	1.8 mGy/pulse
Length of table	10 m		10 m		10 m		1 m		1 m	

Source: From Reference [104]. With permission.

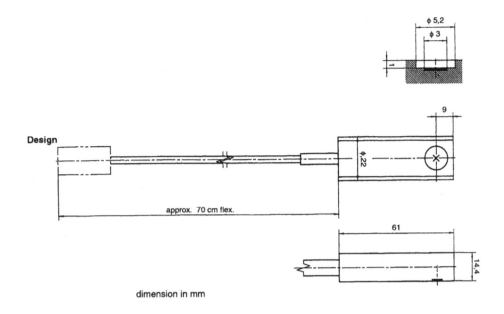

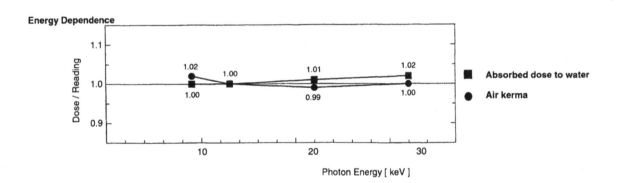

FIGURE 3.111 PTW 23342 ion chamber.

IEC1017-1 and ANSI N42 dose rate response requirements for gamma radiation rate meters. Energy and angular response are being tested.

Figure 3.131 shows the RAM R-200 meter's internal detector measurement accuracy from background (0.001 mSv/h) to 10 mSv/h. Adequate dose rate response from background to 10 mSv/h was obtained with the low-range GM tube ZP1201. The high-range GM tube ZP1313 test shows good dose rate response from 500 μSv/h to 1Sv/h. The meter's software enables a wide range of gamma radiation measurements from background to 1 Sv/h by selecting the appropriate GM and without operator intervention. To avoid continuous GM switching, a large overlapping is provided. Dead time and calibration correction is performed continuously by the meter or external probe's microcontroller.

The RG-40 dose rate response was tested over a large range of gamma fields, from 1 mSv/h to more than 100 Sv/h. This probe operates in a similar way to the RAM R-200 meter by using two GM tubes. The lower-range GM tube 4G60M measures from 1 mSv/h to 10 Sv/h. The higher range GM tube 3G10 measures from 10 mSv/h to a dose rate higher than 100 Sv/h. GM selection switching is performed by the RG-40 microcontroller, according to the internal probe software.

XIII. OTHER IONIZATION CHAMBERS

It has been recommended by several high-energy beam dosimetry protocols [107] that parallel-plate chambers are preferred for electron beam dosimetry, especially for low-energy electron beams. Even in photon beams, parallel-plate chambers have one advantage over cylindrical chambers in that the replacement correction factor P_{repl} is unity. P_{repl} for cylindrical chambers in proton beams is unity only when the chambers are calibrated at d_{max}. The TG21 protocol recommends the use of $P_{repl} = 1$ for parallel-plate chambers having guarded field and internal heights and diameters on the order of 2 and 20 mm, respectively.

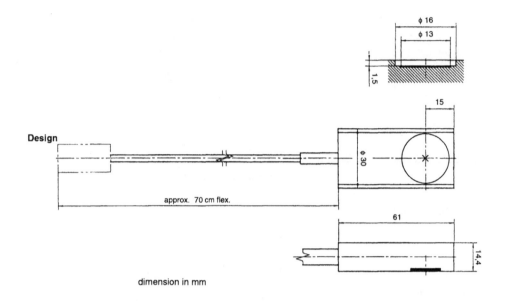

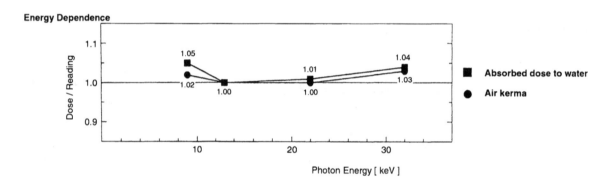

FIGURE 3.112 PTW 23344 ion chamber.

A Shonka model A2 parallel-plate chamber (Exradin Instrumentation, Lisle, IL 60532) is made of conductive plastic. It has a 1-mm-thick build-up, and the overall chamber size is 4.5 cm diameter and 1.7 cm thick. Unlike other commercial parallel-plate chambers, the Shonka chamber takes advantage of using conductive plastic as collecting electrodes, thus eliminating the need to introduce foreign conducting materials to the chamber cavity. One disadvantage of conducting plastic is that the chamber requires a special provision to separate the collecting electrode from the guard.

Solid water has a density of 1.03 gm/cm^3, which is similar to that of water. In addition, it is slightly conductive; therefore, it can be used as electrodes. A major advantage of solid water is that its dosimetric characteristics such as the electron density and the effective atomic number in high-energy photon and electron beams are similar to those of water.

The HK chamber proposed by Kubo [107] produces approximately eight times larger signals than the Attix chamber. The Attix chamber has a 0.025-mm-thin entrance window for build-up dose measurements, whereas the HK chamber has a 4-mm-thick build-up material, thick enough for ^{60}Co beam charge equilibrium, and the guard width of the Attix chamber is 13.7 mm, whereas the HK chamber has a 5-mm guard width.

Figure 3.132 shows a schematic cross-sectional diagram of the HK chamber. The collector (A in Figure 3.132) is insulated from the guard (C) by a 0.5-mm-deep and 25-mm-inner diameter polystyrene cup (B), into which the collector is snugly fit. The thickness of the insulator is 0.25-mm-throughout. The guard is separated from the chamber body by a 0.25-mm-thick polystyrene insert (E). The Attix chamber is schematically shown in Figure 3.133. The front entrance window is a 0.025-mm-thick Kapton impregnated with 5 mg/cm^2-thick graphite. The 12.7-mm diameter collector is made of 0.13-mm-thick polyethylene coated with dag.

The cavity-gas calibration factors for the Attix and HK chambers were determined against the PTW cylindrical chamber using a 20-MeV electron beam, as recommended by the TG21 protocol. The N_{gas} values from the

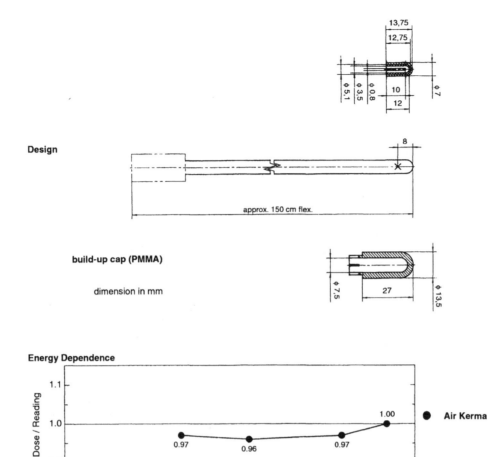

FIGURE 3.113 PTW 23323 ion chamber.

TABLE 3.47
Specifications of the Gammex/RMI Model 449 Chamber

—Dimensions: 6.0 cm O. D. × 1.4-cm height, when removed from its solid-water phantom slab.

—Materials:

Chamber body: Gammex/RMI solid water 457; Nylon screws.

Front wall: 0.025-mm-thick 4.8 mg/cm² conducting Kapton film, operated at ground potential.

Collecting electrode: Thin graphite coating on a polyethylene insulator 0.127 mm thick × 1.27 cm diameter.

—Plate separation (air-gap width): 1 mm.

—Ion-collecting volume: Approx. 0.127 cm³, vented to atmosphere. Guard ring: Outer diameter = 4.0 cm, width = 1.35 cm.

—Electrical leakage: Approx. 3 × 10⁻⁴ A.

—Cable: Low-noise triaxial; BNC male triaxial connector.

—Ion-collecting potential: ±300 V or less, to be applied to collector and guard electrodes through the electrometer's low-impedance terminal.

Source: From Reference 108. With permission.

rest of the beams were normalized to that of the 20-MeV electron beam. The N_{gas} values varied considerably, depending on the type of beams used, i.e., from −1% to almost +3%, contrary to the theory that N_{gas} is constant.

The HK chamber has a cavity volume which is approximately eight times that of the Attix chamber. Therefore, conceptually, the HK chamber is considered to have a statistical advantage in the signal-to-noise ratios over the Attix chamber. The Attix chamber has a three times larger guard width than the HK chamber. Theoretically, this should give a better definition of the collection volume. If the guard widths are inadequate for low-energy electrons, the assumption of $P_{repl} = 1$ will not hold, and N_{gas} at low- and high-electron energies should deviate from each other. This was not observed. Therefore, both chambers are assumed to have adequate guard widths. [107]

Attix a *p-p* chamber [108] is available commercially from Gammex/RMI, as the Model 449 plane-parallel ionization chamber. A schematic diagram is shown in Figure 3.133, and Table 3.47 lists its specifications. Some notes about this chamber include:

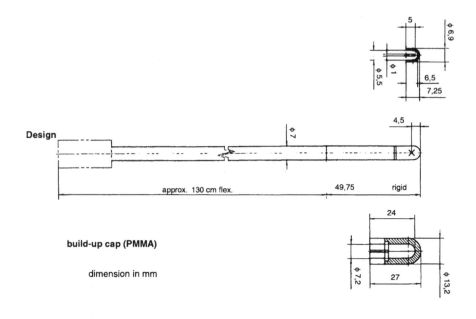

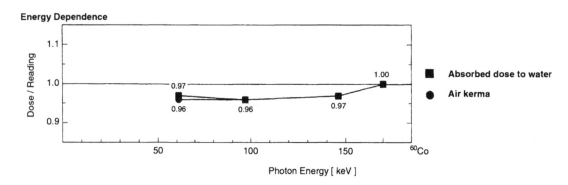

FIGURE 3.114 PTW 31002 ion chamber.

1. It is constructed almost completely of solid water to minimize perturbation of the radiation field when used in a solid-water phantom.
2. The 1-mm plate separation and 4-cm O.D. guard ring allow it to approximate closely the results of an extrapolation chamber without the inconvenience.
3. It is completely guarded, not only against electrical leakage but also against electron in-scattering through the sides of the chamber, for the photon build-up region and for electron beams.
4. The thin Kapton front wall allows measurements to be made as close as 4.8 mg/cm² inside the phantom surface and is easily replaceable by the user if damaged.
5. The polarity effect is less than that of the other *p-p* chambers that have a thin front wall.

Klevenhagen [109] developed a dosimetric method based on an ionization chamber which has an uncalibrated sensitive volume but behaves as a Bragg-Gray cavity in high-energy radiation. The chamber has a variable volume and is constructed from water-similar materials. It can be used in a water phantom directly in a beam of a therapy megavoltage machine under clinical conditions. The chamber allows absorbed dose to be determined from first principles, overcoming many of the problems encountered with conventional dosimetry based on calibrated chambers.

The chamber is shown in Figure 3.134; it was designed to work with small electrode gaps, typically between 0.5 and 5.5 mm, and it is reasonable to assume that the presence of such a small cavity does not disturb the electron fluence in the irradiated medium. Under such circumstances, the ionization volume approximates a thin-air slice surrounded by the medium with minimum perturbation of the electron fluence.

The necessity to calibrate the chamber is overcome by taking advantage of the fact that the mass of air involved in the interactions with the electron beam can be determined from the gradient of a relationship between the measured ionization charge and the gap width between the electrodes. The gradient is all that

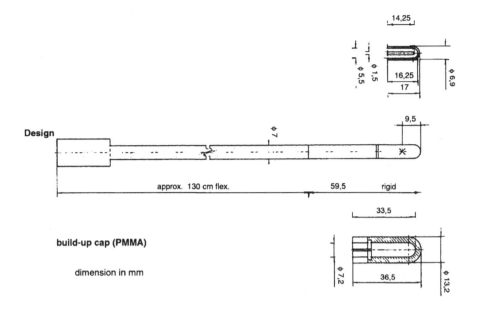

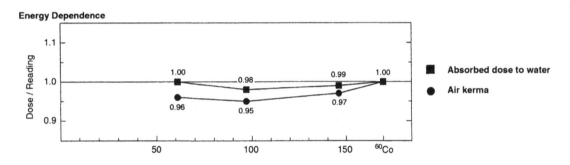

FIGURE 3.115 PTW 31003 ion chamber.

needs to be known since the gradient defines the ionization increment per increment of mass air, which follows from

$$\Delta J_a / \Delta m = (\Delta g_a/\Delta m)(1/A)(1/\rho_a) \quad (3.209)$$

where $\Delta J_a/\Delta m$ is the increment of ionization charge per increment of mass of air, g_a is the gap between electrodes, A is the effective collector area, and p is the air density corrected for ambient conditions.

The material used in construction is Perspex (Lucite), which offers good mechanical stability and electron scattering similar to water. Its plane-parallel geometry is well-suited for electron dosimetry, and its elongated shape ensures that any metal parts such as the stepping motor and driving shaft are kept far enough from the beam to avoid any significant perturbation of the electron fluence. A cross section of the chamber is shown in Figure 3.134.

The collecting electrode (2) of about 22 mm diameter and the surrounding guard ring (1) which has a width of 18 mm were formed on the front face of a mobile cylinder (7) by depositing several layers of a graphite "dag" about

12 m thick. The fine surface finish was obtained by machining off several micrometers from the surface. A clean groove was then cut at about 11 mm radius to provide a uniform electric field and to stop leakage currents from the HV electrode. Its width was sufficient to prevent perturbation of the electron fluence from extending into the sensitive volume. A 2-mm-thick high-purity graphite cap was embedded under the collector at 8 mm depth and was connected to the guard (1) in order to provide a perfect screening for the collector and the signal lead (8), which was guarded right up to the connection with the collector.

The entrance window (4) was made of a polyester film 12.5 m thick, coated with a thin layer of graphite "dag" deposited on its inner surface to create a polarizing electrode connected to the HV lead (5). The estimated thickness of the coating was about 5 m. The polarizing voltage for the chamber was provided from a variable voltage source so that at each electrode spacing the same 150-V mm^{-1} electric field strength was maintained. This was found sufficient to achieve about 99.9% ion-collection efficiency. The stability of the polarizing voltage supply was better than 1%.

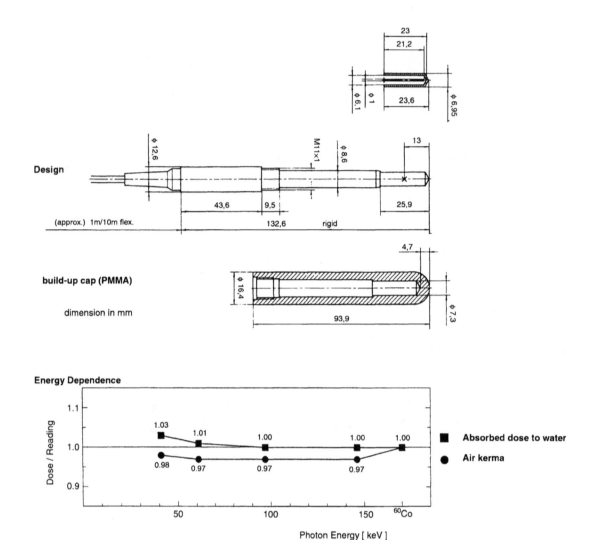

FIGURE 3.116   PTW 30001 ion chamber.

It has been demonstrated that electret ionization chambers (EIC), in which one or both electrodes of an ionization chamber are replaced by very stable electrets, could serve as integrating radiation dosimeters. [110] An external power supply is not necessary since the electret generates the electric field within the chamber volume. Charge carriers produced by photon interaction with the air in the sensitive volume migrate in the field of the electret to the electrodes of opposite sign and cause a decay of the electret charge. If the electric field generated by the electret places the sensitive volume in saturation, the difference in electret surface charge before and after its use is linearly related to the air kerma received. The natural decay rate of the electret surface charge is predictable and is included in the determination of the air kerma received.

Characteristics of a radiation-charged electret dosimeter was described by MacDonald et al. [111] The dosim-eter is based on a parallel-plate ionization chamber with the exception that the collecting electrode is covered by a thin polymer, Teflon or Mylar. During the charging of the dosimeter, ions produced in the sensitive volume by an external radiation source drift in the externally applied electric field and become trapped on the polymer surface, forming an electret. Once the external supply is removed, a field across the sensitive volume is produced by the electret charge, such that during any subsequent irradi-ation, ions opposite in sign to those on the electret sur-face are attracted to the electret and deplete the charge layer in an amount proportional to the air kerma. The remaining charge on the electret is read by an electrom-eter through further irradiation. This technique allows the dosimeter to be simultaneously charged and cali-brated, used in the field, simultaneously discharged and read, and reused again *in situ* without dismantling the dosimeter.

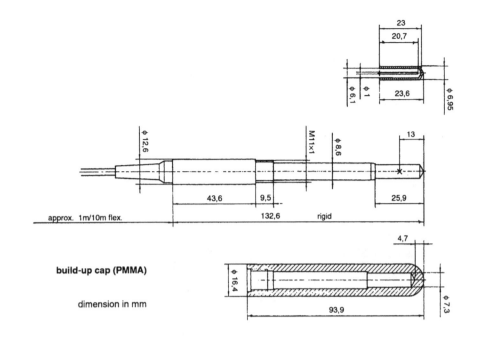

FIGURE 3.117 PTW 30002 ion chamber.

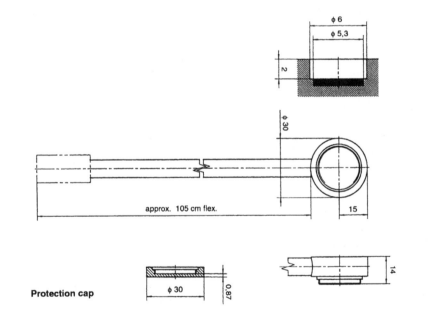

FIGURE 3.118 PTW 23343 ion chamber.

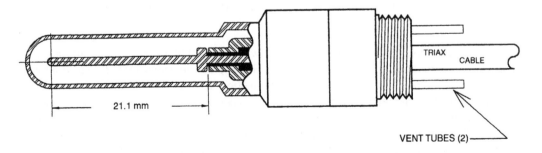

FIGURE 3.119 Extradin model A12 ion chamber.

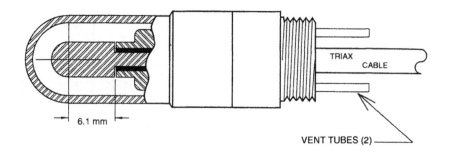

FIGURE 3.120 Extradin model A2, M2, P2, T2 ion chamber.

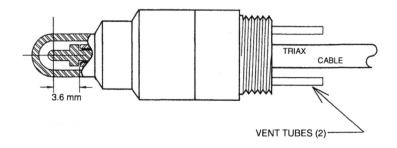

FIGURE 3.121 Extradin model A1, M1, T1 ion chamber.

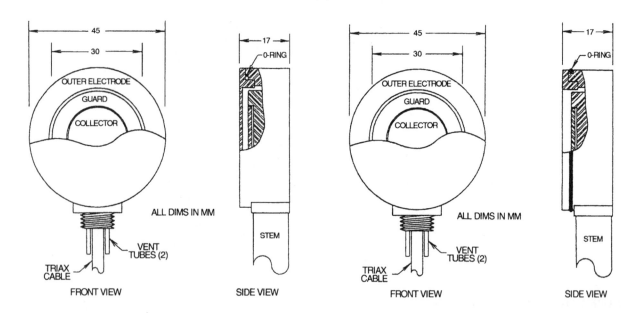

FIGURE 3.122 Extradin model A11, P11, T11 ion chamber.   **FIGURE 3.123** Extradin AIITW, PIITW, TIITW for chamber.

The electret dosimeter resembles a parallel-plate ionization chamber: the polarizing electrode is a thin aluminum foil and the measuring electrode is a metallized polymer foil with the polymer facing the polarizing electrode over an air gap. A schematic diagram of a chamber as connected to charging and discharging measurement apparatus is shown in Figure 3.135. Chambers were constructed with 1.5-mm-thick fiberglass backing, various collector radii c (0.5 to 1.9 cm), guard-ring widths g (1.0 to 19.0 mm), and air gaps a (1 to 9 mm). Small pin connectors were added to allow simple connection to the current measurement equipment and removal from the equipment when the dosimeter was to be used for monitoring. Most of the measurements were performed using a Hewlett Packard Faxitron x-ray system that can operate up to 90 kV_p. Some measurements were also performed on a highly reproducible diagnostic energy unit.

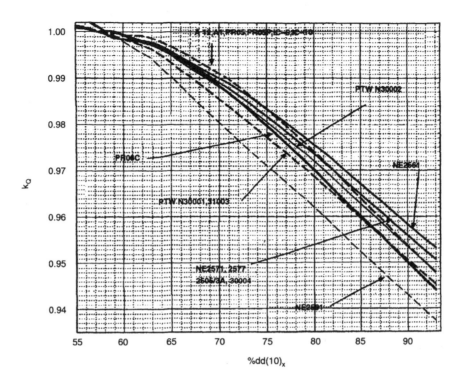

FIGURE 3.124 Values of k_Q at 10 cm depth in accelerator photon beams as a function of $\%dd(10)_x$ for cylindrical ion chambers commonly used for clinical reference dosimetry. (From TG51.)

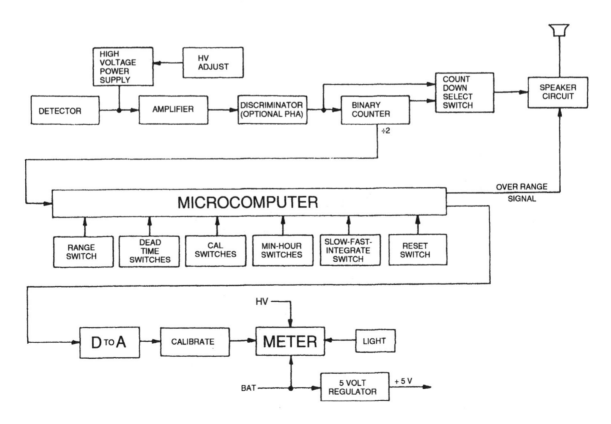

FIGURE 3.125 ASP-1 system block diagram. (From Reference 105. With permission.)

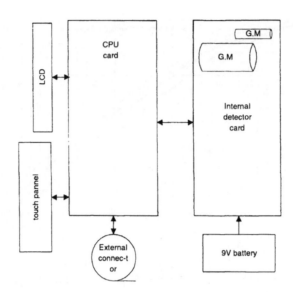

FIGURE 3.126 RAM R-200. Main board layout.

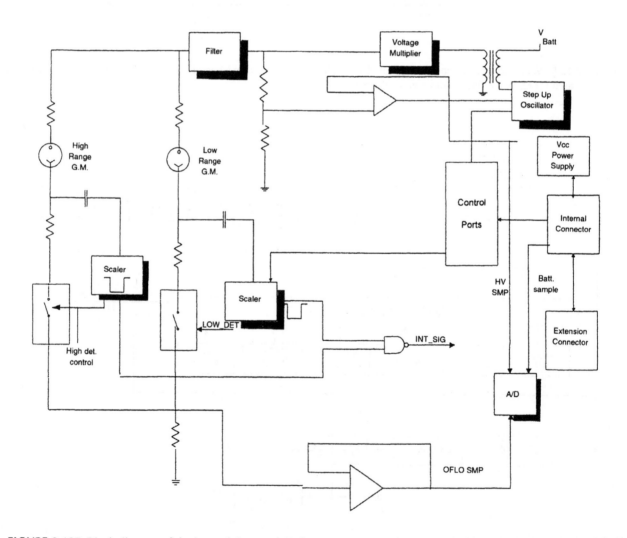

FIGURE 3.127 Block diagram of the internal detector board.

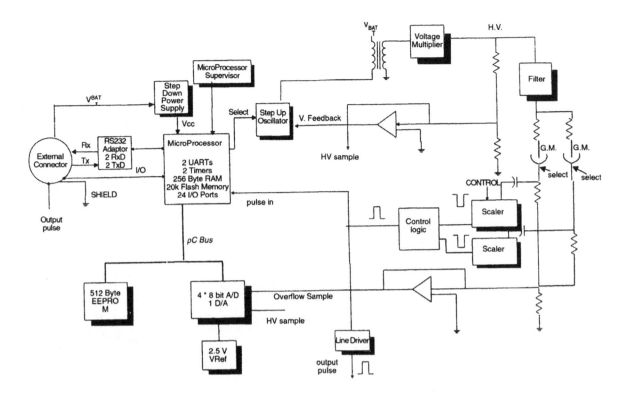

FIGURE 3.128   Block diagram of the external detector board.

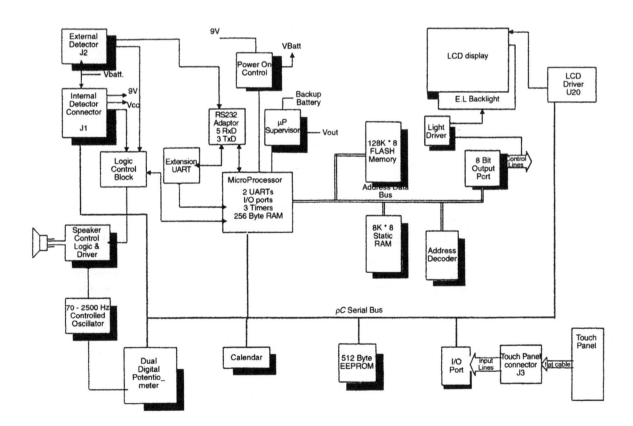

FIGURE 3.129   Block diagram of the CPU board.

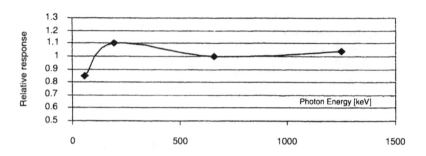

FIGURE 3.130   Energy response of the internal detector.

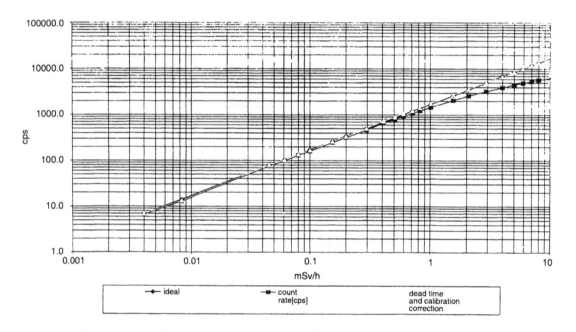

FIGURE 3.131a   RAM R-200 meter–low range GM dose rate linearity.

The surface charge density was measured as a function of air kerma using the radiation discharge method via the electrometer. Typical measurements of charge density removed during discharge over a very large range of air kerma and for various external applied voltages, V_0, are shown in Figure 3.136. These curves first rise linearly with air kerma (or with time for constant air kerma rate) and then saturate as the electret becomes neutral. The maximum level of surface-charge density removed corresponds to the maximum charge density deposited during the charging process.

The dependence of radiation-induced conductivity (RIC) on radiation rate in 254-μm-thick Teflon polyfluoroethylene propylene (FEP) film used in a electret ionization chamber (EIC) was discussed and measured by Mac-Donald and Fallone [112] with regard to radiation dosimetry. RIC permits trapped surface charge to

migrate through the polymer and recombine at the collecting electrode, which results in a reduced EIC charging efficiency and an overestimation of air-kerma in radiation dosimetry.

The EIC does not require an external potential when it is used as a radiation dosimeter because the trapped charges on the electret provide the electric field for the sensitive volume between the collecting and polarizing electrodes. Exposure to radiation produces ions in the air gap which drift towards the electret to neutralize the charge on the electret surface. Within a certain range, the air kerma is related to the difference between the original electret charge and that remaining after the irradiation.

Although RIC affects the charging of electrets only when ionizing radiation is used, when any EIC is used as a dosimeter, irradiation of the chamber allows charge to

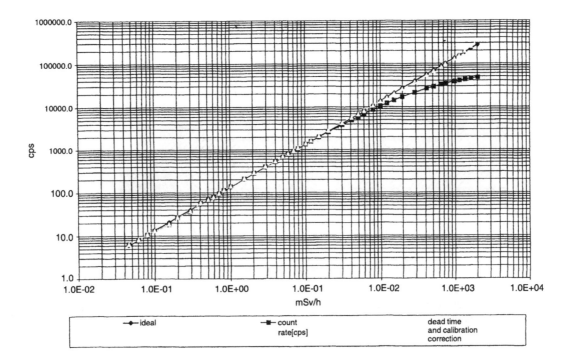

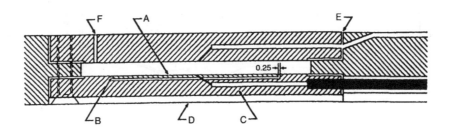

FIGURE 3.132 A cross-sectional view of the HK solid-water parallel-plate chamber. The pancake-shaped cavity size is 2 mm high and 35 mm in diameter. The front-wall thickness is 4 mm. Alphabetical symbols are (A) = 25-mm-diameter collector, (B) = 0.25-mm-thick polystyrene cup to separate the collector from the guard, (C) = 5-mm-wide guard, (D) = 1-mm-thick high-voltage protective polystyrene insulator, (E) = 0.25-mm polystyrene insulator to separate the ground from the chamber body, and (F) = air-communication hole. (From Reference [107]. With permission.)

be depleted from the electret surface because of ion motion through the sensitive volume of the air gap and through the bulk of the electret. The rate-of-charge loss through the electret is time-dependent and nonlinearly dependent on the rate at which the EIC receives air kerma. To understand the limits imposed by RIC on the use of the EIC as a personnel dosimeter, we have studied its dependence on radiation rate both theoretically and experimentally.

The assignment of dose at a point in a medium from a measurement of exposure using an air-equivalent ionization chamber involves a well-known conversion factor (the *f*-factor) that includes the ratio of the average mass energy absorption coefficients (μ_{en}/ρ) for the medium and for air. The *f*-factor is computed as

follows: [113]

$$\bar{f}_{med} = \left(\frac{\overline{W}}{e}\right) \frac{\int [\mu_{en}(E)/\rho]_{med} E\,\phi(E)\,dE}{\int [\mu_{en}(E)/\rho]_{air} E\,\phi(E)\,dE} \qquad (3.210)$$

where $\overline{W}/e$ is the mean energy expended in air per ion formed. In the integrals of Equation (3.210), μ_{en}/ρ is the mass energy absorption coefficient and $\phi(E)$ is the differential fluence spectrum, as a function of the photon energy E, at the point of interest. Because spectra of x-ray beams used in a particular application are often not known, investigators reduce consideration of

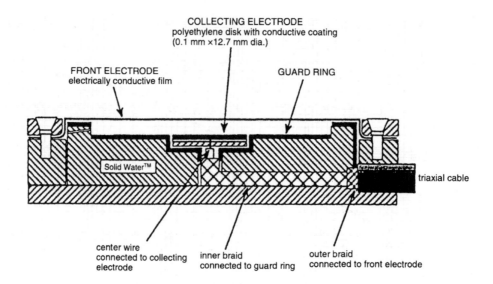

FIGURE 3.133   Schematic cross sectional diagram of the Gammex/RMI model 449 chamber showing its major components. (From Reference 102. With permission.)

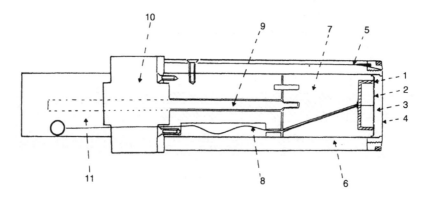

FIGURE 3.134   Schematic diagram of the chamber (not to scale). 1, guard ring; 2, collecting electrode; 3, sensitive volume; 4, entrance window (12-μm-thick polyester film); 5, polarizing voltage lead; 6, external chamber wall (Perspex); 7, mobile cylinder (Perspex) carrying the collector and guard ring; 8, signal lead; 9, screw driving the piston; 10, stepping motor for moving the collector; and 11, connector box. (From Reference [109]. With permission.)

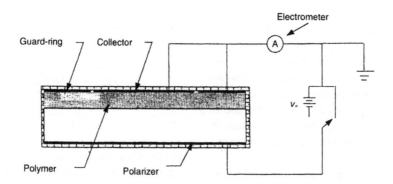

FIGURE 3.135   Schematic of radiation-charged electret ionization chamber and measurement apparatus. The external potential V_0 is applied during the charging mode but removed during the discharging mode. The leads to the external supply and electrometer are removed when the EIC is used as a dosimeter. (From Reference [112]. With permission.)

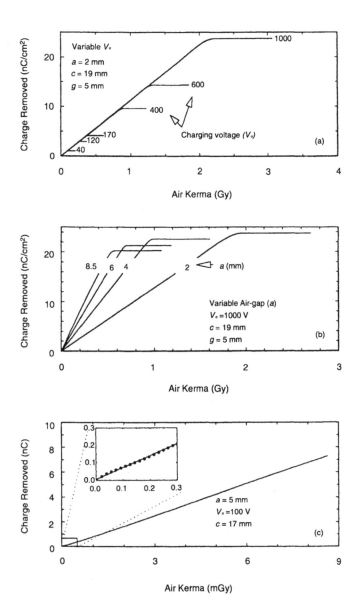

FIGURE 3.136 General dosimeter response for various geometries and V_0, q_n/A (where A is collector area), vs. air kerma for (a) constant air gap a and various initial charging voltages V_0 and (b) constant V_0 and various a. The dosimeter is formed of a 110-μm-thick Mylar electret exposed to ≈ 0.964 mGy s^{-1} (26-keV effective energy x-rays). (c) q_n vs. air kerma for very low air kerma rates of ≈ 0.049 mGys^{-1}. (From Reference [112]. With permission.)

the complete photon spectrum to that of a single photon energy through the use of an equivalent photon energy derived from half-value layer (HVL) measurements according to

$$\mu(\hat{E}) = \frac{\ln 2}{HVL} \qquad (3.211)$$

where $\hat{E}$ is the equivalent photon energy and HVL is the thickness of the first half-value layer. This approach reduces Equation (3.210) to

$$f_{med}(\hat{E}) = \left(\frac{\overline{W}}{e}\right)\frac{[\mu_{en}(E)/\rho]_{med}}{[\mu_{en}(E)/\rho]_{air}} \qquad (3.212)$$

Mono-energetic f-factors in the energy range 1 keV to 1.25 MeV for several tissues, tissue substitutes, and radiation dosimeters are reported in Table 3.48. These conversion factors, in units of Gy R$^{-1} \times 10^{-2}$, can be used in conjunction with a measurement of exposure to estimate absorbed dose at a point in a medium. (The data in this note can also be used, with minor conversions, when the measured quantity is air kerma.)

TABLE 3.48
Mono-Energetic f-factors for Tissues, Tissue Substitutes, and Radiation Dosimeters (Gy R^{-1} × 10^{-2})

E (MeV)	Tissues							Tissue Substitutes				Dosimeters		
	Adipose Tissue	Blood (Whole)	ICRU10b Bone	ICRP23 Bone	ICRU44 Bone	Lung Tissue	Soft Tissue	A-150	B-100	B-110	Water	Alanine	CaF$_2$	LiF
1.00E-03	0.638	0.924	0.836	0.908	0.918	0.923	0.901	0.549	0.780	0.930	0.989	0.744	1.27	1.02
1.50E-03	0.634	0.942	0.858	0.939	0.952	0.944	0.920	0.536	0.796	0.964	1.01	0.743	1.35	1.08
2.00E-03	0.630	0.952	0.871	0.960	0.973	0.953	0.928	0.528	0.809	0.989	1.02	0.742	1.41	1.11
3.00E-03	0.639	1.00	1.28	1.57	1.57	1.02	0.987	0.520	0.830	1.03	1.04	0.741	1.50	1.15
4.00E-03	0.572	0.925	1.20	1.49	1.50	0.939	0.921	0.461	0.760	0.948	0.940	0.663	1.40	1.05
5.00E-03	0.564	0.924	2.78	3.75	3.92	0.938	0.921	0.646	2.63	3.73	0.933	0.654	6.84	1.06
6.00E-03	0.558	0.923	2.96	4.01	4.19	0.938	0.922	0.659	2.82	4.02	0.928	0.647	7.41	1.06
8.00E-03	0.549	0.939	3.24	4.42	4.63	0.937	0.921	0.680	3.12	4.47	0.919	0.636	8.31	1.06
1.00E-02	0.542	0.941	3.45	4.73	4.95	0.936	0.921	0.697	3.35	4.82	0.913	0.628	8.99	1.06
1.50E-02	0.532	0.946	3.82	5.26	5.51	0.934	0.921	0.726	3.74	5.41	0.902	0.615	10.2	1.06
2.00E-02	0.528	0.948	4.04	5.58	5.85	0.933	0.921	0.749	3.99	5.78	0.895	0.608	10.9	1.06
3.00E-02	0.541	0.951	4.20	5.81	6.10	0.932	0.921	0.785	4.18	6.05	0.887	0.613	11.4	1.04
4.00E-02	0.587	0.954	4.01	5.51	5.78	0.934	0.925	0.822	4.01	5.76	0.891	0.645	10.8	1.01
5.00E-02	0.659	0.957	3.51	4.77	4.99	0.939	0.932	0.859	3.52	4.99	0.903	0.700	9.19	0.971
6.00E-02	0.739	0.960	2.91	3.87	4.03	0.945	0.940	0.892	2.92	4.03	0.919	0.761	7.24	0.928
8.00E-02	0.858	0.963	1.95	2.43	2.51	0.955	0.952	0.932	1.96	2.51	0.945	0.853	4.14	0.868
1.00E-01	0.917	0.964	1.45	1.69	1.73	0.961	0.959	0.949	1.46	1.73	0.959	0.899	2.54	0.840
1.50E-01	0.961	0.965	1.07	1.12	1.12	0.964	0.963	0.961	1.06	1.11	0.970	0.934	1.29	0.818
2.00E-01	0.970	0.965	0.982	0.991	0.985	0.966	0.965	0.963	0.978	0.980	0.973	0.941	1.02	0.814
3.00E-01	0.974	0.965	0.944	0.935	0.925	0.966	0.965	0.964	0.940	0.919	0.974	0.944	0.898	0.812
4.00E-01	0.975	0.965	0.936	0.923	0.912	0.966	0.965	0.964	0.931	0.906	0.974	0.945	0.872	0.812
5.00E-01	0.976	0.965	0.933	0.919	0.908	0.966	0.965	0.964	0.929	0.902	0.974	0.946	0.862	0.812
6.00E-01	0.976	0.965	0.932	0.917	0.905	0.966	0.965	0.964	0.927	0.899	0.974	0.946	0.858	0.812
8.00E-01	0.976	0.965	0.930	0.915	0.904	0.966	0.965	0.964	0.926	0.898	0.974	0.946	0.854	0.812
1.00E+00	0.976	0.966	0.930	0.915	0.903	0.966	0.966	0.964	0.926	0.897	0.975	0.946	0.853	0.812
1.25E+00	0.976	0.965	0.930	0.914	0.902	0.966	0.965	0.964	0.925	0.896	0.974	0.946	0.851	0.812

Source: From Reference 113. With permission.

REFERENCES

1. **Khan, F. M. et al.,** *Med. Phys.,* 18, 73, 1991.
2. **Mattsson, L. O. et al.,** *Oncology,* 20, 385, 1981.
3. **ICRU Report 35,** 1984.
4. **IAEA Tech. Rep.,** Series No. 277, 1987.
4a. **Nordic Association of Clinical Physics (NACP),** *Acta Radiol.,* 20, 401, 1981.
5. **Derikum, K. and Roos, M.,** in IAEA Tech. Rep., Series No. 381, 1997.
6. **Andreo, P.,** *Int. J. Radiat. Oncol. Biol. Phys.,* 19, 1233, 1990.
7. **Medin, J. et al.,** *Phys. Med. Biol.,* 40, 1161, 1995.
8. **IAEA Tech. Rep.,** Series No. 381, 1997.
9. **Laitano, R. F. et al.,** *Phys. Med. Biol.,* 38, 39, 1993.
10. **Nystrom, H. and Karlsson, M.,** *Phys. Med. Biol.,* 38, 311, 1993.
11. **Nilsson, B. et al.,** *Phys. Med. Biol.,* 42, 2101, 1997.
12. **IPEMB,** Thwaites, D. I. (Chair) et al., *Phys. Med. Biol.,* 41, 2557, 1996.
13. **AAPM Task Group 21,** *Med. Phys.,* 18, 73, 1991.
14. **AAPM Task Group 39,** *Med. Phys.,* 21, 1251, 1994.
15. **Vatnitsky, S. M. et al.,** *Med. Phys.,* 23, 25, 1996.
16. **Schulz, R. J.,** *Med. Phys.,* 18, 73, 1991.
17. **Vynckier, S.,** *Rad. Oncol.,* 32, 174, 1994.
17a. **Mattsson, L. O.,** Comparison of water calorimeter and ionization chamber, CCEMRI (I), 85, 1985.
17b. **Kubo, H.,** *Radiother. Oncol.,* 4, 275, 1985.
18. **Seuntjens, J. et al.,** *Phys. Med. Biol.,* 38, 805, 1993.
19. **Ma, C-M. and Nahum, A. E.,** *Phys. Med. Biol.,* 38, 1993.
20. **Seuntjens, J. et al.,** *Phys. Med. Biol.,* 33, 1171, 1988.
21. **Podgorsak, E. B. et al.,** *Med. Phys.,* 25, 1206, 1998.
22. **Seuntjens, J. and Verhaegen, F.,** *Med. Phys.,* 23, 1789, 1996.
23. **Ma, C-M. and Nahum, A. E.,** IAEA-SM-330/5, 371, 1994.
24. **Van der Plaetsen, A. et al.,** *Med. Phys.,* 21, 37, 1994.
25. **Tailor, R. C. et al.,** *Med. Phys.,* 25, 496, 1998.
26. **Rosenow, U. F.,** *Med. Phys.,* 20, 739, 1993.
27. **Rogers, D. W. O.,** *Med. Phys.,* 20, 319, 1993.
28. **Ma, C-M. and Nahum, A. E.,** *Phys. Med. Biol.,* 40, 63, 1995.
29. **Weber, L. et al.,** *Phys. Med. Biol.,* 42, 1875, 1997.
30. **Lam, K. L. and Ten Hake, R. K.,** *Med. Phys.,* 23, 1207, 1996.
31. **Khan, F. M. et al.,** *Int. J. Radiat. and Biol. Phys.,* 35, 605, 1996.
32. **Rocha, M. P. O. et al.,** *Phys. Med. Biol.,* 38, 793, 1993.
33. **Nath, R. and Schulz, R. J.,** *Phys. Med. Biol.,* 8, 85, 1981.
34. **Rawlinson, J. A. et al.,** *Med. Phys.,* 19, 641, 1992.
35. **Gerbi, B. J. and Khan, F. M.,** *Med. Phys.,* 24, 873, 1997.
36. **Takata, N. and Matiullah.,** *Phys. Med. Biol.,* 36, 449, 1991.
37. **Zankowski, C. and Podgorsak, E. B.,** *Med. Phys.,* 25, 908, 1998.
38. **Boag, J. W. and Wilson, T.,** *Br. J. Appl. Phys.,* 3, 222, 1952.
39. **Ritz, V. H. and Attix, F. H.,** *Rad. Res.,* 16, 401, 1962.
40. **Takata, N.,** *Phys. Med. Biol.,* 39, 1037, 1994.
40a. **Burns, J. E. and Rosser, K. E.,** *Phys. Med. Biol.,* 35, 687, 1990.
41. **Havercroft, J. M. and Klevenhagen, S. C.,** *Phys. Med. Biol.,* 38, 25, 1993.
42. **Boag, J. W.,** in *The Dosimetry of Ionizing Radiation,* Kase, K. R. et al., eds., Academic Press, Inc., New York, 1987.
43. **Almond, P. R.,** *Med. Phys.,* 8, 901, 1981.
44. **Geleijns, J. et al.,** *Med. Phys.,* 22, 17, 1995.
45. **Das, I. J. and Akber, S. F.,** *Med. Phys.,* 25, 1751, 1998.
46. **Boutillon, M.,** *Phys. Med. Biol.,* 43, 2061, 1998.
47. **Boag, J. W.,** *Int. J. Radiat. Phys. Chem.,* 1, 267, 1969.
48. **Burns, D. T. and McEwen, M. R.,** *Med. Biol.,* 43, 2033, 1998.
49. **Piermattei, A. et al.,** *Med. Biol.,* 41, 1025, 1996.
50. **Almond, P. R.,** *Med. Phys.,* 8, 901, 1981.
51. **Weinhous and Meli,** *Med. Phys.,* 11, 846, 1984.
52. **Biggs, P. J. and Nogueira, I. P.,** *Med. Phys.,* 26, 2107, 1999.
53. **Geleijns, J. J. et al.,** *Med. Phys.,* 22, 17, 1995.
54. **Johansson, B. et al.,** *Phys. Med. Biol.,* 42, 1929, 1997.
55. **Hummel, A.,** *Advances in Radiation Chemistry,* Vol. 4, Wiley, New York, 1974.
56. **Onsager, L.,** *Phys. Rev.,* 54, 554, 1938.
57. **Wittkämper, F. W. et al.,** *Phys. Med. Biol.,* 37, 995, 1992.
58. **Kappas, C. et al.,** *Med. Phys.,* 24, 1797, 1997.
59. **Rogers, D. W. O. and Bielajew, A. F.,** *Phys. Med. Biol.,* 35, 1065, 1990.
60. **Nilsson, B. et al.,** *Phys. Med. Biol.,* 41, 609, 1996.
61. **Hant, M. A. et al.,** *Med. Phys.,* 15, 96, 1988.
62. **Aget, H. and Rosenwald, J.,** *Med. Phys.,* 18, 67, 1991.
63. **Havercroft, J. M. and Klevenhagen, S. C.,** *Phys. Med. Biol.,* 39, 299, 1994.
64. **Gerbi, B. J. and Khan, F. M.,** *Med. Phys.,* 14, 210, 1987.
65. **Van Dyk, J. and MacDonald, J. C. F.,** *Rad. Res.,* 50, 20, 1972.
66. **Williams, J.A. and Agarwal, S. K.,** *Med. Phys.,* 24, 785, 1997.
67. **Nilsson, B. and Montelius, A.,** *Med. Phys.,* 13, 191, 1986.
68. **Ramsey, C. R. et al.,** *Med. Phys.,* 26, 214, 1999.
69. **Harder, D.,** *Biophysik,* 5, 157, 1968.
70. **Roos, M. et al.,** *Med. Physik.,* 24, 1993.
71. **Wittkämper, F. W. et al.,** *Phys. Med. Biol.,* 36, 1639, 1991.
72. **Johansson, K. A. et al.,** *Proc. Symp. On National and International Standards of Radiation Dosimetry,* IAEA SM 222/35, 243, 1978.
73. **Reft, C. S. and Kuchnir, F. T.,** *Med. Phys.,* 26, 208, 1999.
74. **Zoetelief, J. et al.,** *Phys. Med. Biol.,* 35, 1287, 1990.
75. **Zoetelief, J. et al.,** *Phys. Med. Biol.,* 34, 1169, 1989.
76. **Donaldson, J. R. and Tryon, P. V.,** *User Guide to STARPAC,* NBS/R86-3448, 1987.
77. **Reft, C. S. and Kuchnir, F. T.,** *Med. Phys.,* 18, 1237, 1991.
78. **Khan, F. M.,** *Med. Phys.,* 18, 1244, 1991.

79. Cunningham, J. R. and Sontag., M. C., *Med. Phys.*, 7, 672, 1980.
80. Ma, C-M. and Nahum, A. E., *Phys. Med. Biol.*, 40, 45, 1995.
81. Almond, P. R. et al., *Med. Phys.*, 21, 1251, 1994.
82. AAPM Task Group 21, *Med. Phys.*, 10, 741, 1983.
83. Gastorf, R. et al., *Med. Phys.*, 13, 751, 1988.
84. Rogers, D. W. O., *Med. Phys.*, 19, 889, 1992.
85. Rogers, D. W. O., *Med. Phys.*, 19, 1227, 1992.
86. Murali, V. et al., *Phys. Med. Biol.*, 38, 1503, 1993.
87. Huq, M. S. et al., *Med. Phys.*, 20, 293, 1993.
88. Reft, C. S. et al., *Med. Phys.*, 21, 1953, 1994.
89. Huq, M. S. et al., *Med. Phys.*, 22, 1333, 1995.
90. Almond, P. R. et al., *Med. Phys.*, 22, 1307, 1995.
91. Reynaert, N. et al., *Phys. Med. Biol.*, 43, 2095, 1998.
91a. Russell, K. R. and Ahnesjo, A., *Phys. Med. Biol.*, 41, 1007, 1996.
92. Aukett, R. J. et al., in *AAPM Proc. No. 11, Kilovolt X-ray Beam Dosimetry for Radiotherapy and Radiobiology,* 1997, 196.
93. Ma, C-M. et al., *Phys. Med. Biol.*, 39, 1593, 1994.
94. Wickman, G. and Nystrom, H., *Phys. Med. Biol.*, 37, 1789, 1992.
94a. Burlin, T. E. and Chan, E. K., Solid state and chemical radiation dosimetry, IAEA, 393, 1967.
95. Johansson, B. and Wickman, G., *Phys. Med. Biol.*, 42, 133, 1997.
95a. Greening, J. R., *Phys. Med. Biol.*, 9, 143, 1964; 9, 566, 1965.
96. Mie, G., *Ann. Phys. Lpz.*, 13, 857, 1904.
97. Wickman, G. et al., *Med. Phys.*, 25, 900, 1998.
98. Hubbell, J. H. and Seltzer, S. M., NISTIR Report 5632, U.S. Department of Commerce, 1995.
99. Dasu, A. et al., *Phys. Med. Biol.*, 43, 21, 1998.
100. Wickman, G. and Holmstrom, T., *Med. Phys.*, 19, 637, 1992.
101. Johansson, B. et al., *Phys. Med. Biol.*, 40, 575, 1995.
102. *PTW-FREIBURG Ionization Chamber Catalog,* 1999.
103. *Radiation Oncology Source Book, MED-TEC Catalog,* 1999.
104. *Radiological Dosimetry, NE-Technology Precision Dosimetry Catalog,* 1999.
105. *Analog Smart Portable Technical Manual,* Eberline Model ASP-1, 1997.
106. Wangrovich, U., in: *Proc. 20th Conf. of the Nuclear Societies in Israel,* 1999, 73.
107. Kubo, H., *Med. Phys.*, 20, 341, 1993.
108. Attix, F. A., *Med. Phys.*, 20, 735, 1993.
109. Klevenhagen, S. C., *Phys. Med. Biol.*, 36, 239, 1991.
110. Gross, B., *Topics in Applied Physics*, 33, 217, 1987.
111. MacDonald, B. A. et al., *Phys. Med. Biol.*, 37, 1825, 1992.
112. MacDonald, B. A. and Fallone, B. G., *Phys. Med. Biol.*, 40, 1609, 1995.
113. Schauer, D. A. and Links, J. M., *Med. Phys.*, 20, 1371, 1993.
114. Klerenhagen, S. C., *Phys. Med. Biol.*, 36, 1013, 1991.

4 Thermoluminescent Dosimetry

CONTENTS

I. INTRODUCTION

Thermoluminescent dosimetry is used in many scientific and applied fields such as radiation protection, radiotherapy clinic, industry, and environmental and space research, using many different materials. The basic demands of a thermoluminescent dosimeter (TLD) are good reproducibility, low hygroscopicity, and high sensitivity for very low dose measurements or good response at high doses in radiotherapy and in mixed radiation fields. LiF is used for dose measurements in radiotherapy since the effective atomic number of 8.3 is close to that of water or tissue. Lithium tetraborate is more tissue-equivalent than LiF, but it is deliquescent (absorbs moisture from the atmosphere) and its stored signals fade rapidly. Its use is therefore only worthwhile for x-rays, where the closeness of its effective atomic number of 7.3 to tissue outweighs the disadvantages. Calcium sulphate has an effective atomic number of 15.6 and is therefore much less tissue-equivalent, but its effective atomic number is quite close to that of bone. It is very sensitive and therefore can be used for protection dosimetry. Calcium fluoride has an effective atomic number of 16.9 and is also used for protection dosimetry, as it is also very sensitive.

TLDs are relative dosimeters and therefore have to be calibrated against absolute dosimetry systems such as a calibrated ion chamber. A ^{60}Co gamma source is generally used. Due to their small size, TLDs are convenient for dose-distribution measurements in medicine and biology.

The use of TLDs in electron-beam dosimetry is inherently more complicated than its use in photon dosimetry since, for each incident electron beam energy, one obtains a different dose response. As the electron beam penetrates into material, it gradually loses its energy so that both the dose and energy vary with depth in material, making an accurate dose measurement with TLD more difficult.

Fading is an important phenomenon when dosimeters are used for environmental or personal monitoring, which involves reading after long periods of irradiation.

Investigation of Linear Energy Transfer (LET) dependence of TL dosimeters indicates that for all commonly used materials, the sensitivity (TL response/absorbed dose) decreases with increasing LET. TLD-100 exhibits a supralinear dose response above about 2 Gy.

Dosimetry of mixed neutron-gamma radiation is usually performed with a pair of ^{6}LiF and ^{7}LiF dosimeters exhibiting various thermal neutron sensitivities. The TL glow in the ^{6}LiF TLD is primarily generated by the densely ionized secondary alpha particles produced in the dosimeter via a ^{6}Li$(n,\alpha)^3$H reaction. The common assumption that TLD-700 (99.99% ^{7}LiF) has a negligible thermal neutron response relative to its photon response can lead to errors. The TL in a TLD-700 is created by the ^{7}Li$(n,\gamma)^8$Li reaction ($\sigma_{th} = 0.033$ b) as well as the energy transferred to dosimeter material via elastic and inelastic scattering of the impinging neutrons.

By peak analysis of the glow curve of CaF$_2$, it is possible to separate the low and high LET components of a mixed radiation field.

An LiF:Mg,Cu,P dosimeter is useful for low-dose measurement because of its high sensitivity compared to the TLD-100.

The use of computerized glow-curve analysis (GCA) methods has become a normal practice. The application of deconvolution-based GCA methods to complex curves provides information on the parameters of each individual peak.

Knowledge of backscatter factor (BSF) for low-energy x-ray is essential in diagnostic radiology as well as radiotherapy dosimetry to determine the absorbed dose at the surface of the patient.

For proton and alpha particles, a predominant contribution to total stopping power comes from the electronic stopping power due to inelastic collision with electrons.

Electron spin resonance (ESR) investigation of TL samples can, in some cases, help identification of an ongoing mechanism of luminescent centers

II. LIF:MG, TI TL DOSIMETER

LiF:Mg,Ti (TLD-100), a nearly tissue-equivalent material of good sensitivity, is perhaps the best choice for radiation dosimetry. It has been commonly used for medical purposes because it is a nearly tissue-equivalent material (effective atomic number of 8.2 compared to 7.4 for tissue). The weakest characteristic of TLD-100 for radiotherapy dosimetry is accuracy. The existence of fading and sensitivity changes also complicates the situation. The TL response of TLD-100 is not linear for large doses and its efficiency depends on the radiation field. Studies have shown a dependence of the TL efficiency and the supralinearity on the photon energy.

The energy correction factor of LiF TLD calibrated in ^{60}Co γ-rays and used for measurements in electron beam is rather confusing. Some show no energy dependence relative

to ^{60}Co γ-rays, while others show correction factors in the range 1.03–1.1.

An important use of LiF is dosimetry of mixed neutron-gamma radiation. It is usually performed with a pair of ^{6}LiF and ^{7}LiF dosimeters taking advantage of the difference in cross section of the two Li isotopes.

Muniz et al. [1] studied LiF TLD-100 performance for mailed dosimetry in radiotherapy, using glow-curve analysis for TL evaluation and reusable chips. The use of such a glow-curve analysis method permitted determination of the evolution of each peak during storage at room temperature. Delayed TL measurements, including controlled time delays between the steps of the TL cycle, were employed for that determination. Storage took place in a room whose temperature was controlled only during working hours and set at 22°C. During the night the temperature dropped in winter to 15°C and rose in summer to 28°C.

A computer code developed in CIEMAT for the analysis of complex glow curves, resolving the individual peaks, was employed in the experiment. The program fits the experimental points using the first-order kinetics expression for a single peak.

Figure 4.1 shows the kind of result provided by the computer program for a typical prompt LiF TLD-100 glow curve as produced by the preparation treatment employed. The measurement identification and the computing time

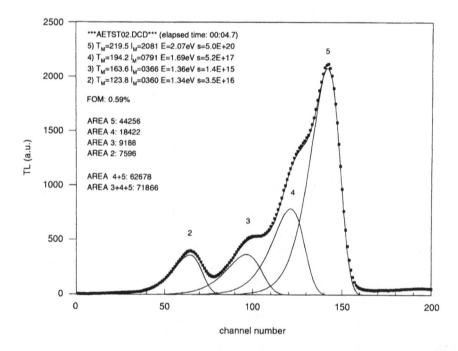

FIGURE 4.1 An analyzed TLD-100 glow curve: squares, experimental points; continuous lines, fitted glow curve and resolved individual peaks. The curve identification, the time required for the analysis in seconds, and the fitted peak parameters are presented. (From Reference [1]. With permission.)

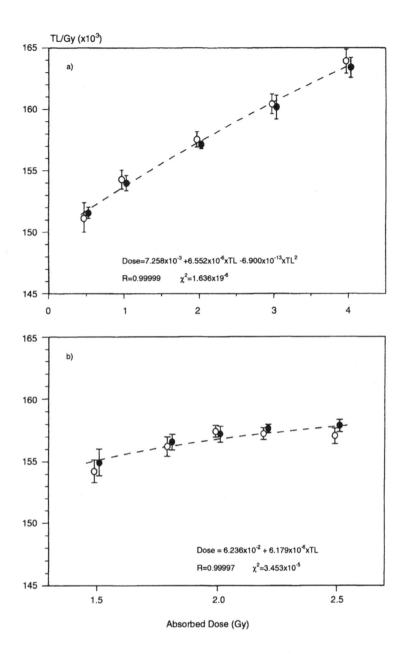

FIGURE 4.2 TL dependence on absorbed dose, expressed as TL per unit dose against dose, (a) in the range 0.5–4.0 Gy and (b) in a reduced range, 1.5–2.5 Gy. Two independent sets of results (open and closed symbols) are presented, obtained from the same group of five detectors. Circles, group mean value; uncertainty bars, two standard deviations of the mean value. In each panel the fitted dose against TL algorithm for the two sets of data is presented: in (a), quadratic dependence, and in (b), linear dependence. Each pair of open and closed points corresponds to the same absorbed dose value. (From reference [1]. With permission.)

can be seen in the figure with the fitted characteristic parameters for each peak (2–5): the temperature of the maximum, T_m; the maximum intensity, I_m; the activation energy, E; and the frequency factor, s. The areas of the different peaks, and the sums of some of them (5 + 4 or 5 + 4 + 3) are also given.

Figure 4.2 presents the results obtained in two tests on the dose against TL dependence near the 2 Gy reference dose using a single group of detectors. Between these tests

the group was employed several times for different types of measurement. Figure 4.2a shows this dependence in the range 0.5–4.0 Gy and Figure 4.2b shows the same dependence but in a reduced range, 1.5–2.5 Gy. In Figure 4.2 the results are expressed as TL Gy^{-1} against absorbed dose in gray instead of the more usual dose against TL plot, in order to appreciate better the dependence on absorbed dose. The uncertainty bars correspond to two standard deviations (mean value) of the group results at each dose

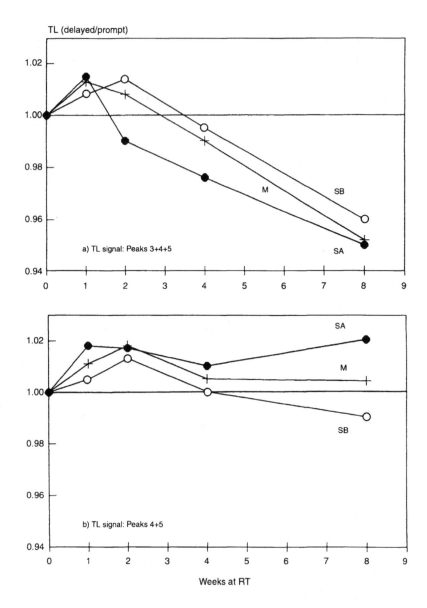

FIGURE 4.3 TL evaluation during storage at room temperature, integrating under several peaks. (From Reference [1]. With permission.)

value. Apart from the closeness of the data obtained in the two tests, a certain supralinearity can be appreciated in Figure 4.2a. A higher TL Gy^{-1} value is obtained at 4.0 Gy than at 0.5 Gy. Each set of data can be well-fitted by a quadratic expression:

$$Dose(\text{Gy}) = A + B\text{TL} + C\text{TL}^2 \qquad (4.1)$$

The quadratic term accounts for supralinearity, so C is always negative and very small. The broken line represents the TL Gy^{-1} against absorbed dose variation as deduced from the fitted quadratic expression.

Figure 4.3 presents the TL results at 2 Gy obtained after different storage intervals normalized to the corre-

sponding prompt values. Data are presented for two integration regions, including three peaks (3 + 4 + 5) (Figure 4.3a), or only two (4 + 5) (Figure 4.3b). Open symbols refer to storage before irradiation (SB) and closed ones refer to storage after irradiation (SA). Crosses are for the groups irradiated in the middle of storage (M). In the figure the different evolution of the SA and SB groups can be appreciated. For the three-peak integration criterion, the groups irradiated at the beginning of storage (SA) show a lower TL signal than the groups irradiated at the end (SB) of the same storage interval. This may be interpreted in terms of fading, affecting filled traps during storage. Nevertheless, the situation is more complex as both groups SA and SB tend to decrease when the storage duration increases, and obviously fading cannot affect empty traps (thc SB groups). Hence, there is also evidence of trap effects. The evidence is stronger if the integration region is limited to peaks 4 and 5 only, omitting peak 3. In this case, the SA groups (filled traps during storage) systematically give higher TL values than the SB groups (empty traps during storage) and also higher values than the reference values in the prompt glow curves. Clearly, this behavior suggests that the dominant processes affecting peaks 4 and 5 are trap processes, causing an increase in sensitivity, and not charge leakage or fading processes. Peak 3, lying at lower temperatures than peaks 4 and 5, is clearly affected by fading (and also by trap effects).

A general-purpose computer code thermoluminescent detector simulator (TLD-SIM) was developed by Branch and Kearfott [2] to simulate the heating of any TLD type using a variety of conventional and experimental heating methods.

In order to calculate the TL intensity, it is first necessary to calculate the temperature T as a function of position and time, which is given by the second order partial differential heating equation:

$$\rho C \frac{\partial T(x,y,z,t)}{\partial t} = -\lambda \nabla^2 T(x,y,z,t) + S \quad (4.2)$$

where ρ is the density, C is the heat capacity, λ is the thermal conduction coefficient, and S is the heat source term for internal heat generation. The Laplacian is numerically evaluated using a second-order linear approxima-

tion, and the above equation simplifies to:

$$\Delta T = -\frac{\lambda}{\rho C} \Delta t \nabla^2 T + \frac{q(x,y,z,t)}{\rho C} \Delta t \quad (4.3)$$

where q is the heating term for laser heating, given by

$$q(x,y,z,t) = \mu(1 - r_f)I_0 e^{-\mu z} \quad (4.4)$$

Here μ is the absorption coefficient, r_f the reflectivity, I_0 the laser power, and z the depth into the chip. It should be noted that $q = 0$ for all other types of heating.

Once the temperature is determined for each volume element, the glow intensity is then calculated by the first order Randall-Wilkins model: [3]

$$I(x,y,z,t) = s_m D(x,y,z) \left[\exp\left(-\frac{E_m}{kT(x,y,z,t)} \right) \right]$$
$$\times \exp\left[-\int_t s_m \exp\left(-\frac{E_m}{kT(x,y,z,t)} \, dt \right) \right]$$
$$(4.5)$$

Here S_m is the frequency factor of oscillation of the particular trap, E_m is the trap depth for that trap, $D(x,y,z)$ is the depth-dose profile, and k is the Boltzmann constant. Trap data for LiF is given in Table 4.1. [2]

TABLE 4.1
Values for Input Trap and Thermophysical Parameters

Parameter	Value	Units
Density	2.4	1/cm
Heat capacity	0.431	cal/(gK)
Thermal conduction coefficient	0.007	cal/(s cm K)
Absorption coefficient	4.0	1/cm
Reflectivity	0.0	none
Free convection heat transfer coefficient	2.80×10^{-4}	cal/(s cm^2 K)
Ambient temperature	300	K
Chip thickness	0.38	cm
Boltzmann constant	8.62×10^{-5}	eV/K
Energy depth of 2nd LiF trap	1.13	eV
Energy depth of 3rd LiF trap	1.23	eV
Energy depth of 4th LiF trap	1.54	eV
Energy depth of 5th LiF trap	2.17	eV
Frequency factor of 2nd LIF trap	6.10×10^{11}	1/s
Frequency factor of 3rd LIF trap	4.00×10^{11}	1/s
Frequency factor of 4th LiF trap	7.30×10^{13}	1/s
Frequency factor of 5th LIF trap	4.00×10^{19}	1/s

Source: From Reference [2]. With permission.

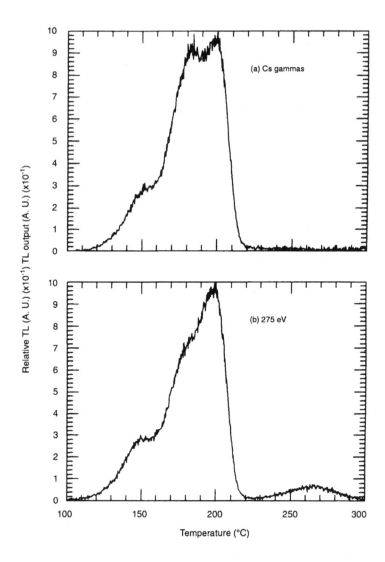

FIGURE 4.4 Glow curves of LiF crystals exposed to ^{137}Cs γ-rays (a) and to 275-eV photons (b). Glow curves following 2550–400 eV irradiations were similar to those shown for 275 eV. Data are normalized at the dosimetric peak at 200°C. (From Reference [4]. With permission.)

LiF:Mg,Ti (TLD-100) extruded ribbons and crystals were exposed by Carrillo et al. [4] to monoenergetic photons of 275-2550 eV energy to determine their potential usefulness as radiation dosimeters for radiobiology experiments. The undesirable effect of air annealing increased with decreasing photon penetration in the dosimeter. Under certain experimental conditions, UV radiation produced anomalous bleaching of high-temperature traps. The crystals and the chips presented a supralinear response. Supralinearity factors were determined to be on the order of 1.5 for crystals and 1.7 for chips.

Extruded chips were $3.2 \times 3.2 \times 0.9$ mm in size. Crystal samples had varying sizes as a result of the cleaving procedure; however, they were cleaved to be as similar in size as possible to the chips. The entire crystal was annealed at 400°C in air for 1 h, after which time the

crystal was removed from the oven and placed on an aluminum block to reach room temperature. Then the crystal was placed in an 80°C oven for 24 h. After this the crystal was cleaved.

Figure 4.4 presents glow curves for a TLD-100 crystal exposed to 193 nJ of ^{137}Cs γ-rays and to a monoenergetic photon beam of 275 eV energy. For ^{137}Cs this exposure (about 1 Roentgen) results in an absorbed dose in the crystal of 8.14 mGy. Glow curves for 2550, 1500, 1200, 900, 730, 600, 500, and 400 eV exposures with 193 nJ of total energy deposited had the same shape as the one for 275 eV. The glow curves were normalized to the same height for the main peak at 200°C. The total energy deposited in each case was also 193 nJ. Photons of 275 and 730 eV energy are the least penetrating ones in LiF.

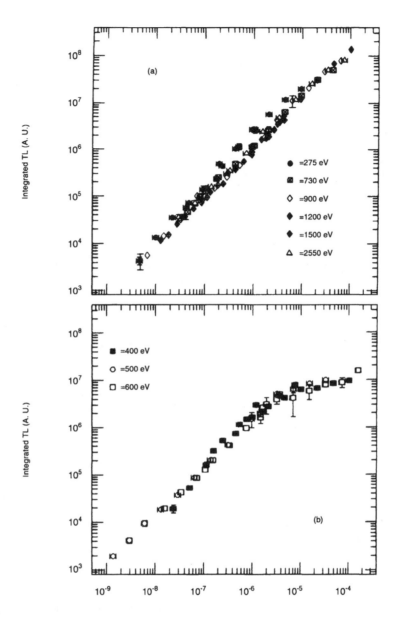

FIGURE 4.5 Integrated TL response of TLD-100 crystals against energy deposited. Irradiation energy as marked. (a) employed the mylar or Be window, (b) employed a thin Al window. (From Reference [4]. With permission.)

Figure 4.5 shows the TL response as a function of total energy deposited in the He-annealed crystals for the photon energies used. The TL response in both cases corresponds to the integrated area of the glow curve between 130°C and 215°C, while the total energy deposited was calculated by using the photodiode measurement. The integrated TL response was a linear function of energy deposited for 275, 730, 900, 1200, 1500, and 2550 eV photons. However, for exposures with photon beams of 400, 500, and 600 eV, a "saturation-like" effect is noticed.

The effect of the annealing atmosphere was found to be crucial for reproducible TL output and glow curves. Three annealing treatments were tested by Carrillo et al. Figure 4.6 shows the glow curves of crystals annealed in

helium and in air for 1 and 23 h, when exposed to 275-eV photons. The adverse effect of an air anneal is clearly seen as an overall reduction in the height of the peaks, mainly the dosimetric 200°C peak. For crystals annealed in air for 23 h, the reduction is even more pronounced. Similar adverse behavior for the air-annealed crystals and chips was noticed for all other photon energies.

The quality dependence of LiF TLD in megavoltage photon beams with qualities from ^{60}Co γ-rays to 25-MV x-rays has been studied experimentally by Mobit et al. [5] against ion chamber measurements and theoretically by Monte Carlo simulation using the EGS4 Monte Carlo code system. The experimental findings are that the energy dependence of 1-mm thick TLD-100 (micro-rods and chips) on average decreases slowly from 1.0 for ^{60}Co

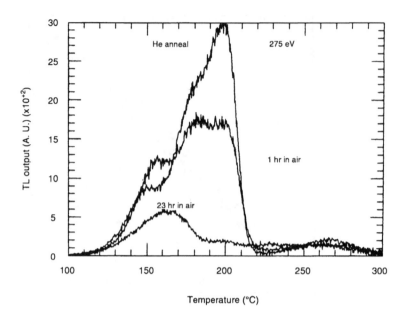

FIGURE 4.6 Glow curves of He- and air-annealed LiF crystals exposed to 275 eV photons. (From Reference [4]. With permission.)

γ-rays to 0.989 ± 1.3% for 6-MV x-rays (TPR$_{10}^{20}$ = 0.685) and to 0.974 ± 1.3% for 25-MV x-rays(TPR$_{10}^{20}$ = 0.800) relative to ^{60}Co γ-rays. The Monte Carlo results vary from 0.991 ± 0.9% for 6-MV x-rays to 0.978 ± 0.8% for 25-MV x-rays. Differences between chips and micro-rods were negligible, and there was no difference in the energy dependence between TLDs irradiated in water or Perspex (PMMA).

When a detector is placed in a medium exposed to ionizing radiation in order to measure dose, it forms a cavity in that medium. The cavity is generally of different atomic number and density from the medium. Cavity theory gives the relation between the dose absorbed in the medium (D_{med}) and the average absorbed dose in the detector or cavity ($\overline{D}_{cav}$):

$$D_{med} = \overline{D}_{cav} f_{med, cav} \qquad (4.6)$$

where $f_{med,cav}$ is a factor that varies with energy, radiation type, medium, size, and composition of the cavity. For a cavity that is small compared with the range of the electrons incident on the cavity in electron and photon beams, the Bragg-Gray relation applies:

$$f_{med,cav} = s_{med, cav} \qquad (4.7)$$

where $S_{med,cav}$ is the average mass stopping power ratio of the medium to the cavity. For a cavity that is large compared to the range of electrons incident on it in a photon beam, the dose in the medium can be obtained from the

mass energy–absorption coefficient ratio of the medium to the cavity material:

$$f_{med,cav} = \left(\frac{\mu_{en}}{\rho}\right)_{med, cav} \qquad (4.8)$$

where (μ_{en}/ρ) is the ratio of the mass energy–absorption coefficients, medium to cavity, averaged over the photon energy fluence spectrum present in the medium. This expression completely neglects any perturbation effects or interface effects that may occur by the introduction of the detector material into the uniform medium.

Burlin proposed a general cavity theory for photons for all sizes which approaches the Spencer-Attix theory in the small-size limit and the ratio of the mass energy–absorption coefficient for very large cavities. According to this theory,

$$f_{med, cav} = ds_{med, cav} + (1 - d)\left(\frac{\mu_{en}}{\rho}\right)_{med, cav} \qquad (4.9)$$

where d is a weighting factor which gives the contribution to the total dose of medium-generated electrons, and $(1 - d)$ is the contribution to the total dose from electrons generated by photon interaction in the cavity.

The quality dependence factor or response (F_{co}^{X}) is defined as

$$F_{Co}^{X} = \frac{TL(X)/D_{med}(X)}{TL(Co)/D_{med}(Co)} \qquad (4.10)$$

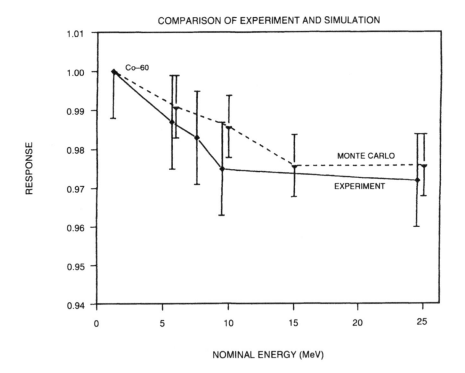

FIGURE 4.7 Comparison of Monte Carlo simulations and experimental determination of the quality dependence factor for LiF TLD. (From Reference [5]. With permission.)

where $TL(X)/D_{med}(X)$ is the light output (TL) per unit dose in a medium for the beam quality X of interest. $TL(Co)/D_{med}(Co)$ is the light output per unit dose in the same medium for ^{60}Co γ-rays. If D_{LiF} is the dose to TLD material, then assuming D_{LiF} is directly proportional to the light output $TL(X)$ at any X, one can write

$$F_{Co}^{X} = \frac{(D_{med}/D_{LiF})_{Co}}{(D_{med}/D_{LiF})_{X}} \qquad (4.11)$$

The assumption of proportionality between the light output and the dose to the TLD ignores the possibility of sensitivity changes in the TLD which depend on interactions between trapping centers. Thus, supralinearity and LET effects will not be accounted for. However, for megavoltage doses under 1 Gy, these are unlikely to be a problem. The energy correction factor to correct for the under-response of a photon beam (X) relative to ^{60}Co γ-rays is

$$f_{Co}^{X} = 1/F_{Co}^{X}$$

where $_{chip}F_{Co}^{X}$, $_{rod}F_{Co}^{X}$ are the quality dependence factor of chips and micro-rods, respectively, and $_{ave}f_{Co}^{X}$ is the average energy correction factor of chips and micro-rods of the same thickness.

Figure 4.7 shows the comparison between Monte Carlo simulation and experimental determination of the quality dependence factor F_{Co}^{X}. The agreement is within 0.7%.

Comparisons have been made for micro-rods, with water as the irradiation medium. In all cases, the Monte Carlo results show a lower deviation from unity of the energy dependence factor compared to the experimental values.

The size of the cavity (thickness of TLD) is critical to whether it behaves as a small cavity, an intermediate, or even a large cavity. To investigate the effect of cavity size on the energy correction factor, simulations were carried out by Mobit et al. with micro-rods of diameter 1, 2, and 5 mm. Figure 4.8 shows the results of such simulations. There is no significant difference between 1-mm and 2-mm diameter micro-rods. The deviation from unity is, however, smaller with 5-mm diameter micro-rods. The results seem to suggest that thicker LiF TLD are better for photon dosimetry than 1-mm diameter LiF TLD, provided that high spatial resolution is not required, e.g., at the edges of the field.

The mass collision stopping power ratio and the mass energy–absorption coefficient ratio are shown in Figure 4.9.

The energy correction factor of LiF TLDs in megavoltage electron beams was calculated by Mobit et al. [6] The experiments show that the energy correction factor of 1-mm thick TLD-100 has an average for both rods and chips which varies from 1.036 ± 1.3% (1 SD) for 4-MeV electron beams to 1.021 ± 1.3% (1 SD) for 20-MeV electron beams for measurement performed at d_{max} in PMMA (Perspex). The results of the Monte Carlo simulations were within 0.6% of the experimental results and ranged from 1.041 ± 0.9% (1 SD) for 2-MeV electrons to 1.028 ± 0.8% (1 SD)

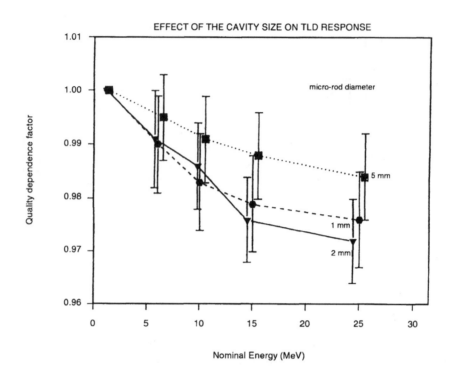

FIGURE 4.8 Quality dependence of LiF TLD rods of different diameter calculated by Monte Carlo simulations. (From Reference [5]. With permission.)

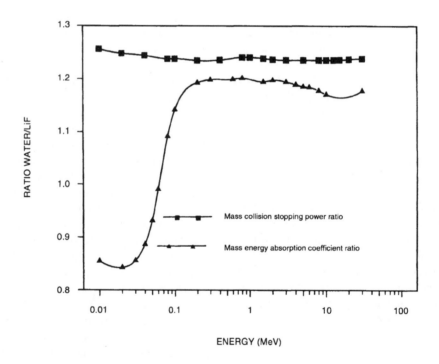

FIGURE 4.9 Variation with energy of the mass energy–absorption coefficient and the mass collision stopping power ratio of water to LiF. (From Reference [5]. With permission.)

for 20-MeV electron beams. Differences in the energy correction factors between rods and chips of the same thickness were negligible. The energy correction factors changed by up to 4% for irradiation of TLD at depths other than at d_{max} for a 5-MeV mono-energetic electron beam.

The energy correction factor is defined as

$$f_{Co}^E = \frac{TL(Co)/D_{med}(Co)}{TL(E)/D_{med}(E)} \qquad (4.12)$$

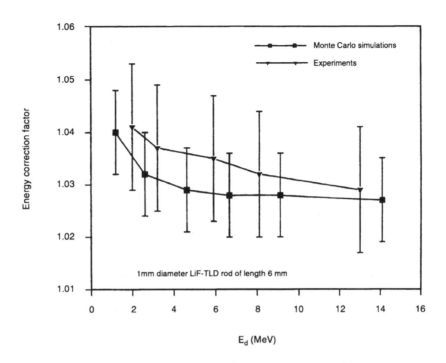

FIGURE 4.10 A comparison of Monte Carlo simulations and experimental determination of the energy correction factor of LiF TLDs. (From Reference [6]. With permission.)

which is the ratio of the light output (TL) per unit dose in the medium for ^{60}Co γ-rays to the light output per unit dose in the medium for an electron beam energy (E). If $\overline{D_{\text{LiF}}}$ is the average dose to LiF TLD material and assuming that $\overline{D_{\text{LiF}}}$ is directly proportional to the light output TL(E) at any E, then

$$f_{\text{Co}}^{E} = \frac{(\overline{D_{med}}(E)/\overline{D_{\text{LiF}}}(E))}{(\overline{D_{med}}(\text{Co})/\overline{D_{\text{LiF}}}(\text{Co}))} \quad (4.13)$$

The energy correction factor can be used to determine the dose as follows:

$$\overline{D}_{med}(E) = TL(E)\frac{\overline{D}_{med}(\text{Co})}{TL(\text{Co})}f_{\text{Co}}^{E} \quad (4.14)$$

The quality-dependence factor F_{Co}^{E} is $1/f_{\text{Co}}^{E}$.

Figure 4.10 shows the comparison of the energy correction factor determined by experiments and Monte Carlo simulations. There is agreement between experiments and Monte Carlo simulations within 1.3% .

To investigate the effect of cavity size on the energy correction factor, Monte Carlo simulations were carried out with rods of diameter 1, 2, and 5 mm. Figure 4.11 shows the results of these simulations. As the thickness of the rod is increased, there is an increase in the energy correction factor; that is, the deviation from unity increases. There is an increase of about 2% in the energy correction factor of 2-mm diameter rods compared with that for 1-mm diameter at high electron energy. At low energy this difference can

be up to 5%. There is a significant difference in the energy correction factor curve among the 1, 2, and 5-mm diameter rods over the entire energy range.

Thermoluminescent characteristics of LiF:Mg,Ti from three manufacturers were compared by Fairbanks and DeWerd. [7] All were chips of size 3 mm × 3 mm × 1 mm. Fifty TLDs of this material were obtained from each of the three manufacturers, Solon Technologies, Inc. (Solon), Victoreen, Inc. (Victoreen), and Teledyne Isotopes (Teledyne). The TLDs were compared on the basis of their precision, sensitivity, thermoluminescent (TL) signal as a function of energy, and linearity of TL signal as a function of exposure.

The Solon and Victoreen chips provided good overall standard deviations, 7.9% and 7.0%, respectively. But the Teledyne standard deviation was nearly twice that of the others, 13.8%. The annealing procedure causes a further spread in the TL signals from a given set of chips, and because an anneal was initially performed, the Solon and Victoreen chips may have been within the mentioned specifications.

All groups showed an over-response at low energies followed by a flat response at higher energies, as shown in Figure 4.12a. The TL signal values have been normalized to remove sensitivity differences and show the correspondence among the manufacturers. Figure 4.13 shows TL signal as a function of exposure.

The dose response of LiF:Mg,Ti (TLD-100) exposed to 15 and 35 kVp (8.0 ± 0.1 and 8.1 ± 0.1 keV effective energy, respectively) x-rays and ^{60}Co γ-rays has been measured by

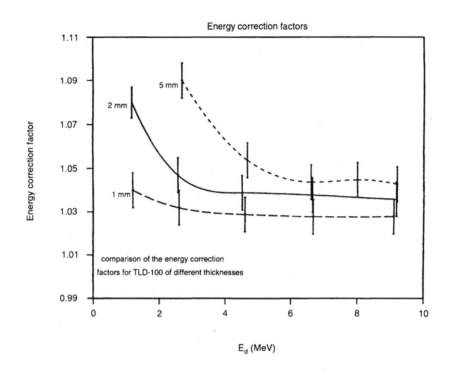

FIGURE 4.11 Energy correction factors of LiF TLD rods of different diameters calculated by Monte Carlo simulations. (From Reference [6]. With permission.)

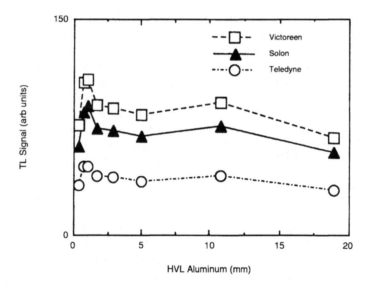

FIGURE 4.12 TL signal as a function of half-value layer (HVL) for each group of TLDs showing an over-response at low energies and a flat response at higher energies. (From Reference [7]. With permission.)

Gamboa-de Buen et al. [8] in the dose interval from $(1.2-5.4) \times 10^3$ Gy for x-rays, and from 0.14 to 850 Gy for γ-rays. In both cases the total TL signal and glow-curve peaks 3 to 9 show supralinearity. The supralinearity function $f(D)$ is similar for both x-ray beams, except for peak 8, where a 30% difference is observed.

Although the onset and degree of supralinearity in the response of LiF:Mg,Ti to photons is dependent on the batch material, the experimental conditions (heating rate and annealing procedure), and the glow peak temperature (McKeever et al. 1995), it has been established that the degree of supralinearity increases when the photon energy increases . The supralinearity increases with the glow peak temperature. Jain and Ganguly [9] have reported a lower degree of supralinearity for peak 8 than for peak 7.

The glow curves were deconvoluted into peaks 3 to 9 (see Figure 4.14a) with the Harshaw/Filtrol software

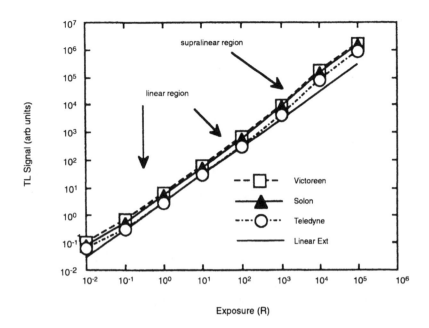

FIGURE 4.13 TL signal as a function of exposure, all three groups showing linearity between 100 mR and 100 R and supralinearity thereafter. A linear extension is provided to make this more apparent. Nonlinearity below 100 mR may be attributed to uncertainty in reader and chip noise. Cessation of supralinearity above 10,000 R may be the beginning of chip damage. (From Reference [7]. With permission.)

that uses the Podgorsak approximation to the Randall-Wilkins first-order kinetics peak shape. In all deconvolutions, the width of peaks 6 to 9 were kept constant (in order to reduce the number of free parameters, since a typical glow curve does not uniquely determine all 21 parameters which describe the 7 peaks) and the position (temperature) and height of all peaks were adjusted in each fit of the glow curves. In the deconvolution of the x-ray glow curves, the width of peak 4 was also fixed. Table 4.2 shows the average values and standard deviations found for the deconvolution parameters. The area under the glow curve, after background and peak 2 subtraction with the deconvolution software, will be referred to as the total TL signal.

Figure 4.14a shows a glow curve after irradiation with x-rays and its deconvolution into peaks 3 to 9. Figures 4-14b and c show a series of glow curves for different doses after x- and γ-ray exposure, respectively. The curves have not been normalized, and the evolution of their shape as a function of the dose indicates a strong dose dependence of the relative contributions from the individual peaks. Figure 4.15 shows the TL response for the total TL signal and individual peaks as a function of dose for both types of radiation. In general, the response follows a linear-supralinear-sublinear behavior.

The supralinearity function $f(D)$, defined as

$$f(D) = \frac{F(F)/D}{F(D_0)/D_0} \qquad (4.15)$$

TABLE 4.2
Glow-Curve Deconvolution Parameters for 8.1-keV x-rays (35 kVp) and ⁶⁰Co γ-Rays

Peak number	8.1 keV x-rays		⁶⁰Co γ-rays	
	Temperature (°C)	FWHM (°C)	Temperature (°C)	FWHM (°C)
3	161 ± 6	38 ± 4	156 ± 2	32 ± 2
4	188 ± 2	20.0	186 ± 1	21 ± 1
5	213 ± 2	26 ± 2	212 ± 1	25 ± 1
6	237 ± 4	40.4	234 ± 1	40.4
7	276 ± 2	40.4	274 ± 2	40.4
8	305 ± 6	32.0	307 ± 1	32.0
9	337 ± 2	32.0	333 ± 3	32.0

Source: From Reference [8]. With permission.

where $F(D)$ is the TL signal per irradiated unit mass at dose D, and D_0 is a low value of the dose such that $F(D_0)$ is linear, has been calculated for the total TL signal and for peaks 5 to 8, and it is shown for 8-keV x-rays in Figure 4.16a and for ⁶⁰Co γ-rays in Figure 4.16b.

The results of a series of experiments performed to determine some fading characteristics of TLD-100 (Harshaw Chemical Co.) were presented by Vasilache et al. The experiment was performed using two batches of dosimeters that were irradiated with x-rays simultaneously and with the same dose. The results indicated that, after the low-temperature peaks had faded out, the response of

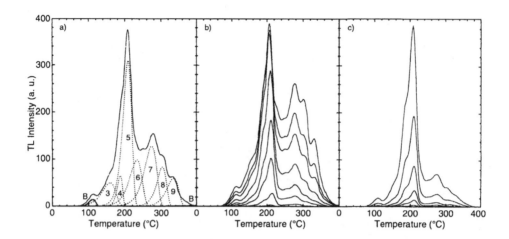

FIGURE 4.14 (a) TLD-100 glow curves after irradiation with x-rays at 1.8×10^3 Gy. Dotted curves show the deconvolution into peaks 3 to 9. Peak 2 and the high-temperature emission (b) were subtracted from the measured glow curve by the deconvolution software. (b)TLD-100 glow curves after irradiation with 8-keV x-rays for different doses. In increasing order: 1.5×10^1, 7.4×10^1, 2.3×10^2, 4.5×10^2, 8.9×10^2, 1.8×10^3, 2.7×10^3, and 5.4×10^3 Gy. (c) TLD-100 glow curves after irradiation with ^{60}Co γ-rays at different doses. In increasing order: 1.7×10^1, 4.3×10^1, 9.2×10^1, 1.7×10^2, 4.3×10^2, and 8.5×10^2 Gy. (From Reference [8]. With permission.)

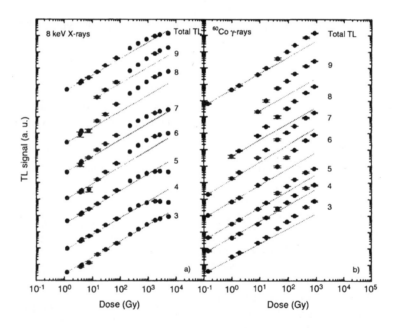

FIGURE 4.15 Response curves of TLD-100 exposed to (a) 8-keV (35-kVp) x-rays and (b) ^{60}Co γ-rays, for the total TL signal and for peaks 3 to 9. The dotted lines correspond to linear responses. The measurements have been arbitrarily displaced for clarity. (From Reference [8]. With permission.)

the dosimeters exposed to light became quite stable, with a very low fading. The signal of the dosimeters kept away from light continued to fade to a lower level than the signal of the dosimeters exposed to light.

To analyze the evolution of each glow peak, the glow curves were deconvoluted using an improved method of gradients. To fit the experimental curves, the well-known Podgorsak approximation of the Randall-Wilkins equation was used:

$$I(T) = I_m \exp\left[1 + \frac{T - T_m}{\omega} - \exp\left(\frac{T - T_m}{\omega} \right) \right] \quad (4.16)$$

along with the relationship between ω and T_m:

$$\beta/\omega = s \times \exp(-\omega T_m) \quad (4.17)$$

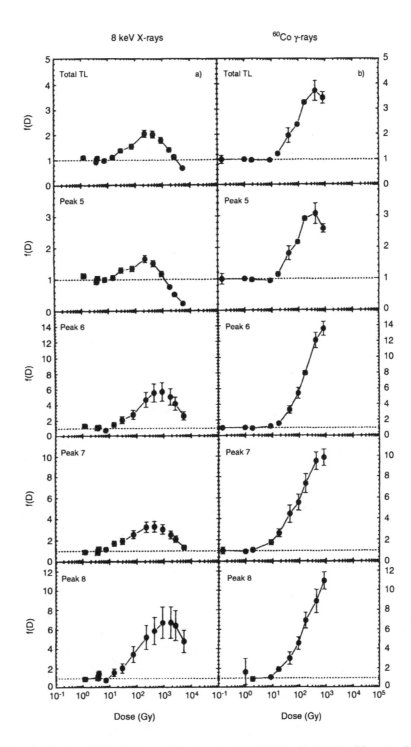

FIGURE 4.16 Supralinearity function $f(D)$ of the total TL signal and peaks 5 to 8 of TLD-100 exposed to (a) 8 -keV (35-kVp) x-rays and (b) ^{60}Co γ-rays. Symbols show the measurements and curves as a guide to the eye. (From Reference [8]. With permission.)

where $I(T)$ is the peak intensity at the temperature T, I_m is the intensity of the maximum, T_m is the temperature of the maximum, β is the heating rate, and s is the attempt to escape frequency of the electrons in the traps. The full width at half-maximum of the peak is given by:

where E is the depth of the energy trap and k is the Boltzman constant.

The results of the first experiment are presented in Figure 4.17. The best fit led to the following two functions for the fading factor: [10]

$$\omega = (kT_m^2)/E \qquad (4.18)$$

$$f_1(\tau) = -0.06\tau^{-0.19} + 1.191$$

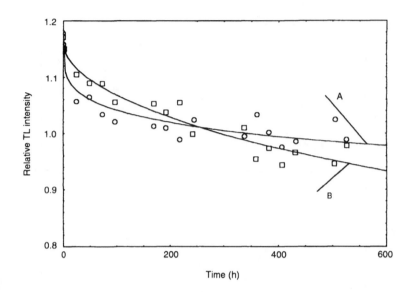

FIGURE 4.17 The fading of the TL signal for TLD-100 detectors irradiated with *bremsstrahlung* x-rays. The continuous lines represent the best fit for the two sets of experimental points. Line A ($\circ$), exposed to light, $(ax^b) + C$, error: 7.200×10^{-3}, $a = 6.151 \times 10^{-2}$, $b = 1.939 \times 10^{-1}$, $C = 1.191$. Line B ($\square$), stored in darkness, $a \exp(-cx^{-d})$, error: 7.492×10^{-3}, $a = 1.162$, $b = 7.326 \times 10^{-3}$, $c = 5.312 \times 10^{-1}$. (From Reference [10]. With permission.)

for the dosimeters that were exposed to light, and

$$f_d(\tau) = 1.16 \exp(-0.007\,\tau^{0.53})$$

for the dosimeters that were kept in the dark. In both cases, the fading factor is given by $f_i(\tau) = I(\tau)/I(\tau = 0)$, where $I(\tau)$ is the integral TL emission at the time τ.

The second experiment (UV post-irradiation) led to a drastic change in the shape of the TL glow curve after the exposure to UV light, by comparison with the glow curve of the TLD-100 detectors irradiated with x-rays only (Figure 4.18a and b). As can be seen, peaks 4 and 5 are now resolved and, furthermore, peak 5 now shows a higher fading than peak 4.

Pre-irradiation annealing treatment at 150°C reduces the sensitivities of peak 4 (197°C) and peak 5 (209°C), whereas the intensity of peak 2 (108°C) is enhanced. The TL intensities of peak 7 (271°C) and peak 10 (441°C) remain practically unaffected (Bhatt et al.). [11]

Figure 4.19 shows an example of a deconvoluted curve. The area under each peak of the deconvoluted curves is being used for the calculation of sensitization factors and intensities.

Figure 4.20 shows a typical glow curve for virgin as well as for 150°C, 6-h pre-irradiation annealed samples. For 150°C, 6-h pretreated samples, the intensity of peak 2 increases by a factor of 1.82, whereas the intensity of TL peak 5 is reduced by a factor of 36. Another TL peak at 235°C (peak 6) appears due to this temperature treatment. TL peaks 7 and 10 are affected only slightly by this treatment.

It is seen that for the 150°C pre-annealing treatment, both peaks 4 and 5 decrease, possibly due to precipitation of phase 6 LiF:MgF$_2$. It may be noted that optical bleaching and thermal annealing investigations have shown convincingly the interrelationship between TL peaks 4 and 5 and the 310-nm absorption band. Thus, the two peaks are expected to behave in a similar fashion during the pre-irradiation annealing treatment.

The glow-curve characteristics of a single a crystal LiF:Mg,Ti, grown by the Bridgman method, were studied by Weizman et al. [12] as a function of Ti concentration (3–14 ppm Ti). The glow curves were deconvoluted into component glow peaks using mixed-order (MO) model kinetics. The TL sensitivities and activation energies of the glow peaks were studied in both post-irradiation annealed and unannealed samples. In glow curves following a 165°C/15-min anneal to remove peaks 2–4, the activation energy of peak 5 was observed to increase from 1.65 ± 0.1 eV at the lowest Ti concentrations to 1.9 ± 0.1 eV from 8–14 ppm Ti concentration. Deconvolution of the glow curve is shown in Figure 4.21.

The sensitivity of the various peaks (2–5) as a function of Ti concentration is illustrated in Figure 4.22. The dependence of the TL sensitivity of peak 5 on the square of the Ti concentration would seem to indicate that the TC/LC complex may be the result of the incorporation of two Ti(OH)$_n$ or O complexes coupled independently to two of the dipoles participating in the formation of the Mg-Li$_{vac}$ trimer.

Thermoluminescence and optical absorption of γ-ray- and α-particle-irradiated LiF:Mg,Ti single crystals have been studied by Bos et al. [13] The optical absorption (OA) bands were measured during a heating regime which

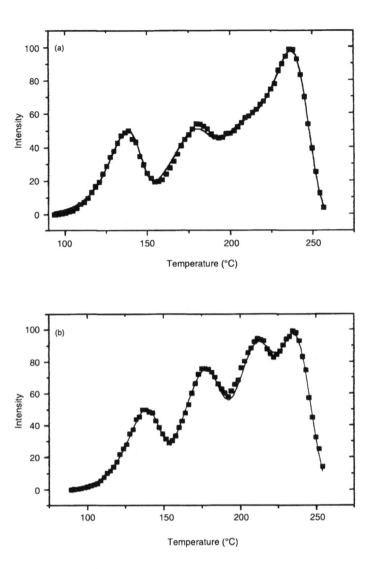

FIGURE 4.18 Glow curves for detectors irradiated only with x-rays (a) and exposed to UV after irradiation with x-rays (b). Both detectors were read immediately after irradiation. The continuous line represents the best fit obtained by computerized glow-curve deconvolution. (From Reference [10]. With permission.)

mimics the glow-curve readout. This technique yields, in principle, a more reliable correlation between glow peaks and OA bands than the conventional procedure of sample-annealing using the l_m-l_{stop} technique. Computerized analysis was used to resolve the OA spectra into Gaussian bands. In f-irradiated samples, apart from the known 5.5-eV band, a previously unnoticed 4.5-eV band has been recognized which may be caused by defects that are active as competitors in the recombination stage.

Optical absorption bands induced by ionizing radiation arise due to changes in the trapping stage only, while a TL glow curve is the result of both trapping and recombination processes. Thermal treatments drastically alter the absorption bands. A commonly used technique to study the effects of thermal treatments is to subject the sample to a pulse-annealing sequence; i.e., the sample is heated to one temperature, then cooled and measured, subsequently heated a second time to a higher temperature, and so on. The record-

ing of the TL glow curve, however, is done by heating the sample just one time without quenches and reheating in between. In LiF:Mg,Ti there are many strongly overlapping OA bands.

Figure 4.23 shows two thermoluminescence glow curves of LiF:Mg,Ti after irradiation at room temperature with 4.5-MeV and 30-keV α particles. The glow curves (above 200°C) show great similarity, from which it is concluded that in both cases the same trapping centers are involved. This is plausible since the stopping powers do not vary very much (7.4×10^2 eVcm^{2}g^{-1} for 4.5-MeV α particles and 5.8×10^2 eVcm^{2}g^{-1} for 30-keV α particles). Figure 4.24 shows the optical absorption spectra of LiF:Mg,Ti at different temperatures following irradiation with gamma-rays and 30-keV α particles. The two radiation qualities produce different spectra. At 50°C the OA band at 5.0 eV (F center) is recognized. It is seen that after α irradiation, the maximum of this band shifts to a

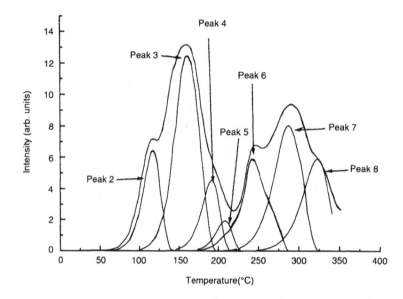

FIGURE 4.19 Typical deconvoluted TL glow curve for LiF TLD-100, annealed at 400°C for 1 h, followed by 150°C, 6-h treatment and irradiated with 1 kGy at 77 K. The peak positions are numbered according to the usual nomenclature. (From Reference [11]. With permission.)

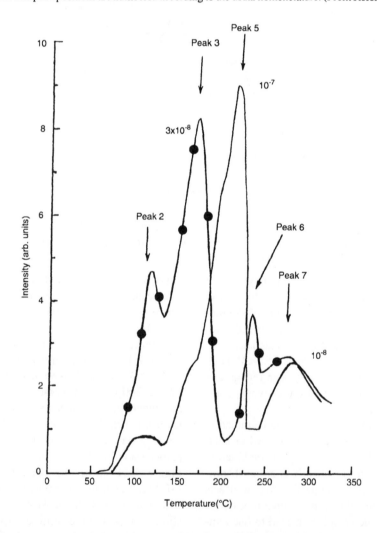

FIGURE 4.20 Typical TL glow curves for LiF TLD-100. Virgin (—) and 400°C for 1 h, followed by 150°C, 6-h pre-annealing (--). The numbers 10^{-7}, 3×10^{-8}, etc. refer to the current ranges of the DC amplifier. (From Reference [11]. With permission.)

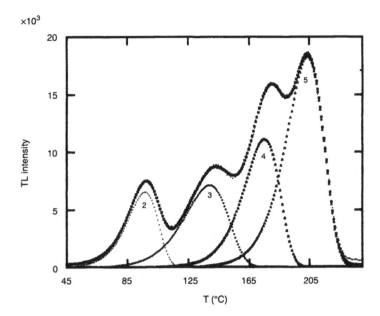

FIGURE 4.21 Deconvoluted glow curve of LiF:Mg,Ti containing 80 ppm Mg and 10 ppm Ti. Analysis of this particular glow curve resulted in the following activation energies: peak 2 (1.3 eV), peak 3 (1.1 eV), peak 4 (1.6 eV), and peak 5 (1.4 eV). (From Reference [12]. With permission.)

lower energy. After α irradiation, the 4.50-and 5.54-eV OA bands are not present; their absence suggests that they are not active during x-irradiation and, therefore, one can expect a linear behavior of OA bands as a function of dose.

Figure 4.25 displays the glow curves plotted for LiF and its mixtures with Li_2CO_3 (nonluminescent material). The shape of the curve obtained for the composite material applied for dosimetry coincides with that observed for the thermoluminescent phosphor. However, the thermoluminescence output (i.e., the values of thermoluminescence signal per unit of dose) appear to be somewhat lower for mixtures due to the presence of additives to phosphor.

Various LiF thermoluminescence dosimeters have been investigated by Osvay and Deme [15] for selective assessment of low and high LET radiation for space dosimetry purposes. The aim of the investigations was to compare the selectivity of the glow peaks of the Harshaw TLD-100, the Polish LiF:Mg,Ti (type MTS-N) and the Austrian LiF-F dosimeters using alpha, gamma, and thermalized neutron radiations. Comparing TL responses using glow-curve analysis of the peaks 4–5 (180–210°C) and on the high-temperature peaks 6–7 (240–270°C) generated by high LET radiation, it was found that separation of radiations with different linear energy transfer (LET) was the most promising in the case of Polish LiF:Mg,Ti (type MTS-N) dosimeters.

Figure 4.26 shows the glow curves of TLD-100, LiF (MTS-N) and LiF-F after 6-mGy gamma irradiation. The gamma sensitivities were calculated from the mean value of the TL responses. The sensitivity of LiF (MTS-N) was found to be two and a half times higher compared with TLD-100, and LiF-F was found to have nearly the same sensitivity as TLD-100. The standard deviation of dose readings for gamma-rays was about 5%.

The glow curves of TLD-100, LiF (MTS-N), and LiF-F dosimeters irradiated by 24-mGy (volume-averaged) alpha dose are shown in Figure 4.27. The glow curves demonstrate the gamma sensitivity and the sensitivity to 5-MeV alpha particles of the LiF dosimeters investigated. Comparing the TL responses, the following can be stated:

1. The glow peak of the gamma-irradiated LiF-F single crystal dosimeter was found at a higher temperature compared to TLD-100 and MTS-N LiF dosimeters; the glow peak temperature of LiF-F is about 240°C (Figure 4.26).
2. The effect of high LET alpha irradiation is considerable for each type of dosimeter investigated, but the qualitative change on the structure of the glow curve was found to be the most significant for the LiF (MTS-N) dosimeter.
3. The LiF-F crystal does not show an explicitly high temperature peak using alpha irradiation in the dose range 1–40 mGy.

Comparing the glow curves of Figure 4.28, the following can be stated. First, the main peak (peaks 4–5) is shifted to lower temperature values for an initial irradiation made by neutrons in comparison with the reversed condition, i.e., with an initial gamma exposure. Second, in addition to this effect, there is a reduction in TL response for an initial irradiation with neutrons.

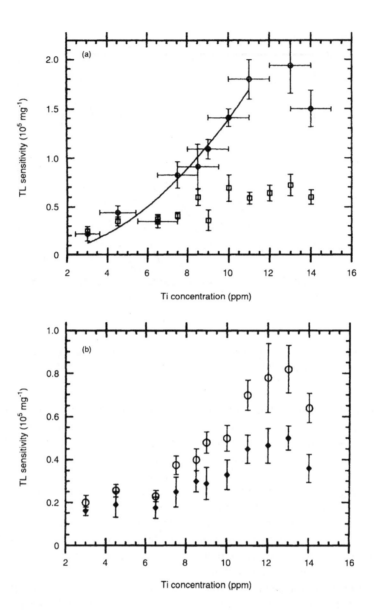

FIGURE 4.22 TL sensitivities of peaks 2–5 as a function of Ti concentration: (a) peak 5 (O); peak 4 (□); (b) peak 3 (O); peak 2 (◆). An estimate of the error in Ti concentration is shown in Figure 4.22a, as well as a quadratic fit (solid line) to the sensitivity of peak 5 as a function of Ti concentration. (From Reference [12]. With permission.)

Miniature LiF:Mg,Ti (MTS-N) pellets, of diameter 1–2 mm and thickness 0.5 mm, specially designed for dosimetry in proton radiotherapy, have been produced. [16] The influence of dopant composition, activation method, and cooling rate on the dose-LET response of these TL detectors was tested. It appears that these dosimetric characteristics are governed mainly by the Mg-dopant, and supralinearity and efficiency for high LET radiation are highest for samples with the lowest content of magnesium.

Two methods of introducing activators into bulk LiF were tested. The first method (denoted as A), based on co-precipitation, is typically used for producing large amounts of LiF:Mg,Ti (in batches of more than 100 g). The second one (method B) exploits a high-temperature treatment to

introduce dopants. Unlike method A, the latter is particularly well-suited for producing a number of small samples originating from a larger batch of raw LiF.

The dose response was described by the linearity index $f(D)$:

$$f(D) = \frac{I(D)/D}{I(D_0)/D_0} \quad (4.19)$$

where $I(D)$ is the TL signal after exposure to a dose D and D_0 is the reference dose. In the work of Bilski et al., $D_0 = 1$ Gy. The dose response of all samples after gamma irradiation was found to be supralinear. After annealing in the PTW oven, a significant increase of $f(D)$ was observed (see examples in Figure 4.29a; error bars in all figures represent

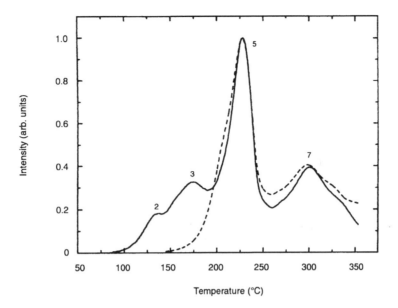

FIGURE 4.23 Glow curves of LiF:Mg, Ti single crystal after irradiation with 4.5-MeV α particles (dotted line) and after implantation with 30-keV He ions (continuous line). Both curves are normalized at the top of glow peak 5. Heating rate is 3°C s^{-1}. (From Reference [13]. With permission.)

1 SD). All samples prepared using method B were particularly susceptible to cooling conditions. In detectors produced using method A, the differences in $f(D)$ after different cooling rates were much smaller—in some detectors no difference was seen, even after the highest dose. The Harshaw TLD-100 detectors showed practically no variation in the value of $f(D)$ after the different cooling rates.

Supralinearity was found to be strongly dependent on the concentration of magnesium. Figure 4.30 shows the linearity index measured at a dose of 15 Gy [$f(D) = 15$ Gy] relative to the concentration of Mg. It can be seen that $f(D)$ increases as the amount of Mg decreases. Figure 4.30 also illustrates the effect of the method of activation on supralinearity: samples prepared using method B show a significantly higher level of supralinearity.

Figure 4.29b presents some examples of the measured proton dose-response curves. The most striking result was obtained for the 120-ppm Mg and 13-ppm Ti sample (method B, standard concentrations), which showed no supralinearity. A similar effect was observed in a few other samples prepared using method B. It is somewhat surprising that after proton exposures, B samples are less supralinear than A samples, while after gamma irradiation the reverse appears to be true. However, as results for samples B are based on rather scant data, they should be treated lightly. Other curves in Figure 4.29b represent data obtained for an A sample with a standard concentration of dopants, for TLD-100, and for the A sample showing the highest supralinearity.

For LET calibration, irradiations were carried out at the Joint Institute for Nuclear Research (JINR) in Dubna with

fluoride ions in the energy range from 65 to 275 MeV amu^{-1} and carbon ions with energies from 100 MeV amu^{-1} up to 3650 MeV amu^{-1}. [17] Some glow curves of TLD-600 after absorption of different radiations are shown in Figure 4.31. The glow curves are normalized to equal height of peak 5. The increase of high-temperature TL emission with increasing LET of absorbed radiation can be seen.

The data obtained in the LET region up to 90 keVμm^{-1} are plotted in Figure 4.32. The graph shows that the LET dependence of the parameter HTR in all three dosimeter types is quite similar. In the range up to 30 keVμm^{-1} in tissue, there are no statistically significant deviations between TLD-100, TLD-600, and TLD-700, and the increase of HTR with LET is very steep, followed by a saturation region where LET is greater than 30 keVμm^{-1}. The relation between HTR and LET for TLD-600 for all irradiations carried out is shown in Figure 4.33. The graph shows that there is an increase of HTR with LET, up to a LET of about 180 keVμm^{-1}.

Extruded LiF ribbons (3.1 $\times$ 3.1 $\times$ 0.9 mm^3) and rods (6 $\times$ 1 $\times$ 1 mm^3) are commonly used TL dosimeters for clinical dosimetry in radiotherapy. The dose distribution in these crystals was investigated by Korn et al. [18] in a 6-MV x-ray beam using smaller LiF TL dosimeter types. In the investigations with small cubes assembled in form of ribbons and rods, it was found that a higher dose was deposited in the center of the ribbons and rods. Accordingly, it was found that TL dosimeters in close contact with each other increase their respective reading.

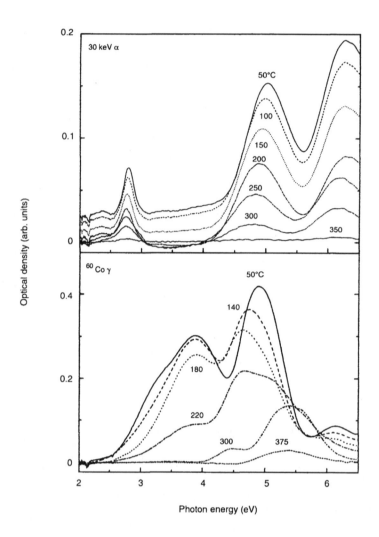

FIGURE 4.24 Optical absorption spectra of a LiF:Mg,Ti single crystal after (a) gamma-ray (D_γ = 780 Gy) and (b) 30-keV α-particle irradiation, measured during heating at a constant heating rate of the sample of 3°C s⁻¹. From all plotted spectra, a spectrum at a high temperature (400°C for the γ-irradiated and 360°C for the α-irradiated) has been subtracted. (From Reference [13]. With permission.)

Figure 4.34a shows the dose response of the ribbons and rods as a function of time after they have first been used. Newly purchased Victoreen ribbons have a higher sensitivity than those produced by Solon/Harshaw. However, Harshaw material seems to have a superior sensitivity after prolonged use. The error bars (±1 SD) in Figure 4.34 show the variability of the dose response between different crystals of one type. No significant difference could be found between the variabilities of chips from different manufacturers.

Figure 4.34b shows the TL response as a function of the number of heating cycles. Each cycle includes annealing, exposure, preread annealing, and readout of the crystals. The dose received per temperature cycle varied between 2 cGy and 100 cGy. Assuming an average of 50 cGy per cycle, one can calculate a total dose administered to the crystals of 50 Gy after 100 cycles.

Figure 4.35 shows the dose response of XTC TL dosimeters stacked together in the geometries illustrated in the figures. The chips were exposed to 100 MU in a 10 × 10 cm² field with 6-MV x-rays. The measured dose in each chip is given as a function of physical depth in LiF. The reading of a single normal ribbon in each geometry is shown by a horizontal line.

It can be seen in Figures 4.35a and b that the effective point of measurement in a normal ribbon is not at the geometric center of the crystal for dose determinations at the surface (geometries A and B). In the 6-MV build-up region, the effective point of measurement is located at about 0.4-mm depth in the ribbon, while its geometric center is at 0.44-mm depth. If one takes the relative electron density of LiF into account, 0.4-mm physical thickness is equivalent to an effective point of measurement about 0.9 mm below the patient's skin for in vivo dosimetry. Readings taken at this depth give a reasonable dose estimate for the deeper blood vessels, where some late damage to the skin originates (ICRP 1991). However, normal TLD ribbons are illustrated for the determination

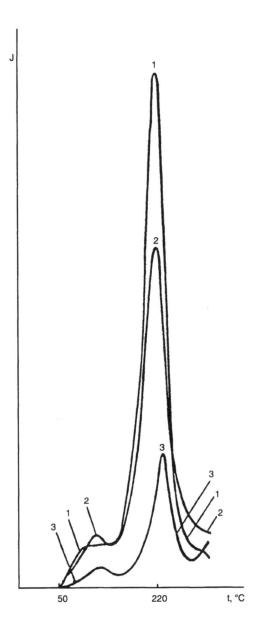

FIGURE 4.25 The glow curves for LiF (1) and for compositions containing 35% (2), and 65% (3) of Li_2CO_3 (nonluminescent material). (From Reference [14]. With permission.)

of the dose to the basal cell layer located at 0.05-mm depth, where major damage to the skin may occur.

Figure 4.35c shows the relative dose measured in 8 XTC dosimeters stacked together, one on top of each other. This results in a stack of 1.12-mm height, which was positioned at 1.5-cm (d_{max}) depth in solid water.

Several dosimetry intercomparisons for whole-body irradiation of mice have been organized by the European Late Effects Project Group (EULEP). [19] These studies were performed employing a mouse phantom loaded with LiF thermoluminescent dosimeters. In phantom, the energy response of the LiF TLDs differs from free in air, due to spectral differences caused by attenuation and scatter of x-rays. Monte Carlo calculations of radiation transport

were performed to verify the LiF TLD energy response correction factors in phantom relative to free in air for full scatter conditions and to obtain energy response correction factors for geometries where full-scale conditions are not met. For incident x-rays with HVLs in the 1 to 3.5-mm Cu range, the energy response correction factor in phantom deviates by 2 to 4 percent from that measured free in air. The energy response correction factors obtained refer to a calibration in terms of muscle tissue dose in phantom using ^{60}Co gamma-rays. For geometries where full scatter conditions are not fulfilled, the energy response correction factors are different by up to about 3 percent at maximum from that at full scatter conditions. The dependence of the energy response correction factor as a function of the position in phantom is small, i.e., about 1 percent at maximum between central and top or bottom positions.

Meigooni et al. [20] investigated the dependence of sensitivity and linearity of the TLD response to the flow of nitrogen gas in the TLD reader at low-dose level. The investigations were performed using small and large LiF TLD (TLD-100, Harshaw) chips. The differences in the physical properties of the TLDs are encountered by using chip-factor correction factors, C_{ij}, obtained from the ratio of each TLD response, $TL_{i,j}$, to the mean response, TL_{mean}, when the whole batch is irradiated to the same dose (e.g., 100 cGy) as

$$C_{ij} = \frac{TL_{ij} - TL_{BKG}}{TL_{mean}} \qquad (4.19)$$

where

$$TL_{mean} = \frac{1}{N} \sum_{i=1, j=1}^{N} [(TL_{ij} - TL_{BKG})] \qquad (4.20)$$

and where TL_{BKG} is the background (reading of several unexposed chips) and N is the total number of chips in the batch. In dosimetry of an unknown radiation field using several TLD chips, a mean or net response, TL_{net}, can be calculated from the responses of the individual chips, TL'_{ij}, that were exposed to the same dose, with correction for the background TL'_{BKG} and chip factors C_{ij}, as follows

$$TL_{net} = \frac{\frac{1}{n} \sum_{i=1, j=1}^{n} [(TL'_{ij} - TL'_{BKG})/C_{ij}]}{F_{lin}} \qquad (4.21)$$

where F_{lin} is correction for the nonlinearity of the TLD response as a function of absorbed dose. F_{lin} is defined here as the ratio of the measured TLD response to the predicted value for the same absorbed dose by linear

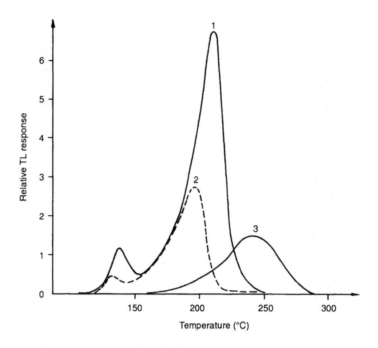

FIGURE 4.26 Thermoluminescence glow curves of TL dosimeters. The dosimeters were irradiated with a gamma dose of 6 mGy. Key to curves: 1, LiF (MTS-N); 2, TLD-100; 3, LiF-F. (From Reference [15]. With permission.)

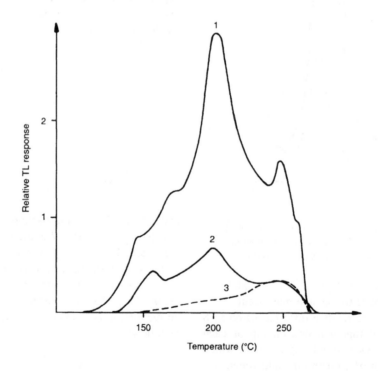

FIGURE 4.27 Thermoluminescence glow curves of TL dosimeters. The dosimeters were irradiated with a volume-average alpha dose of 24 mGy. Key to curves: 1, LiF (MTS-N); 2, TLD-100; 3, LiF-F. (From Reference [15]. With permission.)

extrapolation of the values corresponding to the doses between 50 and 100 cGy.

The effects of nitrogen flow on the TLD responses were measured by Meigooni et al. as a function of absorbed dose from the gamma-rays of ^{137}Cs and x-rays of a 4-MV linear

accelerator. These effects were determined by comparison of the responses of several chips (at least eight chips) that were exposed to the same dose but read partially with and partially without nitrogen gas flow in the TLD reader. For measurements with the ^{137}Cs source, two slabs of Solid Water phantoms

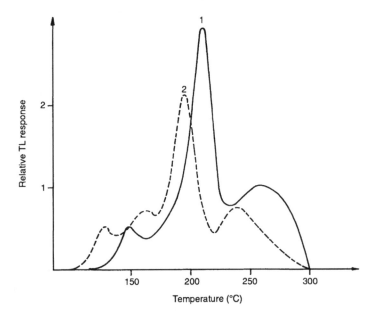

FIGURE 4.28 Thermoluminescence glow curves of a LiF (MTS-N) dosimeter irradiated with a gamma dose of 1 mGy + a neutron dose of 5 mSv (curve 1) and a neutron dose of 5 mSv + a gamma dose of 1 mGy (curve 2).(From Reference [15]. With permission.)

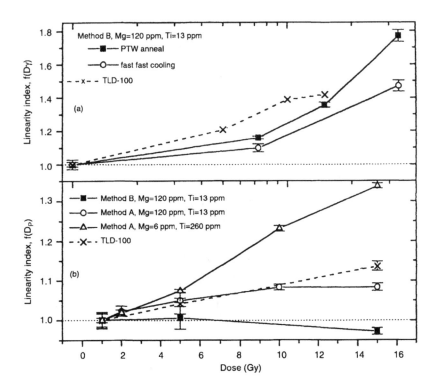

FIGURE 4.29 Linearity index measured for selected samples and TLD-100 after exposures to: (a) ^{137}Cs gamma -rays, (b) modulated proton beam-detectors placed in the middle of extended Bragg peak (PTW annealing). (From Reference [16]. With permission.)

($20 \times 20 \times 4$ Cm^3) were accurately machined to accommodate four TLD chips at each radial distance, ranging from 0.5 to 10 cm relative to the source center.

Figure 4.36a shows the large TLD responses exposed to doses ranging from 0.7 to 30 cGy using the gamma-rays of a ^{137}Cs source, normalized to the value for the largest dose (i.e., 22 cGy). These data were obtained by exposing the TLD placed at a fixed distance from the source center along the transverse bisector of the source. The error bars in this figure reflect the range of absolute dispersion between the individual TLD responses that were exposed to the same dose. This figure also shows that for

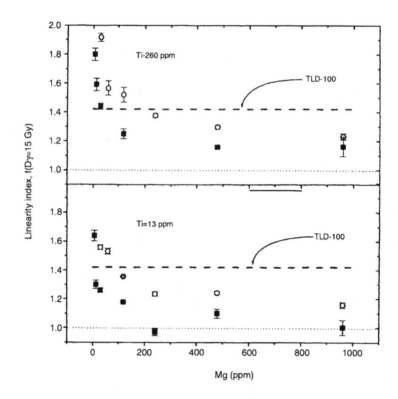

FIGURE 4.30 Values of linearity index for gamma dose $D = 15$ Gy plotted against Mg concentration (PTW annealing). (■) Method A, (O) method B. (From Reference [16]. With permission.)

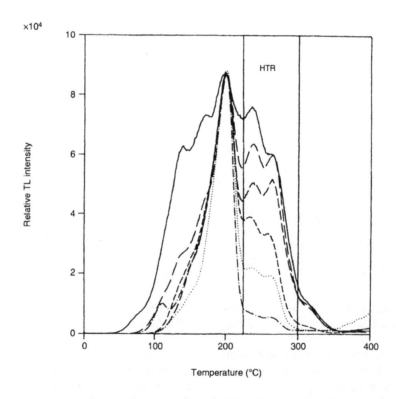

FIGURE 4.31 Glow curves from TLD-600 after irradiation with different radiations. Peak 5 is normalized to equal height. — ^{241}Am alpha, 174 keV μm^{-1}; --- thermal neutrons, 130 keVμm^{-1}; --• ^{19}F ions, 102 keV μm^{-1}; ---• ^{19}F ions, 30 keVμm^{-1}; •••• ^{12}C ions, 12.8 keVμm^{-1}; ----- ^{60}Co gamma. (From Reference [17]. With permission.)

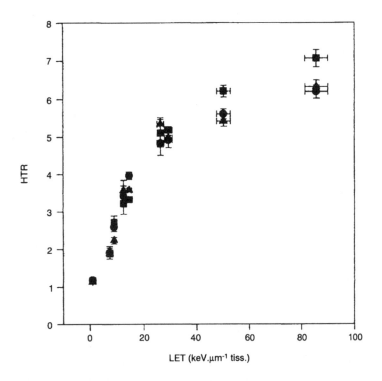

FIGURE 4.32 High-temperature ratio HTR plotted against LET of absorbed radiation for TLD-100 (●), TLD-600 (■), and TLD-700(▲). (From Reference [17]. With permission.)

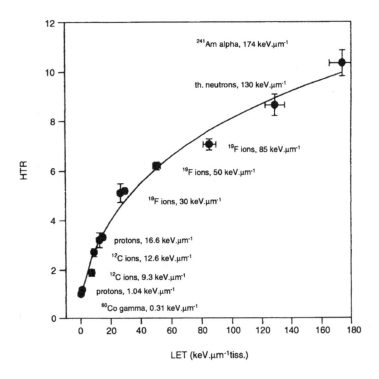

FIGURE 4.33 High-temperature ratio HTR plotted against LET of absorbed radiation for TLD-600. (From Reference [17]. With permission.)

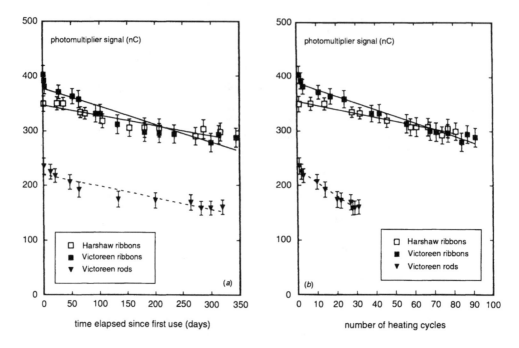

FIGURE 4.34 (a) Photomultiplier signal for normal TLD ribbons and rods exposed to 100 cGy as a function of the time elapsed since their first use. The TL signal is given as the charge collected in the photomultiplier during readout. (b) Photomultiplier signal for normal TL ribbons and rods exposed to 100 cGy as a function of the number of heating cycles. Each cycle includes annealing, exposure, preread annealing, and readout of the crystals. The TL signal is given as the charge collected in the photomultiplier during readout. (From Reference [18]. With permission.)

doses less than 5 cGy, the responses were dispersed by as much as a factor of 2. This dispersion was reduced to less than 5% when the TLDs were read with nitrogen flow in the TLD reader. Moreover, this figure shows a profound increase (about a factor of 2) in the mean TLD response, with doses less than 10 cGy. This large nonlinearity has been eliminated by using nitrogen gas. Similarly, Figure 4.36b shows the effect of nitrogen gas flow in the TLD reader on the responses of small chips exposed to ^{137}Cs gamma-rays. As with the large chips, there was a large (about a factor of 2) dispersion among the individual responses of small chips when nitrogen gas did not flow through the TLD reader. However, the mean responses of these TLD are linear to within $\pm 10\%$. Although nitrogen gas reduced the dispersion to less than $\pm 5\%$, it did not affect the linearity of the TLD response.

Figure 4.37a shows the effect of nitrogen gas on the relative sensitivity or linearity of large TLD responses when they were exposed to 4-MV x-ray beam in the dose range of 1–700 cGy. The relative sensitivity is defined as the ratio of the sensitivity (TL/cGy) for a given dose to the sensitivity for 50 cGy. The absorbed doses were measured using an ADCL-calibrated PTW ion chamber in a polystyrene phantom.

Figure 4.37b demonstrates the effect of nitrogen gas on the relative sensitivity of the small TLD measured response when exposed to 4-MV x-ray beam in the dose range of 1–700 cGy. Dose measurements were made in

the same fashion as for Figure 4.37a. These results indicate no variation of TLD linearity and sensitivity due to nitrogen gas flow, which agrees with the data in Figure 4.36(b). However, the large standard deviation (about a factor of 2) in the TLD response at low doses was reduced to less than 5% by using nitrogen gas.

Surface dose measurements were performed by Korn et al. [21] in the 6-MV beam of a medical linear accelerator with LiF thermoluminescence dosimeters, using a solid water phantom. TLD chips (surface area 3.17×3.17 cm^2) of three different thicknesses (0.230, 0.099, and 0.038 g/cm^2 were used to extrapolate dose readings to an infinitesimally thin layer of LiF.

Figure 4.38 shows the dose build-up in a solid water phantom measured with TLD chips of three different thicknesses. The depth in Figure 4.38 is given in g/cm^2. Assuming the active point of measurement to be in the center of the each LiF chip, half of the chips thickness (in g/cm^2) was added to the solid water depth. The chips were placed free-lying on the surface of a solid water slab and covered by additional layers of solid water to give the depth shown in Figure 4.38.

Figures 4.39 a and b show the dose build-up in LiF TLDs of different thicknesses stacked one on top of each other at the surface of solid water in two different geometries. The TLDs were exposed to 100 mu in a 10×10-cm^2 field of 6-MV x-rays. The results given in Figure 4.39a are for thin chips stacked in air on the surface, and the ones

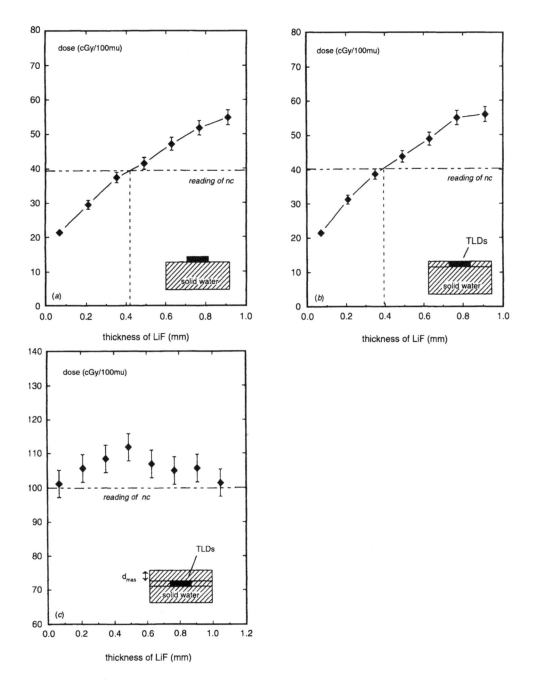

FIGURE 4.35 (a) Dose measured with extra thin ribbons stacked one on top of each other on the surface of a solid water phantom. The thickness of the stack of seven extra thin chips is 0.98 mm. The reading of a normal ribbon (thickness 0.89 mm) in the same geometry is depicted as a horizontal line. (b) Dose measured with extra thin ribbons stacked one on top of each other in the surface of a solid watre phantom. The top XTC is at the same level as the surface of the phantom. The measurement geometry is depicted in the insert. The reading of a normal ribbon in the same geometry is shown as a horizontal line. (c) Dose measured in eight extra thin dosimeters stacked one on top of each other at 1.5 cm (d_{max}) depth in solid water. The readings are normalized to 100 for the reading of a single extra thin ribbon positioned at d_{max}. The reading of a normal ribbon in the same geometry shown in the insert is depicted as a horizontal line. This is by definition 100 cGy per 100 MU at d_{max}. From Reference [18]. With permission.

in Figure 4.39b are for TLDs stacked in solid water with the top chip being flush with the solid water surface. The increased side scatter in the second arrangement leads to a steeper dose build-up in the chips, shown in Figure 4.39b. The measured dose in each chip is given as a function of physical depth in LiF in mg/cm². Error bars at selected values indicate the range of uncertainty (± 2 SD) for the two types of TLDs used.

Lithium fluoride thermoluminescent dosimeter chips (^{6}LiF:Ti,Mg) were irradiated with alpha particles from an

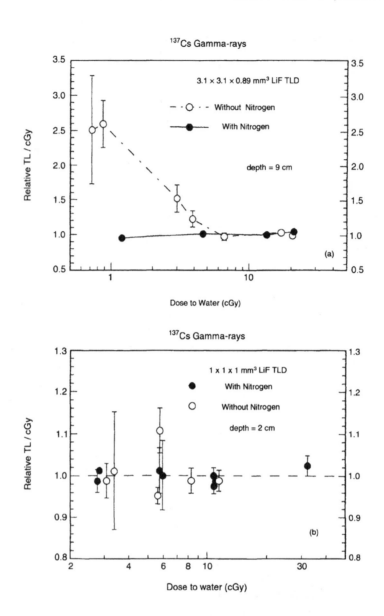

FIGURE 4.36 Effect of nitrogen gas flow on the standard deviation and linearity of large (a) and small (b) LiF TLD chips exposed to the γ-rays of a 10-mg Ra Eq ^{137}Cs tube. The relative sensitivity is defined here as the ratio of the sensitivity (TL/cGy) for a given dose to the sensitivity of the TLDs with the absorbed dose of 22 cGy for large chips and 35 cGy for small chips and located at the same source-to-detector distance. (From Reference [20]. With permission.)

[241]Am source. [22] The energy of the alpha particles was degraded with thin mylar foils of different thicknesses. The net glow curve is shown in Figure 4.40.

The TL glow curves were deconvoluted using the first-order TL kinetics model according to the "Podgorsak approximation" method:

$$I(T) = I_m \exp[1 + (E(T - T)k^{-1}T_m^2)$$
$$- \exp(E(T - T_m)k^{-1}T_m^{-2})] \quad (4.22)$$

where T = temperature (K), I = intensity of the TL glow signal at the temperature T, I_m = the maximum height of the glow peak, E = trap depth of the glow peak (eV),

T_m = temperature at the maximum peak height (K), and k = Boltzmann constant 8.62×10^{-5} eV/K. The trap depth E (eV) is a linear function of the maximum peak temperature T_m; therefore, the following approximation is valid:

$$E \approx 25kT_m \quad (4.23)$$

The responses of extruded TLD-600 and TLD-700 ribbons (Solon Technologies, Ltd., Solon, OH), $3.175 \times 3.175 \times 0.889$ mm³, were calibrated by McParland and Munshi [23] against the thermal neutron fluence in a moderated neutron field created in a tank of water ($60 \times 60 \times 40$ cm³) containing a [252]Cf spontaneous fission source.

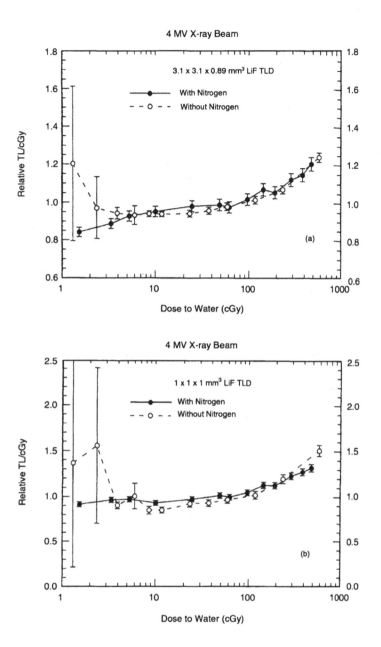

FIGURE 4.37 Effect of nitrogen gas flow on the linearity of large (a) and small (b) LiF TLD. The relative sensitivity is defined as the ratio of the sensitivity (TL/cGy) for a given dose to the sensitivity for 50 Gy. (From Reference [20]. With permission.)

The source intensity was 2.1×10^7 neutrons s^{-1}, and the TLDs were placed 15 cm from the source.

The thermal neutron response of the TLD-700 was initially investigated by comparing the response of the bare TLD-700 in the moderated neutron/photon flux with that with the dosimeter encapsulated in one of a number of thermal neutron-absorbing shields. The absorbers considered were TLD-600 (LiF with 95.62% ^{6}Li and 4.38% ^{7}Li), cadmium, and indium. A natural tin shield was also used. The response of the TLD-700 was unaffected by the placement of the TLD-600 absorber about it, whereas, in contrast, cadmium, which has a thermal neutron absorption

cross section about 2.5 times that of ^{6}Li, elevated the TLD-700 response by a factor of almost 2. The indium led to an even greater increase in the TLD-700 response, whereas the tin shielding did not cause a significant elevation.

TLD-700 chips were irradiated [24] with the fast neutrons from ^{241}Am/Be and ^{252}Cf sources, energy-degraded neutrons from the ^{241}Am/Be source moderated by light water, and gamma-rays from a ^{137}Cs source. The net area under the deconvoluted high-temperature ($\approx 300°C$) peak of neutron-irradiated chips was observed to increase with the average neutron energy. The high-temperature part (250–340°C) of the TL glow curve was deconvoluted using

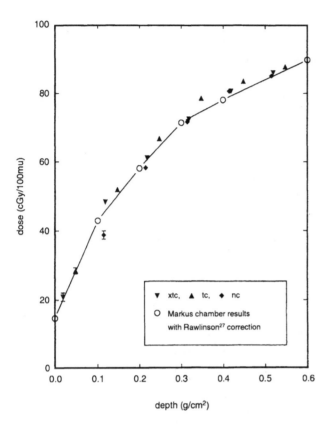

FIGURE 4.38 Dose build-up in solid water for a 6-MV x-ray beam, field size 10 × 10 cm². Three TLD chips of different thicknesses (nominal thickness nc: 0.89 mm, tc: 0.39 mm, and xtc: 0.14 mm) were placed at different depths in solid water. The thickness of the chips was taken into account, assuming the active center of the chip to be in its physical center. Error bars show the range of uncertainty (±2 SD) for the first chip of each type. (From Reference [21]. With permission.)

the first-order TL kinetics model according to the "Podgorsak approximation" method (Equation (4.22)). [25]

All glow curves were normalized to the highest peak height at $T_m \approx 220°C$ (Figure 4.41). It is apparent that the area under the deconvoluted high-temperature glow peak, $A(HT)$, increases with the average neutron energy. It is evident that the high-temperature (≈300°C) peak of TLD-700 dosimeters is sensitive to gamma-rays as well, but to a lesser extent than to fast neutrons.

III. LIF:Mg, Cu, P DOSIMETER

The development of high sensitivity TLD by doping LiF crystals with Mg, Cu, and P was first done by Nakajima et al. [25] The sensitivity of the new TLD was more than 20 times higher than that of LiF:Mg,Ti. Wu et al. [26] showed that LiF:Mg,Cu,P (LiF(MCP) maintains its sensitivity during prepared reuse cycles. The TL characteristics of LiF:Mg,Cu,P include, in addition to high sensitivity, almost

flat photon energy response, low fading rate, and linear dose response. The sensitivity and glow-curve shape are both dependent on the maximum readout temperature and the pre-irradiation annealing parameters. Short low-temperature annealing of 165°C for 10 s prior to readout is capable of removing most of the low-temperature peaks. As a result, there is very little fading up to two months or more at room temperature. High-temperature annealing at 400°C results in irreversable elimination of the main dosimetric peak and causes some increase in the high-temperature peaks.

Advantages of LiF:Mg,Cu,P include high sensitivity as compared to LiF:Mg,Ti, almost flat photon energy response, low fading rate, and linear dose response. The lack of supralinearity at higher dose levels is particularly useful for accident dosimetry and eliminates the source of error usually associated with the application of supralinearity corrections. The main drawbacks are still the relatively high residual signal and the loss of sensitivity for high-readout temperatures. LiF:Mg,Cu,P is interesting in low-dose measurements due to its high sensitivity and its good tissue equivalence.

The glow curve of LiF(MCP) consists of several overlapping glow peaks. The main peak at approximately 220°C, known as peak 4, is the one used for dosimetry applications (the "dosimetric peak"). The rest of the glow curve consists of a low-temperature part in the range of approximately 70–160°C (peaks 1, 2, and 3), and a high-temperature peak at approximately 300°C (peak 5). There is evidence that the glow curve of this material is even more complicated where peaks 4 and 5 are each composed of two overlapping peaks.

The high sensitivity, combined with its tissue equivalence, is the main advantage of this material in personal dosimetry applications. The sensitivity of LiF(MCP) is approximately 25 times higher than that of LiF:Mg,Ti (TLD-100). It is important to note, however, that the measured sensitivity depends not only on the TL properties of the material itself, but also on the spectral response of the light detection system. Both LiF(MCP) and LiF:Mg,Ti have the same effective atomic number (8.2) and could therefore be expected to have a similar photon energy response in reality. The over-response of LiF:Mg,Ti at 30 keV is approximately 35% (relative to 662 keV), as compared to only 6% for LiF(MCP).

The sensitivity of LiF:Mg,Cu,P was studied by Furetta et al. [27] as a function of the annealing temperature and of the repeated cycles of annealing-irradiation-readout. A fading study was carried out over a period of 40 days with the purpose of checking the stability of the stored dosimetric information as a function of different annealing temperatures. 10 LiF:Mg,Cu,P phosphors were cycled 10 times according to the following sequence: annealing, irradiation, readout.

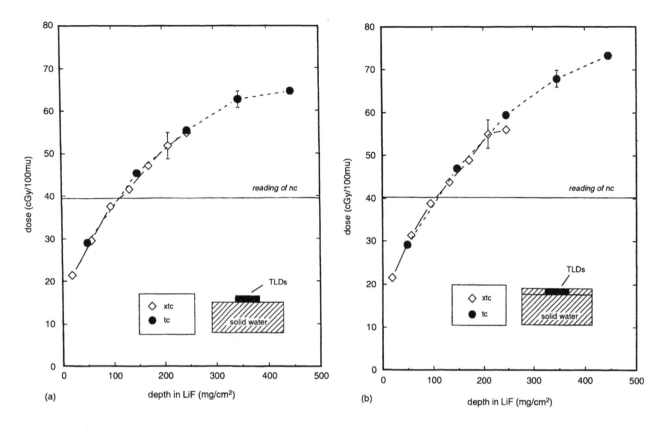

FIGURE 4.39 (a) Dose buildup in LiF as determined with extra-thin (xtc) and thin (tc) TLD chips stacked on top of each other on the surface solid water. The depth is given as the physical depth of the center of each chip. The reading of a single normal TLD chip (nc) placed on the surface is indicated by a horizontal line. (b) Dose build-up in LiF as determined with extra-thin (xtc) and thin (tc) TLD chips stacked on top of each other in solid water. The highest chip is Bush with the solid water surface. The depth is given as the physical depth of the center of each chip. The reading of a single normal TLD chip (nc) at the surface is indicated by horizontal line. (From Reference [21]. With permission.)

Figure 4.42 shows the TL emissions obtained after various thermal procedures. Figure 4.42a (annealing at 220°C) displays the dominance of peak 3 over peak 4, which appears as a shoulder on the descending part of peak 3. At 240°C (Figure 4.42b), the TL emission appears as usually observed: peak 4 becomes the main peak and peak 3 is now a shoulder on the ascending side of the fourth peak. For temperatures larger than 240°C, peak 3 tends to become smaller and smaller, so that the main peak becomes narrower (Figure 4.42c, at 270°C). It was noted that the shape of the TL emission, through the 10 repeated cycles, remains unchanged for each temperature. The plots of the peak 3 and 4 TL integrals, for each temperature and as a function of the repeated cycles, are reported in Figure. 4.43. The TL responses were normalized to that relative to the first cycle, at 240°C. The best reproducibility is obtained at 240°C. At the beginning, the TL emission at 220 and 230°C is a little larger than at 240°C, but in the next cycles, the decrease of the TL is obvious and reaches the 30% level in the tenth cycle at 230°C. The situation is worst for temperatures larger than 240°C. Temperatures higher than 240°C could produce a sort of trap disactivation which increases with the number of cycles. In this sense, a high annealing temperature (> 240°C) seems to produce a progressive quench of the TL emission.

Figure 4.44 shows the plots obtained in a fading experiment over 40 days of storage. All the results are normalized to the TL response of peaks 3 + 4 obtained after annealing at 240°C for "zero" days (reference value). The stability of the TL response is perfect at 240°C annealing, while a decrease is observed at any other temperatures. It is obvious that the trap stability is compromised at temperatures different from the reference value. The maximum loss is observed after 40 days for the group initially annealed at 270°C: the remaining TL is only 43% of the reference TL value.

The response of LiF:Mg,Cu,P thermoluminescence dosimeters to high-energy electron beams used in radiotherapy was investigated by Bartolotta et al. [28] They found that LiF:Mg,Cu,P phosphor is a suitable candidate for quality control of in vivo dosimetry in electron-beam therapy. The TL chips (4.5 mm in diameter, 0.8 mm thick) used were produced by Radiation Detector Works (Beijing) and are commercially known as GR-200A.

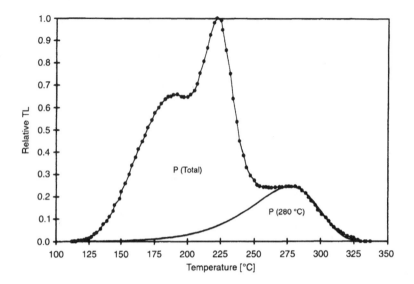

FIGURE 4.40 The glow curve of a TLD-600 chip irradiated with non-degraded alpha particles (E_a = 5.5 MeV) from a 560-Bq ^{241}Am source. The glow curve was recorded with a data logger at a sampling rate of 2 readings per second. The high temperature peak $P(280°C)$ was deconvoluted using the "Podgorsak approximation" of the first-order TL kinetics model. (From Reference [22]. With permission.)

Figure 4.45 shows the TL signal of the GR-200A dosimeters vs. absorbed dose in the entire investigated dose range. Each data point corresponds to the mean of nine readings. The error bars represent the overall uncertainty of the mean to a level of confidence of 95%.

The sensitivity of GR-200A dosimeters to electrons was found to be about 13% less than that of ^{60}Co gamma-rays, in agreement with similar results already known for LiF:Mg,Ti (TLD-100) dosimeters. Variance analysis of the data showed the dependence on electron-beam energy relative to each other to be nil or buried under the measurement uncertainty. When used for dose measurements in high-energy electron beams, it is adequate to calibrate the LiF:Mg,Cu,P dosimeters with any one electron-beam energy without need for energy correction, but if they are calibrated with a ^{60}Co beam, an appropriate energy correction must be applied.

Mailed dosimetry for radiotherapy is one of the most demanding applications of thermoluminescence dosimetry (TLD). Ideally, the uncertainty of a mailed system should be comparable to that of the ionometric methods employed for the calibration of the radiation beams, i.e., around 1% (1σ), and certainly lower than 5%. [29]

The simple dose dependence of GR-200 is in contrast to the supralinear dependence of TLD-100. The proportionality between TL and dose found for GR-200 is in favor of this material, as calibration can be simplified requiring fewer calibration points than with TLD-100.

Very good reproducibility was found for the individual sensitivity factor of every dosimeter. Throughout the experiment and for 20 groups of dosimeters, the worst value found for the standard deviation of 10 determinations of the individual sensitivity factor of a dosimeter was 0.65%. Values of 0.2–0.4% were the most frequent.

Sets of dosimeters were stored at room temperature (18–24°C) for periods of 7, 15, 30, and 60 days. Every set was composed of three groups of five GR-200 dosimeters irradiated to a dose of 2 Gy at different points of the storage—at the beginning, in the middle, and at the end, designated SA (storage after irradiation), M, and SB (storage before irradiation), respectively. [29] Peak 4, the main peak in the GR-200 glow curve, increases in intensity during the first days of storage, stabilizing afterwards. Peak 3, a weak peak, decreases monotonically with time at room temperature. The evolution of the peaks reveals that peak 4 experiences mainly trap effects, leading to an increase in the number of traps available after storage. Peak 3 presents the two effects, trap effect (decreasing the number of traps) and fading. A clear anticorrelation was observed in the evolution of peaks 3 and 4, rendering the sum of these two peaks a more constant dose estimator than peak 4 alone for delayed measurements. Taking these two peaks together, the response change is limited to only 2–4% relative to the prompt sensitivity.

An investigation of the thermal stability of LiF:Mg, Cu,P (GR-200) compared to the more traditional LiF:Mg, Ti (TLD-100) at 40°C and 70°C was presented by Alves et al. [30] Samples of both varieties of dosimeters were stored irradiated or un-irradiated in order to evaluate the relative importance of the temperature/storage-induced effects on either traps or trapped charges. The measured glow curves were analyzed using the deconvolution programs developed at CIEMAT. These techniques allow the detailed characterization of the evolution pattern followed

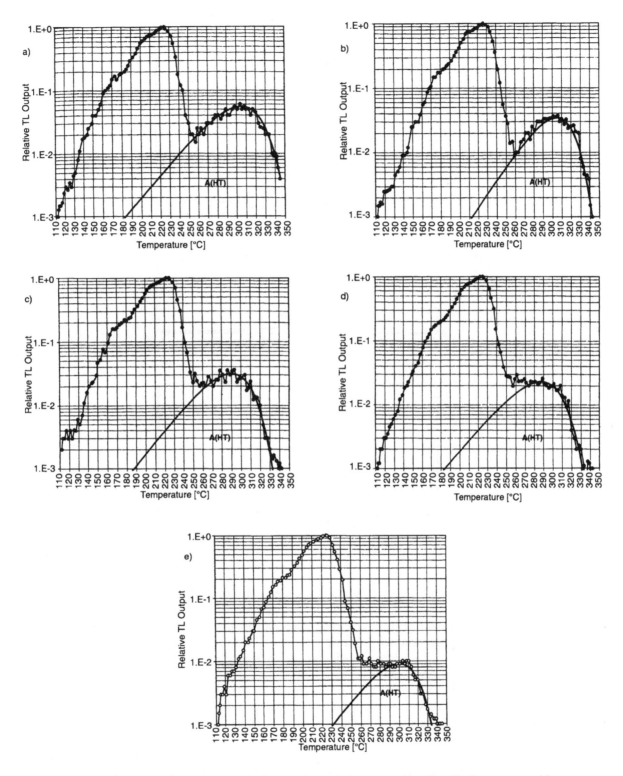

FIGURE 4.41 Glow curves of TLD-700 dosimeter encapsulated in cadmium box and irradiated by fast neutrons with average energy of 5.1 MeV (a), 4.0 MeV (b), 3.2 MeV (c), and 1.9 MeV (d), as well as 662-keV gamma-rays (e) from a ^{137}Cs source. Each data point represents the average value of thermoluminescence output from five readouts from the same TLD chips at a particular temperature. The high-temperature (250–340°C) region of the glow curve was deconvoluted using the "Podgorsak approximation" of the first-order TL kinetics model. The glow curves are normalized to the highest value of the TL output at $T \approx 220$°C. The area under the deconvoluted peak is represented by $A(HT)$. (From Reference [24]. With permission.)

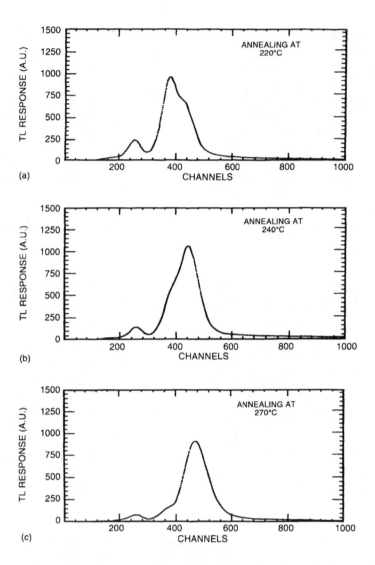

FIGURE 4.42 TL emissions obtained after various thermal procedures of annealing. (From Reference [27]. With permission.)

by each individual peak during the experiment. The results contained confirm that, for both varieties of LiF phosphor, the process affecting their respective main dosimetric peaks is not lading, understood as the spontaneous release of trapped charges. On the contrary, the observed variations in the TL response should be addressed to the modifications experienced by the trap system during storage. However, some slight differences between the evolution of TLD-100 and GR-200 have been found in this comparison. [30]

The decrease of the TL yield observed after exposure periods to environmental radiation has been traditionally interpreted as due to the spontaneous leakage of trapped charges, usually called Randall-Wilkins fading.

The reading cycles used were linear at 7°C s^{-1} in a N$_2$ atmosphere up to 300°C for TLD-100 and, in the case of GR-200, up to a maximum temperature placed 10°C above the detected position of peak 4. Once reached (around 240°C), this temperature was maintained for 5 s. In this way, better

reproducibility with GR-200 has been consistently obtained, reducing the tendency of this material to decrease in sensitivity with reuse to almost negligible levels. Natural cooling down inside the reader was always allowed, meaning a rather rapid and reproducible cooling. No additional annealing other than the readout itself was employed before reuse with both materials.

Figure 4.46 presents the evolution with storage time at 40°C of the relative areas of each individual peak, TLD-100 on the left-hand side and GR-200 on the right-hand side. Closed symbols represent the evolution of irradiated dosimeters (SA, storage after irradiation) and open symbols dosimeters stored un-irradiated (SB, storage before irradiation). The peaks of TLD-100 (as those of GR-200) present a different evolution with time. While the area of peak 5 relative to the prompt area presents a slow tendency to increase independently for irradiated and un-irradiated dosimeters, the area of peak 4, after a small increase, starts to diminish after the fifth day, reaching saturation in 15 days.

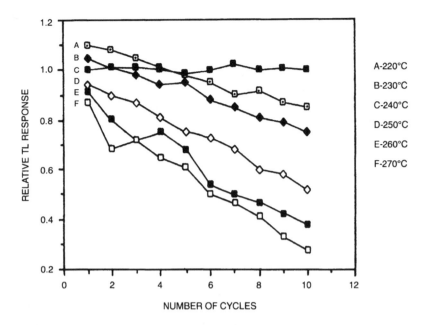

FIGURE 4.43 Peak 3 and 4 TL integrals for each annealing temperature and as a function of repeated cycles. (From Reference [27]. With permission.)

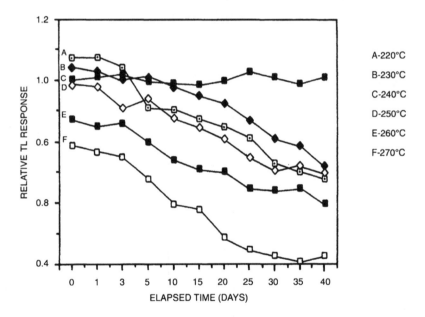

FIGURE 4.44 Plots obtained in fading experiments over 40 days of storage. (From Reference [27]. With permission.)

Peak 3, on the contrary, decreases quickly from the first day to the tenth and then mildly moves also to a saturation level. In the case of GR-200, the area of peak 4 also shows a slow tendency to increase from the beginning of the experiment, whereas peak 3 diminishes even more quickly than the corresponding peak 3 in TLD-100. Peak 4 seems to reach a saturation level at approximately the same time as peak 3 and also presents a softer decrease to a saturation level.

The same situation described for storage at 40°C occurs at 70°C but on a shorter time scale; see Figure 4.47. Peak 5 in TLD-100 and peak 4 in GR-200 present an initial quick rise, observed in the first few hours. Peak 4 in TLD-100 also decreases more rapidly to a saturation level attained earlier than at 40°C. The decrease of peak 3 in both LiFs is also far more rapid than at 40°C, and in GR-200 no peak 3 is detected after two days for SA detectors. This appears to prevent the growth of peak 4, since as soon as peak 3 disappears, peak

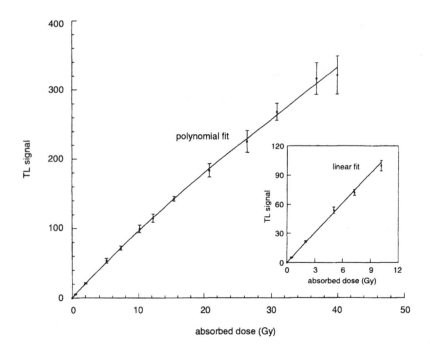

FIGURE 4.45 The TL signal (arbitrary units) vs. absorbed dose for the GR-200A dosimeters irradiated with the 15-MeV electron beam. The error bars represent the standard error (95% confidence level). The full line is the interpolation polynomial curve. The inset shows the linear fit up to 10 Gy. (From Reference [28]. With permission.)

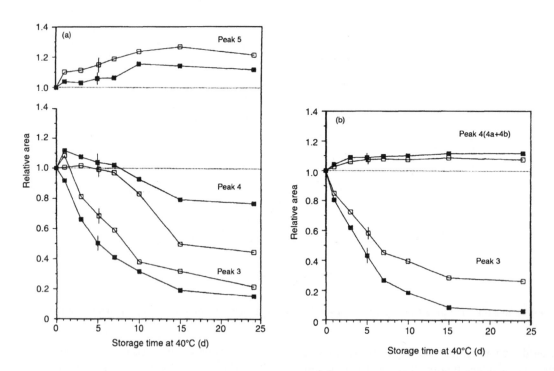

FIGURE 4.46 (a) Individual evolution of peaks 5, 4, and 3 of TLD-100. (b) Individual evolution of peaks 4(a + b) and 3 of GR-200 after different storage intervals at 40°C. Closed symbols represent storage after irradiation (SA) and open symbols represent storage before irradiation (SB). Error bars are typical 3σ values. (From Reference [30]. With permission.)

4 seems to stabilize. For all the peaks studied, the same ordered behavior for filled and empty traps is preserved at 70°C.

A fully automatic computer program for the deconvolution of LiF:Mg, Cu, P glow curves was described by Gomez Ros et al. [31] This program permits the subtraction of the residual contribution of the high-temperature peaks, producing the best fitted values for the kinetic parameters and the areas of the dosimetric peaks.

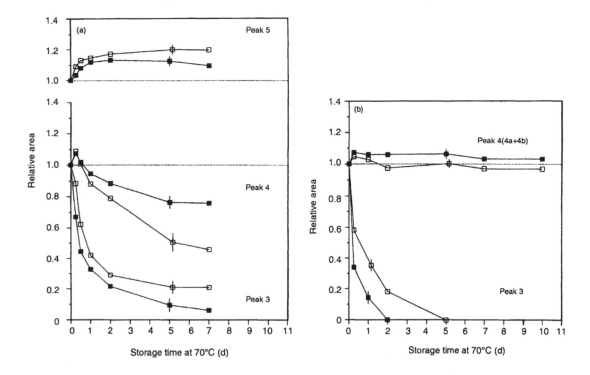

FIGURE 4.47 (a) Individual evolution of peaks 5, 4, and 3 of TLD-100. (b) Individual evolution of peaks 4(a + b) and 3 of GR-200 after different storage intervals at 70°C. Closed symbols represent storage after irradiation (SA) and open symbols represent storage before irradiation (SB). Error bars are typical 3σ values. (From Reference [30]. With permission.)

Figure 4.48a presents some LiF:Mg,Cu,P glow curves (ii) obtained with a linear/plateau heating profile (i), together with the corresponding second readout measurements (X10) (iii). Only peaks appearing during the linear part of the heating cycle can be separated by glow-curve fitting, but the contribution of the highest temperature signal cannot be ignored, and it is one of the most important problems in practical dosimetry with LiF:Mg,Cu,P, especially at low doses. A procedure to subtract the residual signal is illustrated in Figure 4.48b, in which the residual contribution of the high-temperature peaks to the region where dosimetric peaks appear is approximated by a straight line. This "linear background" is identified and subtracted in each glow curve, producing a net TL curve where the individual peaks can be fitted.

The analytical expression used to fit the glow peaks depends on the kinetic model considered for the processes involved. In the case of first-order kinetics the differential equation describing the variation of the peak intensity $I(t)$ with time is:

$$I(t) = -\frac{dn(t)}{dt} = s - \exp\left(-\frac{E}{kT(t)}\right)n(t) \quad (4.24)$$

where $n(t)$ is the trapped charges density, E is the activation energy, s is the frequency factor, k is Boltzmann's constant, and $T(t)$ is the temperature. The solution when

a linear heating profile $T(t)=T_0 + \beta_t$ is applied can be written as:

$$I(t) = I_M \exp\left(\frac{E}{kT_M} - \frac{E}{kT}\right)\exp\left\{\frac{E}{kT_M}\exp\left(\frac{E}{kT_M}\right)\right.$$

$$\left. \times \left[E_2\left(\frac{E}{kT_M}\right) - \left(\frac{T}{T_M}\right)E_2\left(\frac{E}{kT}\right)\right]\right\} \quad (4.25)$$

where T_M and I_M are the temperature and intensity of the maximum, respectively, and E_2 is the second exponential integral function defined as :

$$E_2(Z) = \int_1^{+\infty} \frac{e^{-zt}}{t^2} dt \quad (Z > 0) \quad (4.26)$$

for positive values of z, in particular for $z = E/kT_M$ and $z = E/kT$. Function $E_2(z)$ cannot be evaluated analytically, but a very accurate rational approximation can be used for arguments greater than 1 if we define the function

$$\alpha(z) = e^z E_2(z)$$

$$= 1 - \frac{a_0 + a_1 z + a_2 z^2 + a_3 z^3 + z^4}{b_0 + b_1 z + b_2 z^2 + b_3 z^3 + z^4} + \in (z)$$

$$(4.27)$$

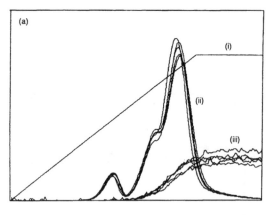

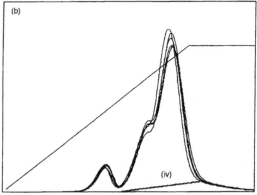

FIGURE 4.48 Background subtraction procedure: (i) heating profile, (ii) LiF:Mg,Cu,P glow curves, (iii) second readout (X10), and (iv) linear approximation of the residual signal in the region of dosimetric peaks. (From Reference [31]. With permission.)

with
$a_0 = 0.2677737343$; $a_1 = 8.6347608925$; $a_2 = 18.059016973$; $a_3 = 8.573328740$; $b_o = 3.9584969228$; $b_1 = 21.099653082$; $b_2 = 25.632956149$; $b_3 = 9.5733223454$
where an upper bound for the error is:

$$|\in(z)| < 2 \times 10^{-8} \quad if \quad (z > 1) \quad (4.28)$$

Then one may rewrite Equation 4.25 as:

$$I(t) = I_M \exp\left(\frac{E}{kT_M} - \frac{E}{kT}\right)\exp\left\{\frac{E}{kT_M}\left[\alpha\left(\frac{E}{kT_M}\right) - \left(\frac{T}{T_m}\right)\right.\right.$$

$$\left.\left. \times \exp\left(\frac{E}{kT_M} - \frac{E}{kT}\right)\alpha\left(\frac{E}{kT}\right)\right]\right\} \quad (4.29)$$

where only algebraic operations should be performed. Applying the condition (4.28) to the arguments of $\alpha(z)$, the error introduced in Equation 4.29 by the approximation of Equation 4.27 is negligible for temperatures (in K) $T < 1.16 \times 10^4 E$, where E is the activation energy (in eV).

Glow-curve analysis of long-term stability of LiF:Mg, Cu, P was compared by Duggan and Korn [32] to LiF:Mg, Ti. Over a six-month period, the long-term stability of LiF:Mg, Cu, P (GR-200A, Beijing, China) was compared with LiF:Mg, Ti (GR-100, Beijing, China). Annealing at 220°C instead of the pre-set temperature of 240°C caused an increase in the contribution from peak 3 in LiF:Mg, Cu, P, causing the material to be more susceptible to fading and peak-area interchanges. The response over time was fitted to a simple exponential of the form:

$$R = R_{t=\infty} + (1 - R_{t=\infty})\exp\left(-t\frac{\ln 2}{T}\right) \quad (4.30)$$

with $R_{t=\infty}$ as the response at day ∞, t as time (days), and T as the "half-life" of the response (days).

Two peaks were fitted for GR-200 and one (SC) or three (FC) peaks were fitted for GR-100, depending on the cooling regime. Program D is given by:

$$I(T) = As \exp\left(\frac{-E}{kT}\right)\exp\left[\frac{-skT^2}{\beta E}\exp\left(\frac{-E}{kT}\right)\right.$$

$$\left. \times \left(0.9920 - 1.602\frac{kT}{E}\right)\right] \quad (4.31)$$

where A = area (counts); s = frequency factor (s^{-1}); E = activation energy (eV); I = intensity (counts s^{-1}); k = Boltzmann's constant (eV K^{-1}); T = temperature (K); β = heating rate (K s^{-1}).

From Experiment A, "fading" of LiF:Mg,Cu,P over six months, for a slow and fast cool-down, were 10% and 6%, respectively, compared with 10% and 14% for LiF:Mg,Ti. From Experiment B, sensitivity changes of LiF:Mg,Cu,P over six months, for a slow and fast cool-down, were 3% and 7%, respectively, compared with 3% and 19% for LiF:Mg,Ti.

Evolution of individual peak area over time in each experimental arm A and B, is shown in Figures 4.49a and b for LiF:Mg,Cu,P and in Figures 4.50a and b for LiF:Mg,Ti. The peak fit for LiF:Mg,Cu,P (FC) was not as good as with the rest of the study, and it is suspected that first-order kinetics approximations may not be appropriate. The total peak area, made up almost entirely of peak 4, was very stable for this case. For LiF:Mg,Ti, for both cool-down regimes, the sum of all peaks, (3 + 4 + 5) and (4 + 5), remained unchanged within experimental uncertainty.

From this study it is clear that the mechanism responsible for the decrease in TLD response is almost entirely sensitivity changes caused by unstable unfilled trap (reactions between defects) and cannot be explained by fading (leakage of trapped charges) alone.

The detection limit (L_D) and the determination limit (L_Q) in personal dosimetry were evaluated by Muniz et al. [33]

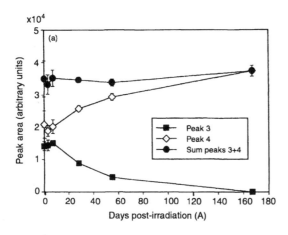

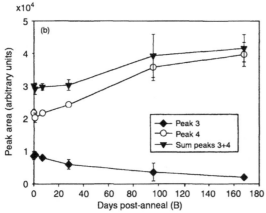

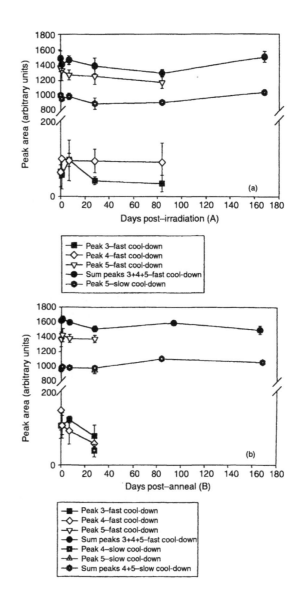

FIGURE 4.49 Evolution of individual peak area over time for GR-200A (SC) for storage (a) post-irradiation (experiment A) and (b) pre-irradiation (experiment B). Peaks were characterized by Equation 4.31. (From Reference [32]. With permission.)

The reading cycles were linear at 7°C s^{-1} in a N$_2$ atmosphere. LiF:Mg,Ti was heated up to 300°C, while LiF:Mg,Cu,P was heated up to a maximum temperature 10°C higher than the detected position of the main dosimetric peak (peak 4), after which this temperature was maintained for 5 s. For both materials, natural cooling inside the reader was always used, resulting in a rather rapid and reproducible cooling. Second readouts were performed immediately after the first so that the zero-dose signal could be estimated for the conventional analysis method. No annealing other than the reading cycle was used for the two materials.

The detection limit, L_D, for any quantity that is measured with an analytical process is the smallest amount of that quantity that can be detected at a specified confidence level. The determination limit, L_Q, specifies the minimum dose that can be measured with a given predefined precision. Hirning [34a] deduced expressions for L_D and L_Q that are given by:

$$L_D = \frac{2(t_n s_b + t_m^2 s_\mu^2 K_b)}{1 - t_m^2 s_\mu^2} \quad (4.32)$$

FIGURE 4.50 Evolution of individual peak area over time for GR-100 for storage (a) post-irradiation (experiment A) and (b) pre-irradiation (experiment B). A comparison is shown between both cooling regimes. Peaks were characterized by Equation 4.31. (From Reference [32]. With permission.)

$$L_D = \frac{k_Q^2 s_\mu^2 K_b + \sqrt{[k_Q^4 s_\mu^4 K_b^2 + k_Q^2 s_b^2 (1 - k_Q^2 s_\mu^2)]}}{1 - k_Q^2 s_\mu^2} \quad (4.33)$$

where t_n and t_m are one-sided Student-t factors for sample sizes of n un-irradiated and m irradiated detectors at the 95% confidence level; s_b is the sample standard deviation of the radiation background measurement, K_b; s_μ is the relative standard deviation (variation coefficient) of the irradiated dosimeters; and k_Q is the desired precision (the inverse of the maximum relative standard deviation).

The expression for the variance of a TLD reading irradiated to a total air kerma K_t is given as a function of

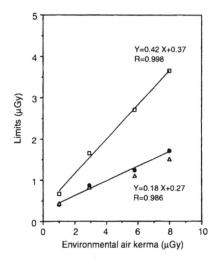 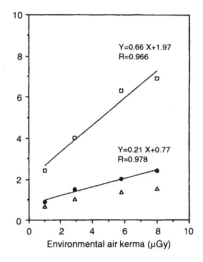

FIGURE 4.51 Results obtained with glow-curve analysis (left-hand side) and conventional analysis (right-hand side) of the glow curves of LiF:Mg,Cu,P. Filled circles, detection limit; open squares, determination limit; open triangles, $3s_b$ values. Linear regression fits are also presented. (From Reference [33]. With permission.)

the variance of null readings $(\sigma)^2$ and of the relative variance observed for high air kerma $(\sigma_\mu)^2$:

$$\sigma_t^2 = \sigma_n^2 + \sigma_\mu^2 K_t^2 \qquad (4.34)$$

If glow curves are to be analyzed with simplified glow-curve analysis (GCA) methods, there is no need for second readouts or for the use of undosed detectors. This leads to a simplification of the above expression, as σ_n does not exist. Then Equation 4.34 becomes:

$$\sigma_t^2 = \sigma_\mu^2 K_t^2 \qquad (4.35)$$

It should be noted that the elimination of σ_n in Equation 4.34 does not mean that the uncertainty of the zero dose signals individual evaluation is zero, since it has an obvious stochastic character. It simply reflects the fact that GCA does not need zero-dose detectors or second readouts.

According to the above, Equation 4.35 with the complex σ_μ was considered in the deduction of the limits for the GCA method. Using sample statistics, L_D and L_Q may be given by: [33]

$$L_D^{GCA} = \frac{t_n s_b}{1 - t_m s_\mu} + \frac{t_m s_\mu}{1 - t_m s_\mu} K_b \qquad (4.36)$$

$$L_D^{GCA} = \frac{k_Q s_\mu}{1 - k_Q s_\mu} K_b \qquad (4.37)$$

which are applicable for GCA methods.

Figure 4.51 shows the results for both L_D and L_Q, obtained with GCA methods on the left-hand side and with conventional analysis on the right-hand side, for LiF:Mg,Cu,P. Values of $3s_b$, obtained with either method, are also presented in the figures, and it can be observed that the use of this parameter for the detection limit underestimates it when the conventional approach is used, while it is quite accurate in the GCA approach. As expected, GCA-produced limits are always lower and the limits for LiF:Mg,Cu,P are also lower than the limits for LiF:Mg,Ti.

Similar to LiF:Mg,Ti, LiF:Mg,Cu,P is available with different thermal neutron sensitivities, depending on the concentrations of LiF:Mg,Cu,P, i.e., 7.5%, 95.6%, and 0.07%, corresponding to TLD-100H, 600H, and 700H, respectively. For use in personal dosimetry, the TLD pellets are encapsulated in thin FEP-coated film and mounted in a standard Harshaw aluminum substrate to form a TLD card. The TL response of this material is extremely sensitive to the readout temperature. [34]

A reader which uses a linear gas-heating technique that combines the advantages of non-contact gas heating and linear ohmic heating was described by Moscovitch. [34] The reader incorporates a linear time-temperature-controlled hot gas-heating technique. The reader can use either nitrogen or air for heating the TL elements. Heating rates may be in the range of 1 to 50°C s^{-1}. The temperature is sensed by individual thermocouples across the end of each nozzle and is sent to a heater control board which compares the measured temperature with that called for by the user-defined heating profile (temperature as a function of time). It then adjusts the current in the heating tubes to maintain the temperature of the gas within ±1°C of the specified level. The TL-emitted light is measured using several photomultiplier tubes (PMT) that are thermoelectrically cooled

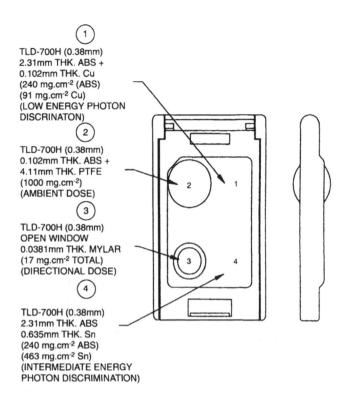

TLD-700H (0.38mm)
2.31mm THK. ABS +
0.102mm THK. Cu
(240 mg.cm⁻² (ABS)
(91 mg.cm⁻² Cu)
(LOW ENERGY PHOTON
DISCRINATON)

TLD-700H (0.38mm)
0.102mm THK. ABS +
4.11mm THK. PTFE
(1000 mg.cm⁻²)
(AMBIENT DOSE)

TLD-700H (0.38mm)
OPEN WINDOW
0.0381mm THK. MYLAR
(17 mg.cm⁻² TOTAL)
(DIRECTIONAL DOSE)

TLD-700H (0.38mm)
2.31mm THK. ABS
0.635mm THK. Sn
(240 mg.cm⁻² ABS)
(463 mg.cm⁻² Sn)
(INTERMEDIATE ENERGY
PHOTON DISCRIMINATION)

FIGURE 4.52 Type 8855 dosimeter. (From Reference [35]. With permission.)

to 12°C. The PMT signal is accumulated via the charge integration technique.

The reader calibration factor for element position i, RCF_i, is defined as follows:

$$RCF_i = \langle Q \rangle_i / L$$

where $\langle Q \rangle_i$ is the average measured charge for that position when a set of calibration dosimeters is exposed to a known quantity of radiation L.

The sensitivity of LiF(MCP) is approximately 25 times higher than the sensitivity of TLD-100, with a batch homogeneity of 8% (one standard deviation). Once encapsulated in a TLD card, however, the relative sensitivity drops to 10. The reason for this sensitivity decrease is that the heating applied to the chip during manufacturing (the encapsulation process) increases the temperature to 280–290°C for short periods of time. This results in permanent loss of sensitivity of the phosphor. [34]

A new environmental TLD dosimeter badge and dose-computation algorithm based on LiF:Mg,Cu,P was described by Perry et al. [35] LiF:Mg,Cu,P, with its high sensitivity, tissue equivalence, energy independence, and low fading characteristics, is a natural choice for environmental dosimetry. The badge consisted of a card and a plastic holder. The card contained four LiF:Mg,Cu,P elements encapsulated in Teflon. The elements were all 3.2 mm square and 0.4 mm thick. The badge was symmetrical and used four fillers to discriminate low- and high-energy

photons and to determine the directional dose equivalent, $H'(0.07,\alpha)$ and ambient dose equivalent $H^*(10)$.

The Type 8855 dosimeter is shown in Figure 4.52.

The filtration-covering element number 1 consists of 91 mg cm⁻² Cu + 240 mg cm⁻² ABS plastic and is used for low-energy photon discrimination. Element number 2 filtration consists of 1000 mg cm⁻² PTFE+ABS plastic and is used for ambient dose measurement. Element number 3 has 17 mg cm⁻² of Mylar plus PTFE encapsulation and is used for directional dose measurement. Element number 4 filtration consists of 464 mg cm⁻² Sn + 240 mg cm⁻² ABS plastic and is used for intermediate energy photon discrimination. The card identification is visible through a red filter window located in the center of the front side of the holder.

The graph in Figure 4.53 represents the photon energy response of each of the elements in the dosimeter referenced to ¹³⁷Cs for pure fields. The results show that with light filtration, such as is the case in element 3, the energy response of the dosimeter is relatively flat. Combined with the response of elements with heavier filtration, ratios of elements will give useful information as to what energy is being measured in a blind test situation.

IV. CAF:Tm (TLD 300) DOSIMETER

TLD-300 dosimeters have been employed in high LET radiotherapeutic fields generated by fast neutron, negative pion, and heavy ion beams. By proper peak height analysis

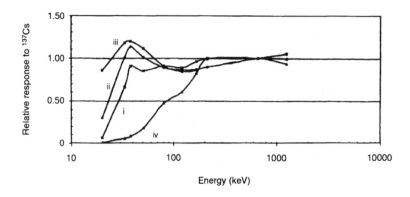

FIGURE 4.53 Photon energy response of 8855 dosimeter. Elements i–iv indicated on curves. (From Reference [35]. With permission.)

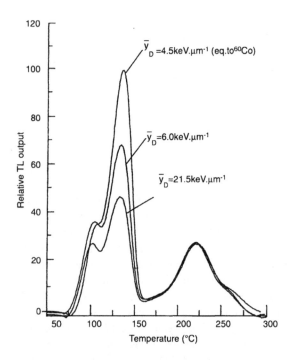

FIGURE 4.54 TLD-300 glow curves taken at different positions along the depth-dose distribution with equal dose; $\bar{y}_D$ values taken from the microdosimetric spectra taken at that position. (From Reference [36]. With permission.)

of the two main peaks of CaF$_2$:Tm glow curves after such irradiations, it is possible to separate the high and low LET content of the radiation field. Therefore, the distribution of the biological equivalent dose can be determined in a single irradiation. The method was extended to a pretherapeutic proton field of 175-MeV maximum kinetic energy by Hoffmann et al. [36] It turned out that in such proton fields, the sensitivity of the high temperature peak is constant; thus, the dose can be determined to an accuracy of ±4%. The sensitivity of the low temperature peak decreases linearly to 50% as a function of $\bar{y}_D$, such that the average linear energy can be determined to an accuracy of ±5 keV μm^{-1}.

The shape of the glow curve changes considerably when the TLD-300 dosimeters are irradiated at different positions along the depth dose. Figure 4.54 shows glow curves taken with the same dose at 0.5, 17, and 19.5-cm depth, which corresponds to 90 values of 4.5, 6.0, and 21.5 keV μm^{-1}. Whereas the sensitivity of the high-temperature peak P2 remains constant, the sensitivity of the low-temperature peak P1 decreases rapidly with increasing LET. The decrease is steeper than measured with negative pions, due to the different shape of the microdosimetric spectra caused by secondary particles. It is, however, similar to the LET dependence found in He^{2+} beams because of the similarity of the spectra.

Peak 3 (P3) of TLD-300 was studied by Angelone et al. [37] as a low-dose TL dosimeter. Measurements showed that it is possible to reach, for the detection limit (L_D), a value of 1.0 ± 0.3 μGy. A reading cycle which allows peak 3 to be read and the TLD-300 to be annealed in the reader allows at least 9-fold reuse of TLD-300 before annealing at 400°C for 1 h is necessary. The residual TL signal was 0.5% of the TL signal of the irradiated dosimeter. A slight decrease of the P3 TL signal with the number of irradiation/reading cycles was also observed.

Since the glow curve of TLD-300 shows two well-separated peaks, the so-called peak 3 (P3) and peak 5 (P5) (the latter being sensitive to high LET particles), a reading cycle able to read peak 3 alone was studied. After several tests the following cycle was adopted: preheating at 80°C lasting 5°C s^{-1}; heating ramp of 5°C s^{-1} for 34 s, up to a maximum temperature of 240°C; quick temperature rise (25° C s^{-1}) up to 400°C; and annealing lasting 15 s (Figure 4.55). The last step allows for P5 to be annealed in the reader. The in-oven annealing procedure used in this work is 1 h at 400°C.

A series of nine cycles (irradiation/preheating/reading) were repeated, irradiating each time the same group of 15 TLDs. The TL residual signal did not exceed 0.5% of the last TL signal reading. It was also found that there is a slight decrease of the P3 response as the number of cycles increases. After nine cycles, a 4% decrease of the

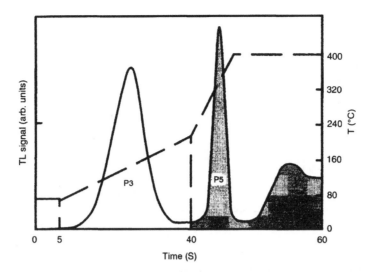

FIGURE 4.55 Typical TL signal for TLD-300 irradiated at 1 mGy and read with the heating cycle reported in the text. The temperature profile is also reported (right axis). (From Reference [37]. With permission.)

last measured signal, with respect to the first TL signal, was observed.

Important parameters are the critical level (L_c) and the detection limit (L_D). L_c is the signal that provides a confidence level of $1-\alpha$ that the result is not due to a background fluctuation, while, as stated by Hirning [34a], L_D is the smallest amount of signal that can be detected at a specified confidence level.

The results for L_c and L_D are reported in Table 4.3. The results clearly indicate that peak 3 of TLD-300 has a very good performance as a low-dose detector. The measured values are only twice that of LiF-600H and GR-206 and a factor five above that of LiF-700H and GR-207, which make peak 3 suitable for most low-dose measurements.

The TL response of CaF$_2$:Tm to 0.7-MeV protons, 5.3-MeV particles, and 120-MeV ^{12}C ions has been investigated by Buenfil et al. [38] The area of peak 3 showed a linear response as a function of fluence for all the fluences studied, while peaks 5 and 6 followed a linear-supralinear behavior. The departure from linearity occurs at high doses, above 10^2 Gy.

Figure 4.56 shows a typical TLD-300 glow curve after low LET irradiation (in this case, ^{60}Co gamma-rays) and its deconvolution into peaks numbered 3 to 6. Figure 4.57 shows a typical glow curve after exposure to heavy charged particles (in this case, 5.3-MeV α particles). The differences between the glow curves displayed in Figures 4.56 and 4.57 explain the interest in TLD-300 as a dosimeter for mixed field situations, where low and high LET contributions need to be separated.

Hoffmann and Prediger [39] found that peak 5 has constant sensitivity to gamma-rays and particles having LET below about 30 keV m^{-1}. Higher LET radiation is less effective than gamma rays in activating peak 5. Buenfil et al. confirmed their results since the highest LET

TABLE 4.3
Measured Critical Level (L_c) and Detection Limit (L_D), in μGy, for the TLDs used by Angelone

Detector	$L_c(\mu Gy)$	$L_D(\mu Gy)$
TLD-300	0.5	1.0
TLD-700H	0.1	0.2
TLD-600H	0.25	0.50
GR-207	0.15	0.30
GR-206	0.26	0.52

Source: From Reference [37]. With permission.

radiation in their data (α particles and ^{12}C) had about half the peak 5 sensitivity than the one measured for protons. Concerning peak 3, Hoffmann and Prediger reported a continuous loss of relative sensitivity as a function of the LET.

Loncol et al. [40] studied the peak 3 (150°C) and peak 5 (250°C) response of CaF$_2$:Tm (TLD-300) to neutron and proton beams, to analyze the effect of different radiation qualities on the dosimetric behavior of the detector irradiated in phantom. The study was extended to an experimental ^{12}C heavy ion beam (95 MeV/nucleon). The sensitivities of peaks 3 and 5 for neutrons, compared to ^{60}Co, varied a little with depth. A major change of peak 5 sensitivity was observed for samples positioned under five leaves of the multi-leaf collimator. While peak 3 sensitivity was constant with depth in the unmodulated proton beam, peak 5 sensitivity increased by 15%. Near the Bragg peak, peak 3 showed the highest decrease of sensitivity. The ratio of the heights of peak 3 and peak 5 decreased by 70% from the ^{60}Co reference radiation to the ^{12}C heavy-ion beam.

Figure 4.58 shows the glow curves of two particular samples positioned at the extreme depths in the phantom

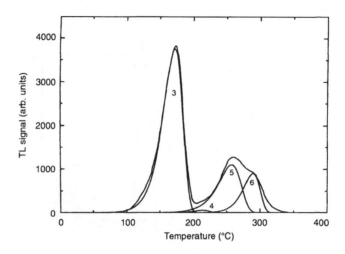

FIGURE 4.56 Typical TLD-300 glow curve for low LET irradiation (^{60}Co γ irradiation). Deconvolution has been applied to separate the curve into peaks 3 to 6. (From Reference [38]. With permission.)

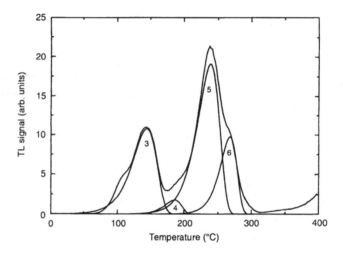

FIGURE 4.57 Typical TLD-300 glow curve for high LET irradiation (^{241}Am α particles). Deconvolution has been applied to separate the curve into peaks 3 to 6. (From Reference [38]. With permission.)

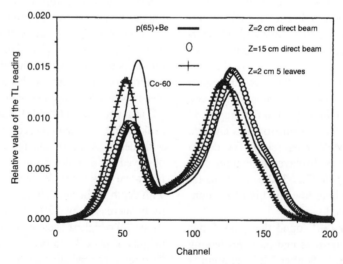

FIGURE 4.58 A glow-curve comparison for samples positioned at different depths in water, after irradiation in p(65) + Be neutron beam and in ^{60}Co beam. (From Reference [40]. With permission.)

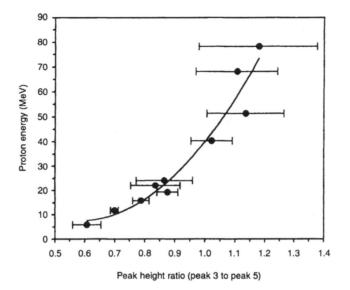

FIGURE 4.59 Proton energy as a function of the peak height ratio (peak 3 to peak 5) for the unmodulated proton beam. The experimental data were fitted with a second-order polynomial. (From Reference [40]. With permission.)

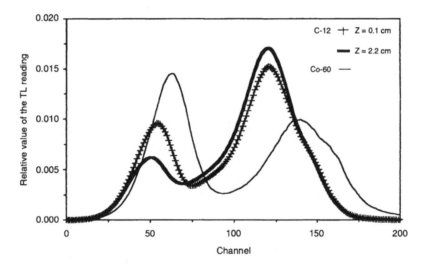

FIGURE 4.60 A glow curve comparison for samples positioned at different equivalent depths in water after irradiation in the ^{12}C ion beam and in the ^{60}Co beam. (From Reference [40]. With permission.)

(2 and 15 cm) irradiated by the direct beam. The glow curve issued from another chip irradiated in a cobalt beam was also plotted. All the curves were normalized to a unit area to allow the comparison between them. From these curves, no major change of sensitivity was observed for peaks 3 and 5 at different positions in the direct neutron beam.

The glow curves of samples positioned at different depths in the phantom irradiated by the unmodulated beam showed that the height of peak 3 decreased relative to that of peak 5 as the TLD position became deeper.

Figure 4.59 shows the proton energy as a function of the peak height ratio in the unmodulated proton beam. These results can be approximated with a second-order polynomial fit. This curve can be used in order to estimate the mean energy in the spread-out Bragg peak.

Figure 4.60 compares the glow curves of samples positioned at the initial plateau (0.1 cm) and at the Bragg peak (2.2 cm) of the unmodulated beam. The glow curve shows an even more pronounced peak height change than for the neutron and proton beams if compared to that for the ^{60}Co beam. As in the proton beam, the height of peak 3 decreased relative to that of peak 5 as the TLD position approached the Bragg peak. Beyond the Bragg peak (2.5 cm), the height of peak 3 increased again to 0.80, as in the proton beam.

The glow-curve structure and kinetics parameters of CaF$_2$:Tm (TLD-300) were investigated by Jafarizadeh [41] at a test dose of ^{90}Sr β rays. A new peak at 68°C, called 1a, was recognized in the complex structure of the initial part of the glow curve. By applying a thermal

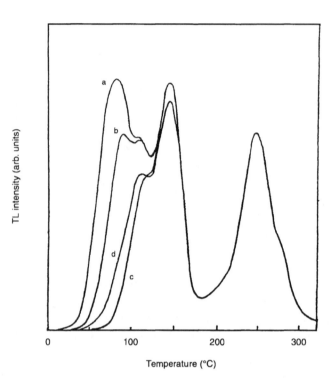

FIGURE 4.61 Glow curves of CaF$_2$:Tm exposed for 10 min to a test dose of 30 mGy of β rays , at a heating rate of 2°C s^{-1}: A, immediately; B, after 1 h; C, 24 h after exposure of 60 mGy; and D, immediately after an exposure of 80 mGy but a long exposure duration time of 16 h. (From Reference [41]. With permission.)

bleaching method using a heating rate of 2°C s^{-1}, seven glow peaks (1a to 6) were separated at a temperature range of 33 to 314°C.

Figure 4.61 shows the glow-curve structure of this phosphor exposed for 10 min to a test dose of 30 mGy of β rays, recorded at a heating rate of 2°C s^{-1} (a) immediately, (b) after 1 h, (c) after 24 h of post-irradiation time, and (d) immediately at a test dose of 80 mGy but for a long exposure duration of 16 h. Seven glow peaks, located at temperatures 68(1a), 80(1), 106(2), 145(3), 191(4), 243(5), and 273(6)°C, were obtained at a heating rate of 2°Cs^{-1}.

The solution of the general order (GO) model formula of May and Partridge [42] for a constant temperature (isothermal decay) leads to Equation 4.38:

$$I^{(1/b)-1} = A + Bt \qquad (4.38)$$

where A and B are constants and t is the decay time.

The phototransferred thermoluminescence (PTTL) of sintered pellets of natural CaF$_2$ from Brazil was compared with that of CaF$_2$ doped with Ce^{3+} and/or Dy^{3+} by Nakamura et al. [43] It is shown that the excitation spectrum provides a convenient way to find an optimum condition of preheat treatment of sintered CaF$_2$ after UV irradiation. The sintered CaF$_2$ containing a small amount of Ce^{3+} 0.2 mol% showed a strong glow peak around

90°C after UV irradiation. On co-doping with Dy^{3+} in CaF$_2$:Ce^{3+}, the prominent glow peak shifts to 270°C. The peak intensity shows a linear response to the UV irradiation dose.

Figure 4.62 shows the glow-curve pattern of natural CaF$_2$, having been irradiated with x-rays to 25 Gy (a) and UV light of 1.2 J (b). Although the TL intensity of UV irradiation is smaller than that of x irradiation, the 260°C peak is prominent compared with low-temperature peaks. The inset shows the 260°C peak-intensity dependence on the UV irradiation time over a wide range as $t = 0.8$ to 103 min.

The 3.65-eV (340-nm) emission reveals a prominent excitation peak at 4.00 eV (310 nm) for S-1 after heating at 700°C for 5 min (Figure 4.63).The peak intensity decreases after subsequent x irradiation (curve B). A decreasing amount of Ce^{3+} may correspond to the TL intensity, since the Ce^{3+} ion must return to the initial state through a recombination with its counterpart. The Ce^{3+} peak height saturates around 310°C, and then the peak height decreases with temperatures again (inset of Figure 4.63).

Sintered CaF$_2$ was prepared containing a small amount (0.2 mol%) of Ce^{3+} (sample S-2). From the excitation spectrum, the optimum temperature of S-2 was found to be 320°C. A strong glow peak at 90°C was observed after heating at 320°C and subsequent UV irradiation (solid curves in Figure 4.64b).

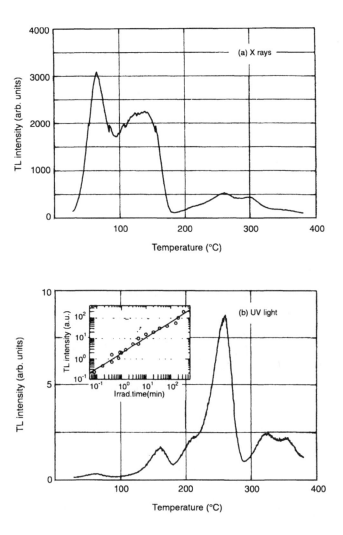

FIGURE 4.62 TL glow curves of as-received natural CaF_2. (a) x irradiation of 25 Gy. (b) UV irradiation of 1.2 J. Inset shows a relation between 260°C peak intensity and irradiation time of UV light. (From Reference [43]. With permission.)

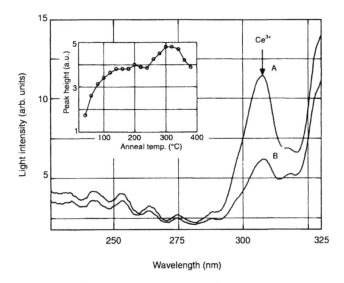

FIGURE 4.63 Excitation spectra of 340-nm emission of Ce^{3+} in sintered $CaF_2:CeO_2$. A: before; B: after x irradiation. Inset shows the Ce^{3+} peak height dependence on the treatment temperature after x irradiation. (From Reference [43]. With permission.)

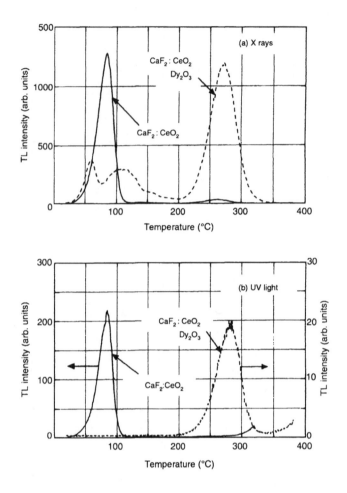

FIGURE 4.64 TL glow curves of S-2 (solid curves) and S-3 (dashed curves): (a) after x irradiation of 25 Gy, (b) after UV irradiation of 1.2 J. (From Reference [43]. With permission.)

TABLE 4.4
Density and Atomic Number of the Materials used in the Determination of the Quality Dependence of TL Materials

TL Material	Effective Atomic Number	Density (g cm⁻³)
Water	7.50	1.00
LiF	8.27	2.64
LiF:Mg:Ti (TLD-100)	8.27	2.64
LiF:Mg (0.18%):Cu (0.0024%):P(2.3%)	8.67	2.64
$Li_2B_4O_7$	7.31	2.44
CaF_2	16.87	3.18
CaF_2:Mn (3%)	17.12	3.18
$CaSO_4$	15.6	2.96

Source: From Reference [44]. With permission.

The TL emission spectra for both x and UV irradiation show a broad band with a peak around 380 nm which is believed to be caused by Ce^{3+} coupled with O^{2-} centers, but no emission bands at 2.64 eV (470 nm) or 2.18 eV (570 nm) have been observed for S-2. From the solid curves of Figure 4.64a and b, it is found that the glow peak of 2 is highly sensitive for UV irradiation as compared to S-1, but the peak temperature is too low to use it for practical purposes. [43]

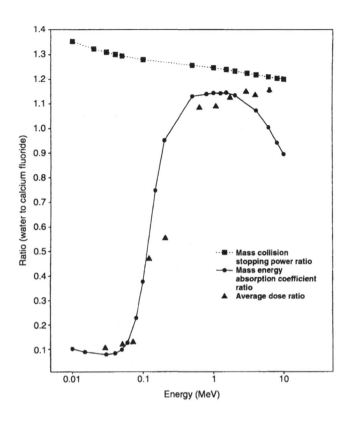

FIGURE 4.65 Calcium fluoride TLD discs, 0.9 mm thick, in MV and kV x-ray beams: comparison of the Monte Carlo derived average dose ratio, water to CaF$_2$, with the mass collision stopping power and mass energy–absorption coefficient ratios, as a function of the mean photon energy. (From Reference [44]. With permission.)

V. THEORETICAL ASPECTS AND MONTE CARLO CALCULATIONS OF TL DOSIMETRY

A Monte Carlo simulation of the quality dependence of different TL materials was performed by Mobit et al. [44] in the form of discs 3.61 mm in diameter and 0.9 mm thick, in radiotherapy photon beams relative to ^{60}Co γ-rays. The TL materials were CaF$_2$, CaSO$_4$, LiF and Li$_2$B$_4$O$_7$. It was found that there was a significant difference in the quality dependence factor derived from Monte Carlo simulations between Lif and LiF:Mg:Cu:P but not between CaF$_2$ and CaF$_2$:Mn. The quality dependence factors for Li$_2$B$_4$O$_7$ varied from 0.990 $\pm$ 0.008 (1 sd) for 25-MV x-rays to 0.940 $\pm$ 0.009 (1 sd) for 5-kV x-rays relative to ^{60}Co γ-rays. For CaF$_2$ the quality dependence factor varied from 0.927 $\pm$ 0.008 (1 sd) for 25 MV x-rays to 10.561 $\pm$ 0.008 (1sd) for 50-kV x-rays. For LiF TLD, there was no significant dependence on the field size or depth of irradiation in the kilovoltage energy range.

Table 4.4 lists the TLD materials simulated, with their effective atomic numbers and densities.

The quality dependence factor, F_{Co}^Q, is defined as

$$F_{Co}^Q = \frac{(\bar{D}_w/\bar{D}_{TLD})_{Co}}{(\bar{D}_w/\bar{D}_{TLD})_Q} \qquad (4.39)$$

where $\bar{D}_{TLD}$ is the average absorbed dose in the TL material and $\bar{D}_w$ is the average absorbed dose in water of the same volume placed at the same point in the uniform water phantom. The assumption used in deriving Equation (4.39) is that the light output is directly proportional to the absorbed dose in the TL material. Figure 4.65 shows the mass energy–absorption coefficient ratio, the mass collision stopping power ratio, and the Monte Carlo calculated average dose ratio of water to CaF$_2$ from 50-kV to 25-MV x-rays. All these ratios are plotted against the mean energy on the phantom surface.

Figure 4.66 shows the average dose ratio of water to LiF obtained by Monte Carlo simulations of kilovoltage and megavoltage photon beams. The mass energy–absorption coefficient ratio and the mass collision stopping power ratio of water to LiF are also shown. The quality dependence factor for LiF TLD in kV x-rays relative to ^{60}Co γ-rays ranges from 1.36 $\pm$ 0.01 for 50-kV x-rays to 1.03 $\pm$ 0.01 for 300-kV x-rays.

Monte Carlo simulations with the EGS4 code system have been performed by Mobit et al. [45] to determine the quality dependence of diamond TLDs in photon beams ranging from 25-kV to 25-MV x-rays and also in megavoltage electron beams. It has been shown that diamond TLDs in the form of discs of thickness 0.3 mm and diameter 5.64 mm show no significant dependence on the incident energy in clinical electron beams when irradiated close to d_{max}, but require an energy correction factor of 1.050 ± 0.008 compared with diamond TLDs irradiated in ^{60}Co γ-rays. The correction factor increases with depth of irradiation and this effect is greater for thicker detectors. The Monte Carlo predicted sensitivity in x-ray beams is constant within 2.5% over the energy range 250 kV to 25 MV. However, the sensitivity decreases by about 60% for 25 kV x-rays compared with ^{60}Co γ-rays.

Stones with controlled amounts of selected impurities can be used with considerable success as detectors of ionizing radiation of all types. Selected stones can be used as thermoluminescent dosimeters (TLDs) or radiation dosimeters in dc mode.

Figure 4.67 shows the comparison between the average dose ratio of water to diamond and the mass energy–absorption coefficient ratio of water to carbon. The two curves are very similar, but in the kilovoltage energy range the dose ratio of water to diamond curve is slightly lower than that for the mass energy–absorption coefficient ratio curve. This could be due to perturbation effects of diamond detectors in kilovoltage photon beams or an effective point of measurement effect, as the dose is averaged over a greater effective volume for the diamond detector (about 3.5 equivalent thicknesses since the density of diamond is 3.5 times greater than that of water).

Figure 4.68 shows a comparison between experimental measurements of Nam et al. [46] and Mobit et al. [45]

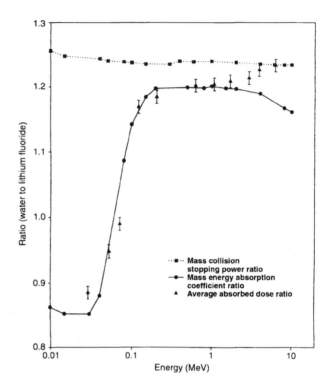

FIGURE 4.66 Lithium fluoride TLD discs, 0.9 mm thick, in MV and kV x-ray beams: comparison of the Monte Carlo derived average dose ratio, water to LiF, with the mass collision stopping power and mass energy–absorption coefficient ratios, as a function of the mean photon energy. (From Reference [44]. With permission.)

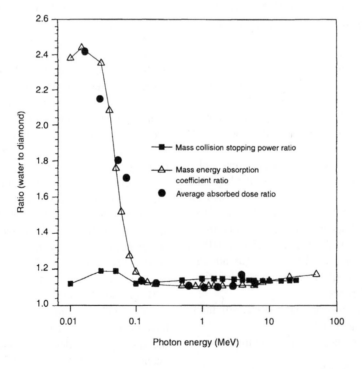

FIGURE 4.67 A comparison of the Monte Carlo derived average dose ratio with the mass collision stopping power ratios and the mass energy–absorption coefficient ratios for water to diamond (carbon). (From Reference [45]. With permission.)

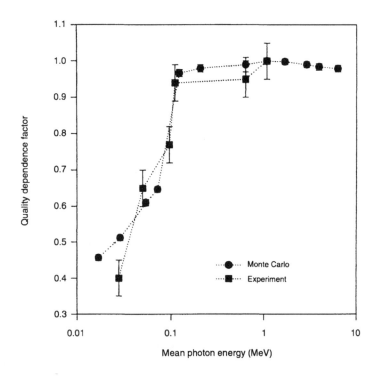

FIGURE 4.68 A comparison of the Monte Carlo derived quality dependence factor of diamond TLDs with experiment as a function of photon quality (kV x-rays). (From Reference [45]. With permission.)

Monte Carlo simulations in photon beams ranging from 25-kV to 25-MV x-rays. The comparisons have been made for TLDs irradiated at the surface of a phantom. In the megavoltage energy range where the dosimeters were irradiated inside a phantom, the difference between Monte Carlo simulation and measurement is less than 1.0%, while in the kilovoltage energy range there is a disagreement of up to 10%. The experimental determination for kilovoltage x-ray beams by Nam et al. was in air and not in a phantom, and this alone can account for a difference of more than 10%.

LET effects are not accounted for by the Monte Carlo simulations. There is also an uncertainty in the energy spectra between experiments and Monte Carlo simulations, and this can be significant, especially for low-energy kV x-rays. Diamond TLDs calibrated in ^{60}Co γ-rays require a correction factor of 1.047 ± 0.008 to correct for the under-response of diamond TLDs in electron beams compared to ^{60}Co γ-rays.

The response of TL dosimeters irradiated with electron beams in a water-equivalent phantom depends not only on the dose but also on the electron energy and dosimeter volume. For a constant electron dose, the TLD response falls with decreasing energy and increasing TL dosimeter volume. [47]

Holt et al. [48] derived a semi-empirical relationship which, similar to the Burlin approach, incorporates the finite size of the dosimeter but succeeds in predicting the TLD sensitivity variations with electron energy quite well. They express the dose, D_m, in the medium as a function of the dose, D_c, in the cavity through the standard relationship

$$\frac{D_m}{D_c} = \frac{\phi_m}{\phi_c} \frac{S_m}{S_c} \qquad (4.40)$$

where ϕ_m and ϕ_c are electron fluences and S_m and S_c, are mass stopping powers in the medium m and cavity c, respectively. Holt et al. then made an assumption that the cavity causes a perturbation of the energy spectrum, i.e., $\phi_m \neq \phi_c$. The ratio ϕ_m/ϕ_c was defined as the cavity perturbation factor K_E and approximated by the following expression:

$$K_E = \frac{\phi_m}{\phi_c} \frac{D_m}{D_c} [S_c^m]^{-1} = fS_c^m + (1 - f) \quad (4.41)$$

where S_c^m signifies the ratio of mass stopping powers S_m/S_c and f represents the fraction of the electron energy entering the cavity and absorbed in the cavity. Similar to Burlin theory, fS_c^m relates the electron spectrum building to its maximum value in the cavity, while $(1 - f)$ represents the decay in the cavity of the electron spectrum which was generated in the medium. In Equation (4.41) K_E is the dose correction factor for the TLD response to electrons relative to ^{60}Co gamma-rays. Thus, $1/K_E$ represents the calculated relative TLD response as

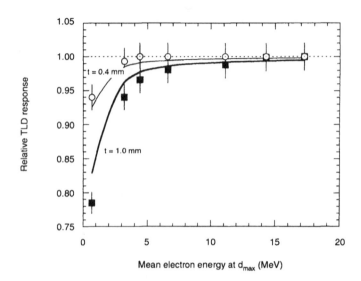

FIGURE 4.69 Relative response of the 0.4- and 1.0-mm thick TLD normalized to the response per unit dose in a ^{60}Co beam. (From Reference [41]. With permission.)

follows:

$$\frac{1}{K_E} = \frac{D_c(E)}{D_m(E)} S_c^m(E) = \frac{D_c(E)}{D_m(\gamma)} S_c^m(E)$$

$$= \frac{D_c(E)}{D_m(\gamma)} \left(\frac{S_c^m(E)}{S_c^m(\gamma)} \right) \approx \frac{D_c(E)}{D_c(\gamma)} = \frac{\eta Q_c(E)}{\eta Q_c(\gamma)} = \frac{Q_c(E)}{Q_c(\gamma)}$$

$$(4.42)$$

where the electron dose to the medium $D_m(E)$ is related to the ^{60}Co gamma-ray dose $D_m(\gamma)$ through Bragg-Gray cavity theory and η is the thermoluminescence calibration factor relating the TL signal Q to the dose. The stopping power ratios $S_c^m(E)$ and $S_c^m(\gamma)$ differ by less than 0.5% over the range of electron energies used in our experiment. Thus, $S_c^m(E)/S_c^m(\gamma)$ is assumed equal to 1 in Equation (4.42)

Figure 4.69 shows the measured relative responses for the 0.4- and 1.0-mm-thick TL dosimeters irradiated at the depth of dose maximum (d_{max}) in the polystyrene phantom, with electrons in the energy range between 0.68–20.9 MeV. The 0.68-MeV points were obtained by irradiating the TL dosimeters on the polystyrene phantom surface with the Sr-Y ophthalmic applicator; all other points were obtained by irradiations with a linac. The relative responses of the TL dosimeters are normalized to the response per unit dose, measured with ^{60}Co gamma-rays. The relative TLD responses are plotted as a function of the mean electron energy at d_{max} phantom. The TLD response per unit dose depends strongly on electron energy and thickness of TL dosimeters.

The CEPXS/ONEDANT code package [49] was used by Deogracias et al. [50] to produce a library of depth-dose profiles for monoenergetic electrons in various materials

for energies ranging from 500 keV to 5 MeV in 10-keV increments. The various materials for which depth-dose functions were derived include lithium fluoride (LiF), aluminum oxide (Al_2O_3), beryllium oxide (BeO), calcium sulfate ($CaSO_4$), calcium fluoride (CaF_2), lithium boron oxide (LiBO), soft tissue, lens of the eye, adipose, muscle, skin, glass, and water. Material data sets were fit to five polynomials, each covering a different range of electron energies, using a least-squares method.

The fifth-order polynomials were of the form:

$$D(x, E) = ax^5 + bx^4E + cx^4 + dx^3E^2 + ex^3E$$
$$+ fx^3 + gx^2E^3 + hx^2E^2 + ix^2E$$
$$+ jx^2 + kxE^3 + lxE^2 + mxE + nx + oE^3$$
$$+ pE^2 + qE + r \qquad (4.43)$$

In the polynomial expression, $D(x,E)$ is the depth dose, x represents the depth in centimeters, E is the energy in keV, and $a – r$ are the polynomial coefficients.

Figure 4.70 shows the comparison between the simulated data and the polynomial fit for an energy of 500-keV. The polynomial fit for the 500-keV energy does not correlate as well as the fits for the higher energies. This is due to the extreme increase in the dose-to-depth ratio in the lower energies, which the polynomial equation cannot adequately compensate. The 500-keV fit is the worst-case fit for the entire energy range in all materials. Figure 4.71 shows the comparison at an energy of 2700 keV.

Figure 4.72 is an example of the depth-dose curve generated from a beta particle spectrum. The energy range for all the spectrums begins at 500 keV, and the endpoint energy of the spectrum is varied from 1300 to 3200 keV.

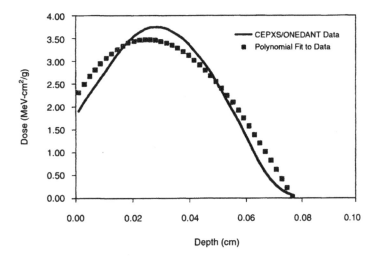

FIGURE 4.70 Depth-dose curve fit comparison between the polynomial dose equation and the CEPXS/ONEDANT data for 500-keV electrons in LiF. (From Reference [50]. With permission.)

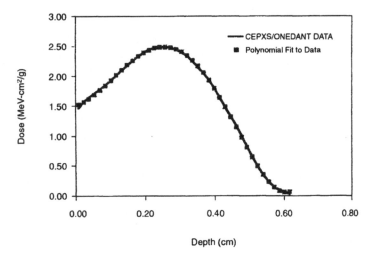

FIGURE 4.71 Depth-dose curve fit comparison between the polynomial dose equation and the CEPXS/ONEDANT data for 2700-keV electrons in LiF. (From Reference [50]. With permission.)

The shape of the curve is based on a complex interaction between the beta spectrum, polynomial dose equation, and the energy range of the spectrum. The increase in energy range produces an increase in dose for all depths up to the range of the highest-energy beta particle.

Thermoluminescent dosimetry is a relative method; that is, it is necessary to initially calibrate the detector in a known radiation field, usually a ^{60}Co source. In most practical situations, the gamma spectra are different from those of primary ^{60}Co gamma rays, because of scattering in irradiated media, and it is important to know the effect of the gamma spectrum on the dosimeter response.

The energy dependence of the response of various TL detectors for ^{60}Co gammas degraded in water was investigated by Miljanic and Ranogajec-Komor. [51] They irradiated several types of thermoluminescent dosimeters (LiF:Mg,Ti; Al$_2$O$_3$:Mg,Y; and CaF$_2$:Mn) at

different depths in a water phantom placed at a distance of 2.5 m from a panoramic ^{60}Co source. Detectors were encapsulated in Plexiglas holders with a wall thickness of 0.5 cm. Reference dosimetry was carried out using a Fricke dosimeter and an ionization chamber.

For ^{60}Co photons, TL detectors are "intermediate"-size detectors. In this intermediate size range, the absorbed dose in the medium is calculated according to the Burlin general cavity theory, which includes cavities of all sizes.

The mean absorbed dose in the detector (D_d) in relation to the absorbed dose at a specified point in the surrounding medium (D_m) is given by

$$D_d/D_m = f = ds_{d,n} + (1-d)(\mu_{en}/\rho)_{d,m} \quad (4.44)$$

where $s_{d,n}$ and $(\mu_{en}/\rho)_{d,m}$ are, respectively, the ratios of mass collision stopping powers and mass energy absorption

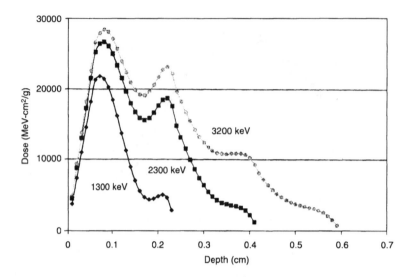

FIGURE 4.72 Depth-dose curves generated from normalized and scaled Y-90 beta particle spectrum source for varying energy ranges in LiF. (From Reference [50]. With permission.)

coefficients of the detector and the medium, and d is a weighting factor which takes into account the size of the detector. It is given by

$$d = (1 - e^{-\beta g})/\beta g \qquad (4.45)$$

where β is the effective mass attenuation coefficient for electrons and g is the average path length through the detector. Burlin originally used the following expression for g:

$$g = 4(V/S),$$

where V is the volume and S is the total surface area.

The ratios of mass energy absorption coefficients for different TLDs and water, weighted over energy spectra given by Seltzer [52], for scattered components and for the whole spectrum are shown in Figure 4.73

The experimental results of dose distribution measurements with different TL dosimeters are shown in Figure 4.74. Doses in water from the results of the Fricke dosimeter were calculated.

EGS4 Monte Carlo simulations have been performed by Mobit et al. [53] to examine general cavity theory for a number of TLD cavity materials irradiated in megavoltage photon and electron beams. The TLD materials were LiF, $Li_2B_4O_7$, CaF_2, and $CaSO_4$ irradiated in Perspex, water, Al, Cu, and Pb phantoms. For megavoltage photon beams, this has been done by determining the dose component $(1 - d)$ resulting from photon interactions in the cavity compared with the dose component resulting from photon interactions in the phantom material (d) by Monte Carlo simulations and analytical techniques. The results indicate that the Burlin exponential attenuation technique can overestimate the dose contribution from photon interactions in a 1-mm thick LiF

cavity by up to 100%, compared with the Monte Carlo results for LiF TLDs irradiated in a water or Perspex phantom. However, there is agreement to within 1% in the quality dependence factor, determined from Burlin's cavity theory, Monte Carlo simulations, and experimental measurements for LiF and $Li_2B_4O_7$ TLDs irradiated in a Perspex or a water phantom. The agreement is within 3% for CaF_2 TLDs. However, there is disagreement between Monte Carlo simulations and Burlin's theory of 6 and 12% for LiF TLDs irradiated in copper and lead phantoms, respectively. The adaptation of Burlin's photon cavity theory and other modifications to his photon general cavity theory for electrons have been shown to be seriously flawed. [53]

Mobit et al. presented cavity theory in the following way. When a detector is placed in a medium exposed to ionizing radiation in order to measure dose, it forms a cavity in that medium. The cavity is generally of different atomic number and density to the medium. Cavity theory gives the relation between the dose absorbed in a medium (D_{med}) and the average absorbed dose in the cavity decay.

$$D_{med} = \bar{D}_{cav} f_{med,cav} \qquad (4.46)$$

where $f_{med,cav}$ is a factor that varies with energy, radiation type, medium, size, and composition of the cavity. For a cavity that is small compared with the range of the electrons incident on it in a photon beam, the Bragg-Gray relation applies and $f_{med,cav}$ is given by the average mass collision stopping power ratio of the medium to the cavity ($s_{med,cav}$), generally evaluated according to Spencer and Attix. For a cavity that is large compared with the electrons incident on it, $f_{med,cav}$ is given by the ratio of the average mass energy absorption coefficient of the medium to the cavity material.

The Burlin cavity theory equation (4.44) requires a value to be provided for d which represents the proportion

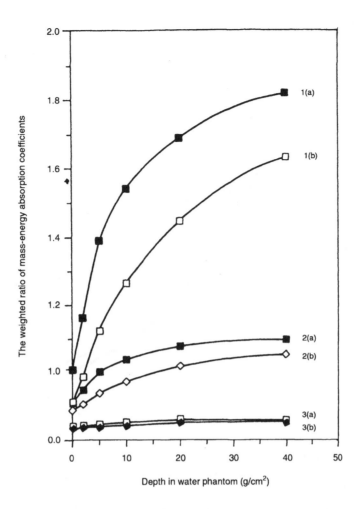

FIGURE 4.73 The weighted ratios of mass energy absorption coefficients for different TLDs and water at different depths in a water phantom (evaluated from Seltzer [52]); curves 1, 2, and 3 are for CaF$_2$, Al$_2$O$_3$ and LiF, respectively, (a) only for the scattered component of the spectra and (b) for the whole spectra. (From Reference [51]. With permission.)

of the dose in the cavity that is due to secondary electrons generated in the medium. This can be calculated directly using the Monte Carlo method, and the values can be compared to the theoretical predictions. Burlin proposed that d could be estimated by assuming that electrons incident on the cavity are exponentially attenuated, so that d is given by

$$d = \frac{\int_0^g e^{-\beta x}\, dx}{\left(\int_0^g dx\right)^{-1}} = \frac{(1 - e^{-\beta g})}{\beta g} \qquad (4.47)$$

where g is the path length of electrons entering the cavity and β is the effective mass attenuation coefficient of electron fluence penetrating the cavity. In order to be able to compare the Monte Carlo calculations of d with the Burlin theory, values of β and g are required. Values for g range from $4V/S$, where V is the volume of the cavity and S is its surface area, to $1.539t$ for a TLD chip, where t is the thickness of the chip. For a cylinder 6 mm long and 1 mm in diameter, the predicted value of g would thus range

from 0.92 to 1.54 mm, although the latter would be an overestimate for a cylinder. Values of β for LiF in a ^{60}Co beam vary from 8.5 cm^{2}g^{-1} [54] through 13.4 cm^{2}g^{-1} [55] to 15.9 cm^{2}g^{-1} [56]. The value of β from Chan and Burlin is based on the expression

$$e^{-\beta R_{CSDA}} = 0.01 \qquad (4.48)$$

where R_{CSDA} is the continuous slowing down approximation of electron range. The value of Paliwal and Almond is derived from

$$\beta_{LiF} = 14/E_{max}^{1.09} \qquad (4.49)$$

where E_{max} is the maximum calculated energy of the secondary electrons, which for ^{60}Co γ-rays is 1.04 MeV. Further discussion can be found in Horowitz. [57]

The quality dependence factor F_{Co}^X is defined as [53]

$$F_{Co}^X = [TL(X)/D_{med}(X)]/TL(Co)/D_{med}(Co) \qquad (4.50)$$

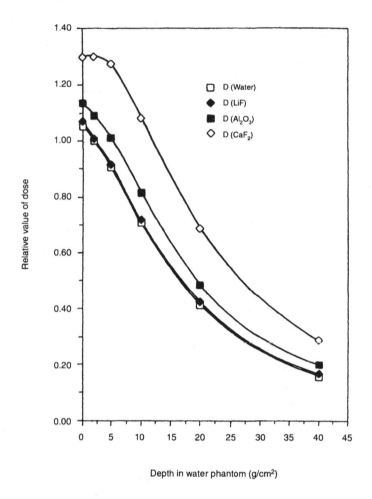

FIGURE 4.74 Results of dose distribution measurements with different types of TL dosimeters as a function of depth in water phantom. Dose to water was evaluated from the measurements using the Fricke dosimeter. (From Reference [51]. With permission.)

where $TL(X)/D_{med}(X)$ is the light output (TL) per unit dose in a medium for the beam quality X of interest. $TL(Co)/D_{med}$ (Co) is the light output per unit dose in the same medium for ^{60}Co γ-rays. In calculating the value of F_{Co}^X using either the Monte Carlo method or cavity theory method, the assumption is made that $TL(X)$ is directly proportional to the absorbed dose in the TL material.

Assuming that $TL(X)$ is directly proportional to the average absorbed dose in the $TLD(\bar{D}_{TLD})$, the dose to the medium in an x-ray beam is obtained from

$$D_{med}(X) = \bar{D}_{TLD}[TL(Co)/D_{med}(CO)]/F_{CO}^X \quad (4.51)$$

for TLDs calibrated in ^{60}Co γ-rays.

The Burlin factor ($f_{w,LiF}$) is given by

$$f_{w,LiF} = D_w/\bar{D}_{LiF} = [ds_{w,LiF} + (1-d)(\bar{\mu}_{en}/\rho)_{w,LiF}]$$

$$(4.52)$$

The Burlin general cavity equation was evaluated for calcium fluoride (CaF_2) and lithium borate ($Li_2B_4O_7$) discs

of thickness 0.9 mm and diameter 3.61 mm. The results of the calculations are displayed in Figures 4.75 and 4.76 for $Li_2B_4O_7$ and CaF_2 respectively.

For lithium borate, the Burlin general theory gives a lower estimate of the deviation from unity of the quality dependence factor by a maximum of 1% compared with the Monte Carlo calculated results. For CaF_2 the maximum difference is 3% for 25-MV x-rays. This difference may be attributable to the cavity perturbation effects of CaF_2 in photon beams. Cavity perturbation effects are larger for CaF_2 than either $Li_2B_4O_7$ or LiF. This means that the secondary electrons produced from photon interactions in the medium are perturbed more by the CaF_2 than either the LiF or the $Li_2B_4O_7$ cavities. Despite this, the agreement between Burlin cavity theory and Monte Carlo simulation for both LiF and CaF_2 is still within 3% for the entire energy range studied. [53]

Figure 4.77 shows that the quality dependence of LiF TLDs in megavoltage photon beams does not depend significantly on the size of the LiF TLD materials, at least for the LiF TLD sizes simulated.

A model was developed by Woo and Chen [58] to enable dose determination for thermoluminescent

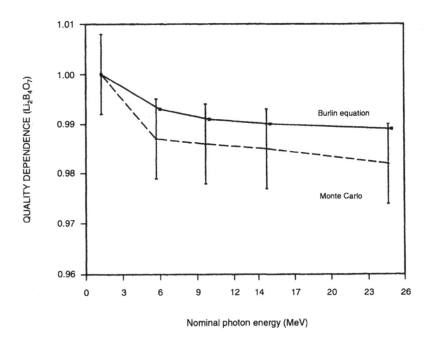

FIGURE 4.75 A comparison of the quality dependence factor derived from Monte Carlo simulation and from the Burlin cavity equation for a 3.61-mm diameter and 0.9-mm thick disc of $Li_2B_4O_7$ in megavoltage photon beams. (From Reference [53]. With permission.)

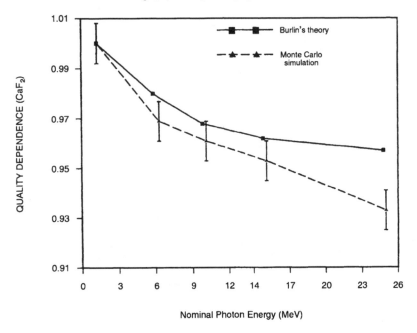

FIGURE 4.76 A comparison of the quality dependence factor derived from Monte Carlo simulation and the Burlin cavity equation for a 3.61-mm diameter and 0.9-mm thick disc of CaF_2 in megavoltage photon beams. (From Reference [53]. With permission.)

dosimeters immersed in a radioactive solution such as used in radioimmunotherapy. For low-energy beta emitters used in such therapy, the size of the dosimeter results in a much lower light output than when irradiated with an external ^{60}Co beam to the same dose as delivered to the medium. The model takes the size of the dosimeter into account and hence allows calculation of the dose in the actual medium.

The dose determined is the dose averaged over the volume of the TLD. If the dose distribution is not uniform across the TLD, as in the case of low-energy electrons whose range is comparable to the TLD thickness, it would be more useful to relate the total light output to the integrated dose or total energy absorbed by the dosimeter. For low-energy beta radiation with range comparable to the TLD size, the dose so determined will be lower than the dose to the medium, due to the displacement effect.

In the case of beta sources, the calculation is based on the fact that all the beta energy is absorbed and the

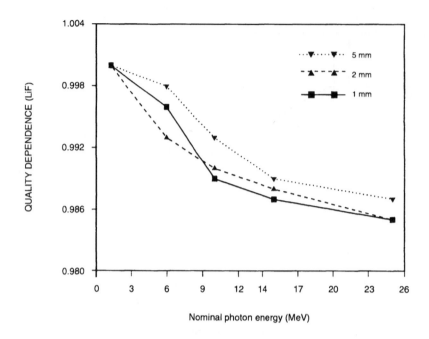

FIGURE 4.77 The quality dependence of LiF TLDs evaluated according to the Burlin cavity expression for an LiF disc of 3.61-mm diameter and different thicknesses in megavoltage photon beams. (From Reference [53]. With permission.)

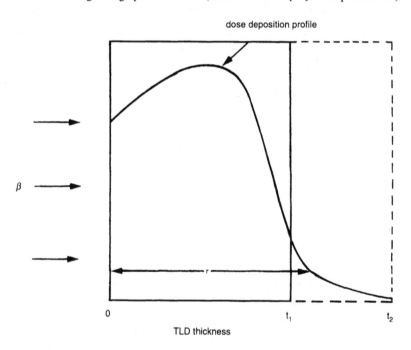

FIGURE 4.78 If the beta range r is large compared to the TLD thickness, as in t_1, then energy is absorbed across the TLD. If the thickness is considerably larger, as in t_2, then only a fraction of the TLD is involved in energy absorption. Dosimeters of thicknesses t_1 and t_2 will give roughly the same total amount of light. The average dose hence determined will be lower for the thick dosimeter. (From Reference [58]. With permission.)

dose is uniform within the solution, except for the region within the beta range to the edge of the container. The equation for the equilibrium dose rate at a point within the solution is given by

$$\text{Dose rate} = 2140 \, C\bar{E} \quad cGy/h, \qquad (4.53)$$

where C is the specific activity of the solution in mCi/g and $\bar{E}$ is the average energy of the emitted beta particle in MeV.

The problem of TLD response dependence on the finite size of the TLD, for both external and internal beta irradiations, can then be summarized in Figure 4.78. where the TLD is irradiated by a beam of beta (β) particles.

For a TLD immersed in a radioactive solution for a specified time, if E_{tot} represents the total energy absorbed within the region R occupied by the TLD in the absence of the TLD E_{self} denotes the energy absorbed within the region R due to radiation sources within region R, and E_{surr} is the energy absorbed within R due to radiation sources outside region R, then

$$E_{tot} = E_{self} + E_{Surr} \qquad (4.54)$$

When the region R is occupied by the TLD, E_{self} is then absent. The energy absorbed in the TLD is just E_{surr}. Since E_{tot} is the energy absorbed by the TLD when calibrated in an external beam to receive the same equilibrium dose as in the radioactive solution, the relative TLD response, (S_u/S_0) is then given by

$$\frac{S_u}{S_o} = \frac{E_{Surr}}{E_{tot}} = 1 - \frac{E_{self}}{E_{tot}} \qquad (4.55)$$

where S_0 is the TLD response in an external photon beam radiation and S_u is the decreased response in-solution irradiation, so that the decrease in relative TLD response is $[1 - (S_u/S_0)]$ or $(S_0 - S_u)/S_0$. This equation indicates that when E_{self} is large, the fraction of energy contributed to the region from the radioactivity displaced is large, and hence the decrease in relative response is large, and vice versa.

When the range of beta is comparable to R, the fraction of energy absorbed depends strongly on the size and geometry of the region, and a more detailed analysis is required.

For internal geometry dosimetry, the objective is to determine dose at a point in a medium. Restricting this consideration to a point in a solution with uniformly distributed radioactivity, the dose delivered to this point can be determined by two separate methods. In the first method, referred to here as the "direct calibration" method, Equation (4.53) is used to establish a calibration curve of light output vs. dose for a TLD immersed in a radioactive solution and, hence, determine an unknown dose. In the second method, to be referred to as "the displacement model" method, the ^{60}Co calibration curve for the TLD is used. To obtain the same dose, this calibration curve requires a larger light output (LO) than in the first approach. Hence, the LO from the immersed TLD in the unknown solution is corrected upward for the calculated decreased relative response, E_{surr}/E_{tot}; i.e., LO is divided by the ratio E_{surr}/E_{tot} to obtain the corrected LO and then the dose. Thus, there are two independent methods of determining the dose, with the displacement model giving a confirmation of the measured dose as well as providing a basic understanding of the decreased response. The results of measurements and the calculations using the displacement model are summarized in Figure 4.79.

An analytical Monte Carlo simulation code has been used by Karaiskos et al. [59] to perform dosimetry calculations around an ^{192}Ir high-dose-rate brachytherapy source utilized in the widely used microSelectron afterloaded system. Measurements of anisotropy functions using LiF TLD-100 rods have been performed in a polystyrene phantom to support the Monte Carlo calculations.

The active source consisted of a 0.60 mm in diameter by 3.5 mm long cylinder of pure iridium metal, within which the radioactive material was uniformly distributed. It was encapsulated in stainless steel and welded to a steel cable.

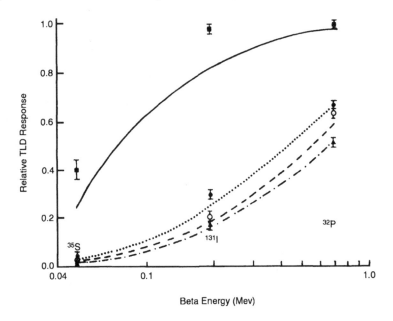

FIGURE 4.79 Comparison of relative TLD response between direct calibration (symbols) and displacement model (lines): rod–(○),(---); chip–(▲), (.–.–.); disc–(●).(···); ultrathin disc–(■), (—). (From Reference [53]. With permission.)

The capsule is 5 mm long and 1.1 mm in overall diameter and its distal end is a steel cap with a 0.55-mm radius. The dose calculation formalism proposed by AAPM Task Group 43 has been followed. Dose rate $\dot{D}(r, \theta)$, in medi-um at point (r, θ), where r is the distance in cm from the active source center and θ is the polar angle relative to the longitudinal axis of the source, is expressed as:

$$\dot{D}(r, \theta) = S_k \Lambda \frac{G(r,\theta)}{G(r_0,\theta_0)} F(r,\theta)g(r) \qquad (4.56)$$

where S_k is the source air-kerma strength in units of U (1 U= 1 μGym^{2}h^{-1} = 1 cGy cm^{2}h^{-1}). Λ is the dose rate constant defined as the dose rate in the medium per unit source strength at radial distance $r_0 = 1$ cm along the transverse axis, and $\theta_0 = \pi/2$, so that

$$\Lambda = D^{\&}(r_0,\theta_0)/S_k \qquad (4.57)$$

$G(r,\theta)$ is the geometry function in units of cm^{-2} which accounts for the variation of relative dose due only to the spatial distribution of activity within the source, ignoring photon absorption and scattering in the source structure. At distances greater than approximately 2–3 times the active source dimension (i.e., for the given source greater than approximately 1 cm), the geometry function is well-approximated (within 1%) by the inverse-square law used in point-source approximation. $g(r)$ is the dimensionless radial dose function which accounts for photon attenuation and scattering in the medium and encapsulation along the transverse axis ($\theta_0 = \pi/2$) and is by definition unity at $r_0 = 1$ cm, i.e., $g(1) = 1$, so that

$$g(r) = \frac{\dot{D}(r,\theta_0)G(r_0,\theta_0)}{\dot{D}(r_0,\theta_0)G(r,\theta_0)} \qquad (4.58)$$

$F(r,\theta)$ is the dimensionless dose anisotropy function which accounts for photon attenuation and scattering at any polar angle θ, relative to that for $\theta_0 = \pi/2$, and is by definition unity at $\theta_0 = \pi/2$; i.e., $F(r,\pi/2) = 1$, so that

$$F(r, \theta) = \frac{\dot{D}(r,\theta)G(r,\theta_0)}{\dot{D}(r,\theta_0)G(r,\theta)} \qquad (4.59)$$

An equivalent dose calculation formalism in terms of the tissue attenuation factor, $T(r)$, is also used in many computer-assisted brachytherapy treatment-planning systems:

$$\dot{D}(r, \theta) = S_k f \frac{1}{r^2} T(r) F(r, \theta) \qquad (4.60)$$

In this expression the product $\Lambda g(r)$ of Equation (4.56) has been replaced by the product $fT(r)$. The factor f is the air kerma in air to water kerma in air conversion factor, which is equal to the ratio of mean mass energy absorption coefficients for water and air. This ratio is almost constant and has a value of 1.11 for energies between 150 keV and 4 MeV. Thus, f is also equal to 1.11 for the ^{192}Ir primary energy range. The factor $T(r)$ is well-approximated by the ratio of water kerma in water to water kerma in air and accounts for the combined atten-uation and scattering in water. It is noted that for the given source, the formalism of Equation (4.60) is valid at radial distances greater than approximately 1 cm, where, as mentioned above, the geometry factor $G(r,\theta)$ is well-approximated by the inverse-square law.

TLDs were placed in small cylindrical TLD polysty-rene receptors (3 mm in diameter and 6 mm long) having a cylindrical hole of 1-mm diameter in their centers to accommodate the dosimeters. A 30 × 30 ×1.1-cm^3 slab of polystyrene, prepared in a milling machine with an accuracy of 0.1 mm, was constructed to accommodate TLD receptors and a plastic catheter (2-mm outer diame-ter), inside which the source was driven. This slab was sandwiched between other identical polystyrene slabs to build a 30 × 30 × 30-cm^3 phantom which approximates the water spherical phantom of 30-cm diameter used in MC calculations. [59]

In Figure 4.80, radial dose functions calculated for spherical water phantoms of different diameters ($d = 10–50$ cm) are presented. The figure demonstrates that phantom dimensions significantly affect the radial func-tions at radial distances near phantom edges, where devi-ations of up to 25% are observed. This effect is due to the reduction of scatter contribution to overall dose at the edges of the phantom and should be taken into account in the case of estimating the dose near the body edges.

Using the MC code and the CSDA electron ranges proposed by ICRU Report 37, the average starting energy of the electrons generated in the TLD material was found to be less than 100 keV over the whole range of measure-ment distances and angles, and the corresponding R was less than 0.006 cm, thus giving the value of factor d less than 0.02. As a result of this, the TLD material behaves like a large cavity in ^{192}Ir γ-rays, and the LiF TLD response can be calculated using the simplified equation:

$$\frac{D_{LiF\ TLD}}{D_{water}} = \left(\frac{\overline{\mu_{en}}}{\rho}\right)_{water}^{LiF}$$
$$= \sum_i E_i \left(\frac{\mu_{en}}{\rho}\right)_{E_i,LiF} \Big/ \sum_i E_i \left(\frac{\mu_{en}}{\rho}\right)_{E_i,\ water} \qquad (4.61)$$

where E_i is the energy of the ith photon in the point of measurement, $(\mu_{en}/\rho)_{E_i,LiF}$ is the mass energy absorption

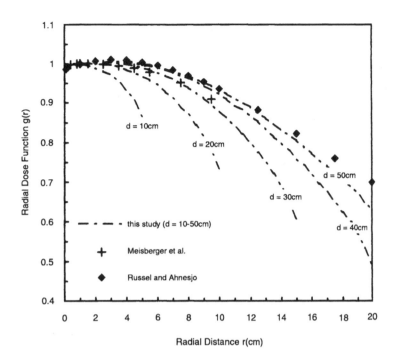

FIGURE 4.80 Radial dose functions, $g(r)$, as a function of radial distance, r, calculated by our code with the ^{192}Ir microSelectron HDR source centered at a spherical water phantom of different diameters ($d = 10$–50 cm). Experimental data of Meisberger et al. performed using a different ^{192}Ir source design centered at a $25 \times 25 \times 30$-cm^3 water phantom, and Monte Carlo calculations of Russel and Ahnesjo performed using the ^{192}Ir microSelectron HDR source centered at an unbounded water phantom, have also been plotted for comparison. (From Reference [59]. With permission.)

coefficient of LiF at energy E_i and $(\mu_{en}/\rho)E_{i,water}$ is the mass energy absorption coefficient of water at this energy E_i.

VI. MISCELLANEOUS TLD

Backscatter factors (BSF) for low-energy x-rays derived from Monte Carlo calculations have been recommended in the UK code of practice for kilovoltage dosimetry published by IPEMB. [60] Coudin and Marinello [61] have measured backscatter factors (BSF) using lithium borate doped with copper (0.1% per mass of copper) in the form of powder (average density 1.30 g cm^{-3}, grain size varying from 74 μm to 175 μm) spread out in thin layers either directly on the surface of the medium or on a mylar sheet for experiments in air. Li$_2$B$_4$O$_7$:0.1% Cu[17] was chosen among different common TL materials due to its relatively flat energy response (Figure 4.81), its high sensitivity per unit volume, its dose-rate independent response, its water equivalence in this energy range, and the fact that its intrinsic response is not influenced by the direction of the beam.

According to the recommendations of the various protocols, the absorbed dose at the surface of a water phantom for low-energy x-rays can be determined from measurement of primary beam kerma and BSF. This quantity is defined as a ratio of water kermas at the phantom surface and free in air in the absence of the phantom. Within the range of 10 to 100 kV, as charged-particle equilibrium is always

achieved and *bremsstrahlung* radiation is negligible, there is little distinction between kerma and dose. Consequently,

$$BSF = \frac{R_{det}^{ph} \times (\mu_{en}/\rho_{det}^{w})_{surf}}{R_{det}^{air} \times (\mu_{en}/\rho_{det}^{w})_{air}} \qquad (4.62)$$

where R_{det} is the reading of the detector (lithium borate or ion chamber) with R_{det}^{ph} and without phantom R_{det}^{air}, and $(\mu_{en}^{r}/\rho_{det}^{w})$ is the ratio of mass energy absorption coefficients of the detector medium to water for photon energy fluence spectra at the phantom surface or in air.

TL measurements have been performed using a very thin layer of 0.4 mm (about three times the average grain size), which is sufficient to ensure the electronic build-up while minimizing the primary photon attenuation.

BSF obtained using Li$_2$B$_4$O$_7$:0.1% Cu are very close (within 1.7%) to the data calculated by Monte Carlo methods, being slightly lower in general.

Figure 4.82 shows experimental results obtained with both the ion chambers and those measured by Klevenhagen with a specially designed flat ion chamber. There is a very good agreement between Klevenhagen's results and the IPEMB data. However, large discrepancies are pointed out with the flat ion chambers used. This is probably due to their different characteristics.

Calibration of Mg$_2$SiO$_4$(Tb)(MSO) thermoluminescent dosimeters for use in determining diagnostic x-ray

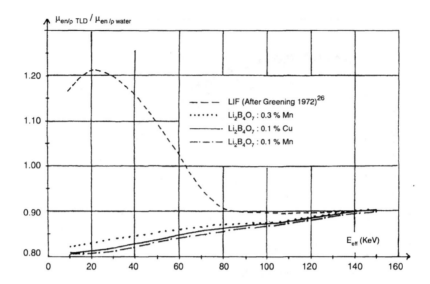

FIGURE 4.81 Ratio of mass energy absorption coefficient for the TL materials to the mass energy absorption coefficient of water as a function of photon energy. The borate curves have been calculated from Hubbells' data. (From Reference [62]. With permission.)

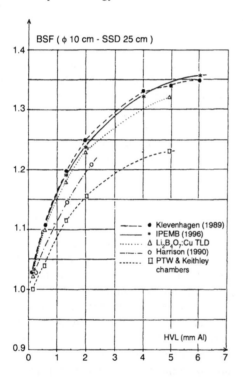

FIGURE 4.82 BSF measured by different authors with different types of detectors compared to the IPEMB's data. (From Reference [61]. With permission.)

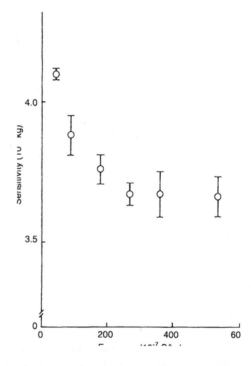

FIGURE 4.83 Sensitivity of the thermoluminescent dosimeters to various exposures. Sensitivity: response of TLD per 1 C/kg exposure; unit of TLD reading is nC. The $Mg_2SiO_4(Tb)$ detectors on a styrofoam block were exposed to x-rays using various exposure times under the following conditions: tube voltage, 100 kVp; filter, 2.5 mm Al; and field size, 20 × 20 cm². Error bars indicate the standard deviations. (From Reference [62]. With permission.)

doses was performed by Kato et al. [62] The results shown in Figure 4.83 demonstrate that the detector sensitivities depend on their exposure in the low-dose region. The sensitivity at 5.16×10^{-6} C/kg (20 mR) was about 10% greater than that at 2.58×10^{-5} C/kg (100 mR). For doses greater than 2.58×10^{-5} C/kg (100 mR), it was quite uniform, within ±1%.

For the x-ray exposures, the tube voltage was altered in 20-kVp increments through a range from 40 to 140 kVp. Exposure times were adjusted so that the exposures were about 2.58×10^{-5} C/kg (100 mR). Other experimental factors were the same as for the experiment described above.

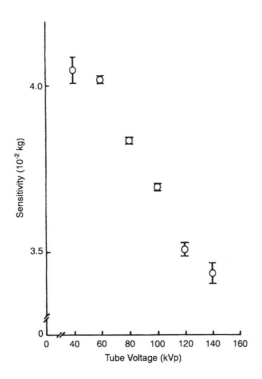

FIGURE 4.84 Sensitivity of the thermoluminescent dosimeters to tube voltages. Sensitivity: response of TLD per 1 C/kg exposure; unit of TLD reading is nC. Exposures were adjusted to approximately 2.58×10^{-5} C/kg (100 mR). The conditions of exposure, excluding total exposures and tube voltage, were the same as those of Figure 4.83. Error bars indicate the standard deviations. (From Reference [62]. With permission.)

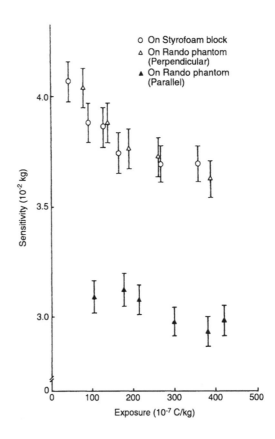

FIGURE 4.85 Sensitivity of the thermoluminescent dosimeters (TLD) in air and on a Rando phantom. Sensitivity: response of TLD per 1 C/kg exposure; unit of TLD reading is nC. The Mg_2SiO_4 (MSO) detectors were exposed to 100-kVp x-rays on a styrofoam block and on a Rando phantom. On the Rando phantom, thermoluminescent sensitivity was determined for 100-kVp x-rays parallel with and perpendicular to the central axis of the cylindrical MSO detectors. Error bars indicate the standard errors estimated from the variation of the TLD sensitivities and the reproducibility of the Shonka-Wyckoff chamber. (From Reference [62]. With permission.)

The results obtained are shown in Figure 4.84. The sensitivity of the MSO detectors used decreased linearly with increasing tube voltage from 60 to 140 kVp. The MSO detector, Mg_2SiO_4(Tb) phosphor, is more sensitive than are other phosphors, such as LiF and $CaSO_4$(Tm). The sensitivity of the MSO detectors depended on their orientation to the central beam. Figure 4.85 shows the sensitivity of the MSO detector.

Experimental evidence has been obtained confirming the significance of deep traps for thermal quenching of luminescence in α-Al_2O_3. [63] The effect of deep traps on thermoluminescence (TL) parameters of the dosimetric peak was first detected in dosimetric α-Al_2O_3 powder and pellets. The presence of deep traps in nominally pure anion-defective crystals of corundum was suggested by indirect evidence. Specifically, the competitive role of deep traps in trapping of carriers was associated with a change in the dosimetric sensitivity when the crystal was exposed to a high-dose irradiation. The state of deep traps affects basic features of the main TL peak at 450 K in anion-defective corundum.

Figure 4.86 illustrates glow curves for anion-defective α-Al_2O_3 crystals excited at 460 K (curve 1), 620 K (curve 2), and 720 K (curve 3) under UV radiation. The data suggest

an important regularity which underlies the proposed model: the occupancy of deep traps increases with the excitation temperature.

Aluminum oxide dosimeters have been shown to exhibit radioluminescence linearly with dose and dose rate to very high values, with minimal visible coloration. [64] Small Czochralski-grown crystals were prepared from high-purity aluminium oxide and chemically pure silicon dioxide. A heating rate of 10°C s^{-1} was used. Al_2O_3:Si has been shown to exhibit a useful log-linear response from 0.1 kGy to at least 10 kGy. In addition, a peak sum ratio method, effectively obviating many calibration problems, has been demonstrated for determining absorbed dose in this dose range. Above 10 Gy, the response is nonlinear, becoming saturated at about 30 Gy.

Figure 4.87 shows the thermoluminescence glow curve for 0.2% silicon doping after exposure at an absorbed dose

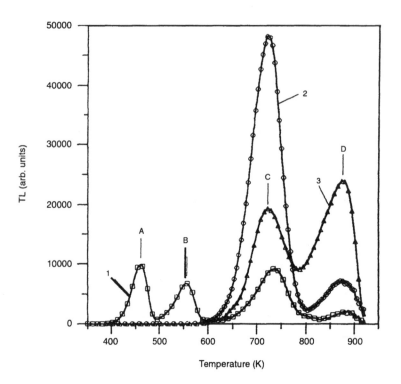

FIGURE 4.86 Glow curves of anion-defective α-Al₂O₃ crystals excited by UV radiation at the heating rate of 2 K s⁻¹. The excitation temperatures were 460 K (1), 620 K (2), and 720 K (3). (From Reference [63]. With permission.)

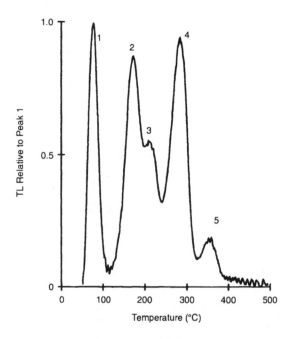

FIGURE 4.87 Glow curve for 0.2% silicon doping of aluminium oxide. The glow curve was measured immediately after a 100-Gy exposure in 1 min. (From Reference [64]. With permission.)

of 100 Gy. Five glow peaks are obvious and well-separated. While the first, appearing at a temperature of approximately 90°C, fades quickly at room temperature, the remaining four are sufficiently stable for use in dosimetry. Figure 4.88

shows the changing shape of the glow curve as absorbed dose increases. Peak 1 is not shown since the time starting from exposure was large with respect to its lifetime. Peak 2 increases in amplitude, peak 3 decreases somewhat, peak 4 decreases dramatically, and, while it is not obvious from the scale of this graph, peak 5 increases. No peak was found to have an amplitude linear with dose. The integral of peak 5 was found to increase nearly linearly with the logarithm of dose, up to a value of 10 kGy.

To characterize the thermoluminescence (TL) of Brazilian topaz, Al_2SiO_4 $(F,OH)_2$, TL emission spectra of as-received samples and gamma-irradiated samples (with 100 Gy, 1 kGy, and 10^6 Gy) submitted or not submitted to a previous annealing procedure were presented by Yukihara et al. [65] All the samples show a broad emission band between 300 and 600 nm for all the glow peaks between 50 and 400°C. In some gamma-irradiated samples, a narrow emission band between 200 and 350 nm was also observed.

Studies were performed using pieces of two large crystals: No. 1, a colorless topaz of unknown origin, and No. 2, a light blue topaz from Marambaia (Minas Gerais, Brazil). From these large crystals, smaller pieces of ~3 × 3 × 1 mm³ (20–60 mg) were obtained. Other topazes were also used for comparison.

Figure 4.89 shows typical emission spectra of unannealed samples of four topazes. Except for topaz 1, the spectra are very broad and similar, ranging from 300 to 560 nm.

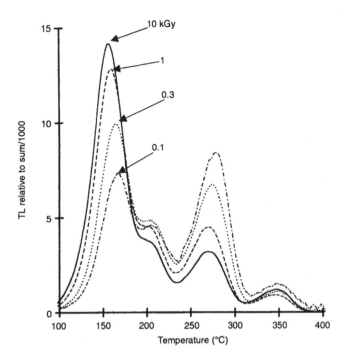

FIGURE 4.88 Change in shape of the glow curve for silicon-doped aluminum oxide with increasing dose. Each glow-curve shape is normalized to the integral for that curve. (From Reference [64]. With permission.)

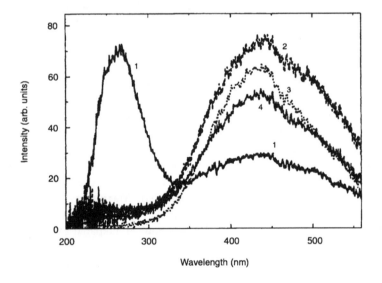

FIGURE 4.89 TL emission spectra of unannealed samples irradiated with 10^3 Gy (No. 2) and 10^6 Gy (Nos. 1, 3, and 4), averaged in the range 10–300°C. (From Reference [65]. With permission.)

LET dependence of Al_2O_3:C and CaF:Mn (TL) dosimeters was investigated by Osvay and Ranogajec-Komor [66], using glow-curve analysis to determine the efficiencies of TL light production for alpha particle and neutron irradiation relative to gamma exposure. No high-temperature peak or significant change of glow-curve shape was found after irradiation by high LET (alpha and neutron) radiation. Gamma sensitivity of Al_2O_3:C compared to CaF:Mn was 6:1. However, the relative TL efficiency for alpha particle irradiation of Al_2O_3:C was a

factor of 10 lower than that of CaF_2Mn. Neutron sensitivity was found to be similar for the two materials.

Figure 4.90 shows the glow curve of Al_2O_3:C and CaF_2:Mn after 1 mGy gamma irradiation. The glow peak of aluminium oxide is at approximately 180°C and of calcium fluoride, it is at 240°C.

Dosimeters were irradiated with alpha particles for 5, 10, 20, 40, and 60 min (3, 6, 12, 24, and 36 mGy). The standard deviation of TL measurements for alpha particles was about 12%. The dose-response curves for the Al_2O_3:C and CaF_2:Mn dosimeters are shown in Figure 4.91. The TL light outputs measured under the peaks of dosimeters irradiated with alpha particles are linear in the dose range investigated.

The glow curves of dosimeters after 3-mGy alpha irradiation can be seen in Figure 4.92. The sensitivity of CaF_2:Mn to 5-MeV alpha particles is a factor of 3 higher than compared to Al_2O_3:C. The peak induced by high LET alpha radiation is split into two components in the dose range investigated.

Figure 4.93 represents the TL output of aluminum oxide to low LET (curve A: 1-mGy gamma) and to high LET (curve B: 1-mSv neutron) radiation. The thermal neutron sensitivity was found to be similar (about 0.04 as compared to gamma irradiation) for both materials.

The thermostimulated luminescence (TSL) of powdered synthetic $CaSO_4$ crystals was studied by Gerome et al. [67] with various types of doping rare earth (RE^{3+}) impurities: Dy^{3+}, Tm^{3+}, Ho^{3+}, Er^{3+} or Gd^{3+}. Defects in anionic SO_4^{2-} groups (SO_x types) were shown to be responsible for the UV lattice emission and the trapping centers, while the RE^{3+} ions are just radiative emission centers.

The origin of this UV emission is generally tentatively explained either (a) by an undesired contamination from radiative impurities, such as Tm^{3+} that has a group of three near-UV lines in the considered spectral range; however, a fine spectral analysis allows one to dismiss such a hypothesis when possible, because these UV emissions are often too weak for good resolution to be possible; or (b) by an emission of the $CaSO_4$ lattice. Similar emissions have been observed in other undoped sulphates ($CaSO_4$, $BaSO_4$, $SrSO_4$, $PbSO_4$, etc).

$CaSO_4$ samples were obtained by a synthesis method. Samples were prepared by dissolving $CaSO_4$ $2H_2O$ in sulphuric acid and vacuum-evaporating the excess acid around 300°C. The oxide (Dy_2O_3, Tm_2O_3, ...) was added for doping. Crystals of undoped and doped calcium sulphate were washed, annealed at 750°C for about 2 h, and sieved to 36–50 μm. Figure 4.94 shows the TSL of $CaSO_4$ undoped or doped with the various RE^{3+} registered through a 340-nm filter, after x irradiation at room temperature.

Figure 4.95 shows the TSL curves of $CaSO_4$:RE^{3+} for the main emission line of each RE^{3+}.

Two new thermoluminescent (TL) phosphors $BaSO_4$:Eu,P and $BaSO_4$:Eu,Na were prepared by Shinde et al. [68] by three different methods, recrystallization, co-precipitation, and sintering at high temperature. The phosphors exhibit variations in the thermoluminescent characteristics and also in their electron spin resonance (ESR) spectra. Both phosphors show changes in their glow-curve shape and TL sensitivity with the method of preparation and with the concentration of co-dopant (sodium/phosphorus). Maximum TL sensitivity ($\approx$5 times that of $CaSO_4$:Dy) was obtained in the case of $BaSO_4$:Eu,Na (0.7, 2.5 mol%), prepared by sintering in air at 900°C.

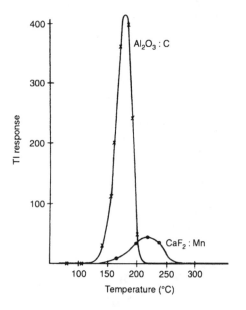

FIGURE 4.90 TL glow curve of gamma-irradiated dosimeters (1 mGy gamma). (From Reference [66]. With permission.)

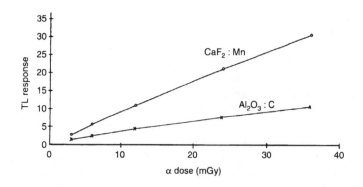

FIGURE 4.91 TL responses as a function of alpha exposure. (From Reference [66]. With permission.)

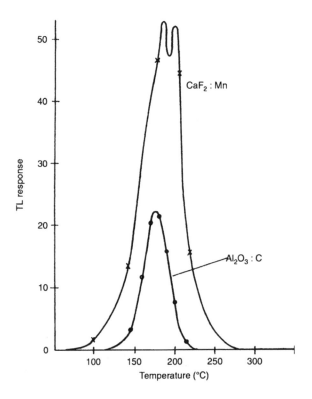

FIGURE 4.92 TL glow curves of alpha-irradiated dosimeters. (From Reference [66]. With permission.)

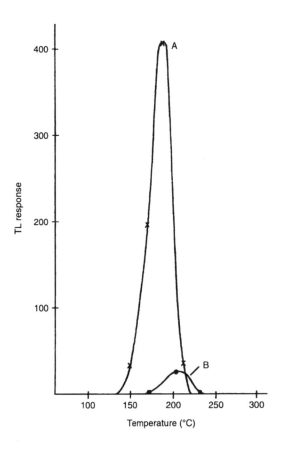

FIGURE 4.93 TL glow curves of Al$_2$O$_3$:C dosimeter irradiated by gamma dose of 1 mGy (curve A) and thermal neutron dose of 1 mSv (curve B). (From Reference [66]. With permission.)

Figures 4.96, 4.97, and 4.98 show the TL glow curves of BaSO$_4$:Eu (0.5 mol%), BaSO$_4$:Eu,P (0.5, 0.2 mol%), and BaSO$_4$:Eu,Na (0.5,0.2 mol %) prepared by three methods: recrystallization (R), co-precipitation (P), and sintering (S). It can be seen from Figure 4.96 that the basic structure of the glow curves of BaSO$_4$:Eu (P) and BaSO$_4$:Eu (S) is similar to that of (Figure 4.98) BaSO$_4$:Eu,Na (P) and BaSO$_4$:Eu,Na (S), respectively. The peak heights are found to change with sodium co-doping. The glow curves of BaSO$_4$:Eu (P) and BaSO$_4$:Eu,Na (P) consist of three glow peaks at about 165, 215, and 270°C, respectively. BaSO$_4$:Eu (S) and BaSO$_4$:Eu,Na (S) show a single broad peak at a relatively high temperature of about 265°C. The glow curve of BaSO$_4$:Eu (R) shows only one peak, at about 170°C, whereas the glow curve of BaSO$_4$:Eu,Na (R) is composed of a main peak at about 165°C and a small peak at 215°C. Figure 4-97 shows the TL glow curves of BaSO$_4$:Eu,P, prepared by three different methods. In this case also, the glow curves are found to change with the method of preparation. This indicates that the glow peak structure is altered when different methods of preparation are used.

Thermoluminescence investigation of Sr$_3$(PO$_4$):Eu^{2+}, a commercially available x-ray phosphor, was reported by Otvos. [69] Its TL emission spectrum is dominated by a single emission band centered at 420 nm, which is characteristic of Eu^{2+}. The glow curve of the Eu^{2+} band has two well-separated peaks. Sr$_3$(PO$_4$): Eu^{2+} seems to

be the best choice for studying thermoluminescence (TL) kinetics and the relation between radioluminescence (RL) and TL.

In Figure 4.99, a TL glow curve can be seen measured in parallel in both emission bands. The irradiation was performed at room temperature (RT) by an ^{90}Sr beta source. The heating rate was 1°C s^{-1}. The glow curve was measured in the two emission regions simultaneously. The figure clearly shows that the blue and red emission bands have completely different glow peak structures. This indicates that not only the luminescence centers responsible for the different emission bands are different, but the charge traps and consequently the charge transfer mechanisms involved are different, as well.

The perovskite-like KMgF$_3$ shows, when properly doped, thermoluminescence (TL) characteristics that make this phosphor a good candidate for ionizing radiation dosimetry. [70] With KMgF$_3$:Yb single crystals, the TL emission shows a sharp peak at 410 nm. This is very advantageous for the instrumentation of TLD analysis. A glow curve is shown in Figure 4.100.

Natural thermoluminescence (TLN) as well as gamma- and beta-induced thermoluminescence (TLI) were studied by Barcena et al. [71] in a set of micas

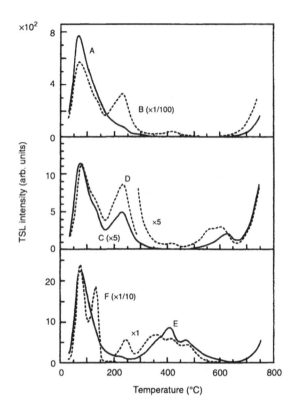

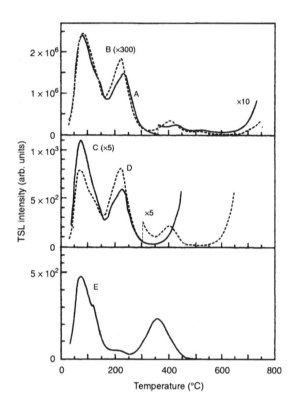

FIGURE 4.94 340-nm TSL response of CaSO$_4$ doped with: curve A, Dy^{3+}; B, Tm^{3+}; C, Ho^{3+}; D, Er^{3+}; and E, Gd^{3+}. Curve F shows the response with undoped CaSO$_4$. (From Reference [67]. With permission.)

FIGURE 4.95 RE^{3+} TSL response of CaSO$_4$ doped with: curve A, Dy^{3+}; B, Tm^{3+}; C, Ho^{3+}; D, E^{3+}; and E, Gd^{3+}. (From Reference [67]. With permission.)

with the general formula (K, Na) X_n AlSi$_3$O$_{10}$(OH,F)$_2$, where X corresponds to Al^{3+}, Mn^{2+}, Mn^{2+}, Fe^{2+}, Fe^{3+}, Mg^{2+}, Ti^{4+}, or Li$^+$, depending on the mica type. These minerals are classified according to crystallographic criteria in dioctahedral (D-micas like muscovite and sericite) and trioctahedral (T-micas like phlogopite and lepidolite). The results provide evidence of the presence of several thermoluminescence (TL) glow peaks induced after laboratory trituration (triboluminescence) of the non-irradiated mica samples.

Figure 4.101 displays the data representing the effects of beta and gamma irradiation on the TL signal for the three mica samples with a lower percentage of scattering (5%) and no fading. The samples used for Figures 4.101a, b, and c were Mos-1, Seri, and Flo-1, respectively. As can be seen, the TLI of the samples is complex, showing the presence of different, complex glow peaks in the range 100 to 450°C. Beta irradiation of muscovite samples was found to induce a prominent glow peak, presumed to be complex, between 150 and 170°C, with a large high temperature signal.

Metal alloy oxides, ceramics, glass, and various papers (carton, filter paper, typewriting paper) have been investigated by Stenger et al. [72] as possible thermoluminescent (TL) detectors or indicators for gamma and electron radiation technology process control. Results

show that there is a good correlation between TL response and absorbed dose in the range 0.1 Gy–15 kGy. Electrochemically oxidized aluminum alloys (E-ALO) can be used in the dose range from 0.1 Gy to 5 kGy. Above 5 kGy they have a non-linear TL response to 50 kGy (Figure 4.102). The useful dose range of ceramics is 100 Gy–8 kGy; the sensitivity of ceramics is similar to that of Al$_2$O$_3$; and the linear range is 100–400 Gy. The dose response curve of glass demonstrates high nonlinear character; glass can store dose information for 10 years.

PERSPEX DOSIMETERS (CHANGE IN LIGHT ABSORPTION)

White Perspex pieces of different thicknesses were examined by Amin et al. [73] for their dosimetric properties in industrial radiation processing. Light absorption of Perspex depends on the dose at which the Perspex is irradiated. The wavelength used for the readout was 314 nm. Pre-irradiation storage under different conditions had no effect on absorbed doses. The post-irradiation studies indicate that the specific absorbance decreases increasingly with storage time. White Perspex of 2-mm thickness could be employed as a suitable dosimeter to measure absorbed dose reproducibly within the common working range of 10–50 kGy.

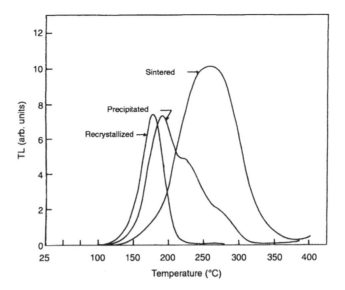

FIGURE 4.96 TL glow curves of BaSO$_4$:Eu (0.5 mol%) prepared by recrystallization, co-precipitation, and sintering methods. (From Reference [68]. With permission.)

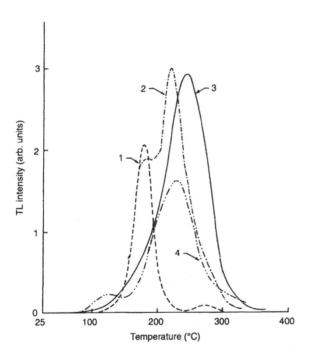

FIGURE 4.97 TL glow curves of BaSO$_4$:Eu,P prepared by: 1, (---) recrystallization; 2, (- . - . -) co-precipitation; 3, (—) sintering; and 4, (-..-..-) CaSO$_4$:Dy. (From Reference [68]. With permission.)

The absorption spectra of both irradiated and un-irradiated local white Perspex materials of two different thicknesses are shown in Figure 4.103. The white Perspex irradiated to 10 kGy gives the absorption spectrum shown, yielding a well-defined peak at 314 nm. The dose-response curves of local white Perspex of two different thicknesses (2 mm and 2.5 mm) and Harwell red Perspex were obtained by irradiating them at different radiation doses, as shown in Figure 4.104. The dose response of white Perspex of 2-mm thickness is better-defined up to 50 kGy, compared with that of 2.5-mm thickness (Figure 4.104). This property may make the former more suitable for use as a dosimeter in the range of 10–50 kGy for industrial radiation processing than the Perspex of 2.5-mm thickness, which is unsuitable for use beyond 35 kGy.

The effect of absorbed dose on the response of a red Perspex dosimeter was studied by Al-Sheikhly et al. [74] Red 4034 Perspex (Batch AW) was irradiated at several doses from 10 to 50 kGy and at three dose rates from 0.83 to 8.6 Gys^{-1} (3.0 to 31 kGy h^{-1}). Although it has been reported that there is only slight temperature dependence of response over the temperature range 20–43°C during irradiation, the effect of elevated temperatures during the period immediately after irradiation may cause errors in dose interpretation related to calibration and storage at ambient laboratory temperatures (e.g., 23°C). The Harwell red 4034 Perspex (Batch AW) dosimeters were kept in hermetically sealed aluminum polymer sachets until read with a Gary Model 219 spectrophotometer.

Figure 4.105 shows the net absorption spectra of red 4034 Perspex (Batch AW), irradiated at a controlled temperature of 22°C, to a series of five doses, at the three indicated dose rates. The vertical "error bar" on each data point shows the estimated random uncertainty (1σ) of 5 replicate samples at each dose and dose rate. The net absorbance per unit thickness (ΔA cm^{-1}) was obtained by placing the un-irradiated sample (having the mean value of absorbance, $\bar{A}_0 = 0.0585$) in the reference beam of the spectrophotometer. The results show that the higher the dose rate,

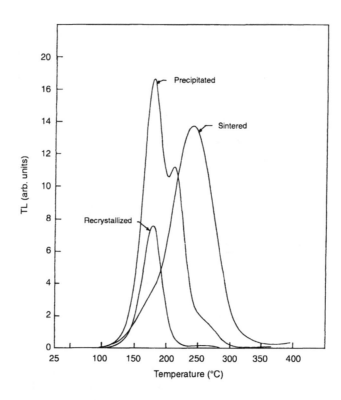

FIGURE 4.98 TL glow curves of $BaSO_4$:Eu,Na (0.5, 0.2 mol%) prepared by recrystallization, co-precipitation, and sintering methods. (From Reference [68]. With permission.)

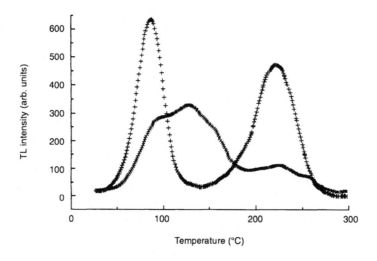

FIGURE 4.99 TL glow curves of $Sr_3(PO_4)$: Eu^{2+} measured after beta irradiation of 0.4-Gy dose at RT. (+) blue emission band (below 450 nm), (x) red emission band (between 500 and 650 nm). (From Reference [69]. With permission.)

the lower the gamma-ray sensitivity of the dosimeter. This finding is at an overlapping higher dose- rate range.

VII. ELECTRON SPIN RESONANCE DOSIMETRY

Electron spin resonance (ESR) is a well-documented technique applied to the study of defects related to paramagnetic impurities in insulators. ESR data can provide definite information on the structure of the dopant surroundings, such as the distribution of charge over the surrounding nuclei, charge and aggregation state of impurity, and characterization of new paramagnetic centers if produced during irradiation.

Splittings between electron energy levels of the defect/impurity is seen due to the interaction of the uncompensated (paramagnetic) electronic spin with crystal fields, nuclear spins, and the applied magnetic field.

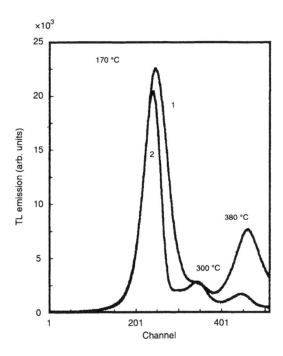

FIGURE 4.100 Glow curves of KMgF$_3$:Yb. (From Reference [70]. With permission.)

A free ion having a resultant electron spin ($\Sigma s = 1/2$), applied magnetic field (H_A) removes spin degeneracy and splits the energy level into two spin non-degenerate levels characterized by $m_s = 1/2$ and $-1/2$. The extent of splitting is $\Delta E = g\mu_B H_A$, where $\mu_B = e\hbar/2m_e c$ is the Bohr magneton and g (Landé g factor) $= 2$ for a free electron. If electromagnetic radiation (microwave) is impressed upon the material, then it will be strongly absorbed (resonance absorption) if its frequency ν is

$$\nu = g\mu_B H_A/h \qquad (4.62)$$

or

$$gH_A = (h/\mu_B)\nu = 7.14 \times 10^{-7} \nu (cgs) \qquad (4.63)$$

As for a free electron, $g = 2$, and if $\nu = 10^{10}$ Hz, then $H_A = 3570$ gauss. It is convenient to use a fixed frequency and vary the magnetic field H_A to observe spin resonance.

Thermal treatment of irradiated (10^6 R) TLD-100 powder and the consequent shape and size of the ESR signal have been utilized [75] to study the defect structures responsible for TL emission. In Figure 4.106, g values for different observed signals are shown. These are determined using the known signal position of DPPH (diphenyl picryl hydrazyl; $g = 2.0036$) as the calibration point (g marker). This is measured experimentally by placing a quartz tube containing DPPH in the second cavity while the irradiated sample powder, in a clean and dry matching quartz tube, is placed in the first cavity of the dual-cavity ESR spectrometer. The difference of the g factor as compared

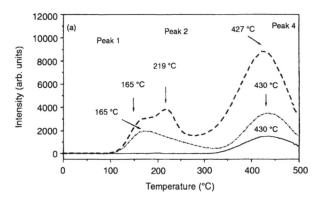

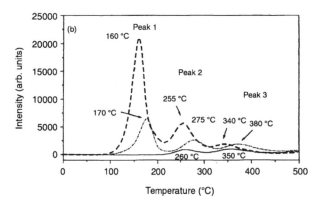

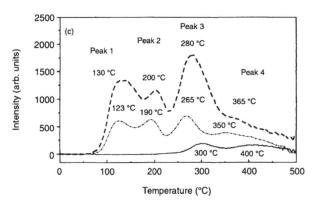

FIGURE 4.101 Thermoluminescence signals of three micas: (a) Mos-1, (b) Seri and (c) Flo-1. The solid line (—) represents the nonirradiated sample, The dash-dotted line (— • —) represents the beta-irradiated sample, and the dashed line (— —) represents the gamma-irradiated sample. (From Reference [71]. With permission.)

to its free-electron value ($g = 2$), i.e., $\Delta g = g_{bound} - g_{free}$) is a measure of the strength of binding of the electron within the defect. The sign of Δg gives the charge state of the defect. The electron gives a negative, and a hole gives a positive, value of Δg. The observation of a broad ESR band is usually related to the presence of centers in a precipitate or colloid form. The precipitates form during the irradiation, a fact which can be verified as the ESR band grows in intensity during the first few irradiation cycles, after which the saturation is exhibited.

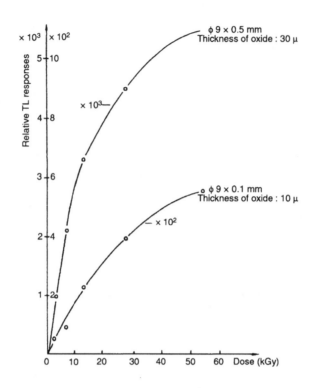

FIGURE 4.102 Dose response curves of E-ALO disks, gamma rate: 120 kGy/h. (From Reference [72]. With permission.)

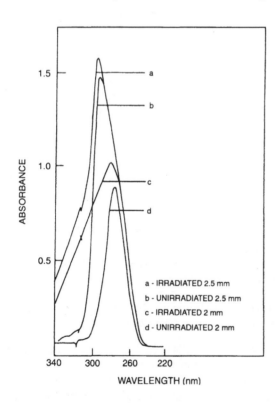

FIGURE 4.103 Absorption spectra of local white Perspex. (From Reference [73]. With permission.)

The dosimetric properties of purified cellulose membrane were investigated by Wieser and Regulla [76] for high-level dosimetry. The response of the cellulose radicals

in terms of absorbed dose in water was determined in the dose range 1–500 kGy. The ESR spectra were analyzed with respect to the radicals' stability under different temperature and humidity conditions.

The use of ESR spectroscopy in the field of dosimetry provides excellent precision and applicability over a wide dose range from 0.1 Gy to about 1 cGy. The ESR quantification of the radiation-induced free radical in, for example, the amino acid alanine has meanwhile proved to be of reference-class quality for high-level dosimetry, i.e., in the range 10 Gy to 0.5 MGy, covering photons and electrons. [76]

Purified cellulose membranes supplied in sheets of 15×30 cm² and 18 μm thickness contain 7.5% water and 12% glycerine. The irradiations were performed with ^{60}Co gamma rays. The dose rate was 15 kGy/h and the temperature in the irradiation chamber was 40°C. The samples were irradiated in the range from 1 kGy to 500 kGy in PMMA containers. The spectrum of a sample measured 4 days after irradiation with 20 kGy is shown in Figure 4.107. It has a total spread of 5.5 mT, and the ratios of the peak amplitudes are approximately 1:3:7:3:1. The central peak is positioned at $g = 2.0027$. Immediately (15 minutes) after irradiation, the amplitudes in the range of ±2 mT around the center of the spectrum are increased by about. 40% and the center is slightly shifted to a higher field ($g = 2.0032$). The peak amplitudes increase linearly with absorbed dose up to about 50 kGy, leading to a saturated level at about 500 kGy (Figure 4.108).

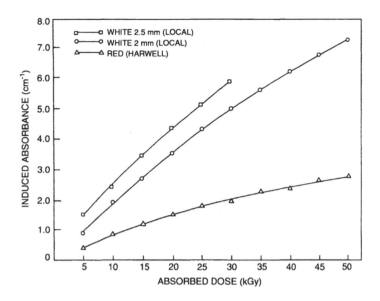

FIGURE 4.104 Dose-response curves of local white and Harwell red Perspex. (From Reference [73]. With permission.)

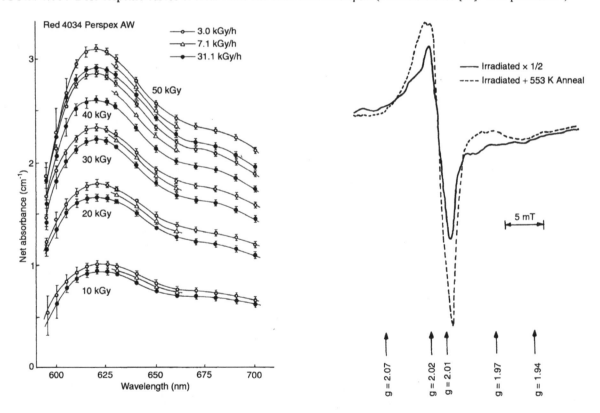

FIGURE 4.105 Net absorption spectra produced by gamma-ray irradiation of red 4034 Perspex, Batch AW, to the indicated doses, at three dose rates (temperature during irradiation: 22°C). (From Reference [74]. With permission.)

FIGURE 4.106 Central ESR signals from LiF (TLD-100) powder after irradiation to 10^6 R (shown at half-intensity) and after annealing at 280°C for 10 min. (From Reference [75]. With permission.)

Of all living tissues in the body, only dental enamel retains indefinitely the history of its radiation exposure. The effect of ionizing radiation on many crystalline materials is to produce free electrons that can be trapped in defects in the crystal lattice. The lifetime of these traps is long and can provide a history of radiation effects.

Both electron spin resonance (ESR) and thermoluminescence have previously been used to examine the trapped electrons in calcified tissues. [77]

When irradiated to high doses, tooth enamel shows a strong signal, which has been investigated previously in dental enamel and hydroxyapatite and identified as the

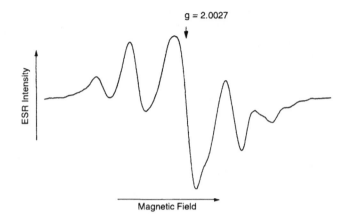

FIGURE. 4.107 The ESR spectrum of cellulose measured at room temperature 4 days after irradiation with 20 kGy. (From Reference [76]. With permission.)

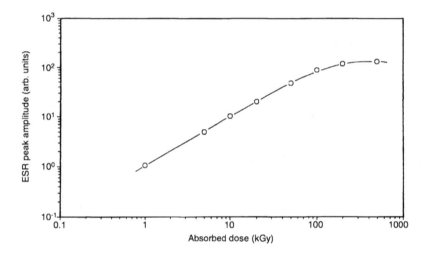

FIGURE 4.108 Dose-response curve of gamma-irradiated cellulose membranes. Prior to irradiation the samples were stored at room conditions. (From Reference [76]. With permission.)

carbonated radical with a *g* of 2.002. At low doses (less than 100 cGy), however, this radiation signal is masked by a signal which is due to the native crystal structure itself and is thought to be due to electrons in deep traps in the crystal at a *g* of 1.996.

Figure 4.109 shows the ESR signals from enamel samples irradiated with doses of up to 10 Gy and the ESR signal of an un-irradiated sample (patient B). A good correlation was found between the strength of the ESR signal and dose up to 20 Gy.

Although ESR dosimetry is a widely accepted means of rapid and early detection of absorbed radiation doses, it does possess two important disadvantages, limited sensitivity and lack of portability. The low sensitivity exists primarily because of the small interaction between the magnetic dipole moment of the trapped electron and the externally applied magnetic field that is required to observe the ESR signal. The magnitude of the associated energy-level transitions is proportional to the product of

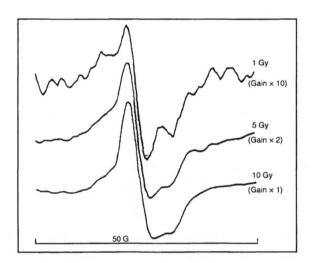

FIGURE 4.109 ESR signals of tooth enamel samples irradiated with doses of ^{60}Co gamma-rays up to 10 Gy. (From Reference [77]. With permission.)

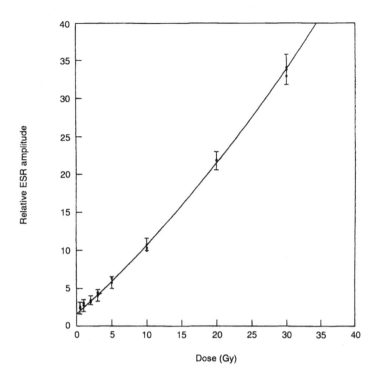

FIGURE 4.110 Dose dependence of the ESR signal amplitude of alanine-polyethylene dosimeters. The full line represents the second-order polynomial fitting function; error bars indicate overall uncertainty (95% confidence level). (From Reference [78]. With permission.)

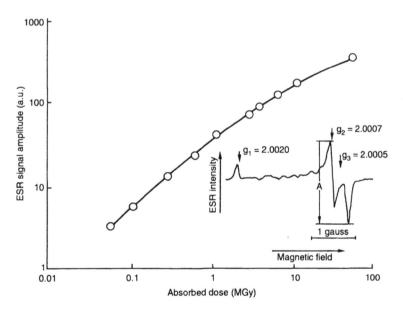

FIGURE 4.111 Response of crystalline quartz powder dosimeter (in terms of total amplitudes of the g_2 and g_3 ESR line signals of the quartz E_1' center) as a function of the gamma-ray absorbed dose .The inset shows the ESR spectrum of the E_1' center in crystalline quartz powder at an absorbed dose of 10 MGy (measurement at ambient room temperature). (From Reference [79]. With permission.)

these two quantities and is only 10^{-6} that of readily observable optical transitions in an atom.

The lack of portability and general applicability of ESR in dental enamel for radiation dosimetry exists for two reasons. First, because of the large externally applied magnetic field strength required to observe the ESR signal,

a wide array of laboratory equipment is used. Second, the large sample size required (typically 0.1 g) necessitates the use of extracted teeth.

The use of electron spin resonance (ESR) dosimetry is also proposed in radiation therapy and accident dosimetry. Experimental investigation preparation procedures of alanine-polyethylene solid state dosimeters (SSD) with optimized characteristics were defined, both modifying the commonly used blend composition and preparation and choosing the best ESR spectra recording conditions. [78]

Dose response to ^{60}Co was studied by Bartolotta et al. [78], irradiating groups of three dosimeters to dose values between 0.5 and 30 Gy. Up to 2 Gy, the signal is almost comparable with BG and no accurate dose measurements are possible. In the 2–30 Gy dose range, ESR amplitude values can be fitted with a second-order polynomial; differences between given and recalculated doses were always less than 4% (Figure 4.110). [78]

SiO$_2$ has been introduced for use at very high doses (10^5 to 5×10^7 Gy) [79] and at relatively high temperatures (up to 200°C). The preferred method of analysis in this case is ESR spectrometry. Figure 4.111 shows the ESR response properties of this system, where the silica material is fine-grain pure crystalline quartz, and free-radical centers consist of oxygen vacancies (E' centers) in the SiO$_2$ lattice. This dosimeter shows an irradiation temperature coefficient of 6.5×10^{-5} % C^{-1}. It can be optimally stabilized by a short heat treatment at 300°C between irradiation and ESR signal analysis.

REFERENCES

1. Muniz, J. L. et al., *Phys. Med. Biol.*, 40, 253, 1995.
2. Branch, C. J. and Kearfott, K. J., *Nucl. Inst. Meth.*, A422, 638, 1999.
3. Randal, J. T. and Wilkins, M. H. F., *Proc. Conf. Roy. Soc. London A*, 184, 366, 1945.
4. Carrillo, R. E. et al., *Phys. Med. Biol.*, 39, 1875, 1994.
5. Mobit, P. N. et al., *Phys. Med. Biol.*, 41, 387, 1996.
6. Mobit, P. N. et al., *Phys. Med. Biol.*, 41, 979, 1996.
7. Fairbanks, E. J. and DeWerd, L. A., *Med. Phys.*, 20, 729, 1993.
8. Gamboa-deBuen, I. et al., *Phys. Med. Biol.*, 43, 2073, 1998.
9. Jain, V. K. and Ganguly, A. K., *Report I-466*, Bhabha Atomic Research Center, 1977.
10. Vasilache, R. A. et al., *Rad. Prot. Dos.*, 85, 183, 1999.
11. Bhatt, B. C. et al., *Rad. Prot. Dos.*, 84, 175, 1999.
12. Weizman, Y. et al., *Rad. Prot. Dos.*, 84, 47, 1999.
13. Bos, A. J. J. et al., *Rad. Prot. Dos.*, 84, 13, 1999.
14. Kalmykov, L. Z., *Med. Phys.*, 21, 1715, 1994.
15. Osvay, M. and Deme, S., *Rad. Prot. Dos.*, 85, 469, 1999.
16. Bilski, P. et al., *Rad. Prot. Dos.*, 85, 367, 1999.
17. Schoner, W. et al., *Rad. Prot. Dos.*, 85, 263, 1999.
18. Korn, T. et al., *Phys. Med. Biol.*, 38, 833, 1993.
19. Zoetelief, J. and Jansen, J. T. M., *Phys. Med. Biol.*, 42, 1491, 1997.
20. Meigooni, A. S. et al., *Med. Phys.*, 22, 555, 1995.
21. Korn, T. et al., *Med. Phys.*, 20, 703, 1993.
22. Mukherjee, B. and Clerke, W., *Nucl. Inst. Meth.*, A361, 395, 1995.
23. McParland, B. J. and Munshi, M., *Nucl. Inst. Meth.*, A322, 308, 1992.
24. Mukherjee, B., *Nucl. Inst. Meth.*, A385, 179, 1997, A361, 395, 1995.
25. Nakajima, T. et al., *Nucl. Inst. Meth.*, 157, 155, 1978.
26. Wu, D. K. et al., *Health Phys.*, 46, 1063, 1984.
27. Furetta, C. et al., *Med. Phys.*, 21, 1605, 1994.
28. Bartolotta, B. et al., *Phys. Med. Biol.*, 40, 211, 1995.
29. Muniz, L. J. and Delgado, A., *Phys. Med. Biol.*, 42, 2569, 1997.
30. Alves, J. G. et al., *Rad. Prot. Dos.*, 85, 253, 1999.
31. Gomez Ros, J. M. et al., *Rad. Prot. Dos.*, 85, 249, 1999.
32. Duggan, L. and Korn, T., *Rad. Prot. Dos.*, 85, 213, 1999.
33. Muniz, J. L. et al., *Rad. Prot. Dos.*, 85, 57, 1999.
34. Moscovitch, M., *Rad. Prot. Dos.*, 85, 49, 1999.
34a. Hirning, C. R., *Health Physics*, 62, 233, 1992.
35. Perry, O. R. et al., *Rad. Prot. Dos.*, 85, 273, 1999.
36. Hoffmann, V. et al., *Rad. Prot. Dos.*, 85, 341, 1999.
37. Angelone, M. et al., *Rad. Prot. Dos.*, 85, 245, 1999.
38. Buenfil, A. E. et al., *Rad. Prot. Dos.*, 84, 273, 1999.
39. Hoffmann, W. and Prediger, B., *Rad. Prot. Dos.*, 6, 149, 1984.
40. Loncol, T. et al., *Phys. Med. Biol.*, 41, 1665, 1996.
41. Jafarizadeh, M. et al., *Rad. Prot. Dos.*, 84, 119, 1999.
42. May, C. E. and Partridge, J. A., *J. Chem. Phys.*, 40, 1401, 1964.
43. Nakamura, S. *Rad. Prot. Dos.*, 85, 313, 1999.
44. Mobit, P. N. et al., *Phys. Med. Biol.*, 43, 2015, 1998.
45. Mobit, P. N. et al., *Phys. Med. Biol.*, 42, 1913, 1997.
46. Nam, T. L. et al., *Med. Phys.*, 14, 596, 1987.
47. Robar, V. et al., *Med. Phys.*, 23, 667, 1996.
48. Holt, J. G. et al., *Phys. Med. Biol.*, 20, 559, 1975.
49. Lorence, L. J. et al., *User Guide to CEPXS/ONEDANT*, SAND 89-1661, SNLA, 1989.
50. Deogracias, E. C. et al., *Nucl. Inst. Meth.*, A422, 629, 1999.
51. Miljanic, S. and Ranogajec-Komor, M., *Phys. Med. Biol.*, 42, 1335, 1997.
52. Seltzer, S. M., *Rad. Phys. Chem.*, 35, 703, 1990.
53. Mobit, P. N. et al., *Phys. Med. Biol.*, 42, 1319, 1997.
54. Chan, F. K. and Burlin, T. E., *Health Phys.*, 18, 325, 1970.
55. Paliwal, B. R. and Almond, P. R., *Health Phys.*, 31, 151, 1976.
56. Burlin, T. E., *Br. J. Radiol.*, 39, 727, 1966.
57. Horowitz, Y. S., *Rad. Prot. Dos.*, 9, 5, 1984.
58. Woo, M. K. and Chen, Z., *Med. Phys.*, 22, 449, 1995.
59. Karaiskos, P. et al., *Med. Phys.*, 25, 1975, 1998.
60. IPEMB Code of Practice, *Phys. Med. Biol.*, 41, 2605, 1996.
61. Coudin, D. and Marinello, G., *Med. Phys.*, 25, 347, 1998.
62. Kato, K. et al., *Med. Phys.*, 18, 928, 1991.
63. Kortov, V. S. et al., *Rad. Prot. Dos.*, 84, 35, 1999.

64. **Lucas, C. A. and Lucas, B. K.,** *Rad. Prot. Dos.,* 85, 455, 1999.

65. **Yukihara, E. G. et al.,** *Rad. Prot. Dos.,* 84, 265, 1999.

66. **Osvay, M. and Ranogajec-Komor, M.,** *Rad. Prot. Dos.,* 84, 219, 1999.

67. **Gerome, V. et al.,** *Rad. Prot. Dos.,* 84, 109, 1999.

68. **Shinde, S. S. et al.,** *Rad. Prot. Dos.,* 84, 215, 1999.

69. **Otvos, N.,** *Rad. Prot. Dos.,* 84, 135, 1999.

70. **Gambarini, G. et al.,** *Rad. Prot. Dos.,* 84, 211, 1999.

71. **Barcena, J. L. et al.,** *Rad. Prot. Dos.,* 84, 289, 1999.

72. **Stenger, V. et al.,** in *Proc. High Dose Dosimetry for Radiation Protection,* IAEA, 1991, 73.

73. **Amin, M. R. et al.,** in *Proc. High Dose Dosimetry for Radiation Protection,* IAEA, 1991, 57.

74. **Al-Sheikhly, M. et al.,** in *Proc. High Dose Dosimetry for Radiation Protection,* IAEA, 1991, 419.

75. **Mahesh, K.,** in *Thermoluminescence in Solids and Its Applications,* Nuclear Technology Publishing, Kent (U.K.), 1989, 159.

76. **Wieser, A. and Regulla, D.,** in *Proc. High Dose Dosimetry for Radiation Protection,* IAEA, 1991, 203.

77. **Pass, B. and Aldrich, J. E.,** *Med. Phys.,* 12, 305, 1985.

78. **Bartolotta, A. et al.,** *Rad. Prot. Dos.,* 84, 293, 1999.

79. **McLaughlin, W. L.,** in *Proc. High Dose Dosimetry for Radiation Protection,* IAEA, 1991, 3.

5 Film Dosimetry

CONTENTS

I. INTRODUCTION

Film dosimetry is attractive due to its high spatial resolution, wide accessibility, and the flexibility to place the film in humanoid phantoms. Also, the short measuring time and the fact that the film dosimetry is intrinsically two-dimensional and integrating in time are appreciated. Film is potentially the ideal detector for determining dose distribution for dynamic beams and for studying the combination of stationary beams treated sequentially. Film dosimetry is widely used to obtain the relative dose distribution of electron and photon beams in water, in plastics, and in inhomogeneous phantom. Film dosimetry in phantoms is advantageous because of high spatial resolution, short treatment unit immobilization time, and 2D information.

Modern film processing units improve the reproducibility and reliability of film dosimetry and make it an attractive method for many applications. Fast film digitizers connected to a computer, equipped with proper evaluation software, allow rapid and accurate analysis of large films in a short time.

The most common setup in relative dose measurements with films is to sandwich a film within a phantom of water-equivalent material with the film plane-parallel to the central axis of the radiation field. With the parallel geometry, two particular precautions must be taken: there must be heavy pressure on the phantom to avoid any air gap on either side of the film, and there must be perfect alignment of the film edge with the surface of the phantom. In addition, the artifacts that result from a thin air layer trapped between the packaging material, the paper spacer, and the film are responsible for the inaccurate dose measurement in the build-up region of the electron depth dose.

Conventional silver halide film has a highly nonlinear photon energy response, especially at low energies. It has radiation interaction properties markedly different from those of tissue. Along with variations introduced by the necessary post irradiation processing, this type of film is extremely difficult to use for accurate analytical dosimetry.

Most radiochromic systems are chemical radiation sensors consisting of solid or liquid solutions of colorless leuco dyes; these become colored without the need for development when exposed to ionizing radiation. Various radiochromic forms, such as thin films, thick films and gels, liquid solutions, and liquid-core waveguides, have been in routine use for dosimetry of ionizing radiation over a wide range of absorbed doses (10^{-2} to 10^6 Gy). Radiochromic film is used for general dosimetry of ionizing radiation in high-gradient areas of electron and photon beams in a wide energy range. The film allowing approximately tissue-equivalent dosimetry has been applied to mapping of dose distribution in brachytherapy.

Radiochromic film (RCF) is of great interest as a planar dosimeter for radiation oncology applications. It consists of a thin, radiosensitive, 7–23-μm thick, colorless leuco dye bonded to a 100-μm-thick mylar base. RCF turns deep blue in color upon irradiation. RCF is approximately tissue-equivalent. GafChromic MD-55 is usable at doses from less than 1 Gy up to 12 Gy when measured at the wavelength of maximum sensitivity (676 nm) and up to about 500 Gy when measured at a wavelength of low sensitivity. The absorption spectra of GafChromic (GC) film contain two peaks with wavelengths in the range 610–680 nm and, thus, the dose-response curve as measured by an optical densitometer or spectrophotometer will be highly dependent on the light source spectrum and sensor material.

II. EXAMPLES OF PHOTOGRAPHIC FILM DOSIMETRY

The use of film as a dosimeter is still limited, due to the various difficulties associated with films such as energy dependence, film orientation, and sensitometric nonlinearity. On the other hand, film is probably one of the best detectors for studying spatial distribution of dose or energy imparted. The dosimetric resolution is limited only by the grain size

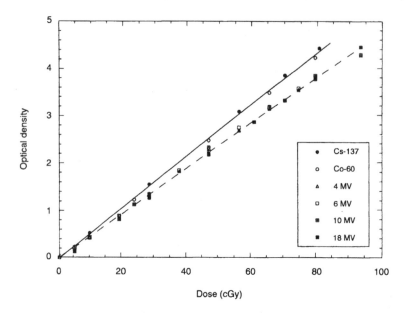

FIGURE 5.1 Characteristic curves of the TVS films for photon beams. (From Reference [1]. With permission.)

and the size of the aperture of an optical densitometer. Commercially available laser-scanning densitometers afford film dosimetry with a resolution of a few microns. Unlike other detectors such as ionization chambers, diode detectors, thermoluminescent detectors (TLDs), scintillation detectors, and diamond detectors where dose information must be recorded before the readout is cleared for another irradiation, radiographic films allow repetitive readouts and provide a permanent record of the dosimetric measurements. Films may be customized in various sizes and shapes to fit any dosimetric application. Due to the relatively small thickness, a film can be treated very close to a Bragg-Gray cavity. The physical flexibility of a film is also suitable for the curved and cylindrical surface dose mapping when other detectors are impractical. In relative dose measurements, the optical density may be taken as proportional to the dose without any correction, since the collisional stopping power ratio of emulsion to water varies slowly with electron energy.

High-energy photon and electron dosimetry was carried out by Cheng and Das using CEA film. [1] The packaging of the CEA films is distinctly different from that of the other films, in that each film is vacuum- sealed in a shiny polyester-made, waterproof packet about 130 μ. thick, permitting film dosimetry to be carried out even in water phantom. Aside from the difference in packaging, the CEA films have a clear polyester film base as opposed to a bluish-dye film base found in the Kodak Readypack XV and XTL films. It is not clear if the difference in film base may have any effect on image quality. The CEA films come in two types: TLF (localization) and TVS (verification), similar to the XTL and XV films of the Kodak Readypack.

Figure 5.1 shows the variation of net optical density with dose for the TVS film for a range of photon energies from ^{137}Cs to 18 MV. The base optical density of the CEA films is on the order of 0.06, compared to about 0.2 for the Kodak films. As shown in Figure 5.1, the linear portion of the characteristic curve covers a range of optical density up to 4.3. This wide range of linearity offers a convenient means in percent depth dose and isodose measurements. It is interesting to note that the film is faster to rays than to *bremsstrahlung* x-rays from a linear accelerator but is independent of energy for each type of radiation. This feature is particularly attractive for high-energy x-rays, as one single sensitometric curve can be used for a wide range of x-ray energies. A straight line is fitted to each of the characteristic curves for rays and x-rays with a regression coefficient of 0.999. For gamma rays, the line of regression is [1]

$$(OD)_{\gamma,\,rays} = 0.054\,07 \times dose, \qquad (5.1)$$

while for x-rays, the line of regression is

$$(OD)_{X,\,rays} = 0.047\,65 \times dose \qquad (5.2)$$

Figure 5.2 compares the characteristic curves for the CEA TVS film and the Kodak XV film for two x-ray energies, 4 and 10 MV. For the CEA TVS film, the linear portion extends over the dose range up to 90 cGy, followed by an abrupt saturation above 90 cGy. The CEA TVS curves have a linear relationship, with a coefficient of regression very close to 1. For the Kodak XV film, on the other hand, the optical density varies curvilinearly with

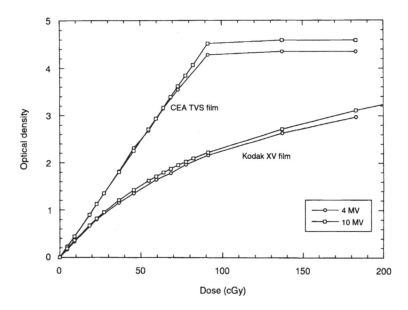

FIGURE 5.2 Comparison of the characteristic curves of CEA TVS film and Kodak XV film for different photon energies. (From Reference [1]. With permission.)

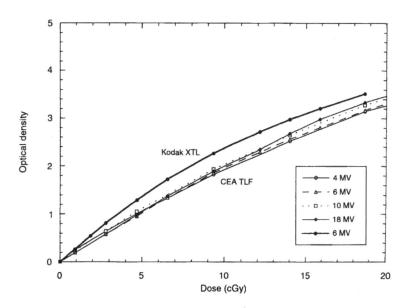

FIGURE 5.3 Characteristic curves of the CEA TLF film for photon energies 4–18 MV. For comparison, characteristic curve of Kodak XTL film is also shown for a 6-MV beam. (From Reference [1]. With permission.)

dose up to 200 cGy. The characteristic curve of the Kodak XV films exhibits a longer tail and a smaller gamma compared to that of the CEA TVS film.

The size of the silver halide crystals is generally smaller in the CEA TVS film than in the Kodak XV film. Indeed, the silver halide crystals in the CEA TVS film are of fairly uniform size and shape, resulting in a linear characteristic curve. On the other hand, the silver halide crystals of the Kodak XV film are of different sizes and shapes, with the largest more than 10 times bigger than the smallest, resulting in a nonlinear characteristic curve. Furthermore, the

CEA TVS film has a higher silver concentration (about 42 g/m²) compared to regular x-ray films (about 7 g/m²).

The characteristic curves for the TLF film over the photon energy range 4-18 MV are shown in Figure 5.3. The film saturates at about 30 cGy, which is considerably higher than the Kodak Readypack XTL film. Unlike the CEA TVS film, the characteristic curves for the CEA TLF films are slightly curvilinear over the range of dose up to 30 cGy. However, if the data is considered only in the range of 0–15 cGy, the curves are all straight lines with a coefficient of regression near unity. The CEA TLF film

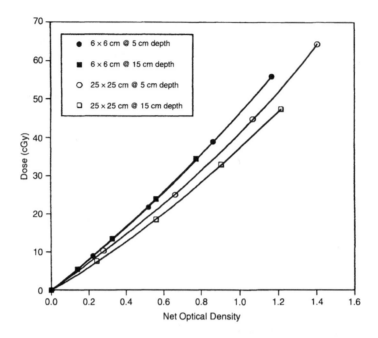

FIGURE 5.4 Film sensitivity dependence upon field size and upon depth. (From Reference [2]. With permission.)

is also energy-independent within the experimental uncertainty.

A method of film dosimetry for high-energy photon beams was proposed by Burch et al. [2] which reduced the required film calibration exposures to a set of films obtained for a small radiation field size and shallow depth (6 cm × 6 cm at 5-cm depth). It involves modification of a compression-type polystyrene film phantom to include thin lead foils parallel to the vertical film plane at approximately 1 cm from both sides of the film emulsion. The foils act as high atomic number filters which remove low-energy Compton scatter photons that otherwise would cause the film sensitivity to change with field size and depth.

High-energy photon beams used in radiation oncology are considered to interact primarily by the Compton scattering process with tissue. However, when film is placed in a tissue-equivalent material, photoelectric interactions associated with the silver atoms in the emulsion cause the film to over-respond relative to the tissue-equivalent material. For the low-energy photons, film dose may be as much as 25 times the tissue dose at the same physical location.

The results for the 6-cm × 6-cm and 25-cm × 25-cm fields shown in Figure 5.4 emphasize the importance of the film sensitivity as a function of field size and depth. As the field size was increased, the dose required to produce a given density on the film was reduced. For sizes smaller than 10 cm × 10 cm, the sensitivity change is small (not shown); and for sizes less than 6 cm × 6 cm, no change in sensitivity occurs. For small fields with less scatter, the change in sensitivity with depth is not apparent.

The effects of foil-to-film separation distance and foil thickness were investigated in order to obtain a single optimum distance-thickness combination, and the results are presented in Figures 5.5 and 5.6. In Figures 5.5 and 5.6, the dose was calculated using data from the 6-cm × 6-cm calibration films. Film/foil separation distances of 0, 0.6, 1.2, and 1.9 cm and lead foil thicknesses of 0, 0.15, 0.30, 0.46, and 0.76 mm were included in the investigation. At 0-cm film/foil separation distance, the curve shows the effect of electrons coming from the lead due to interactions within the foil. This intensification effect exaggerates the shape of the depth-dose curve and the non-uniform film/foil contact is apparent in the data (Figure 5.5). At 0.6 cm the scattered electrons are absorbed in the intervening polystyrene. Further increase in film/foil separation distance produced only minor changes in the calculated dose curve. Figure 5.6 shows the effect of changing foil thickness. A single thickness of 0.15 mm causes a significant decrease in the calculated dose, with only subtle changes as additional layers are added. The best match to ion chamber percentage depth-dose data was observed for lead foil thickness of 0.46 mm with a 1.2 cm film/foil separation distance.

The use of radiographic film for the dosimetry of large (i.e., >10 cm × 10 cm) photon radiotherapy beams is complicated, due to a dependence of film emulsion sensitivity on depth within the phantom. This dependence is caused by a relative increase in depth in the population of lower-energy, scattered photons in the spectrum and the subsequent photoelectric absorption of these photons by the film emulsion. This effect becomes most pronounced with increases in the photon population in the

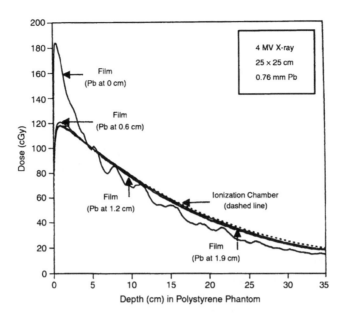

FIGURE 5.5 Lead foils placed adjacent to the film show an exaggerated response at shallow depth and a wavy appearance due to undulation in their surface. Each curve is normalized to 5-cm depth. (From Reference [2]. With permission.)

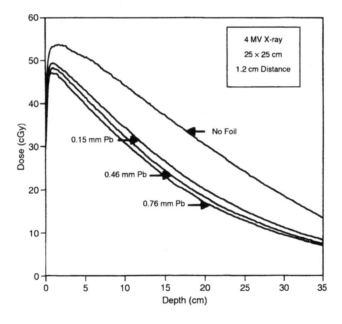

FIGURE 5.6 The effect of foil thickness on the shape of the depth-dose curve. (From Reference [2]. With permission.)

energy region below approximately 400 keV, where the mass attenuation coefficient of emulsion increases rapidly, approximately as $1/(h\nu)^3$. As the low-energy spectral component increases, the tissue equivalence of film emulsion diverges from that of tissue-equivalent materials, due to an approximate Z^3 dependence of the photoelectric interaction cross section. [3]

In order to examine possible dependencies of emulsion exposed by radiosurgical beams, a series of sensitometric curves was established for a range of depths and field sizes. For each exposure a 10-in × 12-in sheet of

Kodak X-Omat V film from a single batch (#194 05 2) was sealed within a light-tight Solid Water (Gammex RMI, Inc.) cassette. The cassette consists of two 2-cm-thick slabs of solid water sealed around three edges by nylon screws and a rubber O-ring. [3]

To minimize possible variations due to film-processing conditions, a Kodak X-Omat RP processor was used by Robar and Clark for which the throughput is very high and quality assurance is performed daily. Developer temperature fluctuated by less than ±0.5°F between processing sessions. Optical density was measured using a scanning densitometer. Base-plus-fog optical density was subtracted from scanned optical densities for each film.

In order to relate optical densities to absolute doses, the dose to water corresponding to each of these depths and for each MU setting was calculated from

$$D(d, A) = \frac{MU PDD(d, A) S_t(A)}{100} \qquad (5.3)$$

where $D(d, A)$ is the dose at depth d for field size A at the phantom surface, MU is the number of monitor units given, $PDD(d,A)$ is the percent depth dose and $S_t(A)$ is the total scatter factor. Both $S_t(A)$ and $PDD(d,A)$ were measured in a water phantom using a p-type silicon electron diode .

Figure 5.7a shows the sensitometric curves obtained for the large (20-cm $\times$ 20-cm) photon beam at depths of 1.0 cm, 10.0 cm, and 20.0 cm. For each depth a curve was fitted to the measured sensitometric data using the single-target/single hit equation

$$OD = OD_{sat}(1 - 10^{-\alpha D}) \qquad (5.4)$$

where OD and D are the measured optical density and given dose, respectively. The saturation density of the film, OD_{sat}, was estimated by delivering a large dose (500 cGy) to a film in phantom, and it was held constant in the fitting algorithm. The fitting parameter α represents emulsion sensitivity and was allowed to vary. For this large field, the curves diverge markedly, indicating that using a sensitometric curve that is not depth-specific would introduce significant error in converting optical density to dose. In order to minimize this error to below approximately 10%, for example, it is necessary to confine the dose range to less than 80 cGy. In contrast, Figure 5.7b shows sensitometric curves corresponding to the same depths for the radiosurgical (2.5-cm-diameter) field. For this small field, the curves agree to within the reproducibility of film development and scanning. By determining the value of α for each curve, film sensitivity was obtained as a function of depth, as illustrated for the 20-cm $\times$ 20-cm and 2.5-cm-diameter fields (Figure 5.8). As expected from the disparity in the sensitometric curves in Figure 5.7a, the film sensitivity for the large (20-cm $\times$ 20-cm) field increases systematically with depth. While small fluctuation of the values of α is apparent for the radiosurgical field, no systematic variation of sensitivity with depth is apparent.

Film dosimetry is most problematic for lower-energy photon beams (e.g., ^{60}Co and 4-MV) due to their lower primary-to-scatter ratio. Figure 5.7 shows the importance of corrections in converting scanned optical density to dose even for the 6-MV beam.

Film is often used for dose measurements in heterogeneous phantoms. In those situations perturbations are produced which are related to the density and atomic

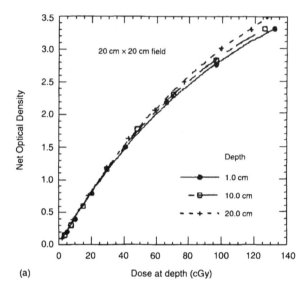

(a)

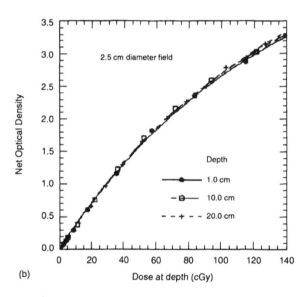

(b)

FIGURE 5.7 The sensitometric curves established experimentally for (a) the 20-cm $\times$ 20-cm radiotherapy field and (b) the 2.5-cm-diameter radiosurgical field, shown for depth in solid water phantom of 1.0 cm, 10.0 cm, and 20.0 cm. (From Reference [3]. With permission.)

number of the phantom material and the physical size and orientation of the dosimeter. Significant differences were observed by El-Khatib et al. between the dose measurements within the inhomogeneity. [4] These differences were influenced by the type and orientation of the dosimeter in addition to the properties of the heterogeneity.

To investigate how the differences in dose measurements are related to the dosimeter configuration, the EGS4 Monte Carlo system, together with the PRESTA algorithm, was used to model the experimental conditions. The percentage depth doses measured in the phantom containing 1-cm bone are shown in Figure 5.9 and are compared to

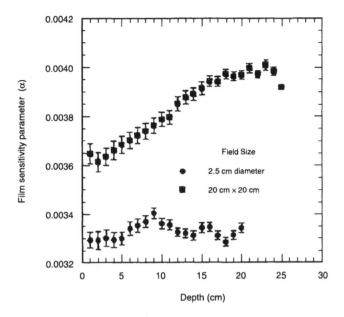

FIGURE 5.8 The curve-fitting parameter α, representing emulsion sensitivity, as a function of depth for the 20-cm $\times$ 20-cm radiotherapy field and the 2.5-cm-diameter radiosurgical field. (From Reference [3]. With permission.)

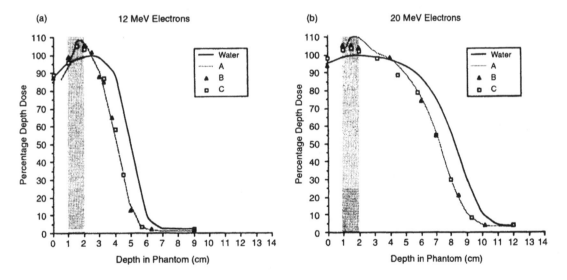

FIGURE 5.9 The percentage depth doses measured in a polystyrene phantom containing 1-cm hard bone at 1-cm depth using (A) film in parallel orientation to the beam, (B) film in perpendicular orientation to the beam, and (C) TLD are shown and compared to the PDD measured in a homogeneous water phantom for electron beams of nominal energy (a) 12 MeV and (b) 20 MeV. (From Reference [4]. With permission.)

the depth doses measured in the homogeneous water phantom. All film measurements were done with the film in the ready pack with the inner paper removed. The doses were normalized to the maximum dose in the homogeneous polystyrene phantom, measured with whatever dosimeter was used to measure the doses at depth. The doses within the bone represent dose-to-unit density material within the bone rather than dose to the bone itself, since a correction by ratio of collisional stopping powers was not done. At several cm beyond the bone, the PDD measured with all dosimeter configurations are identical. The opposite effect

for the two film orientations is observed for the phantom containing the cork (Figure 5.10). The PDD is unaffected in the initial 4 cm of polystyrene and there is greater penetration of the beam in cork. Within the cork the difference measured with the film in different orientations is as much as 6%, and the TLD measurements are lower than the measurements with film. These greater discrepancies are attributed to the fact that the inhomogeneity is much larger and lies in a region of high dose gradient.

A method for the creation of a complete 3D dose distribution from measured data was discussed by

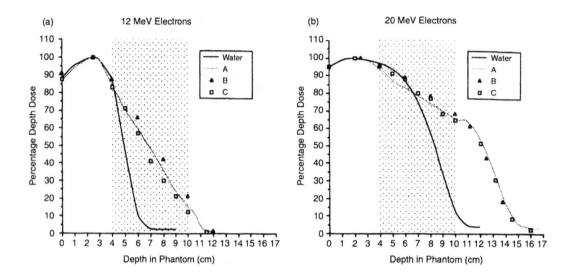

FIGURE 5.10 The percentage depth doses measured in a polystyrene phantom containing 6-cm cork at 4-cm depth using (A) film in parallel orientation to the beam, (B) film in perpendicular orientation to the beam, and (C) TLD are shown and compared to the PDD measured in a homogeneous water phantom for electron beams of nominal energy (a) 12 MeV and (b) 20 MeV. (From Reference [4]. With permission.)

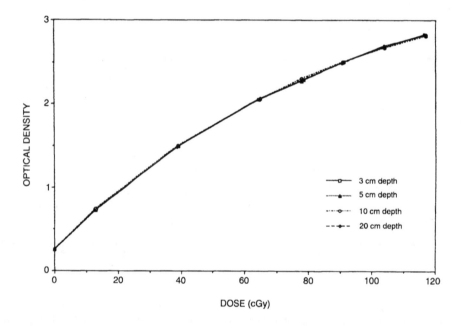

FIGURE 5.11 Film sensitometric curves derived from four sets of BEV films placed at depths of 3, 5, 10, and 20 cm, respectively. (From Reference [5]. With permission.)

Stern et al. [5] It involved the measurement with film of the dose distribution in a series of "beam's-eye-view" (BEV) planes (planes perpendicular to the beam central axis at different depths). Film measurements were made using Kodak Readypack XV-2 film (Eastman Kodak Co., Rochester, NY) sandwiched between sheets of water-equivalent solid. Exposed films were digitized using a laser digitizer with spot size set to 0.42 mm and pixel size 0.45 mm.

Film image data files were next converted into the planning system's standard grayscale image file format.

When necessary, the number of pixels was reduced using nearest-neighbor sampling. The images were entered into the planning system and aligned with the planning system representation of the beam by translating and rotating the coordinate system of the displayed film image. Optical density values were converted to dose values using the appropriate measured film sensitometric curve.

The sensitometric curves obtained at four different depths ranging from d_{max} to 20 cm are plotted vs. absolute dose in Figure 5.11. The largest variation in dose for a given optical density determined from this set of curves is

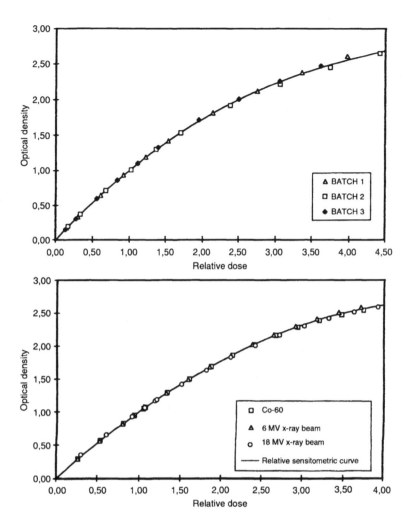

FIGURE 5.12 Relative sensitometric curve for three photon beam qualities (^{60}Co, 6-MV x-rays, 18-MV x-rays). (From Reference [6]. With permission.)

2.3%, at an optical density of approximately 2.3. Variation is less than 1% over most of the measured dose range. This demonstrates that a single sensitometric curve measured at one depth can be accurately used for optical density-to-dose conversion of BEV films at all depths for the experimental situation used.

A feasibility study for mailed film dosimetry has been performed by Novotny et al. [6] The fading of the latent image before film processing is only 3% per month and the normalized sensitometric curve is not modified after a period of 51 days between irradiation and processing. All film experiments have been performed with Kodak Readypack paper envelope films (25.4 cm × 30.5 cm). All films were processed with an automatic processing unit, either an Agfa Gevamatic 1100 (processing time 90 s and temperature 35 ± 1°C) or a modern AGFA Curix 260 (processing time 90 s and temperature 35 ± 1°C). No dependence on the photon energy has been observed in the shape of the normalized sensitometric curve for ^{60}Co, 6-MV, and 18-MV beams (Figure 5.12).

No increase is observed in the fog value for un-irradiated films after mailing or storing for more than 75 days; the average reading is 0.16 ± 0.01 for the two series of films, either processed immediately after irradiation or processed after mailing.

Fading is expressed as the ratio of the optical density of a faded film and the optical density of the reference film, as a function of the time t between film irradiation and film processing. The interval t is equal to 0 for the reference film. The dependence of fading on time t, for films stored in the laboratory, is presented in Figure 5.13. A straight line is fitted through the measured data using the least-squares method. A decrease of about 3% in optical density is observed per month of delay between film irradiation and film processing (corresponding to a given dose of 0.45 Gy and optical density of about 1.3). A smaller decrease of about 1.8% per month in optical density is observed for films irradiated with dose of 1.5 Gy (corresponding to optical density 2.7). These decreases correspond to estimated variations on dose of 3.8% and 3.4%, respectively.

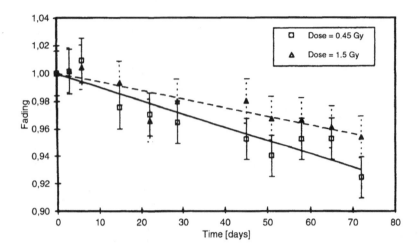

FIGURE 5.13 Decay with time of optical density (fading) of Kodak X-omat V films irradiated with 0.45 Gy or 1.5 Gy. The error bars represent one standard deviation of total uncertainty. (From Reference [6]. With permission).

III. RADIOCHROMIC FILM DOSIMETRY

The introduction of radiochromic dye-cyanide films has resolved some of the problems experienced with conventional radiation dosimeters. The high spatial resolution and low spectral sensitivity of radiochromic films make the dosimeter ideal for the measurement of dose distributions in regions of high dose gradient in radiation fields. The three radiochromic films most commonly used in medical applications have Been GafChromic MD-55 (Nuclear Associates Model No. 37-041), GafChromic HD-810 (Nuclear Associates Model 37-040), and GafChromic DM-1260. The MD-55 films are available in 5″ × 5″ clear sheets and are suitable for dose measurements in the range of 10–100 Gy. Model HD-810 films were available until recently in 8″ × 10″ sheets and were used for measurements of absorbed dose in the range of 50–2500 Gy. Model DM-1260 films were available in 5″ × 5″ rolls and also were used for dose mapping in the range of 50–2500 Gy.

Radiochromic film has significant advantages over silver halide film: it has a relatively flat energy response; it is self-developing so it eliminates variations introduced by the processing step; it is insensitive to visible light, allowing for ease of handling; and the film is fabricated from low-atomic-number materials, so it does not perturb the radiation beam to the same degree as silver halide film. The response of the film has been shown to be independent of dose rate. The uniformity of the film means that no background measurements need to be taken prior to irradiation. The film is sensitive to ultraviolet light with photon energies greater than 4 eV. [7]

Unexposed radiochromic film is transparent with a light blue hue, while film which has been exposed to ionizing radiation is dark blue. The blue coloration means that the broadband light source of a conventional densitometer is attenuated only in the red part of the spectrum. This leads to an overall reduction in the light intensity which is quite small and, thus, the resulting measurements have a low signal-to-noise ratio. Densitometers employing a He-Ne laser, which emits light at 633 nm, close to the peak of about 660 nm in the absorption spectrum of the film, have been used. This improved the sensitivity and the linearity of the dose response. These densitometers can make measurements with resolutions up to 0.3 μm. [7]

The images produced by the scanner were loaded into the software package MATLAB (The Mathworks, Inc., version 4.2c). This allowed for the representation of the image as a matrix which may be manipulated for the purposes of image enhancement, enlargement, and mapping of isodoses or depth doses.

A calibration to convert from raw scanner signal to dose was achieved by placing 12-mm × 30-mm pieces of film on the surface of a 10-cm-thick Perspex phantom of dimensions 20 cm x 20 cm and irradiating them to doses in the range 0–95 Gy.

The net optical density, OD, of a point on the film is given by the equation

$$OD = \log_{10}\left(\frac{S_0}{S}\right) \qquad (5.5)$$

where S_0 is the "dark current," i.e., the scanner signal for an unexposed film, and S is the scanner signal for the film at the point of interest. The optical density depends on the dose D in the following way:

$$\log_{10}\left(\frac{S_0}{S}\right) = aD^\alpha \qquad (5.6)$$

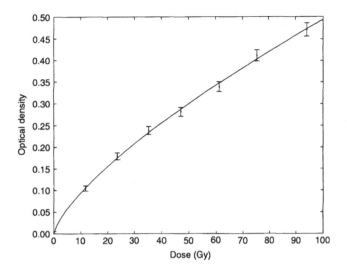

FIGURE 5.14 Curve showing the dependence of optical density on dose. The data points correspond to the optical density of films uniformly irradiated to doses in the range 0–95 Gy, with 55-kV x-rays at the surface of a Perspex phantom. The error bars demonstrate the level of uncertainty in the optical density, based on the uncertainty in measurements of scanner signal. The solid line shows how the model represented by Equation (5.6) fits the experimental data. (From Reference [7]. With permission.)

where a and α are constants. These constants are determined by taking the log of Equation (5.6), resulting in

$$\ln\left[\log_{10}\left(\frac{S_0}{S}\right)\right] = \ln(a) + \alpha\ln(D) \qquad (5.7)$$

A plot of $\ln\left[\log_{10}\left(\frac{S_0}{S}\right)\right]$ against $\ln(D)$ produces a straight line with slope α, intercept $\ln(\alpha)$, and linear regression coefficient $r = 0.999$. A least-squares linear fit to the calibration data yields the constants $\alpha = 0.719$ and $\alpha = 0.0180$. Equation (5.6) is solved for dose and becomes, after substitution of the constants,

$$D = \left[5.56\log_{10}\left(\frac{S_0}{S}\right)\right]^{1.39} \qquad (5.8)$$

The dependence of optical density on dose is shown in Figure 5.14, with Equation (5.6) represented by the solid line. It was shown that the optical density can change by 16% in the 24 h immediately after irradiation, but in the following 24 h the color change is less than 1%. [7]

GafChromic™ Dosimetry Media offers advances in high-dose radiation dosimetry and high-resolution radiography for gamma radiation and electrons. [8] The film consists of either small 1.2-cm × 6-cm strips as individual dosimeters or, for the purpose of large-scale imaging, 0.13-m × 15-m rolls of coated polyester film, or other sizes up to a meter wide and 50 m long. "The sensitive coating is a proprietary sensor layer ≈6 μm thick, and the base is a 100-μm-thick clear polyester film." The film is colorless and "grainless" in that its radiographic image gives a spatial resolution of >1200 lines/mm. Typical radi-

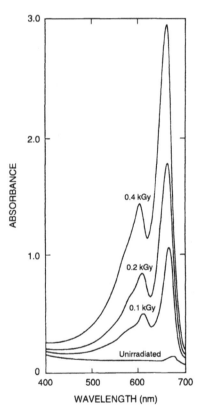

FIGURE 5.15 Absorption spectra of GafChromic™ Dosimetry Media, un-irradiated and irradiated to different doses. (From Reference [8]. With permission.)

ation-induced absorption spectra are shown in Figure 5.15. Doses down to less than 1 Gy can readily be measured by using a special form of the GafChromic™ Dosimetry Media.

The increases in absorbance (ΔA) at four wavelengths are presented in Figure 5.16 as a function of the absorbed dose in water. The four curves are approximately linear functions within the dose range up to 0.5 kGy, but they become nonlinear over the range from 0.5 to 25 kGy (Figure 5.17; McLaughlin et al. [8]).

Figure 5.18 shows hypsochromic shift with storage time for large doses, where measurements have to be made in the near ultraviolet or near-infrared because of unmeasurably high absorbance values in the visible spectrum. The readings on the long wavelength side are more stable than those in the blue and UV part of the spectrum. Long-term stability of the readings of the films at all wavelengths after a one-day initial storage period remains within 4%, when the storage is under controlled laboratory conditions. This long-term stability is observed at all doses from 10 to 1200 Gy.

Figure 5.19 reveals that there are slight shifts to shorter wavelengths of the absorption band maxima as the absorbed dose increases.

Vatnitsky et al. [9] have studied the feasibility of radiochromic film for dosimetry verification of proton Bragg peak stereotactic radiosurgery with multiple beams. High-sensitivity MD-55 radiochromic film was calibrated for proton beam irradiation, and the RIT 113 system was employed for film evaluation.

In order to take into account geometry-related uncertainties, the aperture of each proton beam should include a margin around the target, and the bolus should be "expanded" or "smeared." The choice of margins and bolus smear depends on the treatment technique and varies from center to center.

High-sensitivity MD-55 film consists of a 30-μm polymer sensor layer coated on a 100 μm polyester base and sandwiched between two plastic sheets. 5″ × 5″ film sheets were used and stored in black light-tight plastic envelopes at controlled room temperature (22°C) and humidity (50%). Unpacking, cutting the film to the desired size, irradiation, and evaluation of the films were performed under normal room light. Un-irradiated films were

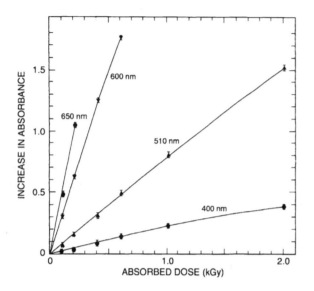

FIGURE 5.16 Curves representing the response of GafChromic™ Dosimetry Media to gamma radiation from a ^{60}Co source, when the absorbance values are measured at the indicated wavelengths with a spectrophotometer, as a function of the absorbed dose in water. The vertical bars on each data point indicate the estimated limits of random uncertainty at a 95% confidence level (2σ). (From Reference [8]. with permission.)

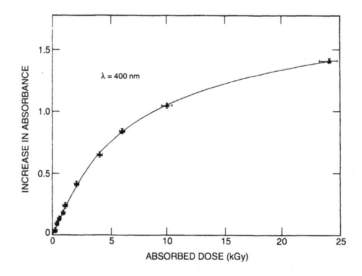

FIGURE 5.17 Response curve for gamma radiation doses up to 25 kGy, when measured with a spectrophotometer at 400-nm wavelength. The vertical bars indicate the 2σ (95% confidence level) uncertainty to the random values of net absorbance at each dose (in water), and the horizontal bars give the corresponding 2σ uncertainty limits of the absorbed dose readings. (From Reference [8]. With permision.)

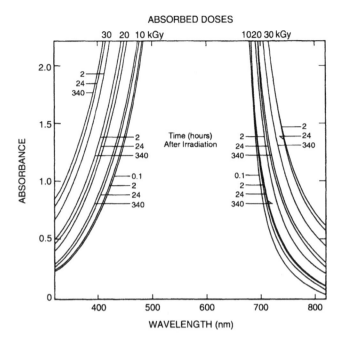

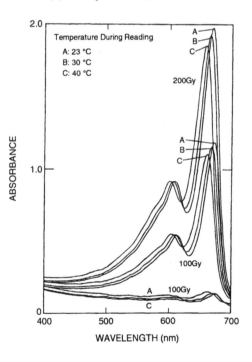

FIGURE 5.18 Changes in the absorption spectra of GafChromic™ Dosimetry Media exposed to high doses of gamma radiation 10, 20, and 30 kGy, and measured at different times after irradiation. (From Reference [8]. With permission.)

handled under the same conditions to control the background optical density. Evaluation of all films was performed 24 h after irradiation.

Irradiated films were processed with the RIT 113 film dosimetry system (RIT Inc., Denver, Colorado). The system configuration includes a digitizer (Argus II) and software for film analysis, which consists of an image-processing platform and various application modules. The Argus II digitizer uses a daylight fluorescent light lamp and CCD (charge coupled device). The CCD's linear sensor has a maximum resolution of 600 horizontal × 1200 vertical dpi (dots per inch). The RIT 113 system is typically used in the following manner. First, calibration films exposed to various doses are scanned in the film digitizer to convert the film to a sequence of pixels whose values represent the optical density at each point on the film. A calibration file is generated to provide the system with the information necessary to convert optical densities on the film to radiation dose levels. Second, subsequent films are exposed to the radiation beam, to qualify the radiation beam characteristics. These films are then digitized and analyzed by the software. [9] Figure 5.20 shows the calibration dose response of the MD-55 film to the irradiation at the center of the spread-out Bragg peak (SOBP) for two beam energies as a function of the absorbed dose to water.

For the five-beam irradiation study, the detector block with radiochromic films was installed in the water phantom so that the intersections between the film planes and CT planes were parallel to the vertical CT axis. Potential errors associated with the angulation of the film were minimized by cutting each film exactly to the shape of the

FIGURE 5.19 Absorption spectra of Gafchromic™ Dosimetry Media, unirradiated and irradiated to the two indicated doses, at indicated temperatures. (From Reference [8]. With permission.)

plastic spacer and aligning the detector block according to labeled marks on the water phantom. The positioning uncertainty of the isocenter projection on each film was estimated to be less than 1 mm.

Figure 5.21 shows calculated and measured dose profiles for the five-beam composite plan in the isocenter

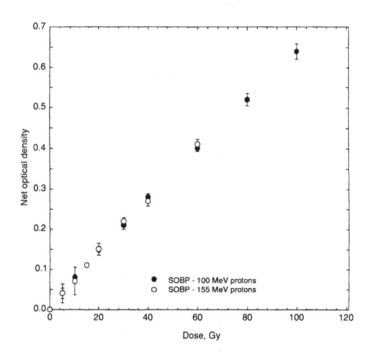

FIGURE 5.20 Calibration response of MD-55 film to proton irradiation at the center of the SOBP. (From Reference [9]. With permission.)

plane (A) and in the plane 3 cm superior to the isocenter (B). The multiple-beam profiles were normalized to a dose of 70 Gy. The comparison presented in Figure 5.21 demonstrates that the treatment plan provided adequate target coverage and that the dose delivered to the reference point (center of the target at isocenter plane) was close to the prescribed dose within accuracy of film dosimetry ($\pm 5\%$). The absolute dose difference between calculated and measured dose values over the target irradiated with multiple beams was also found to be within the uncertainty of the radiochromic film dosimetry ($\pm 5\%$, one standard deviation) for the isocenter plane and off-center planes. The observed difference between measured and calculated profiles can be explained by the dependence of the radiochromic film sensitivity at a very steep distal edge of the Bragg peak.

Zhu et al. [10] have studied the uniformity, linearity, and reproducibility of a commercially supplied radiochromic film (RCF) system (Model MD-55). Optical density (OD) distributions were measured by a helium-neon scanning laser (633 nm) 2D densitometer and also with a manual densitometer. All film strips showed 8–15% variations in OD values, independent of densitometry technique, which are evidently due to nonuniform dispersal of the sensor medium. A double exposure technique was used to solve this problem. The film is first exposed to a uniform beam, which defines a pixel-by-pixel nonuniformity correction matrix. The film is then exposed to the unknown dose distribution, rescanned, and the net OD at each pixel corrected for nonuniformity. The double exposure technique reduces OD/unit dose variation to a 2–5% random

fluctuation. RCF response was found to deviate significantly from linearity at low doses (40% change in net OD/Gy from 1 to 30 Gy).

The film consisted of a thin, radiosensitive (7–23 μm thick) colorless leuco dye bonded to a 100-μm-thick Mylar base. RCF films are colorless before irradiation and turn a deep blue color upon irradiation without physical, chemical, or thermal processing. RCF is approximately tissue-equivalent, integrates simultaneously at all measurement points, and has a high spatial resolution (>1200 lines/mm). It shows a stable, reproducible response if protected from UV light and unstable temperatures and humidity. Since radiochromic dye is an aromatic hydrocarbon, like plastic scintillator, its deviation from tissue-equivalent energy response is much smaller compared to silicon diode detectors and has the same magnitude (but opposite direction) as that of TLD.

To suppress the effects of nonuniform RCF response, a double-exposure densitometry technique was used. Before use, each film was first given a uniform dose of 20 Gy or 40 Gy, and fiducial marks were placed on two opposite corners of the film using a punch with a diameter of 600 μm. After a time interval ranging from 48 to 96 h from irradiation, the film was then scanned and the resultant optical density distribution, $OD_1(i, j)$, was used to form a pixel-by-pixel sensitivity correction matrix, $f(i, j)$.

$$f(i, j) = \frac{OD_1(i, j)}{\langle OD_1(i, j) \rangle} \qquad (5.9)$$

where i, j are the X and Y position indices and $\langle OD_1(i, j) \rangle$ is the mean OD over the entire film. This pre-exposed film

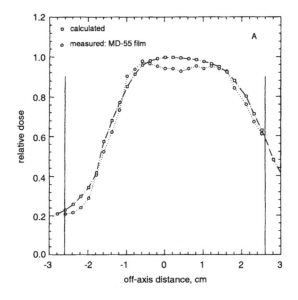

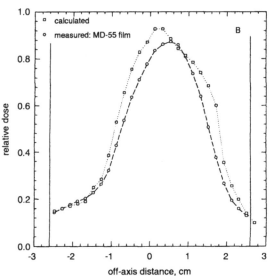

FIGURE 5.21 A comparison of measured and planned dose profiles for five-beam irradiation. Vertical lines represent boundaries of a sliced polystyrene cylinder with films in the detector block. (A), a comparison of measured and planned dose profiles for five-beam irradiation, isocenter plane. (B), a comparison of measured and planned dose profiles for five-beam irradiation, CT plane, 3 cm superior to the isocenter. (From Reference [9]. With permission.)

is then exposed to the unknown dose distribution, held for an interval of 48–108 h, and scanned, yielding the cumulative optical density distribution, $OD_2(i, j)$. The time interval between the OD_1 scan and exposure to the unknown radiation field was highly variable, ranging from a few hours to several weeks. Both OD_1 and OD_2 images are then transferred to the graphics workstation. There, a software package displays the two images, allows corresponding fiducial marks on the two films to be identified, correlates the double- and single-exposed images pixel-by-pixel by

aligning the fiducial marks, and then calculates and displays the net optical density distribution, $\Delta OD(i, j)$ corrected for pixel-to-pixel sensitivity variations:

$$\Delta OD(i, j) = \frac{OD_2(i, j) - OD_1(i, j)}{f(i, j)} \quad (5.10)$$

The corrected net optical density, $\Delta OD(i, j)$, is taken to be the response of the detector to the unknown radiation field. Pixel-by-pixel correction for nonuniformity of film response was performed using the original OD matrix consisting of 50-μm pixels. Films were always scanned with the sensitive side of the RCF emulsion facing the ceiling to prevent the two OD images from being flipped.[10]

The response of RCF with respect to absorbed dose in a 6-MV x-ray beam for both single- and double-exposure techniques was studied by Zhu et al. using the same films and OD distributions as were used for interfilm reproducibility analysis. For each of the dose groups, the mean of the average optical ΔD density of each film was taken, yielding the final mean optical density $\langle \overline{\Delta OD} \rangle$ corresponding to the net dose, ΔD. The deviation of RCF response from linearity was quantitated by means of a relative linearity correction factor F_{lin}, defined as

$$F_{lin}(\Delta OD) = \frac{[\langle \overline{\Delta OD} \rangle / \Delta D]}{[\langle \overline{\Delta OD} \rangle / \Delta D]} \quad \begin{array}{l} \text{At each dose level} \\ \text{At a reference of 30 Gy} \end{array}$$

$$(5.11)$$

Zhu et al. chose the 30-Gy dose level for normalizing F_{lin} to unity because it represents the center of the dose range typically used in their RCF dose measurements; however, its choice is completely arbitrary and affects neither the accuracy nor precision of subsequent dose measurements. The linearity correction factor of RCF response using the Macbeth manual densitometer was defined as

$$F_{lin}(\Delta OD) = \frac{[\langle \overline{OD} \rangle / D]}{[\langle \overline{OD} \rangle / D]} \quad \begin{array}{l} \text{At each dose level} \\ \text{At a reference of 30 Gy} \end{array}$$

$$(5.12)$$

where $\langle \overline{\Delta OD} \rangle$ represents the single-exposure OD averaged over manual readings measured over a 3 × 3-mm grid from each film averaged over the four films in each dose group with background density subtracted. For doses over 20 Gy, the cumulative OD_2 densities were measured and plotted as a function of D_2. To measure the sensitometric curve at lower doses, a series of unexposed films were irradiated to the desired doses and readout as described above.

Figure 5.22 shows a typical OD profile through the center of a 12-cm-long strip, illustrating the sinusoidal behavior of its nonuniform response. The upper panel shows the profile resulting from a uniform dose of 40 Gy, while the center panel illustrates the effect of an additional 20-Gy exposure. This demonstrates that the nonuniformity is a multiplicative phenomenon, independent of the dose delivered. Finally, the lower panel of Figure 5.22 illustrates that nonuniform film response is dramatically reduced by the double-exposure method. An 11% systematic variation of film response is reduced to a random OD fluctuation, which is less than 5%.

Figure 5.23 shows the results of interfilm reproducibility for both the single and double-exposure techniques. For both techniques, the average OD in the central 5×5 block of 50-μm pixels was extracted from each film and then averaged over the four films in each dose group for doses ranging from 1 Gy 230 Gy. For the double-exposure technique, an initial 20-Gy exposure was used to define the relative pixel sensitivity correction. The graphs also show film response as a function of absorbed dose. For the single-exposure technique, the relative standard deviation is as large as 11% at low doses and stabilizes at a value of 4% for doses in excess of 30 Gy. For the double-exposure technique, the relative standard deviation is as high as 9% at low doses and rapidly falls to a plateau of 2% at 30 Gy. This result indicates that for four repeated readings using separately exposed films, the double-exposure technique should be able to characterize radiation fields with an experimental precision of $\pm 2\%$ (95% confidence interval) when exposed to doses of at least 20 Gy.

GafChromic MD-55 was irradiated with ^{60}Co γ-rays by Klassen et al. [11] A double irradiation method was used in which a dosimeter is given an unknown dose followed by a known, calibration dose. It was found that the measured optical density of GafChromic MD-55, as currently fabricated, is affected by the polarization of the analyzing light, an important consideration when using GafChromic MD-55 as a precision dosimeter.

The thin, radiosensitive layers in GafChromic MD-55 are made of a colorless gel containing polycrystalline substituted–diacetylene monomer, which undergoes partial polymerization when irradiated with ionizing radiation such that the blue color of the polymer becomes progressively darker as the dose increases. The ΔOD_u due to the polymer increases roughly linearly with absorbed dose. Figure 5.24a shows the optical absorption spectrum of GafChromic MD-55 from 600 nm to 700 nm, before and after a dose of 6 Gy. Figure 5.24b, with its expanded scale of OD and wavelength, shows the spectrum of an unirradiated film from 700 nm to 720 nm taken with a bandpass of 1.0 nm. The oscillations in the OD are interference fringes which constitute a channel spectrum. [11]

Some of the apparent absorption is actually a loss of light due to reflection, not absorption. Using the equation

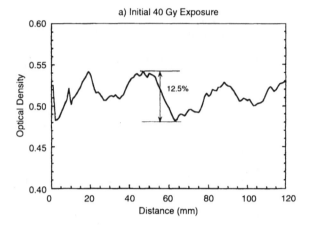

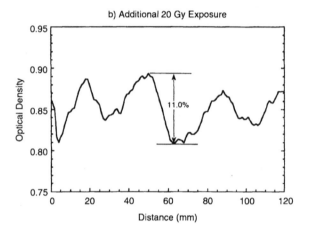

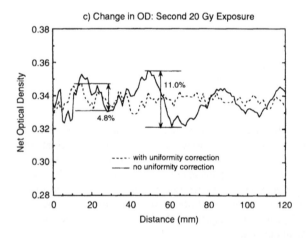

FIGURE 5.22 Central line OD profiles from a single film. Graph (a) shows the results of a single 40-Gy exposure. The middle figure, graph (b), shows the profile resulting from an additional dose of 20 Gy. The lower figure gives the net OD profile [derived by pixel-by-pixel subtraction of graph (a) from graph (b)], with (dashed line) and without (solid line) the pixel-by-pixel uniformity correction derived from the initial 40-Gy exposure. (From Reference [10]. With permission.)

for reflective losses,

$$R = [(\eta_{medium} - \eta_{air})/(\eta_{medium} + \eta_{air})]^2 \qquad (5.13)$$

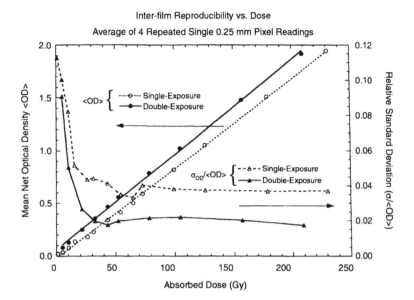

FIGURE 5.23 Dose response curve of RCF to 6-MV x-rays at doses ranging from 1 Gy to 230 Gy for single- and double-exposure techniques plotted on the left-hand scale. Interfilm reproducibility (relative standard deviation of the central effective pixel readings from the four film pieces) is plotted on the right-hand axis of the graph. (From Reference [10]. With permission.)

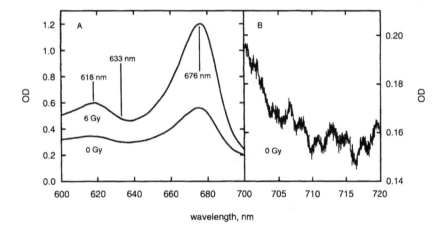

FIGURE 5.24 (a) The optical absorption spectrum of un-irradiated GafChromic MD-55 (lower spectrum) and GafChromic MD-55 several weeks after a dose of about 6 Gy (upper spectrum). The bandpass used was 3.5 nm. Some useful wavelengths are indicated. (b) The spectrum of un-irradiated GafChromic MD-55, taken using a bandpass of 1 nm and displayed at a higher sensitivity than in (a) in order to demonstrate the periodic fluctuations due to interference fringes (channel spectrum). (From Reference [11]. With permission.)

where R is the reflectivity, $100R$ is the percent reflection, η_{medium} is the refractive index of the medium (taken to be 1.5 for Mylar), and η_{air} is the refractive index of air (1.0). Klassen et al. calculated that the reflection from each Mylar/air surface leads to a 4.0% light loss, or 8.0% from the two such surfaces, which corresponds to an OD of 0.036. There are seven layers in GafChromic MD-55, meaning eight reflecting surfaces.

The temperature dependence of OD at 676 nm was measured in two studies using six dosimeters that had received 0, 1.0, 3.5, 6.9, and 14.5 Gy. Readings were taken at seven temperatures between 18.8 and 28.1°C.

By returning to the initial temperature several hours later, it was found that the OD did not change irreversibly during the measurements. The results, displayed for 673, 674, 675, 676, 677, and 678 nm in Figure 5.25, show that at 22°C the temperature dependence is smallest at 675 nm, only 0.0002 OD units for a 0.1 degree change in temperature between 20°C and 24°C. At 24°C, a wavelength of 674 nm has the smallest dependence of OD on temperature. Figure 5.25 shows that poor control of the spectrophotometer temperature can be countered by selecting the wavelength with the smallest temperature dependence.

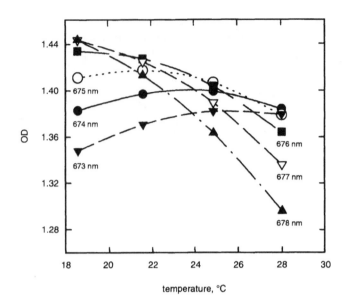

FIGURE 5.25 A plot of the OD of a single dosimeter, which had received 6.9 Gy as read at a variety of temperatures and wavelengths. The measurements were made several months after the irradiation. (From Reference [11]. With permission.)

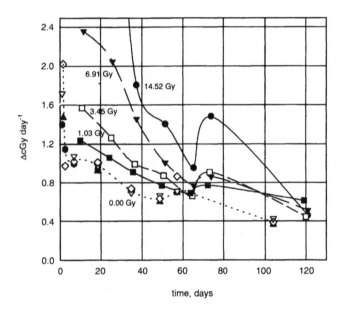

FIGURE 5.26 The rate of change of optical density of un-irradiated and irradiated GafChromic MD-55 dosimeters where ΔOD_u is expressed in terms of dose (assuming that 1 Gy = 0.11 OD units). Time zero for the irradiated dosimeters was the end of the irradiation. The dose to each dosimeter is displayed on the figure. A dotted line is given as a visual aid to the time dependence of the four un-irradiated dosimeters for which time zero was the time they were cut from the sheet. A jump in the rate of change, most easily seen at 65 days for the dosimeter that received 14.52 Gy, occurred over the same dates for all the dosimeters, time zero having different dates for the different dosimeters. (From Reference [11]. With permission.)

The change in OD was measured vs. time for a number of dosimeters, some un-irradiated and others irradiated to a variety of doses up to 14.52 Gy. Over this dose range, 1 Gy corresponds to about 0.11 OD units. Assuming 0.11 OD units equals 1Gy, the rate of change of OD is shown in Figure 5.26 as the equivalent rate of change in cGy day^{-1}. (ΔOD_u is the change in absorbance due to polymer formation.)

GafChromic MD-55 is composed of seven layers as illustrated in Figure 5.27. Either outer Mylar layer could be pulled off the gel by flexing the dosimeter under liquid nitrogen and pulling the Mylar layer off or, alternatively, by soaking the dosimeter in water overnight and pulling it apart at the Mylar/gel interface.

In order to optimize the measurement sensitivity and thus improve precision, Reinstein et al. [12] described a

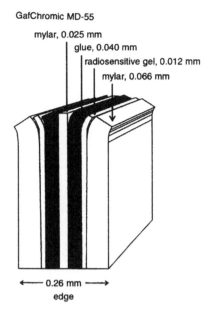

GafChromic MD-55

mylar, 0.025 mm

glue, 0.040 mm

radiosensitive gel, 0.012 mm

mylar, 0.066 mm

←—— 0.26 mm ——→
edge

FIGURE 5.27 A schematic view of the structure of GafChromic MD-55, indicating the seven layers and their thicknesses. The uncertainty in the measurement of the thicknesses was estimated to be about ±0.001 mm for the hard Mylar layers and ±0.002 mm for the soft, gel, and glue layers. (From Reference [11]. With permission.)

method to calculate the dose response curves (net optical density at a given wavelength or spectrum vs. absorbed dose) for different densitometer light sources using measured GC film absorption spectra. Comparison with measurements on the latest version of GC film (Model MD-55-2) using four types of densitometers [He-Ne laser, broadband (white light) densitometer, and two LED (red-light) filtered densitometers] confirm the accuracy of this predictive model. They found that an LED (red-light) source with a narrow band-pass filter centered at 671 nm near the major absorption peak achieves nearly the maximum possible sensitivity (almost four times more sensitive than He-Ne laser, 632.8 nm) and may be suitable for in vivo dosimetry.

For LED light source coupled to a band-pass filter, the resulting intensity as a function of wavelength after having passed through the filter is

$$I_1(\lambda) = I_0(\lambda)10^{-A_{filter}(\lambda)} \quad (5.14)$$

where $I_0(\lambda)$ is the intensity of the light source (LED) at wavelength λ incident on the filter; $I_1(\lambda)$ is the intensity after having passed through the filter; and $A_{filter}(\lambda)$ is the absorbance of the filter at wavelength λ. For the case of the broadband (white light) light source, $I_0(\lambda)$ was determined from the temperature of the tungsten filament, an assumed blackbody radiation spectrum, and, for mathematical convenience, $A_{filter}(\lambda) = 0$ (no filter). By coupling

the light source $I_1(\lambda)$ with the absorption spectra of GafChromic film, the intensity reaching the detector can be written as

$$I_2(\lambda, D) = I_1(\lambda)10^{-A_{GAF(\lambda,D)}} \quad (5.15)$$

where $A_{GAF}(\lambda, D)$ is the absorbance of GafChromic film at wavelength λ, irradiated to a dose D. Optical density as a function of dose $[OD(D)]$ for GC film was calculated by combining the equations above to yield

$$OD(D) = -\log\left\{ \int_0^\infty I_0(\lambda)10^{-A_{filter}(\lambda)}10^{-A_{GAF}(\lambda,D)} \times \right.$$

$$\left. d\lambda \middle/ \int_0^\infty I_0(\lambda)10^{-A_{filter}(\lambda)}d\lambda \right\} \quad (5.16)$$

By subtracting the optical density of an un-irradiated sample of GafChromic film, the net optical density (NOD) as a function of dose was obtained (dose response curve) for each densitometer light source. Note that the model assumes a uniform response by the detector for the range of wavelengths measured. The transmission spectra for the two band-pass filters used in our measurements with the filtered LED densitometers are shown in Figure 5.28. Absorption spectra of GC film exposed to total doses 0, 10, 20, 30, 50, and 100 Gy and measured at 31.9 days post-irradiation are shown in Figure 5.29. The minor and major absorption peaks can clearly be seen. At 30 Gy exposure, these peaks are centered at 614 nm and 674 nm, respectively, with peaks shifting to lower wavelengths with increasing dose.

Figure 5.30a plots the predicted dose response curves at fixed wavelengths 400 nm, 510 nm, 632.8 nm, 650 nm, and 671 nm as calculated from the measured absorption spectra of GC film (MD-55-2). It is observed that the dose response curve measured at 671 nm (near the major peak) is the most sensitive with a DNODI = 14 Gy. The least sensitive dose response curve is at 400 nm (not near either of the absorption peaks), where a net optical density of 1 is not achieved, even for a dose as high as 100 Gy. Of the fixed wavelengths, the dose response curve at 650 nm (on the left shoulder of the major peak) exhibits the greatest linearity, with a DNODI = 34 Gy. Also shown is the dose response curve for a fixed wavelength of 632.8 nm, which can be compared with the measured data from the He-Ne laser densitometer. The predicted and measured results show excellent agreement (DNODI = 56 Gy for both). [12]

The dose response of high-sensitivity GafChromic film to photons from ^{125}I seeds for doses up to 200 Gy

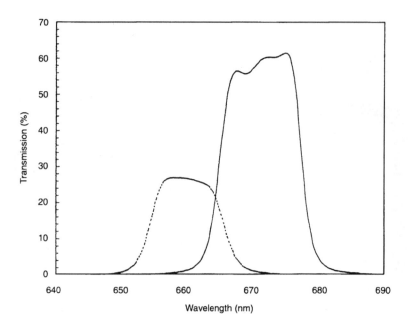

FIGURE 5.28 Transmission spectra for the 660 nm filter (dashed line) and 621 nm filter (solid line) measured using the Du 640 B spectrometer. (From Reference [12]. With permission.)

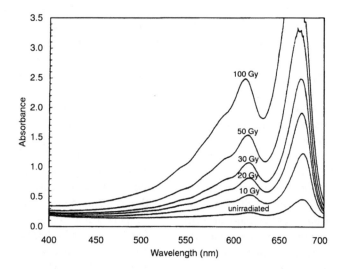

FIGURE 5.29 Absorbed spectra of GafChromic film (MD-55-2) for the dose levels 0, 10, 20, 30, 50, and 100 Gy. The spectra were measured on a Beckman DU 640 B spectrometer. (From Reference [12]. With permission.)

was established by Chiu-Tsao et al. [13] The net optical density was found to be a power function of dose with exponents of 0.858 and 0.997 for the Macbeth and LKB densitometers, respectively. Film sensitivity with the LKB densitometer was about double that with the Macbeth densitometer.

The dose-response curves of the high-sensitivity GafChromic films were plotted for ^{125}I, ^{137}Cs, and ^{60}Co and shown in Figures 5.31a, b, and c, respectively, for doses up to 200 Gy. The curves for the same radionuclide obtained with the LKB and Macbeth densitometers were compared to the figure. The net OD increases with

increasing dose. The values with the LKB were about twice those with the Macbeth, depending on the dose. This means that the film sensitivity measured with red light from the He-Ne laser is higher than that obtained with the broadbeam spectrum. The data obtained with the Macbeth densitometer were fitted to a power function of dose, with exponents of about 0.858, 0.839, and 0.849 for ^{125}I, ^{137}Cs, and ^{60}Co, respectively. As for the LKB densitometer, the curves were linear for ^{137}Cs and ^{60}Co. The curve for ^{125}I, however, was better-fitted to a power function of dose with a power of 0.997, i.e., essentially linear.

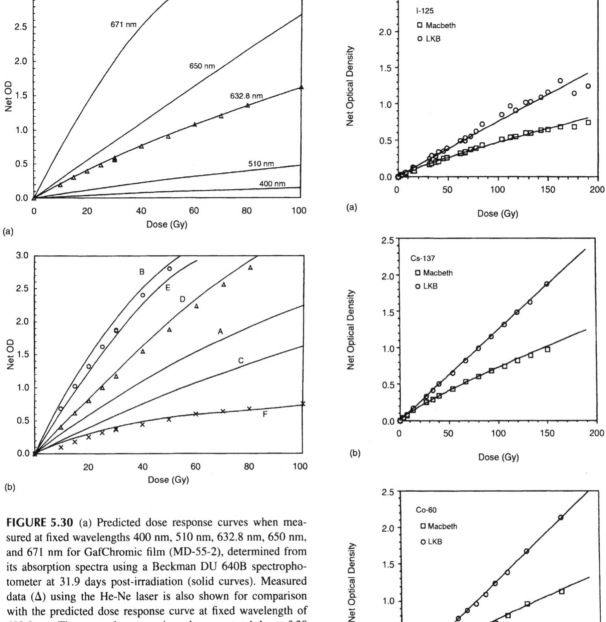

FIGURE 5.30 (a) Predicted dose response curves when measured at fixed wavelengths 400 nm, 510 nm, 632.8 nm, 650 nm, and 671 nm for GafChromic film (MD-55-2), determined from its absorption spectra using a Beckman DU 640B spectrophotometer at 31.9 days post-irradiation (solid curves). Measured data (Δ) using the He-Ne laser is also shown for comparison with the predicted dose response curve at fixed wavelength of 632.8 nm. Three samples were given the same total dose of 30 Gy and used to check for measurement reproducibility. (b) Predicted dose response curves (solid curves) for a tunable light source to the minor peak (A), major peak (B), and for valley between peaks (C), and for complex light sources: an LED light source coupled to a 660-nm band-pass filter (D), an LED light source coupled to a 671-nm filter (E), and a broadband light source (F). Measured dose response curves are also shown for comparison, namely a 660-nm filtered densitometer (Δ), 671-nm filtered densitometer (O) and a broadband densitometer (x). Three samples were given the same total dose of 30 Gy and were used to check for measurement reproducibility. (From Reference [12]. With permission.)

FIGURE 5.31 The dose-response curves of the high-sensitivity GafChromic film for three radionuclides,[125]I, ^{137}Cs, and ^{60}Co, in a, b, and c, respectively. The square and circle symbols are for the reading using Macbeth and LKB densitometers, respectively. The solid curves represent the corresponding function fit. (From Reference [13]. With permission.)

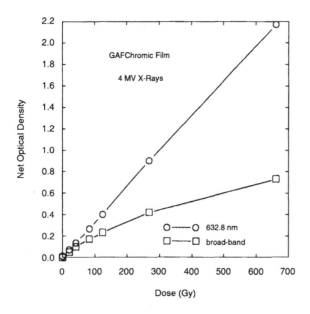

FIGURE 5.32 Dose response curves for radiochromic film. (From Reference [14]. With permission.)

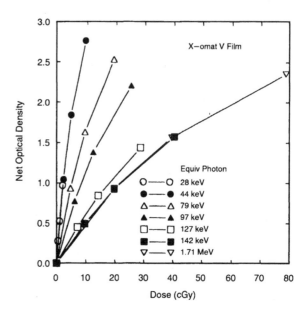

FIGURE 5.34 Family of dose-response curves for silver halide film. Each curve is for one x-ray beam (one equivalent photon energy). (From Reference [14]. With permission.)

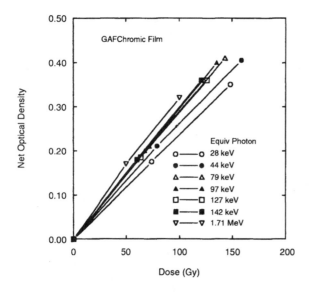

FIGURE 5.33 Family of dose response curves for radiochromic film. (From Reference [14]. With permission.)

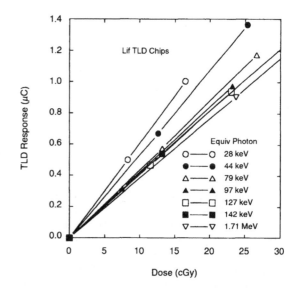

FIGURE 5.35 Family of dose-response curves for LiF TLD chips. Each curve is for one type of x-ray beam (one equivalent photon energy). (From Reference [14]. With permission.)

The dose response of the radiochromic film for 4-MV x-rays is shown in Figure 5.32. [14] The x-rays were produced by a Clinac 4 linear accelerator, and the film was irradiated at a depth of 1.25 cm in an acrylic phantom set up at an SSD of 41 cm. The dose rate was about 900 cGy/min. The lower curve shows the results from optical density measurements using broadband absorption, and the upper curve shows the results of measurements using 632.8-nm light.

A family of dose-response curves for radiochromic film for different photon energies is shown in Figure 5.33. These curves are in the low-dose range for this film, although the

doses are about 250 times the doses used for the silver halide film curves (Figure 5.34) and about 600 times the doses used for the LiF chip curves (Figure 5.35). The curves show linearity of dose response, and the similarity of slopes shows that the dependence on photon energy is slight. The optical densities are for 632.8-nm light.

An "improved film" was developed with higher sensitivity and better uniformity than the model MD-55 films by Meigooni et al. [15] The improved film is a multilayered structure composed of a nominal 30-μm thickness of the

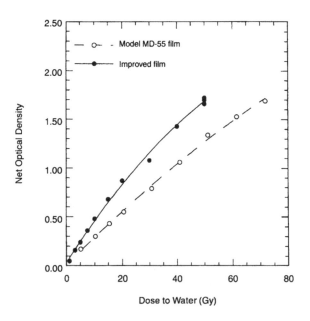

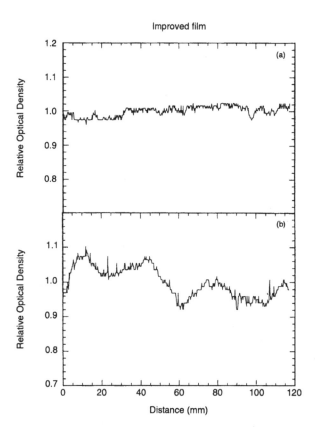

FIGURE 5.36 Sensitometric curves of the improved radiochromic film (solid symbol) compared with that of MD-55 film (open symbol). These results were obtained by irradiating the films to doses ranging from 1–72 Gy of gamma rays from a ^{60}Co teletherapy unit. (From Reference [15]. With permission.)

FIGURE 5.37 Relative optical density of a uniformly irradiated improved radiochromic film (a) measured in one direction along the film and (b) measured along the orthogonal direction. (From Reference [15]. With permission.)

sensitive material, sandwiched between two pieces of 75-μm-thick polyester base material and two pieces of 13 μm laminated material, yielding a total thickness of 206 μm. In contrast, MD-55 film is composed of a layer (23-μm-thick) of sensitive material on one side of a 100-micron polyester base. These films are colorless and transparent before being exposed to ionizing radiation. As they are irradiated, their color changes to blue, due to the polymerization of radiochromic dye. The darkness of the blue color depends on the absorbed dose and can be measured with a laser densitometer with a wavelength ranging from 610–675 nm.

Figure 5.36 shows the measured sensitometric curves of the improved radiochromic film and MD-55 film as a function of absorbed dose. This figure indicates that the improved film requires about 40% less dose than MD-55 film to yield an optical density of 0.5. This difference is reduced to about 30% for an optical density of about 1.5. The combination of all sources of uncertainties in these measurements is estimated to be less than ±5%. This uncertainty includes the reproducibility of the laser scanning system, uncertainty of determination of absorbed dose given to the film, and standard deviation of the optical density measured along a path of about 5 mm in a uniformly exposed film.

The uniformity of the improved radiochromic film and MD-55 film along two orthogonal directions was determined by measuring the optical density of a film irradiated with a uniform radiation field. Figure 5.37a shows very good uniformity (within 4%) along one direction of the

improved radio-chromic film. A similar result was obtained at a distance of 5 cm from the beam axis, along the same direction of the film. However, along the perpendicular direction (Figure 5.37b), a non-uniformity of up to 15% was observed. This non-uniformity is not consistent from one sample of film to another.

Figure 5.38 shows the variation of relative optical density of the improved radiochromic film and MD-55 film as a function of time after irradiation for absorbed dose as shown in the figures.

Evaluation of the influence of irradiation temperature differences on the response of three widely used radiochromic film dosimeters was made by McLaughlin et al. [16] Freestanding nylon base film (mean thickness 48 μm) FWT-60-00, thin-coated (~7 μm) sensor on 100-μm polyester base GafChromic DM-1260 dosimetry media, and freestanding polyvinylbutyral base film (mean thickness 23 μm) Risø B3 were investigated.

The irradiations were made with gamma-rays from two ^{60}Co sources, one at an absorbed dose rate of 12.2 kGy/h and the other at 30.0 kGy/h. During irradiation the films (three films for each irradiation) were held between two electron equilibrium layers of 3-mm-thick polystyrene and sealed at 50% relative humidity in a triple layer pouch of

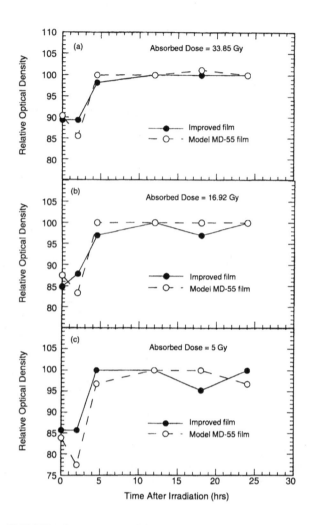

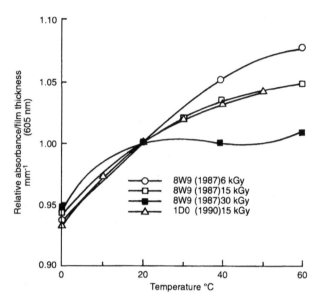

FIGURE 5.39 Temperature dependence, relative to that at 20°C, of the gamma-ray response of two batches of radiochromic film FWT-60-00. Batch 8W9 was irradiated at three absorbed doses (6, 15, and 30 kGy) and batch 1D0 was irradiated at 15 kGy. The absorbances per unit thickness ($\Delta A\ mm^{-1}$) were measured at 605-nm wavelength. (From Reference [16]. With permission.)

FIGURE 5.38 Variation of the optical density with time after irradiation for the improved radiochromic film (solid symbol) compared to MD-55 film (open symbol). These results were obtained by simultaneous irradiation of pieces of each film-type absorbed doses of 5, 16.92, and 33.85 Gy using a ^{60}Co teletherapy unit. (From Reference [15]. With permission.)

30-μm-thick polyethylene–aluminum foil laminate. Each packet was conditioned for 30 min at each irradiation temperature before irradiation, then irradiated, and then stored at room temperature immediately after irradiation.

Figures 5.39, 5.40, and 5.41 show, respectively, the temperature dependence of the gamma-ray responses of the three most recent batches of radiochromic films, FWT-60-00 (batches 8W9-1989 and 1D0-1990), GafChromic 1260 (batch 09031501-1989), and Risø B3 (batch 85-128-1989). In all three cases, it is seen that a different trend occurs at different absorbed doses, particularly at the higher temperatures.

A technique of readout was developed by Kellermann et al. [17] to measure the optical density distributions of the film in purely directed light. This technique implements radiochromic film dosimetry near the film's absorption maximum by using a single-mode top-surface

emitting laser diode (675.2 nm). The effective sensitivity of the film, compared with a helium-neon laser densitometer (632.8 nm), is increased approximately threefold.

Figure 5.42 shows the optical density of the radiochromic film depending on the wavelength of the analyzing beam after irradiation with four different doses. [18] In the helium-neon laser densitometer, the film is scanned with an analyzing beam and the transmission is measured by a photomultiplier. The emission wavelength (632.8 nm) of the analyzing beam lies in the valley of the film's absorption spectrum. Since the absorption spectrum of the radiochromic film contains two peaks centered at approximately 615 nm and 675 nm, it has been suggested [17] that significant improvement in response might be achieved by tailoring the wavelength of the incident densitometer light to one of these peaks. Densitometers with a light-emitting diode (LED) (660 nm, spectral line width >30 nm) and with an LED coupled to a band-pass filter (670 nm, bandwidth 11 nm, GatChromic densitometer) have been presented, but their spectral line width is too broad to match the absorption maximum exactly, and the improvement in response is reduced.

The radiochromic film's sensitive layer consists of microcrystals of radiation-sensitive monomer uniformly dispersed in a gelatin binder. When the microcrystals are exposed to ionizing radiation, polymerization occurs and the polymers alter the crystal color to various shades of blue. The optical density increases continuously with the absorbed dose. The optical density depends neither

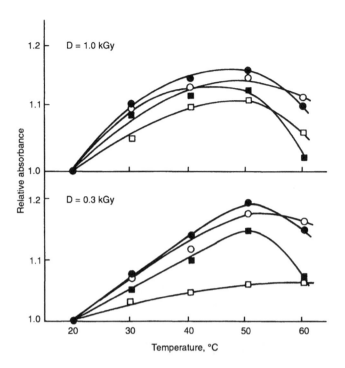

FIGURE 5.40 Temperature dependence, relative to that at 20°C, of the gamma-ray response of radiochromic film GafChromic Dosimetry Media, DM-1260, batch 09031501, irradiated at two absorbed doses. The wavelengths of absorbance measurements: •−650 nm; +−600 nm; X − 510 nm; ◊ −400 nm. (From Reference [16]. With permission.)

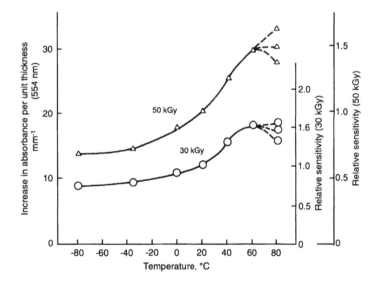

FIGURE 5.41 Temperature dependence of the gamma-ray response of radiochromic film Risø B3, batch 85-128, irradiated at two absorbed doses. The two right ordinates show the response relative to that at 20°C. The absorbances per unit thickness ($\Delta A \ mm^{-1}$) were measured at 554-nm wavelength. (From Reference [16]. With permission.)

on photon energy in the range 0.1 to 20 MeV nor on the energy of secondary electrons in the range 0.01 to 20 MeV.

GafChromic MD-55™ irradiated in a uniform radiation field showed a non-uniformity in optical density of up to 15% in one direction and within 4% in the direction orthogonal to it. This would result in a dose error of ≈10% or ≈2.5%.

A laser diode with an emission wavelength of 675.2 nm was chosen by Kellermann et al. as a light source to transilluminate the radiochromic film. At 675.2 nm, near the film's absorption maximum, a dose gradient in the radiation

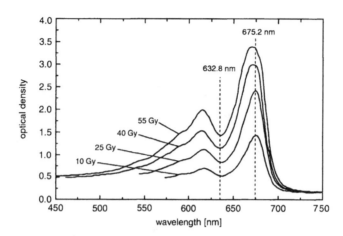

FIGURE 5.42 Optical density of the radiochromic film (Gafchromic MD-55™, Model No. 37-041), depending on the wavelength of the analyzing beam after irradiation with four different doses. [18] The wavelength of the helium-neon, laser (632.8 nm) lies in the valley of the film's absorption spectrum, while the wavelength of the laser diode (675.2 nm) used by Kellermann et al. is near the film's absorption maximum. (From Reference [17]. With permission.)

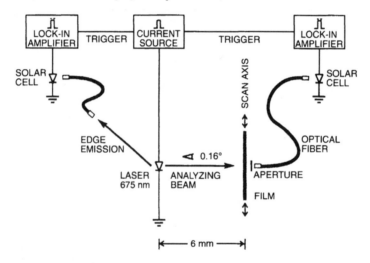

FIGURE 5.43 The schematic measurement setup is sketched. The laser light transmits the film and is sent via an optical fiber to a solar cell, where it is transformed to an electrical signal. This is measured with a lock-in amplifier. The light of the edge emission is used as a reference signal. (From Reference [17]. With permission.)

field leads to a change in optical density that is up to three times larger than the change that is caused by the same dose gradient in the valley (632.8 nm) of the film's absorption spectrum. The laser light propagating in the active region excites a transverse electrically polarized (TE_0) surface mode in the waveguide structure on the top of the laser diode through grating coupling. The surface mode couples both into the vacuum light cone, resulting in top-surface emission with narrow beam divergence, and back to the active region, leading to a gain mechanism and thus to single-mode emission.

The film is fastened (with double-sided adhesive tape) to a stage that allows precise movements along the scanning direction (and orthogonal to it) in 50-μm steps along a distance of 75 mm. The schematic of Figure 5.43 displays the measurement setup.

After irradiation, the optical density of radiochromic film increases logarithmically. Therefore, the optical density measurements with the densitometry system are done four days after the end of irradiation, by which time the changes in optical density are small.

The calibration curve—change in optical density at 675.2 nm vs. radiation dose—is shown in Figure 5.44. The curve rises rapidly, but the upward gradient becomes smaller with higher doses. In comparison, the optical density vs. radiation dose at 632.8 nm is flat but approximately linear. [19] At 675.2 nm, a dose gradient from 0 Gy to 20 Gy in the radiation field leads to a change in optical density that is 3.1 times larger than at a wavelength of 632.8 nm. This ratio decreases to 1.9 with a dose gradient from 20 Gy to 40 Gy.

Film digitizers are, compared with mechanical scanners, much faster, while their spatial resolution is even higher.

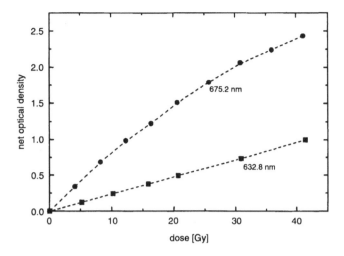

FIGURE 5.44 Calibration curve—change in optical density at 675.2 nm vs. radiation. The curve rises rapidly, but the upward gradient becomes smaller with higher doses. In comparison, optical density vs. radiation dose at 632.8 nm is flat but approximately linear. (From Reference [17]. With permission.)

Most commercially available digitizers are based on either a scanning He-Ne laser beam or a one-dimensional charge-coupled device (CCD) combined with a standard fluorescent lamp. Mersseman and De Wagter [20] investigated a CCD-based digitizer.

The principles of the digitizer are as follows. A broad light beam from the fluorescent lamp (Philips F17T8/TL841) crosses the light-diffusion plate and travels through the scrolling film to be digitized. Two fixed vertical and two movable horizontal shutters subsequently collimate this light beam to a narrow rectangular field that covers the film lengthwise. The transmitted light is subsequently reflected by a stationary mirror, focused by the 40-mm lens and finally detected by the linear CCD element. The CCD detector reads line per line at 300 dots per inch (dpi) over its full length of 35.6 cm (14 in). In order to achieve a 150, 75, and 60 dpi resolution, the film digitizer combines data from nearby pixels. A 12-bit analog-to-digital converter digitizes the signal. Conversion tables (built-in or loaded from the host computer) are allowed to further translate the data. The resulting data matrix is transferred to the host computer (Pentium PC) using an SCSI interface.

Nuclear Associates' GafChromic Dosimetry Media is a radiation-sensitive imaging film designed for use in medical and industrial quality control measurement applications. The media is colorless and grainless and offers a very high spatial resolution (1200 LP/mm), making it invaluable for measuring dose distribution around small brachytherapy sources and stereotactic radiosurgery fields. [21] GafChromic Dosimetry Media MD-55 (Model 37-041) comprises 5″ × 5″ clear film sheets that are suitable for medium-dose measurement applications covering the 1.5 to 50-Gy dose range, when used with the radiochromic densitometer. Higher doses (100 Gy and greater) can be measured using other commercially available densitometers. MD-55 has a higher sensitivity than HD-810 and was developed for use in external-beam radiotherapy and brachytherapy applications. MD-55 is a laminated film composed of two pieces of 2.65-mil polyester base, each with a nominal 15-micron-thick coating.

HD-810 (Model 37-040) films are low-sensitivity 8″ × 10″ clear film sheets suitable for high-dose beam profiling and dose-mapping applications in the 10 to 300-Gy dose range, when used with the radiochromic densitometer. Higher doses (up to 2500 Gy) can be measured using other commercially available densitometers. This media has a nominal 7-micron-thick radiation-sensitive layer on a 3.9 mil polyester base.

GafChromic Dosimetry Media is composed of materials with low atomic numbers, so the film can be considered tissue-equivalent for most radiation therapy applications. Additionally, this material has low sensitivity to ambient room light, which simplifies handling procedures and enhances image stability. Its response is dose rate independent; it is linear dose response and energy independence above 100 keV. [21]

TECHNICAL DATA

MD-55: Polyester Base: 67-μm; Sensitive Layer: 15-μm; Pressure-Sensitive Adhesive Layer: 44.5 μm; Polyester Base: 25 μm; Pressure-Sensitive Adhesive Layer: 44.5 μm; Sensitive Layer: 15 μm; Polyester Base: 67 μm.

HD-810: Sensitive Layer: 7 μm; Adhesive Layer: 1.5 μm; Conductive Layer: 0.05 μm; Polyester base: 99 μm.

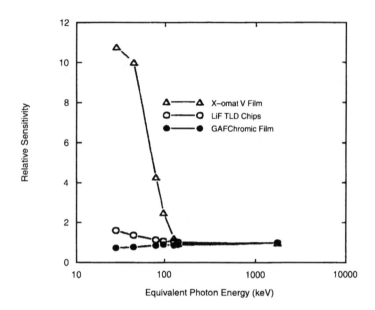

FIGURE 5.45 GafChromic sensitivity to energy composed to that of silver based film on LiF TLD. (From Reference [21]. With permission.)

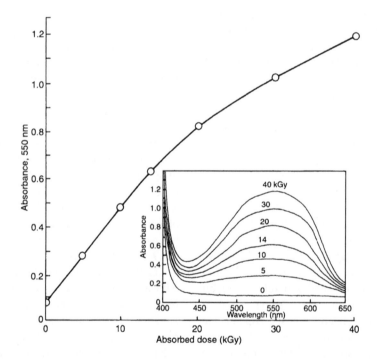

FIGURE 5.46 Response of radiochromic film of tetrazolium chloride in polyvinyl pyrrolidone (in terms of increase in absorbance measured at 522 nm wavelength) as a Junction of the gamma ray absorbed dose. The inset shows the increase in the absorption spectral amplitudes with dose. (From Reference [22]. With permission.)

37-041 MD-55 GAFCHROMIC Dosimetry Media; Package of Five 5″ × 5″ Sheets, Weight: .10 lb (.04 kg). 37-040 HD-810 GAFCHROMIC Dosimetry Media: Package of Five 8″ × 10″ Sheets, Weight: .50 lb (.24 kg)

The GAFCHROMIC sensitivity to energy compared to that of silver based film is shown in Figure 5.45. Another promising thin film polymer dosimeter for high

doses consists of a film or gel cast from an alcohol solution of blue tetrazolium or triphenyl-tetrazolium chloride combined with a polymer matrix material (e.g. polyvinyl alcohol, polyvinyl pyrrolidone), McLaughlin [22]. Figure 5.46 shows the radiation response properties of one example of this film system, in this case with a thickness of 30 μm.

REFERENCES

1. **Cheng, C-W. and Das, I. J.,** *Med. Phys.,* 23, 1225, 1996.
2. **Burch, S. E. et al.,** *Med. Phys.,* 24, 775, 1997.
3. **Robar, J. L. and Clark, B. G.,** *Med. Phys.,* 26, 2144, 1999.
4. **El-Khatib, E. et al.,** *Med. Phys.,* 19, 317, 1992.
5. **Stern, R. L. et al.,** *Med. Phys.,* 19, 165, 1992.
6. **Novotny, J. et al.,** *Phys. Med. Biol.,* 42, 1277, 1997.
7. **Stevens, M. A. et al.,** *Phys. Med. Biol.,* 41, 2357, 1996.
8. **McLaughlin, W. L. et al.,** *Nucl. Inst. Meth.,* A302, 165, 1991.
9. **Vatnitsky, S. M. et al.,** *Phys. Med. Biol.,* 42, 1887, 1997.
10. **Zhu, Y. et al.,** *Med. Phys.,* 24, 223, 1997.
11. **Klassen, N. V. et al.,** *Med. Phys.,* 24, 1924, 1997.
12. **Reinstein, L. E. et al.,** *Med. Phys.,* 24, 1935, 1997.
13. **Chiu-Tsao, S-T. et al.,** *Med. Phys.,* 21, 651, 1994.
14. **Muench, P. J. et al.,** *Med. Phys.,* 18, 769, 1991.
15. **Meigooni, M. F. et al.,** *Med. Phys.,* 23, 1883, 1996.
16. **McLaughlin, W. L. et al.,** in *Proc. High Dose Dosimetry for Radiation Protection,* IAEA, 1991, 305.
17. **Kellermann, P. O. et al.,** *Phys. Med. Biol.,* 43, 2251, 1998.
18. **Ertl, A. et al.,** *Phys. Med. Biol.,* 43, 1567, 1998.
19. **Meigooni, A. C. et al.,** *Med. Phys.,* 23, 1803, 1996.
20. **Mersseman, B. and De Wagter, C.,** *Phys. Med. Biol.,* 43, 1803, 1998.
21. **Nuclear Associates,** *Diagnostic Imaging and Radiation Therapy Catalog,* 1999.
22. **McLaughlin, W. L.,** in *Proc. High Dose Dosimetry for Radiation Protection,* IAEA, 1991, 3.

6 Calorimetry

CONTENTS

I. INTRODUCTION

Calorimetry has been used for a long time as a technique for establishing the absorbed dose. Graphite calorimetry has been used to establish absorbed dose standards for use in radiation therapy. A conversion process is necessary to convert from dose to graphite to dose to water. In other radiation measurement areas, too, calorimetry is recognized as a good approach for establishing absorbed dose standards. The basic assumption is that all (or a known fraction) of the absorbed radiation energy appears as heat, so that the measurement of absorbed dose reduces to a measurement of a temperature change. Most calorimeters developed for the purposes of radiation dosimetry have been constructed from graphite because of the perceived difficulties of working with a liquid system. Graphite has been chosen because its radiation absorption characteristics are similar to those of water, and thermally isolated segments can be machined and configured so as to permit the measurement of absorbed dose to graphite. Given the dose to graphite, the dose to water is obtained using a conversion process. The conversion from dose to graphite to dose to water introduces additional uncertainty.

The main technical difficulty is the problem of constructing a thermally isolated segment in which to measure the temperature change. Any wall which might be used to hold a small mass of water (of the order of 1 g) would significantly perturb the measurement. A calorimetric determination of the dose at a point requires that a thermally isolated element has been arranged so that no significant heat transfer occurs during the irradiation. If ΔT is the measured temperature rise in this element, then the absorbed dose to the material, D_m, is given by

$$D_m = c_m \Delta T/(1 - k_{HD}) \qquad (6.1)$$

where c_m is the specific heat of the calorimetric material and k_{HD} is the heat defect. The heat defect, which can be

positive or negative, is given by

$$k_{HD} = (E_a - E_h)/E_a \qquad (6.2)$$

where E_a is the energy absorbed by the irradiated material and E_h is the energy which appears as heat.

The approach to equilibrium temperature in time and space is governed by the heat equation. The effect of conductive heat transfer on the measured temperature change can be estimated using the equation

$$\partial T/\partial t = \alpha \nabla^2 T \qquad (6.3)$$

where T is the temperature and t is the time. The thermal diffusivity, α, is equal to $\kappa/\rho c$ where κ is the thermal conductivity, ρ is the density, and c is the specific heat. Beyond the maximum of the depth-dose curve, the variation of the dose with depth, $D(z)$, can be represented approximately by

$$D(z) = D_0 e^{-\mu(z - z_o)} \qquad (6.4)$$

where D_0 is the dose at depth Z_0 and μ for ^{60}Co γ-rays is approximately 0.05 cm^{-1}.

II. WATER CALORIMETER

When a large water calorimeter is irradiated by a beam directed vertically downward, apart from the build-up region, the temperature will in fact decrease with increasing depth. Thus, at depths well beyond the peak of the depth-dose distribution, the liquid will be stable with respect to convection. Domen [1] constructed a water calorimeter in which no effort would be made to thermally isolate the volume element in which the dose was to be measured. The essential features of Domen's calorimeter are shown

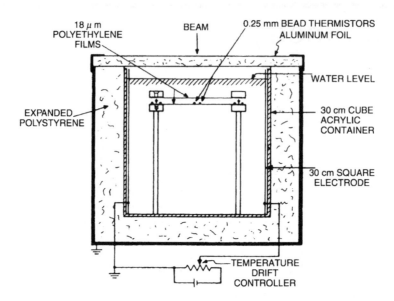

FIGURE 6.1 Schematics of Domen's absorbed dose to water calorimeter. The large mass of motionless water was thermally isolated from the environment. The thermistors were fixed in position and protected from the water by polyethylene film. The large electrodes at opposite sides of the container were used to control temperature drifts. (From Reference [2]. With permission.)

in Figure 6.1. The calorimeter consisted of a 30-cm cube of once-distilled water. During irradiation, the water was assumed to be motionless. Although the water was thermally isolated from the environment, it was not sealed against the exchange of gases with the atmosphere. The water temperature at a depth of 5 cm was measured using two thermistors sandwiched between thin polyethylene films. The calorimeter was irradiated with a ^{60}Co γ-ray beam directed vertically downward.

For low-linear-energy-transfer (LET) radiation, about 70% of the energy is deposited in spurs. In water, the primary spur products are highly reactive, but some may escape from the region of the spur before reacting. Others may undergo reactions within the spur to give rise to stable products which diffuse throughout the liquid. For a given LET, each species which escapes from the spur can be assigned a yield, G, which is independent of the composition of the dilute aqueous solution.

For water, the species which escape from the spurs have been identified and their G-values have been measured for a wide range of LET. Figure 6.2 shows how the G-values of the eight spur products change with LET. As the LET increases, adjacent spurs begin to overlap, so that reactions within the spurs become more important. The result is that fewer of the highly reactive species, such as $e_{\bar{a}q}$ and OH, escape from the region of the spur before reacting.

Fleming and Glass [3] constructed an absorbing element which consisted of discs of aluminum and plastic, each thick enough to fully stop the proton beam. The discs were in good thermal contact with each other and thermally isolated from the environment, so that the temperature of the assembly could be measured. If the same total energy was delivered by the proton beam to either the

aluminum or the plastic, any difference in the temperature rise must be due to a difference in the heat defect of the two materials.

Selbach et al. [4] measured the heat defect of water but used low-energy (17–30 kV) x-rays. Figure 6.3 shows the experimental arrangement used by Selbach et al. [4] The water-filled absorber was 30 mm in diameter and 15 mm long. A mu-metal disc 1 mm thick was immersed in the water and its position along the axis of the radiation field could be varied by using magnets. The disc was thick enough to be almost totally absorbing, and measurements were made with the disc either completely forward or completely back. The reference material used by Selbach et al. was mu-metal, an alloy consisting of 75% Ni, 18% Fe, 5% Cu, and 2% Cr by weight, and they assumed that its heat defect was zero. Assuming no heat-loss corrections and total absorption of the x-ray beam, Equation (6.2) can be used to show that the heat defect of water is [2]

$$k_{HD}^{w} = 1 - (\Delta T_{w}/\Delta T_{\mu m}) \qquad (6.5)$$

where ΔT_{w} and $\Delta T_{\mu m}$ are the measured temperature changes of the composite absorber.

Ross et al. [5] have described a technique for measuring the heat defect of water which is based on the total absorption of low-energy (1–5 MeV) electrons. They compared the temperature rise induced in water by the electron beam with the temperature rise caused by a known amount of electrical energy. Any difference (after making any necessary corrections) must be due to the heat defect of the water used.

The essential features of their apparatus are as follows. The electron-beam energy was determined with an

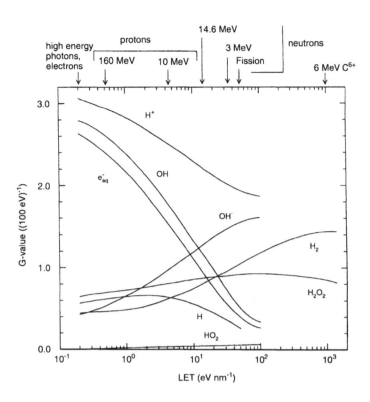

FIGURE 6.2 *G*-values of various spur products as a function of the track-averaged LET. The smooth curves summarize the trends in the measured data, except for H$^+$ and OH$^-$. (From Reference [2]. With permission.)

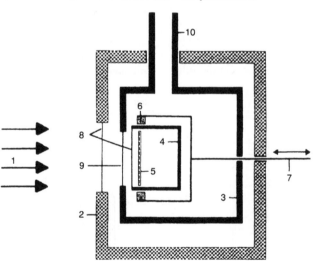

FIGURE 6.3 Section through the measuring system of Selbach et al. [4] The principal components are identified by numbers as follows: 1, x-ray beam; 2, calorimeter casing; 3, vacuum container; 4, water-filled absorber; 30 mm in diameter and 15 mm long; 5, mu-metal disc which can be moved through the water using magnets; 6, permanent magnets; 7, assembly for changing the location of the mu-metal disc; 8, plastic foils; 9, beryllium window; 10, vacuum container. (From Reference [2]. With permission.)

uncertainty of 0.2% using a magnetic spectrometer. The charge delivered to the absorber was measured with an uncertainty of 0.3% using a toroidal monitor. The beam entered the water through a Mylar window 23 μm thick. The water was stirred during both irradiation and electrical heating in order to minimize the effects of temperature gradients. The temperature rise was measured for both modes of heating and the temperature rise per unit input energy

calculated. The heat defect was then obtained from

$$k_{HD}^w = 1 - (\Delta T_r / \Delta T_e) \qquad (6.6)$$

where ΔT_r and ΔT_e are the temperature changes per unit energy for radiation and electrical heating, respectively.

A calorimeter in which water quality was carefully controlled was developed by Ross et al. [5]; it is shown in Figure 6.4. The water was purified by passing it through a particulate filter, an ion-exchange column, a 5 nm filter, and an ozonizer. Finally, the water was doubly distilled and stored in quartz containers. Approximately 100 ml of water was held in a thin-walled glass vessel which was thermally isolated from the environment. The water was stirred, and the temperature rise was measured using thermistors. Facilities were provided for bubbling the water with various gases and for sealing the vessel from the atmosphere. If the measured temperature change was ΔT, then

$$\overline{D}_w = [c_m\Delta T/(1 - k_{HD}^w)]k_{FM} \qquad (6.7)$$

The factor k_{FM} accounts for the effect of the glass in contact with the water.

Domen [7] has modified his original calorimeter design so that water quality is carefully controlled. Figure 6.5 shows the revised calorimeter in which the water in the vicinity of the measuring thermistors is enclosed in a clean glass vessel. The water is prepared using a filter, deionizer, and organic absorber. The rest of the water in the calorimeter does not need to be of high quality, since any heat defect in it will not affect the temperature measurement in the vicinity of the thermistors. The diameter of the glass vessel is about 33 mm, and its length is such that it holds 90 cm³ of water. The thickness of the glass wall is about 0.3 mm. The diameter of the vessel is a compromise based on two considerations. If the diameter were too small, excess heat generated in the glass by the radiation field would affect the measured temperature rise. On the other

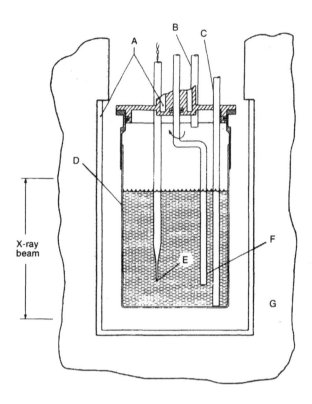

FIGURE 6.4 Cross-section view of Ross's calorimeter. The water volume is 5 cm in diameter and 5 cm high. A: channel for temperature-regulated water; B: outlet tube for gas flow; C: glass filling and bubbling tube; D: thin-walled (0.040 g cm⁻²) Pyrex calorimeter vessel; E: one of two thermistors; F: glass stirring paddle; G: styrofoam insulator. (From Reference [2]. With permission.)

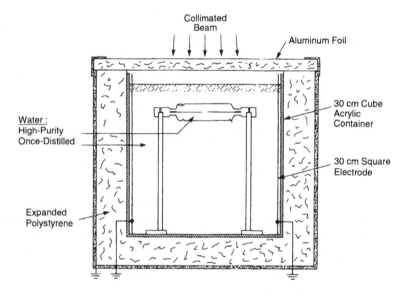

FIGURE 6.5 Main features of Domen's sealed water calorimeter. The main body of the calorimeter consists of a large mass of motionless water thermally isolated from the environment. The measuring thermistors are located in a sealed glass vessel in which the water quality is carefully controlled. With this design, the heat defect of the water at the point of measurement is known, and the vessel also acts as a convective barrier, potentially permitting measurements in horizontally directed radiation fields. (From Reference [2]. With permission.)

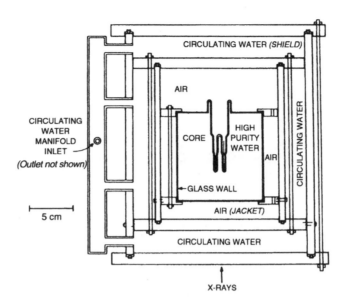

FIGURE 6.6 Cross section of the water calorimeter. Omitted from this drawing are two Teflon-seated glass valves and a 3-cm³, nitrogen-filled expansion chamber, located at the distal end of the core. (From Reference [8]. With permission.)

hand, if the diameter were too large, the vessel would not serve as a convective barrier. For the diameter chosen, Domen showed that the effect of excess heat should be less than 0.1% 60 s after a 60-s irradiation. Furthermore, he estimated that the glass should act as an effective convective barrier, permitting measurements to be made in a horizontally directed electron beam.

The determination of the dose to water using a water calorimeter is absolute in the sense that it does not require the application of radiation-dependent parameters such as stopping power ratios, replacement corrections, ion collection efficiency, ⁶⁰Co exposure calibration factors, etc. Indeed, the determination requires only the specific heat of water, and the calibration of the calorimeter depends upon an accurate thermometer. If the thermal defect of water is zero, i.e., all absorbed energy is converted to heat, then the temperature change of the water multiplied by the specific heat is equal to the absorbed dose. (The thermal defect is $(T_e - T_0)/T_e$, where T_e is the expected temperature rise and T_0 is the observed temperature rise of an irradiated material.) [8]

Figure 6.6 shows a cross section of the calorimeter with x-rays incident upon it from below.

The core was filled with high-purity water that has dissolved oxygen removed by bubbling ultra-high-purity nitrogen through it for several hours. To eliminate the possibility of convection currents, the core is maintained at 4.0°C by circulating refrigerated water through the jacket that surrounds it.

Calibrations are done with the thermistors in the glass capillary tubes of the core, and the core is filled and submerged in the reservoir of a refrigerated circulator. To minimize the effects of ambient temperature variations, this apparatus is placed in a 4°C cold room for the period of one week that it takes to complete a calibration. The resistance of the thermistors is determined using the same Wheatstone bridge. Water temperatures in the 2–10°C range and thermistor power levels in the range 5–200 microwatts are routinely used in the calibration procedure.

In order to investigate experimentally the overall correction factor for a cylindrical ionization chamber, water calorimetry was used by Seuntjens et al. [9] An important limitation of water calorimeters open to impurities is the heat defect arising from chemical reactions induced by the radiolysis of water. Using different types of closed-vessel calorimeters containing water with well-defined additives, agreement between experiments and model calculations of the relative heat defect caused by the radiolysis of water was obtained. [9] Using high-purity deoxygenated water, Schultz et al. obtained agreement between water calorimetry and ionization chamber dosimetry.

A schematic drawing of the calorimeter is shown in Figure 6.7. The calorimeter essentially consisted of a cylindrical PMMA tube with a length of 15 cm, a diameter of 4 cm, and a wall thickness of 0.5 mm, containing high-purity water suspended in a large water phantom. For its construction, the different parts of the cylinder were glued together using acrylate glue flushed with air for several hours, rinsed numerous times with high-purity water, and pre-irradiated up to several kGy in order to reduce influences of the heat defect.

The calorimeter tank was a 30-cm × 30-cm × 30-cm (inner dimensions) double-walled phantom with 4-cm wall thickness and a PMMA front window (8.25-mm thickness). In the bottom of the calorimeter, a small magnetic stirrer is built in for agitation of the water. The water temperature is measured with two small PT100 resistance probes. [9]

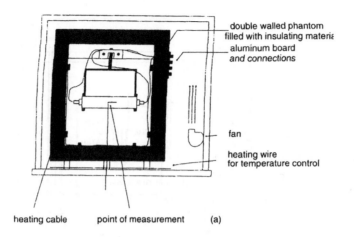

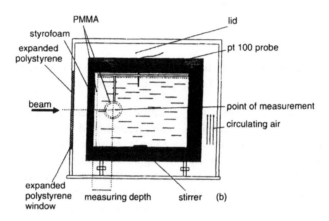

FIGURE 6.7 Schematic drawing of the water calorimeter and enclosing box: (*a*) front view; (*b*) side view. (From Reference [9]. With permission.)

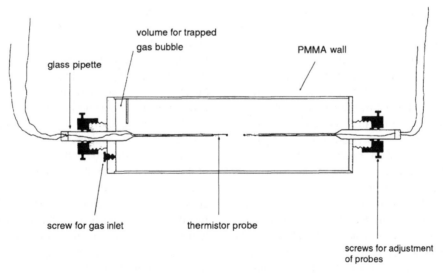

FIGURE 6.8 Schematic drawing of the calorimeter detector vessel (PMMA) with thermistor probes. The vessel can be bubbled with various gases to establish various chemical systems. (From Reference [9]. With permission.)

The water temperature can be adjusted using insulated heating cable mounted in the phantom. The calorimeter core (calorimeter detector vessel) was suspended in the water phantom and adjusted using a tiny spacer between the inner side of the phantom front window and the detector vessel so as to set the point of measurement (the center of the thermistors) at a geometrical depth of 5 cm from the outside of the phantom PMMA window. The calorimeter core is

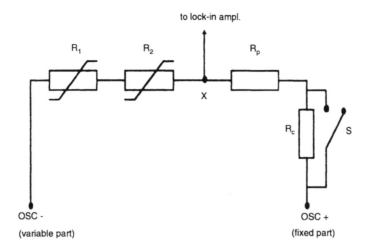

FIGURE 6.9 Schematic diagram of bridge circuit. (From Reference [9]. With permission

shown schematically in Figure 6.8. The temperature increase is measured using thermistor probes. The calorimeter is irradiated by a horizontal beam with a field diameter of at least the total length of the detector vessel (15 cm).

The thermistor probes essentially consisted of small commercial glass thermistor probes (Thermometrics, Inc., type P20, outer glass diameter 0.5 mm, length 5 mm), glued into the end of small glass capillaries using epoxy. The thermistors were serially connected to the one-arm bridge circuit, shown in Figure 6.9.

The thermistor power level was varied in the range 1–200 μW for ten temperature points in the range 15–30°C. Because the temperature was varied over more than 10 K and the relation between $\ln R$ and $1/T$ is slightly nonlinear the fit

$$\ln R = a + b/T + c/T^2 \qquad (6.8)$$

was used.

Due to the low dose rates, long irradiation times (typically 300 s) are required to obtain a sufficiently large signal. For this purpose, corrections have to be applied for heat loss (or gain) at the point of measurement by the non-uniformity of the dose distribution. These corrections are also needed due to excess heat from non-water materials caused by their lower thermal capacity and their different radiation absorption characteristics.

Palmans et al. [10] reported water calorimeter operation in an 85-MeV proton beam and a comparison of the absorbed dose to water measured by ionometry with the dose resulting from water calorimetric measurements. The results showed that pure hypoxic water and hydrogen-saturated water yielded the same response with practically zero heat defect, in agreement with the model calculations. The absorbed dose inferred from these measurements was then compared with the dose derived from ionometry by applying the European Charged Heavy Particle Dosimetry

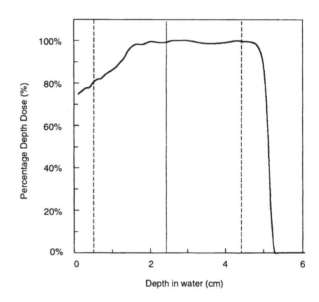

FIGURE 6.10 Percentage depth-dose curve in water for the modulated 85-MeV proton beam. The full vertical line shows the position of the measuring point; the dashed lines show the upstream and downstream limits of the calorimeter detection vessel. (From Reference [10]. With permission.)

(ECHED) protocol. Restricting the comparison to chambers recommended in the protocol, the calorimeter dose was found to be 2.6% ± 0.9% lower than the average ionometry dose.

For 85-MeV protons, the range in water amounts to 50.5 mm, thereby irradiating the entire cylindrical core of the calorimeter. The percentage depth-dose curve is shown in Figure 6.10.

Figure 6.11 shows a schematic side view of the water calorimeter. The calorimeter consists of a 30-cm cubic water phantom, stabilized at 4°C in order to remove any concern related to convection. The water temperature is measured using calibrated platinum resistor probes at different positions in the phantom. The temperature increase

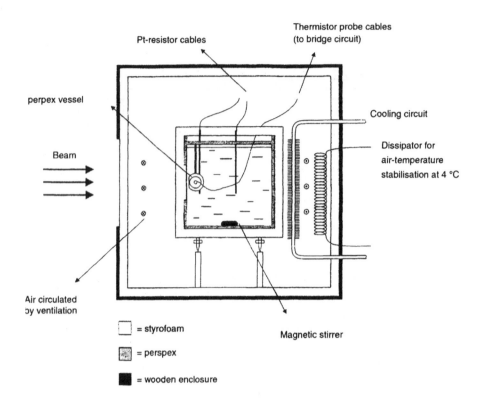

FIGURE 6.11 Schematic drawing of the water calorimeter used by Palmans et al. (side view). (From Reference [10]. With permission.)

due to radiation is measured in the center of a 4-cm-diameter acrylic vessel (14-cm length, 0.5-mm wall thickness) using glass thermistor probes (0.5 mm in diameter at the probe end), each of 24 kΩ at 4°C.

In water calorimetry, dose to water is immediately derived from the temperature change by multiplication with the specific heat capacity (c_w) of water at the measuring temperature, which is known with high accuracy. Specifically, we considered correction factors for conductive heat losses (k_c) and the chemical heat defect (h). The effect of the thermistor probe ends (0.5-mm diameter) and the vessel wall (0.5-mm PMMA) on the dose at the measuring point has been studied in detail for medium-energy x-rays and high-energy photons, where it has been found to be smaller than 0.2% $\pm$ 0.1%. Therefore, considering the scatter properties of protons compared to photons, the effect is probably very small for protons. The correction factor (k_{sc}) for this effect for protons has, therefore, been ignored and an uncertainty of 0.2% has been assigned. By measuring the overall fractional thermistor resistance change of the two thermistors and applying the following equation [10],

$$D_w = c_w \times \frac{\overline{\Delta R}}{R} \times \bar{S}^{-1} \times k_c \times k_{sc} \times \frac{1}{1 - h'} \quad (6.9)$$

absorbed dose to water was derived. $\bar{S}$ represents the average thermistor sensitivity which results from the calibration against standard thermometry. The average

fractional thermistor resistance change $\overline{\Delta R}/R$ of the two thermistors was derived from linear extrapolations of pre-irradiation drift and post-irradiation drift to mid-run and measurement of the bridge output voltage change ("irradiation-run"). This voltage change is converted into an overall fractional bead resistance change by increasing the opposite bridge arm with a known resistor and measuring the output voltage change ("calibration-run"). Corrections of approximately 0.05% for change in thermistor power dissipation, and therefore excess heat, are made for both irradiation and calibration runs.

The heat defect h is defined as

$$h = \frac{E_a - E_h}{E_a} \quad (6.10)$$

where E_a represents the energy absorbed and E_h is the energy appearing as a temperature rise. For low LET radiation and for water containing well-defined impurities, it has been shown that the heat defect can be calculated based on the primary yields (G-values = number of molecules formed due to an energy deposition of 100 eV) of the species produced 10^{-7} s after the passage of the secondary electron through water, and assuming the species are homogeneously distributed throughout the solution. The method is based on the calculation of product yields in the irradiated solution long (i.e., seconds to minutes) after the passage of the radiation using reaction kinetics and long after evaluation of the overall energy balance using published enthalpies of formation. [10]

Absorbed dose to water at the point of measurement in the calorimeter was derived from the ion chamber reading M using the ECHED code of practice

$$D_w = MN_kC_p \qquad (6.11)$$

where N_k is the air-kerma calibration factor for a ^{60}Co photon beam and C_p is an overall dose conversion factor given by

$$C_p = (1 - g) \times A_{wall} \times k_m \times \left[\left(\frac{S}{\rho}\right)_{air}^{water}\right] \times \frac{(W_{air}/e)_P}{(W_{air}/e)_C} \qquad (6.12)$$

where the stopping power ratio of water to air is taken from ICRU Report 49 [17] for the effective proton energy $(E_P)_{eff}$ at the point of measurement. For $(W_{air}/e)_p$ the value of 35.2 J/C is recommended in the protocol. For consistency, the other chamber-dependent factors in the calibration beam (A_{wall} and k_m) are taken from the IAEA code of practice for all chambers. Here A_{wall} is a correction factor for scattering and attenuation in the chamber wall, and k_m is a correction factor for the non-air equivalence of the chamber construction materials. For the comparison with water calorimetry, values for A_{wall} (equal to IAEA values for the chambers used) and k, recommended in the ECHED protocol, were used, yielding doses deviating by no more than 0.4% for the same chamber reading.

Figure 6.12a shows a schematic side view of the water calorimeter positioned in its enclosure. The calorimeter tank consists of a 30-cm × 30-cm × 30-cm water phantom, thermally isolated by polystyrene foam. Isolation of the enclosure allows temperature stabilization of the air surrounding the calorimeter phantom. The operating temperature is 4°C, in order to remove any concern related to convection. The water temperature is continuously measured with calibrated Pt-resistor probes inserted at different positions in the water phantom. The temperature increase due to irradiation is measured at the center of a cylindrical vessel using small thermistors that are embedded in the tip of small glass probes (Figure 6.12c). Two probe types are shown, one in which the thermistor bead is epoxied in the end of the probe tip and one where the bead is sealed in a small glass rod. Both types have been used for the ^{60}Co measurements, but only the second type has been used in the 5-MV and 10-MV photon beams. No significant difference in calorimeter response has been observed due to the use of the two different types. This is also confirmed by excess heat calculations. Figure 6.12b shows the cylindrical glass vessel and the positioning of the glass probes. The vessel contains deionized and three-times-distilled water and has a diameter of 4 cm, a length of 14 cm, and a wall thickness varying from 0.3 mm in

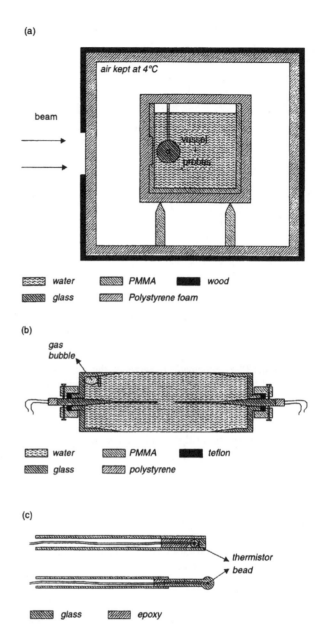

FIGURE 6.12 Schematic design of the Gent sealed water calorimeter: (a) phantom and enclosure, (b) glass vessel with position mechanism for thermistor probes, and (c) tip of the thermistor probes. (From Reference [11]. With permission.)

the central part corresponding to the central axis of the beam to 1.6 mm at the lateral edges. The chemical water systems inserted into the vessel for the present study are Ar-saturated pure water and H$_2$-saturated pure water. [11]

Absorbed dose to water is derived from the temperature increase due to irradiation at the location of the thermistors, ΔT, as

$$D_w = c_w\Delta T k_c k_{sc} k_{dd}\frac{1}{1 - h} \qquad (6.13)$$

where c_w is the specific heat capacity of water at the measuring temperature. The temperature change ΔT is measured

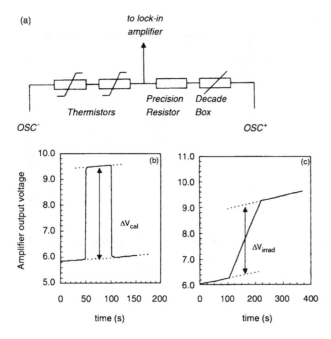

FIGURE 6.13 (*a*) Schematic drawing of the a.c. measuring bridge and typical time evolution of the bridge output voltage during (*b*) a "calibration run" and (*c*) an "irradiation run." (From Reference [11]. With permission.)

using the measuring bridge that is shown schematically in Figure 6.13a. Two thermistors are serially connected in one arm of the a.c. bridge, and in the opposite arm a fixed high-precision resistor and a variable resistor are connected to compensate the bridge. The out-of-balance voltage is amplified by a lock-in amplifier. Figures 6.13b and 6.13c show a typical calorimeter run in the 10-MV photon beam with first a "calibration run" and thereafter an "irradiation run." The procedure used to determine the voltage change at mid-run is shown. The calibration of the bridge output voltage is performed in a "calibration run" by adding a well-known resistor to the bridge circuit (opposite arm as thermistors) and determining the corresponding out-of-balance voltage. This calibration is then used to derive the average fractional thermistor resistance change $\Delta R/R$ due to irradiation from the voltage change at mid-run, ΔV, during an "irradiation run." On both calibration and irradiation runs, corrections for changes in self-heating of the thermistors and for nonlinearity of the bridge response are applied. ΔT is calculated as $\Delta T = \Delta R/RS$, where S is the average thermistor sensitivity resulting from calibration of the thermistors against standard thermometry. [11]

III. GRAPHITE CALORIMETER

Potentially, water calorimeters are most accurate, but they are delicate systems and require considerable care to control the purity of the water and the effects of radiochemical reactions occurring in it. The standard based on the graphite

calorimeter is improving continuously. Graphite calorimeters are stable and are very accurate. However, suitable conversion procedures are necessary for establishing the absorbed dose to water. A particular problem in graphite calorimetry is gap effect and its corrections.

The primary standard of absorbed dose to water established at ENEA for the ^{60}Co gamma-ray quality is based on a graphite calorimeter and an ionometric transfer system. [12] The graphite calorimeter is of the Domen type. The gap configuration of the ENEA calorimeter is shown in Figure 6.14. The four calorimeter bodies are enclosed in a 1-cm-thick PMMA housing with 152-mm external diameter. The external housing is radially surrounded by a close-fitting 300-mm-diameter graphite annular body. It is possible to increase the thickness of this annular body by adding posterior 300-mm-diameter full backscattering graphite plates and additional graphite front plates to increase the measurement depth in graphite. The minimum measurement depth (distance of the central plane of the core from the calorimeter front surface) is 0.88 g cm^{-2}. The calorimeter is thermoregulated at a temperature of about 27°C, with a stability better than 5×10^{-4}°C during a typical measurement run. The jacket has approximately the same heat capacity as the core, which is a disc of 20-mm diameter and 2.75-mm thickness. In two holes radially drilled in the core edge, two measuring thermistors are embedded, thus doubling the sensitivity during measurements. Additional thermistors are in the other calorimeter bodies (one thermistor for each body) for temperature monitoring. Electrical heaters are embedded in all of the four bodies of the calorimeter. The heaters are manually operated to inject heat amounts into each body for the calorimeter electrical calibration and for controlling the heating or cooling drifts in the different bodies. [12]

To transfer the dose from graphite to water, a thick-walled (TW) ionization chamber is used. This is a cylindrical homogeneous chamber with graphite wall sufficiently thick (0.5 g cm^{-2}) to allow electron equilibrium at the ^{60}Co gamma radiation.

In each gap of the calorimeter there are three different regions contributing in a different way to the variation of the radiation fluence in the core. These regions are those in front of (anterior gap), behind (posterior gap), and around (annular gap) the core. The effects of these three gap regions were evaluated, separately but a single correction was determined for the same region of all three gaps.

For conversion of the absorbed dose from graphite to water, the TW chamber was irradiated in graphite phantom at the same depth, $z(P_g)$, where the absorbed dose to graphite, D_g, was known by absolute measurement with the calorimeter. A chamber calibration factor, N_g, in terms of absorbed dose to graphite, was then obtained as

$$N_g = D_g(P_g)/Q(P_g) \tag{6.14}$$

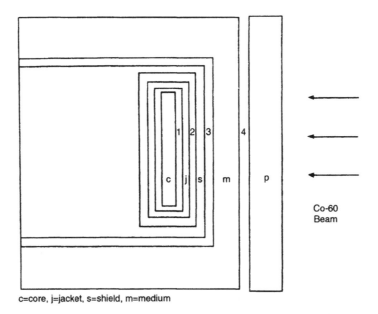

c=core, j=jacket, s=shield, m=medium

FIGURE 6.14 The schematic configuration of the gaps in the ENEA graphite calorimeter. The gaps around the calorimeter core, c, are those denoted 1–3. Gap 4 is between the calorimeter and the frontal external plate, p, used to vary the measurement depth. The gap width is 0.5 mm for all of the anterior gaps and for the annular gaps 1 and 2, 1 mm for the annular gap 3, and 0.65 mm and 1.2 mm for the posterior gaps 1 and 2, respectively. (From Reference [12]. With permission.)

where $Q(P_g)$ is the ionization charge due to the absorbed dose D_g at P_g. The chamber was then irradiated in the water phantom at a depth $z(P_w)$, where the photon fluence differential in energy was expected to be the same as that existing at the reference depth $z(P_g)$ in the graphite phantom. The "corresponding points" P_w and P_g were determined according to the scaling rule derived from the photon fluence scaling theorem

$$D_g(P_w) = Q(P_w)N_g \quad (6.15)$$

where $D_g(P_w)$ is the absorbed dose to graphite at the center of a small mass of graphite placed in water with its center in P_w; $Q(P_w)$ is the corrected charge reading due to the absorbed dose D_g at P_w.

The existence of transient electron equilibrium conditions (TCPE) in that mass of graphite allows us to express the absorbed dose given by Equation (6.15) also as

$$D_g(P_w) = \Psi_g(P_w)(\mu_{en}/\rho)_g \beta_g \quad (6.16)$$

where $\Psi_g(P_w)$ is the photon energy fluence at the center of the mass of graphite placed in water at P_w, $(\mu_{en}/\rho)_g$ is the mass energy absorption coefficient in graphite averaged over the photon energy spectrum, and β_g is the ratio of the absorbed dose to the collision part of kerma at the center of the small mass of graphite.

Similarly, the absorbed dose to water at P in undisturbed water can be written as [12]

$$D_w(P_w) = \Psi_w(P_w)(\mu_{en}/\rho)_w \beta_w \quad (6.17)$$

where the symbols have an analogous meaning to those in Equation (6.16), with water in place of graphite. In particular, $\Psi_w(P_w)$ is the photon energy fluence in undisturbed water at P_w.

By the ratio of Equations (6.16) and (6.17), one then obtains

$$D_w(P_w) = D_g(P_w)[(\mu_{en}/\rho)_w/(\mu_{en}/\rho)_g](\beta_w/\beta_g)$$
$$\times \Psi_m/\Psi_g \quad (6.18)$$

Combining Equations (6.16) and (6.17) will give

$$D_w(P_w) = Q(P_w)N_g(\bar{\mu}_{en}/\rho)_g^w \beta_g^w \Psi_g^w \quad (6.19)$$

where the notation x_b^a represents the ratio $(x)_a/(x)_b$. [12]

An electron beam graphite calorimeter has been developed by Burns and Morris [13] covering the dose range from kilogray levels down to doses of 1 Gy, at dose rates down to 5 Gy/min. The thermistor-bridge system is calibrated by reference to three triple-point temperatures, using a platinum resistance thermometer as a transfer device. The uncertainty in the measurement of the absorbed dose to graphite core is estimated lobe $\pm 1\%$ at the 95% confidence level.

The calorimeter system is shown schematically in Figure 6.15.

The calorimeter core is a disc of graphite 50 mm in diameter and 7 mm in thickness. A hole, 1.5 mm in diameter and 7 mm deep, is drilled radially into the mid-plane

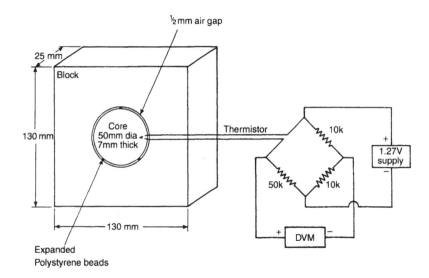

FIGURE 6.15 Schematic diagram of calorimeter system. The calorimeter block is completely enclosed in 25 mm of expanded polystyrene. (From Reference [13]. With permission.)

of the core to accommodate a glass bead thermistor with a nominal resistance of 50 kΩ at 20°C. Thermal contact with the graphite is improved by using a small amount of zinc oxide–based heat sink compound around the thermistor bead. This disc is inset into the surface of a rectangular block of graphite approximately 130 × 130 × 25 mm deep and is thermally isolated from the block by a nominal 0.5-mm air gap. Small expanded polystyrene beads separate the disc from the block. The dimensions of the block were chosen to be sufficiently large to ensure the complete scattering of 15-MeV electrons into the core, but at the same time they are small enough to give relatively uniform heating of the rectangular block, in area and in depth, in a scattered electron beam from the NPL linear accelerator (linac). The core and surrounding block are completely enclosed in 25 mm of expanded polystyrene, which has been shown to have no measurable effect on the response of the calorimeter. If this assembly is allowed to reach equilibrium in an environment which is temperature controlled to ±200 mK, any temperature drift of the core is linear over a period of tens of minutes and is always less than 0.5 mK/min. Dose rates down to about 3 Gy/min, and doses below 1 Gy, have been measured without the need for further temperature control. In particular, no vacuum system or controlling heaters are necessary. [13]

A description was given by DuSautoy [14] of the UK primary standard graphite calorimeter system. The calorimeter measures absorbed dose to graphite for photon radiations from ^{60}Co to 19-MV x-rays and is the basis of the NPL therapy-level absorbed dose to water calibration service. Absorbed dose to graphite from the photon calorimeter has been compared with three other standards: an ionization chamber and cavity theory, for ^{60}Co gamma radiation; the NPL electron calorimeter, for 12–14-MeV

electron beams; and the BIPM ^{60}Co absorbed dose standard. The calorimeter has been called the photon-beam calorimeter to distinguish it from the electron-beam calorimeter. This calorimeter has been the primary standard for the NPL calibrations in absorbed dose to water and the chemical dosimetry services since 1988. The mean absorbed dose to carbon (graphite) in the core of the calorimeter D_c is determined by measuring the energy deposited by ionizing radiation, E_R, in the core of mass m, according to the definition of absorbed dose given by

$$D_c = \frac{E_R}{m} \qquad (6.20)$$

For a dose of 1 Gy, the energy absorbed, E_R, raises the temperature of the core by about a millikelvin. This temperature rise is measured with a resolution of a few tens of microkelvin by a thermistor embedded in the core.

Figure 6.16 shows the vertical cross section through the axis of the assembled calorimeter and gives the dimensions in more detail. The core is placed near the front so that electron beam measurements can be made. The minimum measurement depth is about 1 g cm^{-2}. Two thermistors are included in the core; one is used as a sensor and the other as a heater for the electrical calibration.

The core resistance measurement and heating circuits are shown in Figure 6.17. Initially the second jacket heating is regulated to produce a constant second jacket temperature, and the core and first jackets are allowed to reach thermal equilibrium. The core, first jacket, and second jacket bridges are then balanced by adjusting the balancing resistor (e.g., R_M in the core circuit). Before a measurement, the heating power in the second jacket is fixed at a

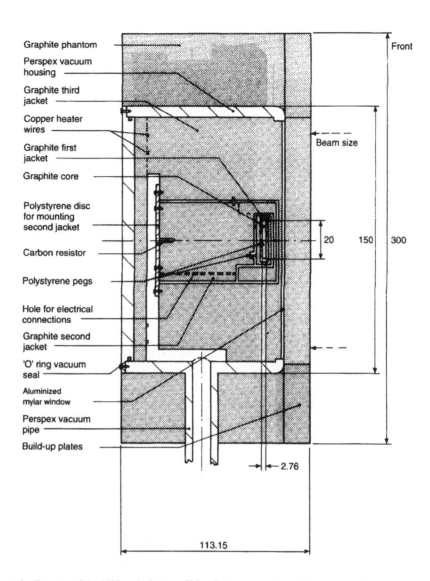

Graphite phantom

Perspex vacuum
housing

Graphite third
jacket

Copper heater
wires

Graphite first
jacket

Graphite core

Polystyrene disc
for mounting
second jacket

Carbon resistor

Polystyrene pegs

Hole for electrical
connections

Graphite second
jacket

'O' ring vacuum
seal

Aluminized
mylar window

Perspex vacuum
pipe

Build-up plates

Front

Beam size

20   150   300

2.76

113.15

FIGURE 6.16 Schematic diagram of the NPL calorimeter. Side-view cross section. Not to scale. Dimensions in mm. (From Reference [14]. With permission.)

level which gives steady increase in temperature at a rate of approximately 12 μK min^{-1}.

Figure 6.18 shows a typical measurement sequence. The bridge amplifier output voltages of the core and two jackets are recorded for about 300 seconds (A to B). The calorimeter is then irradiated (B to E), reducing the resistance of the sensing thermistors embedded in the core and two jackets and increasing their bridge output voltages. Irradiations last for about 80 seconds. Midway through the irradiation time (at C), the switch, S1, is closed, reducing the resistance of that arm of the core bridge by a small fixed amount (i.e., 0.1 Ω). Simultaneously, the balancing resistances of the first and second jacket bridges are reduced to offset their bridge output voltages below their balance points. The irradiation continues to heat the calorimeter, increasing the bridge output voltages. The irradiation is stopped (at E) when the core bridge output voltage returns to the pre-heating balance voltage (at B).

The output voltages of all the bridges then level off and are recorded for a further 300 s (E to F) after the heating.

The dose the core would have received if it had been entirely graphite, D_c, can be calculated from [14] as

$$D_c = E_R / \left(m_g + \sum_{i=1}^{n} m_i \frac{D_i}{D_g} \right) \qquad (6.21)$$

where m_g is the mass of graphite, m_i is the mass of the non-graphite impurity identified by the subscript i, n is the number of non-graphite impurities, D_g is the average absorbed dose in graphite, and D_i is the average absorbed dose in the impurity labeled i. The ratio of the doses D_i/D_g is sufficiently near unity and the impurities are so small for this calorimeter that the effective mass of the core m was taken (for all beams and energies) as $m = m_g + \sum_{i=1}^{n} m_i$, i.e., 1.5043 $g \pm 0.05\%$.

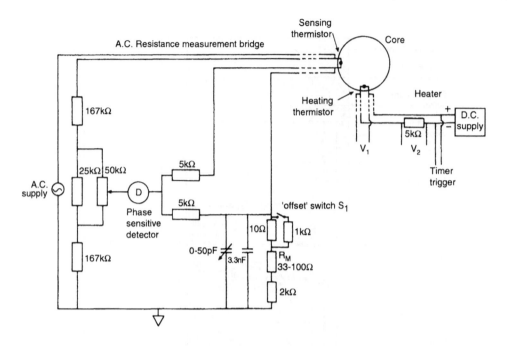

FIGURE 6.17 Schematic diagram of electrical circuits. (From Reference [14]. With permission.)

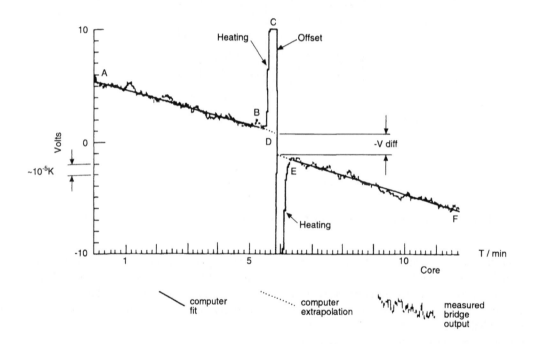

FIGURE 6.18 Plot of bridge output voltage against time. (From Reference [14]. With permission.)

IV.   OTHER MATERIAL CALORIMETERS

The thermal defects of A-150 plastic and graphite referenced to aluminum were determined by Schulz et al. [15] for 800-keV protons scattered by a 2-μm nickel foil (mean transmitted energy = 550 keV). Composite cores of Al/A-150, graphite/A-150, and Al/graphite which could be irradiated from one side or the other were employed. The temperature increase of a core caused by 30–100 s of irradiation (3–6 nA of proton beam current) was detected by two thermistors mounted in opposite legs of a Wheatstone bridge. The thermal defect of A-150 plastic was determined to be 0.0421 ± 0.0036 (SE) referenced to aluminum and 0.0402 ± 0.0034 referenced to graphite. The thermal defect of graphite referenced to aluminum is 0.0043 ± 0.0034. No change in the thermal defect of A-150 plastic was detected for accumulated doses up to 8×10^5 Gy.

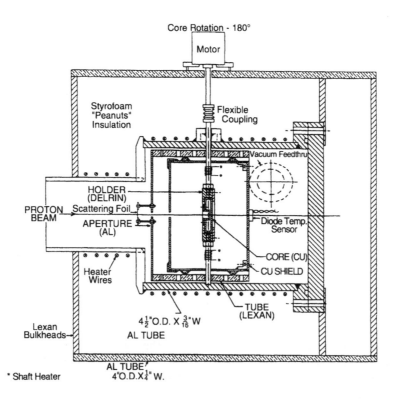

A key consideration in the application of tissue-equivalent materials, solids, or liquids to absorbed-dose calorimeters is the extent to which energy absorbed from ionizing radiations is converted to heat. Specifically, the thermal defect of a material is defined as the expected temperature change minus the observed temperature change, the difference then divided by the expected temperature change; positive thermal defects indicate endothermicity, while negative thermal defects indicate exothermicity. [15]

A flat, cylindrical absorber (the core), one side of which is the reference material, is mounted so that it can be rotated through 180° (about a diameter passing through its mid-thickness) and irradiated from one side or the other. As the temperature rise caused by irradiation depends on the heat capacity of the composite core, a temperature rise caused by irradiation of the test material which differs from that for the reference materials is indicative of a thermal defect. A cross-sectional drawing of the thermal-defect calorimeter is shown in Figure 6.19, and a more detailed drawing of the core support and A-150/ aluminum core is shown in Figure 6.20. Each core has three holes in its periphery to accept three thermistors, two of which were connected to opposite legs of a Wheatstone bridge and the third used as an ohmic heater for system testing. Because the thermal diffusivity of A-150 plastic (2.7×10^{-3} cm^2 s^{-1}) is much lower than aluminum

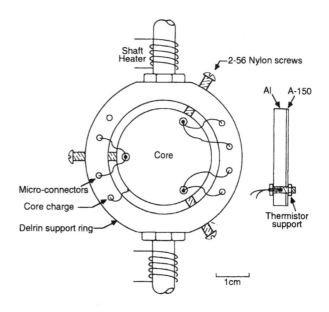

FIGURE 6.20 The A-150 plastic/aluminum core and the core support assembly. (From Reference [15]. With permission.)

(4.1 cm^2 s^{-1}) or graphite (4.6 cm^2 s^{-1}), the A-150 absorber was made about 0.4 mm thick.

The core assembly was surrounded by an insulated copper shield which is connected in series to an electrometer and a high-voltage power supply. With this arrangement,

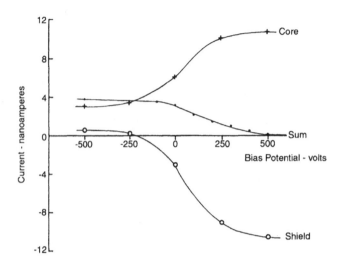

FIGURE 6.21 The effect of shield bias on the current collected by the core and shield. (From Reference [15]. With permission.)

the bias applied to the shield can be varied between ±500 V and the shield charge recorded. The entire calorimeter is maintained at 300 K by electrical heating wires wrapped around its body. The calorimeter is suspended at the center of a 25-cm OD aluminum tube.

The thermal defect was calculated from

$$\frac{(d/Q_{ref} - d/Q_{test})}{d/Q_{ref}} \qquad (6.22)$$

where d is the chart recorder deflection and Q is the sum of core and shield charges.

The effect of shield bias potential on the charge collected from the core and shield was tested with the aluminum/graphite core, and the results for aluminum using potentials in the range ±500 V are shown in Figure 6.21. Data similar to those shown in Figure 6.21 were also measured for graphite and, although different values for the sums of core and shield charges were obtained for zero and positive shield bias, essentially the same values as for aluminum were obtained when the shield was negative. All of the thermal-defect data reported below were obtained with negative 500 V applied to the shield.

An experimental setup was described by Brede et al. [16] which allows the calorimetric determination of the heat defect of solids and fluids relative to that of gilded copper, which was used as the reference material. The calorimeter operated in a quasi-adiabatic mode and was suitable for protons, deuterons, α particles, and other heavy ions with energies above 5 MeV. Corrections have been made for secondary electron emission and also for heat losses due to the temperature gradient on the surface of the calorimeter core.

The heat defect, h, describes that fraction of the absorbed energy which does not result in a temperature change of the absorber. It is defined by the equation

$$h = (W_a - W_h)/W_a \qquad (6.23)$$

where W_a represents the energy absorbed and W_h is the energy that appears as heat.

In mixed neutron-photon fields, the absorbed dose to water is generated by secondary electrons and charged particles such as protons, deuterons, α particles, and recoil oxygen nuclei. The various particles produce markedly different ionization densities along their tracks, and this can be described by their linear energy transfer, LET. Knowledge of the LET dependence of the heat defect in water is therefore a prerequisite for the water calorimeter, in order to reduce the main component of the uncertainty in water calorimetry and to apply a correction factor in mixed neutron-photon fields.

The material under investigation—in this case A-150 plastic or water—was contained in a cylindrical canister of gilded copper with an inner diameter of 15 mm and a length of 5 mm and which has an eccentric beam stop (see Figure 6.22). Two different A-150 plastic absorbers from different production lots were investigated. All samples had the same geometrical dimensions—a cylindrically shaped probe, 3.5 mm in thickness and 15 mm in diameter. They were pressed into the canister. A 7-μm-thick indium-sealed molybdenum entrance window allowed charged particles to enter the core.

The calorimeter (Figure 6.22) comprised a core, inner and outer jacket, and vacuum chamber which were thermally and electrically insulated from one another and operated in an adiabatic mode. The temperature increase, ΔT_i, (i indicates either A-150 or water) measured on the

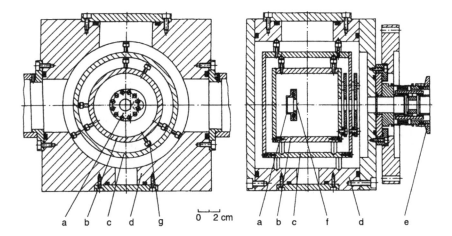

FIGURE 6.22 Core (a); inner jacket (b); outer jacket (c); vacuum chamber (d); flange connection to the beam line (e); entrance window (f); and screws (g) to align the core with nylon cords (not shown). (From Reference [16]. With permission.)

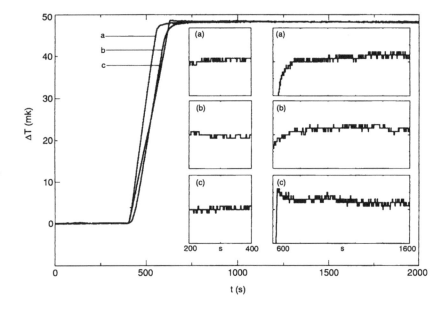

FIGURE 6.23 Time (t) plotted against temperature (ΔT) response of the calorimeter. Beam stop: A-150 plastic (a), water (b), and gilded copper (c). The inserts on the left-hand side show the initial temperature drifts; the inserts on the right-hand side show the response after the end of irradiation. The temperature scale of all inserts is 2 mK. (From Reference [16]. With permission.)

core surface is given by

$$\Delta T_i C (1 - h_i)^{-1} = Q_b U_b \qquad (6.24)$$

where C is the total heat capacity of the calorimeter core, h_i is the heat defect in either A-150 plastic or water, Q_b is the total beam charge deposited in the core, and U_b is the voltage which is used to accelerate the particles; i.e., the product $Q_b U_b$ is the total energy imparted in the absorber.

The temperature measurements of the core and jackets were made with cylindrically shaped thermistors (NTC) 0.35 mm in diameter and 0.5 mm in length (type 6331 from Philips) and were connected to

thin copper wires 0.06 mm in diameter and 30 cm in length.

The calorimetric measurement started with an analysis of the initial drift for about 400 s. After an irradiation time of 200 s, the system equilibrated during a time of approximately 500 s. During the next 500 s, the final temperature drift was recorded (see Figure 6.23). The temperature increase was determined from the linear region of the initial and final drifts and was normalized to the beam charge collected from the core. The measurements were carried out alternately with gilded copper and water or A-150 plastic as the beam stop. The quantity ($\Delta T_i/\Delta T_m$) was determined from sequential measurements in order to reduce systematic errors. [16]

REFERENCES

1. **Domen, S. R.,** *J. Res.,* NBS 87, 211, 1982.
2. **Ross, C. K. and Klassen, N. V.,** *Phys. Med. Biol.,* 41, 1, 1996.
3. **Fleming, D. M. and Glass, W. A.,** *Rad. Res.,* 37, 316, 1969.
4. **Selbach, H. J. et al.,** *Metrologia,* 24, 341, 1993.
5. **Ross et al.,** *Metrologia,* 29, 59, 1992.
6. **Schulz, R. J. et al.,** *Med. Phys.,* 14, 790, 1987.
7. **Domen, S. R.,** *J. Res. Natl. Inst. Stand. Tech.,* 99, 121, 1994.
8. **Schulz, R. J. et al.,** *Med. Phys.,* 18, 1229, 1991.
9. **Seuntjens, J. et al.,** *Phys. Med. Biol.,* 38, 805, 1993.
10. **Palmans, H. et al.,** *Med. Phys.,* 23, 643, 1996.
11. **Palmans, H. et al.,** *Phys. Med. Biol.,* 44, 647, 1999.
12. **Guerra, A. S. et al.,** *Phys. Med. Biol.,* 41, 657, 1996.
13. **Burns, D. T. and Morris, W. T.,** in *Proc. High Dose Dosimetry for Radiation Protection,* IAEA, 1991, 123.
14. **DuSautoy, A. R.,** *Phys. Med. Biol.,* 41, 137, 1996.
15. **Schulz, R. J. et al.,** *Phys. Med. Biol.,* 35, 1563, 1990.
16. **Brede, H. J. et al.,** *Rad. Prot. Dos.,* 70, 505, 1997.
17. **ICRU49,** stopping powers and ranges for protons and alpha particles, 1973.

7 Chemical Dosimetry

CONTENTS

The major advancement of the last decade in chemical dosimetry is the development of the gel dosimeter. This is actually an application of a physical phenomenon for better utilization of the well-known Fricke dosimeter. The present chapter mentions in brief some of the contributions made to the use of the already known chemical dosimeters. Chapter 9 is devoted to gel dosimetry.

When dose is measured in water irradiated by an electron beam, the absorbed dose measured with the plane-parallel chamber in the polystyrene phantom can be corrected to obtain the dose to water, according to the equation

$$D_{water}(d_{water}) = D_{med}(d_{med})[\bar{S}/\rho]_{med}^{water} \phi_{med}^{water} \qquad (7.1)$$

where the depth in water d_{water} is related to the depth in the medium d_{med} by

$$d_{water} = d_{med} \times \rho_{eff} = d_{med}(R_{50}^{water}/R_{50}^{mef}) \qquad (7.2)$$

where $[(\bar{S}/\rho)]_{med}^{water}$ is the ratio of the mean unrestricted mass collision stopping power in water to that in the medium; ϕ_{med}^{water} is the fluence factor, i.e., the ratio of electron fluence in water to that in the solid phantom; R_{50} is the depth of the 50% ionization; and P_{eff} is the effective density of the medium and is discussed in the AAPM TG 25 protocol, where recommended values of P_{eff} and ϕ_{med}^{water} are given.

An independent method of absolute dosimetry is the use of the ferrous sulphate dosimeter. The Fricke solution can be prepared by combining 1 mmol/l ferrous ammonium sulphate with 1 mmol/l sodium chloride and 0.4 mol/l sulfuric acid in double distilled water. [1] When the solution is irradiated, the ferrous ions Fe^{2+} are oxidized by radiation to ferric ions Fe^{3+}. The oxidation of the ferrous ions is directly proportional to the absorbed dose. The ferric-ion concentration can be measured by the change in the absorbance using a spectrometer.

The absorbed dose to the dosimeter solution, D_{fricke}, can be calculated according to the equation

$$D_{fricke} = \frac{\Delta A_t}{\rho l \in_{mt} G_{t'}} \qquad (7.3)$$

where ΔA_t is the increase in the absorbance due to irradiation at a temperature t during the spectrophotometric measurement; ρ is the density of the Fricke solution ($\rho = 1.024 \times 10^3$/kg/m^3); l is the length of the light path in the photometer cell (l is generally 0.01 m); $\in_{mt}$ is the molar absorption coefficient for the ferric ions at temperature t; and $G_{t'}$ is the radiation chemical yield of ferric ions at the irradiation temperature t'. For all the electron beam energies, an $\in_m G$ value of 352×10^6 m^{-2} kg^{-1} Gy^{-1} was adopted from ICRU Report No. 35 [9] at an irradiation temperature t' of 25°C and a spectrophotometer measurement temperature t of 25°C. A correction for the temperature t differing from 25°C was made according to the equation

$$\in_{mt} G_{t'} = \in_{m25°C} G_{25°C}[1 + k_1(25 - t)] \times [1 + k_2(25 - t')] \qquad (7.4)$$

where the temperature coefficient k_1, is approximately $0.0007°C^{-1}$ and k^2 is $0.0015°C^{-1}$. Using the mass stopping power ratio $S_{W,Fricke} = 1.004$ to convert the dose to the ferrous sulphate solution, D_{Fricke}, to the dose to water, D_{water}, Equation 7.3 becomes

$$D_{water} = \frac{\Delta A_t}{\rho l \in_{mt} G_t}$$
$$= \frac{278.54 \Delta A_t}{[1 + 0.007(25 - t)][1 + 0.0015(25 - t')]} \qquad (7.5)$$

The ratios of the values of the absorbed dose to water determined with the Exradin chamber, calibrated using all three different calibration methods and the Fricke dosimetry, as a function of the mean energy at depth, E_z, are shown in Figure 7.1.

As mentioned above, gel dosimetry is discussed in Chapter 9. Because of the strong affiliation of that technique to chemical dosimetry, an example is given here. The nuclear magnetic resonance (NMR) longitudinal relaxation

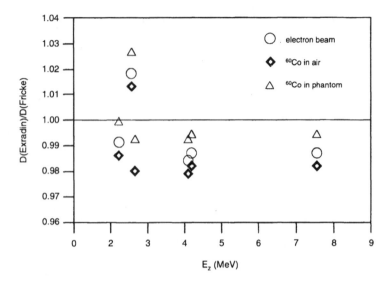

FIGURE 7.1 The ratio of the values of the absorbed dose to water determined with the Extradin chamber, calibrated using three different calibration methods and Fricke dosimetry. (From Reference [1]. With permission.)

rate R_1 dose-response characteristics of a ferrous-sulphate-doped chemical dosimeter system (Fe MRI) immobilized in a gelatin matrix were explored by Hazle et al. [2] Samples containing various concentrations of the $FeSO_4$ dosimeter were irradiated to absorbed doses of 0–150 Gy. R_1 relaxation rates were determined by imaging the samples at a field strength of 1.5 T (^1H Lamor frequency of 63.8 MHz). The response of the system was found to be approximately linear up to doses of 50 Gy for all $FeSO_4$ concentrations studied (0.1–2.0 mM). Changing concentrations in the range of 0.1–0.5 mM affected both the slope and intercept of the dose response curve. For concentrations of 0.5–2.0 mM, the slope of the dose-response curves remained constant at approximately 0.0423 $s^{-1}Gy^{-1}$ in the dose range of 0–50 Gy.

The gels were prepared by mixing 5% gelatin by weight with 75% of the total desired volume of water under constant heating and stirring. This particular gelatin was used because of its high bloom rating (higher bloom ratings indicate stronger or more viscous gels), allowing for lower concentrations of the gel to achieve an acceptable thickness. De-ionized water was used to avoid contamination with other paramagnetic species. The inorganic components were mixed in 25% of the desired total volume at or below room temperature. The resulting final concentrations (for the total volume) were 0.1–2.0 mM $FeSO_4$, 1 mM sodium chloride and 0.5 M sulphuric acid. After the gelatin had been allowed to melt completely at 55°C for at least 15 min, heating was discontinued and the solution containing the inorganic components was added. The gel was then allowed to cool with continual mixing for 30–60 min. After the nominal cooling period, the gel was transferred into 12-mm-diameter polypropylene tubes and allowed to harden at room temperature for

at least 1 h. At this point the gels had congealed to a semi-solid state and did not require refrigeration. After the samples were positioned in the tank, the tank was filled with tap water at room temperature and the gels were allowed to equilibrate for approximately 15 min to the ambient temperature of the tap water (typically 22–24°C). The water tank served to maintain constant temperature and to reduce magnetic susceptibility artifacts at the external interface of the gel samples. Tap water was used in the tank (R_1 typically 0.333 s^{-1}) rather than paramagnetically dopted water (R_1 typically 1.25 s^{-1}), as is the usual case in MRI studies, so that the ambient water surrounding the samples would have low signal intensity for all but the images obtained using extremely long repetition times. [2]

The NMR experiments were carried out using a 1.5-T Signa whole-body scanner. The longitudinal relaxation rates were determined using seven single-slice images of the same spatial location acquired with different repetition times by saturation recovery analysis. Typical repetition times were 100, 300, 600, 1000, 1500, 3000, and 6000 ms. Single-slice images were obtained in lieu of multi-slice images to avoid any out-of-plane saturation effects. R_1 was calculated using a three-parameter fit of the equation below, using the standard region-of-interest software provided with the system:

$$M(T_R) = M_0[1 - (1 - \cos\alpha)e^{-T_R R_1}] \quad (7.6)$$

where $M(T_R)$ is the magnetization observed at a repetition time T_R, M_0 is the equilibrium magnetization ($T_R = 0$) and α is the read-pulse flip angle (nominally 90°).

The general dose response of the dosimeter was determined by irradiating gels of 0.0, 0.5, 1.0, and 2.0-mM

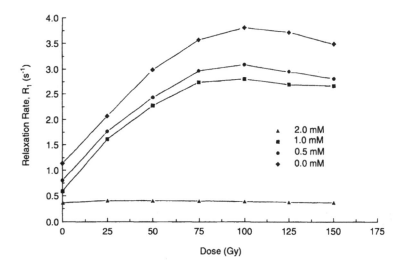

FIGURE 7.2 The dose-response curve for a gel concentration of 5% and $FeSO_4$ concentrations of 0.0, 0.5, 1.0, and 2.0 mM are shown for dose up to 150 Gy. The usable range seems to be 0–50 Gy. (From Reference [2]. With permission.)

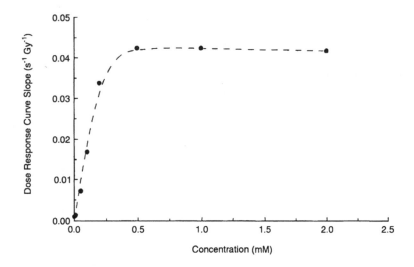

FIGURE 7.3 The slopes of the dose-response curves plotted as a function of initial $FeSO_4$ concentration to demonstrate the effect of concentration on dose-response sensitivity. In the range of 0 mM to about 0.5 mM, increasing the $FeSO_4$ concentration results in increasing sensitivity. Concentrations in the range of 0.5 to 2.0 mM have approximately constant sensitivity (average = 0.0423 $s^{-1}Gy^{-1}$). (From Reference [2]. With permission.)

$FeSO_4$ to doses of 0–150 Gy. A single sample was irradiated for each dose level. The results of this experiment (Figure 7.2) suggest that the linear range of the system for the concentrations considered is 0–50 Gy, irrespective of $FeSO_4$ concentration. The useable range could be extended to 100 Gy if higher-order response terms are acceptable.

To demonstrate the effect of concentration on sensitivity, the slopes of the response curves were plotted against concentration (Figure 7.3). Increasing $FeSO_4$ concentration affects not only the initial R_1 but also the slope of the dose-response curves for concentrations up to about 0.5 mM. [2]

The mean absorbed dose in the detection volume of Fricke dosimeters can be written

$$\langle D_f \rangle = \Delta(OD)/\rho_f l \varepsilon G(Fe^{3+}) \qquad (7.7)$$

where $G(Fe^{3+})$ is the radiation chemical yield of the oxidation of ferrous ions into ferric ions, ε is the molar extinction coefficient of ferric ions at the considered wavelength, ρ_f is the density of Fricke solution, l is the light pathlength through the spectrophotometer cell, and $\Delta(OD)$ is the increase in optical density of the solution. Direct dose measurements require knowledge of all the above parameters.

The transfer method uses the ratio of absorbed doses in the detection volume of the transfer dosimeter in both phantoms. After simplification, this ratio reduces to the ratio of instrument responses, either ionization currents or increases in optical density. In particular, the constants W_{air} or $\in G(Fe^{3+})$ and their associated uncertainties are eliminated.

Chauvenet et al. [3] discussed the ratio of dose absorbed in the detecting material vs. the dose to the wall. The dosimeter is made of a detection volume V, filled with material "det" surrounded by a wall of material "wall." The fraction of dosimeter response due to electrons arising in the detection volume is noted α_{det}, and the fraction of dosimeter response due to electron arising in the dosimeter wall, is α_{wall}. The correction factor for the perturbation of photon energy fluence at point C, caused by the replacement of medium "med" by material "wall" in the volume of the wall, is denoted $_{wall}(\psi_c)_{med,wall}$; the same correction factor caused by the replacement of medium "med" by material "det" in the detection volume is denoted $_{med}(\psi_c)_{med,det}$. Similar symbols are used for stopping power and absorption coefficient ratios. The corresponding absorbed-dose to collision-kerma ratios are denoted $\beta_{med,wall}$ and $\beta_{med,det}$. Using these notations, the absorbed dose in medium "med" at point C without dosimeter, $D_{med}(C)$, and the mean absorbed dose in the detection volume V of the dosimeter, $\langle D_{det}\rangle$ are then related by [3]

$$
\begin{aligned}
D_{med}(C) = {} & \langle D_{det}\rangle\,_{wall}(\psi_c)_{med,wall,det}(\psi_c)_{med,det} \\
& \times [(1 - \alpha_{wall} - \alpha_{det})S_{med,det} \\
& + \alpha_{wall}(\beta\bar\mu_{en}/\rho)_{med,wall}S_{wall,det} \\
& + (\beta\bar\mu_{en}/\rho)_{med,det}]
\end{aligned}
\tag{7.8}
$$

Equation (7.8) is applied to two identical Fricke dosimeters with glass walls. The optical density increase caused by the production of Fe^{3+} ions under irradiation was measured by spectrophotometry at a wavelength of 304 nm, relative to a blank realized with a non-irradiated sample. The measured optical densities are corrected for irradiation temperature (t_G) and reading temperature (t_e), according to the following formula:

$$
\Delta(OD)
$$
$$
= \frac{\Delta(OD)_{uncorrected}}{(1 + 0.0069(t_e - 20))(1 + 0.0012(t_G - 20))}
\tag{7.9}
$$

Correction factors for photon fluence perturbation $_{wall}(\psi_c)_{med,wall}$ and $_{det}(\psi_c)_{med,det}$ are evaluated by derivation of the ratio of transmissions through the zone corresponding to the wall or to the detection volume filled with appropriate materials. For this purpose, the following parameters are introduced: (1) the effective attenuation coefficient μ', (2)

the effective mean wall thickness x, and (3) the effective mean distance d between the front face of the wall-detection volume interface and the median plane of the detection volume perpendicular to the beam axis. From these considerations, the following relations can be derived:

$$
_{wall}(\psi_c)_{med,wall} = \exp(-\mu'_{med}x)/\exp(-\mu'_{wall}x)
\tag{7.10}
$$

$$
_{det}(\psi_c)_{med,det} = \exp(-\mu'_{med}x)/\exp(-\mu'_{det}x)
\tag{7.11}
$$

Coefficients α_{det} and α_{wall} can be expressed as follows, with reasonably good approximation:

$$
\alpha_{det} \approx 1 - (\langle u_{det}\rangle/\langle t_{det}\rangle)
\tag{7.12}
$$

$$
\begin{aligned}
\alpha_{wall} & \approx (1 - (u_{wall}/t_{wall}))(\langle u_{det}\rangle/\langle t_{det}\rangle) \\
& \approx [1 - (u_{wall}/t_{wall})](1 - \alpha_{det})
\end{aligned}
\tag{7.13}
$$

where u_{wall} is the ratio of absorbed doses to the wall at the outer interface and at the inner interface; t_{wall} is the photon energy fluence transmission through the wall; $\langle u_{det}\rangle$ is the mean value of the ratios of absorbed doses to material "det" at each point of V and at the wall interface, averaged over the whole detection volume V; and $\langle t_{det}\rangle$ is the mean value of the photon energy fluence transmission through "del" from the wall interface to each point of volume V, averaged over the volume V. According to the exponential decay and build-up assumption of electron fluence (Burlin [10]) or, more exactly, of energy deposition, one can write

$$
u_{wall} = \exp(-\mu_{e,wall}x')
\tag{7.14}
$$

$$
\langle u_{det}\rangle = [1 - \exp(-\mu_{e,det}y')]/\mu_{e,det}y'
\tag{7.15}
$$

where $\mu_{e,wall}$ and $\mu_{e,det}$ are effective attenuation coefficients for electron energy deposition in materials "wall" and "det," respectively; x' is the mean wall thickness for electrons arising in medium "med", and y' is the mean effective thickness of volume $V(y' = 2d)$.

The EGS4 Monte Carlo calculation for a 1-mm-thick Pyrex wall yielded $\alpha_{det} = 0.89$ and $\alpha_{wall} = 0.10$. [3] Therefore,

$$
D_w(C)/\langle D_f\rangle = fP_{wall} = 1.0037(10)
\tag{7.16}
$$

Chauvenet et al. obtained

$$
D_w(C)/\langle D_f\rangle = 1.0026(24)
\tag{7.17}
$$

Dose conversion and wall correction factors for Fricke dosimetry in high-energy photon beams were calculated by Ma and Nahum [4], using both an analytical general cavity model and Monte Carlo techniques. The conversion

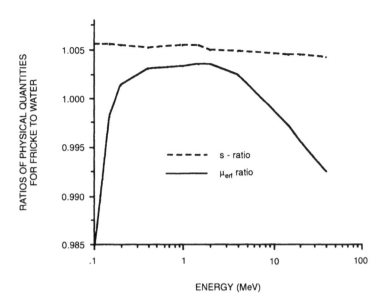

FIGURE 7.4 The photon mass energy absorption coefficients for water to the Fricke dosimeter solution. The dashed line shows the electron mass stopping power ratio for water to the Fricke dosimeter solution. (From Reference [4]. With permission.)

factor is calculated as the ratio of the absorbed dose in water to that in the Fricke dosimeter solution with a water-walled vessel. The wall correction factor accounts for the change in the absorbed dose to the dosimeter solution caused by the inhomogeneous dosimeter wall material. It is shown that Fricke dosimeters in common use cannot be considered to be "large" detectors and, therefore, "general cavity theory" should be applied in converting the dose to water. It is confirmed that plastic dosimeter vessels have a negligible wall effect. The wall correction factor for a 1-mm-thick Pyrex-walled vessel varies with incident photon energy from 1.001 ± 0.001 for a ^{60}Co beam to 0.983 ± 0.001 for a 24-M ($TPR_{10}^{20} = 0.80$) photon beam. (TPR_{10}^{20} is the ratio of absorbed dose at 20-cm depth to that at 10 cm—"tissue-phantom ratio"). This implies that previous Fricke measurements with glass-walled vessels should be re-evaluated.

For high-energy photon (and electron) beams, the absorbed dose to water averaged over the volume occupied by a wall-less Fricke dosimeter, D_w, can be obtained as

$$D_w = f D_f \qquad (7.18)$$

where f is the absorbed dose conversion factor and D_f is the dose calculated according to Equation (7.3) for a wall-less Fricke dosimeter; Equation (7.18) effectively defines f. For the case of a non-water-walled vessel with Fricke dose D_f', one can write

$$D_f = P_{wall} D_f' \qquad (7.19)$$

which defines P_{wall} as the ratio of the absorbed dose to the Fricke solution obtained with a wall-less dosimeter to that

with the non-water-walled dosimeter. Putting Equations (7.18) and (7.19) together, D_w is given by

$$D_w = f P_{wall} D_f' \qquad (7.20)$$

Instead of using two factors in determining the dose to water, a general conversion factor F_p can be used; i.e.,

$$F_p = f P_{wall} \qquad (7.21)$$

For megavolt photons, the ratio of the mass energy absorption coefficients for water to the Fricke dosimeter solution, $(\mu_{en}/\rho)_{water,f}$, decreases with increasing photon energy (see Figure 7.4). The dose conversion factor derived from general cavity theories should be somewhere between $(\mu_{en}/\rho)_{water,f}$ and $s_{water,f}$. The fractional dose resulting from the electrons generated by photon interactions in the dosimeter solution decreases with increasing incident photon energy (see Figure 7.5).

EGS4 Monte Carlo calculations show for megavoltage radiotherapy photon beams, and experimental measurements confirm with an accuracy of 0.2%, that glass- or quartz-walled vials used in Fricke dosimetry increase the dose in the Fricke solution. [5] This is caused mainly by increased electron scattering from the glass, which increases the dose to the Fricke solution. For plastic vials of similar shapes, calculations demonstrate that the effect is in the opposite direction, and even at high energies it is much less (0.2% to 0.5%).

ICRU Report 35 [9] recommended the use of plastic dosimeter vessels for Fricke dosimetry in the determination of absorbed dose in a phantom irradiated by high-energy photon and electron beams. Plastic vessels have

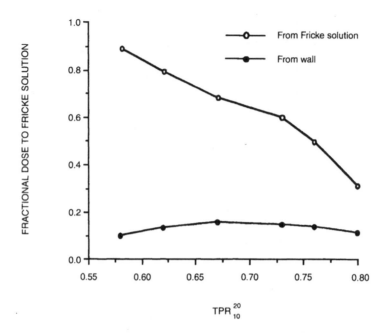

FIGURE 7.5 The fractional doses resulting from electrons generated by photon interactions in either the Fricke dosimeter solution or in the wall material. The remaining contribution is from the medium. The dosimeter is 1.354 cm in diameter and 5.5 cm in length. The wall material is Pyrex glass and the thickness is 1 mm. (From Reference [4]. With permission.)

linear attenuation coefficients and stopping-power values somewhat less than those for water, and this compensates for the somewhat greater values in the ferrous sulphate solution. The main purpose of using plastic vessels is to minimize the perturbation effects on the electron scattering introduced by the presence of the Fricke dosimeter in the water phantom. However, a great deal of care is required with plastic vessels because of storage effects, i.e., chemical effects on the ferrous sulphate solution when stored in the plastic container. For glass vessels, the perturbation and cavity theory effects are potentially larger. Burlin and Chan [11] showed both experimentally and theoretically that the cavity-theory effects could be as large as 7% with small volumes of Fricke solution in thick-walled silica vessels in a ^{60}Co beam. This is because in the normal, i.e., large volume detector, one considers the Fricke solution to be a photon detector in which all the dose is delivered by electrons starting in the Fricke solution and, hence, $D_{med} = D_f(\bar{\mu}_{en}/\rho)_{Fricke}^{med}$, where D_{med} is the dose to the medium and D_f is the dose to the Fricke solution. In the small volume detector, one has an electron detector in which the dose is delivered by electrons starting in the walls; hence, Bragg-Gray cavity theory applies and one has $D_{med} = D_f(\mu_{en}/\rho)_{Fricke}^{wall}(\mu_{en}/\rho)_{wall}^{med}$.

Results for the calculated absorbed-dose conversion factors relating the absorbed dose in the Fricke solution to the absorbed dose in the homogeneous water phantom at the same point are presented in Figure 7.6.

Figure 7.7 presents summaries of the calculated wall-correction factors P_{wall} for the standard coin-shaped,

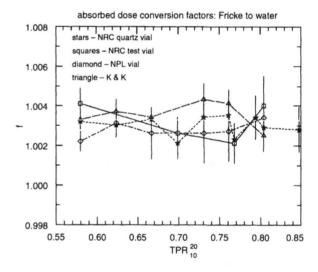

FIGURE 7.6 Absorbed-dose conversion factors, f, for converting dose to Fricke solution into dose to water for Fricke vials of various shapes as described in the text. Stars, NRC standard vial shape; squares, NRC test vial shape; diamonds, NPL vial shape with no air gaps; triangles, long circular vial shape to model Kwa and Kornelsen vials. (From Reference [5]. With permission.)

quartz-walled Fricke vials used at NRC. These corrections are substantial and must be taken into account. To facilitate their use, these data have been fit to a simple linear expression in terms of TPR_{10}^{20}:

$$p_{wall} = 1.0478 - 0.08223(TPR_{10}^{20}) \qquad (7.22)$$

The data show an apparent departure from a linear relationship near $TPR_{10}^{20} = 0.76$. This may be associated with the change in the reference depth from 5 to 7 cm at this point, or it may indicate that TPR_{10}^{20} is not a good beam quality indicator for this correction, or it may just be a statistical fluctuation.

The polyethylene walls reduce the dose to the Fricke solution.

The NRC vials are made of quartz because they were designed for use in a calibration service in which the vials were shipped to the clinic for measurements and hence

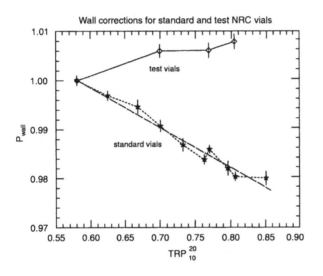

FIGURE 7.7 Wall-correction factors p_{wall} for converting absorbed dose to Fricke solution measured in the standard quartz-walled vials and the polyethylene-walled test vials to dose in wall-less detectors. The dashed line shows the values from Equation (7.22). (From Reference [5]. With permission.)

must not suffer from chemical storage effects. In view of the correction factors, it may be more appropriate in-house to use plastic-walled vials.

If electron transport is not included in the Monte Carlo calculation, the effect of the front and back wall is only a 0.27% reduction in dose; i.e., although the front wall attenuates the primary beam by 0.55%, the additional dose from photon scatter changes the net effect to 0.27%. [5]

Ma and Nahum [6] presented Fricke-to-water dose conversion and wall correction factors for Fricke ferrous sulphate dosimetry in high-energy electron beams. The dose conversion factor has been calculated as the ratio of the mean dose in water to the mean dose in the Fricke solution with a water-walled vessel, and the wall correction factor accounts for the change in the Fricke dose due to the presence of the non-water wall material. The results show that for a Fricke dosimeter of 1.354-cm diameter and 5.5-cm length, the dose conversion factor is nearly constant at 1.004 (within 0.1%) for electron energies of 11–25 MeV if the dosimeter is placed at the depth of maximum dose, but it can vary by a few percent if the dosimeter is placed on the descending portion of the depth-dose curve. The wall corrections are smaller than 0.3% for 0.1-cm-thick polystyrene vessels. For 0.1-cm-thick Pyrex glass vessels, the wall correction factor varies from 0.989 for 11 MeV at 2.75-cm depth to 0.997 for 25 MeV at 7-cm depth. This confirms recent experimental findings that Fricke doses obtained with glass-walled vessels were up to 1% higher than those with polystyrene-walled vessels.

The wall correction factor p_{wall} accounts for any change in the absorbed dose in the Fricke dosimeter solution due to the presence of the non-water-equivalent

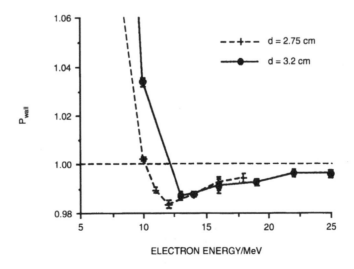

FIGURE 7.8 The Monte Carlo–calculated p_{wall} for a Fricke dosimeter of 1.354-cm diameter and 5.5-cm length at a depth of either $d = 2.75$ cm or $d = 3.2$ cm in water irradiated by monoenergetic electron beams of 10×10 cm^2 field size. The wall thickness is 0.1 cm and the wall material is Pyrex. The Monte Carlo statistical uncertainties are shown as error bars on the curves. (From Reference [6]. With permission.)

wall material. The perturbation of the electron fluence in the phantom by the non-water-equivalent wall depends on the combined effect of the difference in mass stopping power, scattering power, and density between the wall and phantom material. For a Pyrex glass wall, for instance, the mass stopping power is lower than that of water, but it has a much greater density (2.23 g cm^{-3}) and scattering power compared to water. The glass-wall effect will therefore mainly result from the high density and scattering power of the wall material. On the other hand, plastic wall materials such as polystyrene generally have smaller stopping powers and/or scattering powers and hence may cause effects in the opposite direction.

It is seen in Figure 7.8 that the high density of the Pyrex glass causes more rapid fall-off of the dose with depth and therefore reduces the dose in the Fricke dosimeter solution.

Aqueous coumarin was investigated by Collins et al [7] as a possible dosimeter for radiation therapy applications. Coumarin-3-carboxylic acid in aqueous solutions converts, upon irradiation, to the highly fluorescent 7-hydroxy-coumarin-3-carboxylic acid. The intensity of the fluorescence signal is linearly proportional to the number of the formed 7-hydroxy-coumarin-3-carboxylic acid molecules, which in turn is proportional to the radiation-absorbed dose. The system exhibits nearly linear behavior with dose, in the range of 0.1 to 50 Gy.

Concentrations in the range of 10^{-3} to 10^{-5} M were prepared by diluting the appropriate amount of coumarin in distilled water and heating the solution. Buffer (PBS pH 7.4) was added at the end. The 7-hydroxy-coumarin-3-carboxylic acid is easily soluble in water and heating was not necessary.

Coumarin solutions were irradiated in polystyrene or polymethylmethacrylate (PMMA) 1-cm-path-length cuvettes (4.5-ml volume, 1-mm wall thickness) at room temperature with either a ^{137}Cs gamma-ray source (dose rate approximately 1 Gy/min) or a Varian Clinac 2100C linear accelerator (nominal dose rates 0.8, 1.6, 2.4, and 4 Gy/min) in air or water. The absorption spectra of coumarin (10^{-4} M), 7-hydroxy-coumarin (10^{-4} M), and irradiated coumarin (10^{-4} M, dose of 250 Gy) are shown in Figure 7.9.

Fluorescence spectra of 7-hydroxy-coumarin (10^{-7} M), coumarin (10^{-4} M) and irradiated solution of coumarin (10^{-4} M, dose of 10 Gy) are shown in Figure 7.10 in the range of 350 to 600 nm. The 7-hydroxy-coumarin and the irradiated coumarin spectra were generated under 388-nm excitation. The coumarin emission was excited with 330 nm (coumarin excited at 388-nm shows negligible fluorescence). The spectra of the 7-hydroxy-coumarin and the irradiated coumarin exhibit the same emission peak centered at 450 nm. The un-irradiated coumarin solution exhibits a weak luminescence maximum at 410 nm. Both the 450-nm emission band in hydroxy-coumarin and the 410-nm

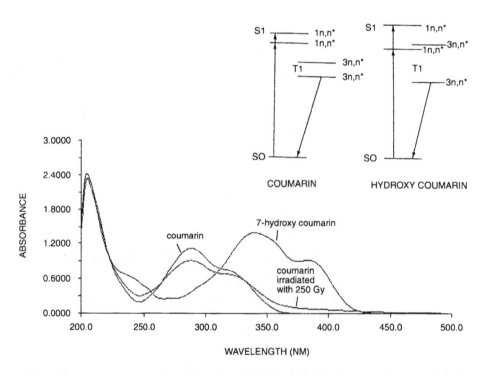

FIGURE 7.9 Room-temperature absorption spectra of coumarin-3-carboxyl acid (10^{-4}M), 7-hydroxy-coumarin-3-carboxyl acid (10^{-4}M), and coumarin-3-carboxyl acid (10^{-4}M) irradiated with 250 Gy. (From Reference [7]. With permission.)

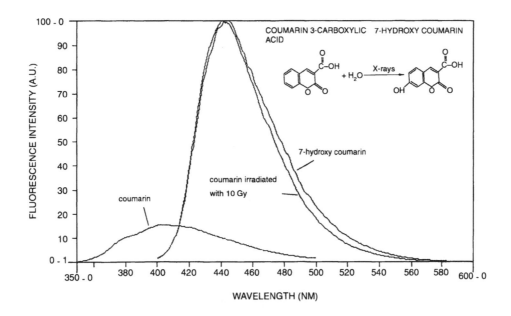

FIGURE 7.10 Room-temperature fluorescence of coumarin-3-carboxyl acid (10^{-7}M) and coumarin-3-carboxyl acid (10^{-4}M) irradiated with 10 Gy. (From Reference [7]. With permission.)

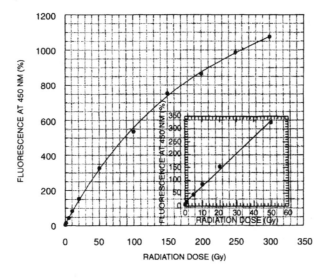

FIGURE 7.11 Fluorescence of irradiated 10^{-4} M aqueous coumarin-3-carboxylic acid vs. radiation-absorbed dose. The excitation was at 400 nm, the emission at 450 nm, and the resolution at 5 nm. (From Reference [7]. With permission.)

band in coumarin are assigned to a transition from the triplet $^3\pi\pi^*$ state to the ground state. [7]

Figure 7.11 depicts the fluorescence of 10^{-4} M coumarin solution irradiated in air with a linear accelerator 6-MV pulsed photon beam at a nominal dose rate of 4 Gy/min in the dose range of 0 to 300 Gy. The solution was irradiated and measured in PMMA containers with a volume of 3 ml. The measured signal was compared to the signal of a standard solution of 10^{-7} M 7-hydroxy-coumarin.

Gupta et al. [8] gave the absorption spectra of the ferric-xylenol orange complex and discussed the effect of acidity on the absorption maxima in the presence of alanine,

glutamine, and valine. The irradiated alanine and glutamine were read by five different methods:

1. Amino acid dissolved in 9.8 mL of aerated sulphuric acid. 0.1 mL each of xylenol orange and ferrous ions were added one after the other.
2. Amino acid dissolved in 9.8 mL of aerated sulphuric acid. Then 0.1 mL each of ferrous ions and xylenol orange were added one after the other; xylenol orange was added after the reaction with ferrous ions was over.
3. Amino acid dissolved in 10 mL of aerated FX solution.

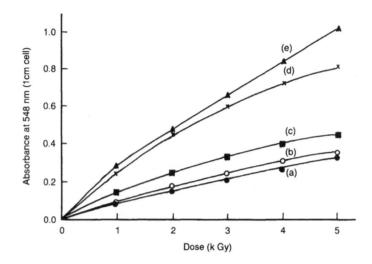

FIGURE 7.12 Oxidation of ferrous ions by irradiated alanine. (From Reference [8]. With permission.)

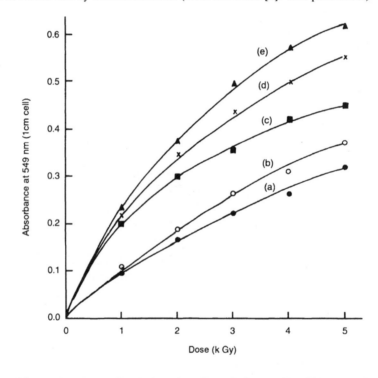

FIGURE 7.13 Oxidation of ferrous ions by irradiated glutamine. (From Reference [8]. With permission.)

4. Amino acid dissolved in 9.8 or 9.9 mL of aerated ferrous sulphate solution, and 0.2 or 0.1 mL of 0.01 mmol dm^{-3} xylenol orange was added later.

5. Amino acid was dissolved in 9.8 or 9.9 mL of oxygenated ferrous sulphate solution, and 0.2 or 0.1 mL of 0.01 mmol dm^{-3} xylenol orange was added later.

Figures 7.12 and 7.13 show that the oxidation of ferrous ions for method 3 is almost double that for method 1. The oxidation increases further for methods (4) and (5).

Glutamine powder was irradiated to 50-kGy and 5-kGy doses. Twenty mg of irradiated powder was dissolved in 5 mL of 0.033 mol dm^{-3} aerated sulphuric acid and 5mL of FX solution containing 0.4 mmol dm^{-3} of ferrous ammonium sulphate; 0.2 mmol dm^{-3} of xylenol orange in 0.033 mol dm^{-3} sulphuric acid was added later (method b). Ten measurements were done using methods (3) and (6). Figure 7.14 shows the oxidation of ferrous ions by the two methods for valine. [8]

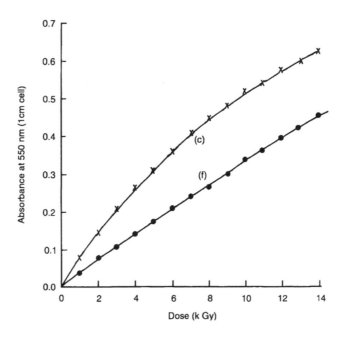

FIGURE 7.14 Oxidation of ferrous ions by irradiated valine (methods 3 and 6). (From Reference [8]. with permission.)

The oxidation of ferrous ions by the dissolution of amino acids in aqueous aerated acidic solution can be written as follows:

$$\dot{R}_1H + O_2 = R_1H\dot{O}_2$$

$$\dot{R}_1H + Fe^{3+} = R_1 + Fe^{2+} + H^+$$

$$\dot{R}_2H + Fe^{2+} + O_2 = products + Fe^{3+}$$

$$R_2H\dot{O}_2 + Fe^{2+} = Fe^{3+} + products$$

where $\dot{R}_1H$ and $\dot{R}_2H$ are two free radicals formed on the radiolysis of amino acids. There is a competition between reactions (1) and (2). Here R_1 is a stable product. Some free radicals oxidize ferrous ions directly. This competition kinetics gives the following relationship:

$$1/[A] = C_1/[D] + C_2 \qquad (7.23)$$

where A is the absorbance of the ferric-xylenol orange complex in the presence of the amino acid, D is the dose, and C_1 and C_2 are constants.

REFERENCES

1. **Xu, Z. et al.,** *Med. Phys.,* 23, 377, 1996.
2. **Hazle, J. D. et al.,** *Phys. Med. Biol.,* 36, 1117, 1991.
3. **Chauvenet, B. et al.,** *Phys. Med. Biol.,* 42, 2053, 1997.
4. **Ma, C-M. and Nahum, A. E.,** *Phys. Med. Biol.,* 38, 93, 1993.
5. **Ma, C-M. et al.,** *Med. Phys.,* 20, 283, 1993.
6. **Ma, C-M. and Nahum, A. E.,** *Phys. Med. Biol.,* 38, 423, 1993.
7. **Collins, A. K. et al.,** *Med. Phys.,* 21, 1741, 1994.
8. **Gupta, B. L. et al.,** in *Proc. High Dose Dosimetry for Radiation Protection,* IAEA, 1991, 327.
9. **ICRU 35,** Radiation dosimetry: electron beam with energy 1–50 MEV, 1984.
10. **Burlin, T. E.,** *Br. J. Radiol.,* 39, 727, 1966.
11. **Burlin, T. E. and Chan, F. K.,** *Int. J. Appl. Radiat. Isot.,* 20, 767, 1969.

8 Solid-State Dosimeters

CONTENTS

I. DIAMOND DOSIMETER

Diamond detectors are attractive for high-resolution megavoltage photon and electron beam measurements because of their small size and near tissue equivalence ($Z = 6$ compared to $Z_{eff} = 7.42$ for soft tissue). Diamond detectors work as solid-state ionization chambers. The absorption of ionizing radiation induces a temporary change in the electrical conductivity of the diamond through the production of electrons and positive holes that have sufficient energy to be free to move through the crystal. They have dimensions and spatial resolution similar to commonly used silicon diode detectors, and they are almost tissue-equivalent. Other advantages include high sensitivity, low leakage current, and high resistance to radiation damage. Diamond detectors can be considered resistive elements where, for a given bias voltage, the current is zero with no radiation and increases almost linearly with dose rate. For a given dose rate, current is nearly proportional to bias voltage. The response of the diamond detector has been shown to be nearly independent of the incident photon energy, since the mass attenuation coefficient ratio of water to carbon is nearly constant.

In a pure crystal, recombination quickly neutralizes the ion pairs produced. Nitrogen atoms are generally present as an impurity in the diamond crystals. After bonding covalently with the carbon atom, the nitrogen atom is left with an unpaired electron, which has a high probability of jumping to the conduction band. The ionized impurity created acts as a trap for electrons generated during electron-hole pair production. The free electrons captured in the ionized trap produce polarization, thus reducing the electric field generated by the external $+100$-V bias. As the diamond detector is irradiated, its response initially decreases due to an increase in the polarization effect. The detector response is finally stabilized when an equilibrium trap population is attained.

The suitability of a natural diamond detector with a special contact system for the measurement of relative dose distributions in selected radiotherapy applications was studied by Vatnitsky and Jarvinen. [1] A natural diamond plate of 0.2–0.4-mm thickness and with the special contact system was positioned inside a cylindrical plastic capsule at a depth of 0.2 mm for detector Dl, 0.3 mm for detector D2, and 1.0 mm for detector D3; see Figure 8.1. The material of the capsule was PMMA (Dl, D2) and polystyrene (D3). The front window of detector Dl was transparent, while the windows of detectors D2 and D3 were opaque to light. The detector bias was 250 V (Dl), 150 V (D2), and 100 V (D3). After switching on the high voltage, a pre-irradiation of the detectors up to a dose of about 2 Gy (Dl) and 3Gy (D2, D3) was performed to settle the response of the diamond to a stable level.

The variation of the stopping-power ratio (water/material) as a function of electron energy for the three detector materials is shown in Figure 8.2.

Diamond detectors are radiosensitive resistors whose conductivity varies almost in proportion to dose rate and is almost independent of bias voltage for a constant dose rate. At the recommended bias of $+100$ V, and also at $+200$ V, the detector is operating with incomplete charge collection due to the electron-hole recombination time being shorter than the maximum time for an electron to be collected by the anode. As dose rate is varied by changing FSD or depth (changing dose per pulse), detector current and dose rate are related by the expression $i \propto D^{\Delta}$, where Δ is approximately 0.98. This manifests itself in an overestimate in percentage depth dose of approximately 1% at a depth of 30 cm when compared to ionization chamber results. The dose rate dependence is attributed to the reduction in recombination time as dose rate increases. [2]

The recombination rate in a pure crystal (when an equilibrium number of free electrons is established) is proportional to the square root of the rate of ion-pair production

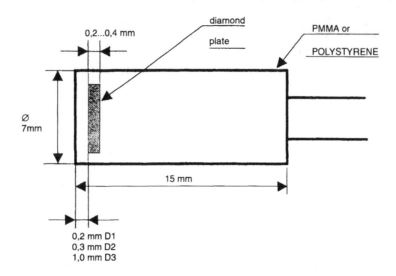

FIGURE 8.1 Construction of the diamond detectors. (From Reference [1]. With permission.)

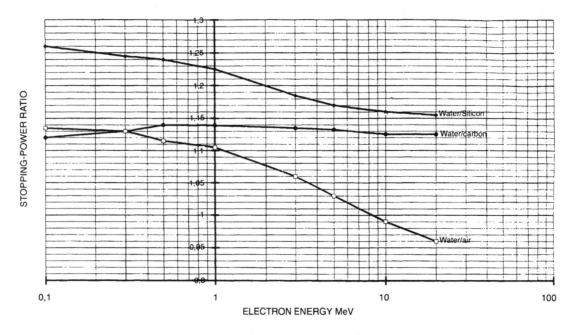

FIGURE 8.2 Ratios of mass collision stopping powers for water and the three detector materials (ICRU 37 [33]). (From Reference [1]. With permission.)

and, hence, to the dose rate. This is due to the increase in the probability of recombination with the number of vacant holes. Charge-collection efficiency therefore decreases with dose rate, making a pure crystal unsuitable for dosimetry. If impurities are present, metastable states are introduced which trap many electrons that would otherwise recombine with holes. If the number of electrons in traps is large, the proportional increase in the number of vacant holes can be almost independent of dose rate. This means that the recombination rate, and hence the efficiency of charge collection, is almost independent of the rate of ion-pair production, giving an almost linear increase in detector signal with dose rate. (Hoban [2])

For a diamond crystal to be suitable for radiation dosimetry, some impurities are necessary in order for the signal to increase linearly with dose rate, but excess impurities will cause the diamond to be insensitive and to suffer polarization effects. Diamond crystals suitable for dosimetry are of type IIa, meaning that they are almost transparent to ultraviolet light. The degree of transparency increases with a reduction in the concentration of nitrogen impurities, which provides a basis for choosing crystals with a low nitrogen concentration.

Due to the rapid rate of electron-hole recombination, a very high bias voltage is required for complete charge collection to occur.

The current measured using a Nuclear Enterprises 2570 electrometer (with bias supplied externally) for bias voltages of 20–200 V, at a dose rate of 2.0 Gy min⁻¹, is shown in Figure 8.3. The current plotted is the charge collected for a dose of 1.0 Gy divided by the irradiation time of 30 s.

The charge collection efficiency is the ratio of detector current to the rate at which charge is produced in ionization (gain factor). The rate of charge production in the crystal is readily calculated from knowledge of the dose rate D, density r, and volume V of the sensitive element, and the energy w required to produce an ion pair:

$$\frac{dQ}{dt} = \frac{DV\rho e}{w} \qquad (8.1)$$

The density of diamond is 3.5 g cm⁻³, the volume of the crystal was 1.4×10^{-3} cm³, and w is approximately 16 eV, so at a dose rate of 2.0 Gy min⁻¹ (3.3×10^{-5} J g⁻¹s⁻¹), the charge production rate is 1.0×10^{-8} C s⁻¹. From Figure 8.3, the detector current at 2.0 Gy min⁻¹, with a bias of 100 V, is 5.8×10^{-9} A, which gives a gain factor of 0.58.

Dose-rate dependence measurements were made with the RK ionization chamber, diode, and diamond. Results are plotted in Figure 8.4, where it is seen that the relative response of the diamond appears to be the same as the ionization chamber for dose rates up to approximately 1.5 Gy min⁻¹ and then decreases steadily as the dose rate increases. In contrast to the findings of Rikner [3], the diode shows a relatively large increase in over-response as dose rate increases.

The suitability of diamond detectors has been investigated by Seuntjens et al. [4] to measure relative central-axis depth kerma curves in medium-energy x-rays by comparing them to the NE2571 cylindrical ionization chamber. To investigate their limit of application in terms of energy dependence, two diamond detectors were first calibrated in terms of air-kerma free in air using free-air ionization chambers for several low-and medium-energy x-ray qualities. Energy dependence correction factors were used to convert depth signal curves measured with a diamond detector into depth kerma curves and to compare them to curves measured using the well characterized NE2571 ionization chamber. The results show that the diamond detector is directly suitable for relative dosimetry of medium-energy x-rays with qualities higher than or equal to 100 kV (HVL 4.8-mm Al) to within 2% depth dose. For lower x-ray qualities (down to 80 kV, HVL 2.6-mm Al), relative energy dependence correction factors

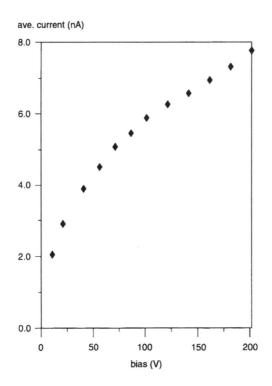

FIGURE 8.3 Average diamond detector current in nA vs. bias voltage for a dose rate of 2.0 Gy min⁻¹. Current is the stable electrometer reading in nC for a dose of 1.0 Gy divided by the irradiation time of 30 s. (From Reference [2]. With permission.)

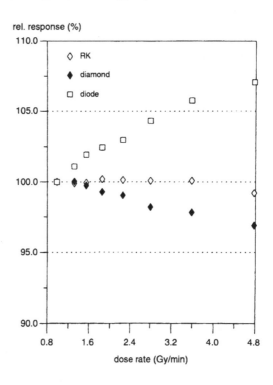

FIGURE 8.4 Response of diamond, diode, and RK ionization chamber as average dose rate varies from 0.98 to 4.77 Gy min⁻¹ by changing FSD from 140 cm to 60 cm. Dose per pulse varies from 0.079 to 0.387 mGy. Values on the graph are the ratio of each detector reading to Farmer ionization chamber reading, normalized to 100% at 0.98 Gy min⁻¹. (From Reference [2]. With permission.)

of at most 12% of the kerma value are necessary at a depth of 10 cm when normalized to the response near the phantom surface.

The diamond detector types used by Seuntjens were from PTW (type Riga). The radiation-sensitive region of the diamond detector is a low-impurity natural diamond plate of thickness specified between 0.1 and 0.3 mm, sealed in a cylindrical polystyrene housing of diameter 7.3 mm. The surface of the diamond is 1 mm beneath the outer surface of the housing. As part of the irradiation procedures, for each energy, the detectors were first pre-irradiated up to an accumulated dose of at least 3 Gy as determined by studying their response stability. Two diamond detectors (PTW type Riga) were involved in relative air-kerma calibrations against free-air chambers.

Figure 8.5 shows the response curves normalized to 1 at the radiation quality with average photon energy of 121.9 keV. The response exceeds unity by up to 6% for average photon energy of 80 keV, presumably due to the high Z materials involved in biasing the diamond chip. Below 80 keV it gradually drops off, down to 65% of the response at 29 keV, primarily due to photon attenuation in the high-density chip. The differences between the two PTW detectors in the decreasing portion of the response curve are due to differences in chip thicknesses. For the purpose of evaluating the quality correction factor, the responses of both detectors were averaged. Also shown in Figure 8.5 (dashed-dotted line) are the ratios of mass energy absorption coefficients graphite to air, normalized at the same energy (121.9 keV).

A PTW Riga diamond detector has been evaluated by Heydarian et al. [5] for use in electron beam dosimetry, by comparing results with those obtained using a diode

(Scanditronix p-Si) and an ionization chamber (Scanditronix RK). The directional response of the diamond at 6 MeV and 15 MeV is more uniform than that of the diode, but for both detectors there is a dip in response when the beam axis is perpendicular to the detector stem. Spatial resolution of the diode detector, measured beneath a 2-mm-wide slit, is slightly better than that of the diamond detector with detector stems both parallel and perpendicular to the beam axis. Diamond and diode depth-dose curves both agree well with corrected ionization chamber results at 15 MeV, while at 6 MeV the diamond is in better agreement. This indicates that the diode provides better spatial resolution than the diamond for measuring profiles, but the diamond is preferable for low-energy depth-dose measurements.

Silicon diode detectors have the advantage of a small, high-density, sensitive volume and thus have a high spatial resolution. A significant disadvantage in principle, however, is the non-water-equivalence of the silicon. As Figure 8.6 shows, the silicon:water collision-mass stopping power ratio is not constant with electron energy, especially at energies below 5 MeV (ICRU 37 [33]). In addition, separate diode detectors are required for electron and photon do-simetry.

The diamond detector is attractive because of its near water-equivalence and its high spatial resolution. As Figure 8.6 shows, the carbon:water stopping power ratio remains approximately constant over the energy range 1–20 MeV, which implies an advantage of diamond detectors over silicon diodes for electron dosimetry. Figure 8.6 also shows the increase in the air:water stopping power ratio with energy caused by the density effect.

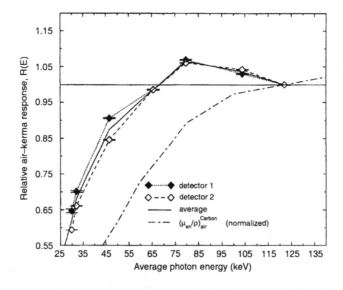

FIGURE 8.5 Relative air-kerma response of PTW diamond detectors measured against free-air ionization chambers. All results were arbitrarily normalized to unity at an average photon energy of 121 keV. (From Reference [4]. With permission.)

The sensitive region of the diamond detector is a low-impurity natural diamond plate of thickness 0.33 mm and volume 1.4 mm³, sealed in a cylindrical polystyrene housing of diameter 7.3 mm. An operating bias of +100 V (recommended by PTW) is applied through 0.05–0.6-μm gold contacts and 50-nm silvered copper wire. The surface

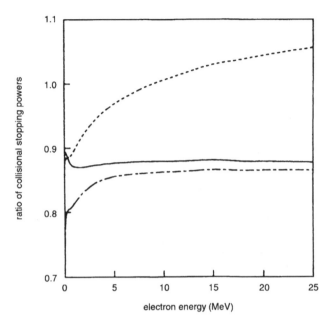

FIGURE 8.6 Collision stopping power ratios of carbon, air, and silicon to water as a function of energy (from ICRU Report 37 [33]). —, carbon:water; ----, air:water; — • —, silicon:water. (From Reference [5]. With permission.)

of the diamond plate is 1.0 mm beneath the end of the housing. In Figure 8.7 a comparison of the longitudinal cross sections of the diamond and diode detectors is shown. Physical and operating parameters of the diamond detector are given in Table 8.1. It should be noted that the mode of operation of a diamond detector is as a resistive element, where conductivity is proportional to dose rate.

The 6 MeV results are shown in Figure 8.8a, where the readings at 0° are normalized to 100%. It can be seen that the minimum in the diode detector response curve (at approximately 100°) is less than that for the diamond detector, while the maxima (at approximately 50°) are equal. Results for a 15-MeV beam are shown in Figure 8.8b, where again the minimum in the diode response is less than that for the diamond. Note that for both detectors,

TABLE 8.1

Physical Parameters of the Diamond Detector (as Supplied by PTW).

Main impurities	Nitrogen and boron ($<10^{19}$ atomis cm^{-3})
Sensitive volume	1.4 mm³
Sensitive area	4.3 mm²
Thickness of sensitive volume	0.33 mm
Operating bias	+100 V ($\pm$1%)
Dark current	$<10^{-12}$ A
Sensitivity to ^{60}Co radiation	1.75 × 10^{-7} C Gy^{-1}
Pre-irradiation dose	$\leq$5 Gy

From Reference [5]. With permission.

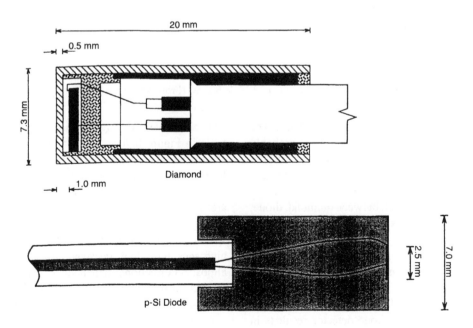

FIGURE 8.7 Longitudinal cross sections of the diamond detector and Scanditronix p-Si diode detector. The +100 V bias is applied to the diamond through gold contacts on the diamond surface. (From Reference [5]. With permission.)

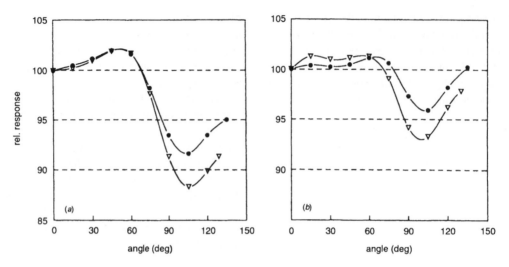

FIGURE 8.8 Directional response of the diamond and diode detectors for (a) 6 MeV and (b) 15 MeV. •, diamond; ∇, diode. (From Reference [5]. With permission.)

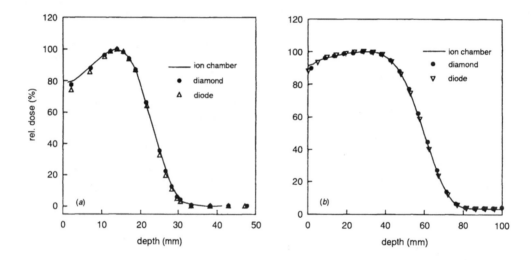

FIGURE 8.9 Depth-dose curves as measured with the diamond and diode detectors in the parallel orientation, compared with corrected ionization chamber results for (a) 6-MeV and (b) 15-MeV beams. (From Reference [5]. With permission.

the curves are flatter at 15 MeV than at 6 MeV. The variation in response with incident direction is probably due to changing interface effects [3], where a varying degree of delta-ray equilibrium is reached within the sensitive volumes (ICRU 22 [34]).

In Figure 8.9a, a comparison between diamond, diode, and ionization chamber results is shown for 6 MeV, where the diamond and diode were both orientated parallel to the beam. The effective points of measurement for the diamond and diode are 1.00 mm and 0.5 mm beneath the end of the detector housings, respectively.

Depth-dose curves for 15 MeV are shown in Figure 8.9b, where it can be seen that the diode detector result is in closer agreement with the diamond detector and ionization chamber results than is the case for 6 MeV; this is because the silicon-water stopping power ratio becomes almost

constant above 5 MeV (see Figure 8.6). Note that the depth scale in Figure 8.9b is smaller than that of Figure 8.9a; any deviation from the ionization chamber curve is therefore more noticeable for the 6-MeV curve.

Characteristics of the PTW/RIGA diamond detector are: [6]

The diamond detector has the advantage of excellent spatial resolution, low energy and temperature dependence, high sensitivity, high resistance to radiation damage, and almost no directional dependence. The diamond detector is used for relative dosimetry, and it is typically used for dose and dose-rate measurements in high-energy photon and electron beams, where the fields are very small or have steep gradients.

Technical Details of the diamond detector are as follow:

- Photon energy range: 0.08–20 MeV
- Electron energy range: 4–20 MeV
- Dose-rate range: 0.05–30 Gy/min
- Linearity of response: ±2%
- Pre-irradiation dose: 5–10 Gy
- Operating bias: +100 V ± 1%
- Sensitivity to ^{60}Co radiation: 0.5–5.0×10^{-7} C/Gy (Figure 8.10)
- Dark current: $<5 \times 10^{-12}$ A
- Charge collection time: $<10^{-8}$ s
- Sensitive area: 3–15 mm^2
- Thickness of sensitive volume: 0.1–0.4 mm
- Sensitive volume: 1–6 mm^3
- Radiation resistance: $>10^5$ Gy
- Outer probe diameter: 7.3 mm
- Weight incl. cable and connector: approx. 170 g
- Directional dependence: less than 2% in the range of 0° to 170° in a water phantom for depths larger than d_{max} (d_{max} = depth of dose maximum) (Figure 8.13)

^{60}Co and 15 MeV electron depth dose distributions are shown in Figures 8.11 and 8.12, respectively.

Since the diamond detector uses a naturally grown diamond, the exact dimensions of the sensitive volume slightly differ. The exact data of each individual probe are specified in the calibration certificate.

The absorbed dose to muscle from proton beam charge measurements, using the ^{60}Co absorbed dose calibrated detectors and proton beam quality correction factor was proposed by Vatnitsky et al. [7] According to modified cavity theory, the diamond detector can be considered a carbon cavity in a water phantom when exposed in a proton beam. The absorbed dose proton beam quality correction factor for the diamond detector, $k_{Qp}(DD)$, can be calculated by

$$k_{QP}(DD) = \frac{W_{pr,carbon}/e}{W_{\gamma,carbon}/e} \times \frac{[(S/\rho)^m_{carbon}]_{pr}}{[(\bar{L}/\rho)^m_{carbon}]_{cobalt}}$$

$$\times \frac{(p^{polyst}_{carbon})_{pr}}{(p^{polyst}_{carbon})_{cobalt}} \tag{8.2}$$

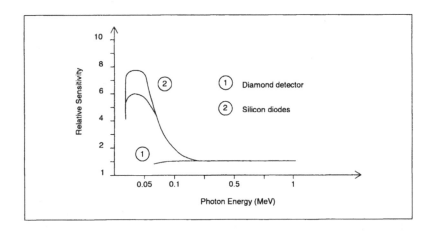

FIGURE 8.10 Relative photon sensitivity. (From Reference [6]. With permission.)

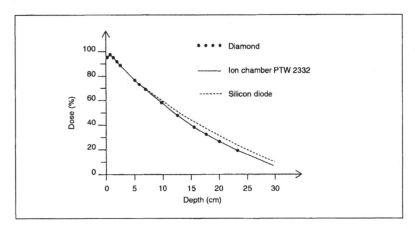

FIGURE 8.11 ^{60}Co depth dose curve. (From Reference [6]. With permission.)

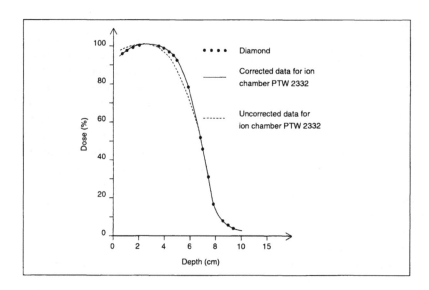

FIGURE 8.12 15-MeV electron depth dose curve. (From Reference [6]. With permission.)

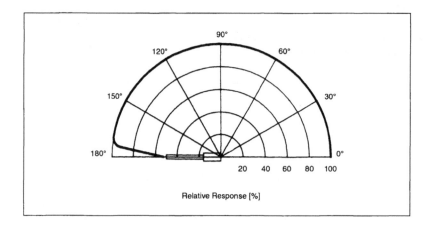

FIGURE 8.13 Directional response. (From Reference [6]. With permission.)

They assumed that the energy to produce an electron-hole pair in a diamond detector is independent of the energy and type of the particles. A second assumption is that the diamond detector can be considered a thin disk-shaped detector. In a proton beam, the scattering effect can be considered negligible since the large mass of a proton results in similar in-scattering and out-scattering from the detector's volume. Secondary electrons produced in proton-electron collisions have an average range on the order of several microns. It was assumed that all of the ionization in a diamond plate is produced by these electrons and some primary protons. There is a fraction of delta-rays produced in proton-electron interactions with a range comparable to the pathlength through the detector's volume, but the contribution of this component to the total ionization is small compared with secondary electrons and primary protons. It is estimated that perturbation effects for the diamond detector in proton beams can be considered to

be at least smaller than in an electron beam, with their ratios safely assumed equal to unity. The uncertainty for this assumption was estimated at 0.5%, resulting in a total uncertainty for the perturbation factors ratios of -0.7%.

With the two above-stated assumptions, Equation (8.2) can be rewritten for diamond detectors that have ^{60}Co absorbed dose to polystyrene calibration factors as follows:

$$k_{QP}(DD) = \frac{[(S/\rho)^m_{carbon}]_{pr}}{[(\bar{L}/\rho)^{polyst}_{carbon}]_{cobalt}} \tag{8.3}$$

The mass energy absorption ratio of carbon to water is nearly constant over a wide energy range, making the diamond detector nearly tissue-equivalent. The directional dependence of the radiation response of the diamond detector for ^{60}Co 6-MV and 18.MV photon beams was found by Rustgi [8] to be more uniform than that of the diode. The spatial resolution of the diamond detector, as measured by penumbra width, is slightly less than that of the diode detector but clearly superior to that of the 0.14-cm³ ionization chamber. The diamond detector, with high radiation sensitivity and spatial resolution, is an excellent choice as a detector in photon fields with high dose gradients such as brachytherapy and radiosurgery. [8]

A commercially available diamond detector (PTW) was evaluated by Rustgi. The diamond detector had a small measuring volume of 1.9 cm³ with a sensitive area of 7.3 mm² and a plate thickness of 0.26 mm. As shown in Figure 8.7, the detector is enclosed in a cylindrical plastic housing with the natural diamond plate displaced 1 mm below the front circular face of the housing. For optimum operation, a bias of +100 V from an external source was applied to the detector, as recommended by the manufacturer. The conductivity induced in the diamond plate by irradiation is directly proportional to the dose rate. The charge produced in the diamond plate was measured with a Keithley 35614 electrometer with its electronic bias grounded. In order to stabilize the response of the diamond detector, it was irradiated to a dose of 500 cGy, as suggested by the manufacturer.

After stabilizing the response of the diamond detector by irradiating it to a dose of 500 cGy in a solid water phantom, the maximum variation of the diamond detector response was found to be less than 0.5% over an 8-h period. The radiation sensitivity of the diamond detector was measured to be approximately 2.2×10^{-9} C cGy^{-1}, compared to 1.0×10^{-9} C cGy^{-1} for the photon diode under identical irradiation geometries. The diode detector has a sensitive volume of 0.3 mm³, compared to 1.9 mm³ for the diamond detector.

The mass energy absorption coefficients of carbon, air, and silicon relative to water as a function of photon energy are shown in Figure 8.14. The mass energy–absorption coefficient ratio of carbon to water changes by less than 7%, compared to 42% for the silicon to water ratio in the 0.1–20 MV energy range. Measurements made with a silicon diode detector require energy-dependent correction factors to convert diode response to dose.

Energy and dose rate dependence of a diamond detector in the dosimetry of 4–25-MV photon beams has been measured by Laub et al. [9] The diamond detector was a low-impurity natural diamond plate of thickness 0.032 cm and volume 0.003 cm³. It was connected to a Unidos Universal Dosimeter (PTW- Freiburg) with an applied detector bias of +100 V. Dose measurements were carried out in an MP3-water phantom (PTW-Freiburg). The diamond detector was positioned on the central axis of a photon beam, entering the surface of the water phantom perpendicularly.

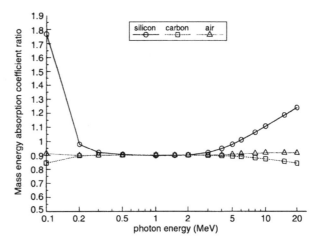

FIGURE 8.14 Mass energy attenuation coefficients of carbon, air, and silicon relative to water as a function of photon energy. (From Reference [8]. With permission.)

The diamond detector's response turned out to be slightly decreasing with increasing dose rate. This is actually to be expected from the theory of radiation-induced conductivity in an insulator, which explains the decrease in response as a consequence of a very short electron-hole recombination time. The alteration of detector current with increasing dose rate can be approximated by the empirical expression

$$i = i_{dark} + R \times \dot{D}^{\Delta} \tag{8.4}$$

where i is the detector current and $\dot{D}$ is the dose rate. The parameter i_{dark} indicates dark-current influences and is consequently nearly zero. R is a fitting parameter for the response of the diamond detector, and Δ is one for the slight sublinearity of response. In this examination, Δ turned out to be 0.963 ± 0.005.

The response of the diamond detector, indicated by the fitting parameter R, was established at different levels of photon energy. The detector's response did not change significantly at any stage; consequently, the diamond detector shows no energy dependence within the covered energy range of 4–25 MV (Figure 8.15). Although this is to be expected due to the near tissue equivalence of carbon, the contact material of the detector might introduce an element of energy dependence. Heydarian et al. [5], exposing a diamond detector to a 6- and a 15-MeV electron beam, could not find any energy dependence requiring correction. According to Laub results, this holds true for high-energy photon beams, as well. The error bars inserted in Figure 8.15 principally correspond to variations in the dose rate delivered by the accelerator. The diamond detector was also found to possess a superior spatial resolution and a high sensitivity (about 4.1×10^{-7} C-Gy^{-1}). [9]

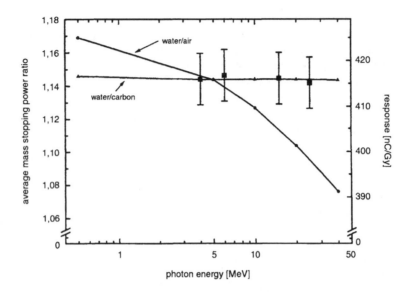

FIGURE 8.15 Response of diamond detector at photon energies of 4, 6, 15, and 25 MV (▮, right axis). For comparison, the average mass stopping power ratio of water/carbon and of water/air (left axis) is drawn in. (From Reference [9]. With permission.)

Using a diamond detector for the dosimetry of brachytherapy sources was investigated by Rustgi. [10] A high-activity ^{192}Ir source was selected for this purpose. The dosimetric characteristics measured included the photon fluence anisotropy in air, transverse dose profiles in planes parallel to the plane containing the HDR source, and isodose distributions. The "in-air" anisotropy of the photon fluence relative to seed orientation was measured at 5 and 10 cm from the source.

A type 60003 diamond detector manufactured by PTW Freiburg was used by Rustgi. The radiation-sensitive region of the diamond detector had a volume of 1.8 mm^3. The low-impurity diamond crystal of unknown shape had a thickness of a 0.26 mm. The crystal was enclosed in a plastic housing, with the surface of the crystal displaced 1 mm below the housing front surface. All measurements were made with an external bias of +100 V, as recommended by the manufacturer, to achieve near-total ion-pair collection.

The EGS4 Monte Carlo code has been used by Mobit and Sandison [11] to investigate the response of a PTW/diamond detector irradiated in both clinical and monoenergetic megavoltage electron beams ranging in energy from 5 to 20 MeV. The sensitive volume of the PTW/diamond detector simulated has a thickness of 0.4 mm and a diameter of 4.4 mm. The results show that the PTW/diamond detector has a constant response (within 1.0%) in electron beams if irradiated at depths closed to d_{max}, and its response is almost independent of irradiation depth or incident electron energy (within 3%). The encapsulation of the bare diamond detector with low-Z epoxy and polystyrene wall material does not affect its response in electron beams.

Figure 8.16 shows the schematic of the simulated PTW/diamond detector (model T60003) designed to measure relative dose distributions in high-energy photon and electron beams. The sensitive volume of the PTW/diamond detector is a disk made from natural diamond (density 3.51 g/cm^3) with a radius ranging from 1.0 to 2.2 mm and a thickness ranging from 0.2 to 0.4 mm.

When the PTW/diamond detector is irradiated in a water phantom at given point p, the absorbed dose to the water phantom at this point $D_w(p)$, following the Spencer-Attix cavity equation and the AAPM-TG21 protocol, is as follows:

$$D_w(p) = \bar{D}_{PTW} \times s_{w,dia}^{SA} \times P_{cav}^{PTW} \times P_{wall}^{PTW} \times N_{PTW} \qquad (8.5)$$

where $\bar{D}_{PTW}$ is the average absorbed dose in the sensitive volume, of the PTW/diamond detector, $s_{w,dia}^{SA}$ is the Spencer-Attix mass collision stopping power ratio of water to diamond (graphite), P_{wall}^{PTW} is the perturbation correction factor due to the non-water equivalence of the encapsulating wall material of the PTW/diamond detector, and P_{cav}^{PTW} is the perturbation correction factor resulting from the insertion of the diamond-sensitive volume into the water medium, which can be considered as the perturbation due to the departure of the wall-less (bare) detector from non-ideal Bragg-Gray (Spencer-Attix) conditions. Here, P_{cav}^{PTW} in the TG21 formulation is regarded as the product of two perturbation correction factors, the electron fluence perturbation correction factor and the dose gradient perturbation correction factor.

Since the sensitive volume of the PTW/diamond detector is very small (6.1 mm^3), one can assume that the

TABLE 8.2
Monte Carlo–Determined Variation with Energy of the Average Absorbed Dose Ratio of Water to the Sensitive Volume of the PTW/Diamond Detector in Monoenergetic Electron Beams

E_O (MeV)	Depth (cm)	E_d (MeV)	$\overline{D}_w/\overline{D}_{PTW}$
5	1.2	2.2	1.145 ± 0.002
10	2.7	4.1	1.146 ± 0.002
15	3.6	6.7	1.149 ± 0.002
20	5.4	7.0	1.148 ± 0.002

The electron beam size is a square of dimension 10 cm at the phantom surface, and the incident electron beam is assumed to be parallel and nondivergent. The mean energy at depth (E_d) was calculated following the IAEA method.

From Reference [11]. With permission.

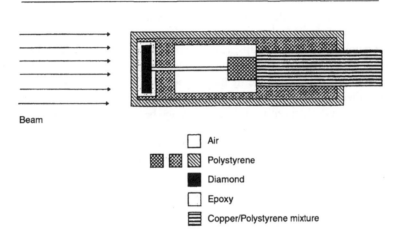

Beam

□ Air

▨ ▩ ▨ Polystyrene

■ Diamond

□ Epoxy

▤ Copper/Polystyrene mixture

FIGURE 8.16 Composition of the PTW/diamond detector (the simulated geometry).(From Reference [11]. With permission.)

absorbed dose in the water phantom at point p is equal to the absorbed dose to the water phantom averaged over the same volume as the cavity. This means $D_w(p) = \overline{D}_w$, and Equation (8.5) can be written as

$$\overline{D}_w/\overline{D}_{PTW} = s_{w,dia}^{SA} \times P_{cav}^{PTW} \times P_{wall}^{PTW} \qquad (8.6)$$

where $\overline{D}_W$ is the average absorbed dose in water of the same volume, as the wall-less diamond detector placed at the same point p is the uniform water phantom.

The PTW/diamond detector response depends on the average absorbed dose ratio of water to diamond ($\overline{D}_w/\overline{D}_{PTW}$), and this ratio can be used to determine if the PTW/diamond detector response depends on the depth of irradiation or the incident electron beam energy. Table 8.2 shows the average absorbed dose ratio of water to the sensitive volume of the PTW/diamond detector ($\overline{D}_w/\overline{D}_{PTW}$) for incident monoenergetic electron beams.

For wall-less detectors like TLDs, the deviation of the absorbed dose at a point p in a uniform medium, from the

Spencer-Attix cavity equation, can be expressed as:

$$P_{med,cav} = \frac{\overline{D}_{med}/\overline{D}_{cav}}{s_{med,cav}^{SA}} \qquad (8.7)$$

where $p_{med,cav}$ is the perturbation correction factor, ($\overline{D}_{med}/\overline{D}_{cav}$) is the average absorbed dose ratio of the medium to cavity, and $s_{med,cav}^{SA}$ is the Spencer-Attix mass collision stopping power ratio. However, for an encapsulated diamond detector like the PTW/diamond detector, the perturbation effects are from two different sources: (1) the encapsulation and (2) the perturbation due to the sensitive volume of the detector (air for ion chambers and diamond for the PTW/diamond dosimeter) not acting as a perfect Bragg-Gray cavity [Equation (8.7)]. The effect of the encapsulation, denoted by P_{wall}^{PTW}, can be determined by comparing the PTW/diamond response in a water phantom to the response of a nonencapsulated diamond detector of identical sensitive volume (identical dimensions) irradiated at the same depth with the same incident electron energies. [11]

II. MOSFET RADIATION DOSIMETER

MOSFET is a sandwich-type device consisting of a p-type silicon semiconductor substrate separated from a metal gate by an insulating oxide layer. The advantages of MOSFET devices include being a direct reading detector with thin active area ($<25\mu m$) and having a small size. The signal can be permanently stored and is dose-rate independent. Dose enhancement was observed for photon energies below 40 keV, where the photoelectric effect is dominant. With adequate filtration, uniform response was reported for photon energies above 80 keV. The sensitivity and linearity of MOSFET devices is greatly influenced by the fabrication process and the voltage-controlled operation during and after irradiation.

When a metal-oxide-semiconductor (MOS) device is irradiated, three mechanisms within the silicon dioxide layer predominate: the build-up of trapped charge in the oxide; the increase in the number of interface traps; and the increase in the number of bulk oxide traps. Electron-hole pairs are generated within the silicon dioxide by the incident-ionizing radiation. Electrons, whose mobility in SiO_2 at room temperature is about 4 orders of magnitude greater than holes, quickly move toward the positively biased contacts. Depending on the applied field and the energy and kind of incident particle, some fraction of electrons and holes will recombine. The holes that escape initial recombination are relatively immobile and remain behind, near their point of generation.

A negative bias applied to the FET gate causes a positive charge to build-up in the silicon substrate. This build-up of charge allows current to pass through the silicon substrate from the source to the drain terminals. The gate voltage necessary to allow conduction through the MOSFET is known as the threshold voltage.

When the MOSFET is exposed to ionizing, radiation, electron hole pairs are formed in the oxide insulation layer. The junction potential between the device layers, or an applied positive potential to the gate, causes the electrons to travel to the gate while the holes migrate to the oxide silicon interface, where they are trapped. These trapped positive charges cause a shift in the threshold voltage, since a larger negative voltage must be applied to the gate to overcome the electric field of the trapped charges to achieve conduction. The threshold voltage shift is proportional to the radiation dose deposited in the oxide layer, and this relationship is the basis for using MOSFETs as dosimeters. Irradiated MOSFETs have been stored under zero bias condition for over ten years with approximately 1% loss of signal.

The efficiency of charge trapping or effective sensitivity of the MOSFET depends on the thickness of the oxide layer and the bias potential applied across the layer during irradiation. For a given oxide layer thickness, the sensitivity can be controlled by the bias potential applied between the gate and substrate during irradiation, since large electric fields applied across the oxide layer result in more rapid separation of the electrons from the holes.

Positive-ion diffusion toward the oxide-silicon interface results in a negative shift of the threshold voltage, while phonon-induced (thermal) release of trapped charges, as well as charge annihilation by electrons tunneling from the silicon to the oxide layer (nonthermal), results in annealing or a positive shift of threshold voltage. These signal drifts become particularly apparent during low dose rate, long-time irradiations. In order to obtain meaningful dosimetric information from a MOSFET dosimeter under low dose rate conditions, the drift effects must be characterized and deconvoluted from the threshold voltage data.

Prototype miniature dosimeter probes have been designed, built, and characterized by Gladstone et al. [12], employing a small, radiation-sensitive metal oxide semiconductor field effect transistor (MOSFET) chip. It was used to measure, in vivo, the total accumulated dose and dose rate as a function of time after internal administration of long-range beta particle radiolabeled antibodies and in external high-energy photon and electron beams. The MOSFET detector is mounted on a long narrow alumina substrate to facilitate electrical connection. The basic MOSFET probe design is shown in Figure. 8.17.

A plot of the sensitivity vs. bias voltage is shown in Figure 8.18. A line is drawn through the data points, showing that the sensitivity of the MOSFET detector varies logarithmically with bias potential between 1.5 and 9 V. The sensitivity does not change as a function of total absorbed dose from 1 to 1000 cGy or dose rate from 0.195 cGy/h to 400 cGy/min.

Figure 8.19 is a plot of the threshold voltage vs. temperature for our MOSFET probe. The threshold voltage for this MOSFET type is linear with temperature, with a slope of 7.4 mV/°C. This linear temperature shift is easily characterized and corrected for by adding an offset to account for the temperature of the MOSFET at the time the threshold voltage is measured. If pre- and post-irradiation measurements are made at the same temperature, no correction is needed.

The drift rate of the threshold voltage is observed to be linear with $\ln (t/t_0)$ (t is time after irradiation and t_0 is normalization time of 1 min) at times greater than 150 min, with the rate of drift increasing with increasing temperature.

A direct-reading semiconductor dosimeter has been investigated by Soubra et al. [13] as a radiation detector for photon and electron therapy beams of various energies. The operation of this device is based on the measurement of the threshold voltage shift in a custom-built metal oxide–silicon semiconductor field effect transistor (MOSFET). This voltage is a linear function of absorbed dose. The extent of the

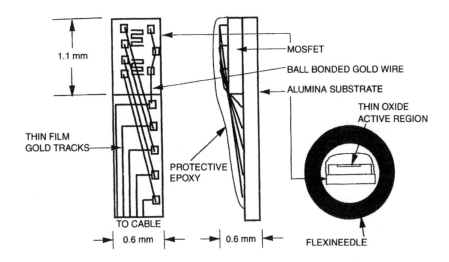

FIGURE 8.17 Miniature MOSFET radiation detector probe design. The MOSFET chip is mounted on an alumina substrate. Gold wire is used to electronically connect the MOSFET to the leads of the substrate. A cable is connected to the end of the substrate not shown. The probe is inserted into a plastic flexineedle for protection before insertion into tissue. (From Reference [12]. With permission.)

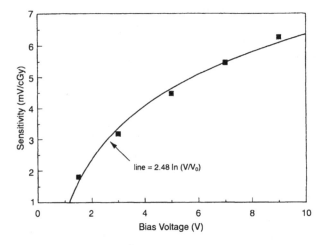

FIGURE 8.18 Sensitivity vs. bias voltage. The sensitivity increases with increasing bias voltage.(From Reference [12]. With permission.)

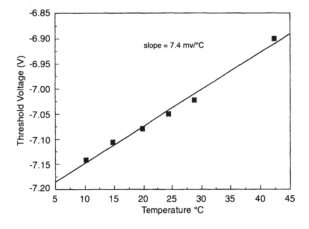

FIGURE 8.19 Threshold voltage vs. temperature. The magnitude of the threshold voltage decreases with increasing temperature. (From Reference. [12] With permission.)

linearity region is dependent on the voltage-controlled operation during irradiation. Operating two MOSFETS at two different biases simultaneously during irradiation will result in sensitivity (V/Gy) reproducibility better than $\pm 3\%$ over a range in dose of 100 Gy and at a dose per fraction greater than 20×10^{-2} Gy. The modes of operation give this device many advantages, such as continuous monitoring during irradiation, immediate reading, and permanent storage of total dose after irradiation.

The basic MOSFET structure is depicted in Figure 8.20. The type shown is known as a *p* channel enhancement MOSFET, which is built on a negatively doped (*n*-type) silicon substrate. Two terminals of the MOSFET, called the source (*s*) and the drain (*d*) are situated on top of a positively doped (*p*-type) silicon region. The third terminal shown is the gate (*g*). Underneath the gate is an insulating silicon dioxide layer, and underneath this oxide layer is the *n* silicon substrate. The region of the substrate immediately below the oxide layer is known as the channel region. When a sufficiently negative V_G (gate voltage) is applied (with reference to the substrate), a significant number of minority carriers (holes in this case) will be attracted to the oxide-silicon surface from both the bulk of the silicon and the source and drain regions. Once a sufficient number of holes have accumulated there, a conduction channel is formed, allowing an appreciable amount of current to flow between the source and the drain (I_{ds}). Figure 8.21 shows how the magnitude of current varies with gate voltage.

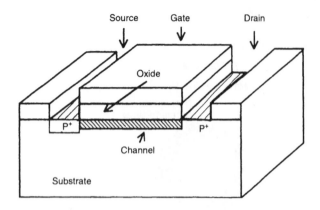

FIGURE 8.20 Schematic cross section of a *p*-channel MOSFET showing the oxide (SiO₂), the substrate (Si), the source, the gate, and the drain. (From Reference [13]. With permission.)

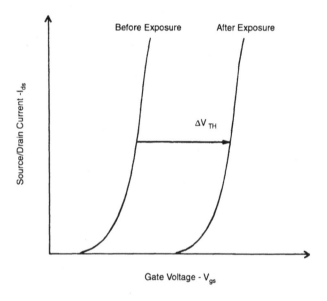

FIGURE 8.21 Typical MOSFET source to drain current I_{ds} vs. gate voltage V_{gs}. The difference in the gate voltages V_{gs} needed to attain a predetermined current flow I_{ds}, before and after radiation, is equal to the threshold voltage shift (ΔV_{TH}). (From Reference [13]. With permission.)

Threshold voltage (V_{TH}) is defined as the gate voltage needed to attain a predetermined current flow (I_{ds}).

Typically commercial MOS transistors have 200- to 800-Å-thick gate oxides. To maximize the MOSFET sensitivity to ionizing radiation, a relatively thick gate oxide is preferred. However, there is a limitation to how thick a gate oxide can be grown. Growth of extremely thick oxide layers (>1 μm) requires growth times of more than 100 h at temperatures of 1000°C. These oxidation times are not feasible and, in addition, thick oxide will tend to be highly stressed, leading to dislocations at the Si-SiO₂ interface, resulting in fast surface states and threshold instability. [13]

The response of the single MOSFET detector (ΔV_{TH}) as a function of accumulated dose will exhibit a nonlinear region at high dose levels. The main reason for this phenomenon is the accumulation of the radiation-generated positive charge (holes) in the oxide traps, which effectively reduces the magnitude of the electric field produced by the positive gate bias between the positive-charge region and the gate.

The MOSFET is approximately 200 μm in diameter and consists of a 0.5-μm Al electrode on top of a 1-μm SiO₂ and 300-μm Si substrate. Results for percentage surface dose measured by Butson et al. [14] were within ±2% compared to the Attix chamber and within ±3% of TLD extrapolation results for normally incident beams. Percentage surface dose for 10 × 10-cm and 40 × 40-cm field size for 6-MV x-ray at 100-cm SSD using the MOSFET was 16% and 42% of maximum, respectively. Factors such as its small size, immediate retrieval of results, high accuracy attainable from low applied doses, and recording of its dose history make the MOSFET a suitable in vivo dosimeter where surface and skin doses need to be determined.

Dose in the build-up region varies considerably with changes in parameters such as beam energy, SSD, patient geometry, angle of incidence, field size, and the use of blocks, block trays, and wedges and boluses. Dose deposited at the surface and in the build-up region has two main contributors:

1. In-phantom scatter, where x-rays are incident on the phantom and interact with electrons via mainly the Compton effect and, to a lesser extent, photoelectric and pair production at 6-MV x-ray energy, which, in turn, deposit their energy along their path; and
2. Electron contamination from electrons produced outside the phantom and incident on the phantom to deposit their dose. These electrons are produced by x-rays interacting with material in and on the machine head, such as flattening filters, ion chambers, collimators, block trays and wedges, plus the air column between the source and phantom. These electrons will be referred to as electron contamination.

Skin dose measurements represent a greater technical challenge because of the small distances to superficial layers; hence, any suitable detector should have minimal build-up due to encapsulation.

The experimental configuration for the MOSFET consisted of a scientific grade (stock item) n-type MOSFET device from which Butson et al. have removed the nickel casing, leaving the bare crystal intact and exposed, as shown in Figure 8.22. This was easily achieved using pliers to ease off the cap that was connected at three positions around its base to the device's main body. We

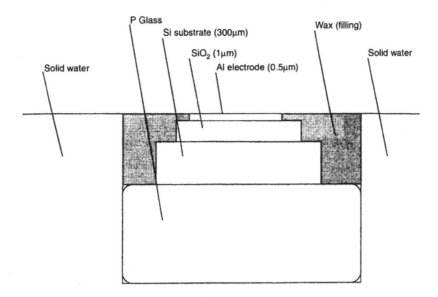

FIGURE 8.22 Physical design of unencapsulated MOSFET. (From Reference [14]. With permission.)

refer to this state as the unencapsulated MOSFET. During irradiation, the MOSFET was connected to a 6-V bias voltage supplied by a DC battery with a circuit, as shown in Figure 8.23.

The unencapsulated MOSFET has a few advantages over the TLD:

1. Immediate retrieval. Dose can be ascertained accurately within minutes after irradiation with as low as 10 cGy of applied dose.
2. MOSFETs can be used again immediately. They do not require an anneal process after irradiation; TLDs do require such a process.
3. Inexpensive support equipment. MOSFETs require only a small power source and voltmeter for data retrieval.
4. MOSFETs retain their dose history. Information is not lost during the retrieval process.

All of the above properties and their accuracy make the unencapsulated MOSFET an ideal in vivo dosimeter for surface and skin dose measurements.

Rosenfeld et al. [15] was concerned with the role of the package of MOSFETs used in measurements of gamma dose in mixed gamma-neutron fields, in high energy *bremsstrahlung*, and in soft x-ray fields. It is shown that a kovar cap should be avoided for dosimetry applications in the presence of strong thermal neutron fluences. In regions of strong electronic disequilibrium, the "bare" or unencapsulated MOSFET is a unique tool for surface dose measurements and Monte Carlo model validation. For depths where electronic equilibrium exists (i.e., $x >$ depth of D_{max}), the MOSFET package is not critical. For low-energy X-ray fields, the energy dependence of the dose

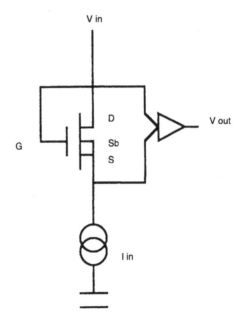

FIGURE 8.23 MOSFET circuit diagram. (From Reference [14]. With permission.)

enhancement factor (DEF) will differ, depending upon whether the irradiation is performed in free-air geometry or on the surface of a phantom.

The study of the neutron response of MOSFET dosimeters was initiated primarily for the dosimetry of mixed gamma-neutron fields. It has previously been shown that such MOSFETs are much less sensitive (~100 times) to fast neutron tissue dose than to gamma tissue dose. The investigations have been carried out in TO-5 and TO-18 kovar packages exposed to 3-MeV and 15-MeV fast neutron fields and cadmium filtered

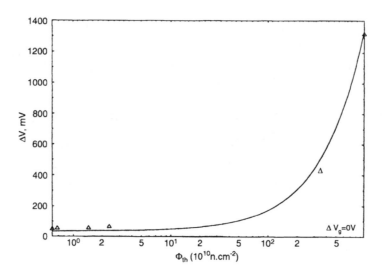

FIGURE 8.24 Change in MOSFET threshold voltage as a function of thermal neutron fluence ($V_g = 0$ V, passive mode). (From Reference [15]. With permission.)

reactor beam spectra with an average neutron energy of 1 MeV.

The MOSFET dosimeter is unique due to the small size of its active dosimetric volume of SiO_2. This makes the MOSFET very attractive for dosimetry in regions of electronic disequilibrium. Such fields occur in radiation hardness testing on pulsed *bremsstrahlung* facilities (~10^9 Gy(Si)/s) and also on medical linear accelerators having end energies in the range 6–25 MeV. Different approaches are required for tissue dosimetry in soft x-ray spectra compared to high-energy *bremsstrahlung* fields. The appropriate MOSFET packaging will differ between these cases.

The Ukrainian n-MOSFET with a 1-μm-thick oxide layer and aluminum gate and encapsulated in a standard kovar package has been investigated in the MOATA research reactor thermal (TC-10) facility.

To study the role of different thermal neutron fluences on the threshold shift while maintaining the gamma and fast neutron fluences approximately constant, Rosenfeld et al. surrounded the TO-18 packaged MOSFETs with additional ^{6}LiF/epoxy encapsulation. Each of these lithiated covers was cylindrical in shape and of uniform thickness. To maintain constant gamma irradiation conditions, MOSFETs were exposed at the same point in the irradiation channel and for the same period of time but with different thicknesses of ^{6}Li encapsulation. The change in threshold voltage of the MOSFETs as a function of thermal neutron fluence is shown in Figure 8.24.

The model of the MOSFET follows the geometry shown in Figure 8.25. The gold-plated leads and kovar base were represented as a homogeneous mixture of gold and nickel. Similarly, the silicon oxide layer and aluminum gate electrode were considered a homogeneous mixture of silicon oxide and aluminum. These approximations are standard techniques in MCNP modeling. The thickness of the SiO_2

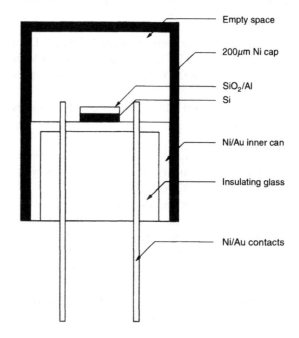

FIGURE 8.25 Schematic diagram of the geometry used in the Monte Carlo model of the MOSFET. (From Reference [15]. With permission.)

layer was increased from its real value of 1 μm to 450 μm in the model to get better statistics.

Figure 8.26 shows the role of MOSFET packaging on the relative dose depth distribution in the phantom build-up region for 6-MeV *bremsstrahlung*. The MOSFET encapsulated in a TO-18 package showed a strong dose enhancement phenomenon in the region of strong electronic disequilibrium. At the interface between the air and the solid water, the encapsulated MOSFET yielded an overestimation of the dose by four times relative to Attix chamber measurements.

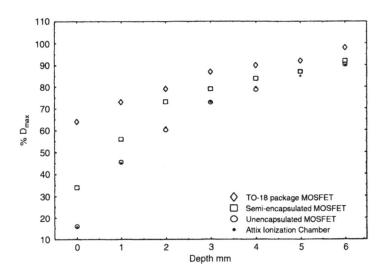

FIGURE 8.26 Depth dose distributions measured using a MOSFET in the build-up region of a solid water phantom irradiated with a 10 × 10-cm² field from a 6-MeV *bremsstrahlung* source. (From Reference [15]. With permission.)

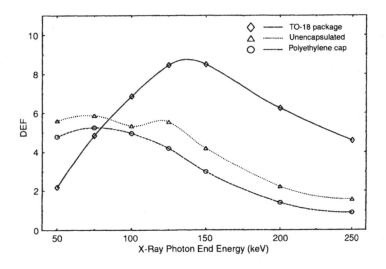

FIGURE 8.27 Dose enhancement factors for four differently packaged MOSFETs in free-air geometry. (From Reference [15]. With permission.)

The semi-encapsulated MOSFET showed a 2.5 times overestimation of dose for the same comparison. The "bare" MOSFET, however, was in agreement with the Attix chamber to more than 1% for the entire 15-mm build-up region.

Figure 8.27 illustrates the DEF for a thick-oxide n-MOSFET in three different packages. The energy axis shows the end energy of the x-ray spectrum, and the DEF is normalized to the sensitivity observed for an end energy of 6 MeV. All irradiations were performed with the beam normally incident on the top surface of the MOSFET can or die.

MOSFET integral dosimetry based on charge build-up in the SiO_2 layer, when exposed to radiation was studied by Rosenfeld et al. [16]. Microdosimetry using silicon micro-volumes based on spectroscopy of charge in a *p-n*

junction, is a further step in the characterization of mixed radiation environments. For that characterization LET must be measured. That is the most serious problem encountered in the data interpretation, i.e., the determination of the average chord length of the sensitive volume. This quantity is required for the conversion of pulse height spectrum to LET spectrum.

The *n*-MOSFET chip (*p*-Si substrate, doping level 1.35×10^{15} cm^{-3}, drain surrounded by 60-μm length gate, drain-active area 1.5×10^{-2} mm²) originally was used in this study in a TO-18 package from which the cover was removed. Either ^{210}Po or ^{241}Am α source was placed inside the vacuum chamber, with the MOSFET chip located immediately above it. The measurements were made in two modes. In the first mode (Figure 8.28a), the source and gate

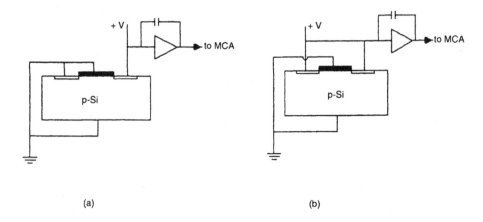

(a)   (b)

FIGURE 8.28 Connections for pulse-height measurement using the MOSFET (a) with only the drain connected to the PA and (b) with source and drain both connected to the PA. (From Reference [16]. With permission.)

were grounded and a charge sensitive preamplifier (PA) was connected to the n^+-p drain junction reverse biased at 18 V. In the second mode (Figure 8.28b), both the source and the drain were connected to the PA and the gate was grounded.

Dose response of various radiation detectors to synchrotron radiation was studied by Korn et al. [17] Monoenergetic x-rays from synchrotron radiation offer the unique opportunity to study the dose response variation with photon energy of radiation detectors without the compounding effect of the spectral distribution of x-rays from conventional sources. The variation of dose response with photon energies between 10 and 99.6 keV was studied for two TLD materials (LiF:Mg,Ti and LiF:Mg,Cu,P), MOSFET semiconductors, radiographic film, and radiochromic film. MOSFET detectors and the radiographic film were found to over-respond to low-energy x-rays by a factor of up to 7 and 12, respectively, while the radiochromic film underestimated the dose by approximately a factor of 2 at 24 keV. The TLDs showed a slight over-response with LiF:Mg, Cu,P, demonstrating better tissue equivalence than LiF:Mg,Ti (maximum deviation from water less than 25%).

The increasing recombination effects with decreasing photon energy result in a reduced sensitivity. Other contributing factors are the predominance of the photoelectric effect in SiO_2 in comparison with tissue for photon energies less than 100 keV, a dose enhancement effect due to the metal-SiO_2 interfaces used in the MOSFET devices, and the packaging of the chip.

The MOSFET dosimeter system used in Kron's study was developed by the radiation physics group at the University of Wollongong (Australia). The n-MOSFET detectors, which were obtained from the Ukraine, feature a 1-μm-thick silicon oxide layer with an aluminum gate electrode of 0.5-μm thickness. For the experiments, the detector was placed in a solid water plate of approximately 20×40 mm^2 and a thickness of 4 mm.

The gate of the MOSFETs was biased during irradiation with 5 V. They were evaluated using a battery-operated

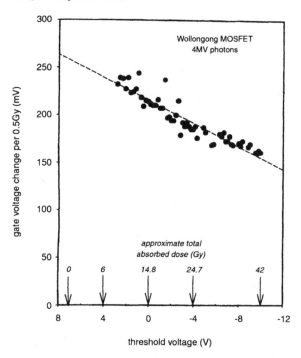

FIGURE 8.29 Variation of response with total absorbed dose in MOSFET detector, A, exposed to 4-MV photons from a medical linear accelerator (Varian Clinac 600 C/4 MV). The change required in gate voltage to compensate for the charge accumulated due to 0.5 Gy absorbed dose was used as dose response. It is shown as a function of total absorbed dose and the actual MOSFET total gate voltage required to maintain a constant current between source and drain. The bias applied during irradiation was 5 V. The broken line shows the best linear fit to the experimental data ($r^2 = 0.877$). (From Reference [17]. With permission.)

portable reader which was designed for measurements of the threshold voltage in the range of $+10$ V to -10 V. The stability of the threshold voltage assessment was within 2 mV.

Figure 8.29 shows the reduction of dose response in the Wollongong MOSFET, A, as a function of total gate voltage.

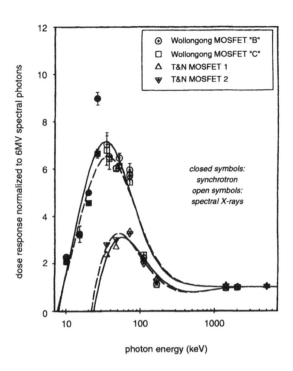

FIGURE 8.30 Variation of the dose response with effective radiation energy for MOSFET detector. The response is normalized to 6-MV spectral x-rays. (From Reference [17]. With permission.)

The total gate voltage is related to the total absorbed dose that is also indicated on the horizontal axis. The reduction of dose response, expressed as gate voltage change per absorbed dose, is a linear function of the total absorbed dose. The detector response is reduced by approximately 3% for a gate voltage change of 1 V. This function was used to correct the readings of the other Wollongong MOSFET detectors.

Figure 8.30 shows the variation of dose response with photon energy relative to water for MOSFET detectors. The two Wollongong-type detectors exhibit a maximum of dose response relative to water around 35 keV. The difference between the two detectors in the absolute dose maximum of 7.1 and 6.5, respectively, approaches statistical significance.

For comparison, the results obtained in spectral radiation beams with the Thomson and Nielson [17] MOSFET detectors are also plotted in Figure 8.30. The reduction of dose response with total absorbed dose was found to be less than 1% for a gate voltage change of 1 V, which is considerably smaller than the one observed in the Wollongong MOSFETs. The dose response for different radiation energies was corrected for this variation.

As can be seen in Figure 8.30, the Thomson and Nielson detectors also exhibit a smaller variation of dose response with photon energy, which is most likely due to the use of a 2-mm resin on the detector. The over-response was found to be on the order of a factor of 3, just above an effective photon energy of 50 keV.

The sensitive element of a MOSFET detector is a silicon oxide layer underneath the aluminum transistor gate. The geometry of the sensing regions of the silicon dioxide is slightly different for the two MOSFET devices used by Rosenfeld et al. [18] In the case of the TOT500 p-MOSFET device, the gate oxide has a serpentine shape, and in the case of the n-MOSFET device it was a rectangular ribbon shape with an outer boundary dimension of about 200 × 200 μm^2 and a thickness of about 1 μm for both detectors.

The change in MOSFET threshold voltage, which is proportional to the absorbed radiation dose, was measured under constant current conditions. The n-channel MOSFET chip was mounted inside a kovar package with the lid removed. RADFET has a different topology than the n-channel MOSFET and was mounted on a plastic board under an epoxy bubble. Both MOSFET and RADFET detectors were irradiated with a gale bias voltage of 5 V to increase sensitivity and improve linearity of response.

The n-MOSFET dosimeters were exposed in air, i.e., without any build-up material, to the microbeam in "normal" and "edge-on" orientations, and the p-MOSFET dosimeter was placed in "edge-on" orientation only. The dose delivered to MOSFET per single irradiation shot in free-air geometry was 17 monitor units (MU), which corresponds to a maximum threshold voltage shift in the center of the beam of about 150 mV. The results (Figure 8.31) prove that the spatial resolution of a MOSFET dosimeter is indeed affected by the orientation of the detector in the beam, the resolution being superior for the "edge-on" mode. The results demonstrate that the responses of n-MOSFET and RADFET in "edge-on" mode are almost independent of package geometry and material for irradiation in free-air geometry.

Electron fluence was calculated by Rosenfeld et al. [19] from the dose measured in the water at a distance of 1 m from the virtual source. The fluence at the point of irradiation was calculated using the inverse-square law. The electron fluence and dose in Si was measured independently using the Hamamatsu photodiode at the point of irradiation from the relation

$$D(Si) = \frac{I_d w_{Si} t_{ir}}{e \rho_{Si} V_{Si}} \qquad (8.8)$$

where, I_d = photodiode current
w_{Si} = 3.62 eV
t_{ir} = irradiation time
e = 1.6×10^{-19} C
ρ_{Si} = density of Si
V_{Si} = sensitive volume of the photodiode (1 × 1 × 0.3 cm³).

A LINAC electron beam was used for monitoring of the Hamamatsu photodiode response. The degradation of

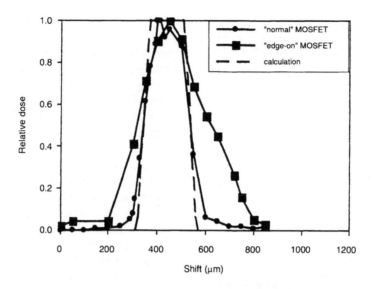

FIGURE 8.31 Comparison of transverse radiation dose profile across the 200-μm-wide microbeam measured by the MOSFET detector in both "normal" and "edge-on" orientations. (From Reference [18]. With permission.)

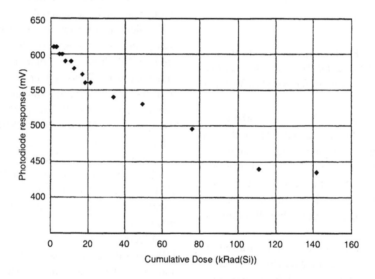

FIGURE 8.32 Hamamatsu photodiode response to a 20-MeV electron field as a function of absorbed dose. (From Reference [19]. With permission.)

the photodiode response in terms of cumulative dose (Si) to the 20-MeV electron field is shown in Figure 8.32. It can be seen that after a dose of 50 krad, which is the ionization dose expected within the BELLE SVD per year in a worst-case scenario, the response is degraded by ~15%.

In Figure 8.33, the reverse current measured under full depletion conditions before and after electron irradiation, with $\Phi_e \sim 10^{12}$ e/cm^2, is plotted as a function of reverse voltage. Following irradiation, the reverse current increased by a factor of ~13. The reverse current was monitored as a function of time to determine the effects of room temperature annealing. Over a period of two months, the reverse current was seen to decrease by 35% from the value measured 5 hrs after electron irradiation.

All measurements were made at a constant temperature of 20°C.

The radiation damage of the detector is measured by an increase in the reverse bias current per unit volume of fully depleted detector, according to Equation 8.9:

$$\frac{I - I_0}{Vol} = \alpha_p \Phi_p \qquad (8.9)$$

where α is a current damage parameter which is dependent on the type and energy of radiation

I_0 = reverse current before irradiation
I = reverse current after irradiation
Vol = full depletion volume of the detector.

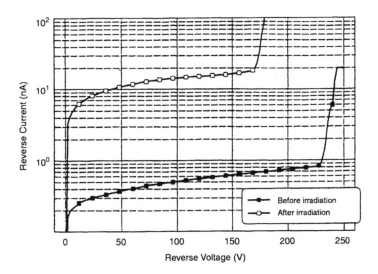

FIGURE 8.33 Detector reverse current measured before and after 20-MeV electron irradiation at a fluence. $\Phi_e = 9.2 \times 10^{11}$ cm^{-2}. (From Reference [19]. With permission.)

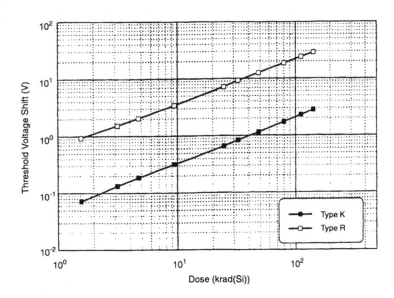

FIGURE 8.34 Calibration curves for the RADFETs in terms of dose in Si. (From Reference [19]. With permission.)

To obtain calibration curves for the RADFET system under high-energy electrons in terms of dose in Si, four such devices were exposed to the 20-MeV electron field, together with the Si photodiode and PIN dosimeter diodes. The shift of the threshold voltage under a constant current corresponding to the thermal stable point was measured. All measurements were done during the 3-hour irradiation. The spread in the change in threshold voltage (ΔV_{TH}) as a function of the dose in silicon (D(Si)) was found to be less than 3% for different RADFETs from the same batch. Figure 8.34 shows the calibration curves for both "R" and "K" type RADFETS in the 20-MeV electron field in terms of dose in Si. The analytical expression derived from the calibration curves for the "K" type

RAFET is $D = 40.23 \times (\Delta V_{TH})$, where D is the dose measured in krad(Si) and ΔV_{TH} is the voltage shift in volts. For the "R"-type RADFETs, $D = 1.81 \times (\Delta V_{TH})$.

III. P-TYPE SEMICONDUCTOR DOSIMETER

MOSFETs have a disadvantage of a response which is dependent upon the incident photon energy, particularly at low energies. In recent years, p-FET dosimeters have been developed with specially-grown thick-gate oxides which have a large number of oxide traps. Sensitivities of >10 mV/rad(Si) have been achieved. The sensitivity to radiation can be enhanced by applying a large positive bias during radiation, which forces more of the positive oxide charge to the interface. The sensitivity to temperature can be minimized by operating the p-FET with a current at the temperature-independent point. The main limitation in using silicon semiconductor detectors is effects caused by radiation damage. A nonlinear response with dose rate has proved to be serious for n-type detectors when pre-irradiated to 10 kGy or more in 20-MeV electrons. A sensitivity drop after pre-irradiation is another effect which is more pronounced for n-type detectors compared to p-type detectors of the same resistivity. A third effect of radiation damage is the sensitivity increase with increasing temperature. [20]

A Pantak HP-320 quasi-monoenergetic x-ray unit was used by Edwards et al. [21] to determine the response of two Thomson and Nielson TN-502RD MOSFETs, a Scanditronix EDP-10 semiconductor (build-up cap 10 mm: tissue equivalence), a BDD-5 semiconductor (build-up cap 4.5 mm: tissue equivalence), and a LiF:Mg:Ti TLD over the energy range 12–208 keV. The sensitivity of each detector was normalized to the value produced by exposure to 6-MV x-rays. The maximum relative sensitivities of the two MOSFET detectors were 4.19 ± 0.25 and 4.44 ± 0.26, respectively, occurring at an incident x-ray energy of 33 keV. The maximum relative sensitivity of the Scanditronix EDP-IO of 2.24 ± 0.13 occurred at 65 keV, and for the EDD-5, it was 7.72 ± 0.45 at 48 keV. The TLD produced a maximum relative sensitivity of 1.31 ± 0.09 at 33 keV. Compared with available data based on heteroenergetic x-ray sources, these measurements have identified a more representative response for each detector to low-energy x-rays.

Scanditronix p-type semiconductor detectors were investigated (Scanditronix, Uppsala, Sweden). These included the EDD-5 and the EDP-IO, both connected to a DPD-510 direct patient dosimeter. The 0.9-mm sensitive volume of the EDD-5 detector is encapsulated by Perspex, which provides an effective tissue-equivalent depth of 4.5 mm. This detector is designed primarily for critical organ measurements outside the main beam. The 0.9-mm^3 sensitive volume of the EDP-IO detector is surrounded by a 0.75-mm-thick stainless steel cap (providing a tissue-equivalent depth of 10 mm) and is adhered to a Perspex plate (1-mm thickness). This detector is designed for measurements inside the main beam generated by 4–8 MV linear accelerators.

A correction factor U_d for each detector d was calculated to allow for the small difference between the SSD to the center of its sensitive volume (J_d) and the SSD to a rigid surface:

$$U_d = \left(\frac{0.4}{J_d}\right)^2 \qquad (8.10)$$

The sensitivity $S_{x,d}$ of a detector at each x-ray energy

relative to that produced by 6-MV x-rays was calculated from:

$$S_{x,d} = \frac{L_d M_d (R_{x,d} - b_d)}{k_x t_x A_x F_x U_x} \qquad (8.11)$$

where $R_{x,d}$ is the detector response, b_d is the response to background radiation, k_x is the kerma rate recorded by the NE2550 Protection Level Secondary Standard with a beam current of 1 mA and SSD of 1 m, and t_x is the exposure time. The corrected response per unit air-kerma rate was normalized to the detector's response per unit air kerma rate from a 6-MV linear accelerator by multiplying by two previously determined calibration factors. The first factor L_d converted each detector's response to the absolute dose rate, measured at the maximum depth dose in a water phantom irradiated at an SSD of 1 m with a 10-cm × 10-cm field from a 6-MV linear accelerator. The second factor M_d converted this to the corresponding air-kerma rate and included any effects due to attenuation within the detector and from backscatter.

The variation in relative sensitivity of each detector with x-ray energy is illustrated in Figure 8.35. Relative to its response to a 6-MV linear accelerator beam, the sensitivity of each detector increased as the x-ray energy decreased until it reached a maximum value, and then it decreased with further decrease in x-ray energy.

On-chip p-FETs were developed by Buehler et al. [22] to monitor the radiation dose of n-well CMOS ICs by monitoring threshold voltage shifts due to radiation-induced oxide and interface charge. The design employs closed-geometry FETs and a zero-biased n-well to eliminate leakage currents. The FETs are operated using a constant current chosen to greatly reduce the FETs' temperature sensitivity. The dose sensitivity of these p-FETs is about -2.6 mV/krad(Si) and the off-chip instrumentation resolves about 400 rad(Si)/bit. When operated with a current at the temperature-independent point, it was found that the pre-irradiation output voltage is about -1.5 V, which depends only on design-independent silicon material parameters. The temperature sensitivity is less than 63 μV/°C over a 70°C temperature range centered about the temperature insensitive point.

The schematic cross section of the device shown in Figure 8.36 indicates that the n-well and source are separated so that they can operate at slightly different biases required by the operational amplifier, U_1. This allows all the forced current to flow through the p-FET channel. Drain-to-well leakage is shunted to ground. Grounding the n-well is a departure from normal CMOS circuit operation, where the n-well is normally connected to VDD.

In operation, all terminals of the p-FET are operated near ground except the drain, which operates near $V0 = -1.5$ V. During irradiation, the device is biased in the off state. The two RL resistors are used to bleed off any charge

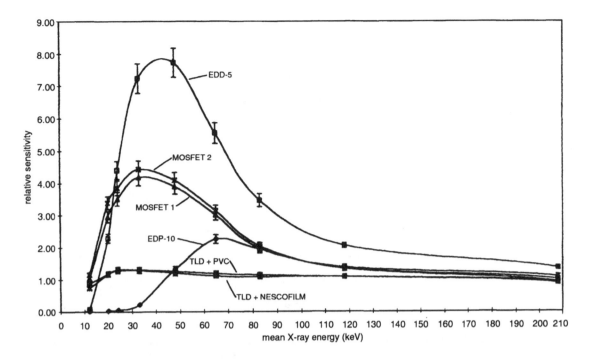

FIGURE 8.35 The relative sensitivity $S_{r,d}$ (response per unit air-kerma rate) normalized to that produced by 6-MV x-rays of two MOSFET p-type semiconductor diodes (EDP-10 and EDD-5) and two LiF:Mg:Ti TLD (one in PVC, one in Nescofilm™) to low-energy quasi-monoenergetic x-rays (error bars are ± 1 SD). (From Reference [21]. With permission.)

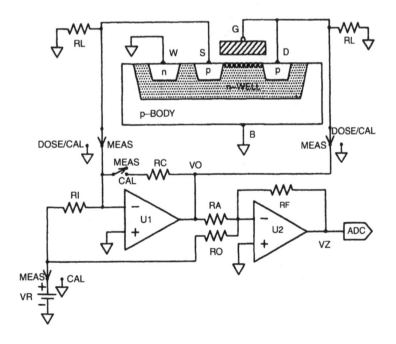

FIGURE 8.36 p-FET total dose circuitry. (From Reference [22]. With permission.)

remaining on the p-FET. The instrumentation is calibrated by positioning the switch to CAL, shown in Figure 8.36. This replaces the p-FET with calibration resistor RC.

The threshold voltage is described by:

$$VT = VT_0 + VT_T(T - T_0) + VT_D \times D \quad (8.12)$$

where D is the dose, VT_0 is the threshold voltage at T_0 and $D = 0$, $VT_T = \partial VT/\partial T|_{T \rightarrow T_0}$ and $VT_D = \partial VT/\partial D|_{D \rightarrow 0}$. The temperature and dose dependence of β (transconductance) is given by:

$$\beta = \beta_0(T/T_0)^{-n} + \beta_D \times D \quad (8.13)$$

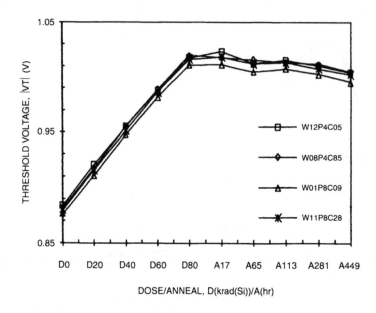

FIGURE 8.37 Four 1.2-μm CMOS p-FET threshold voltage dose/anneal responses. (From Reference [22]. With permission.)

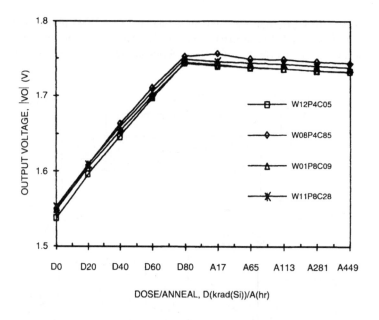

FIGURE 8.38 Four 1.2-μm CMOS p-FET output voltage dose/anneal responses determined at the current, ID_m. (From Reference [22]. With permission.)

where β_0 is β evaluated at T_0 and $D = 0$, $\beta_D = \partial\beta/\partial D|_{D \to 0}$, n characterizes the mobility temperature dependence, and T is the absolute temperature. $\beta = KP \times W_e/L_e$, $KP = \mu C_0$. $W_e = W - DW$ and $L_e = L - DL$ are the effective channel width and length, respectively, μ is the zero field channel mobility, C_0 is the gate oxide capacitance per unit area.

The VT values, plotted in Figure 8.37, show a high degree of linearity during irradiation and a slight recovery with anneal. The group average slope of the VT vs. dose curve during irradiation is $VT_D = -1.698 \pm 0.038$ mV/krad(Si). The shift in VT with radiation is consistent with the build-up of positive oxide charge and interface states. The slight recovery of VT

during room-temperature anneal is consistent with the slight loss of oxide charge. The interface state density is stable, as seen by the flat response of the mobility during anneal.

The output voltage for the four p-FET samples is shown in Figure 8.38. This plot was obtained from data sets. The data shows a nearly linear rise in V_0 with dose and a slight annealing effect.

Brucker et al. [23] present the results of studies of dose enhancement in dual and single dielectric pMOSFET dosimeters for various package and die designs. Eight different MOSFET designs and package types were investigated over a photon energy range from 14 to 1250 keV. Packages filled with silicon grease, aluminum oxide, or

DUAL-IN-LINE CERAMIC PACKAGE   TO–18 KOVAR PACKAGE

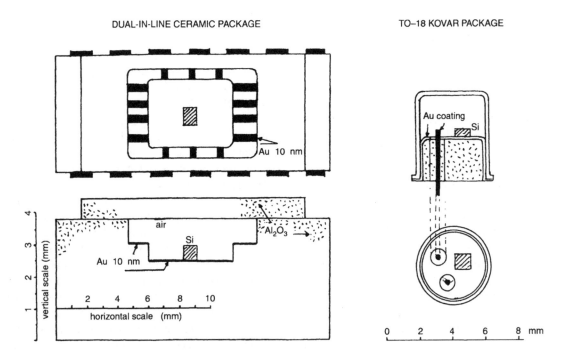

FIGURE 8.39 Cross-sectional and top views of dual-in-line ceramic and TO-18L kovar can packages. (From Reference [23]. With permission.)

paraffin eliminated the contribution of back scatter to the enhanced dose. These modifications allowed measurements of the usual dose enhancement at the aluminum or polysilicon gate-silicon nitride (dual dielectric devices) or silicon dioxide interfaces (single dielectric parts) and at the silicon nitride–silicon dioxide interface. In addition to the primary peak in the DEF (dose enhancement factor) curve vs. energy at 45.7 keV, there is a second peak at about 215 keV. This peak might be due to enhancements at the interfaces of a MOSFET. These interface effects were small in the single-insulator parts in standard ceramic packages and significantly larger in the dual-insulator devices. The effects were reduced by filling the packages with the materials as previously described. The geometry of the package, the size of the air gap between the die's surfaces, and the lid of the package impacts the value of the DEF.

The three-dimensional models used by the code were those in Figure 8.39. The figure shows a cross-sectional and top view of the dual-in-line ceramic and TO-18L kovar can packages. The gold coatings on the bonding pads and in the well or cavity of the ceramic package are indicated. This diagram actually represents the REM 501A part type. The other REM devices did not have the coating in the cavity.

Figure 8.40 shows a comparison of results obtained with air- and grease-filled TO-5 kovar cans containing #3 transistors. Figure 8.41 contains plots that compare the DBF values for Sandia #1 parts in ceramic and kovar packages. Results for air- and grease-filled ceramic packages are also shown.

A dose-rate-independent p-MOS dosimeter for space applications was discussed by Schwank, et al. [24] A dual dielectric p-MOS dosimeter (RADFET) has been designed

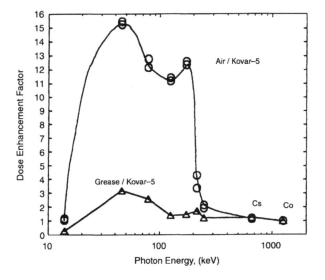

FIGURE 8.40 Dose enhancement vs. energy for Sandia #3 in TO-5 kovar cans with and without grease in package. (From Reference [23]. With permission.)

at Sandia. The RADFET consists of a thermally grown oxide and a CVD deposited nitride. With a negatively applied bias, holes generated in the SiO_2 transport to and are trapped at the SiO_2/Si_3N_4 interface, producing a measurable threshold-voltage shift. Because holes are trapped away from the Si/SiO_2 interface, hole neutralization by tunneling and interface-trap build-up are minimized, resulting in little fade or annealing of the RADFET output response. RADFETs were irradiated at dose rates from 0.002 to 50 rad(Si)/s, with biases from -5 to -20 V.

RADFETs were also annealed for times up to 10^7 s at temperatures up to 100°C. Within experimental uncertainty, no difference in RADFET output response at a given bias was observed over the dose rate range examined and for 25°C anneals. At an anneal temperature of 100°C, only a 20% decrease in RADFET output response was observed.

To monitor total dose, RADFETs were characterized for the threshold-voltage shift, ΔV_{TH}, and the change in source-to-drain voltage, ΔV_{DS}, necessary to maintain a constant source-to-drain current, I_{DS}.

The RADFETs were designed at Sandia and fabricated at Micrel Semiconductor Corporation. The structure of the RADFET is shown in Figure 8.42. The RADFETs used by Schwank et al. consisted of a 60-nm thermal oxide grown on a (100) n on n^+ expitaxial substrate, and a low-pressure CVD 100-nm silicon nitride layer deposited on top of the

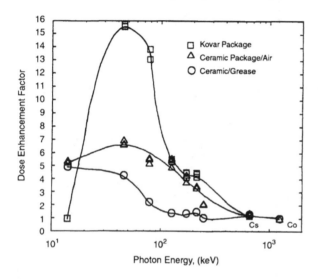

oxide. All RADFETs were p-channel transistors with metal gates and were taken from the same wafer. The wafer-to-wafer and lot-to-lot variation in RADFET response is not known, so other wafers would have to be calibrated independently. Transistors were packaged in 24-pin DIP packages.

p-MOS dosimeters with a single dielectric have been fabricated using a variety of techniques. Most of these measure the threshold-voltage shift induced by positive charge trapping in the oxide. For thermal oxides, most of the trapped charge occurs in the oxide near either the top gate-Si/SiO$_2$ (negative gate bias during irradiation) or bottom Si/SiO$_2$ (positive gate bias during irradiation) interface. If holes are trapped near the top interface, the measured threshold-voltage shift is very small, due to a small charge moment arm. For this reason, standard p-MOS dosimeters are normally operated with a positive applied bias. Unfortunately, for positive applied biases, interface traps will be generated at the Si/SiO$_2$ interface and trapped holes can be neutralized by thermal emission and tunneling processes. The time dependence of interface-trap build-up and the neutralization of holes will result in fade of the dosimeter's output characteristics. [24]

The radiation-induced threshold voltage shift, ΔV_{TH}, of Sandia's dual dielectric RADFET is given by:

$$\Delta V_{TH} = \frac{L_{ox}}{\varepsilon_{ox}\varepsilon_0}\left[\frac{q}{L_{ox}}\int_0^{L_{ox}}(L_{ox}-x)n_h^+(x)\,dx\right]$$

$$+ \left[\frac{L_T - L_{ox}}{\varepsilon_{Si_3N_4}\varepsilon_0}\int_0^{L_{ox}}qn_h^+(x)\,dx\right] \quad (8.14)$$

where ε_{ox} and $\varepsilon_{Si_3N_4}$ are the dielectric constants for SiO$_2$ ($\varepsilon = 3.9$) and Si$_3$N$_4$ ($\varepsilon = 7.5$), respectively, ε_0 is the permittivity of free space, L_{ox} is the thickness of the oxide layer, L_T is the total dielectric thickness (SiO$_2$ + Si$_3$N$_4$),

FIGURE 8.41 Illustrates effect of package and filling in reducing DBF for Sandia #1 in TO-18L and ceramic FLTPAC. (From Reference [23]. With permission.)

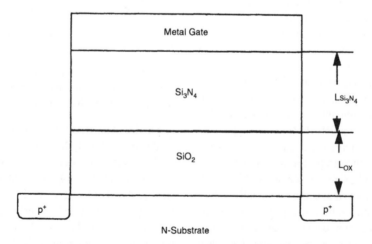

FIGURE 8.42 RADFET transistor structure. (From Reference [24]. With permission.)

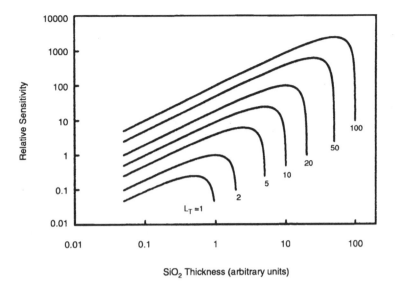

FIGURE 8.43 Relative RADFET sensitivity for the case where holes are trapped near the SiO$_2$/Si$_3$N$_4$ interface for oxide thicknesses from 0.05 to 100 (arbitrary units) and total insulator thicknesses, L_T, from 1 to 100 (arbitrary units). (From Reference [24]. With permission.)

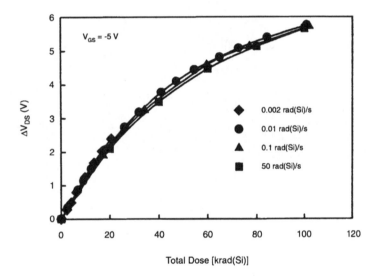

FIGURE 8.44 Δ V_{DS} vs total dose for RADFET. (From Reference [24]. With permission.)

and $n_h(X)$ is the distribution of trapped holes in the SiO$_2$ layer (or net positive charge if both electrons and holes are trapped). x is measured relative to the metal gate/Si$_3$N$_4$ interface.

Figure 8.43 is a plot of calculated relative radiation sensitivity vs. SiO$_2$ thickness for several values of insulator thickness (SiO$_2$ + Si$_3$N$_4$).

At very low temperatures (~100°K), holes generated in the SiO$_2$ will be immobile and the distribution of holes throughout the SiO$_2$ is uniform. Under these conditions, the radiation sensitivity is proportional to

$$S \propto \frac{L_{ox}^2}{2\varepsilon_{ox}} + \frac{L_{ox}(L_T - L_{ox})}{\varepsilon_{Si_3N_4}} \qquad (8.15)$$

The radiation response of the RADFETs for dose rates from 0.002 to 50 rad(Si)/s is shown in Figure 8.44. This figure is a plot of ΔV_{DS} for a constant source-to-drain current (~4 mA) for RADFETs irradiated at room temperature to 100 krad(Si) with a constant gate-to-source bias, V_{GS}, of −5 V.

For low total dose levels, the output response of the RADFETs is linear with total dose. Figure 8.45 is a plot of ΔV_{DS} for RADFETs irradiated to 20 krad(Si) at dose rates from 0.002 to 50 rad(Si)/s with $V_{GS} = -5$ V. Over this dose range, for all dose rates, the dose is given by

$$\text{Dose[rad(Si)]} = 8300 \times \Delta V_{DS} \qquad (8.16)$$

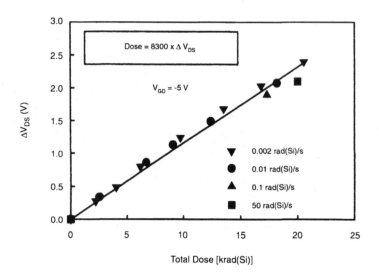

FIGURE 8.45 ΔV_{DS} vs. total dose for RADFETs irradiated to 20 krad(Si) at dose rates from 0.002 to 50 rad(Si)/s with $V_{GS} = -5$V. (From Reference [24]. With permission.)

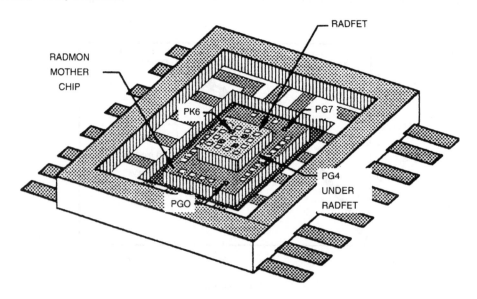

FIGURE 8.46 RADFET mounted on the RADMON mother chip in a 16-pin flat package. Three radiation sensitivities are achieved: PK6 has highest sensitivity, PG0 and PG7 have moderate sensitivity, and PG4 has lowest sensitivity, being shielded by the RADFET. (From Reference [25]. With permission.)

where ΔV_{DS} is given in volts. At higher total doses, >20 krad(Si), the RADFET output response becomes sublinear (see Figure 8.44) as the trapped charge in the insulator begins to reduce the internal field in the oxide. As the oxide electric field is lowered, the amount of electron-hole recombination increases. The increase in electron-hole recombination will result in fewer trapped holes in the insulator. [24]

A stacked p-FET dosimeter consisting of a RADMON mother chip with three p-FETs and multiplexer and an attached RADFET has been developed by Buehler et al. [25] for the STRV-2/MWIR detector. Calibration of the dosimeter using an Am-241 source indicates that the RADFET is about 20 times more sensitive to radiation than the RADMON. This dosimeter is expected to measure radiation dose from rads to megarads.

The dosimeter, seen in Figure 8.46, consists of a thin-oxide RADMON mother chip with three p-FETs, a multiplexer, and other devices. A thick-oxide RADFET is attached to the mother chip. The device operates on the principle that radiation passing through the gate oxide region of the p-FET creates hole-electron pairs. Since the oxides have mainly hole traps, radiation-induced positive charge is trapped in the oxide. This build-up of positive charge changes the threshold voltage of the device, which is interpreted in terms of radiation dose. A cross-section view is shown in Figure 8.47.

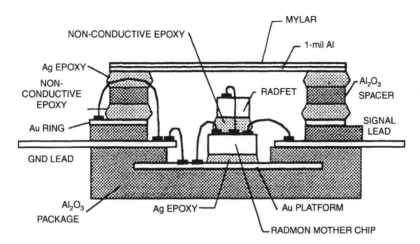

FIGURE 8.47 Cross section of the MWIR dosimeter package showing the wire bonds that ground the Au ring and aluminum-coated mylar lid. (From Reference [25]. With permission.)

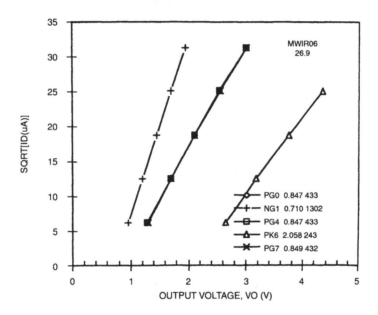

FIGURE 8.48 Typical current-voltage characteristics for the five FETs located on the MWIR. Note that the FET characteristics for PG0, PG4, and PG7 overlap. The numbers in the key are *VT* in volts and β in $\mu A/V^2$. (From Reference [25]. With permission.)

Room temperature current-voltage measurements are shown in Figure 8.48 for the five FETs in this experiment. These curves were modeled using the saturation drain current expression:

$$ID = \left(\frac{\beta}{2}\right)(VO - |VT|)^2 \qquad (8.17)$$

where *VO* is the FET output voltage, $\beta = KP \times W/L$ is the transconductance, $KP = \mu C_0$, *W* is the channel width, *L* is the channel length, μ is the channel mobility, and C_0 is the gate oxide capacitance per unit area. The threshold voltage, *VT*, is determined from the extrapolation of the curves shown in Figure 8.48 to zero drain current using a

least-squares fitting technique. The β values were determined from the slope of the characteristics shown in Figure 8.48.

Semiconductor detectors based on *p*-type silicon but with different doping levels have been investigated by Grusell and Rikner. [26] It was shown that a *p*-type detector with a low doping level and high resistivity showed a nonlinear dose rate response if it were radiation-damaged in a high-energy photon beam, which contains neutrons. By increasing the doping level, it was shown that a detector with a resistivity of 0.2 Ωcm stayed linear after pre-irradiation in radiation fields from high-energy electrons, photons, and protons. Other parameters did not show any changes of clinical importance at the different doping levels.

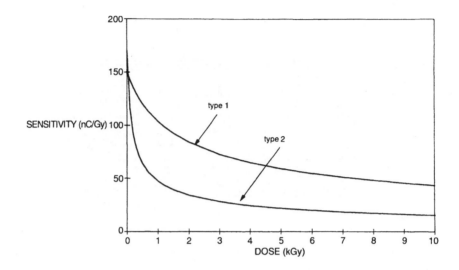

FIGURE 8.49 The sensitivity as a function of accumulated dose of 20-MeV electrons for the two types of detectors. From Reference [26]. With permission.)

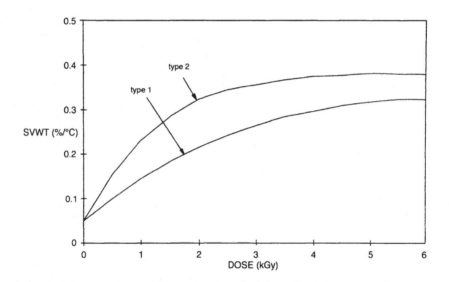

FIGURE 8.50 The sensitivity variation with temperature (SVWT) as a function of accumulated dose of 20-MeV electrons for the two types of detectors. (From Reference [26]. With permission.)

Detectors with two different resistivities were designed. Type 1 had a resistivity of 10 Ωcm, and detectors of type 2 had a resistivity of 0.2 Ωcm. The ionization volume was about 0.3 mm^3, with a front surface of about 4 mm^2. The detectors were encapsulated in a water-resistant epoxy resin (Araldite D). During the measurements they were in the short-circuit mode, connected to an integrating electrometer with a very low and stable input offset voltage.

The reduction in sensitivity as a function of pre-irradiation with 20-MeV electrons is shown in Figure 8.49 for the two detector types. As can be seen, the detectors made on low-resistivity silicon (type 2) show a faster decrease in sensitivity compared to the detectors of type 1.

Figure 8.50 shows the sensitivity variation with temperature (SVWT) as a function of pre-irradiation dose with 20-MeV electrons for the two detectors. Detectors of type 2 showed a slightly faster variation with accumulated dose than detectors of type 1. The saturation value of about 0.4% per °C was, however, the same for the two types of detectors.

The sensitivity of a cylindrical p-type silicon detector was studied by Piermattei et al. [27] by means of air and water measurements using different photon beams. A lead filter cap around the diode was used to minimize the dependence of the detector response as a function of the brachytherapy photon energy.

The silicon diode EDD-5, designed as a direct patient-dose monitor for ⁶⁰Co irradiation, manufactured by Scanditronix, is a *p*-type delector with a build-up of 5 mm for ⁶⁰Co γ-ray beams. The geometrical features (10 mm in length, 5 mm in diameter) are shown in Figure 8.51. The cylindrical symmetry of the EDD-5 gives an angular dependence ≤2%, while the sensitivity increases by 0.4% per °C in the temperature range 20°C–40°C.

Figure 8.52 shows the sensitivity of the EDD-5 silicon diode as a function of the photon beam energy normalized to the value obtained for the ⁶⁰Co γ-ray beam. The uncertainty of the sensitivity values was estimated to be within 4% (2σ).

A mean sensitivity, $\bar{S}(r)$, of the unshielded and shielded EDD-5 diode at distances *r*, between 1 and 10 cm, from a point source in water, was calculated as the sum of the diode sensitivities weighted by the absorbed dose in water: [27]

$$\bar{S}(r) = \sum_i \Phi(E_i,r)E_i\left(\frac{\mu_{en}(E_i)}{\rho}\right)_{water}$$
$$\times S(E_i) \Big/ \sum_i \Phi(E_i,r)E_i\left(\frac{\mu_{en}(E_i)}{\rho}\right)_{water} \quad (8.18)$$

where $\Phi(E_i,r)$ are at distance *r* from the source in water; the relative photon fluences at energy E_i, $(\mu_{en}(E_i)/\rho)_{water}$, are the mass energy absorption coefficients in water for photon energy E_i; and $S(E_i)$ are the diode sensitivities given by the solid lines, shown in Figure 8.52, which connect the experimental data.

IV. OTHER SOLID-STATE DOSIMETERS

VeriDose solid-state diode dosimeters are commercially available solid-state diode dosimeters. [28] Their characteristics are as follows:

- hemispherical shape improves isotropic response and reduces angular and field-size dependencies
- waterproof design with appropriate build-up for all clinical photon and electron energies
- flat bottom permits secure, easy placement on the patient
- color-coded for ease of identification
- dose rate independent
- responds to photons and electrons

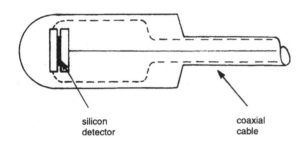

FIGURE 8.51 Scheme of the EDD-5 diode. The effective thickness of the sensitive volume, approximately 60 μm, is located in a drop shape-encaptulation with outer dimensions 10 mm in length and 5 mm in diameter. (From Reference [27]. With permission.)

silicon detector

coaxial cable

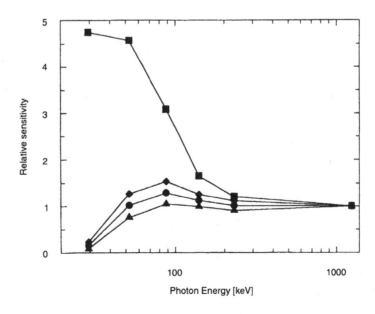

FIGURE 8.52 Relative sensitivity of an EDD-5 diode, unshielded (■) and shielded, with a cylindrical closed-ended lead cap, 0.15 (◆), 0.22 (●), and 0.25 mm (▲) thick, as a function of the photon energy. Data are normalized to the ⁶⁰Co γ-ray beam. Continuous lines connect experimental data. (From Reference [27]. With permission.)

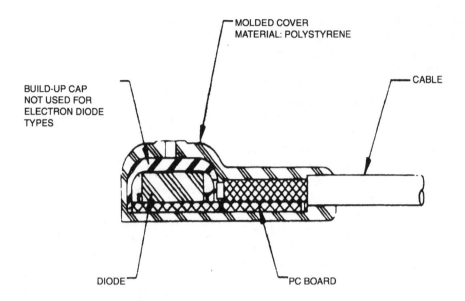

FIGURE 8.53 Cross section of a VeriDose diode detector. (From Reference [27]. With permission.)

- responds to dose rates of 1.0 cGy/min to 1000 cGy/min
- ^{60}Co x-ray beams, pulsed (linear-accelerator) x-ray beams, and electron beams.
- optimized for use with all Nuclear Associates patient dose monitors and high-quality medical-grade ionization-chamber electrometers.

Nuclear Associates VeriDose diode detectors are solid-state, silicon-based radiation detectors that utilize a *p-n* junction. These rugged diodes are encased within an FDA-approved polystyrene material. A low-noise coaxial cable is used to connect the diode to an electrometer. When attached to an electrometer, these diodes provide enhanced sensitivity and instantaneous response time.

VeriDose diodes are constructed using a "parallel-plate" geometry with planar electrodes opposing each other at a given spacing. They are platinum-doped *n*-type diodes; their nominal sensitivity is 1.5 nC/cGy; and their sensitivity volume is 0.25 mm^2. Figure 8.53 shows a cross section of the diode assembly.

A high-precision patient dosimetry method has been developed by Heukelom et al. [29], based on the use of *p*-type diodes. Entrance as well as exit dose calibration factors have to be determined under reference irradiation conditions. A set of correction factors must be available.

The measurements were made using a set of seven *p*-type diodes (type EDP-20, Therados). Each diode was encapsulated by a hemispherical build-up cap of 2.2-mm stainless steel and 2.8-mm epoxy, which together is equivalent to approximately 2 cm water for 8-MV x-rays.

Dose calibration factors, $N_D^{entrance}$ and N_D^{exit}, defined as D^{ref}/R^{ref}, were measured under reference conditions for each diode. For each diode, the variation of the calibration

factors has therefore been investigated systematically as a function of phantom thickness, field size at the isocenter, SSD, the presence of wedges, and temperature. The influence of these variations in irradiation conditions on the calibration factor can be expressed as a correction factor, C_i. The subscript *i* refers to the actual irradiation condition, e.g., $C_{thickness}$, $C_{field size}$, C_{SSD}, C_{wedge} and C_{temp}. The dose value at the entrance or exit point, determined from the diode reading, can be expressed as

$$D_{diode} = R_{diode} N_D \prod_i C_i \qquad (8.19)$$

in which the C_i values are independent of each other.

In Figure 8.54, $C_{temperature}$ is plotted against the temperature of the water present in a Perspex container on which the diodes were positioned. The temperature dependence has been determined for two integrated dose values, representing the situation after about 0.5 year and 1.5 years of clinical use. The variation in sensitivity of the diodes as a function of the dose per pulse is indicated in Figure 8.55, again for two integrated dose values.

The sensitivity of all diodes as a function of the mean primary photon beam energy, normalized to the value for the 8-MV photon beam, is indicated in Figure 8.56 for both the entrance- and exit-side geometries. Mean photon energies were derived from mass attenuation coefficients. These were obtained for the low-energy photon beams from half-value layers measured in copper. For the megavoltage photon beams, mass attenuation coefficients were estimated from percentage depth-dose values for a small field size.

The energy dependence of the response of silicon detectors was determined first by the increase in cross section of the photoelectric effect in silicon relative to

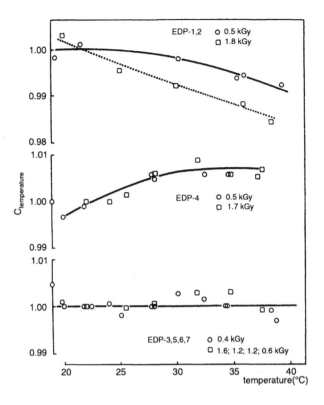

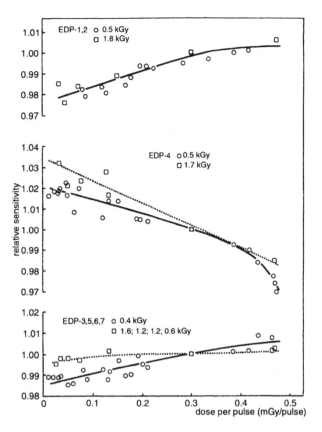

FIGURE 8.54 The correction factor $C_{temperature}$ plotted against the temperature of water in a Perspex container on which the diodes were positioned. $C_{temperature}$ was determined for different values of the total dose delivered to these diodes. (From Reference [29]. With permission.)

FIGURE 8.55 The relative sensitivity of the diodes as a function of the dose per pulse for different values of the total dose delivered to these diodes. (From Reference [29]. With permission.)

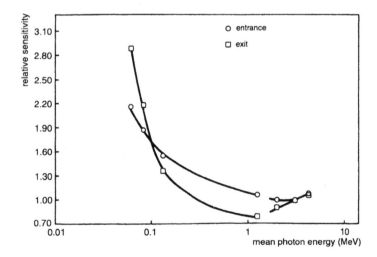

FIGURE 8.56 The relative sensitivity of the diodes plotted against the mean photon energy as determined at the entrance (O) and exit (□) sides of the rod phantom. (From Reference [29]. With permission.)

water for decreasing photon energy and to pair production for high energies.

A variety of high-purity silicon grown on the basis of different manufacturing technologies by Khivrich et al. [30] was exposed to gamma irradiation (up to a dose of 10^8 rad(Si)) and to neutron irradiation (up to a fluence of 10^{15} n/cm²). Observation was made of the conduction type and carrier concentration as a function of dose. The conversion point (n-Si to p-Si) of gamma-irradiated silicon was found to vary over 2 orders of magnitude of

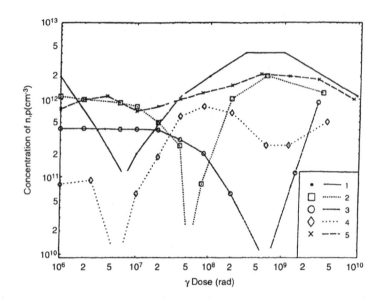

FIGURE 8.57 Charge carrier concentration vs gamma dose (in air) for high purity-Si from different manufacturers with different initial charge carrier concentrations in cm^{-3} n-Si: curve (1) 2.5 × 10^{12} cm^{-3} (NTD), (2) 1.05 × 10^{12} cm^{-3} (ZTMK), (3) 5 × 10^{11} cm^{-3} (producer 2), (4) 6 × 10^{10} cm^{-3} (producer 3); p-Si: (5) 6 × 10^{11} cm^{-3} (ZTMK). (From Reference [30]. With permission.)

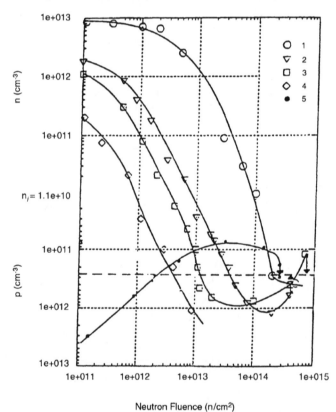

FIGURE 8.58 Charge carrier concentration in high-purity silicon with efferent initial resistivity in kΩcm vs. neutron fluence, n-Si: curve (1) 0.39 kΩcm, (2) 2.5 kΩcm, (3) 5 kΩcm, (4) 40 kΩcm; p-Si:(5) 3.2 kΩcm. (From Reference [30]. With permission.)

gamma dose for different manufacturers of high-purity silicon, independent of the initial carrier concentration. A systematic study of the radiation hardness of high-purity silicon allowed the development of silicon detec-tors working under harsh radiation environments and operating over a wide range of dose. High-purity silicon PIN diodes were calibrated using an epithermal neutron beam to determine whether response in terms of 1-MeV(Si)

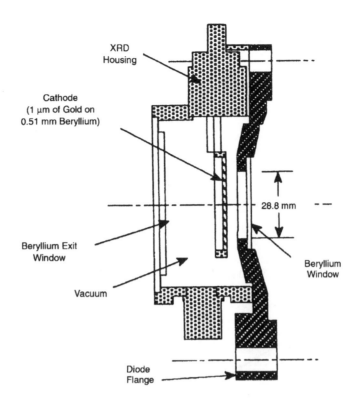

FIGURE 8.59 SEMIRAD x-ray diode illustration. (From Reference [31]. With permission.)

neutrons was independent of the calibration spectrum used.

Figure 8.57 shows the change of charge carrier concentration (electrons or holes) and the conductivity type vs. gamma dose in air from a ^{60}Co source. All of the high purity n-Si samples investigated showed conversion to p-type. However, there does not exist for these particular samples a correlation between the conversion dose and the initial charge carrier concentration, as is the case for silicon manufactured using the same technological process but with different initial concentrations of charge carriers. The conversion dose for n-Si with $\rho \sim 20$ kΩcm of producer 3 is two orders of magnitude less than for n-Si with $\rho \sim 9$ kΩcm of producer 2; see Figure 8.57 curves 3, 4.

The hole concentration in high-purity p-type silicon does not change significantly (Figure 8.57, curve 5). This corresponds to the small contribution of boron to the creation of stable radiation defects.

High-purity n- and p-type silicon was investigated under fast neutron irradiation. Behavior similar to high-purity silicon under gamma irradiation was observed; n-Si converted to p-Si (Figure 8.58, curves 1–4), with a flattening-off of the hole concentration of the order of 10^{12} cm^{-3}. In contrast to gamma irradiation, a strong correlation between the initial resistivity of n-Si and the neutron fluence required for conversion of n- to p- type has been observed. The p-type high resistivity Si has similar fast

neutron fluence dependence to n-Si after conversion to p-type (Figure 8.58, curve 5).

Bellem et al. [31] described the method used to calibrate a large-volume continuous-wave x-ray radiation test chamber. Three detector types—x-ray vacuum diodes (XRD), silicon PIN diodes, and PMOS FETs—were used to measure spectral intensity and dose deposition in silicon devices.

The x-ray vacuum diode, the Secondary Electrons Mixed Radiation Dosimeter or SEMIRAD (referred to as the XRD), is an evacuated chamber constructed to collect secondary electron emissions from a gold film cathode within the chamber. The XRD is ideally suited to measure a constant high-intensity x-ray beam with radiation intensities at energies up to 160. The XRD was used to quantify the incident spectrum. A cross-sectioned illustration of the XRD is shown in Figure 8.59.

The PIN diodes (Quantrand 010PIN025) have a nominal area of 10 mm^2 and an effective collection depth of about 25 μm. The diodes provide a direct readout of the ionization-induced current, which is proportional to the dose rate, and can accurately measure dose rates up to 200 krad(Si)/min.

The PMOS FETs used were manufactured by the AT&T Allentown Microelectronics Facility. The devices were fabricated using polysilicon gate SiO_2:Si structures with a hardened gate oxide under 20 nm thick. A test strip

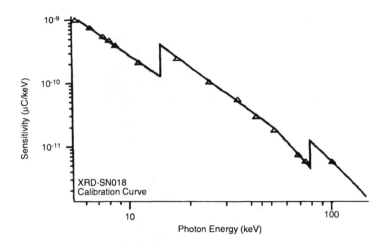

FIGURE 8.60 SEMIRAD x-ray diode (XRD-SN018) sensitivity vs. photon energy. (From Reference [31]. With permission.)

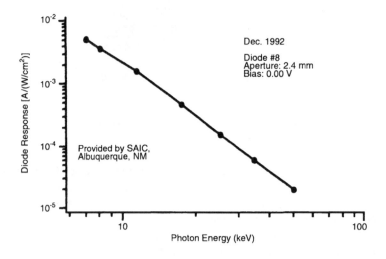

FIGURE 8.61 ARACOR PIN diode (CS 1003) sensitivity vs. Photon Energy. (From Reference [31]. With permission.)

containing a variety of gate lengths and widths was available on-chip. Four transistors were selected with a gate width of 30 μm and gate lengths of 1.50, 1.75, 2.00, and 5.00 μm.

The device response was then calculated by integrating a calculated or measured incident spectrum with the experimentally measured detector sensitivity. A computer program was written to integrate the product of these two quantities; that is,

$$i_d = A_a \int_0^{E'} \phi'(E) S(E) \, dE \qquad (8.20)$$

where i_d = diode current (μA)

E' = spectral energy distribution end point energy (keV)

A_a = aperture area (cm²)

$S(E)$ = diode sensitivity data $\left(\dfrac{\mu C}{\text{kev}}\right)$

$\phi'(E)$ = incident spectral energy distribution

$$\left(\frac{\text{keV}}{\text{keV} \times \text{cm}^2 \times \text{s}}\right)$$

The XRD was calibrated from 5.41 keV (K-line for Cr) up to 101 keV (K-line for V) by measuring the current response. A plot of the sensitivity data obtained for the XRD calibration is shown in Figure 8.60. The calibration errors are estimated at less than 5% over the energy range from 5 keV to 140 keV, with the errors increasing to about 10% as the energy approaches 160 keV. Below 5 keV, errors are introduced in the sensitivity due to the M-lines of Au; however, this energy will have no effect in our calculations since the photons with energy below 5 keV are absorbed by the intervening air.

The PIN diode was calibrated at the SAIC/DNA facilities from 6.40 keV (the K-line for Fe) to 52 keV (the K-line

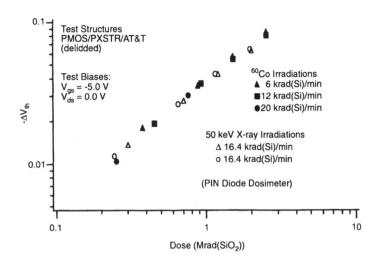

FIGURE 8.62 Threshold voltage shift of PMOS FETs with x-Ray and ^{60}Co data overlaid. (From Reference [31]. With permission.)

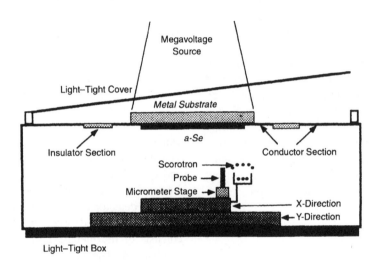

FIGURE 8.63 Schematics of the detector and read-out unit (From Reference [32]. With permission.)

for Yb). The PIN diode response is shown in Figure 8.61. This diode was found to be an accurate dosimeter for measuring dose rate in silicon. Due to the potential for PIN degradation due to oxide charge build-up from leakage by the collimator shield, the PIN is limited to spectral energies less than 50 keV. In theory, however, the PIN diode responds to all incident radiation. Therefore, to determine dose rates for endpoint energies above 50 keV, a method which combines measured and calculated results was used.

The rate that energy is absorbed by a material (i.e., the dose rate) is given by: [31]

$$\dot{D}_{E'} = K\int_0^{E'} \phi'_{E'}(E)\mu_a(E)\,dE \qquad (8.21)$$

where $\dot{D}_{E'}$ = dose rate for a spectral energy distribution with end-point energy E'

K = unit conversion factor

E' = endpoint energy

$\phi'_{E'}(E)$ = spectral energy distribution with endpoint energy E'.

$\mu_a(E)$ = energy-dependent mass absorption coefficient.

An overlay comparison of x-ray data is shown in Figure 8.62. The maximum variation between data points in Figure 8.62 is less than 10%. For example, the slope of the 50-keV x-ray data is 31.5 mV/Mrad, while the slope of the 6-krad/min ^{60}Co data is 32.2 mV/Mrad. This correlation is excellent for these types of measurements over the broad range of photon energies being considered. [31]

A feasibility study has been performed by Falco et al. [32] on metal/amorphous selenium detectors for megavoltage portal imaging. The metal plates of the detectors were positioned facing the incident 6-MV and ^{60}Co photon spectra. The detectors consisted of various thicknesses

(0.15 mm, 0.30 mm, and 0.50 mm) of amorphous selenium (a-Se) deposited on metal plates of varying thicknesses: aluminum (2.0 mm), copper (1.0 mm and 1.5 mm), stainless steel (0.9 mm), or glass (1.1 mm). The detectors were charged prior to irradiation by corona methods, and the portal images were subsequently digitized after irradiation with a noncontact electrostatic probe. The sensitivity of the detectors to dose, electric field across the a-Se layer, metal plate type, and a-Se thickness was studied. An increase in electric field increases the sensitivity (gradient of the a-Se surface voltage vs. dose curve) and dynamic range of the resultant image. However, increase in a-Se thickness, although also increasing the sensitivity, decreases the dynamic range.

A layer of *a*-Se is deposited on a metal (conductor) substrate or plate. The formation of a latent electrostatic image on an *a*-Se photoconductor surface takes two steps: (1) the *a*-Se is charged (e.g., corona charging) in the dark to achieve a uniform charge distribution on its surface; and (2) a patient-modulated photon beam forms a latent image by the local centralization of the uniform charge distribution. The extent of this local neutralization is proportional to the number of electron-hole pairs produced by the irradiation per unit volume of the *a*-Se layer. Therefore, the varying intensity across the radiation beam exiting the patient will result in a corresponding charge-distribution pattern on the *a*-Se surface. In complete darkness, the *a*-Se layer is a charge capacitor, and a uniform charge distribution σ on the *a*-Se surface will result in a potential difference V across the layer:

$$V = \frac{\sigma d}{\varepsilon} \qquad (8.22)$$

where d is the thickness of the *a*-Se layer and ε is the permittivity of *a*-Se. X-ray irradiation on the detector will create electron-hole pairs within the *a*-Se and these charges will drift under the influence of the electric field ($E = V/d$) toward opposite surfaces to neutralize the initial charge on the *a*-Se surfaces. The decrease of surface charge $\Delta\sigma$ is proportional to the radiation dose ΔD and inversely proportional to the average energy $W_{\pm}$ required to generate and collect one electron-hole pair in *a*-Se:

$$\Delta\sigma\alpha - \frac{\Delta D}{W_{\pm}} \qquad (8.23)$$

With some mathematical manipulation, the voltage difference between the a-Se layer and the metal substrate as a function of radiation dose is: [32]

$$V(D) = V_0\left[1 - \frac{\alpha d^{1/3}}{\varepsilon V_o^{1/3}} \times D\right]^3 + V_{bias} \qquad (8.24)$$

where $\alpha((\text{Coulomb/cGy})(\text{volt})^{-2/3}/(\text{m})^{4/3})$ is a proportional constant characterizing the photoconductor's sensitivity to megavoltage photons. It is defined as the sensitivity parameter. V_0 is the initial voltage potential difference on the metal *a*-Se detector before irradiation and equals ($V_{grid} + |V_{bias}|$), where V_{grid} is the potential on the *a*-Se surface deposited by corona charging and $|V_{bias}|$ is the absolute value of the negative bias voltage applied to the metal plate. The voltage $V(D)$ drops to V_{bias} (i.e., total discharge) at dose

$$D_m = \frac{\varepsilon}{\alpha} \times \left(\frac{V_0}{d}\right)^{1/3} \qquad (8.25)$$

which thus defines the dynamic range. Substituting Equation (8.25) into Equation (8.24), we obtain

$$V(D) = V_0\left(1 - \frac{D}{D_m}\right)^3 + V_{bias} \qquad (8.26)$$

The sensitivity can be determined from the steepness of the slope of the discharging curve:

$$\frac{dV}{dD} = -\frac{3\alpha d^{1/3} V_0^{2/3}}{\varepsilon} \times \left(1 - \frac{D}{D_m}\right)^2 \qquad (8.27)$$

which becomes less steep as the radiation dose increases. The dynamic range and the sensitivity depend on α, d, and V_o, and are investigated by measuring the discharging curve of the *a*-Se layer by megavoltage irradiation.

The detector and read-out unit shown in Figure 8.63 are a self-contained system within a light-tight box which consists of the metal/*a*-Se detector and a 2D scanning mechanism. The scanning mechanism has attached to itself a scorotron and a noncontact electrostatic probe. The 2D scanning mechanism allows 1D scanning of the scorotron for corona charging of the *a*-Se surface prior to irradiation and 2D scanning of the probe in a raster fashion for read-out of the *a*-Se surface immediately after irradiation. Two-dimensional scanning, precise to 2-μm increments, is provided by a two-dimensional servomotor-operated motion stage. The stage rests below the metal/*a*-Se detector and faces the *a*-Se layer. A commercial package on a 486 PC controls the motion of the 2D stage and the acquisition of data from the probe. Communication of control and data is provided through a GPIB line from the PC located outside the room to the self-contained system which always remains inside the room.

The detector composed of the glass substrate has one side of the glass covered with *a*-Se and the other side covered with a 1-μm-thick layer of an indium-tin-oxide conductor. The physical densities of glass, Al, Se, SS, and Cu are 2.2 g/cm³, 2.7 g/cm³, 4.3 g/cm³, 7.5 g/cm³, and 8.9 g/cm³, respectively. [32]

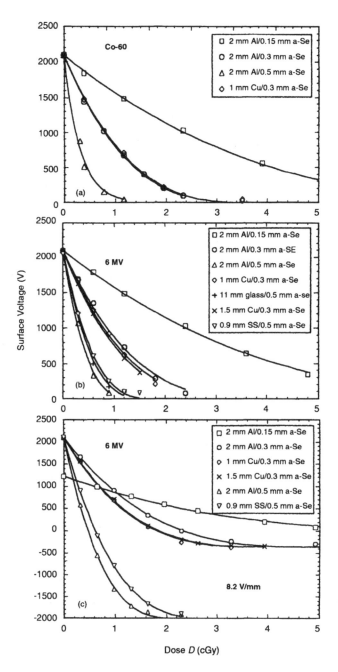

Surface Voltage (V)

Dose D (cGy)

FIGURE 8.64 Radiation discharge curves for different metal plate/a-Se detectors measured with the (a) ^{60}Co spectrum, (b) 6-MV spectrum using V_{grid} = 2100 V and V_{bias}= 0, and (c) radiation discharge curves for detectors having the same electric field E = 8.2 V/μm across the a-Se layer. The same E was obtained by applying V_0 = 1230 V, 2460V, and 4100V for the detectors with a-Se layer thicknesses of 0.15 mm, 0.30 mm, and 0.50 mm, respectively.(From Reference [32]. With permission.)

Figure 8.64a shows the sensitivity curves for four detectors (0.5 mm, 0.3 mm, 0.15 mm a-Se on 2-mm Al, and 0.3 mm a-Se on 1-mm Cu) irradiated with the ^{60}Co spectrum. The uncertainties of all the values in Figure 8.64a are

less than 0.9%; thus, the error bars are too small to be shown.

There is no significant difference between the sensitivity curves of the detectors with Al (0.54 g/cm^2) or Cu (0.89 g/cm^2) metal plates and which have the same a-Se thicknesses. The relative α values were calculated by the Monte Carlo method. The D_m values produced the best fit of Equation (8.26) to the data in Figure 8.64a. The corresponding correlation coefficients R are at least 0.999 for all fittings. Once D_m is obtained, the parameter a is then calculated from Equation (8.25).

Detector sensitivity was also measured for the 6-MV spectrum. In Figure 8.64b, the discharge curves for seven detector types are shown. The uncertainties of all the values in the figure are less than 1.2%.

The sensitivity curve of the metal/a-Se detectors was measured with constant electric field, and results are shown in Figure 8.64c. The largest values for V_{grid} and $|V_{bias}|$ used in our experiments were 2100 V and 2000 V, respectively. Since the resultant electric field (E) across the largest a-Se thickness (0.50 mm) was 8.2 V/μm, this constant value of E was used for all studies. For the a-Se thicknesses less than 0.50 mm, smaller values of V_{grid} and V_{bias} were required to achieve the same E. Thus, potentials V_0 of 1230 V, 2460 V, and 4100V were placed on detectors with a-Se layer thicknesses of 0.15 mm, 0.30 mm, and 0.50 mm, respectively. [32]

REFERENCES

1. **Vatnitsky, S. and Jarvinen, H.,** *Phys. Med. Biol.,* 38, 173, 1993.
2. **Hoban, P. W. et al.,** *Phys. Med. Biol.,* 39, 1219, 1994.
3. **Rikner, G.,** *Acta Radiologica Oncol.,* 24, 71, 1985.
4. **Seuntjens, J. et al.,** in AAPM *Proc. No. 11, Kilovolt X-Ray Beam Dosimetry for Radiotherapy and Radiobiology,* 1997, 227.
5. **Heydarian, M. et al.,** *Phys. Med. Biol.,* 38, 1035, 1993.
6. *PTW-Freiburg Ionization Chamber Catalog,* 1999.
7. **Vatnitsky, S. et al.,** *Med. Phys.,* 22, 469, 1995.
8. **Rustgi, S. N.,** *Med. Phys.,* 22, 567, 1995.
9. **Laub, W. U. et al.,** *Med. Phys.,* 24, 535, 1997.
10. **Rustgi, S. N.,** *Phys. Med. Biol.,* 43, 2085, 1998.
11. **Mobit, P. N. and Sandison, G. A.,** *Med. Phys.,* 26, 839, 1999.
12. **Gladstone, D. J. et al.,** *Med. Phys.,* 21, 1721, 1994.
13. **Soubra, M. et al.,** *Med. Phys.,* 21, 567, 1994.
14. **Butson, M. J. et al.,** *Med. Phys.,* 23, 655, 1996.
15. **Rosenfeld, A. B. et al.,** *IEEE Trans. Nucl. Sci.,* 42, 1870, 1995.
16. **Rosenfeld, A. B. et al.,** *IEEE Trans. Nucl. Sci.,* 43, 2693, 1996.
17. **Korn, T. et al.,** *Phys. Med. Biol.,* 43, 3235, 1998.
18. **Rosenfeld, A. B. et al.,** *IEEE Trans. Nucl. Sci.,* 46, 1774, 1999.

19. **Rosenfeld, A. B. et al.,** *IEEE Trans. Nucl. Sci.,* 46, 1766, 1999.
20. **Van Dam, G. et al.,** *Radiother. Oncol.,* 19, 345, 1990.
21. **Edwards, C. R. et al.,** *Phys. Med. Biol.,* 42, 2383, 1997.
22. **Buehler, M. G. et al.,** *IEEE Trans. Nucl. Sci.,* 40, 1442, 1993.
23. **Brucker, G. J. et al.,** *IEEE Trans. Nucl. Sci.,* 42, 33, 1995.
24. **Schwank, J. R. et al.,** *IEEE Trans. Nucl. Sci.,* 43, 2671, 1996.
25. **Buehler, M. G. et al.,** *IEEE Trans. Nucl. Sci.,* 43, 2679, 1996.
26. **Grusell, E. and Rikner, G.,** *Phys. Med. Biol.,* 38, 785, 1993.
27. **Piermattei, A. et al.,** *Med. Phys.,* 22, 835, 1995.
28. **Nuclear Associates,** *Diagnostic Imaging and Radiation Therapy Catalog,* 1999.
29. **Heukelom, S. et al.,** *Phys. Med. Biol.,* 36, 47, 1991.
30. **Khivrich, V. I. et al.,** *IEEE Trans. Nucl. Sci.,* 43, 2687, 1996.
31. **Bellem, R. D. et al.,** *IEEE Trans. Nucl. Sci.,* 41, 2139, 1994.
32. **Falco, T. et al.,** *Med. Phys.,* 25, 814, 1998.
33. **ICRU 37,** Stopping powers for electrons and positrons, 1984.
34. **ICRU 22,** Measurement of low-level radioactivity, 1972.

9 | Gel Dosimetry

CONTENTS

I. INTRODUCTION AND BASIC CONCEPTS

In 1984, Gore et al. [1] proposed that the radiation-induced changes in the well-established Fricke solution could be probed with nuclear magnetic resonance (NMR) relaxation measurements rather than using the conventional spectrophotometry. Gore et al. realized that the NMR spin-lattice and spin-spin relaxation rates ($1/T_1$ and $1/T_2$, respectively) of the dosimeter are related to the amount of Fe^{3+} present in the Fricke solution and, since these relaxation parameters govern the contrast of MR images, it was possible to observe the radiation-induced changes in Fricke phantoms by MRI. This initial work indicated that there was a possibility for a 3-D dosimetry based on MRI and ferrous sulphate (Fricke) dosimeters. Following the initial proposal, a number of groups reported on MR-based radiation dosimetry. Most imaging applications used ferrous sulphate solutions incorporated into gel matrices (Fricke-gels) to stabilize the spatial integrity of the dosimetry. Various gel matrices were investigated for Fricke-gel MR-based dosimetry, including gelatin, agarose, and sephadex. Each system had its advantages and limitations. A few papers have suggested that dose distributions can be determined directly from MR images based on signal intensity and calibration curves. However, the dosimeters require high doses, typically 10–40 Gray (Gy), for radiation-induced changes to be readily observed by NMR. The Fe ions diffuse even in the gel matrices and the spatial information eventually is destroyed. The time between the start of irradiation to the end of the dose measurement should be no more than about two hours.

It was well-known that irradiating polymers or macromolecules could alter their molecular dynamics and structure. This made polymer systems good candidates for MRI-based dosimetry. For many polymers, the changes occurred only at very high doses ($>10^4$ Gy). However, by choosing a suitable polymer or polymer-solvent system, the dose range could be extended to a range more extensive than that offered by Fricke dosimeters (e.g., 50 cGy to 100 Gy with aqueous polyacrylamide).

While the relaxation times had been shown to be very sensitive probes of the viscosity and molecular weight of polymer systems, the changes in the relaxation times of irradiated systems were not expected a priori to be related to dose in a simple manner.

Polymer-gel dosimeters present many advantages over Fricke-gel systems (principally since there is significantly less diffusion of polymers within the gel matrix so that, theoretically, the radiation-induced changes maintain their spatial integrity indefinitely). There are still a number of features (cost, duration of polymerization reactions after irradiation, monomer toxicity) which seem less attractive.

Radiation-induced polymerization of polymer-gel dosimeters is clearly visible. Experience with optical changes in gel dosimeters has initiated a new realm in gel dosimetry, with dose measurement using optical techniques and image reconstruction in two and three dimensions. [2] Most of the reactions involving the free radicals initially produced by the radiation are very rapid and are essentially complete within a microsecond. However, subsequent reactions of non-radical products, or of large free radicals on polymer chains, may be quite slow. Figure 9.1 is an example of the agarose/ferrous system. [3]

NMR RELAXATION

Nuclear magnetic resonance (NMR) methods have been very useful in the study of the structure, composition, and molecular dynamics of various materials. Of particular interest are NMR relaxation or relaxometry studies of radiation chemical dosimeters such as the ferrous sulfate–doped or Fricke gels and polymer gels. The radiation-induced changes in the solutes of these aqueous dosimeters affect the relaxation properties of the water protons (hydrogen nuclei) constituting most of the magnetization signal studied using proton NMR or magnetic resonance

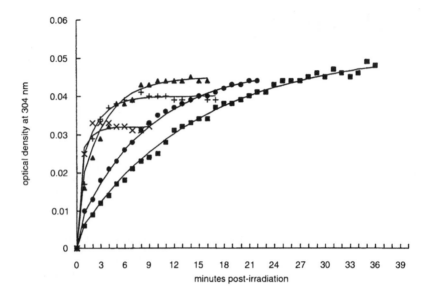

FIGURE 9.1 Optical density of agarose/ferrous gels after 1.87 Gy of gamma irradiation. 0.1 mM Fe^{2+}(■), 0.2 mM Fe^{2+} (●), 0.3 mM Fe^{2+}(▲), 0.5 mM Fe^{2+}(+), 0.9 mM Fe^{2+}(×). (From Reference [3]. With permission.

imaging (MRI) methods. Models show that the NMR dose response of the gel dosimeters is governed by two mechanisms: the chemical response of the gel to radiation and the response of NMR parameters to radiation products. NMR dose response models allow for the absorbed radiation dose to be determined from fundamental physical variables rather than a calibration of the dosimeter's response. [4]

If the material is irradiated with a time-dependent magnetic field or radio frequency (rf) energy, such as rf pulse sequences, the equilibrium magnetization may be perturbed such that the magnetization becomes time-dependent and possesses components transverse to the external magnetic field. Once the perturbing rf energy is removed, the magnetization begins to relax to its equilibrium value and direction. The longitudinal and transverse magnetizations relax at the characteristic longitudinal and transverse rates R_1 and R_2, which are simply the inverse of the relaxation times T_1 and T_2.

The macroscopic equilibrium magnetization **M** (bold type represents a vector quantity) of a material placed in an external magnetic field **H** ($= 0,0,H_0$)), oriented by convention along the z-axis, results from the superposition of the magnetic moments of individual nuclei in the material. Only nuclei with a non-zero spin possess a nuclear magnetic moment, $\mu = \gamma \mathbf{I}$, where γ is the gyromagnetic ratio specific to a particular nucleus, 2π, and $\mathbf{I}(= (I_x, I_y, I_z))$ is the quantum mechanical spin operator. In MRI gel dosimetry the nuclei of interest are mainly the hydrogen nuclei or protons on the water and macromolecules in the gels. The following discussions is limited to these nuclei of spin 1/2.

The overall magnitude and direction of $\mathbf{M}(=(0,0, M_0))$ depend on the behavior of individual nuclear magnetic moments or spins interacting with the magnetic field. The

spins may be classically described as precessing about the z-axis at the Lamor frequency of $\omega_0 = \gamma H_0$ and aligned with or against **H**. Both orientations represent a different spin state with a different energy level. Spins aligned along the field are in the lower energy state of $E = -1/2\hbar\gamma H_0$, and those aligned against the field are in the higher energy state of $E = -1/2\hbar\gamma H_0$. This description of spin behavior concurs with that determined quantum-mechanically by solving the equation of motion of μ,

$$\frac{d\mu(t)}{dt} = \frac{i}{\hbar}[\mu(t), H_0] \tag{9.1}$$

where $H_0 (= \mu \cdot \mathbf{H} = -\gamma\hbar H_0 I_z)$ is the Zeeman Hamiltonian representing the interaction between the spin and the magnetic field. Taking the expectation value of the solution for μ (t) also yields an oscillatory time dependence for the $\mu_{x,y}(t)$ component and a constant μ_z component aligned with or against **H**.

For a material composed of an ensemble of spins, the components of the net macroscopic magnetization are the average of the expectation values of the corresponding μ components, $\langle \bar{\mu}_i \rangle$, multiplied by the number of spins, N:

$$M_i = N \langle \bar{\mu}_i \rangle \tag{9.2}$$

where i designates the axis ($i = x, y, z$). At equilibrium the phase of the oscillating $\mu_{x,y}$ component for individual spins varies randomly over all spins; i.e., there is no phase coherence between the $\mu_{x,y}$ of the spins. Thus, the average $\langle \mu_{x,y} \rangle$ is zero, and no net transverse magnetization $M_{x,y}$ is observed. The average $\langle \mu_z \rangle$ depends on the number of spins directed along the magnetic field and the number directed against. Since more spins are aligned along the direction of the magnetic field ($E = -\frac{1}{2}\gamma\hbar H_0 < 0$) than against it ($E = -\frac{1}{2}\gamma\hbar H_0 > 0$), M_z is greater than zero. The overall equilibrium magnetization of the ensemble of spins is thus given by $\mathbf{M} = (0, 0, M_0)$.

A system of nuclear spins placed in an external homogeneous static magnetic field will possess an equilibrium magnetization $\mathbf{M} = (0, 0, M_0)$. If the material is irradiated with a time-dependent magnetic field or radio frequency (rf) energy oscillating at the Larmor frequency, ω_0, resonant absorption of the rf energy occurs and perturbs the equilibrium magnetization such that $\mathbf{M}$ becomes time-dependent, and $M_z \neq M_0$ and $M_{xy} \neq 0$. For a given field H_0, different types of nuclei, each characterized by a different γ, will have different Larmor frequencies and, therefore, each type can be selectively irradiated; i.e., only protons will absorb rf energy of 25 MHz in a 0.6-T field. Once perturbed, the magnetization tends to return to or relax to its equilibrium value $\mathbf{M} = (0, 0, M_0)$.

The perturbation and relaxation dynamics of the macroscopic magnetization can be described by the phenomenological Bloch equations:

$$\frac{dM_z}{dt} = \gamma(\mathbf{M} \times \mathbf{H})_z + \frac{M_0 - M_z}{T_1} \quad \text{and} \quad \frac{dM_{x,y}}{dt}$$

$$= \gamma(\mathbf{M} \times \mathbf{H})_{x,y} - \frac{M_{x,y}}{T_2} \quad (9.3)$$

where T_1 and T_2 are the relaxation time constants characterizing the relaxation of the respective magnetization components. The first terms on the right-hand side of the two expressions in Equation 9.3 describe the motion of the macroscopic magnetization in the presence of an applied field, $\mathbf{H}$.

The field applied during MR imaging or NMR experiments typically has the form $\mathbf{H} = (H_1\cos(\omega_0 t), -H_1\sin(\omega_0 t), H_0)$, where the z term is the static field and the x and y terms are the perturbing transverse fields of the rf energy pulses of magnitude H_1. During the application of short pulses, usually only microseconds long, little magnetization relaxation occurs and the magnetization dynamics may be approximated by ignoring the second right-hand terms in Equation 9.3. In this case the magnetization in a frame of reference rotating about the z-axis at ω_0 ($= \gamma H_0$) becomes:

$$M'_y = M_0\sin(\gamma H_1 t) \quad \text{and} \quad M'_z = M_0\cos(\gamma H_1 t) \quad (9.4)$$

where the prime designates the rotating reference frame, and H_1 is arbitrarily taken to point along the x-axis. Thus, applying a transverse rf field will cause $\mathbf{M}$ to rotate away

from the z-axis at a frequency $\omega_1 = \gamma H_1$. The angle of the rotation of $\mathbf{M}$ is determined by $\theta = \gamma H_1 \tau_p$, where τ_p is the duration of the rf pulse. This angle is used to describe the rf energy pulses in the pulse sequences that manipulate the magnetization during MR imaging or NMR experiments. At a more fundamental level, M_z is affected by the coupling between the rf energy and the spins, since the coupling creates non-zero transition probabilities between the eigenstates. As a result, the absorption of rf energy can affect the populations of the eigenstates and hence the average $\langle \mu_z \rangle$. The M_{xy} is also affected by the coupling, as the coupling also creates phase coherence in the time dependence of the average $\langle \mu_x \rangle$ and $\langle \mu_y \rangle$, such that the average magnetic moments gain magnitude.

Once the perturbing rf energy is removed, the Bloch equations reduce to:[4]

$$\frac{dM'_z}{dt} = \frac{M_0 - M'_z}{T_1} \quad \text{and} \quad \frac{dM'_{x,y}}{dt} = -\frac{M'_{x,y}}{T_2}$$

$$(9.5)$$

such that only the relaxation terms remain. The longitudinal magnetization M'_z grows exponentially to its equilibrium value M_0 with a characteristic time T_1, the longitudinal relaxation time, whereas the transverse magnetization $M'_{x,y}$ decays exponentially to zero with a characteristic time T_2, the transverse relaxation time.

There are four water proton groups in a Fricke gel dosimeter: the bulk water protons, the water protons hydrating the Fe^{2+} and Fe^{3+} ions, and the water protons hydrating the gel. The exchange of the water protons between all four groups determines the overall observed relaxation. Rapid exchange has been observed for the overall R_1 and R_2 of water in Fricke gels containing gelatin, but not for the R_2 of water in gels containing agarose. [4]

Within the limit of fast exchange, the longitudinal magnetization recovery is characterized by a single exponential and hence by a single overall water spin-lattice relaxation rate R_1. This rate is simply the sum of the inherent rates, each weighted by the fraction of water protons p^i in their respective groups, as follows:

$$R_1 = p^{3+}R_1^{3+} + p^{2+}R_1^{2+} + p^{gel}R_1^{gel}$$
$$+ (1 - p^{3+} + p^{2+} - p^{gel})R_1^{water}$$
$$= k^{3+}(R_1^{3+} - R_1^{water})[Fe^{3+}] + k^{2+}(R_1^{2+} - R_1^{water})$$
$$\times [Fe^{2+}] + k^{gel}(R_1^{gel} - R_1^{water})[gel] + R_1^{water}$$
$$= r_1^{3+}[Fe^{3+}] + r_1^{2+}[Fe^{2+}] + r_1^{gel}[gel] + R_1^{water}$$

$$(9.6)$$

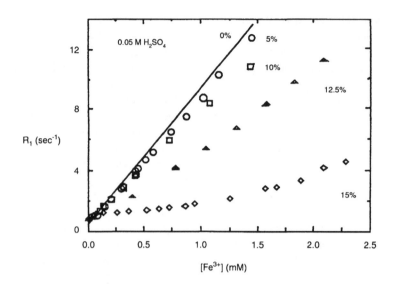

FIGURE 9.2   The spin-lattice relaxation rate dependence on the ferric-ion concentration for the different gel concentrations (% by weight) labeled on the graph and 0.05-M sulfuric acid. (From Reference [4]. With permission.)

where the r_i^j are the relativities of the respective proton groups and specify the ability of a solute to enhance spin-lattice relaxation of water protons. Equation 9.6 indicates that under a fast exchange regime, R_1 varies linearly with the concentration of ions. The spin-lattice relaxation rate dependence on ferric concentration is shown in Figure 9.2.

Equation 9.6 may be expressed as follows by noting that $[Fe^{2+}] + [Fe^{3+}] = [Fe^{2+}]_0$:

$$R_1 = (r_{eff}^{3+} - r^{2+})[Fe^{3+}] + [Fe^{2+}] + R_1(0Gy) \quad (9.7)$$

where

$$R_1(0Gy) = (r^{2+}[Fe^{2+}]_0 + r^{gel}[gel]_0 + R_1^{water})$$

The final expression for the R_1 dose response is

$$R_1 = \left\{ (r_{eff}^{3+} - r^{2+})G(Fe^{3+})\frac{10\rho}{eN_A} \right\}D + R_1/(0Gy) \quad (9.8)$$

where N_A is Avogadro's number, ρ is the density in kg/liter, e is the number of Joules per electron volt, and $G(Fe^{3+})$ is the chemical yield for Fe^{3+} in number of ions per 100 electron volts, given by

$$G(Fe^{3+}) = \frac{[Fe^{3+}]eN_A}{10\rho D} \quad (9.9)$$

The single spin-spin relaxation rate R_2 observed for water and monomer protons suggests that the protons on the bulk water, monomer, and water hydrating the polyacrylamide gelatine and monomer are all under fast exchange. Under the fast-exchange regime, the water and monomer R_2 is the sum of the inherent water R_2 weighted by the fraction of water proton in the respective group:

$$R_2 = p^p R_2^p + p^g R_2^g + (1 - p^p - p^g)R_2^b$$
$$= k^p(R_2^p - R_2^b)[p] + \{p^g R_2^g + (1 - p_g)R_2^b\} \quad (9.10)$$

where the superscripts are: b = bulk water, p = polyacrylamide, and g = gelatin. $k^p = p^p/[p]$ is the fraction of water-proton-hydrating polymer per weight fraction of polymer in the dosimeter. Equation 9.10 illustrates how an increase in R_2 with dose results from an increase in polymer concentration. Using the fast exchange model for the water in a PAG gel dosimeter, the following dose response model results: [4]

$$R_2 = \{r^p G^p\}D + R_2(0Gy) \quad (9.11)$$

where r^p is the polymer relaxivity, $G^p = [p]/D$ is the polymer yield in units of percent weight fraction of polymer formed per Gy, and $R_2(0Gy)$ is the spin-spin relaxation rate of an un-irradiated PAG gel dosimeter.

One of the most important qualities possessed by a dosimeter gel is that it forms both phantom and detector. The advantages to be expected with gel dosimetry compared with conventional dosimeters may thus be summarized in the following properties: [5]

- independence on radiation direction, radiation quality, and dose rate for conventional clinical beams
- absorbed dose integration in the dosimeter (of utmost importance for dynamic treatment and multiple beams)

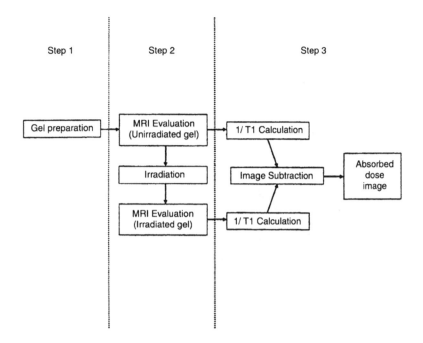

Step 1 | Step 2 | Step 3

FIGURE 9.3 A schematic description of the procedure used for the FeGel system, including the background subtraction. (From Reference [5]. With permission.)

- evaluation of a complete volume
- potential for true three-dimensional dosimetry utilizing a high spatial resolution
- possibility to measure in a phantom that is equivalent to anatomical soft tissue

The contributions to uncertainty of the different steps included in FeGel relative absorbed-dose measurements (Figure 9.3) can be evaluated according to the guidelines given by the International Organization for Standardization recommendations. The uncertainty is divided into two categories, according to the way in which its numerical value was estimated: [5]

- type A uncertainty determined by statistical methods
- type B uncertainty determined by other means, e.g., estimated or obtained from a calibration certificate, etc.

In step 1 (Figure 9.3), the gel is produced, and if a linear and spatially uniform $1/T_1$ dose response is assumed, the uncertainty introduced by this step is negligible. The next steps include MRI acquisition (step 2, Figure 9.3) and background subtraction (step 3, Figure 9.3) to obtain the final absorbed dose image.

The local water proton NMR relaxation rates ($R_1 = 1/T_1$ and $R_2 = 1/T_2$) in the gel increase in proportion to the absorbed radiation dose. This is believed to be caused by the increase of the population of those water molecules which are bound to the surfaces of the rigid polymer particles. [6] These bound water molecules exchange their protons with the polymer and with the bulk water molecules. The relaxation rates that can be measured from MRI images are averaged over a voxel and represent the bulk water relaxation. Figure 9.4 shows R2-dose calibration curve.

Figure 9.5 may indicate that the polymer gel dosimeter is independent of photon energy and dose rate; it is true only within certain limits for dose rate. Different gel formulations, and especially different monomers, have great effect on dose rate dependence. The "supersensitive" BANG-3 formulation shows independence of dose rate up to a certain dose, above which the curves which represent higher dose rates begin to saturate (Figure 9.6).

High LET radiations saturate the polymer gels at lower doses than lower LET radiations. However, the extent of this effect depends on the choice of the monomers and the response modifiers. BANG-1 formulation (bis + acrylamide + gelatin + water) did not show measurable LET dependence, whereas the BANG-3 series (gelatin + methacrylic acid + water) showed a significant effect. [6]

A formulation of a tissue-equivalent polymer gel dosimeter for the measurement of three-dimensional dose distributions of ionizing radiation has been developed by Maryanski et al. [7] It is composed of aqueous gelatin infused with acrylamide and N,N'-methylene-bisacrylamide monomers and made hypoxic by nitrogen saturation. Irradiation of the gel, referred to as BANG, causes localized polymerization of the monomers, which, in turn, reduces the transverse NMR relaxation times of water protons.

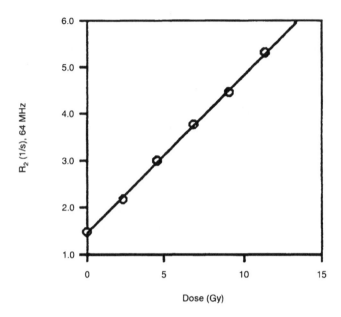

FIGURE 9.4 R_2 (dose) calibration. (From Reference [6]. With permission.)

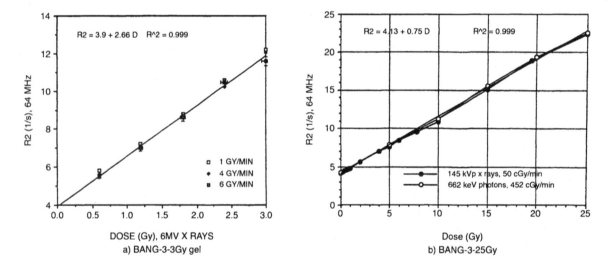

FIGURE 9.5 a), b) Dose response for BANG-3-type gels irradiated with photons at low to medium dose rates: a) 3Gy, b) 25Gy. (From Reference [6]. With permission.)

The dose dependence of the NMR transverse relaxation rate, R_2, is reproducible (less than 2% variation) and is linear up to about 8 Gy, with a slope of 0.25 s^{-1}Gy^{-1} at 1.5 T. Magnetic resonance imaging may be used to obtain accurate three-dimensional dose distributions with high spatial resolution. Since the radiation-induced polymers do not diffuse through the gelatin matrix, the dose distributions recorded by BANG gels are stable for long periods of time and may be used to measure low-activity radioactive sources. Since the light-scattering properties of the polymerized regions are different from those of the clear, non-irradiated regions, the dose distributions are visible, and their optical densities are dependent on dose. [7]

The formation of cross-linked polymers in the irradiated regions of the gel increases the NMR relaxation rates of neighboring water protons. Therefore, the radiation-induced polymerization in polymer gels plays a role similar to that of the radiation-induced oxidation of ferrous ions in Fricke gels, with four important advantages. First, in polymer gels the spatial distribution of NMR relaxation rates, which reflects the distribution of dose, is stable and does not change with time. Second, Fricke gels have an intrinsically high electrical conductivity. Consequently, the RF field is strongly attenuated. The polymer gel dosimeters do not contain ionic species and show insignificant RF attenuation. Third, polymerized regions can be seen visually. Last, polymer gels

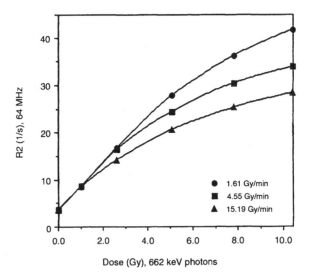

FIGURE 9.6 Dose response saturation and dose rate dependence at higher dose rates in BANG-3 gel formulation. (From Reference [6]. With permission.)

are considerably more sensitive to radiation than the Fricke gels.

Three BANG gels were prepared by Maryanski et al. Gels I and II were contained in tissue culture flat flasks made of polystyrene, and gel III was prepared in a rectangular Lucite box. Each gel was irradiated by 4×4-cm^2 fields using 6-MV x-rays produced by a Varian 2100C linear accelerator at a dose rate of 4 Gy min^{-1}. The following doses were delivered at d_{max}: 5, 10, 15, 20 Gy to gel I; 2, 4, 6, 8, 10 Gy to gel II; 1, 2, 3, 4, 5, 6 Gy to gel III. Gels I and II were irradiated at an ambient room temperature. Gel III was irradiated at 0°C, immediately after being removed from an ice-and-water bath. All gels were stored in a refrigerator at 4°C immediately after irradiation.

Figure 9.7 shows values of transverse relaxation rates (R_2) for gels I, II, and III as a function of dose. The pooled data (Figure 9.7a) show that the dose response was highly reproducible over a wide range of doses. For doses below 8 Gy, the dose response is well-fitted by a straight line; Figure 9.7b shows the individual straight-line fits for each gel. For example, a relaxation rate of 2.0 s^{-1} corresponds to doses of 2.83, 2.92, and 2.90 Gy in the three different gels, a variation of less than 2% from the mean value.

The signal in a spin-echo MR image produced by a dose D is given by

$$S(TE) = S(0)e^{-R_2(D)TE} \qquad (9.12)$$

where S is the signal at echo time TE.

Radiation dose distributions in three dimensions from tomographic optical density scanning of polymer gels were described by Gore et al. [8] The dosimetric data stored within the gels were measured using optical tomographic densitometry. The dose-response mechanism relied on the production of light-scattering micro-particles, which result from the polymerization of acrylic comonomers dispersed in the gel. The attenuation of a collimated light beam caused by scattering in the irradiated optically turbid medium was directly related to the radiation dose over the range 0–10 Gy. An optical scanner has been developed which incorporates a He-Ne laser, photodiode detectors, and a rotating gel platform.

The intensity of a monochromatic light beam passing through the medium is attenuated exponentially under narrow-beam conditions. If $\mu(x, y)$ denotes the optical attenuation coefficient per unit length in a section of the object, then the intensity exiting the sample at position x is $I(x)$ when the incident intensity is I_0:

$$I(x) = I_0 e^{-\int \mu(x,y)\,dy} \qquad (9.13)$$

A schematic diagram that illustrates the operation of the prototype device is shown in Figure 9.8.

After acquisition, the projection data are transferred to an image reconstruction program.

Progress in the development of polymer gel dosimetry using MRI was reported by Maryanski et al. [9] The dose distribution image produced in the tissue-equivalent gel by radiation-induced polymerization and encoded in the spatial distribution of the NMR transverse relaxation rates (R_2) of the water protons in the gel is permanent. Maps of R_2 are constructed from magnetic resonance imaging data and serve as a template for dose maps, which can be used to verify complex dose distributions from external sources or brachytherapy applicators. The integrating, three-dimensional, tissue-equivalent characteristics of polymer gels make it possible to obtain dose distributions not readily measured by conventional methods. An improved gel formulation (BANG-2) has a linear dose response that is independent of energy and dose rate for the situations studied to date.

The so-called BANANA (acronym based on Bis, Acrylamide, Nitrous Oxide, And Agarose) and BANG (acronym based on Bis, Acrylamide, Nitrogen, and Gelatin) formulations of polymer gels differ in their gel matrices, which are agarose and gelatin, respectively; 1% by weight agarose in BANANA gel was replaced by 5% by weight gelatin in BANG gel. This resulted in a lower background R_2 ($= 1/T_2$, transverse NMR relaxation rate of the water protons) for the nonirradiated gel, and a more transparent medium in which the irradiated region is clearly visible.

An improved polymer-gel formulation was developed, containing 3% N,N′-methylene-bisacrylamide (referred to as bis), 3% acrylic acid, 1% sodium hydroxide, 5% gelatin, and 88% water, where all percentages are by weight (see Table 9.1). This gel differs from BANG mainly in the substitution of acrylic acid for acrylamide. Henceforth, it

TABLE 9.1

Chemical Composition of the BANG-2 Gel

Component	Formula	Weight Fraction in the Gel
Gelatin	$(C_{17}H_{32}N_5O_6)_x$	0.05
Acrylic acid	$CH_2CHCOOH$	0.03
Bis	$(CH_2CHCONH)_2CH_2$	0.03
Sodium hydroxide	NaOH	0.01
Water	H_2O	0.88

From Reference [9]. With permission.

will be referred to by the acronym BANG-2. Dissolved oxygen, which inhibits free-radical polymerization reactions, is removed from the mixture by passing an inert gas such as nitrogen through it when it is above the gelling temperature and prior to sealing the vessel.

In all preparations gelatin type A (acid-derived), approximately 300 Bloom (a gel strength indicator), was used. The water was from an ion-exchange purifier, and the nitrogen gas contained less than 100 ppm oxygen.

To establish the tissue equivalence of the gel, its physical density was determined by weighing the gel and measuring its volume at room temperature, one day after gelation.

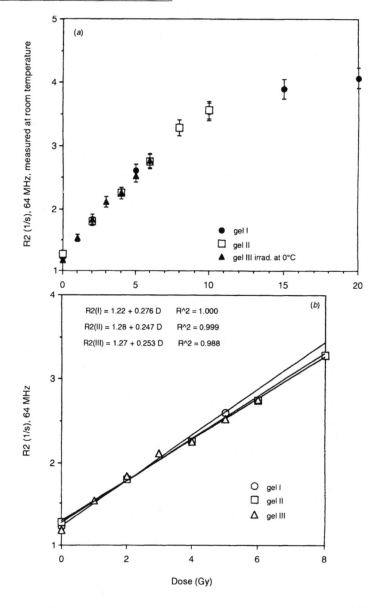

FIGURE 9.7 (*a*) The dose dependence of the water-proton NMR transverse relaxation rate R_2, combining the data from three separately prepared BANG gels, over the range 0–20 Gy. The error bars indicate 5% relative error. (*b*) The dose dependence of the water-proton NMR transverse relaxation rate R_2, measured in three separately prepared BANG gels, over the range 0–8 Gy. The separate linear fits were $R_2(I) = 1.22 + 0.276D$, $r^2 = 1.000$; $R_2(II) = 1.28 + 0.247D$, $r^2 = 0.999$; $R_2(III) = 1.27 + 0.253D$, $r^2 = 0.988$. (From Reference [7]. With permission.)

TABLE 9.2

Comparison of Elemental Composition (Weight Fractions are Denoted as W_k), Electron Densities, and Average Atomic Numbers for BANG-2 Gel, Human Muscle Tissue, and Water.

Material	w_C	w_H	w_N	w_O	w_{Na}	ρ (kg m^{-3})	ρ_e ($\times 10^{29}$ m^{-3})	ρ_e/ρ ($\times 10^{26}$ kg^{-1})	$\bar{Z}^a$
BANG-2	0.0564	0.1051	0.0135	0.8173	0.0058	1030	3.42	3.32	7.14
Muscle Tissue	0.1230	0.1020	0.0350	0.7289	0.0008	1040	3.44	3.31	6.93
Water	0.00	0.1111	0.00	0.8889	0.00	1000	3.34	3.34	7.22

^{a}Calculated as $\bar{Z} = \Sigma_k w_k Z_k$.

Source: From Reference [9]. With permission.

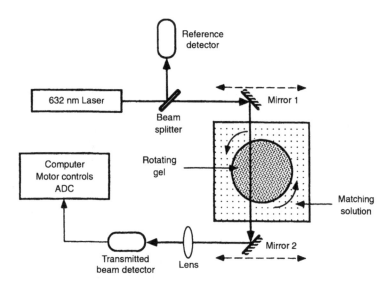

FIGURE 9.8 A schematic diagram of the prototype scanner. The mirrors translate left to right to obtain projections of the gel optical attenuation. Between each translation, the gel is rotated by a second stepping motor. (From Reference [8]. With permission.)

These measurements were repeated several times. The measured density was 1.03 ± 0.005 g/cm^3. The atomic composition, electron density, and average atomic number of the BANG-2 gel of human muscle tissue, and of water are compared in Table 9.2.

All gels used were held in glass bottles, flasks, or beakers, so as to obviate the problem of oxygen contamination that occurs in containers made from oxygen-permeable plastics.

II. APPLICATION OF FERROUS IONS

Ferrous sulphate gel analyzed by relaxation time measurements with NMR-imaging 3D determinations of absorbed dose was discussed by Gambarini et al. [10] The dose-response curve slope is about 0.2 s^{-1} Gy^{-1}, and the G-factor turns out to be ~185 ions per 100 eV of absorbed energy.

The spin-echo images were obtained employing a multiple spin-echo sequence. Since the spin-spin relaxation rate $1/T_2$ proved to have a dose sensitivity higher than the spin-lattice relaxation rate $1/T_1$, only the value of $1/T_2$ was generally evaluated. The transverse relaxation rate was determined by a multiple spin-echo sequence with 16 echoes; the echo times were $T_E = (28 + 20n)$ ms where $n = 0,...,15$, and the repetition time was $T_R = 2.5$ s. The T_2 values were calculated utilizing a one-exponential fit with a nonlinear, least-squares, three-parameter algorithm. The matrix size was 256×256 mm^2, and the voxel size was $10 \times 1.2 \times 1.2$ mm^3. The results of the fit, as can be seen from Figure 9.9, support the validity of the one-exponential analysis and confirm the mono-exponential trend of the magnetization recovery process in agar, or other gels, doped with paramagnetic ions.

The set-up for the gel preparation completely eliminates steam loss and makes reproducibility of the operations possible. This equipment consists of a cylindrical Pyrex container whose cover is supplied with an opening for the thermometer, a water-cooled coil for continuous

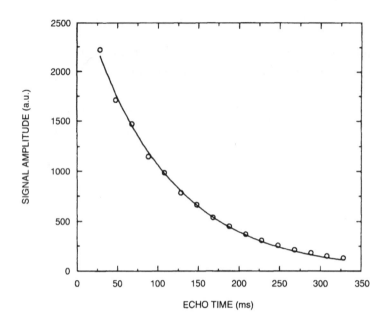

FIGURE 9.9 The spin-echo amplitude taken from measurements in a $FeSO_4$ agarose gel, with a one-exponential three-parameter fit. (From Reference [10]. With permission.)

steam recovery, and a Pyrex tube with a fritted glass ending to bubble oxygen into the gel. The container is put into a Si oil bath, over an oven whose power supply is controlled by a thermometer plunged into a the Si oil. In the oven plate a magnetic stirrer with an independent power supply is incorporated. A view of the apparatus is shown in Figure 9.10.

From the dose-response result, one can estimate the G value of this dosimeter.

$$G = (\Delta R/D)9.64 \times 10^9/\rho(R_b - R_a)$$
$$\times (Fe^{3+} \text{ ions per 100 eV}) \qquad (9.14)$$

where R_a is the relaxation-rate increase per unit concentration of ferrous ion, R_b is the relaxation-rate increase per unit concentration of ferric ion, and ρ is the density (kg m^{-3}).

The resulting density of the doped gel was $\rho = 1076$ kg m^{-3}. From the experimental results of Prasad et al. [22] for relaxation rates vs. ion concentrations, one obtains

$$(R_b - R_a) = (10.2 - 0.4)s^{-1}\text{mM}^{-1}$$
$$= 9.8 \times 10^3 s^{-1}\text{M}^{-1} \qquad (9.15)$$

Assuming this value, the ferrous-ion yield becomes

$$G = 183 \ (Fe^{3+} \text{ ions per 100 eV})$$

The diffusion problem in the traditional Fe(II/III) agarose gel system employed in MRI studies of radiation

dosimetry was studied by Balcom et al. [11] The diffusion coefficient of Fe(III) in 1% agarose gel at pH 1.1 is $D = 2.7 \pm 0.3 \times 10^{-6}$ cm^2 s^{-1}. The diffusion coefficient of Fe(II) is $D = 3.3 \pm 0.5 \times 10^{-6}$ cm^2 s^{-1}. Measurement of the diffusion coefficients permits simulation of the MRI-signal intensity from phantoms with model radiation dose distributions.

The basis for the investigated diffusion measurement is the linear relationship between spin-lattice relaxation rates and paramagnetic concentration. The paramagnetic may be either Fe(II) or Fe(III). The H relaxation rate of water protons in a gel is a first-order process governed by Equation (9.16):

$$1/T_1 = a + bC \qquad (9.16)$$

where C is the concentration of the paramagnetic ion, b is the molar relaxivity, and a is the inverse natural lifetime of water in this environment. The values of a and b are determined for each paramagnetic species through a systematic calibration.

The signal intensity at any point in the profile, ignoring instrumental factors, is given by:

$$S = \rho \exp(-TE/T_2)[1 - 2\exp(-t_d/T_1)] \qquad (9.17)$$

where ρ is the proton density, TE is the echo time, and T_2 is the spin-spin relaxation time of the water. If the delay t_d after the first 180° pulse is set equal to $(\ln 2 \times T_1)$, the signal from that point in space will be zero. Signal variation along the profile is primarily determined by the spatial variation of T_1 in the last term of Equation (9.17).

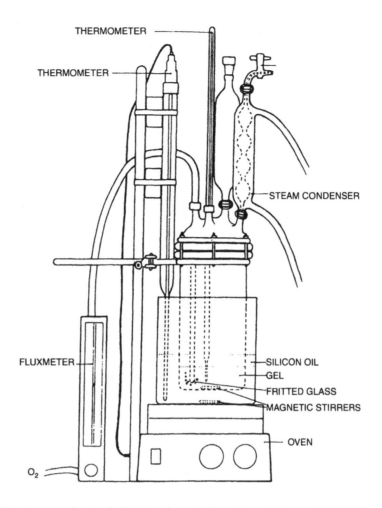

THERMOMETER

THERMOMETER

STEAM CONDENSER

FLUXMETER

SILICON OIL

GEL

FRITTED GLASS

MAGNETIC STIRRERS

OVEN

O_2

FIGURE 9.10 A view of the apparatus for the dosimeter gel preparation. (From Reference [10]. With permission.)

The spatial variation of concentration is governed by the one-dimensional diffusion equation

$$\frac{\partial C}{\partial t} = D \frac{\partial^2 C}{\partial x^2} \qquad (9.18)$$

The solution to Equation (9.18) for free diffusion into a semi-infinite slab from a constant concentration reservoir is given by

$$\frac{C(x,t)}{C_0} = Erfc \left[\frac{x}{(4Dt)^{1/2}} \right] \qquad (9.19)$$

The ratio C/C_0 is the normalized concentration, D the diffusion coefficient, x the spatial position, and t the elapsed time of the experiment. The initial boundary conditions are $C = C_0$ for $x < 0$ and $C = 0$ for $x > 0$. The boundary conditions require that for the length of the experiment the reservoir stay at a fixed concentration and be well-mixed.

Figure 9.11 shows the concentration variation with position for the two initial boundary conditions at times

0, 1, and 24 h. It is seen that the initial boundary has a significant influence on the curves only at very short times. After 8 h, the Fe(III) concentration was 25% and 75% of the initial value at a distance of ±0.27 cm from the step function boundary and a distance of ±0.34 cm from the center of the broad boundary. After only 24 h, the corresponding values were ±0.47 cm and ±0.51 cm, respectively. These two values are strikingly similar, considering the difference between the initial conditions of the two systems. After 24 h, for the case of Fe(III) in agarose gel phantoms, the details of the initial boundary are unimportant, compared to the effects of diffusion in determining the variation of concentration with position.

NMR relaxation times T_1 and T_2 of agarose and Fricke agarose gels have been measured by Luciani et al. [12] in the range 17–51 MHz. The analysis of the spin-echo curves indicates multi-exponential behavior, characterized by three components, at all the examined frequencies. The relative T_2 values, ranging from a few to a hundred milliseconds, can be attributed to different species of water molecules present in the gel. Two of these components are characterized by relaxation rates R_2^a and R_2^b more sensitive than R_1 to gamma irradiation, with the sensitivity

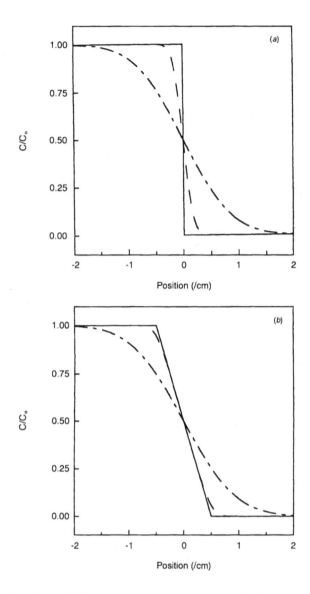

FIGURE 9.11 Simulation of the integrated diffusion equation at times 0 (—), 1 h (---), and 24 h (—•—) for (a) step function and (b) extended initial boundary conditions. The plot is generated with $D = 2.7 \times 10^{-6}$ cm^2 s^{-1} and the initial boundary is centered about $x = 0$ cm. (From Reference [11]. With permission.)

S being $S(R_1) = 0.066$ s^{-1} Gy^{-1}, $S(R_2^a) = 0.088$ s^{-1} Gy^{-1}, $S(R_2^b) = 0.17$ s^{-1} Gy^{-1}. The three T_2 values decrease as a function of frequency, but no gain in dose sensitivity is obtained by changing the working frequency in the examined range. It was possible to estimate the irradiation yield from three independent parameters, R_1, R_2^a, and R_2^b.

Preparation of the agarose gels was as follows: agarose powder, Sigma Chemicals, type VII, was dissolved in triple-distilled water, then boiled to obtain a clear solution. It was subsequently oxygenated by bubbling the gas through it for 30 min and allowed to cool to 36°C before the Fricke solution was added. All steps after boiling were conducted under constant stirring. Volumes of about 70 μl were then poured into 5-mm-diameter NMR tubes and allowed to gel at room

temperature. The final concentrations were 1.5-mM ferrous ammonium sulphate, 1-mM sodium chloride, 50-mM sulphuric acid, and 1% agarose by weight. [12]

FERRIC AGAROSE GEL

The diffusion of ferric ions produced by irradiation in a dosimeter gel, consisting of a ferrous sulphate solution and agarose gel, has been studied by Olsson et al. [13] The purpose of the study was to measure the diffusion coefficient of ferric ions in the gel, to investigate how the diffusion of ferric ions affects a dosimetric image with time, and to evaluate when it is necessary to correct for this distortion. The diffusion coefficient of ferric ions in the gel was found to be 1.91×10^{-2} cm^2 h^{-1} $\pm$ 5%. It was shown that the dose image obtained with an MR scanner deteriorates due to diffusion. This deterioration can be predicted with the aid of the measured diffusion coefficient. It was concluded that if the MR measurements ($1/T_1$ image) of a typical depth-dose distribution are carried out within 2 hours of irradiation, the diffusion will not have a significant effect on the results.

Among the dosimeters used to measure ionizing radiation, the dosimeter gel has recently been recognized as a new and interesting tool for the determination of absorbed dose distributions. The dosimeter consisted of a ferrous sulphate solution mixed with a gel, usually agarose gel. The radiolysis induced by the radiation produces ferric ions in an amount proportional to the absorbed dose. The concentration of ferric ions is also proportional to the inverse of the proton spin-lattice relaxation time, $1/T_1$, which can be measured by magnetic resonance imaging (MRI). To maintain the spatial distribution of the produced ferric ions over a period of time, sufficient to allow measurements in an MR scanner, the ferrous sulphate solution is incorporated within a gel.

The diffusion of medium-sized solutes in uncharged dilute gels, such as the diffusion of ferric ions in agarose, can be regarded as normal aqueous diffusion in the presence of an inert polymer matrix. The main effect of the polymer matrix is to increase the path length traveled by a diffusing solute. A number of equations predicting the ratio of the diffusion coefficient in a gel, D, to the corresponding aqueous diffusion coefficient, D_{aq}, have been proposed. The equation of Mackie and Meares,

$$D/D_{aq} = (1 - \Phi_p)^2/(1 + \Phi_p)^2 \qquad (9.20)$$

where Φ_p is the polymer volume fraction of the gel , has been shown to agree well with experimental data. This equation is an attractive alternative to experiments, provided that an accurate value of D_{aq} is available. Unfortunately, the value of the aqueous diffusion coefficient D_{aq} for the ferric ion, valid for the present electrolyte mixture, could not be found in the literature. The experimental approach of determining D was thus chosen since a reliable and precise technique of measuring diffusion coefficients in gels was available. [13]

The dosimeter gel is a mixture of ferrous sulphate solution and agarose gel. In the dosimeter gel, the final concentrations of the ingredients were: ferrous ammonium sulphate, 1.0 mM; sodium chloride, 1.0 mM; sulphuric acid, 50 mM; and for SeaPlaque and SeaGel, 1.25 and 0.25% by weight, respectively.

A diaphragm diffusion cell was used, which consists of two compartments (A and B) separated by a gel diaphragm of known thickness, l, and area, A. Each compartment contains a well-stirred solution of known volume, V_A and V_B, respectively. The solution in compartment A initially contains the solute being studied, while solution B is pure solvent. A pseudo-steady-state diffusional solute flux, N, through the diaphragm is attained after an initial time delay, t_0. Measurement of the solute concentrations, C_A and C_B, is then started and continued for a period of approximately $4t_0$ or longer. Assuming one-dimensional diffusional flux, perfect mixing, and negligible mass-transfer resistance at the gel-liquid interface, the diffusional flux is given by

$$N = (1 - \Phi_p)D\left(\frac{C_A - C_B}{l}\right) \qquad (9.21)$$

where $(1 - \Phi_p)$ is identified as the void fraction of the gel that is accessible to the diffusing solute. Equation (9.21) is simply the well-known Fick law, integrated over the diaphragm from $x = 0$ to $x = 1$, assuming constant D and a porous medium. However, C_A and C_B are not constant, and Equation (9.21) is not a suitable "working equation." With the aid of a material balance and the assumption of a pseudo-steady state, a convenient Equation (9.22) is obtained. Plotting the experimental values of the left-hand side as a function of time t, the diffusion coefficient is obtained from the slope of the straight line: [13]

$$\ln\left(\frac{C_A - C_B}{C_{A_0} - C_{B_0}}\right) = -\frac{(1 - \Phi_p)DA}{l}\left(\frac{1}{V_A} + \frac{1}{V_B}\right)(t - t_o) \qquad (9.22)$$

Here, C_{A_0} and C_{B_0} are the concentrations of chambers A and B, respectively, at t_0. Many authors simply ignore the very existence of $(1 - \Phi_p)$ or implicitly assume it to be unity. While dubious in many cases, the latter assumption is justified for the present case since Φ_p is very low and the size of the diffusing ferric ion is several orders of magnitude smaller than the "pore radius" of the gel.

A standard procedure for an accurate estimate of a $1/T_1$ map is to calculate it from a set of spin-echo images with different repetition times. The total scanning time for this procedure may be more than one hour. When diffusion is the parameter to be studied, an hour may be a rather long acquisition time, as the diffusion may be significant during this time. Therefore, single images, in which the echo-amplitude

(or MR signal) can be approximated as linear with $1/T_1$ (which also means linear with respect to the absorbed dose) are used. An absorbed dose of 30 Gy yields T_1 values in the region of 650 ms and, for non-irradiated gel, 1700 ms (at 1.0 T); T_2 for the gel is approximately 80 ms. In a spin-echo sequence with a short repetition time ($TR = 100$ ms) and echo-time ($TE = 23$ ms), the MR signal is approximately linearly dependent on $1/T_1$ (or the absorbed dose). The relative signal can be calculated from [13]

$$S = e^{-TE/T_2}[1 - 2e^{-(TR - TE/2)/T_1} + e^{-TR/T_1}]/$$
$$(1 + e^{-TR/T_1}e^{-TR/T_2}) \qquad (9.23)$$

A value of $D = 1.91 \times 10^{-2}$ cm² h⁻¹ ($\pm 5\%$) was obtained from three identical but separate experiments. This can be compared with the result of Schulz et al. [14] They used a method of unsteady-state diffusion into agarose slabs and obtained $D = 1.58 \times 10^{-2}$ and 1.83×10^{-2} cm² h⁻¹ for sulphuric acid concentrations of 25 and 12.5 mM, respectively. The temperature at which their measurements were carried out was not given. [13]

Gambarini et al. [15] described a technique to obtain three-dimensional (3D) imaging of an absorbed dose by optical transmittance measurements of phantoms composed of agarose gel in which a ferrous sulphate and xylenol orange solution were incorporated. The analysis of gel samples was performed by acquiring transmittance images with a system based on a CCD camera provided with an interference filter matching the optical absorption peak of interest. The proposed technique for 3D measurements of an absorbed dose is based on the imaging of phantoms composed of sets of properly piled up gel slices.

Gel containing ^{10}B in the amount typically accumulated in tumors for BNCT was analyzed by Gambarini et al. [16] The isodose curves were obtained from NMR analysis of a phantom of borated gel after irradiation in the thermal column of a nuclear reactor.

In NMR analysis, a good result is achieved if the dosimeter response R is defined as the difference between the relaxation rate $(1/T)$ measured in the irradiated sample and that measured, at the same time, in an un-irradiated sample from the same gel preparation:

$$R = (1/T)_{irr} - (1/T)_{blank} \qquad (9.24)$$

The highest sensitivity has been obtained with the following gel composition.

- Ferrous sulphate solution: 1mM $Fe(NH_4)$-$6H_2O$, 50 mM H_2SO_4 in the amount of 50% of the final weight
- Agarose SeaPlaque: $[C_{12}H_{14}O_5(OH)_4]$ in the amount of 1% of the final weight
- Highly purified water: H_2O in the amount of 49% of the final weight

The sensitivity of this gel is better than that of the standard Fricke dosimeter; in the γ-ray field of a ^{137}Cs biological irradiator delivering a dose rate of 0.14 Gys^{-1}, the dose-response curve, obtained by measuring the transverse relaxation rates $(1/T_2)$ with a Somatom Siemens imaging analyser operating at 1.5 T, 63 MHz, shows good linearity up to $\approx$40 Gy, with slope equal to 0.2 s^{-1}Gy^{-1}.

The additional absorbed dose due to 40 μgg^{-1} of ^{10}B has been evaluated by means of the relation:

$$D = 1.602 \times 10^{10} \sigma FNE\phi(\text{Gy}) \qquad (9.25)$$

where

D is the absorbed dose in Gy,
$\sigma = 3.837 \times 10^{-21}$ cm^2 is the reaction cross section,
$F = 4 \times 10^{-5}$ is the ^{10}B fraction by weight,
$N = 6 \times 10^{22}$ is the number of atoms per gram of ^{10}B,
$E = 2.28$ MeV is the energy released per event, and
Φ = the thermal neutron fluence in ncm^{-2}.

For the absorbed dose per unit fluence:

$$D = 3.364 \times 10^{-12}(\text{Gy cm}^2)$$

Gels infused with the Fricke solution limit the diffusion rate and can be scanned in 3D using magnetic resonance imaging (MRI). While most of the development of MRI dosimetry has been with Fricke solution infused in protein gelatin, more promising results have been obtained in a new polymer gelatin system. However, this MRI approach has several drawbacks. This technique requires access to MRI facilities in reasonable proximity to the irradiation facilities. With the current sensitivity and signal-to-noise ratio of MRI, good dosimetric images have been produced only for thick image slices through the gel. The chemical preparation of the acrylamide polymer is potentially very toxic and also requires that the gel be free of oxygen in order for the reaction to proceed. [17]

Optical methods have been used to map 2D dose distributions in slices of gel containing ferrous sulphate–benzoic acid–xylenol orange (FBX) solution. Optical computed tomography (OCT) reconstruction techniques have been introduced to map 3D dose distributions in cylindrical and spherical volumes of polymer gel material. The change in optical attenuation of the acrylamide gels is due mainly to optical scattering, and this will ultimately limit dose sensitivity accuracy.

Kelly et al. reported the use of radiochromic gels in order to improve safety, convenience, and the performance of optical CT dosimetry. The optical densities of radiochromic compounds like the FBX solution show an increase with absorbed dose when irradiated with x-ray photons and the change is due to absorption rather than scattering of light. FBX was chosen over standard Fricke solution because it could be probed using light of longer wavelength ($\lambda = 543.5$ nm) in order to reduce Rayleigh scattering and thereby increase our signal-to-noise ratio.

The radiochemistry of the FBX liquid dosimeter has been shown to have properties very similar to the standard liquid Fricke solution. When the ferrous ion (Fe^{2+}) is irradiated, it oxidizes to form the ferric ion (Fe^{3+}), and the corresponding complex with xylenol causes a change in absorption, which has a broad peak centered at a wavelength of 540 nm (Figure 9.12). The radiation-induced change in the absorption spectrum is known to be a linearly proportional to the radiation dose and can be probed easily using a helium-neon laser beam at a wavelength of 543.5 nm.[17]

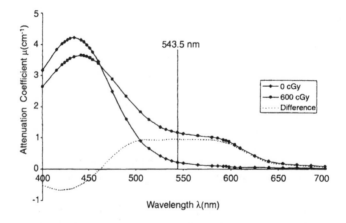

FIGURE 9.12 Change in FBX-gel absorption spectrum due to irradiation. The attenuation of the un-irradiated gel (0 cGy curve) and the irradiated gel (600 cGy curve) were measured using water as the reference. The difference between the two curves gives the change in the absorption coefficient due to the irradiation, consisting of a broad peak centered near 540 nm. The vertical line indicates the frequency of the He-Ne laser used to probe the FBX-gel system. (From Reference [17]. With permission.)

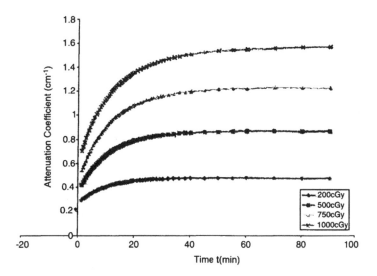

FIGURE 9.13 Graph showing the reaction completion time for FBX-gel irradiated to a dose of 1000 cGy. The graph clearly indicates that at least a 1-h post irradiation time is required before the dose distribution can be scanned optically. (From Reference [17]. With permission.)

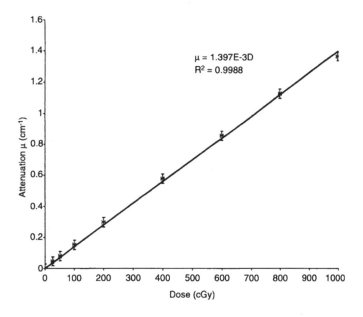

FIGURE 9.14 Plot of the reconstructed attenuation coefficient vs. dose for FBX-gelatin. The error in the dose corresponds to an uncertainty of ~2%, while the error in the reconstructed attenuation is a fixed value of ~0.03. (From Reference [17]. With permission.)

Figure 9.13 shows the reaction completion time of the exposed FBX-gel. Samples were irradiated in methacrylate cuvettes (4.5 ml, 1.0-cm pathlength) to doses of 200, 500, 1000, and 2000 cGy. Immediately following the irradiation, the absorption of the samples was measured as a function of time with a spectrophotometer. The temperature of the cuvettes was maintained at 10°C by placing the cuvettes in a thermal reservoir filled with double-distilled water. In Figure 9.13, the 200-cGy sample reached quasi-saturation and stabilized after ~40 min. The attenuation of the 1000-cGy sample, on the other hand, was still noticeably changing after 60 min.

The dose response curve for the preparation of FBX-gelatin is plotted in Figure 9.14. Cylinders of gel were irradiated with a 4.0-cm × 4.0-cm field of ^{60}Co gamma-rays.

The optical characteristics of a ferrous benzoic acid xylenol orange in gelatin (FBXG) have been studied by Bero et al. [18] over the wavelength range 300–700 nm as a function of radiation dose. The un-irradiated gel exhibits a strong absorption peak at 440 nm; with increasing dose, this peak starts to reduce in intensity, while a new broad peak centered at 585 nm begins to appear. Using ^{60}Co gamma-rays, the absorption coefficients for these two peaks were found to vary linearly with dose up

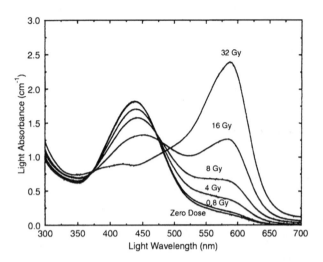

FIGURE 9.15 FBXG optical absorbance spectra for samples irradiated to 0.8, 4, 8, 16, and 32 Gy. (From Reference [18]. With permission.)

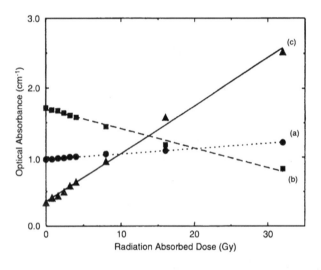

FIGURE 9.16 FBXG dose response curves at different wavelengths: (a) at 304 nm, (b) at 440 nm, (c) at 585 nm. (From Reference [18]. With permission.)

to at least 30 Gy with slopes of -0.028 cm^{-1}Gy^{-1} (440 nm) and 0.069 cm^{-1}Gy^{-1} (585 nm). The NMR response of FBXG gels was found to be marginally reduced compared to the standard Fricke dosimeter in gel form, and the NMR technique was much less sensitive than the optical read-out method.

The FBXG dosimeter has two main constituents: 75% by volume is the gel component, and the modified Fricke solution makes up the rest. The gels were prepared by mixing 5% by weight gelatin with "Milli-Q" (highly purified) water; the mixture was heated to 45°C under constant stirring until the gelatin powder was completely dissolved. The FBX solution consisted of 0.5 mM ferrous ammonium sulphate, 25 mM sulphuric acid, 1mM benzoic acid, and 0.1 mM

xylenol orange (sodium salt). The benzoic acid crystals were put into the ion-indicator solution and heated until they dissolved, and then this mixture was added to the inorganic components; all these concentrations, are expressed relative to the final FBXG volume.

Figure 9.15 shows the optical absorption spectra for FBXG exposures of 0–32 Gy. The most prominent features of these data are two broad peaks at 440 and 585 nm; the 440 nm peak, which is a feature of the un-irradiated gel, consistently reduces with increasing dose and has almost disappeared at 32 Gy. By contrast, the 585-nm absorbance increases from a very low value at zero dose to become a very strongly defined peak at higher doses. The absorbance at 304, 440, and 585 nm is plotted as a function of dose in Figure 9.16, where it is clear that the 585-nm response gives both the greatest sensitivity and the lowest detectable dose (estimated to be about 0.5 Gy).

NMR measurements of R_1 and R_2 are shown in Figure 9.17 as a function of dose for the conventional Fricke doped gelatin gel and for FBXG. The FBXG R_1, R_2 dose responses are slightly smaller than those of the simple Fricke gel and much smaller than the optical response at 585 nm.

III. APPLICATION OF BANG POLYMER GEL

Optical properties of the BANG polymer gel were discussed by Maryanski et al. [19] The dose-response mechanism relies on the production of light-scattering polymer micro-particles in the gel at each site of radiation absorption. The scattering produces an attenuation of transmitted light intensity that is directly related to the dose and independent of dose rate. For the BANG polymer gel (bis, acrylamide, nitrogen, and gelatin), the shape of the dose-response curve depends on the fraction of the cross-linking monomer in the initial mixture and on the wavelength of light. At 500 nm the attenuation coefficient (μ) increases by approximately 0.7 mm^{-1} when the dose increases from 0 to 5 Gy. The refractive index of an irradiated gel shows no significant dispersion in the visible region and depends only slightly on the dose. The average sizes of the cross-linked particles produced by radiation, as a function of dose, was established. The particle sizes increase with dose and reach approximately the wavelength of red light.

The BANG polymer gels, prepared by Marganski et al., contain 5% by weight gelatin, 89% water, and 6% acrylic monomers acrylamide and bis (i.e., N,N′-methylene-bis-acrylamide), with four different fractions of the cross-linking monomer bis in the monomer feed. Expressed in percentage units (%C) as 17, 33, 50, and 67%C. The gels are irradiated in test tubes to graded doses using 250-kV x-rays, with a dose rate of 3.5 Gy min^{-1}.

Figure 9.18. demonstrates that the refractive index does not increase significantly when a solution of monomers is

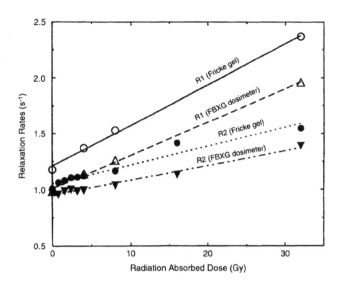

FIGURE 9.17 A comparison between the NMR relaxation rates of FBXG gel and the standard Fricke gel dosimeter; solid lines are for Fricke gel and dashed lines are for FBXG. (From Reference [18]. With permission.)

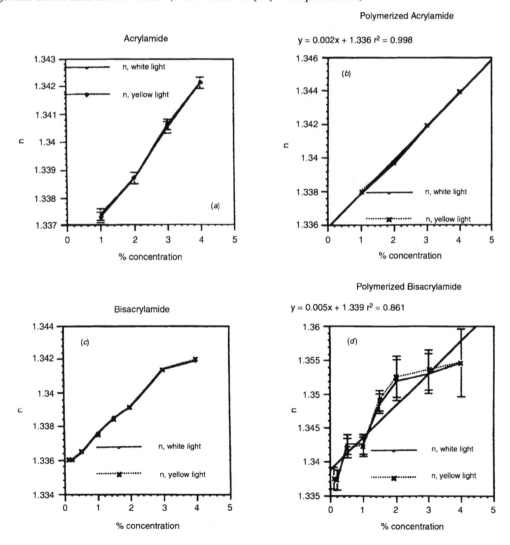

FIGURE 9.18 (a), (b), (c), (d) The concentration dependence of the index of refraction for (a) acrylamide and (c) bis monomer solutions and (b), (d) for corresponding polymers. (From Reference [19]. With permission.)

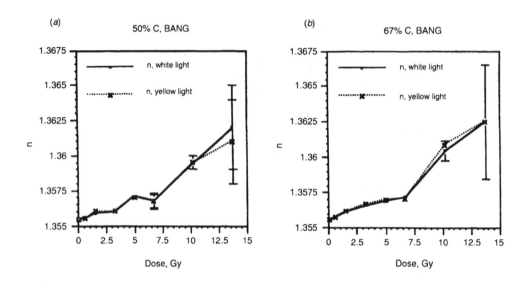

FIGURE 9.19 (a), (b) The dose dependence of the refractive index of polymer gels. (a) 50%C BANG, (b) 67%C BANG. (From Reference [19]. With permission.)

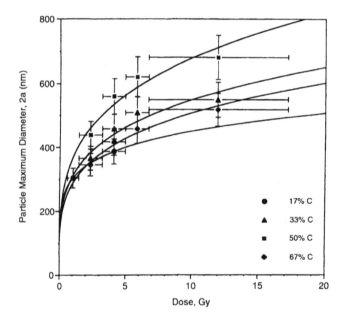

FIGURE 9.20 Maximum sizes of radiation-induced cross-linked polyacrylamide panicles for seven dose intervals. The power law fits are as follows:

17%C	$2a = 304.2D^{0.171}$	$r^2 = 1.00$
33%C	$2a = 312.0D^{0.246}$	$r^2 = 0.91$
50%C	$2a = 363.3D^{0.266}$	$r^2 = 0.91$
67%C	$2a = 288.6D^{0.246}$	$r^2 = 0.97$

(From Reference [19]. With permission.)

polymerized thoroughly by chemical means. This effect can be expected to be even smaller when the polymerization is only fractional, as it is in polymer gels irradiated to low doses. Figure 9.19 shows how the refractive index depends on the dose for two cross-linker fractions. The changes overall do not exceed 0.5% at the maximum dose

of 14 Gy; therefore, it can be expected that refraction in the regions of high-dose gradient should not significantly distort the optical density projections in the optical scanner.

Calculated average upper limits of the particle sizes produced within several dose intervals are shown in Figure 9.20.

The effects of varying the weight fraction (%C) of the cross-linker N,N′-methylene-bisacrylamide (bis) per total amount of monomer (6% w/w), and the NMR measurement temperature, on the dose response of the transverse relaxation rate (R_2) of bis-acrylamide-nitrogen-gelatin (BANG) aqueous polymer gel dosimeters have been investigated by Maryanski et al. [20] The gel samples were irradiated in test tubes with 250-kV x-rays, and the water proton NMR transverse relaxation rates were measured at 0.47 T using a Carr-Purcell-Meiboom-Gill multiecho pulse sequence. Both the dose sensitivity (slope of the linear portion of an R_2-dose response) and the maximum rate at which the R_2-dose response saturated (R_2^{max}) were found to depend strongly on the cross-linker fraction and on the temperature of the R_2 measurement. The dose sensitivity peaked at approximately 50%C and, for this composition, varied from 0.14 s^{-1}Gy^{-1} at 40°C to 0.48 s^{-1}Gy^{-1} at 10°C. The maximum transverse relaxation rates ranged from 0.8 s^{-1} at 33%C and 40°C to 11.8 s^{-1} at 83%C and 5°C. These results suggest that water proton transverse relaxation in the gel is controlled by an exchange of magnetization between the aqueous phase and the semi-solid protons associated with the polymer, and that the latter experience spectral broadening from immobilization, which increases with cross-linking or cooling.

Solutions were prepared in 100-ml quantities. Gelatin (5% w/w), acrylamide, and bis (6% w/w total comonomers) were dissolved in water (89% w/w) at 60°C. The bis weight fraction (%C) per total amount of comonomer was varied from 17%C to 83%C. Two ml of each solution were poured into glass test tubes, which were maintained at 60°C in a water bath. Humidified nitrogen gas was bubbled through each sample for 2 min, using Pasteur micropipettes, and then the test tubes were hermetically sealed with rubber stoppers and refrigerated.

Gelled samples were equilibrated at room temperature (21°C) and irradiated to different doses with 250-kV x-rays filtered through 2 mm Al and with a dose rate of 3.5 Gy min^{-1}. The doses to the gel were assumed to be equal to those measured in the Fricke solution, irradiated to graded doses in identical test tubes. The gel proton NMR transverse relaxation rates ($R_2 = 1/T_2$) were measured 1 day following the irradiation using an IBM Minispec NMR pulsed spectrometer operating at 20 MHz. A multiecho Carr-Purcell-Meiboom-Gill (CPMG) pulse sequence was employed, with pulse spacing $\tau = 1$ ms and with each eighth echo sampled. The amplitude of the last echo was always less than one-fifth of the first. The temperature of the samples was varied from 5°C to 40°C using an automated water heater-circulator.

Figure 9.21a shows dose-response curves obtained from NMR measurements performed at 20°C on gels of varying %C. Figure 9.21b shows the lower-dose region for the same set of data. It is evident from these data that

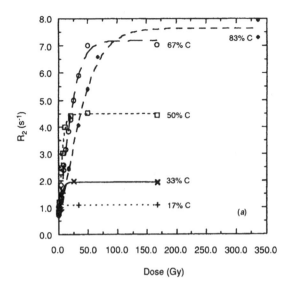

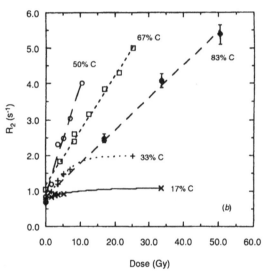

FIGURE 9.21 (a) Dose-response curves obtained from R_2, measured at 20°C in gels with different weight fractions of crosslinker per total comonomer (%C). (b) Lower-dose region of the data from (a). (From Reference [20]. With permission.)

the amount of cross-linker in the monomer mixture significantly affects the R_2 sensitivity (defined as the initial R_2 increment per unit dose) and the dynamic range over which changes in R_2 reflect variations in dose.

Figure 9.22a illustrates the dependence of the R_2 sensitivity on crosslinker content for three different temperatures at which R_2 was measured. It can be seen that the sensitivity peaks at about 50%C and increases with decreasing temperature for any %C.

Figure 9.22b shows the dependence of the dose sensitivity on the temperature of NMR measurement for several different weight fractions of the cross-linker (%C). It is worth noting that this dependence is most pronounced for the same cross-linker fraction at which the sensitivity is at its maximum (about 50%C).

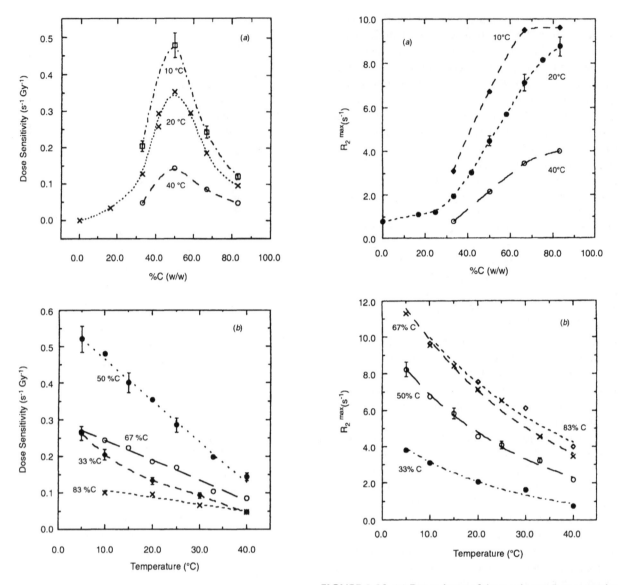

FIGURE 9.22 (a) Dependence of the initial dose-response sensitivity on the cross-linker content for three different temperatures at which R_2 was measured. (b) Dependence of the dose sensitivity on the measurement temperature for several different proportions of the cross-linker. (From Reference [20]. With permission.)

FIGURE 9.23 (a) Dependence of the maximum R_2 measured at a saturation dose (R_2^{max}) on cross-linker fraction for three different temperatures. (b) Dependence of R_2^{max} on temperature for several different values of cross-linker fraction. (From Reference [20]. With permission.)

Figure 9.23a shows the effect of varying the cross-linker fraction on R_2^{max}, the maximum R_2 achievable by a saturation dose, at three different temperatures. It can be seen that R_2^{max} increases with cross-linking and that this dependence increases dramatically above approximately 30%C. In addition, this effect is clearly enhanced by cooling the gels during the NMR measurment.

Figure 9.23b illustrates the dependence of R_2^{max} on temperature for several different values of cross-linker fraction. R_2^{max} appears to be more sensitive to temperature changes as the cross-linker fraction increases.

Figure 9.24 shows how the dose-response curve for a 50%C BANG gel changes with temperature. It is clear that

it is the radiation-induced build-up of the cross-linked polyacrylamide in the gel that enhances the solvent R_2 as well as rendering it more sensitive to temperature changes.

The spatial uniformity of gel sensitivity to radiation was found to depend strongly on the presence of oxygen, which must be eliminated for the gel dosimeter to be of use. The gel formula used by Oldham et al. [21] is shown in Table 9.3.

The 3D gel-image data produced by the NMR scanner is in the form of calculated T_2 relaxation times. This image must be converted to absorbed dose to allow verification of radiotherapy dose distributions. Conversion is achieved by means of a calibration curve, obtained by measuring

TABLE 9.3
Chemical Composition of BANG–gel

Component	Formula	Weight Fraction	Amol
Gelatin	$C_{17}H_{32}N_5O_6$	0.05	402.22
Acrylamide	$CH_2CHCONH_2$	0.06	71.04
Bis	$(CH_2CHCONH)_2CH_2$	0.06	154.1
Water	H_2O	0.83	18

From Reference [21]. With permission.

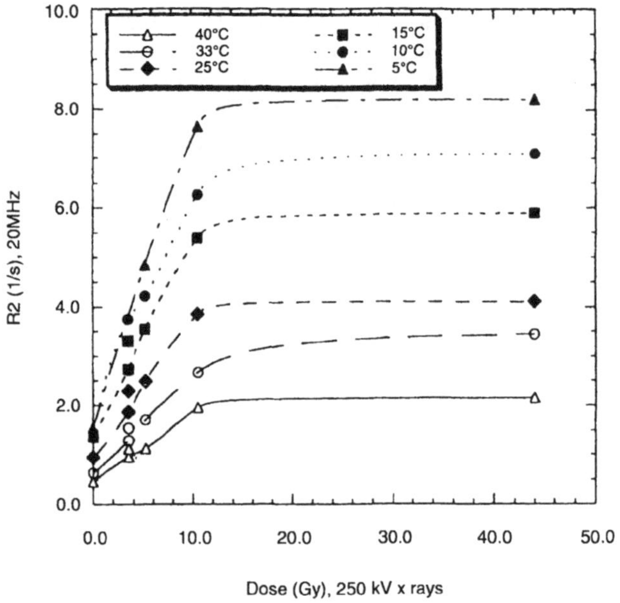

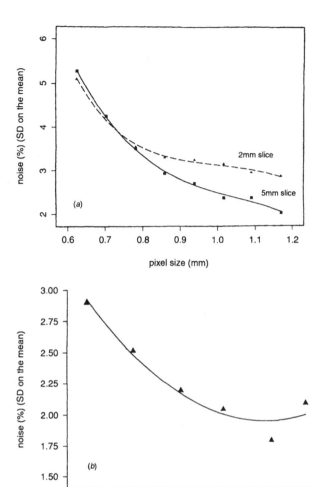

FIGURE 9.24 Effect of temperature on the R_2 dose-response curve of the 50%C BANG. (From Reference [20]. With permission.)

the T_2 relaxivity in regions of the gel that have been irradiated to known doses.

The final step in forming the calibration curve was to determine the dose actually delivered in the regions of interest for which the T_2 values were measured. This determination was performed by film measurement in a simulated geometry. The required quantity is D_m, the absorbed dose delivered by a 2×2-cm², 6-MV radiation field to the measurement plane after it has been attenuated through a 6-mm Perspex plate and a 2-cm layer of gel. D_m was calculated according to

$$D_m = MU(D_f \text{ISC} - \text{BSC}) \qquad (9.26)$$

where D_m is the dose in the measurement plane; MU is the number of monitor units delivered; D_f is the dose per MU measured by a piece of film placed in a Perspex sandwich at the effective radiological path length (d) to

FIGURE 9.25 (a) The relationship between noise (i.e., standard deviation on the mean) in small regions (~1 cm²) in the T_2 (or $R_2 = 1/T_2$) gel image and the pixel size. (b) The relationship between noise (i.e., standard deviation on the mean) in the T_2 (or $R_2 = 1/T_2$) gel image and R_2. High R_2 corresponds to high dose and high polymerization. (From Reference [21]. With permission.)

the measurement plane ($d = 2.38$ cm is the Perspex-equivalent radiological path length of 6 mm of Perspex and 2 cm of gel); ISC is an inverse-square law correction to allow for the fact that the measurement plane is 2 mm further from the source than the film; and BSC is a backscatter correction to account for the reduced backscatter radiation from the nitrogen space compared with solid Perspex. The ISC term is $(102.38/102.6)^2 = 0.9957$. The BSC term for a 2×2-cm² field, from Perspex that is ≥2 cm from the measurement plane, is negligible and was ignored.

Figure 9.25a shows the relationship between the standard deviation (SD) in a 1-cm-diameter region placed centrally in the irradiated region and the pixel size. It is seen that for a 2-mm slice thickness, little improvement in noise is gained by increasing the pixel size above $\approx 0.8 \times$ 0.8 mm^2. The latter therefore represents a good tradeoff between noise and spatial resolution. Figure 9.25b shows the relationship between noise (SD on the mean) and R_2.

The MR-imaging sequence parameters were chosen to minimize the errors for fitting the exponential decay of short T_2 (i.e., high-dose regions ≈ 6–8 Gy, corresponding to an R_2 of ≈ 5–6.0).

REFERENCES

1. **Gore, J. C. et al.,** *Phys. Med. Biol.,* 29, 1189, 1984.
2. **Schreiner, L. J.,** in *Proc. 1st International Workshop on Radiation Therapy Gel Dosimetry,* 1999, 1.
3. **Appleby, A.,** in *Proc. 1st International Workshop on Radiation Therapy Gel Dosimetry,* 1999, 9.
4. **Audet, C.,** in *Proc. 1st International Workshop on Radiation Therapy Gel Dosimetry,* 1999, 31.
5. **Back, S. A. J. and Olsson, L. E.,** in *Proc. 1st International Workshop on Radiation Therapy Gel Dosimetry,* 1999, 47.
6. **Maryanski, M. J.,** in *Proc. 1st International Workshop on Radiation Therapy Gel Dosimetry,* 1999, 63.
7. **Maryanski, M. J. et al.,** *Phys. Med. Biol.,* 39, 1437, 1994.
8. **Gore, J. C. et al.,** *Phys. Med. Biol.,* 41, 2695, 1996.
9. **Maryanski, M. J. et al.,** *Med. Phys.,* 23, 699, 1996.
10. **Gambarini, G. et al.,** *Phys. Med. Biol.,* 39, 703, 1994.
11. **Balcom, B. J. et al.,** *Phys. Med. Biol.,* 40, 1665, 1995.
12. **Luciani, A. M. et al.,** *Phys. Med. Biol.,* 41, 509, 1996.
13. **Olsson, L. E. et al.,** *Phys. Med. Biol.,* 37, 2243, 1992.
14. **Schulz, R. J. et al.,** *Phys. Med. Biol.,* 35, 1611, 1990.
15. **Gambarini, G. et al.,** *Nucl. Inst. Meth.,* A422, 643, 1999.
16. **Gambarini, G. et al.,** *Rad. Prot. Dos.,* 70, 571, 1997.
17. **Kelly, R. G. et al.,** *Med. Phys.,* 25, 1741, 1998.
18. **Bero, M. A. et al.,** *Nucl. Inst. Meth.,* A422, 617, 1999.
19. **Maryanski, M. J. et al.,** *Phys. Med. Biol.,* 41, 2705, 1996.
20. **Maryanski, M. J. et al.,** *Phys. Med. Biol.,* 42, 303, 1997.
21. **Oldham, M. et al.,** *Phys. Med. Biol.,* 43, 1113, 1998.
22. **Parasad, P. V.,** *Rad. Res.,* 128, 1, 1991.

10 Neutron Dosimetry

CONTENTS

I. INTRODUCTION

Neutrons are non-ionizing particles and are thus identified by detection of ionizing particles emitted in neutron interaction. The most common reactions are (n, α), (n, p), (n, γ), and proton recoil. The dosimeters can be gas-filled, scintillators, TLD, solid-state, track detectors, photographic film, and other types. Gas-filled neutron dosimeters are generally tissue-equivalent ionization chambers. The properties required for a neutron dosimeter are high efficiency for neutron detection, constant response per unit kerma in tissue as a function of neutron energy, and the possibility to discriminate against γ-radiation. It is important that the dose or the energy transferred to the dosimeter is similar to that transferred to the human body irradiated, taking into consideration the body composition, such as bone, tissue, etc.

Neutron dosimetry is done for medical purposes as well as in nuclear research laboratories, reactor centers, nuclear power stations, and industry. For routine radiation surveys, the "rem-counters" have practically solved the problem of measuring dose-equivalent rates without much information on neutron spectra. Routine personnel neutron dosimetry is done by TLD badges and pocket ionization chambers. The requirements for accident dosimetry are different from those for routine survey and dosimetry. Superheated drop detectors are useful for environment dosimetry.

Most neutron dosimeters that exist for direct reading or quick, easy evaluation are of the military or civil defense type, such as tissue-equivalent condenser chambers, neutron-sensitive diodes, or radiation elements, suitable for foolproof operation. Accident dosimetry for neutrons is well-established. The important systems are reliable critical alarms and similar dosimetry systems that are partly worn by persons and partly set up on fixed positions in the controlled areas. Fission-track dosimeters and activation-threshold detectors, together with body counting for ^{24}Na by a simple body counter, permit a reasonable quick evaluation of the severity of exposures after an accident.

Standardization of neutron dosimetry procedures has been established by drafting protocols for neutron dosimetry for radio therapy in Europe and the U.S. The use of calibrated tissue-equivalent (TE) ionization chambers with TE gas filling is recommended as the practical method for obtaining the tissue kerma in air and the absorbed dose in a TE phantom. The TE chamber should have a calibration factor for photons applied to it.

Correction factors convert the reading of the chamber to the charge produced in an ideal chamber at a reference temperature and pressure. Also, correction factors account for the finite size of the TE chamber, its build-up cap, and its stem when measurements are made in air. For the conversion of measured charge to absorbed dose, several parameters are necessary, including the average energy required to create an ion pair in the gas, the gas-to-wall absorbed dose conversion factor, the neutron kerma ratio of reference tissue to that in the wall material of the ionization chamber, and the displacement correction factor.

The types of personnel neutron dosimeters commonly used for radiation-protection purposes are neutron film type A (NTA film), thermoluminescent albedo dosimeters (TLD albedo), superheated drop detectors, and fission track detectors. Deficient energy response and sensitivity, fading, use of radioactive material, and other limitations prevent any of these dosimeters from being satisfactory for universal application in a wide variety of radiation environments.

A primary requirement for the personnel neutron dosimetry system is that it be able to meet current regulatory requirements. Neutron dosimeters should be worn if the neutron whole-body dose equivalent exceeds 10% of the γ- plus X-ray dose equivalent. The dosimeter is considered satisfactory for operation in mixed radiation fields if it can measure 1 rem of fast neutrons in the presence of 3 rem of γ-rays with energy above 500 keV. The dosimeter response must simulate the dose-equivalent conversion factor, which varies by a factor of 40 over that range. Radiation weighting factor W_R is given below for various kinds of radiation and for neutrons of various energies.

Type and energy range of radiation	Radiation weighting factor W_R
Photons, all energies	1
Electrons and muons, all energies *	1
Neutrons, energy < 10 keV	5
10 keV to 100 keV	10
> 100 keV to 2 MeV	20
> 2 MeV to 20 MeV	10
> 20 MeV	5
Protons, other than recoil protons, energy > 2 MeV	5
Alpha particles, fission fragments, heavy nuclei	20

* Excluding Auger electrons emitted from nuclei to DNA, for which special microdosimetric considerations apply.

Calculation of the radiation weighting factor for neutrons requires a continuous function; the following approximation can be used, where E is the neutron energy in MeV:

$$W_R = 5 + 17e^{-(\ln(2E))^{\frac{2}{6}}} \qquad (10.1)$$

The rationale for fast neutron therapy (high LET radiation) is based on radiobiological arguments: [1]

a. a reduction in the oxygen enhancement ratio (OER) with high LET radiations; i.e., a selective efficiency against hypoxic cancer cells;
b. a greater selective efficiency against cells in the most radio-resistant phases of the mitotic cycle;
c. a smaller role of the sub-lethal lesions; i.e., less importance of the dose per fraction.

These factors can bring an advantage or a disadvantage, depending on the characteristics of the cancer cell population and of the normal tissues at risk. This raises the problem of patient selection. According to NCI, fast neutron therapy should be the treatment of choice for advanced-stage salivary gland tumors, and surgery should be limited to cases where there is a high likelihood of achieving a negative surgical margin and where the risk of facial nerve damage is high.

A standard A-150 tissue-equivalent or graphite thimble chamber having a standard ^{60}Co calibration factor is the recommended reference dosimeter for clinical proton dosimetry. The mean neutron energy or related parameters such as the half-value thickness (HVT) have also been used (Figure 10.1). The development of the microdosimetric programs has shown that this approach is indeed an oversimplification and that several types of secondary charged particles must be taken into account (Figure 10.2).

In a given neutron beam, the RBE does not vary as a function of depth (Figure 10.2). For proton beams, the observed RBEs are lower, ranging from about 1.0 to 1.2. This range is clinically significant but only slightly larger than the experiment uncertainties on RBE determinations (Figure 10.3). The proton energy decreases with depth and, thus, there is possible increase in RBE.

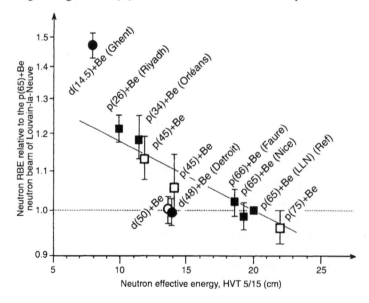

FIGURE 10.1 Variation of RBE on different neutron beams as a function of energy. (From Reference [1]. With permission.)

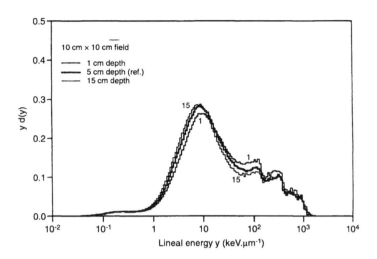

FIGURE 10.2 Comparison of microdosimetric spectra measured at different depths in a phantom on the beam axis of the $p(65)$ + Be neutron therapy beam of Louvain-la-Neuve (the highest energy used in neutron therapy). The curves represent the distributions of individual energy deposition events in a simulated volume of tissue 2 μm in diameter. The parameter y (lineal energy) represents the energy deposited by a single charged particle traversing the sphere, divided by the mean cord length. Four peaks can be identified in the $p(65)$ + Be neutron spectra: the first is at 8 keV μm^{-1} and corresponds to high-energy protons, the second is at 100 keV μm^{-1} and corresponds to low-energy protons, the third is at 300 keV μm^{-1} and is due to a-particles, and the last is at 700 keV μm^{-1} and is due to recoil nuclei. The (rather flat) peak corresponding to the gamma component of the beam is at 0.3 keV μm^{-1}. The differences between the spectra measured at differents depths are relatively small and no significant RBE variation has been detected. (From Reference [1]. With permission.)

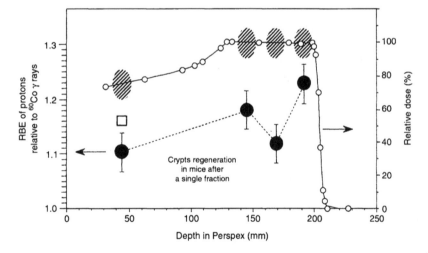

FIGURE 10.3. Variation of RBE with depth in the modulated 200-MeV proton beam produced at the National Accelerator Centre (NAC), Faure (left ordinate, closed circles). The width of the spread-out Bragg peak is 7 cm (from 13 to 20 cm in depth). The error bars indicate the 95% confidence intervals. The open square corresponds to the RBE for the unmodulated beam. The depth-dose curve in the modulated beam is superimposed for comparison (right ordinate, small open circles). The hatched areas indicate the different positions where the mice were irradiated. The width of the shaded areas corresponds to the size of the mice abdomens. (From reference [1]. With permission.)

II. KERMA CALCULATION AND MEASUREMENT

Kerma is defined as the average kinetic energy released in matter (per unit mass) and is the sum of all energy transferred to light-charged particles and residual nuclei in a reaction. The neutron kerma factor is the kerma pro-

duced per unit neutron fluence. The absorbed dose is the energy deposited per unit mass.

Neutron dose is obtained by calculating the kerma. The kerma factor k_f for neutrons of energy E_n, for a given material is defined as the ratio of the neutron kerma K and the neutron fluence. The kerma factor is usually calculated from the average energy transferred to kinetic energy of charged particles in the various types of interactions and the corresponding cross sections. For neutron energy E_n less than 20 MeV, the data can be taken from the ENDF/B-V data file. For higher energies, the data is from relied-on calculations only. Above 15 MeV the number of nuclear reactions increases and substantially contributes to the kerma, with strong uncertainty in the cross section. Erroneous cross sections due to the use of inadequate calculation models, in particular for light elements, leads to large discrepancies in kerma factor calculations. More experimental data are needed at energies above 20 MeV.

The kerma factor k_f can be obtained by direct dose measurement with ionization chamber or proportional counter. Knowing the dose, the kerma K, and the fluence provides the necessary data. A-150 tissue-equivalent plastic and graphite chambers are used for this purpose.

The energy transferred (kerma per neutron per cm³) for an incident neutron of energy E in the laboratory is defined by

$$K(E) = \sum_i N_i \left[\sum_j \bar{\varepsilon}_{ij}(E) \sigma_{ij}(E) \right] \quad (10.2)$$

where i identifies the element, j identifies the type of nuclear reaction, N_i is the number of nuclei of the ith element per gram, and $\varepsilon_{ij}(E)$ is the average amount of energy transferred to kinetic energy of charged particles in a collision whose cross section is $\sigma_{ij}(E)$. For elastic scattering, the kinetic energy (laboratory system) of a recoiling nucleus of mass A relative to that of the neutron is

$$\varepsilon_{i,el}(\theta_{CM}, E) = \frac{2AE}{(A + 1)^2}(1 - \cos\theta_{CM}) \quad (10.3)$$

where θ_{ij} is the angle of scattering of the neutron in the center of mass system.

For inelastic scattering the energy transferred to a recoil nucleus left in an excited state with energy $\bar{E}'$ above the ground is

$$\varepsilon_{i,inel}(\theta_{CM}, E) = \frac{2AE}{(A + 1)^2} - \frac{\bar{E}'}{A + 1} - \frac{2AE}{(A + 1)^2}\cos\theta_{CM}$$

$$\times \left[1 - \frac{(A + 1)\bar{E}'}{AE} \right] \quad (10.4)$$

The kerma can be calculated by

$$K = \sum_j \int n_j(E)E \, dE \quad (10.5)$$

where j is the secondary particle type produced by the neutron interaction, E is the particle energy, and $n_j(E)$ is the intensity of the particle spectrum (particles per gram per neutron per cm²). W is given by

$$\bar{W} = \sum_j \int n_j(E)E \, (dE) \Big/ \sum_j \int n_j(E)[E/W(E)_j]dE \quad (10.6)$$

For elastic scattering, if the neutron fluence intensity is $\phi(E_n)$ and the atomic density in the material is N (atoms/g), the differential spectrum intensity of the recoiling particle of mass A and energy E_2 is

$$n(E_2) = N\phi(E_n)\frac{(1 + A)^2}{4AE_n}\sigma(E_n)_{el}\sum_l (2l + 1)$$

$$\times f_l(E_n)P_l(\cos\theta_{CM}) \quad \text{particles/g.MeV} \quad (10.7)$$

The neutron mass is assumed to be unity, $P_l(\cos\theta_{cm})$ is a Legendre polynomial of degree l, and $f_l(E_n)$ is its coefficient. In the case of elastic scattering by hydrogen,

$$n(E_2) = N\phi(E)\frac{(1 + A)^2}{4AE_n}\sigma_{el}(E_n) M(E_n) \quad (10.8)$$

where

$$M(E_n) = (1 - 1/2\cos\theta_{CM} + 1/2b\cos^2\theta_{CM})/(1 + b/6)$$

and

$$b = 2(E_n/90)^2$$

The total kerma induced by neutrons in the wall material of the PC is derived from

$$K = k_w(r_{m,g})_n(W_n/W_c)(C/m)\int_{h_{min}}^{h_{max}} hn(h) \, dh \quad (10.9)$$

where k_w is the correction factor for the attenuation and scattering caused by the wall, $(r_{m,g})_n$ is the gas-to-wall absorbed dose conversion factor, and W_n is the average W-value (average energy required to produce an ion pair in the gas) for neutron-induced secondaries. W_c is the W-value for the radiation used for calibration, C is the calibration factor which relates pulse height h to energy imparted (the energy deposited in the cavity), m is the mass of the gas, and $n(h)$ is the number of neutron-induced events measured as a function of pulse height h.

Kerma factors for carbon and A-150 tissue-equivalent plastic were determined by Schuhmacher et al. [2] in two nearly monoenergetic neutron beams. Kerma was measured with low-pressure proportional counters (PC) with walls made of graphite and A-150 plastic. The neutron fluence was measured with a proton recoil telescope. The corrections for the kerma contributions from low-energy neutrons in the beams were determined by time-of-flight measurements with an NE213 scintillation detector and with the PCs. For neutron energy of (26.3 ± 2.9)MeV and (37.8 ± 2.5)MeV, the kerma factors for carbon were (35.5 ± 3.0)pGy cm^2 and (42.6 ± 3.9)pGy cm^2, respectively, and for A-150 they were (75.6 ± 5.4) pGy cm^2 and (79.0 ± 5.6) pGy cm^2. At these energies the kerma factor ratio for ICRU muscle tissue to A-150, calculated on the basis of the new kerma factors, is 0.93.

A proton recoil telescope (PRT) was used for the fluence measurements; it was based on the detection of recoil protons produced by elastic scattering in a hydrogen-containing radiator foil. This type of instrument enables the neutron fluence to be determined with uncertainties of only 2.5% to 3% in the neutron energy range from about 1 to 15 MeV and is therefore used as a primary fluence standard at various laboratories. The PRT described by Schuhmacher et al. consisted of a polyethylene (PE) radiator foil (areal density 144.7 mg cm^{-2}) mounted on a 1.0-mm thick Ta backing, two proportional counters, a solid-angle defining aperture, and two solid state detectors, S_1 and S_2 (Figure 10.4). The proportional counters were filled with isobutane at a pressure of 50

kPa. A 1.0-mm-thick Si surface barrier detector was chosen for S at 27 MeV and a 2.5-mm-thick one at 39 MeV, while in both cases S_2 consisted of a 5.0-mm -thick Si(Li) detector operated at room temperature. For the 39-MeV neutron field, a 0.5-mm-thick aluminum absorber was inserted into the PRT between the aperture and S_1, so that protons of the maximum energy were stopped in the silicon detector S_2. The signals from the four detectors were analyzed with a multi-parameter data acquisition system which allowed the protons to be discriminated from other types of charged particles emitted from the radiator foil.

The PRT was used to determine the fluence of neutrons of the nominal energy, i.e., within the main peak. The probability of energy depositions by recoil protons in the various detectors of the PRT for a given spectral neutron fluence at the position of the radiator was calculated with a Monte Carlo program. It simulates the production of recoil protons and their transport through the telescope, including energy loss straggling and angle straggling. The cross sections for n-p scattering were based on experimental results published by Fink et al. [3] These data, which were determined with total uncertainties of less than 2%, are about 5% higher at 27 MeV and 10% higher at 39 MeV than those from a phase shift analysis for the scattering angles which were observed with the PRT. The phase shift was employed for interpolating the measured data in energy and angle.

The tissue absorbed-dose determination depends upon knowledge of the relative rate of charged-particle energy production per unit mass for the tissue and tissue substitute

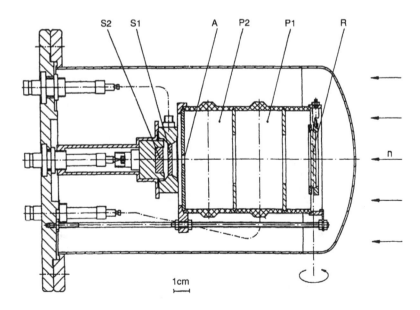

FIGURE 10.4 Schematic drawing of the PRT with radiator (R), "pillbox"-proportional counters (P_1 and P_2), solid-angle defining aperture (A), and semiconductor detectors (S_1 and S_2). The PRT is irradiated from the right. The radiator assembly consists (from right to left) of a graphite disc, a Ta backing, and a PE disc. The assembly was turned by 180 degrees for the background measurement. (From Reference [2]. With permission.)

material, i.e., the kerma or kerma factor ratio. As the kerma ratio is a function of the neutron energy, information about the neutron energy spectrum is needed. Hence, neutron dose determinations using tissue-substitute materials require accurate kerma factor values for all constituent materials for both the tissue and tissue substitute over the entire neutron energy range. [4] Recommended hydrogen kerma factors for neutron energies from 10 to 70 MeV calculated from cross sections obtained from the "VL 40" phase shift analysis are given in Table 10.1. Figure 10.5 shows the information known for both measured calculated neutron kerma factors for carbon as a function of incident neutron energy. Figure 10.6 shows a summary of measured and calculated neutron kerma factors for oxygen as a

function of incident neutron energy. The experimental database for oxygen is limited.

Modern neutron radiotherapy employs energies extending to 70 MeV, while industrial applications such as transmutation and tritium breeding may generate neutrons exceeding energies of 100 MeV. Secondary neutrons produced by advanced proton therapy facilities can have energies as high as 250 MeV. Evaluated nuclear data were presented for C, N, and O by Chadwick et al. [5].

An important consequence of these microscopic cross-section calculations is the determination of integral radiation dosimetry quantities—kerma factors. Figure 10.7 shows total and ejected particle partial kerma factors for C and O. Such detailed information is a direct result of the inclusive modeling technique. Note that the proton kerma increases rapidly with energy and exceeds that for alpha particles at 50 MeV(C) and 42 MeV(O). Another striking consequence is the contribution of deuterons, which exceeds that for alpha particles at higher neutron energies, reaching 25%(C) and 20%(O) of the total kerma at 100 MeV. This rapid increase in proton and deuteron kerma at higher energies due to pre-equilibrium mechanisms was not recognized in the earlier extension to 32 MeV of ENDF/B-VI.

Kerma factors of carbon and A-150 tissue-equivalent plastic have been measured by Schrewe et al. [6] in nearly monoenergetic neutron beams at energies between 45 and 66 MeV. The kerma was measured with low-pressure proportional counters (PC), with the walls consisting either of graphite or A-150 plastic material, and the neutron beam fluence was measured with a proton recoil telescope. The kerma factor corrections were determined from

---

TABLE 10.1
Recommended Neutron Kerma Factor for Hydrogen.

E_n (MeV)	K_f (fGy.m^2)
10.00	45.69
12.38	46.59
15.62	46.93
20.46	46.50
31.42	44.21
46.19	41.11
59.79	38.92
70.00	37.80

The table is designed for linear-linear interpolation. The uncertainty (1 standard deviation) is ±1% at all energies.

From Reference [4]. With permission.

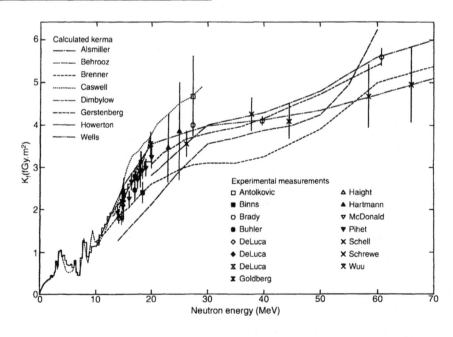

FIGURE 10.5 Measured and calculated neutron kerma factors for carbon as a function of incident neutron energy from 0 to 70 MeV. (From Reference [4]. With permission.)

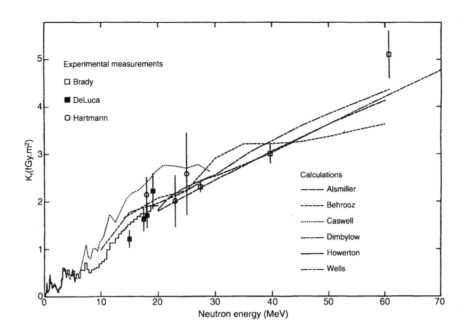

FIGURE 10.6 Measured and calculated neutron kerma factors for oxygen as a function of incident neutron energy from 0 to 70 MeV. (From Reference [4]. With permission.)

time-of-flight measurements with NE213 scintillation detectors and PCs. At neutron energies of (44.5 ± 2.6) MeV and (66.0 ± 3.0) MeV, the kerma factors were (40.9 ± 4.1) pGy cm^2 and (49.5 ± 8.7) pGy cm^2 for carbon and (80.4 ± 6.3) pGy cm^2 and (80.1 ± 8.2) pGy cm^2 for A-150 plastic.

The kerma K produced by monoenergetic neutrons of energy E and fluence Φ may be expressed as $K(E) = \Phi k_f(E)$, where k_f is the kerma factor. Kerma factors may be calculated from the cross sections $\sigma_j(E)$ as the sum of the interactions of type j which produce charged particles of an average energy $\bar{E}_j(E)$, multiplied by the number of target atoms per unit mass n:

$$k_f = K/\Phi = n\sum_j \bar{E}_j(E)\sigma_j(E) \qquad (10.10)$$

Above 20 MeV, however, calculations are much more complex, since the number of reaction channels rises drastically with increasing neutron energy, there is a lack of precise cross-section data and the reaction kinematics are scarcely known, particularly for the multi-particle breakup reactions. Since kerma and absorbed dose are numerically almost equal if the charged secondary particle achieves equilibrium, the kerma was determined by measuring the absorbed dose with a cavity chamber.

The total kerma K_{tot} obtained from the PC measurement can be written as the sum of three portions: the desired kerma of the nominal energy K_0, which refers to the neutron energy range E_2 to E_{max}, and two background contributions K_1 and K_2. K_1 and K_2 are made accessible by the two TOF measurements; K_1 is produced by neutrons of energy $0 - E_1$, and K_2 by $E_1 - E_2$.

The correction factor k_c, defined as the ratio K_0/K_{tot}, can be expressed as:

$$k_c = K_0/K_{tot} = (1 - K_1/K_{tot})/(1 + K_2/K_0) \qquad (10.11)$$

where K_1/K_{tot} is directly obtained from the PC-TOF measurements and K_2/K_0 is calculated from the spectral neutron fluences and kerma factors extrapolated from recent measurements and theoretical predictions. Although the kerma fraction of low-energy neutrons, $(1 - k_c)$, was quite substantial, k_c was determined with a relatively low uncertainty (see Table 10.2).

The ICRU has introduced the quantity Personal Dose Equivalent, $H_p(10)$ (see Chapter 1), as an estimator for the effective dose, E, as the primary limiting quantity for all conditions of penetrating radiation, i.e., radiation-type energy and angle of incidence. It is defined as the dose equivalent in 10 mm depth of an individual. The definition of Personal Dose Equivalent, $H_p(10)$, does not lead to a unique value because it depends on both the individual and the site chosen in which to obtain $H_p(10)$. [7] For the purposes of development, type testing, and calibration of individual dosimeters, the quantity was extended by the ISO to the TE slab phantom. In this case $H_p(10)$ has a unique value for a given radiation field. It serves as some kind of "estimator for the estimator" and shall be sufficiently conservative compared with the effective dose, at least for an irradiation incidence which is restricted to the frontal half-space. Monte Carlo calculations were performed by Leuthold et al. [7] using the MCNP code version 4A. This code is capable of transporting neutrons and photons as well as electrons.

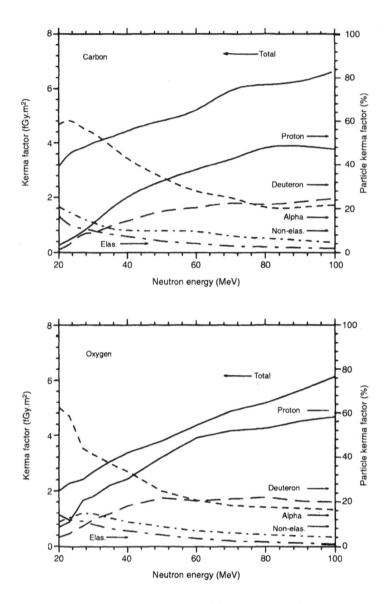

FIGURE 10.7 Calculated carbon and oxygen values for total kerma and for that due to different ejectiles plotted against neutron energy. Total kerma is given on the left ordinate while percentage kerma values by particle type are given on the right ordinate. (From Reference [5]. With permission.)

Calculations were performed for parallel monoenergetic beams of neutrons. The neutron cross-section data library EFF1LIB used in these calculations is based on the ENDF/B-5. The phantoms used for the calculations were the ICRU slab phantom (30 cm × 30 cm × 15 cm), consisting of 4-element ICRU tissue, and the male anthropomorphic phantom ADAM.

For the calculation of the personal dose equivalent in the ICRU slab phantom, a neutron quality factor was used which takes into account the revised $Q(L)$ relationship of ICRP 60 and the new stopping power data for protons and alpha particles in water by ICRU 49. The kerma values applied are taken from ICRU 26. Above 10-MeV neutron energy, the recommended kerma values of White et al. are used. Figure 10.8 shows the quality factor as function of neutron energy.

Calculations of personal dose equivalent, $H_p(10, \alpha)$, in the ICRU slab phantom were performed for energies 0.025 eV, 2 keV, 565 keV, and the high energies 14.8 and 19 MeV in order to see the influence of the above-mentioned difference in the kerma values and quality factor. [7] The values are given in Table 10.3. Figure 10.9 shows the ratio of personal dose equivalent, $H_p(10, \alpha)$, given by IEC to $H_p(10)$ obtained in this work for the five selected neutron energies as a function of the angle of incidence. The differences at low neutron energies may be attributed to the use of the $S(\alpha,\beta)$ cross-section tables for light water, whereas for the calculation of the IEC data, the $S(\alpha,\beta)$ cross sections of polyethelene were applied. The difference at high energies due to the different kerma values used exceeds 20%.

TABLE 10.2
Experimental Results for A-150 Plastic (A) and Carbon (C)

E_p MeV	E_n MeV	M	K_{tot}/Q pGy.nC^{-1}	k_c	E_2 MeV	K_0/Q pGy.nC^{-1}	Φ_0/Q cm^{-2}.nC^{-1}	k_f pGy.cm^2	k_f^C/k_f^A
31.9	26.3(29)	A	1539(58)	0.562(11)	20	865(37)	11.44(64)	75.6(54)	0.469(36)
		C	563(35)	0.721(11)	20	406(26)	11.44(64)	35.5(30)	
43.4	37.8(25)	A	1573(59)	0.493(11)	32	775(34)	19.81(55)	79.0(56)	0.539(45)
		C	691(48)	0.605(11)	32	418(30)	9.81(55)	42.6(39)	
50.0	44.5(26)	A	1799(75)	0.470(10)	38	845(40)	10.52(70)	80.4(65)	0.506(45)
		C	768(59)	0.560(10)	38	430(32)	10.52(70)	40.9(41)	
63.8	58.4(29)	A	1899(195)	0.427(12)	(50)	811(86)	10.50(8)	77.0(10)	0.61(11)
		C	976(134)	0.504(11)	(50)	492(68)	10.50(8)	46.8(74)	
71.2	66.0(30)	A	2030(128)	0.405(10)	58	822(55)	10.26(79)	80.1(82)	0.62(11)
		C	1058(169)	0.480(10)	58	508(82)	10.26(79)	49.5(88)	

The numbers in brackets denote the uncertainty (one standard deviation) of the last digits of the respective values.

From Reference [6]. With permission.

TABLE 10.3
Personal Dose Equivalent, H_p (10, α) in the ICRU Slab Phantom for Five Neutron Energies at Different Angles of Incidence

E_n(MeV)	$H_p(10,\alpha)/\Phi$(pSv.cm^2)					
	0°	15°	30°	45°	60°	75°
2.5×10^{-8}	11.50	10.50	8.87	6.32	3.97	1.62
2.0×10^{-3}	9.69	8.91	7.73	5.86	3.80	1.72
0.565	347	334	321	292	245	117
14.8	472	456	454	449	454	396
19.0	492	478	477	473	486	436

From Reference [7]. With permission.

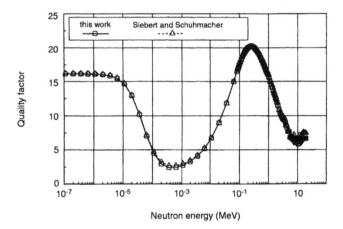

FIGURE 10.8 Quality factor as function of neutron energy in an ICRU tissue mass element. Differences in the high-energy region come from different kerma values used. (From Reference [7]. With permission.)

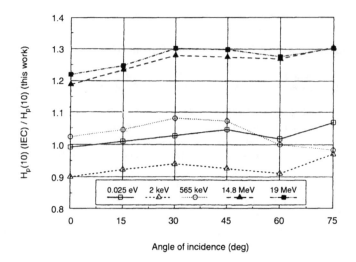

FIGURE 10.9 Ratio of personal dose equivalent, $H_p(10)$, in the ICRU slab phantom given by IEC to $H_p(10)$ obtained for five neutron energies as function of the angle of incidence. (From Reference [7]. With permission.)

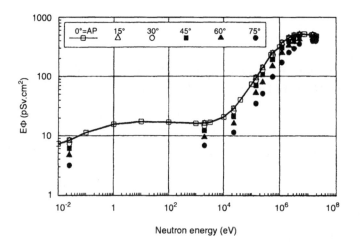

FIGURE 10.10 Effective dose E in the ADAM phantom as function of neutron energy for six angles of incidence to the left side of the phantom. (From Reference [7]. With permission.)

Figure 10.10 shows the result for the effective dose, E, as a function of neutron energy for different angles of incidence to the left side of the phantom. Figure 10.11 gives the ratio of personal dose equivalent to effective dose, $H_p(10,\alpha)/E$. For the personal dose equivalent, the IEC values are used except at 14.8 and 19 MeV.

The personal dose equivalent, $H_p(10)$, was calculated by Hollnagel [8] in the anthropomorphic MIRD phantom. $H_p(10)$ is compared with the primarily limited quantities, the effective dose E, and the effective dose equivalent, H_E, computed in the same Monte Carlo run. The operational quantity, $H_{sl}(10)$, in the ICRU slab was found to approximate $H_p(10)$ for multidirectional incidences.

The personal dose equivalent, $H_p(d)$, as defined in ICRU Report 51, is the dose equivalent (DE) in ICRU soft tissue at an appropriate depth d below a specified point on the body. For strongly penetrating radiation, a depth of

10 mm is frequently employed, ICRU states, in which case it is denoted as $H_p(10)$. The shape of the $H_p(10)$ curves in Figure 10.12 is roughly similar to that of other conversion functions for neutrons; the PA curve is rather low, as expected. The curve for H_E at AP incidence is also displayed, showing that $H_p(10)$ is a conservative estimate for H_E at AP geometry. Figure 10.13 shows that the ratio $H_p(10)/E$ ranges between 0.4 and 1.8 except for the PA incidence, which is an exceptional situation for H_p. In general, $H_p(10)$ underestimates the effective dose E below 50 keV neutron energy. In Figure 10.14 the ratio $H_p(10)/H_E$ exhibits a spread between 1 and 3.4, again except at PA. Consequently, $H_p(10)$ could replace the effective DE H_E as a limiting quantity. However, there exists no advantage in calculating $H_p(10)$ instead of H_E or E, because the degree of complexity of the computations is the same for each quantity.

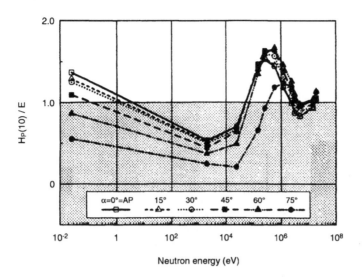

FIGURE 10.11 Ratio of personal dose equivalent, $H_p(10)$, in the ICRU slab phantom to effective dose E in the ADAM phantom for six angles of incidence. The shaded area exhibits the region of non-conservativity. (From Reference [7]. With permission.)

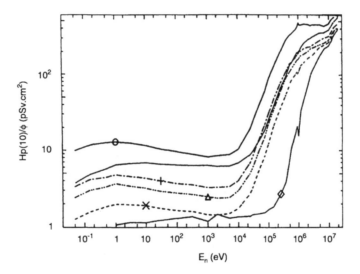

FIGURE 10.12 The personal dose equivalent (DE), $H_p(10)$, in the trunk of the male MIRD phantom for five irradiation geometries: (O) AP, ($\Diamond$) PA, (X) lat, (+) Rot, ($\triangle$) Iso. The dotted curve represents the effective DE, H_E, at AP in the same phantom. (From Reference [8]. With permission.)

The definition of the term "Computational Dosimetry" was presented by Siebert and Thomas [9], interpreted as the sub-discipline of computational physics, which is devoted to radiation metrology. Computational simulations directed at basic understanding and modeling are important tools provided by computational dosimetry, while another very important application is the support that it can give to the design, optimization, and analysis of experiments. The primary task of computational dosimetry is to reduce the variance in the determination of absorbed dose. It is essential for the analysis and organization of the complex body of data needed in dosimetry, and it provides models, evaluations, and concepts for this purpose. The ability to use numerical means to simulate

practical situations is one of the most powerful tools of computational dosimetry: experimental or clinical setups may be represented; particle spectra and dose distributions may be determined; and response functions (or matrices) for instruments and conversion coefficients may be calculated. Computational dosimetry greatly facilitates the optimization of experiments and supplements the analysis of the data. In appropriate cases computational dosimetry can even replace experiments.

The absorbed dose in tissue, D_t, can be estimated from the absorbed dose in A-150 plastic, D_p, with $D_t = \alpha_{t,p} D_p$, where $\alpha_{t,p}$ denotes the A-150 plastic-to-tissue absorbed dose conversion factor. [10] In monoenergetic neutron fields, and under charged particle equilibrium conditions,

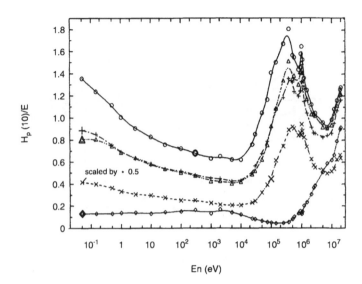

FIGURE 10.13 Ratio of the personal *DE*, $H_p(10)$, to the effective dose *E*, both in phantom ADAM for five neutron incidences. Symbols as in Figure 10.12 apply. (From Reference [8]. With permission.)

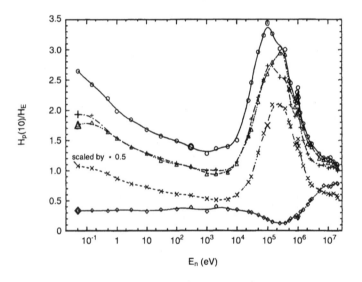

FIGURE 10.14 Ratio of the personal *DE*, $H_p(10)$, to the effective *DE*, H_E, both in phantom ADAM for five neutron incidences. Symbols as in Figure 10.12 apply. (From Refernce [8]. With permission.)

the conversion factor can be calculated from the kerma factors of the individual elements:

$$\alpha_{t,p} = \left(\sum_i t_i K_i\right)\bigg/\left(\sum_i p_i k_i\right) \qquad (10.12)$$

where t_i and p_i denote the elemental mass fraction tissue and A-150 plastic, respectively calculated values of $\alpha_{t,p}$ are shown in Figure 10.15.

Ambient dose equivalent $H^*(d)$ is defined as $H = \overline{Q}D$ at the reference point inside the ICRU sphere, with $\overline{Q}$ given by:

$$\overline{Q} = \frac{1}{D}\int_0^x D_L Q(L)dL \qquad (10.13)$$

where D_L denotes the dose distribution in L and $Q(L)$ is the quality factor. The TEPC (Tissue Equivalent Proportional Counter) can be used to estimate D_L. While the different compositions of the TEPC wall (A-150 plastic) and standard tissue influence D and may be converted with $\alpha_{t,p}$, their influence on D_L requires further consideration.

MCNP Monte Carlo calculations with an extended cross-section data set have been performed by Mares et al. [11] for an anthropomorphic MIRD (ADAM) phantom in the neutron energy range between 20 and 100 MeV. The irradiation conditions varied from 0° (frontal incidence) up to 75° to the left side of the MIRD phantom. For the calculation of organ dose equivalent $H_{T,q}$ (the index q

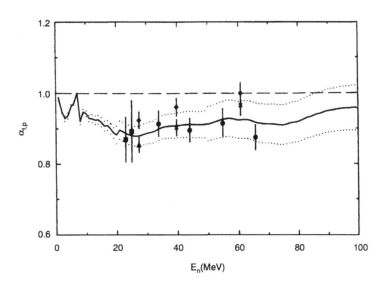

FIGURE 10.15 Absorbed dose conversion factor $\alpha_{t,p}$ for A-150 plastic to ICRU standard tissue. The solid line is based on evaluated kerma factor data. The dotted lines correspond to a 68% confidence level. The data points are derived from experimental kerma factors of C and O that have been published for $E_n > 20$ MeV. The symbols refer to experimental works cited elsewhere. (From Reference [10]. With permission.)

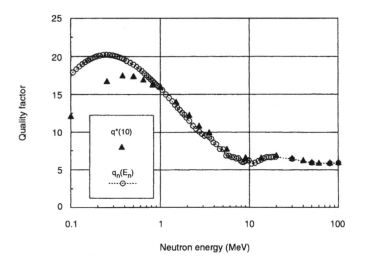

FIGURE 10.16 Neutron quality factor and $q^*(10)$ in the ICRU sphere as function of neutron energy. Above 20 MeV, data for $q^*(10)$ of Sannikov et al. [60] are used as an approximation for the quality factor q_n. (From Reference [11]. With permission.)

indicates that the quality factor is applied) a quality factor q_n for neutrons was applied which takes into account the revised $Q(L)$ relationship of ICRP 60 and the new stopping power data for protons and alpha particles in water of ICRU 49. (Mares et al. [11])

$$H_{T,q}(E) = \int q_n(E')k_n(E')\Phi_{T,n}(E',E)\,dE'$$
$$+ \int k_{ph}(E'')\Phi_{T,ph}(E'',E)\,dE'' \qquad (10.14)$$

where

$q_n(E')$ is the quality factor as a function of neutron energy E'

$k_n(E')$ is the fluence-to-kerma conversion factors as a function of neutron energy E', depending on the elemental composition of the tissue in organ T.

$k_{ph}(E'')$ is the fluence-to-kerma conversion factor as a function of photon energy E''

$\Phi_{T,n}(E',E)$ and $\Phi_{T,ph}(E'',E)$ are the neutron and photon fluences in the organ.

For the calculation of the equivalent organ dose H_T (according to ICRP60), the smooth radiation weighting factor function W_R was applied:

$$H_T(E) = w_R(E) \times$$

$$\left[\int k_n(E')\Phi_{T,n}(E', E)\, d\acute{E} + \int k_{ph}(E'')\Phi_{T,ph}(E'',E)\, dE'' \right]$$

(10.15)

Effective dose E and effective dose equivalent H_E were calculated by use of the new organ weighting factors w_T of ICRP 60 where

$$E = \sum_T w_T H_T$$

and

$$H_E = \sum_T w_T H_{T,q}$$

Figure 10.16 shows the quality factor q_n and the ratio $q^*(10) = H^*(10)/D^*(10)$ at 10 mm in the ICRU sphere as a function of neutron energy.

The measurement of neutron fluence-to-kerma conversion coefficients, (commonly called kerma factors) $(k(E))$, was performed by Newhauser et al. [12]; these factors are shown in Figure 10.17.

A comprehensive set of kerma factor values for H, C, O, N, and Si, as well as for tissue and frequently used high-energy particle detectors, was given by Gorbatkov et al. [13] Assuming charged in equilibrium at neutron energies below 100–150 MeV, the absorbed dose rate $(D, \text{Gy s}^{-1})$ can, with satisfactory accuracy, be found from the following relation:

$$D = \int F(E)K_f(E)\, dE$$

(10.16)

where $F(E)$ is the neutron spectrum (neutron cm^{-2} MeV^{-1} s^{-1}) and $K_f(E)$ is the neutron kerma factor (Gy cm^2). The neutron kerma factor is defined as the average kinetic energy released in matter (per unit mass and per unit neutron fluence) and is the sum of all energy transferred to light charged particles and residual nuclei in an elastic or nonelastic nA-reaction. For composite materials, the neutron kerma factor can be expressed as

$$K_f(E) = N_A \sum_i \gamma_i \sigma_i E_i = \sum_i \rho K_{fi}(E) \quad (10.17)$$

where N_A is Avogadro's number, γ_i is the atomic density of the ith isotope, σ_i is the charged particle production cross section ($\sigma_i = \sigma_i^{non} + \sigma_i^{el}$), E_i is the

TABLE 10.4
Neutron Kerma Factors for H, C, O, N, Si (fGy m^2)

E (MeV)	H	C	O	N	Si
15	48.1	2.06	1.31	1.5	1.21
20	47.0	3.10	1.85	2.04	1.53
25	45.85	3.41	2.08	2.46	1.74
30	44.51	3.64	2.37	2.78	2.02
40	42.92	4.0	3.0	3.62	2.51
50	40.75	4.39	3.67	4.44	3.08
60	39.04	4.97	4.4	5.24	3.62
70	37.72	5.49	4.96	5.9	4.16
80	36.74	6.03	5.6	6.58	4.73
90	36.05	6.69	6.3	7.2	5.33
100	35.58	7.54	6.94	8.0	5.86
120	35.23	9.03	8.35	9.34	6.93
150	35.82	11.05	10.2	11.34	8.48

From Reference [13]. With permission.

average kinetic energy of all charged particles produced by monoenergetic neutrons when interacting with a nucleus of the ith isotope, $K_{fi}(E)$ is the neutron kerma factor for the ith isotope, and ρ_i is the ith isotope fraction in the mixture. Comparison of calculated and measured values of neutron kerma for ^{12}C, ^{16}O, and ^{14}N are shown in Figures 10.18–10.20, respectively. Neutron kerma values for some elements are given in Table 10.4 and for some materials in Table 10.5. Calculated values for kerma factor are shown in Figure 10.21.

III. GAS-FILLED NEUTRON DOSIMETERS

Tissue-equivalent proportional counters became the popular neutron dosimeters. The newer dosimeters were developed from the well-known microdosimeters. Still, there are limitations to those dosimeters' applications. The accuracy achievable in radiation-protection dosimetry of mixed neutron-photon radiation fields still needs improvement. The area monitors are heavy in weight because of the amount of moderator needed for fast neutron dosimetry. Also, their dose equivalent response has a strong energy dependence. Measurement of dose due neutrons of energy above 15 MeV is practically impossible with a gas-filled dosimeter. Some later developments and investigations of the gas-filled detectors are discussed in this section.

The original motivation to develop a tissue-equivalent proportional counter (TEPC)-based device came from the fact that a quantity can be derived from a TEPC that is similar to H. H can be written as

$$H = \int Q D_L\, dL$$

(10.18)

with the quality factor Q defined as a function of L and the differential distribution, D_L, of the absorbed dose with

TABLE 10.5
Neutron Kerma Factors for Composite Materials (fGy m²)

E (MeV)	Tissue	CH	NaI	CsI	BaF₂	CeF₃	BGO	LS	PbWO₄
15	6.12	5.6	0.08	0.02	0.46	0.62	0.20	0.36	0.18
20	6.55	6.45	0.21	0.08	0.52	0.69	0.36	0.50	0.27
25	6.66	6.67	0.35	0.16	0.59	0.78	0.44	0.58	0.33
30	6.78	6.75	0.47	0.25	0.68	0.89	0.56	0.69	0.41
40	7.16	6.99	0.77	0.47	0.91	1.15	0.83	0.94	0.62
50	7.52	7.19	1.08	0.72	1.21	1.49	1.13	1.23	0.85
60	7.98	7.59	1.40	0.99	1.50	1.82	1.44	1.54	1.11
70	8.35	7.97	1.72	1.24	1.82	2.18	1.72	1.82	1.35
80	8.82	8.39	2.02	1.49	2.11	2.51	2.01	2.12	1.59
90	9.37	8.95	2.29	1.72	2.38	2.80	2.26	2.40	1.82
100	9.93	9.70	2.57	1.94	2.66	3.12	2.54	2.70	2.06
120	11.2	11.0	3.15	2.42	3.28	3.79	3.12	3.32	2.58
150	12.9	13.0	3.97	3.08	4.08	4.68	3.95	4.16	3.28

From Reference [13]. With permission.

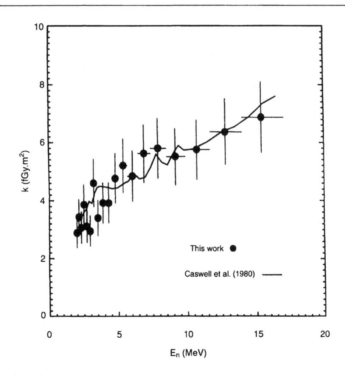

FIGURE 10.17 Neutron fluence-to-kerma coefficients k in A-150 plastic as a function of neutron energy E_n from a proportional counter measurement that was iteratively unfolded. Results are normalized to the value of k at 12 MeV from Caswell et al. [61] (From Reference [12]. With permission.)

respect to L. Lineal energy, y, defined as the ratio of the energy imparted to a volume (to the sensitive volume of the detector) and the average chord length in that volume, is closely related to L. [14] Various procedures can be used to determine the dose equivalent reading of a TEPC, M_H. One method is:

$$M_H = \int q D_y \, dy \qquad (10.19)$$

with a weighting function q which approximates the quality factor Q, e.g., using $q = Q$ and setting $y = L$. A corresponding reading for the mean quality factor can be obtained from

$$M_Q = M_H \bigg/ \int D_y \, dy \qquad (10.20)$$

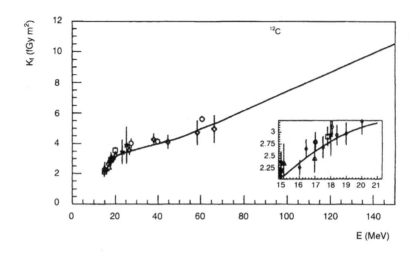

FIGURE 10.18 Comparison between experimental (symbols) and calculated (curve) neutron kerma factors for carbon. (From Reference [13]. With permission.)

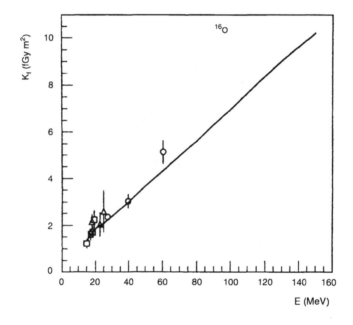

FIGURE 10.19 Comparison between experimental (symbols) and calculated (curve) neutron kerma factors for oxygen. (From Reference [13]. With permission.)

An instrument for the direct measurement of dose equivalent for any type of radiation seems to be feasible if the relation between L and y, or more precisely, between D_L and D_y, are known.

The neutron interaction with a person's body greatly influences the radiation field at a point of interest. In fact, for neutron energies below about 10 keV, H is dominated by the contribution of neutrons thermalized in the body while the contribution of the primary neutrons is negligible. The "operational quantities," such as the ambient dose equivalent, penetrating $H^*(10)$ are therefore defined at a reference point in a 30-cm-diameter spherical phantom.

Each instrument must therefore be adjusted to achieve an energy-independent dose equivalent response R_H.

The energy dependence of three typical detector systems—a thick-walled TEPC (KFA), a thin-walled TEPC (BIO), and a moderator-type dose equivalent rate meter (Leake)—is shown in Figure 10.22. The systems were calibrated in the radiation field of a D_2O-moderated Cf source in terms of the neutron ambient dose equivalent. While R_H varies for Leake by a factor of 20, the factor for the other two systems is 4 and 5, respectively. The reasons for the energy dependence of the TEPC systems are well understood: for the thick-walled TEPC, the low

TABLE 10.6
Technical Properties of TEPC-based Radiation Protection Dosimeters

	CIRCE	CIRCG	EG&G	HANDI	KFA	PNL	REM-402
m (kg)	5	10	0.6	6	7[a]	2	2
V (l)	6	30	0.8	18	10[a]	1.5	6.4
V_s (cm³)	98	27	109	110	270	1	97
T (h)	–	–	40	24	–	240	17
Price[b]	25000	16000[c]	2000	10000[d]	40000	–	5000
Application[e]	S	A	P	A	A	S	A

m = total mass of the system (detector and electronics)

V = total volume of the system (detector and electronics)

V_s = sensitive volume of detector

T = minimum time of operation with internal power supply (if available)

[a]Excluding a personal computer, required for data analysis.
[b]Approximate price in US$.
[c]Excluding the detector.
[d]A small number of prototypes were produced on a net-cost basis.
[e]Main area of application, A: area monitoring, P: personnel monitoring, S: space dosimetry.

From Reference [14]. With permission.

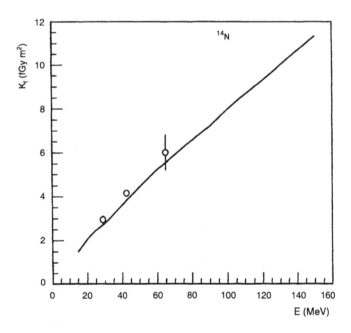

FIGURE 10.20 Comparison between experimental (symbols) and calculated (curve) neutron kerma factors for nitrogen. (From Reference [13]. With permission.)

R_H at about 100 keV is mainly explained by the neutron fluence attenuation in the detector wall, and for the thin-walled TEPC, the low R_H below 50 keV is mainly caused by the insufficient thermalization of neutrons and the short range of the recoil protons produced.

The main properties of systems which have been developed for and applied in radiation protection dosimetry are listed in Table 10.6. The prices listed are approximate values given by the manufacturers and compose the complete systems. Depending on the intended field of application, very differing detector sizes were employed.

A spherical proton recoil proportional counter was used by Weyrauch and Knauf [15] to measure neutron fluences. The response functions of that counter have recently been calculated with sufficient accuracy, taking into account gas amplification and wall effects. Neutron scattering from the

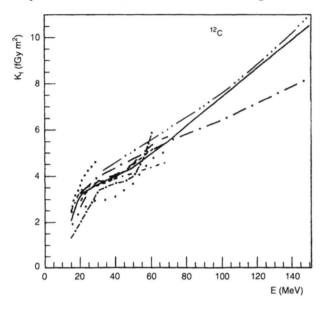

FIGURE 10.21 Calculated neutron kerma factors for carbon as a function of incident neutron energy from 15 to 150 MeV. The figure shows kerma factor values of Gorbatkov et al. (solid curve), the results of Morstin et al. [62] (—·—··—··—··—·), and the results of Savitskaya et al. [63] (—·—·—·). (From Reference [13]. With permission.)

counter walls was estimated using the MCNP code. Significant progress has been made in quantitatively understanding the response of the spherical counter, and it could be shown that including the variation of gas amplification along the counting wire, as a result of the variation of the electric field strength, is sufficient to explain experimental data. A comparison between calculated and measured response functions is shown in Figure 10.23 for an incident neutron energy of 144 keV (filtered reactor beam). In the proton energy range from about 25 keV to 100 keV, the shape of the calculated and measured responses agree on the 0.1% level.

The measured pulse height distribution $P(E_p)$ is related to the spectral neutron fluence $\Phi(E)$ by

$$P(E_p) = N \int_0^\infty R(E_p - E_n, E_n)\sigma(E_n)\Phi(E_n)\, dE_n \quad (10.21)$$

where E_p is the recoil proton energy, E_n is the energy of the incident neutron, N denotes the total number of protons in the counter, and $\sigma(E_n)$ is the neutron-proton total cross section. The response function $R(E_p - E_n, E_n)$ is normalized, so that the total number of recoil protons Z is obtained from

$$Z = \int_0^\infty P(E_p)\, dE_p$$

$$= N \int_0^\infty \sigma(E_n)\Phi(E_n)\, dE_n \quad (10.22)$$

To obtain the absolute fluence distribution, Equation 10.22 should be followed.

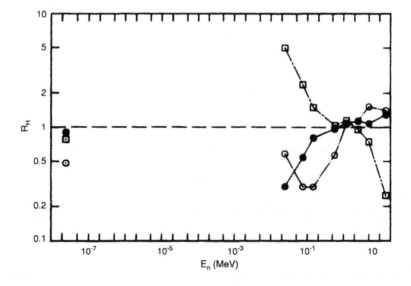

FIGURE 10.22 Ambient dose equivalent response, R_H, for monoenergetic neutrons as a function of neutron energy, E_n, for two TEPC systems (BIO: full circles; KFA: open circles) and a moderator-type neutron dose equivalent meter (Leake: squares). The TEPC systems were calibrated in a ^{252}Cf(D$_2$O) field. The lines serve as eye guides. The results were obtained within the framework of an intercomparison. (From Reference [14]. With permission.)

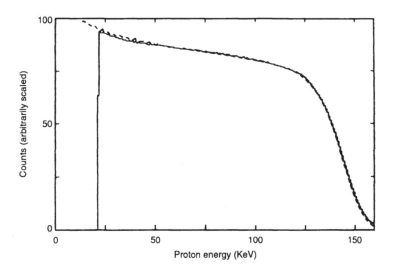

FIGURE 10.23 Measured pulse-height distribution (full line) compared with the calculated response (dashed line). The SP2 counter was exposed to reactor neutrons filtered through 1372 mm Si and 60 mm Ti, which produces a nearly monoenergetic neutron line at 144 keV. Both curves are normalized to unity. (From Reference [15]. With permission.)

If the relative spectral distribution of neutrons $\lambda(E_n)$ is known, Equation 10.22 can be used to obtain the fluence $\Phi(E_n) = \Phi_0 \lambda(E_n)$. In this case, only Φ_0 remains to be determined. Normalizing $\lambda(E_n)$, we obtain

$$\int_0^\infty \lambda(E_n)\, dE_n = 1 \qquad (10.23)$$

and, from Equation (10.22),

$$\Phi_0 = \frac{Z}{N \int_0^\infty \sigma(E_n) \lambda(E_n)\, dE_n} \qquad (10.24)$$

The principle of a multi-element tissue-equivalent proportional counter dosimeter was proposed by Kliauga. [16] It consists of 200 elements (right cylinders, diameter = height = 3.6 mm) in 25 groups. Each element corresponds to a microdosimetric proportional counter, and each group has a common wire anode. The counter radiation cross section is 5.3 times that of a spherical counter with the same volume. The mechanical structure is very complex, as is shown in the corresponding schematic diagram (Figure 10.24).

A multichannel tissue-equivalent proportional counter MC-TEPC dosimeter is based on a microdosimetric tissue-equivalent proportional counter. [17] The sensitivity of this detector is enhanced by increasing the counter radiation cross section. The 8-mm-thick cathode is drilled with some 250 4-mm-diameter holes, and electron swarms, created by interactions of secondary charged particles in the counting gas, are drifted along the holes' axis to 10 high-potential wire anodes by an electrokinetic field. This electric field is obtained by a bias current flowing from the internal to the external race of the cathode.

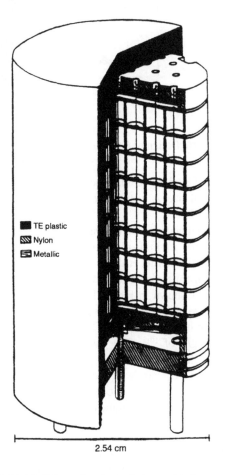

TE plastic
Nylon
Metallic

2.54 cm

FIGURE 10.24 Schematic view of the multi-element tissue-equivalent proportional counter proposed by Kliauga. (From Reference [17]. With permission.)

Electrons are multiplied around and collected by the wire anodes. The sensitivity of the detector is between 5 and 7

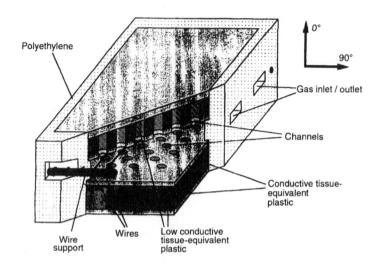

FIGURE 10.25 Schematic view of the multi-channel tissue-equivalent proportional counter (MC-TEPC) developed at the SDOS.(From Reference [17]. With permission.)

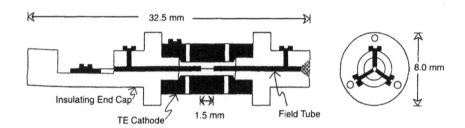

FIGURE 10.26 Design of 0.0027-cm³ TEPC. (From Reference [18]. With permission.)

times that of a multi-cellular and spherical counter of the same external size. A view of the counter interior is shown in Figure 10.25.

Paired miniature tissue-equivalent proportional counters for dosimetry in high-flux epithermal neutrons were described by Burmeister et al. [18] A pair of miniature TEPCs has been constructed with sensitive volumes of 0.0027 cm³. The design is shown in Figure 10.26. The cathode of one TEPC is made of A-150 conducting tissue-equivalent plastic. The cathode of the paired TEPC is made of A-150 loaded with 200 μg/g of ^{10}B. The gas fill is a propane-based tissue equivalent (TE) counting gas. Propane-based tissue-equivalent (p-TE) gas, which consists of 54% propane, 40.5% CO_2, and 5.5% N_2, is one of the two TE gases most commonly used in microdosimetry ionization chambers and in low-pressure proportional counters.

The mean energy required to form an ion pair after complete dissipation of incident particle energy, W, for electrons, protons, alpha particles, and heavy ions (^{12}C, ^{14}N, ^{16}O) in propane-based tissue-equivalent gas was summarized and analytically represented by Bronic. [19] The energy dependence of the W value for electrons can be described as:

$$W(E) = \frac{W_\infty}{1 - (U/E)} \qquad (10.25)$$

Here, W_∞ is the asymptotic high-energy electron W value, E is electron energy, and U is the average kinetic energy of sub-ionization electrons.

The experimental data for heavier particles can be fitted using the function:

$$W(E) = \frac{A}{[\ln(E + B)]^C} + D \qquad (10.26)$$

where E is the energy of the particle in keV amu^{-1} and A, B, C, and D are fitting parameters. W values for electrons in propane-based TE (p-TE) gas were fitted using Equation (10.25) (Figure 10.27).

The W value for He ions shows somewhat stronger energy dependence in propane than in p-TE gas (Figure 10.28). The average ratio of the W values in p-TE gas and those in propane is 1.0, which is lower than the corresponding ratio for electrons and protons (1.07), as well as for heavy ions.

The neutron sensitivity of a C-CO_2 ionization chamber has been experimentally determined by Endo et al. [20] in the neutron energy range of 0.1–1.2 MeV. Eleven k_U data were obtained which ranged from 0.04 to 0.1 as a function of neutron energy. There were two peaks in the spectrum of the k_U factor and their positions corresponded to the resonance energies of the kerma factors

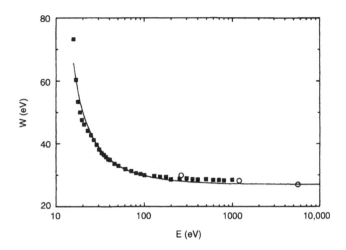

FIGURE 10.27 W value for electrons in propane-based TE gas. Experimental data: (■), Combecher; [64] (○), Krajear Bronic et al.; [65] (—) fitted line, Equation (10.25). (From Reference [19]. With permission.)

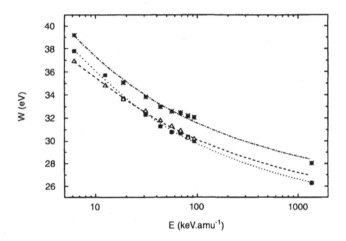

FIGURE 10.28 W value for alpha particles in propane and propane-based TE gas. Experimental data: (■), propane; (●), propane-based TE gas, fitted lines, Equation 10.26; (...), propane; (– –), p-TE, experimental data; (–.–), p-TE, calculated values. (From Reference [19]. With permission.)

of carbon dioxide, of which the energies are about 0.4 and 1 MeV.

The paired-chamber method is the most reliable technique to separately determine kerma values in neutron and gamma-ray mixed fields. Usually, one of the pair is a TE-TE chamber which is made with an A150 plastic wall and filled with tissue-equivalent gas. This chamber determines the sum of neutron and gamma-ray kermas. The other chamber of the pair is a C-CO$_2$ chamber which is made with a graphite wall and is filled with carbon dioxide gas. This chamber essentially determines the gamma-ray kerma except for a small response for neutrons. The neutron kerma is approximated by subtracting the gamma-ray kerma from the sum of the gamma-ray and the neutron kerma. In actuality, the effect of its small neutron sensitivities for the C-CO$_2$ chamber (k_U factor) must be considered, since the chamber has a response of about 0.1–10%, depending on

the neutron energy. [20] The k_U factors are approximately proportional to the gas kerma factor.

The neutron sensitivity of the C-CO$_2$ chamber is denoted by (k_{U-c}). The k_{U-c} factors range from 0.04 to 0.1 and are dependent on the neutron energy, as shown in Table 10.7. The neutron sensitivity of a C-CO$_2$ chamber is roughly proportional to the kerma ratio (K_{CO_2}/K_t) where K_{CO_2} in the gas and K_t in the tissues. It is also proported to the ratio $\overline{W}_{Co}/\overline{W}_n$, where Co means the calibration field of ^{60}Co and n means the neutron field. The cavity is assumed sufficiently large compared to the ranges of the recoil particles. The energy deposited in the gas is due only to charged particles liberated in the gas. The k_{U-c} factor is expressed by

$$k_{U-c} \approx \beta(K_{CO_2}/K_t)(\overline{W}_{Co}/\overline{W}_n) \qquad (10.27)$$

where β is a normalization constant which may be related to the effective kerma component ratio of the graphite wall to the carbon dioxide gas.

Recombination chambers are high-pressure tissue-equivalent ionization chambers, designed and operated in such a way that ion-collection efficiency in the chambers is governed by initial recombination of ions.

TABLE 10.7
The Results for the k_U Factor as a Function of the Neutron and the Proton Energies

Proton energy (MeV)	Neutron energy (MeV)	k_U factor
3.0	1.2	0.095 ± 0.008
		0.100 ± 0.009
2.9	1.1	0.100 ± 0.008
2.8	0.99	0.110 ± 0.009
		0.120 ± 0.010
2.7	0.90	0.086 ± 0.007
2.6	0.78	0.076 ± 0.006
		0.076 ± 0.010
2.5	0.69	0.072 ± 0.006
2.4	0.57	0.072 ± 0.006
		0.074 ± 0.010
2.3	0.48	0.076 ± 0.006
2.2	0.37	0.075 ± 0.009
		0.089 ± 0.010
2.1	0.25	0.045 ± 0.008
2.0	0.11	0.038 ± 0.010
		0.041 ± 0.011

The uncertainty of the k_U factor value is the standard deviation (68% confidence interval) of the mean for several measurements.

From Reference [20]. With permission.

A large chamber of REM-2 type was filled to a pressure of about 1 MPa with a gas mixture consisting of methane and 5% (by weight) nitrogen. [21] The output signal of the chamber is the ionization current (or collected charge) as a function of collecting voltage (Figure 10.29). All the recombination methods require the measurement of the current (or charge) at at least two values of the collecting voltage. The highest voltage should provide the conditions close to saturation; i.e., almost all the ions generated in the chamber cavity should be collected. In practice, the maximum voltage that can be applied to the chamber without causing a considerable dark current or other unwanted effects is used.

Depending on the number of measured points of the saturation curve (see Figure 10.29), one can obtain different information:

a. The ionization current measured at maximum applied voltage is proportional to the absorbed dose D (some small corrections for lack of saturation can be introduced when needed).

b. Measuring at two voltages enables determination of the recombination index of radiation quality, Q_4, that approximates the quality factor. Then the ambient dose equivalent, $H^*(10)$, can be determined as $H^*(10) = DQ_4$.

c. The measurement of ionization current at six collecting voltages and mathematical analysis of the saturation curve enable separation of low-LET and high-LET dose fractions and determination of quality factor and $H^*(10)$.

d. Measuring at 15 or more collecting voltages enables determination of crude microdosimetric spectra terms of dose distributions in LET. The method is based on similar but more detailed mathematical analysis than methods used when

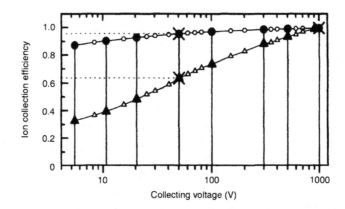

FIGURE 10.29 Example of saturation curves measured with REM-2 recombination chamber. Circles, ^{137}Cs reference gamma radiation; triangles, 2.5-MeV neutrons; crosses show two points needed for determination of ICRP 21 quality factor; solid points are needed for measurements of $H^*(10)$ with ICRP 60 quality factor. and all the points are used for microdosimetric distributions. (From Reference [21]. With permission.)

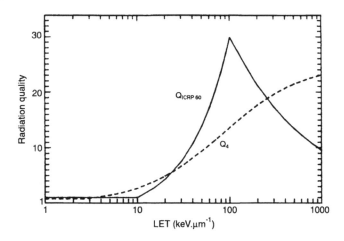

FIGURE 10.30 Dependence of Q_4 and quality factor (ICRP Report 60) on LET. (From Reference [22]. With permission.)

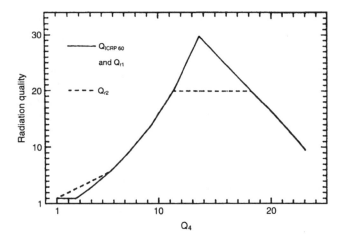

FIGURE 10.31 The relationship between Q_4 and $Q_{ICRP\ 60}$ for radiations with a single value of LET. The same curve is used for the dependence of Q_{r1} on Q_4. Dashed lines represent the differences between Q_{r1} and Q_{r2}. (From Reference [22]. With permission.)

measurements are performed at six points. Somewhat better accuracy can be achieved in determination of $H^*(10)$; however, the time needed for measurements is longer and the mathematical analysis requires some experience. [21]

The recombination index of radiation quality (*RIQ*), denoted as Q_4, is a measurable quantity that directly approximates the radiation quality factor. Determination of Q_4 involves two steps:

1. calibration of the chamber in a reference gamma radiation field from a ^{137}Cs source; during the calibration, the recombination voltage U_R is determined in such a way that the ion-collection efficiency $f_\gamma(U_R) \approx 0.96$; and
2. determination of ion-collection efficiency, $f_{mix}(U_R)$. The investigated radiation field

requires measurements at two voltages, as indicated by crosses in Figure 10.29.

Q_4 is then calculated according to its definition as:

$$Q_4 = \frac{1 - f_{mix}(U_R)}{1 - f_\gamma(U_R)} \qquad (10.28)$$

Because $f_R(u_R) \approx 0.96$,

$$Q_4 = (1 - f_R)/0.04 \qquad (10.29)$$

A recombination chamber was used by Golnik and Zielczynski [22] for determination of Q_4. It was shown that both Q_4 and the ICRP 21 quality factor had a similar dependence on LET. Q_4 is plotted against LET in Figure 10.30 as a dashed line. Since ICRP 60 introduces a new functional relationship between the quality factor and LET (solid line in Figure 10.30), it is not possible to use Q as a direct approximation of the ICRP 60 quality factor. The relationship between the two is shown in Figure 10.31.

The recombination index of radiation quality (RIQ) was determined by Golnik et al. [23] in high-energy neutron beams from the 660-MeV phasotron of the Joint Institute for Nuclear Research, Dubna (JINR). Measurements were performed with a high-pressure ion chamber at two collecting voltages and compared with calculations considering spectra of all secondary and tertiary charged particles.

Measurements of $Q_R(RIQ)$ were performed using a 3.8-cm³ in-phantom recombination chamber of F-l type (Figure 10.32), filled with TE gas up to 500 kPa. The disc-shaped chamber has three parallel-plate tissue-equivalent electrodes, 34 mm in diameter (total diameter of the chamber is 62 mm). The distance between electrodes is 1.75 mm. The sensitivity of the chamber is about 350 nC Gy⁻¹. Its operational dose rate range is from 10^{-3} up to 1.5 Gy s⁻¹. Two collecting voltages, $U_R = 30$ V and $U_s = 800$ V, were sequentially applied to the chamber electrodes. RIQ was determined as:

$$Q_R = \frac{1}{R}\left(1 - \frac{i(30V)}{i(800V)}\right) \qquad (10.30)$$

where i is the ionization current of the recombination chamber related to the monitor reading, $R = 1 - f_\gamma(U_R)$, and $f_\gamma(U_R)$ is the ion-collection efficiency at collecting voltage U_R, determined in the gamma radiation field of the reference ^{137}Cs source. In experiments described here, $R \approx 0.04$.

IV. TLD NEUTRON DOSIMETRY

Using TL detectors for the dosimetry in mixed neutron-gamma fields requires knowledge of their response to neutrons over a broad energy range. For the calculation of the neutron fluence response, the conversion of the energy deposited by the secondary heavy charged particles to the TL reading effect must be considered. For this purpose, the light conversion factors for all secondaries as a function of their energy must be known. The relative light conversion factor is the ratio of the detector reading coming from a mass element where the dose was absorbed as a result of the deposition of the ion. The energy deposition by secondary ions and gamma radiation generated as a result of neutron interaction in the detector material or in the surrounding material (detector holder, radiator) is of fundamental importance.

Thermoluminescent materials such as LiF(TLD-100), ^{7}LiF(TLD-700, ^{6}LiF(TLD600), and CaF$_2$(TLD-300) show curves (*TL* intensity as a function of temperature, T) in which the different peaks exhibit a different dependence on the LET of the radiation. This allows the separate determination of photon and neutron doses in a mixed (n,γ) radiation field with the two-peaks method.

The thermal neutron response of any *TL* materials depends on the capture cross section of the elements of those materials. Boron and lithium have large cross sections for thermal neutrons. All neutron reactions in LiF used for the determination of the intrinsic response component are summarized in Table 10.8.

The energy deposited yields fluence-to-kerma conversion factors. Figure 10.33 shows the intrinsic fluence-to-kerma conversion factor for different materials. While in the low neutron energy region, the reaction ^{6}Li$(n,\alpha)^3$H dominates; i.e., the different contents of ^{6}Li cause different factors, and for high-energy neutrons, the polyethylene reaction has the similar contribution. The additional K/Φ component from the 1-mm-thick

TABLE 10.8
Selected Neutron Interactions in LiF Materials

Reaction	Q value (MeV)
Elastic scattering ^{6}Li	—
Inelastic scattering ^{6}Li	-1.47128
^{6}Li$(n,\alpha)^3$H	4.786
^{6}Li$(n, p)^6$He	-2.7336
Elastic scattering ^{7}Li	—
Inelastic scattering ^{7}Li	-0.477484
^{7}Li$(n,d)^6$He	-7.76382
Elastic scattering ^{19}F	—
Inelastic scattering ^{19}F	-0.11
^{19}F$(n,p)^{19}$O	-4.0363
^{19}F$(n,d)^{18}$O	-5.76892
^{19}F$(n,t)^{17}$O	-7.55613

From Reference [24]. With permission.

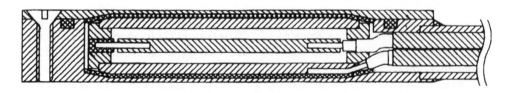

FIGURE 10.32 Cross section of the in-phantom chamber of F1 type (for historical reasons this chamber is sometimes called the KR-13 chamber). (From Reference [23]. With permission.)

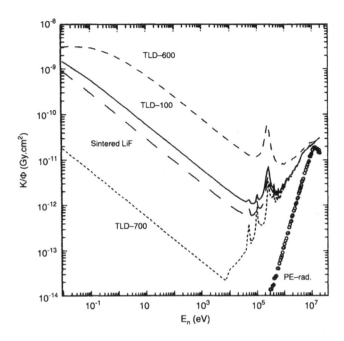

FIGURE 10.33 Kerma factors for LiF TLDs. Additional kerma in sintered LiF TLD caused by a 1-mm thick polyethylene radiator. (From Reference [24]. With permission.)

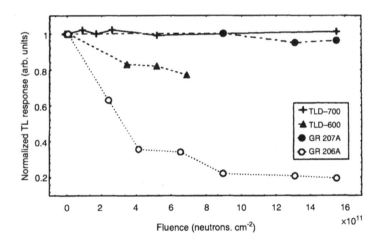

FIGURE 10.34 Comparison between commercial ^{6}Li-enriched and depleted LiF for γ-ray irradiation after thermal neutron exposure at different fluences. TL intensities are normalized to the response after a 1 Gy-γ irradiation. (From Reference [25]. With permission.)

PE radiator to the sintered LiF detector is included in Figure 10.33.

^{6}LiF-enhanced dosimeters exposed to high fluences of thermal neutrons undergo irreversible radiation damage, preventing their subsequent utilization. ^{6}LiF-depleted dosimeters do not show such an effect, but in mixed fields of high fluxes with thermal neutrons, they do not provide a simple way of discriminating among the contributions of the field components. The responses to thermal neutrons and to γ-rays of some TL single crystals have been investigated by Gambarini et al. [25] LiF:Mg,Cu,P and LiF:Cu^{2+} single crystals have shown a much higher sensitivity to thermal neutrons than to γ-rays.

Owing to the high cross section of ^{6}LiF for the reaction with thermal neutrons, ^{6}Li$(n,\alpha)^3$H, LiF dosimeters of different isotopic composition, exposed in thermal neutron fields, present different sensitivities and different shapes of glow curve. With regard to thermal neutrons, ^{6}LiF:Mg,Cu,P chips have shown a response about nine times higher than that of ^{6}LiF:Mg,Ti, linear in the low-dose range.

In Figure 10.34, the reduction factor of the response to γ-rays of TLDs exposed to thermal neutrons and afterward annealed is shown as a function of the total nth fluence under previous exposure. The results show that ^{6}LiF chips, after exposure to high fluences of thermal neutrons, depart from linearity, moderately for the Mg,Ti-doped phosphors and markedly for the Mg,Cu,P-doped ones; moreover, they reveal irreversible radiation damage which invalidates their new utilization after such exposures. In contrast, ^{6}LiF chips, doped with Mg,Ti or Mg,Cu, P, do not show such radiation damage in the fluence-investigated interval.

The different LET dependences of the low- and high-temperature glow peaks of CaF$_2$:Tm thermoluminescent material (TLD-300) allow the determination of neutron and gamma dose simultaneously.

To calibrate TLD-300 dosimeters, the individual sensitivities of the two glow peaks to neutrons and photons are experimentally determined. A calibration method was proposed by Kriens et al. [26] The response R_i' of the glow peak i to the mixed n-γ field relative to its sensitivity to the γ-rays used for calibration is given by:

$$R_i' = k_i D_N + h_i D_\gamma \qquad (10.31)$$

where $k_i =$ sensitivity of the glow peak i to neutrons, relative to its sensitivity to the γ-rays used for calibration, and $h_i =$ sensitivity of the glow peak i to photons in the mixed n-γ field, relative to its sensitivity to the γ-rays used for calibration.

With the assumption that the sensitivities of the TLD-300 glow peaks to the photons in the mixed n-γ field are equal to their sensitivities to the γ-rays used for calibration ($h_i = h_2 = 1$), the relative neutron sensitivities of the two peaks are given by: [26]

$$k_{1,2} = \frac{(R_{1,2}' - D_\gamma)}{D_N} \qquad (10.32)$$

Dividing Equation (10.31) by the total dose $D_N = D_T + D_\gamma$ yields:

$$\frac{R_{1,2}'}{D_T} = k_{1,2} + (h_{1,2} - k_{1,2})\frac{D_\gamma}{D_T} \qquad (10.33)$$

The linear regression of multiple values R_1'/D_T or R_2'/D_T and D_γ/D_T determines k_1 or k_2 and h_1 or h_2, if D_T and the ratios D_γ/D_γ are known for multiple measuring conditions. The variation of D_γ/D_T is performed by three procedures: different phantom depths, different field sizes, and additional exposure to ^{60}Co γ rays.

For all calibration methods used, the mean value of the relative sensitivities of the two TLD-300 glow peaks were found to be $k_i = (0.078 \pm 0.029)$ and $k_2 = (0.215 \pm 0.047)$ for fast neutrons, as well as $h_1 = (1.02 \pm 0.36)$ and $h_2 = (1.02 \pm 0.26)$ for photons of the mixed n-γ beam, comparable to recent measurement.

The relative neutron responses of both peaks in TLD-300 chips were found by Hoffmann et al. [27] to be 0.10

and 0.32. Using this method, various dose distributions of neutron and gamma dose in a water phantom were measured and compared with the results of GM counter measurements and Monte Carlo dose calculations.

The normalized TLD readings for corresponding gammas and neutrons of the low-and high-temperature glow peaks should be different according to [27]

$$P_{1N} = \frac{P_1}{(P_1)_{60\text{Co}}} = D_\gamma + (\varepsilon_1 K)D_n \qquad (10.34)$$

$$P_2 N = \frac{P_2}{(P_2)_{60\text{Co}}} = D_\gamma + (\varepsilon_2 K)D_n \qquad (10.35)$$

where $K =$ kerma ratio CaF$_2$/H$_2$O $= 0.27$ at 14 MeV. From these equations, the neutron, gamma, and total dose distribution are:

$$D_n = \frac{P_{2N} - P_{1N}}{(\varepsilon_2 K) - (\varepsilon_1 K)} \qquad (10.36)$$

$$D_\gamma = \frac{\dfrac{\varepsilon_2 K}{\varepsilon_1 K}P_{1N} - P_{2N}}{\dfrac{\varepsilon_2 K}{\varepsilon_1 K} - 1} \qquad (10.37)$$

$$D_{total} = D_n + D_\gamma \qquad (10.38)$$

where ε_1 is the sensitivity of the low-temperature peak and ε_2 is the sensitivity of the high-temperature peak. The larger the difference $(\varepsilon_2 - \varepsilon_1)$ and the higher the kerma ratio K, the higher the accuracy of the dose determination.

The CaF$_2$:Tm (TLD-300) material was used as chips of size 3 mm $\times$ 3 mm $\times$ 0.9 mm. Before irradiation, the chips were annealed at 400°C for 1 h. Each dosimeter was calibrated individually by ^{60}Co γ irradiation at an exposure level equivalent to 130 mGy (13 rad) in muscle tissue. [27]

After neutron irradiation, the glow curves show a very different shape as compared to those induced by γ irradiation (see Figures 10.35 and 10.36). Due to the high LET component, the response of the low-temperature peak 1 is largely reduced, whereas the response of peak 2 remains relatively high. The peak height ratio therefore reduces from about 3:1 for gammas to 1.1:1 for free-air neutron irradiations.

The resulting neutron responses can then be derived from Equations 10.34 and 10.35 to be

$$\varepsilon_2 K = 0.32, \qquad \varepsilon_1 K = 0.10$$

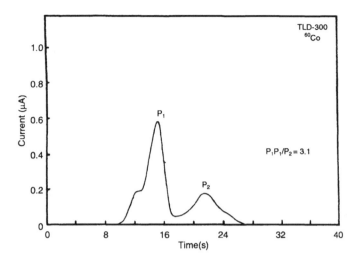

FIGURE 10.35 Glow curve of a TLD-300 dosimeter after ^{60}Co irradiation. (From Reference [27]. With permission.)

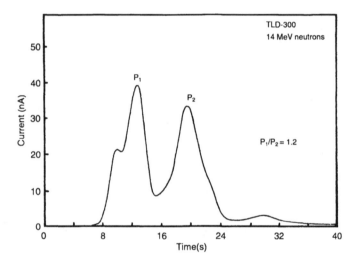

FIGURE 10.36 Glow curve of a TLD-300 dosimeter after irradiation in a 14-MeV neutron beam. (From Reference [27]. With permission.

The dose is then calculated according to

$$D_n = \frac{P_{1N} - P_{2N}}{0.22} \qquad (10.39)$$

$$D_\gamma = \frac{3.2 P_{1N} - P_{2N}}{2.2} \qquad (10.40)$$

$$D_{total} = D_n + D_\gamma$$

where P_{1N} and P_{2N} are the normalized TL readings of both peaks, measured as peak heights after irradiation in the neutron field.

TLD-300 detectors were used by Angelone et al. [28], allowing the neutron and gamma contribution to the total nuclear heating to be separated by using the two-peak method. The so-called peak 3 and peak 5 show different sensitivities to gamma and neutrons. TLD-300 was chosen because of its sensitivity to low gamma-ray dose and its high $Z_{eff}(Z_{eff} = 16.2)$.

A 14-MeV neutron generator produces up to 1.0×10^{11} s^{-1} 14-MeV neutrons by using the *D-T* fusion reaction. The total measured doses ranged from 4.9×10^{-19} Gy per neutron up to 1.9×10^{-14} Gy n^{-1}, which covers the whole linear response range for TLD-300.

The TLD-300 dosimeters were calibrated in a secondary standard gamma radiation field using a ^{60}Co source calibrated at 0.5% up to 3.5%, depending on the dose level. The calibration ranged from 10 μGy up to 35 Gy. Measuring the response of the two main peaks of TLD-300, it is possible to separate the neutron and gamma absorbed doses. The corresponding equation, accounting for the fact that the phosphors are embedded in stainless steel, is:

$$R_v = h_v D_\gamma^* + k_v C_n D_n \qquad (10.42)$$

TABLE 10.9
Data and Parameters Used to Solve Equations 10.43 and 10.44

Depth (cm)	$\bar{E}_n$ (MeV)	f_5	$\langle k_3 \rangle$	$\langle k_5 \rangle$	R_3 (Gy)	R_5 (Gy)
2.50 SS	7.40	1.65	1.09×10^{-1}	3.73×10^{-1}	9.38	17.75
9.51 SS	3.14	1.21	7.33×10^{-2}	2.58×10^{-1}	2.78	3.73
16.52 SS	1.93	1.08	6.24×10^{-2}	2.22×10^{-1}	1.09	1.21
23.40 SS	1.43	1.0	5.76×10^{-2}	2.06×10^{-1}	4.56×10^{-1}	4.85×10^{-1}

The quoted uncertainty is from 6% up to 8% for R_3 and from 7% up to 12% for R_5.

Source: From Reference [28]. With permission.

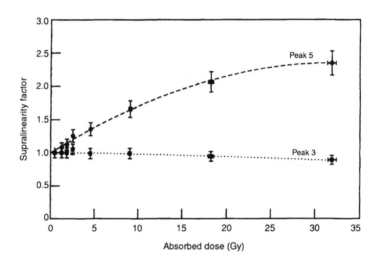

FIGURE 10.37 Measured supralinearity factor (f_ν) for peak 3 and peak 5 of TLD-300 plotted against absorbed dose (Gy). (From Reference [28]. With permission.)

where R_ν is the measured TLD response, normalized to a ^{60}Co gamma-ray source, for peak ν ($\nu = 3$ or 5), and k_ν and D_n are the neutron sensitivities (for peak ν) referring to the ^{60}Co gamma-ray and the neutron absorbed dose (in SS), respectively. C_n is the neutron-kerma ratio between the TLD-300 and the surrounding material (SS), D_γ^* is the gamma-ray absorbed dose (in SS), and h_ν is the photon sensitivity of peak ν in the mixed n-γ field, relative to the γ-rays used for calibrating the detectors.

The supralinear zone for peak 5 arises above 2 Gy. A study of the supralinear behavior of our TLD-300 detectors was carried out and the results are summarized in Figure 10.37.

Solving Equation 10.42, we get for D_γ and D_n:

$$D_\gamma = \frac{\left(R_5 - \frac{k_5}{k_3} R_3 \right)}{\left(f_5 - f_3 \frac{k_5}{k_3} \right)} \qquad (10.43)$$

$$D_n = \frac{(f_5 R_3 - f_3 R_5)}{C_n(f_5 k_3 - f_3 k_5)} \qquad (10.44)$$

Values are gathered in Table 10.9.

Dosimetry in a mixed n-γ field was measured by Piters et al. [29] The measured glow curves were analyzed according to the glow curve superposition method and according to the two-peak method. The peak heights were obtained by fitting the tops of the peaks to a Gaussian function. Figure 10.38 shows the results of a fit with the glow curve superposition method. The peaks used for the two-peak method are indicated by arrows. Corrected and uncorrected glow curve fits are shown in Figure 10.39. The temperature shifts dT_a and dT_b are on the order of 6°C.

The neutron sensitivities of the total response (k_T) as well as of separate peaks 3 (k_3) and 5 (k_5) on the glow curve were measured by Angelone et al. [30] for CaF$_2$:Tm (TLD-300) thermoluminescent dosimeters. The TLD-300 were encapsulated in A-150 TE plastic and located at different depths in the water phantom. The neutron sensitivities

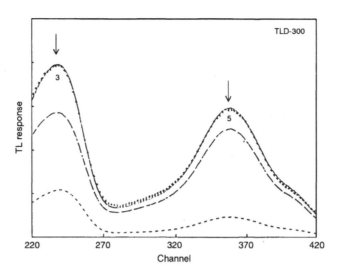

FIGURE 10.38 Glow curve of TLD-300 (+) after an irradiation in a mixed (n,γ) radiation field with actual dose $D_\gamma = 10$ mGy and $D_n = 92$ mGy ($E_n = 6$ MeV). The composed glow curve (solid line) is a superposition of the calibration curves I_a multiplied by $A = 0.085$ (dotted curve) and I_b multiplied by $B = 0.32$ (broken curve). The arrows indicate the peaks used in the two-peak method. Results for superposition method: $D_\gamma = 9.4$ mGy, $D_n = 92$ mGy. Results for the two-peak method: $D_\gamma = 9.6$ mGy, $D_n = 91$ mGy. (From Reference [29]. With permission.)

of the TLDs were derived by comparison with the results obtained with the twin detector method.

It is possible to use TLD-300 to separate the neutron and gamma dose, even for a complex neutron spectrum, provided that the neutron sensitivities of peak 3 and peak 5 are obtained by a proper weighting procedure of the monoenergetic data, using the calculated neutron spectra as weighting functions. This method requires the neutron spectrum to be known, and it is applicable for neutron energies lower than 25 MeV, since above this energy no data for the neutron sensitivity of TLD-300 are available.

TLD-300 shows a typical glow curve characterized by two main peaks, designated as peak 3 and peak 5 (Figure 10.40). The two peaks show different responses to different LET particles.

Angelone et al. found that there is a linear relationship between the ratio of the area under peak 3 to that under peak 5 (A_3/A_5) and the ratio of the gamma dose to the total dose (D_γ/D_T) (Figure 10.41).

Thermal neutron responses (TNR) of some thermoluminescence dosimeters (TLDs) were investigated by Matsumoto [31] for the application mixed-field dosimetry of thermal neutrons and gamma-rays. The TNR values of Mg_2SiO_4:Tb (MSO-S) and Mg_2SiO_4:Tb+LiF (MSO-LiF) elements were determined to be 3 mGy and 2.9 Gy, respectively, for a thermal neutron fluence of 10^{10} n cm^{-2}.

Figure 10.42 shows the relationship between the thermal neutron fluence and the TLD response. It is clear that TL materials having large cross sections, such as boron, lithium and gadolinium, are capable of a large thermal neutron response. However, Mg_2SiO_4:Tb and BeO also have small responses for thermal neutrons.

V. SUPERHEATED DROP NEUTRON DOSIMETER

The superheated drop detector (SDD) consists of thousands of superheated drops dispersed in a small vial of gel which vaporize upon exposure to high LET radiation, thereby providing a directly observable indication of neutron dose. This detector possesses high sensitivity to neutrons and insensitivity to high-energy photons and electrons, making it suitable for the determination of neutron dose-equivalent rates around high-energy photon and electron radiotherapy beams.

The principle of a bubble dosimeter is the following. A great number of very small droplets of a liquid with a low heat of vaporization are incorporated into an organic gel or polymer. For a specific temperature and pressure, the liquid droplets are superheated in a metastable state. This unstable equilibrium can be destroyed by an external energy supply.

At the interaction point between droplet and high LET particle, such as a neutron or heavy charged particle, the liquid droplet becomes vaporized as a visible gas bubble. The overall absorbed dose is correlated to the number of bubbles. If the gel viscosity is low, bubbles can reach the surface and increase the upper gas volume. The corresponding dose can be derived by measuring the gas volume thus produced with a simple optical scaling system. A sound is emitted for each bubble being created and the bubbles are counted in real time by an electro-acoustic transducer (Figure 10.43). The dose rate corresponds to bubble frequency.

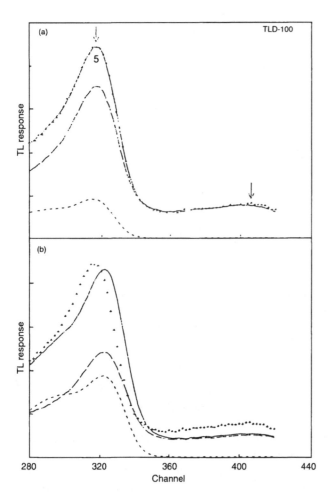

FIGURE 10.39 Glow curve of TLD-100 (+) irradiated in a mixed (γ,n) radiation field. $D_\gamma = 3.5$ mGy and $D_n = 98.5$ mGy (1 MeV) analyzed (a) with and (b) without corrections in the heating profiles of the calibration curves. The dolled and broken curves are the contributions of the calibration curves I_a and I_b to the total glow curve (solid line). Results for (a): $D_\gamma = 3.6$ mGy, $D_n = 92.7$ mGy. Results for (b): $D_\gamma = 4.8$ mGy, $D_n = 63.7$ mGv. (From Reference [29]. With permission.)

Bubble dosimeters generally have low sensitivity to photon radiation.

The superheated drop detector (SDD), invented by Apfel et al. [32], consists of a collection of superheated drops suspended in a gel. Its operation is based on the same principle as that of the bubble chamber, namely, the initiation of vapor bubbles by energetic ions in superheated liquids. The volume of vapor or the acoustic waves given out when superheated drops vaporize serve as a measure of neutron exposure. Since the drops are kept in a "perfectly smooth" container, i.e., another liquid, the sample can be maintained in a superheated state for a long period of time. The repressurization procedure needed in the conventional bubble chamber is avoided, and when one drop boils explosively, it does not trigger boiling in adjacent drops. Therefore, each drop in the SDD is a continuously sensitive miniature bubble chamber. It rep-

resents stored mechanical energy which is released when triggered by radiation. The operation of the SDD does not require any power source. Using appropriate materials, it can operate at room temperature and have a near dose-equivalent response. Direct readability, simplicity of preparation, and low cost make it potentially useful in the measurement of neutron dose-equivalence for patients undergoing high-energy x-ray or electron radiotherapy.

The elemental composition of an SDD by weight is 9.6% hydrogen, 23.4% carbon, 66.3% oxygen, 0.24% fluorine, and 0.46% chlorine, and its physical density is 1.14 g cm^{-3}, very similar to that of tissue-equivalent liquid used in neutron dosimetry.

The SDD is available in several forms, from a pen dosimeter to an electronic bubble event counter. The detector employed by Nath et al. [33] was a 4-cm³ glass vial filled with a gel into which approximately 20,000 drops were mixed. Each drop was about 65 μm in diameter and upon vaporization expanded to about 500 μm. The superheated drop material was a halocarbon (CCl_2F_2), otherwise known as Freon 12.

A graduated pipette was fitted onto the vial cap to read the gel volume displaced by neutron-induced bubbles. After screwing the cap fitted with the pipette to the vial, the vials were inverted to trap the rising bubbles in the gel and prevent them from rising to the surface and escaping through the indicator pipette, thereby causing erroneous volume readings.

The neutron dose equivalent vs. volume relationship was fitted by least squares with an equation of the form:

$$\text{Dose Equivalent(mSv)} = c_1 \ln(1 - \text{volume(cm)}^3/c_2)$$

$$(10.45)$$

where c_1 and c_2 are the fitting parameters determined from absolute calibration. The calibration curve was linear up to 5.0 mSv (500 mrem), with a cumulative maximum usable dose equivalent of about 50 mSv (5000 mrem).

The sound produced by bubble formation was recorded by transducers that sense the accompanying pressure pulse. The active survey meter (ASM) electronically discriminates against spurious noise and vibration. Several devices based on the technology described have been designed. Some are active counters acoustically detecting the sound pulses emitted when drops vaporize. Recent versions perform pulse shape analysis, record the exposure time history, and provide dose rate estimates. Others are passive, integrating meters typically employed in personnel monitoring. Two types of passive dosimeters exist: they are based either on the optical counting of the bubbles, which are permanently trapped in a stiff polymer matrix, or on the measurement of their total volume, which is, in this case, inferred from the expansion of a soft gel they are dispersed in.

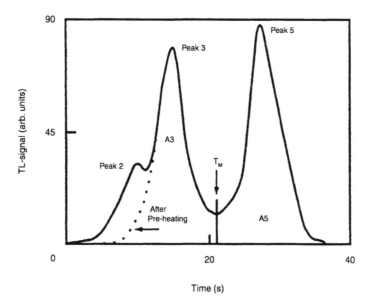

FIGURE 10.40 TL signal vs. reading time for TLD-300 irradiated with neutrons. The three main peaks before and after the preheating cycle are shown. The two areas corresponding to peak 3 and peak 5 under the glow curve are separated at T_M, the time at which the minimum between the two peaks occurs. (From Reference [30]. With permission).

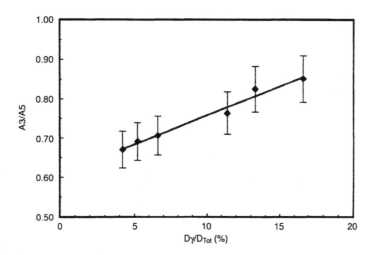

FIGURE 10.41 Plot of the area ratio of peaks 3 and 5 vs. the measured (D_γ/D_T) ratio. (From Reference [30]. With permission.)

The approach of D'Errico et al [34] relies on a triplet of the passive, volumetric emulsions having different degrees of superheat. These permit the detection of neutrons above distinct energy thresholds and provide a three-bin energy spectrum as well as an estimate of the total neutron dose equivalent.

Three halocarbons were chosen by d'Errico et al.: dichlorotetrafluorocethane (R-114), octafluorocycloblitane (C-318), and dichlorofluoromethane (R-12). At 30°C, the selected operating temperature, R-12 emulsions respond from thermal to fast neutrons while R-114 and C-318 present effective energy thresholds of 1.0 and 5.5 MeV, respectively. At 30°C, all these emulsions are insensitive to γ-and x-rays, at least below the energy thresholds

for photo-nuclear reactions—generally around 15 MeV for the low-Z detector constituents.

In order to avoid saturation effects due to the intense and pulsed nature of the radiation field, passive custom-made superheated emulsions (SE) with volumetric readout were employed in the photo-neutron measurements. The dosimeters consist of ~4.5-ml vials (14-mm inner diameter, 30-mm height) connected to graduated pipettes (Figure 10.44). The bottom 1.5 ml of each vial is the active part filled with neutron-sensitive emulsion, i.e., tens of thousands of ~60-μm droplets suspended in a glycerin-based gel, while the top contains inert gel only. When vapor bubbles form, they displace a commensurate volume of gel into the pipette, thus providing an immediate, though passive, response. [34]

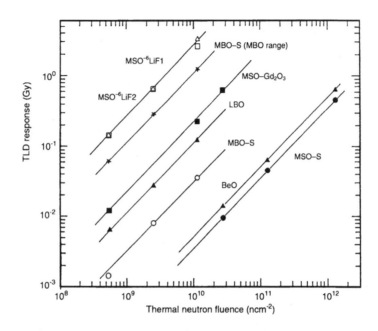

FIGURE 10.42 Relationship between TLD response and thermal neutron fluence. (From Reference [31]. With permission.)

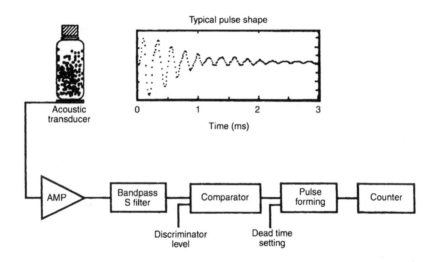

FIGURE 10.43 Principle of the electro-acoustic bubble dosimeter. (From Reference [17]. With permission.)

The detection efficiency, or fluence response (number of bubbles per incident neutron per square centimeter), of all SE types as a function of operating temperature and neutron energy was determined with electronic instrumentation recording the drop vaporization events acoustically. Active bubble counting was possible in low-intensity metrology fields and was chosen for the calibrations since it allows for higher precision compared to the passive volumetric readout.

The relative fluence responses of the three emulsions used in this study are shown in Figure 10.45, plotted as a function of neutron energy. These curves present thresholds corresponding to the minimum neutron energies that can trigger the nucleation. The general rule

they reflect is that the higher the superheat of the liquid, the lower the minimum energy that neutrons, or rather their charged secondaries, must impart to the drops in order to nucleate their evaporation. In particular, the emulsions of halocarbons C-318 and R-114 require the deposition of hundreds of keV per micrometer and are nucleated only by fast neutrons, via the recoils generated inside or next to the superheated drops. In fact, bubble nucleation occurs only when sufficient energy is deposited within an extremely localized region of the superheated drop. While charged recoils generate trails of microscopic vapor cavities inside the droplets, only those exceeding a minimum critical size can develop into macroscopic bubbles. When this size is reached,

TABLE 10.10
Chemo-Physical Data for Superheated Emulsions ICRU-Recommended TE Composition and TE Liquid

	Elemental composition by weight percent					
Material	H	C	O	N	F, Cl	Density (g cm⁻³)
SE	8.8	28.2	62.5	...	0.5	1.2
ICRU TE	10.1	11.1	76.2	2.6	...	1.0
TE liquid	10.2	11.1	76.1	2.6	...	1.1

From Reference [34]. With permission.

1 cm

FIGURE 10.44 Passive, integrating superheated drop (bubble) detector with volumetric readout for the measurement of high neutron dose rates. (From Reference [34]. With permission.)

the expansion becomes irreversible and the whole droplet evaporates.

Figure 10.45 shows that emulsions of halocarbon R-12 (CCl_2F_2) are also sensitive to thermal and intermediate neutrons. This occurs as the droplets contain chlorine and are sufficiently superheated to be vaporized by the exoergic $^{35}Cl(n,p)^{35}S$ capture reaction, which releases 615 keV and creates densely ionizing 17-keV sulfur ions. Although R-l 14 ($C_2Cl_2F_2$) has two chlorine nuclei in its molecule, it is not superheated enough to be nucleated by the energy deposition pattern of the reaction products. Thus, neither R-l 14 nor C-318 (C_4F_8) emulsions are sensitive to thermal neutrons in the temperature range of interest, as was verified at the PTB reactor. [34]

The normalized experimental and calculated fluence response curve for R-12 is plotted in Figure 10.45 against

the "kerma-equivalent-factor" $k_f Q_n$. The latter is the average tissue kerma per unit neutron fluence weighted by the quality factor (QF) of the neutron secondary particles.

Neutron dose-equivalent measurements were performed by Loye et al. [35] in and around an 18-MV x-ray beam using superheated drop detectors and phosphorous pentoxide (P_2O_5) powder. The neutron capture reaction $^{31}P(n,\gamma)^{32}P$ is significant only for thermal neutrons, and the $^{31}P(n,p)^{31}S$ reaction is sensitive to fast neutrons above 0.7 MeV. The activation products ^{31}Si and ^{32}P are pure β emitters with half-lives of 2.62 h and 14.28 d, respectively. Five ml of a solution of 0.32 gm of P_2O_5 powder per ml of distilled water were placed in each scintillation vial. The samples were irradiated with approximately 40 Gy of x-ray dose. Two ml of solution from each of the irradiated vials were transferred to fresh scintillation vials, and 15 ml of the liquid scintillation counter (LSC) cocktail Insta-gel Plus™ were added to each of these vials. The vials were counted for 1 min each, 8–10 times, and were corrected for decay between counts. ^{32}P activity for the thermal neutrons was measured the next day after sufficient decay of ^{31}Si.

The superheated drop detectors used were Neutrometer-HD™ (SDD 100) purchased from Apfel Enterprises, Inc. The response of these detectors is dose-equivalent from thermal-energy (0.025 eV) to fast-energy (66 MeV) neutrons. Each detector has thousands of droplets (approximately 20,000) of Freon 12 suspended in a gel contained in a 4-cm³ vial.

The drops expand from roughly 65 to 500 μm and the expansion of the bubbles displaces the gel into the pipette. The volume displaced is equated with dose equivalence (DE) by Equation 10.45. For the detectors used by Loye et al, C_1 equals -28.5 and C_2 equals 0.685, as calibrated by Apfel Enterprises, Inc.

Seitz's theory [36] suggests that when a heavy charged particle slows down moving through a liquid, its kinetic energy is transferred as thermal energy to extreme small regions (temperature spikes) through the intermediary of δ rays. The intense heating induces localized boiling,

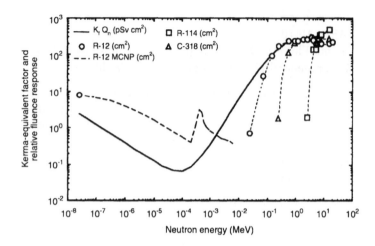

FIGURE 10.45 Kerma equivalent factor, $k_f Q_n$, plotted as a function of neutron energy vs. the relative fluence responses of the superheated emulsions (all normalized to the kerma factor plateau): R-12 (O, experimental, and —,calculated), C-318 (△), and R-114 (□). (From Reference [34]. With permission.)

creating trails of microscopic vapor cavities which develop into macroscopic bubbles when the density of energy deposition is high enough. It is assumed that a spherical vapor cavity of radius r, embedded in a liquid of surface tension σ and vapor tension p_v, expands indefinitely when $(p_v - p_e) > 2\sigma r$, where p_e is external pressure. Therefore, a critical radius $r_c = 2\sigma/\Delta p$ defines the discriminator between growing bubbles and those collapsing under the action of external forces. Various expressions of the formation energy for such critical bubbles have been proposed, such as: [37]

$$W = \left(4\pi r_c^2 \sigma - \frac{4}{3}\pi r_c^3 \Delta p\right) + \frac{4}{3}\pi r_c^3 \rho_v h_{fg}$$
$$+ 2\pi\rho_1 r_c^3 \dot{r}^2 + F$$
$$= W_{Gibbs} + H + E_{wall} + F \qquad (10.46)$$

where ρ_v is vapor density, ρ_1 is liquid density, h_{fg} is latent table heat of vaporization, $\dot{r}$ is vapor wall velocity, W_{Gibbs} is the minimum reversible work required for bubble formation or Gibbs free energy, H is the vaporization energy, E_{wall} is the kinetic energy imparted to the liquid by the motion of the vapor wall, and F is the energy imparted to the liquid during the growth of the bubble by the viscous forces. Neglecting the last two terms and substituting for $r_c = 2\sigma/\Delta p$:

$$W = \frac{16\pi\sigma^3}{3\Delta p^2}\left[1 + \frac{2p_v}{\Delta p}\left(1 + \frac{h_{fg}}{R^*T}\right)\right] \qquad (10.47)$$

where R^* is the gas constant. It may be immediately observed that when the degree of superheat Δp increases,

both the critical radius and the minimum formation energy decrease. Values of k between 1 and 13 have been proposed for the expression $E_{min}=kr_c(dE/dx)$. Although this semi-empirical approach allows for an estimate of the measured bubble formation energy in superheated liquids, it is far from reflecting the physics of the bubble nucleation.

The flow fields, as a function of time t and radial distance r, of a viscous, heat-conducting, and compressible fluid subject to the singular initial condition of a sudden energy deposition are then governed by five fluid dynamic equations: three conservation equations (mass, momentum, and energy), one equation of state, and its associated specific internal energy equation (treating the medium as a Horvath-Lin fluid). These are sufficient to solve for the five unknowns: temperature, T; pressure, p; velocity, u; specific volume, v; (density $\rho = 1/v$); and specific internal energy, e. It is noteworthy that a critical radius equal to 2/3 of Seitz's had already been shown to improve consistency with the experimental findings. [37]

The correlation existing between density of energy transfer and efficiency of radiation-induced vaporization was examined by d'Errico et al. [38] by determining the response to thermal neutrons of dichlorofluoromethane (CCl_2F_2) emulsions as a function of temperature. These emulsions contain ^{35}Cl that undergoes the exoergic capture reaction $^{35}Cl(n,p)^{35}S$ with low-energy neutrons. Of the reaction products, the proton receives 598 keV and deposits it over 13 μm, a relatively long range compared to the critical radius, whereas the sulphur ion receives 17 keV and deposits it within 43 nm.

Measurements carried out at carefully controlled temperatures between 10°C and 40°C showed that the sensitivity to thermal neutrons is thermodynamically inhibited below 15°C, while it rises sharply above ~22°C (Figure 10.46). The responses of octafluorocyclobutane (C_4F_8) and dichlorotetrafluoroethane ($C_2Cl_2F_2$) are shown in

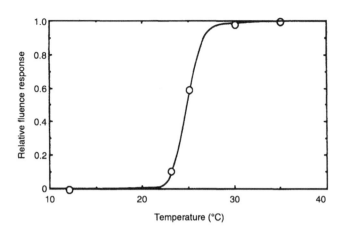

FIGURE 10.46 Relative fluence response of a dichlorofluoromethane emulsion to thermal neutrons as a function of operating temperature. (From Reference [37]. With permission.)

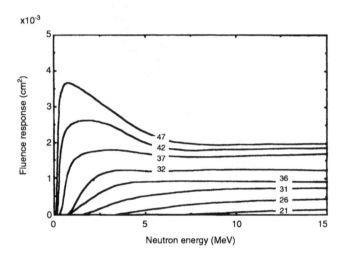

FIGURE 10.47 Fluence response of octafluorocyclobutane (left-hand side) and dichloroletrafluoroethane (right-hand side) emulsions as a function of degree of superheat (reported on each curve) and neutron energy. (From Reference [37]. With permission.)

Figure 10.47: two sets of four curves are reported, each one identified by the corresponding degree of superheat ΔT (expressed in Kelvin above the boiling temperature). It is evident that the higher the superheat, the lower the energy that neutron-charged secondaries must impart to the droplets in order to nucleate their evaporation.

Figure 10.47 also documents the progressive appearance of a maximum in the response of highly superheated emulsions. This is correlated with the arising detection efficiency for neutron-recoiled protons entering the drops from the surrounding hydrogenated gel.

Design of neutron dosimeters based on superheated drop (bubble) detectors was described by d'Errico et al. [38] Designs rely on the double transducer scheme but adopt a comparative pulse-shape analysis of the signals. Pulses from the two piezoelectric transducers are amplified and fed to rectifier detectors producing their envelopes, which are then processed by analog/digital converters

(Figure 10.48). When the signal from the SDD vial exceeds an adjustable threshold and presents the typical decay pattern of a bubble pulse, the two digitized tracks are compared. The detector pulse is accepted only when its shape clearly differs from the signal in the noise channel.

In order to minimize the temperature dependence of the response, an effective and elegant solution was proposed and is now implemented in passive devices. In a sealed vial, a low-boiling-point liquid is introduced over the free surface of the sensitive emulsion (Figure 10.49). The vapor tension of this liquid varies with ambient temperature, and the pressure thus applied to the detector compensates for the temperature effects. In fact, the neutron detection efficiency depends on the overall degree of superheat of the emulsion, i.e., on the combination of pressure and temperature values. The dependence on these quantities is illustrated in Figure 10.50, showing

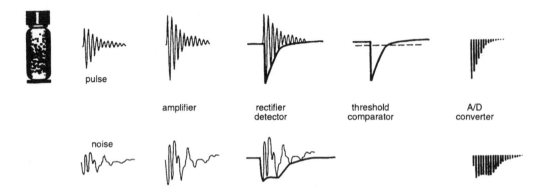

FIGURE 10.48 Schematic of an active bubble counter based on the acoustical recording of bubble vaporizations. (From Reference [38]. With permission.)

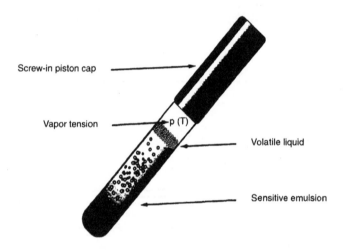

FIGURE 10.49 Passive bubble detector with re-pressurization cap and temperature compensation system. (From Reference [38]. With permission.)

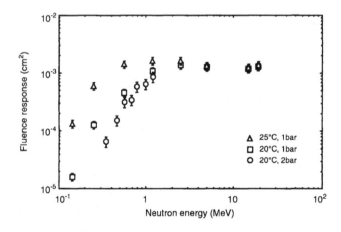

FIGURE 10.50 Fluence response of a dichlorofluoromethane emulsion as a function of pressure, temperature, and neutron energy. (From Reference [38]. With permission.)

that the increase in response of a dichlorofluoromethane detector warmed up from 20 to 25°C is almost exactly counterbalanced by raising the operating pressure from 1 to 2 bar. In dosimeters with pressure compensation, the variation of sensitivity per degree Celsius is reduced from 5 to about 1%.

A Radiation Assessment Instrument for Space Applications (RAISA) based on bubble detectors has been

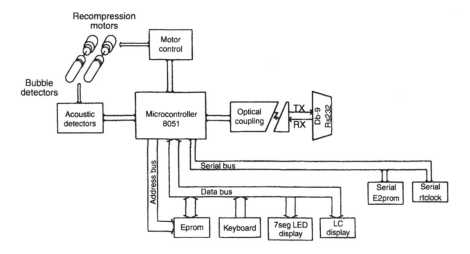

FIGURE 10.51 Block diagram of main components of RAISA. (From Reference [40]. With permission.)

developed by Ing et al. [40] to meet the needs for radiation monitoring for long-term space missions. RAISA uses two bubble detectors, but at any time only one of the detectors is used as the radiation sensor. When a maximum number of bubbles has accumulated in this detector, a microprocessor turns on the second detector and recompresses the first detector. In addition to instantaneous display of ambient dose rate, RAISA records information in memory every second.

Measurements of space radiation with bubble detectors have been done using specially formulated BD-PNDs. Although the BD-PND can be packaged so that its contents remain unchanged (chemically and physically) for many years, this detector cannot be used continuously for years since the number of bubbles continues to accumulate with radiation exposure and the detector can accommodate only a maximum number of bubbles before filling the entire detector. Thus, it is necessary that the bubbles be recompressed (or "cleared") once a specified number of bubbles (say, 300) has accumulated. Also, it is generally desirable to clear the bubbles after a certain period of time because the accumulated bubbles continue to grow (very slowly) after their initial formation and oversize bubbles are more difficult to recompress. For normal applications, the bubble detector is recycled (i.e., read and recompressed) every few days during use. The reading is usually done with an automatic reader which counts the total number of bubbles in the detector using sophisticated image-processing software.

Figure 10.51 shows a block diagram of all the components that make up RAISA. The acoustic detectors provide bubble-explosion detection and external-noise sensing to ensure that no external noise can be mistaken as a bubble explosion. The acoustic bubble detectors consist of piezoelectric sensors, charge-sensitive preamplifiers, analog switches for selecting bubble detectors, level comparators, and output timers. Serial Eprom is a non-volatile memory

capable of storing up to 15,700 bubble events; this information includes the time at which the bubble event was recorded. This data can be retrieved using any computer via serial communication.

The properties of neutron bubble detectors (type BD-100 R*) have been investigated by Schulze et al. [41] with fast neutrons from D-T and Am-Be sources. The experimentally determined responses of the bubble detectors have shown considerable fluctuations (up to 50%). The response after repeated recompressions has been examined. The detectors proved to be insensitive to gammas.

The bubble detectors (type BD-100 R) consist of a small test tube (overall length 80 mm, active length 45 mm, diameter 15 mm) filled with an elastic clear polymer. Interspersed in this polymer are superheated freon droplets. Recoil protons may be produced by neutron interaction with the polymer. If these protons strike such a droplet, it may vaporize. This vaporized droplet remains trapped as a visible bubble in the polymer. Recharging is accomplished by pressurizing the bubble detectors considerably above the vapor pressure of the freon gas mixture, thereby reforming the bubbles to liquid droplets.

Figure 10.52 shows the deviation of the dose equivalent measured with the bubble detectors as a function of the recompression cycles. The values from bubble detectors of the two low-sensitivity classes, as well as of the two high-sensitivity classes, were combined for greater clarity. In particular, the bubble detectors of the low-sensitivity classes showed large deviations when used for the first time and the reading was significantly too high (mean + 150%). After the third recompression (i.e., the fourth irradiation), all bubble detectors showed a 40% to 50% higher dose equivalent than that applied, but the variations were still large.

Superheated drop compositions have been employed by Apfel et al. [42] in two passive (non-electronic) direct reading detectors for neutron monitoring of personnel

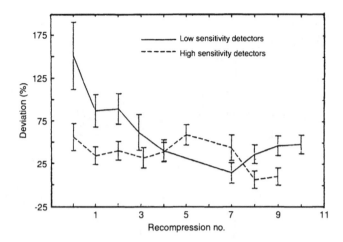

FIGURE 10.52 Deviation of the dose equivalents measured with bubble detectors from the dose equivalent as a function of the recompression cycles. Irradiation was done with an Am-Be source. (From Reference [41]. With permission.)

and in accelerator applications. The Neutrometer™ is a pen-sized tube partially filled with a superheated drop composition. Neutrons impinge on the superheated drops, causing them to vaporize suddenly. The resulting bubbles push a piston up a calibrated tube, giving a direct indication of the neutron dose equivalent. The Neutrometer™ can be reused for several successive exposure periods, with each exposure adding an increment to the total accumulated dose. The Neutrometer™-HD is designed for higher dose ranges. Vials containing the superheated drop composition are placed at various locations in the irradiation area, such as near a high-energy medical accelerator. Bubbles that are nucleated by neutrons force an indicator liquid up a calibrated pipette. This displacement gives the neutron dose equivalent. The spatial dependence of the dose equivalent can therefore be measured immediately, without recourse to activation foils, scintillation powders, or other similar systems that require extensive post-processing. Tests to characterize the performance of these detectors suggest that they are nearly dose-equivalent, that batch-to-batch uniformity is good, and that they meet many of the demands required in their respective applications.

The personnel neutron dosimeter, Neutrometer-100, is 12.5 cm tall and weighs 10 g. The round glass tube which holds the superheated drop composition is held in a triangular plastic holder. The bottom section holds approximately 0.5 ml of aqueous gel composition with approximately 8000 drops of 100-μm diameter. Atop this composition is a gel piston, a white disc (which is what is measured against the calibrated scale), and a small amount of gel which acts as a lens so that the scale is magnified. The SDD-100 drop material has been characterized in the energy range 0.2 to 14 MeV for thermal energies and at 2, 24, and 144 keV. The energy response is shown in Figure 10.53a, compared with the fluence to ambient dose-equivalent conversion factor. In Figure 10.53b, two modifications have been made. The albedo

response of the detector has been estimated, comparing it with the fluence to individual dose-equivalent, penetrating, conversion factor. The contribution due to n-capture gamma-rays has been subtracted, from this factor, showing better agreement with the estimated albedo response based on measurements.

It was found that the displacement of the piston, d in mm, fits the following depletion equation ($\pm$10%):

$$d = 65(1 - e^{-0.27D}) \qquad (10.48)$$

where D is the neutron dose equivalent in mSv. The practical limit of the neutrometer was 57 mm, corresponding to 8 mSv. The smallest calibration mark on the scale corresponded to 0.9 mm for a dose equivalent of 50 μSv (5 mrem). The calibration factor used in the results was:

$$\text{Dose equivalent (mSv)} = -21.5\ln(1 - 1.37\ V)$$

where V is the volume displacement of indicator liquid in the pipette, in ml.

VI. SOLID-STATE NEUTRON DOSIMETER

Solid-state detectors were very limited in their use for neutron detection. The main reasons were: (a) neutron radiation damage to the crystal made the detector useful only for a short period of usage (a small number of total counts) and (b) solid-state detectors were relatively too expensive for uses as described in (a). In the last few years, the manufacturing technique of solid-state devices has developed tremendously. Many different kinds of devices are available on the market for a low price. Radiation damage to the device is not a problem when neutron dosimetry is the concern, because the dose is measured by the damage caused to the detector. A change in resistivity or leakage current in the device is a measure of the

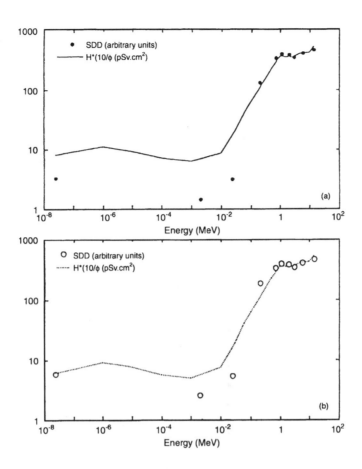

FIGURE 10.53 (a) The experimental results of exposure of SDD-100 material to monoenergetic neutrons from RARAF (Nevis Labs), Van de Graaff, and the US National Institutes of Standards and Technology research reactor. The solid line is adapted from information provided in the ICRP Report 60. (b) Data points have been modified to include estimated albedo response and then compared to fluence to individual dose-equivalent, penetrating, conversion factor, from which n-capture gamma-ray contribution has been subtracted. (From Reference [42]. With permission.)

dose absorbed in it. These are just a few considerations of the great change that has taken place in solid-state dosimetry. Some of the more specific developments are described below.

An electronic neutron dosimeter based on a combination of a polyethylene converter with a double diode (Canberra CD-NEUT-200-DBL) has been studied by Fernandez et al. [43] The dosimeter has been irradiated by monoenergetic neutron beams of energies from 73 keV to 2.5 MeV, with dose equivalents between 0.3 and 5.7 mSv. The differential method has been applied to separate the neutron response from the gamma contribution.

A dosimeter arrangement composed of a 40-μm-thick layer of polyethylene converter followed by a double-diode detector has been used. The double-diode detector (CD-NEUT-200-DBL) consists of two Canberra diodes, with an effective area of 2 cm and a bulk resistivity of about 500 Ω cm, located on the opposite sides of a single silicon block. The depleted zone of each of the two diodes is, in the actual configuration, 30 μm deep for a polarization tension of 10 V. These depleted zones are separated

by 222 μm of Si. Ortec preamplifiers and a Canberra multichannel analyzer with amplifier are used and connected to a Toshiba T-180 portable computer for data acquisition, as indicated in Figure 10.54. Data analysis is performed using the Ortec Maestro II software.

Figure 10.55 shows the simulated dose-equivalent response R_H for normally incident neutrons as a function of neutron energy for this dosimeter configuration and several polyethylene converter thicknesses, together with the experimental values obtained. The dosimeter response may become flatter if the polyethylene converter thickness is reduced, as deduced from the figure. Once this thickness is selected, it is obvious from the results obtained that it is necessary to reduce as much as possible the energy cutoff introduced in order to increase the response to 144, 250, and 570-keV neutrons.

Superheated emulsions and silicon diodes were studied by Vareille et al. [39] as radiation sensor for personal and area monitors. The detectors were analyzed with respect to their neutron sensitivity and their overall suitability for practical dosimetry. The Classical Differential

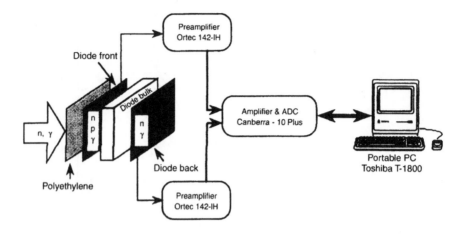

FIGURE 10.54 Schematic diagram of the experimental setup. (From Reference [43]. With permission.)

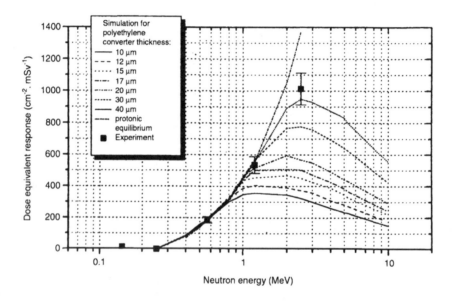

FIGURE 10.55 Experimental and calculated dose-equivalent response R_H for normally incident neutrons as a function of neutron energy. (From Reference [43]. With permission.)

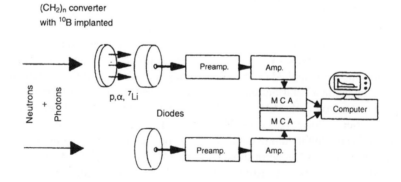

FIGURE 10.56 Block diagram of the Classical Differential Method. (From Reference [39]. With permission.)

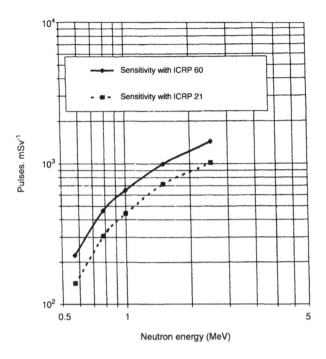

FIGURE 10.57 Fluence response of the CDM coupled diodes as a function of neutron energy (normalized to 1 cm²). (From Reference [39]. With permission.)

Method (CDM) was studied, employing two identical *p-n* junctions, one bare and one covered by a ¹⁰B-loaded polyethylene converter (typically 35 μm thick). In the CDM (Figure 10.56), the difference of the two diode counts is taken, as it should theoretically be free of electronic background and *r* contribution. Unfortunately, the two diodes are never truly identical and, to avoid the γ sensitivity, it was necessary to set a threshold at 600 keV, despite the removal of most metal parts and the reduction of the depleted layer to 10 μm. In the 1–4 MeV neutron energy range, the sensitivity in terms of individual dose equivalent falls between 0.5 and 1 pulse per μSv (Figure 10.57). Thermal neutrons are also detected, due to the ¹⁰B converter.

A method using one PN junction covered with a hydrogenous converter (¹⁰B-loaded) was proposed by Bordy et al. [44] for individual neutron dosimetry. This method is based on a pulse-shape analysis to discriminate the photon signal from the neutron signal. It allows drastic reduction of the photon sensitivity (by a factor of 1000). By applying a neutron correction factor to the low-energy events, the gap in neutron sensitivity for intermediate energy can be partly filled. Lead shields used to surround the detector allow the remaining photon sensitivity to be decreased by a factor of two. An especially designed hydrogenous moderator placed at the top of the detector allows the neutron sensitivity to be increased by a factor of two for 250-keV neutron energy.

The device (Figure 10.58) consists of 3 parts: (1) the detector and preamplifier (Intertechnique PSC 762); (2) the amplifier step and the pulse shape analyzer, all made of commercial NIM standard board; and (3) three multichannel analyzers that record the spectra corresponding to Sol, So2, and So3. The detector consists of (a) a Canberra PIPS diode (300-μm thickness, bulk resistivity: 600 Ω cm, sensitive area: 1 cm²) covered with (b) a polyethylene converter ¹⁰B-loaded (5 × 10¹⁵ bore, 30 μm thick). Conductive rubber rings (c) and an aluminum plate (d) are used to apply a reverse-bias voltage to the junction for 12 V. The depleted layer is about 30 μm thick. Two printed circuits (e) join the different pieces together.

The rise time measurements are carried out with a time amplitude converter (TAC). The rise time is measured between a "start" and a "stop" signal related to the event being measured. The amplitude of the logic pulses generated by the TAC is related to the interval between these two signals. The "start" signal is generated when the analog input pulse crosses a low-level threshold. This marks the arrival time of the event. The "stop" signal is produced by a timing single channel analyzer when the peak of the input pulses is detected; this technique uses the zero-crossing of a bipolar pulse related to the input pulse. Thus, the rise time is virtually independent of the input signal amplitude.

A dip in neutron sensitivity is noticed around 250 keV (Figure 10.59). An attempt has been made to obviate this problem by applying a weighting factor k_w and adding hydrogenous moderator. The moderator allows the sensitivity to be increased around 0.25 MeV (Figure 10.59), but this sensitivity remains low.

A real-time personal neutron dosimeter has been developed by Nakamura et al. [45] The dosimeter contains two neutron sensors, a fast neutron sensor and a slow neutron sensor, which are both *p*-type silicon semiconductor detectors contacted with two different radiators of polyethylene and boron. The neutron detection efficiencies of these sensors were measured in a thermal neutron field and monoenergetic neutron fields from 8 keV to 22 MeV.

Figure 10.60 shows the cross-sectional views of the fast and slow neutron sensors. The fast neutron sensor is a 10 × 10-mm² *p*-type silicon crystal, on which an amorphous silicon hydride is deposited. The slow neutron sensor is also a 10 × 10-mm² *p*-type silicon on which a natural boron layer is deposited around an aluminum electrode to detect α and Li ions from the ¹⁰B(n,α)⁷Li reaction. Both sensors are in contact with 80-μm-thick, polyethylene radiators to produce recoil protons from the H(n, n)p reaction. The slow neutron sensor has some sensitivity for fast neutrons but is mainly used to measure neutrons with energies less than 1 MeV, while the fast neutron sensor measures neutrons with energies in the MeV region.

The detector is operated by applying the opposite bias of +15 V, and a depletion layer of 60 μm thickness is generated under the amorphous silicon hydride, which can fully absorb α, ⁷Li, and recoil proton energies but absorbs

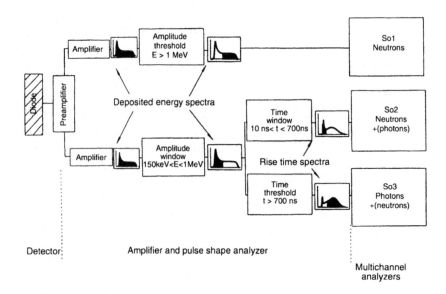

FIGURE 10.58 Block diagram of the experimental setup. (From Reference [44]. With permission.)

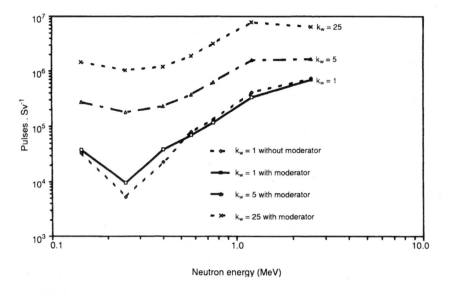

FIGURE 10.59 Dose-equivalent response obtained with and without moderator and by using the weighting factor k_w. (From Reference [44]. With permission.)

only a part of the gamma-ray energy in order to discrim-inate easily between neutron output pulse heights from gamma-ray output pulse heights. By using such pulse height discrimination, the contribution of the gamma-ray dose to the neutron dose can be suppressed within 1% under a 1 Sv h^{-1} gamma-ray mixed field.

The energy response of the dosimeter to neutrons was measured in a monoenergetic neutron field ranging from 8 keV to 15 MeV and in a 22-MeV quasimonoenergetic neutron field. Detection efficiency was obtained by sum-mation of all neutron pulses beyond the bias level and division of the sum by the neutron fluence, and the results were expressed as counts per n cm^{-2}. The bias level for gamma-ray pulse cutoff was determined to be at a position

corresponding to the ambient gamma-ray dose equivalent of 1 Sv h^{-1} in free air. The responses of fast and slow neutron sensors to fluence as a function of neutron energy are shown in Figure 10.61, together with the calculated response using the MORSE-CG Monte Carlo code. An ideal response of the neutron dosimeter must be close to the fluence-to-dose equivalent conversion factor (drawn as a dotted line in Figure 10.61) over a wide range of neutron energy but, as seen in Figure 10.61, for neutron energies from 100 keV to 1 MeV. However, practically speaking, this underestimation can be compensated for by the slight overestimation in the lower-energy region (espe-cially in the thermal energy region), considering that the actual neutron field has a continuous neutron spectrum,

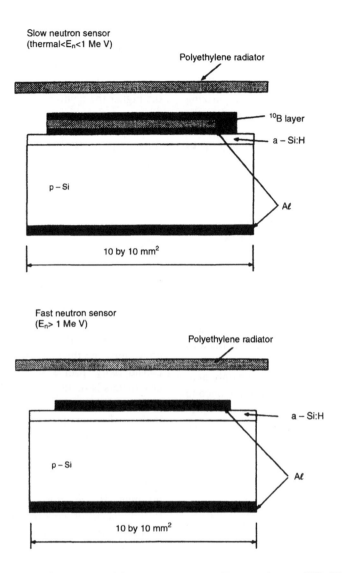

FIGURE 10.60 Cross-sectional view of the slow and fast neutron sensors. (From Reference [45]. With permission.)

usually containing an amount of thermal neutron components.

A combined active/passive personal neutron dosimeter was proposed by Luszik-Bhadra et al. [46] It consists of a position-sensitive silicon diode which is covered on four different positions by different thermal and fast neutron absorbers, moderators, and converters—including a passive CR-39 detector.

A sketch of the dosimeter configuration under test is shown in Figure 10.62. For a first feasibility study, it is assumed that a single position-sensitive silicon diode can be used. In front of the silicon diode (315 μm thick, 3-cm² sensitive area), a batch of thermal and fast neutron converters was placed at 1 mm distance. The thermal neutron converters (section (c) in Figure 10.62) consist of three LiF layers behind different shieldings. The positions are denoted by "Albedo", "Closed" and "Front." The fast neutrons are detected separately in position 4. Fast neutron converters and moderators are different sheets of hydrogen-

containing plastics. They consist of a polyethylene sheet 0.5 mm thick (with the evaporated LiF layers), followed by two CR-39 sheets, each 0.7 mm thick. The one is used as a passive CR-39 detector and the other one serves as a CR-39 designed converter.

The registration of heavy charged panicles by means of MOS breakdown counters was suggested by Tommasino et al.[47] A neutron sensor consists of a MOS thin-film capacitor in combination with a fissile radiator (Figure 10.63). The operation voltage applied to the MOS element must be chosen such that spontaneous breakdowns through the silicon dioxide layer can be neglected. Incident neutrons generate fission fragments within the radiator which enter the silicon dioxide. Along their paths secondary charge carriers are produced in high concentration, which gives rise to a local degradation of the breakdown field strength. The result is electric breakdowns at the sites of the fission fragment tracks which can be counted as voltage pulses by using a simple electronic circuit. The pulse rate z is proportional to the momentary neutron flux density in time-dependent neutron fields or to the corresponding derived dosimetric quantities. In the course of each breakdown, a hole is created in the silicon dioxide and in the aluminum top electrode by vaporizing material. The total number of vaporization spots on the sensor surface is an additional measure of the fluence of incident neutrons—similar to the number of etched tracks in solid-state nuclear track detectors. (Dorschel et al. [48])

VII.   BONNER SPHERE AND OTHER PORTABLE NEUTRON MONITORS

Bonner sphere consists of a central detector, sensitive to thermal neutrons, and a moderator sphere. A ^{3}He proportional counter is often used as the central detector. A knowledge of the Bonner sphere response as a function of incident neutron energy is required to interpret Bonner sphere measurements. The response of a Bonner sphere is defined as the ratio of yield over neutron fluence.

A Bonner sphere spectrometer (BSS) is the only neutron monitor which allows the spectral neutron fluence to be measured in a wide energy range, from thermal up to about 20 MeV. The ambient dose-equivalent value can be

calculated from the spectrum by applying the corresponding fluence to dose-equivalent conversion function. The responses of the spheres are functions of the neutron energy and moderator diameter. The larger the sphere diameter, the higher the neutron energy where the response reaches its maximum.

To achieve an isotropic response, a portable monitor for measuring $H^*(10)$ should ideally consist of spherical detecting elements; an orthospherical detector surrounded by a spherical or an orthospherical converter is convenient from a practical point of view. This is generally the case for rem counters where the detecting element is a proportional or GM counter and the converter is a polyethylene sphere.

To yield an isodirectional response with respect to H_p, an individual dosimeter for measuring $H_p(d)$ should have a flat converter and a flat active detector volume. This is usually the case, for example, for dosimeters based on the use of a nuclear emulsion, a solid-state detector (e.g., CR-39 foils), or a silicon diode covered with hydrogenous converters, the thickness of which, however, often varies considerably from $d = 10$ mm, and none of the above dosemeters is really isodirectional. This can also be seen in Figure 10.64, where the ideal angular response of an individual dosimeter with respect to fluence for various neutron energies is compared with experimental results for CR-39 dosimeters. [49]

The response functions of four widely used Bonner sphere spectrometers (BSS) with an LiI scintillator and different ^{3}He detectors were calculated by Kralik et al. [50] by means of the three-dimensional Monte Carlo neutron transport code, MCNP, taking into consideration a detailed description of the detector setup; they were then compared with experimental calibration data. In order to obtain agreement between the calculated response functions and the measured calibration data, different scaling factors had to be applied to the response functions calculated for the various sphere diameters, and even the shape had to be adjusted in order to fit the thermal neutron response measured (see Figure 10.65). Better results were obtained by means of the three-dimensional neutron transport Monte Carlo code, MCNP, although details of the detector setup were still neglected.

Response functions of Bonner sphere spectrometers were calculated by Sannikov et al. [51] for neutron energies from 10 MeV to 1.5 GeV by the Monte Carlo high-energy transport code, HADRON. Calculations were made for two types of thermal neutron detectors inside the Bonner spheres: ^{3}He proportional counter and ^{6}LiI scintillation detector. The results obtained are compared with calculations using the MCNP and the LAHET Monte Carlo codes.

The MCNP code was extended up to 100 MeV using the neutron cross sections for hydrogen and carbon from the

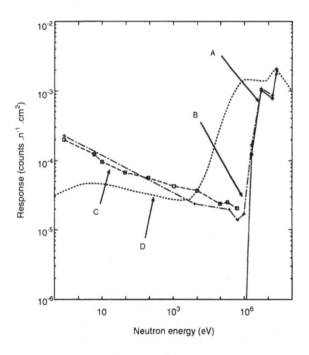

FIGURE 10.61 Dosimeter response to fluence as a function of neutron energy with the ambient dose-equivalent conversion factor described in ICRP 74. Key to lines: A (— — — —) fast neutron sensor, B (·—·—·) slow neutron sensor, C (----) calculated response, D (····) ambient dose-equivalent conversion factor. (From Reference [45]. With permission.)

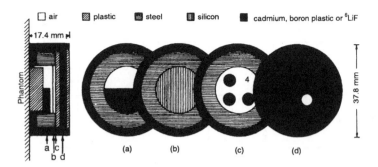

FIGURE 10.62 Sketch of the dosimeter setup (a) the albedo window, (b) the silicon diode, (c) the front side of the plastic converter with three LiF layers, and (d) the cadmium sheet with front window. The four positions on section (c) are, counterclockwise, (1) "Albedo," (2) "Closed," (3) "Front," and (4) "Fast" without ^{6}Li converter. (From Reference [46]. With permission.)

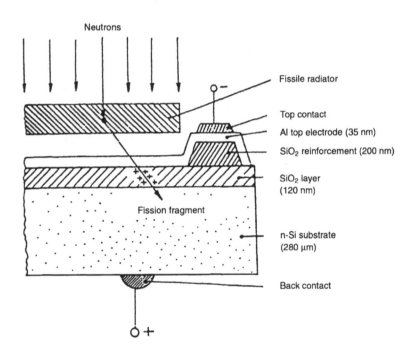

FIGURE 10.63 Scheme and operation principle of a MOS breakdown sensor. (From Reference [48]. With permission.)

LA 100 library. The high-energy transport code HADRON is based on the cascade-exciton model of nuclear reactions. This model includes a cascade stage (modified version of the Dubna cascade model), a pre-equilibrium stage (exciton model), and an equilibrium evaporation stage. Low-energy neutrons are transported by the Monte Carlo code FANEUT. The results for low neutron energy responses of the ^{3}He BSS were found to be in a good agreement with the MCNP data.

The LAHET code, developed at Los Alamos National Laboratory, is based on the HETC high-energy transport code of ORNL. This program has different options for simulation of inelastic interactions of hadrons with nuclei: Bertini and the ISABEL cascade models. At the de-exci-

tation stage of nuclear reaction, two versions of the pre-equilibrium exciton model or the Fermi breakup model may be used. The transport of the low-energy neutrons is calculated by the code HMCNP, which is a modified version of MCNP.

The results of calculations performed by the three Monte Carlo codes are shown in Figure 10.66 for several Bonner spheres. The data obtained by the MCNP and HADRON programs agree within the limits of 15% below 100 MeV. This is not the case for the LAHET data, which are much higher as a rule.

The more pronounced differences for small spheres in Figure 10.66 may be explained by the large contribution of low-energy neutrons from the first interaction of

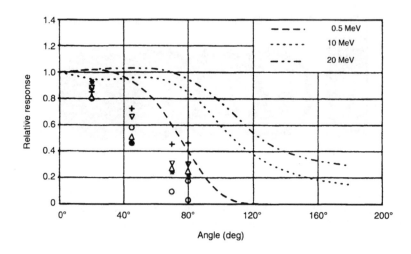

FIGURE 10.64 Angular dependence of the response of individual dosimeters with respect to fluence for various neutron energies. Ideal theoretical curves (approximated by $H'(10)/\Phi$ and experimental results with CR-39 neutron dosimeters. Key: (+) 14 to 16 MeV, (∇) 3 to 5 MeV, (Δ) 1.3 to 1.6 MeV, (*) 0.45 to 1.1 MeV, (o) 0.13 to 0.4 MeV. (From Reference [59]. With permission.

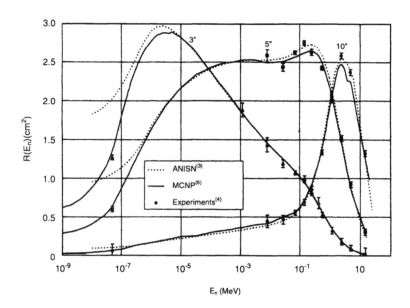

FIGURE 10.65 Comparison of calculated response functions $R(E_n)$ of the PTB-C BSS for the 3″, 5″. and 10″ spheres with corresponding experimental data. (From Reference [50]. With permission.

neutrons with carbon nuclei. In the case of large spheres, this effect is masked by averaging the processes of neutron production and absorption over several collisions.

The full set of the ^{3}He BSS smoothed-response functions is shown in Figure 10.67 by solid curves from 1 MeV to 1500 MeV (MCNP below 10 MeV and HADRON above 10 MeV). These responses include only ^{3}He$(n,p)^3$H events with energy depositions from 0.2 to 0.9 MeV. The dashed curves in Figure 10.67 show the total responses, including all events above 0.2 MeV.

Measurements of Bonner sphere response functions in reactor-filtered neutron beams were intercompared by

Siebert et al. [52] with calculations, and the influence of the moderator's mass density is studied in calculations. The ^{3}He content of the proportional counter used as the central detector has been experimentally determined and roughly confirmed by calculations.

The central detector is a spherical ^{3}He proportional counter, Centronic Type SP 90, 3.2 cm in inner diameter and with a stainless steel wall 0.5 mm thick. The moderator spheres are made of polyethylene $(CH_2)_n$ and are 7.62 cm, 11.43 cm, 15.24 cm, and 20.32 cm in diameter with a mass density, ρ, of 0.95 $\pm$ 0.005 g cm^{-3}. The fittings around the detector also consist of $(CH_2)_n$, their mass

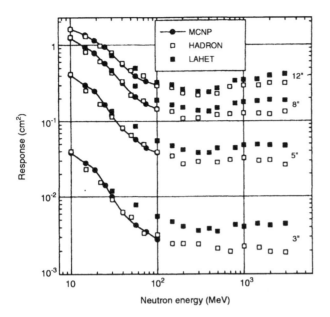

FIGURE 10.66 Comparison of the ³He BSS responses calculated by the Monte Carlo codes MCNP, HADRON, and LAHET. (From Rference [51]. With permission.)

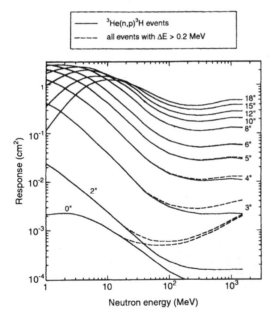

FIGURE 10.67 Response functions of the ³He BSS with (dashed curves) and without (solid curves) contribution of secondary charged particles from high-energy hadrons calculated by the codes MCNP (below 10 MeV) and HADRON (above 10 MeV). (From Reference [51]. With permission.)

density being 0.944 ± 0.003 g cm⁻³. Reactor filtered beams with nominal energies 24 keV and 144 keV serve as the neutron source.

Figure 10.68 shows smoothed Bonner sphere (BS) responses as calculated for four sets of parameters as a

function of moderator diameter. The moderator mass density is 0.95 g cm⁻³. The response increases with higher pressure. The peak position for the higher energy is shifted to larger diameters.

$$P_{He} = 200 \text{ kPa}, \ E_n = 144 \text{ keV} \text{ ------}$$
$$P_{He} = 172 \text{ kPa}, \ E_n = 144 \text{ keV} \text{ --- . ---}$$
$$P_{He} = 200 \text{ kPa}, \ E_n = 24 \text{ keV} \text{ - - -}$$
$$P_{He} = 172 \text{ kPa}, \ E_n = 24 \text{ keV} \text{ ———}$$

A neutron counter which is applicable to spectrometry and dosimetry over a wide energy range has been developed by Toyokawa et al. [53] It gives energy spectra, integral fluences, and dose equivalent of incident neutrons with energies from thermal to 15 MeV. The counter consists of a spherical polyethylene moderator and three slender ³He position-sensitive proportional counters inserted into the moderator. The position-sensitive proportional counters give detection position profiles of neutrons slowed down to thermal energies in the spherical moderator, which gives the above-mentioned information.

A schematic drawing of the counter and the setup used for response calculations is shown in Figure 10.69. The diameter of the spherical moderator is 26 cm. It was estimated using Monte Carlo simulations so that the responses extend up to 15 MeV. The outer diameter and counter are 1 cm and 0.9 cm, respectively. The counting gas is a mixture of ³He (101 kPa) and CF₄ (70 kPa). Because the addition of the CF₄ gas shortens the ranges of the proton and the triton produced in the ³He(np)³H reaction, the proportional counter has a good position resolution, evaluated to be about 0.7 cm in FWHM when the applied voltage is 1500 V. The position of the neutron detection was calculated by the charge division method.

Figure 10.70 shows the calculated position profiles (or the counter responses as a function of the axial position) for various monoenergetic neutrons. In each calculation, 10⁷ neutrons, which corresponds to the neutron fluence of 1.9×10^4 cm⁻² at the center of the moderator (when the moderator is removed), were incident of the counter.

The abscissa of Figure 10.70 shows the axial position in the position-sensitive proportional counter placed along the x-axis (Figure 10.69). Therefore, the neutrons are incident on the counter from the right-hand direction in Figure 10.70. As the energy of the incident neutrons increases, the position profiles spread to the deep region, which is within about ±5 cm of the axial position. The position profiles show the approximate distribution of the thermalized neutrons. It is shown in Figure 10.70 that the distribution is sensitive to the energy of the incident neutrons. In the first step of the feasibility study of the counter, thought of as a neutron spectrometer to be used under various geometric conditions of neutron irradiation, only the responses obtained for the x-proportional counter for the point-source

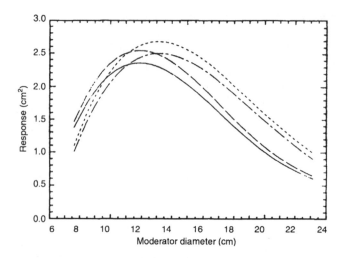

FIGURE 10.68 Smoothed responses of Bonner spheres as a function of the moderator diameter for two partial ³He pressures and incident neutron energies. The moderators consist of polyethylene with a specific mass density of 0.95 g cm⁻³. The diameter of the central detector is 3.2 cm. (From Reference [52]. With permission.)

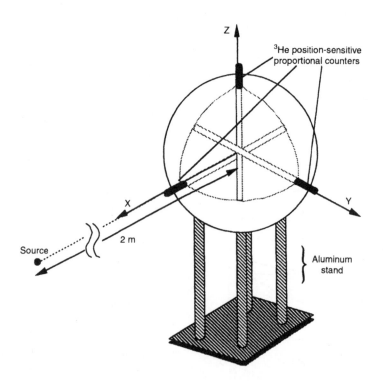

FIGURE 10.69 A schematic drawing of the counter and the source-detector geometry. (From Reference [53]. With permission.)

geometry were examined. The numbers shown at the top of Figure 10.70 are position groups which were introduced to calculate the response at a certain depth from the moderator surface along the x-axis as a function of the neutron energy.

Figure 10.71 shows the counter responses as a function of the incident neutron energy obtained for the x-proportional counter. Each of the curves shows the

detection counts at one of the five position groups, per unit fluence at the center of the moderator, as a function of the incident neutron energy. The energy resolution seems to be rather poor because the response curves overlap considerably.

A rem counter with a response tailored to match the shape of H*(10)/Φ as defined by ICRU and ICRP has been developed by Burgkhard et al. [54] The rem counter LB

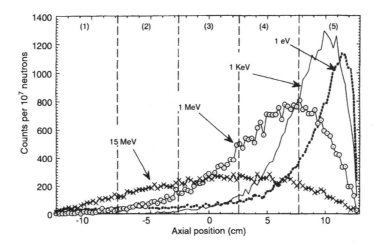

FIGURE 10.70. Detection position profiles calculated with the Monte Carlo simulations or the responses as a function of the axial position for various monoenergetic neutrons. The position profiles are shown for only the x-proportional counter. The numbers indicate the position groups: (1) -13 cm to -7.8 cm, (2) -7.8 cm to -2.6 cm, (3) -2.6 cm to 2.6 cm, (4) 2.6 cm to 7.8 cm, and (5) 7.8 cm to 13 cm, respectively. (From Reference [53]. With permission.)

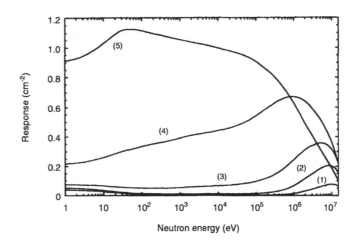

FIGURE 10.71 Calculated responses for the five position groups shown in Figure 10.70 for the x-proportional counter as a function of the incident neutron energy. (From Reference [53]. With permission.)

6411 consists of a polyethylene moderator sphere with a diameter of 25 cm, a central cylindrical ^{3}He proportional counter, and internal Cd absorbers and perforations. The instrument has an integrated high-voltage supply and signal processing and is connected to a microprocessor-controlled portable data logger. Counter features are summarized in Table 10.11.

The original data and the fitting results are shown in Figure 10.72, which demonstrates the quality of the fitted procedure, especially in the relevant high-neutron-energy range. This fluence response has then been used to establish a numerical calibration factor for a broad calibration field, N, defined as the ratio of the ambient dose equivalent, $H^*(10)$, and the value indicated by the rem counter

in the calibration field, M, by integration:

$$H^*(10) = \int_{E_{n-min}}^{E_{n-max}} dE_n h_\phi^*(E_n)\Phi_{E_n}(E_n) \quad (10.49)$$

$$M = \int_{E_{n-min}}^{E_{n-max}} dE_n R_\phi(E_n)\Phi_{E_n}(E_n) \quad (10.50)$$

where E_n, R_ϕ and Φ_{En} are the neutron energy, the evaluated fluence response, and the spectrum, respectively. E_{n-min} and E_{n-max} are taken as 1 meV and 20 MeV, respectively.

TABLE 10.11
Essential Features of the LB 6411

Electronics	Integrated high voltage supply and signal processing, connected to a microprocessor-controlled portable datalogger.
Physical basis	^{3}He proportional counter, moderating polyethylene sphere of 25 cm diameter and Cd absorbers.
Optimization	Search for optimal physical instrument parameters using detailed MCNP modeling and the newest cross sections (European Fusion File, EFF).
Experimental verification	Monoenergetic neutrons from thermal through 19 MeV at the PTB.
New quantity indication	Ambient dose equivalent H*(10) for neutrons according to ICRP 60 from thermal to 20 MeV neutrons.
Low energy dependence	+10% to −30% in the energy range 50 keV up to 10 MeV.
High sensitivity	≈ 3 counts per nSva.

From Reference [53]. With permission.

Figure 10.73 shows the H*(10) response of the LB 6411 and the Leake counter, numerically calibrated in the bare ^{252}Cf spectrum. The LB 6411 shows a three-times-lower over-response for intermediate neutrons and provides a four-times-higher sensitivity than the Leake counter.

The so-called single sphere albedo counter makes use of three active ^{3}He proportional counters as thermal neutron detectors positioned in the center and on the surface (albedo dosimeter configurations) of a polyethylene sphere. The linear combination of the detector readings allows for the indication of different quantities like H*(10), D, and ϕ, and reduces the energy dependence significantly. [55]

The counter makes use of three ^{3}He tubes, one at the center of a PE sphere, designated "c", and two on the surface, one just inside the moderator (albedo detector configuration, designated "i"), front-shielded, and the other just outside the moderator (detector configuration, designated "a", for thermal field neutrons), back-shielded by a boron-plastic absorber (Figure 10.74). Detector positions, size, and thickness of the boron absorber have been optimized to simulate the KfK albedo dosimeter. Small ^{3}He proportional counter tubes of the type Thomson 0.5 NH 1/1K, with an effective height of 10 mm and a diameter of 10 mm combined with a KfK-improved Thomson ACH amplifier, offer a dose rate range from 1 μSv h^{-1} to 100 mSv h^{-1} with a background rate of ≤0.01 μSv h^{-1} and a gamma discrimination factor of 3×10^{-5}. The count rates of the three detectors are indicated simultaneously.

The neutron dose equivalent H*(10) was obtained by a combination of the three detectors response ($M(a)$, $M(i)$, and $M(c)$) (Figure 10.75) and the corresponding weighting factor K.

$$H^*(10) = k_1 M(a) + k_2 M(i) + k_3 M(c) \qquad (10.51)$$

For the estimation of the ambient dose-equivalent, fluence to dose-equivalent conversion factors have been used, based on the new Q-L relationship recommended by ICRP 60. The results in Figure 10.76 show that a sphere diameter of 25 cm would result in a sufficient energy response for H*(10), whereas a diameter of 27 cm would improve the underestimation of the 30-cm sphere in the energy range 0.1 to 1 MeV. The prototype with a diameter of 25 cm is a compromise of energy response and weight of the sphere.

The Andersson-Braun rem counter is widely employed for radiation protection purposes, but its efficiency shows a marked decrease for neutron energies above about 10 MeV. A neutron monitor with a response function extended up to 400 MeV has been achieved by Birattari et al. [56] by modifying the structure of the moderator-attenuator of a commercial instrument.

The Tracerlab NP-I Portable Neutron Monitor Snoopy was chosen as a reference instrument for the design of the extended range version. A moderator-attenuator structure was constructed to house the same BF$_3$ proportion counter. The new structure is obtained by adding a layer of 1 cm of Pb around the boron plastic attenuator (at a cost of about 8 kg of extra weight) and shifting the outer polyethylene moderator 1 cm outward (i.e., its thickness remains unchanged). Figure 10.77 shows a longitudinal cross section of the instrument. The response function of the modified rem counter shows a marked increase at high energies and no changes in the other regions. This modified instrument was called the Long Interval NeUtron Survey meter (LINUS). Figure 10.78 shows its calculated response function; the response of the standard A-B rem counter and the H*(10) curve are also shown for comparison.

The response plotted in Figure 10.78 refers to lateral irradiation of the monitor by a uniform and parallel radiation field, i.e., the conditions under which H*(10) is calculated. A response of 275 s^{-1} per mSv h^{-1} has been adopted. The conversion coefficients from neutron fluence to ambient dose equivalent were taken from Stevenson. [57]

The Model NRD (Eberline [58]) neutron rem detector is a nine-inch-diameter, cadmium-loaded polyethylene sphere with a BF$_3$ lube in the center. This detector has been shown to have an energy response which closely

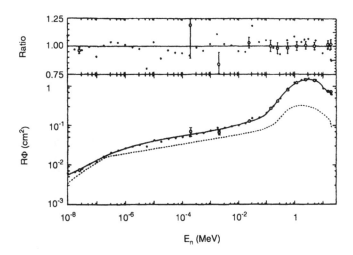

FIGURE 10.72 Lower part: experimental (circles), calculated (dots), and evaluated neutron fluence responses for the LB 6411 (solid line) and the Leake counter (dashed line) as a function of neutron energy. Upper part: response ratios of experimental and calculated values to the filled values. (From Reference [54]. With permission.)

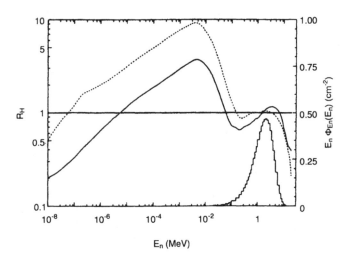

FIGURE 10.73 Ambient dose-equivalent response as function of neutron energy for the LB 6411 (solid line) and the Leake Counter (dashed line), numerically calibrated (Equations 10.49 and 10.50) in a bare ^{252}Cf spectrum (histogram). (From Reference [54]. With permission.)

follows the theoretical dose from neutrons over the energy range from 0.025 eV (thermal) to about 10 MeV. The BF$_3$ tube allows excellent gamma rejection. The detector is identical to that of the Model PNR-4 portable neutron rem counter. The detector specifications are:

1. Detector: BF$_3$ tube in nine-inch, cadmium-loaded polyethelene sphere
2. Plateau: approximately 200 V with a slope of 5% per 100 V
3. Operating voltage: dependent on sensitivity of counter and cable length; typically 1600 to 2000 V.
4. Directional response: within ±10%

5. Energy range: thermal to approximately 10 MeV
6. Gamma rejection: up to 500 R/h, dependent on high-voltage setting
7. Sensitivity: approximately 50 counts per minute (cpm) per mrem/h (3000 counts per mrem)
8. Connector: MHV series coaxial
9. Energy response: see Figure 10.79
10. Size: 9 inches in diameter × 10.5 inches high (22.9 cm diameter × 26.7 cm high)
11. Weight: 13.75 lb (6.3 kg)

Typical neutron plateau and gamma response of the NRD neutron remdetector is shown in Figure 10.80.

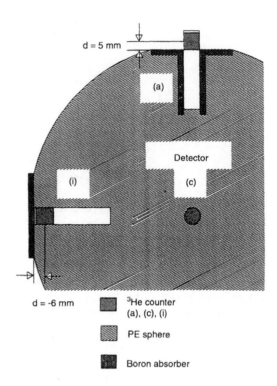

FIGURE 10.74 Cross section of the dose rate meter using ³He detectors in the boron-plastic absorber configurations (i), (a), and (c) in a polyethylene sphere. (From Reference [55]. With permission.)

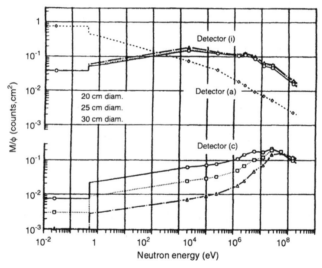

FIGURE 10.75 Energy dependence of the fluence response M/Φ for the ³He detectors for frontal (AP) irradiation of the detector configuration (i) and (a) and for (c) in the PE spheres of different diameters. (From Reference [55]. With permission.)

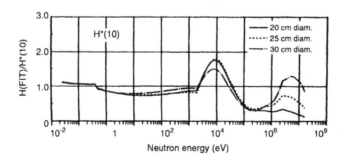

FIGURE 10.76 Energy dependence of the dose equivalent response H(*FIT*)/H*(10) of the counter using different sphere diameters and fits of the detector readings and frontal irradiation, based on the new *Q-L* relationship of ICRP 60. (From Reference [55]. With permission.)

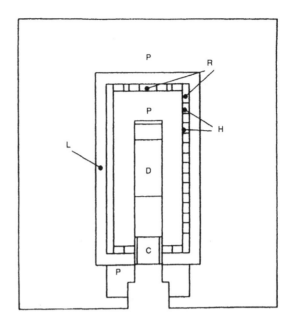

FIGURE 10.77 Longitudinal cross section of LINUS. P, polyethylene; R, boron-doped synthetic rubber; L, lead; H, holes; D, detector; C, connectors. (From Reference [56]. With permission.)

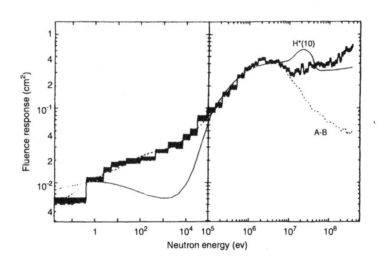

FIGURE 10.78 Response to lateral irradiation of LINUS. The response of a conventional A-B rem counter (the Tracerlab NP-1 Snoopy) and the H*(10) curve are shown for comparison. (From Reference [56]. With permission.)

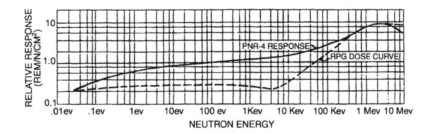

FIGURE 10.79 Neutron energy response curve. (From Reference [58]. With permission.)

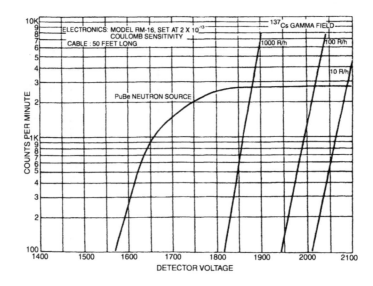

FIGURE 10.80 Typical neutron plateau and gamma response. (From Reference [58]. With permission.)

REFERENCES

1. **Wambersie, A. and Menzel, H. G.,** *Rad. Prot. Dos.,* 70, 517, 1997.
2. **Schuhmacher, H. et al.,** *Phys. Med. Biol.,* 37, 1265, 1992.
3. **Fink, G. et al.,** *Nucl. Phys.,* A518, 561, 1990.
4. **White, R. M. et al.,** *Rad. Prot. Dos.,* 70, 11, 1997.
5. **Chadwick, M. B. et al.,** *Rad. Prot. Dos.,* 70, 1, 1997.
6. **Schrewe, U. J. et al.,** *Rad. Prot. Dos.,* 44, 21, 1992.
7. **Leuthold, G. et al.,** *Rad. Prot. Dos.,* 70, 379, 1997.
8. **Hollnagel, R. A.,** *Rad. Prot. Dos.,* 70, 387, 1997.
9. **Siebert, B. R. L. and Thomas, R. H.,** *Rad. Prot. Dos.,* 70, 371, 1997.
10. **Schrewe, U. J. et al.,** *Rad. Prot. Dos.,* 70, 17, 1997.
11. **Mares, V. et al.,** *Rad. Prot. Dos.,* 70, 391, 1997.
12. **Newhauser, W. D. et al.,** *Rad. Prot. Dos.,* 70, 13, 1997.
13. **Gorbatkov, D. V. et al.,** *Nucl. Inst. Meth.,* A388, 260, 1997.
14. **Schuhmacher, H.,** *Rad. Prot. Dos.,* 44, 199, 1992.
15. **Weyrauch, W. and Knauf, K.,** *Rad. Prot. Dos.,* 44, 97, 1992.
16. **Klianga, P.,** Columbia University internal report, 1992. (see Reference [17].)
17. **Barthe, J. M. et al.,** *Rad. Prot. Dos.,* 70, 59, 1997.
18. **Burmeister, J. et al.,** *Nucl. Inst. Meth.,* A422, 606, 1999.
19. **Bronic, I. K.,** *Rad. Prot. Dos.,* 70, 33, 1997.
20. **Endo, S. et al.,** *Phys. Med. Biol.,* 41, 1037, 1996.
21. **Golnik, N.,** *Rad. Prot. Dos.,* 70, 211, 1997.
22. **Golnik, N. and Zielczynski, M.,** *Rad. Prot. Dos.,* 44, 57, 1992.
23. **Golnik, N. et al.,** *Rad. Prot. Dos.,* 70, 215, 1997.
24. **Hahn, T. et al.,** *Rad. Prot. Dos.,* 44, 297, 1992.
25. **Gambarini, G. et al.,** *Rad. Prot. Dos.,* 70, 175, 1997.
26. **Kriens, M. et al.,** *Rad. Prot. Dos.,* 44, 309, 1992.
27. **Hoffmann, W. and Songsiriritthigu, P.,** *Rad. Prot. Dos.,* 44, 301, 1992.
28. **Angelone, M. et al.,** *Rad. Prot. Dos.,* 70, 169, 1997.
29. **Piters, T. M. et al.,** *Rad. Prot. Dos.,* 44, 305, 1992.
30. **Angelone, M. et al.,** *Med. Phys.,* 25, 512, 1998.
31. **Matsumoto, T.,** *Nucl. Inst. Meth.,* A301, 552, 1991.
32. **Apfel, R. E. and Lo, Y. C.,** *Health Phys.,* 56, 79, 1989.
33. **Nath, R. et al.,** *Med. Phys.,* 20, 781, 1993.
34. **d'Errico, F. et al.,** *Med. Phys.,* 25, 1717, 1998.
35. **Loye, T. et al.,** *Med. Phys.,* 26, 845, 1999.
36. **Seitz, F.,** *Phys. Fluids,* 1, 2, 1958.
37. **d'Errico, F. et al.,** *Rad. Prot. Dos.,* 70, 109, 1997.
38. **d'Errico, F. et al.,** *Rad. Prot. Dos.,* 70, 103, 1997.
39. **Vareille, J. C. et al.,** *Rad. Prot. Dos.,* 70, 79, 1997.
40. **Ing, H. et al.,** *Rad. Prot. Dos.,* 85, 101, 1999.
41. **Schulze, J. et al.,** *Rad. Prot. Dos.,* 44, 351, 1992.
42. **Apfel, R. E., et al.,** *Rad. Prot. Dos.,* 44, 343, 1992.
43. **Fernandez, F. et al.,** *Rad. Prot. Dos.,* 70, 87, 1997.
44. **Bordy, J. M. et al.,** *Rad. Prot. Dos.,* 70, 73, 1997.
45. **Nakamura, T. et al.,** *Rad. Prot. Dos.,* 85, 45, 1999.
46. **Luszik-Bhadra, M. et al.,** *Rad. Prot. Dos.,* 70, 97, 1997.
47. **Tommasino, L. et al.,** *J. Appl. Phys.,* 46, 1484, 1975.
48. **Dorschel, B. et al.,** *Rad. Prot. Dos.,* 44, 355, 1992.
49. **Hankins, D. et al.,** *Rad. Dos. Prot.,* 23, 195, 1982.
50. **Kralik, M. et al.,** *Rad. Prot. Dos.,* 70, 279, 1997.
51. **Sannikov, A. V. et al.,** *Rad. Prot. Dos.,* 70, 291, 1997.
52. **Siebert, B. R. L. et al.,** *Rad. Prot. Dos.,* 44, 89, 1992.
53. **Toyokawa, H. et al.,** *Rad. Prot. Dos.,* 70, 365, 1997.
54. **Burgkhard, B. et al.,** *Rad. Prot. Dos.,* 70, 361, 1997.
55. **Burgkhardt, B. et al.,** *Rad. Prot. Dos.,* 44, 179, 1992.
56. **Birattari, C. et al.,** *Rad. Prot. Dos.,* 44, 193, 1992.
57. **Stevenson, G. R.,** *Dose Equivalent Per Star,* CERN Divisional Report TIS-RP/173, 1986.
58. **Eberline Co.,** *Technical Manual for Neutron Rem Detector Model NRD,* 1999.
59. **Portal, G. and Dietze, G.,** *Rad, Prot. Dos.,* 44, 165, 1992.
60. **Sannikov, A. V. and Savitskaya, E. V.,** *Rad. Prof. Dos.,* 70, 383, 1997.
61. **CasWell, R. L. and Cogne, J. J.,** *Rad. Res.,* 83, 217, 1980.
62. **Morstin, K. et al.,** Nuclear and atomic data for radiotherapy, IAEA, 1987.
63. **Savitskaya, E. W. and Sannikov, A. V.,** *Rad. Prof. Dosin.,* 60, 135, 1995.
64. **Combecher, D.,** *Rad. Res.,* 84, 189, 1980.
65. **Bronic, I. K. et al.,** *Rad. Res.,* 115, 213, 1988.

Index